HYDROGEOLOGY and GROUNDWATER MODELING

SECOND EDITION

HYDROGEOLOGY
and GROUNDWATER
MODELING

SECOND EDITION

Neven Kresic

CRC Press
Taylor & Francis Group
Boca Raton London New York

CRC Press is an imprint of the
Taylor & Francis Group, an informa business

CRC Press
Taylor & Francis Group
6000 Broken Sound Parkway NW, Suite 300
Boca Raton, FL 33487-2742

International Standard Book Number-10: 0-8493-3348-2 (Hardcover)
International Standard Book Number-13: 978-0-8493-3348-4 (Hardcover)

Library of Congress Cataloging-in-Publication Data

Krešić, Neven.
   Hydrogeology and groundwater modeling / Neven Krešić.
      p. cm.
   ISBN-13: 978-0-8493-3348-4
   ISBN-10: 0-8493-3348-2
   1. Hydrogeology--Mathematical models. 2. Groundwater flow--Mathematical models. I. Title.

GB1001.72.M35K75 2006
551.4901'5118--dc22
                                                 2006012881

Visit the Taylor & Francis Web site at
http://www.taylorandfrancis.com

and the CRC Press Web site at
http://www.crcpress.com

To Joanne and Miles

Preface

On various occasions, and for different purposes, I have asked myself what would be the best approach to introduce novices to the mysterious world of groundwater in some (hopefully) captivating way, and then have them interested in what were to follow. Although the word "mysterious" may not sound scientific or engineering-like, groundwater is by any definition mysterious because we cannot see it. As soon as one can see it, it is not groundwater anymore, also by definition. And I suspect that "the following subjects" mentioned above could be rather important to everyone waiting to hear what I had to say. Here are just a few examples: a class of geology seniors some of whom were facing a career-defining decision based, partially, on how successful my hydrogeology lectures were; or a group of attorneys helping themselves and me prepare for an expert witness testimony in front of jury members most of whom, in all probability, never heard of word hydrogeology before; or a state regulator about to endure my explanation regarding a very complex "groundwater fate and transport model" developed for an industrial client, without previously having a single groundwater-related academic course in his life. Not pretending that I have found the right answers (nor that I ever will), here follows one suggestion that might work.

Please find a cheap transparent plastic container of about a gallon or so when you have a chance; cheap, because you will poke a small hole somewhere in the middle of one side of it, and transparent because you will fill it, about two thirds, with some sandy soil or anything similar, and because you will then want to see what happens inside the container when you do the following: please start pouring (relatively slowly) about half a gallon of water into the container until water starts seeping out of the hole. By doing this you will suddenly become a student of hydrogeology as you also try to answer the following questions:

- How long did it take before water (groundwater, in a way) started seeping out of the hole?
- How much water did you have to pour into the container before it happened?
- Where was all that water stored inside the container?
- Why did the water start seeping out anyway?

After the (ground)water starts seeping out, pour the rest of the water all at once and observe what happens next, both inside the container (by looking through the transparent wall) and at the top of the sandy soil. You may be able to see how the completely wet portion of the sandy soil inside the container thickens, rising above the hole level, and how the water at the top of the soil disappears slower than before and eventually even stops disappearing for a while. In case you see some of your sandy soil trying to escape through the hole together with water, have a piece of porous cloth handy to prevent that. As all these events develop, you may find yourself asking a few additional questions about your little creation (hydro-geologists and nonhydrogeologists alike may refer to your creation as a little model of an "aquifer").

Now if you were already a hydrogeologist, or had anything to do with groundwater in your professional life, you may try to entertain the following question (even if you did not create the little aquifer): is it possible to predict (calculate) how much water one needs to pour into the plastic container before it starts seeping out of the hole? And then, if you were really interested in our little aquifer, you may come up with all kinds of strange questions such as:

- How long would it take for water to start seeping out if we pour two ounces of water per minute into the plastic container?
- If, after the water starts seeping out, we pour one ounce of orange juice (without pulp) on the top of the soil opposite of the hole, in a small localized area very close to the edge of the container, is it possible to predict (calculate) if and when the orange juice will start seeping out of the hole as well?
- Is it possible to model all this with computers (mathematically)?

Expanding on the example of our little aquifer, my intention with this book is to provide a foundation for understanding the importance of the above questions as they relate to real world situations, and then to answer many more about groundwater movement, quantity, quality, availability, impacts of various contaminants on groundwater, all illustrated in the second part of the book with problems solved in a "user-friendly way." This book is a logical extension of *Quantitative Solutions in Hydrogeology and Groundwater Modeling*, also published by CRC Press in 1997, which was focused only on solving problems related to groundwater flow (i.e., "physical" hydrogeology, or what one could call "noncontaminant" hydrogeology). This new book includes general chapters on groundwater flow principles, aquifers and aquitards, aquifer vulnerability and sustainable development, groundwater quality and contaminants, and theory of contaminant fate and transport. All relevant groundwater theory is presented at the beginning of the book, with references to the related solved problems included in the second part of the book. Most problems from the first book are included, and there are new examples of groundwater flow modeling, and contaminant fate and transport analyses. This new book, *Hydrogeology and Groundwater Modeling*, is therefore suitable as a textbook for undergraduate and graduate courses in the two subjects. It should also be useful as a reference book for working professionals, reminding them how to solve practical problems using "real" numbers, step by step, or showing them how to do the same with some problems they have to face for the first time.

I would like to thank many readers of the first book, who sent me their suggestions regarding the presented materials. I am grateful to Mr. Fletcher Driscoll of Fletcher Driscoll & Associates, LLC, and Mr. Richard Mandle of Michigan Department of Environmental Quality, for their invaluable advice and time spent in reviewing the contents of both books. I am especially thankful to Mr. Robert Masters of National Ground Water Association for performing the same review, and for his continuing support and friendship. I also thank Dr. Michael Kavanaugh and Richard Brownell of Malcolm Pirnie for their support. Finally, I am very grateful to Mr. James Derouin of Steptoe & Johnson for many challenging questions and his tireless drive to make various subjects in hydrogeology and groundwater modeling more understandable and closer to nongeologists, future hydrogeologists, and others interested in the mysterious world of groundwater.

Neven Kresic
Rixeyville, Virginia

Author

Neven Kresic is a professional hydrogeologist with Malcolm Pirnie, Inc., Washington, D.C., and serves as the company's quality control consultant for groundwater modeling. Dr. Kresic received his BS in hydrogeologic engineering, and MS and PhD in hydrogeology and geology from Belgrade University, in the former Yugoslavia. He is a professional geologist, professional hydrogeologist, and certified groundwater professional with over 20 years of international groundwater and surface water–related consulting, research, and teaching experience.

Dr. Kresic taught at major universities in Europe and the United States, and was the recipient of the Senior Fulbright Scholarship for research at the United States Geological Survey in Reston, Virginia. His areas of expertise include groundwater engineering, development and remediation, karst and fractured rock hydrogeology, and groundwater modeling.

Dr. Kresic served as chair of the Continuing Education Committee of the American Institute of Hydrology, is a founding member of the Ground Water Modeling Interest Group sponsored by the National Ground Water Association, and a permanent member of the Karst Commission of the International Association of Hydrogeologists. Dr. Kresic has authored numerous papers and three books in the areas of hydrogeology and groundwater modeling.

Table of Contents

Part II
Solved Problems in Hydrogeology and Groundwater Modeling

Part I

Basic Concepts in
Hydrogeology and
Groundwater Modeling

1 Hydrogeology and Groundwater Modeling

1.1 GROUNDWATER AND HYDROGEOLOGY

The term hydrogeology (*hydrogéologie* in French) was first created by the French biologist and naturalist Jean-Baptiste Lamarck in 1802, in a publication with the same name, published in Paris by the Museum of Natural History (Lamarck, 1802). Lamarck also introduced the term biology. Although an entire scientific and engineering field, which for its subject has groundwater, owes its name to the French scientist, there seemed to be not much more Lamarck and hydrogeology have shared since. At least not until now, as a global approach to the preservation of both the environment and humankind is being endorsed by more and more people, including most of the governments around the world. Like Werner and Humboldt (also prominent scientists of the time), Lamarck considered nature as a whole, emphasizing the close interconnections between its abiotic and biotic parts. He even declared that there must be one integrative science *physique terrestre* (physics of the Earth), which would be able to embrace the *météorologie* (the study of the atmosphere), *hydrogéologie* (the study of Earth's crust), and *biologie* (the study of living organisms) (Ghilarov, 1998).

Lamarck's idea of Earth's crust is that it is not only embraced by life, but is under its control, is one that periodically emerges in the scientific community, causing various controversies (for example, Vernadsky, 1926, 1929, 1997; Lovelock, 1979). It seems that all similar concepts inevitably contain speculative elements and, what is more important, could not be tested. Only now, the level achieved by empirical science equipped with appropriate techniques of measuring allows us to hope that the relationship of the scientific community to the global ecosystem problem is changing. Recognizing the way that science evolved to this state, we must remember the role of Lamarck as a forerunner of the global approach to science (Ghilarov, 1998).

The material presented in this chapter is, to a large extent, based on the pioneering work of geologists of the United States Geological Survey (USGS) and other American researchers, who established hydrogeology as the mainstream scientific and engineering discipline in the late nineteenth and early twentieth centuries. Together with the contributions of French engineers Darcy and Dupuit, and of Daubree, who was arguably the first to publish a systematic work on groundwater (1887), their observations and quantitative analyses serve as the foundation of modern hydrogeology. References to publications by King (1892, 1899a,b, 1900), Hazen (1892, 1893), Norton (1897), Slichter (1899, 1902, 1905), Fuller (1904, 1905, 1906, 1908), Fuller et al. (1910, 1911), Darton (1902, 1904), Veatch (1906), Ellis (1909), Lee (1912, 1915), and Meinzer (1923a,b, 1927a,b, 1932, 1940), here and elsewhere in the book, are tributes to these fathers of hydrogeology in the United States.

The beginning of systematic hydrogeologic investigations in the United States is explained by Fuller (1905):

> In the earlier years of the Survey no special provision was made for the study of underground waters, although a considerable amount of information was gathered in connection with the investigation of other problems and a number of reports were published. Beginning with 1894 provision was made by Congress for the investigation of underground currents and deep wells, and from 1894 to 1902 a considerable number of special reports on underground waters were prepared under this authority. In the latter year, in order to satisfactorily meet the new demands resulting from the great increase in the use of underground waters in recent years and to develop, specialize, and systemize the work, the investigations relating to underground waters were segregated from the division of hydrography and placed in charge of a distinct organization known as the division of hydro-geology or hydrology. This division is divided into two sections, eastern and western, the first embracing the States east of the Mississippi and those bordering that river on the west, and the second including the remaining, or the so-called 'reclamation' States and Territories and Texas. The two sections have been placed in charge of geologists, Mr. N.H. Darton acting as chief of the western section and the writer as chief of the eastern. The work of the division includes the gathering, filing, and publication of statistical information relating to the occurrence of water in artesian and other deep wells; the gathering and publication of data pertaining to springs; the investigation of the geologic occurrence, from both stratigraphic and structural standpoints, of underground waters and springs; a study of the laws governing the occurrence and flow of subterranean waters and springs, including the investigation of variations due to tidal, temperature, and barometric fluctuations; direct measurements of rate of underflow; detailed surveys of regions in which water problems are of great importance and urgency; and the publication of reports on irrigation, city water supplies, and other important uses of underground waters.

The USGS continued its lead in advancing hydrogeology through the present day. Charles Theis, the founder of quantitative hydrogeology, published his groundbreaking paper on transient groundwater flow to wells in 1935 and gave numerous other contributions such as on anisotropy and heterogeneity, and transport of radionuclides and other contaminants in subsurface. First widely used computer models of groundwater flow, and contaminant fate and transport were developed at the USGS: Konikow and Bredehoeft introduced MOC in 1978, and McDonald and Harbaugh released MODFLOW for public use in 1988 (Konikow and Bredehoeft, 1978; McDonald and Harbaugh, 1988).

The following discussion by Meinzer (1940) illustrates the importance of groundwater as a resource, and is remarkably relevant for understanding the current utilization of water resources in the United States as well:

> One of the greatest achievements in American history has been the progressive development of domestic water supplies of good quality—that is, the water supplies used by the people for drinking, cooking, and laundry and toilet purposes.
>
> In the pioneer period the domestic supplies were obtained chiefly from springs or from shallow dug wells that were generally unreliable and exposed to pollution. In contrast, a very large proportion of the 75,000,000 people who now live in communities that are served by public waterworks have ample and perennial supplies of safe, acceptable water delivered under pressure to the taps in their kitchens, bathrooms, and laundries. Within the communities that have public waterworks, including large cities, there are still a considerable number of people who obtain their domestic supplies from private wells. The supplies from many of these wells are subject to pollution and are otherwise unsatisfactory, and they should for the most part be abandoned in favor of the supplies furnished by the public waterworks. It is estimated that there are in the United States between 3,000 and 4,000 communities of less than 1,000 inhabitants that are

provided with public water supplies. However, according to the 1930 census, the total number of incorporated places having less than 1,000 inhabitants is slightly over 10,000 and their aggregate population is somewhat more than 4,000,000. It may therefore be inferred that there are still several thousand of the smaller incorporated communities, with an aggregate population of perhaps 2,000,000, that do not have public water supplies. Most of the people in these communities have private wells.

According to the 1930 census there are nearly 45,000,000 people who live in rural territory—that is, in homes that are widely distributed on about 6,500,000 farms and in the very small rural communities that are not incorporated. It should be noted that the United States differs from most other countries in that the people engaged in agriculture do not generally live in compact communities but in widely separated homes on the individual farms. It is not generally practicable to supply these homes from public waterworks, but as a rule each farmer has developed a private supply for the household, or in some cases more than one household, on his farm.

By far the largest number of the rural inhabitants obtain their domestic water from private wells, but a considerable number obtain it from springs, and some depend on rain water stored in cisterns, water in irrigation ditches, or other sources. In the hard-water areas it is common practice for a household to have two distinct sources of supply—a well that provides hard but otherwise relatively acceptable water for drinking and cooking, and a cistern that provides soft water for laundry and toilet use.

The large cities obtain their public water supplies chiefly from surface sources, and the small communities chiefly from wells. Of the 25 cities that have a population of more than 300,000 all obtain their supplies from streams or lakes, except that a few of these have supplementary supplies of groundwater. In 1930 the aggregate population of these 25 cities was about 25,000,000, and the total population of all the cities supplied wholly or chiefly from surface sources or from springs is estimated to have been about 55,000,000. The supplies are generally ample, except for some of the smaller cities, and almost without exception the water, after treatment, is of good sanitary quality.

About 6,500 communities, with an aggregate population of about 20,000,000, having public waterworks supplied wholly or chiefly from wells or other structures for recovering groundwater. This number includes 15 of the 68 cities having a population between 100,000 and 300,000; somewhat less than one-third of the 283 cities between 25,000 and 100,000; about one-half of the 1,457 cities between 5,000 and 25,000; and about two-thirds of the approximately 8,000 smaller communities that have public waterworks. The cities between 100,000 and 300,000 that have water supplies derived from wells or comparable structures, named in the order of their population, are as follows: Houston, Memphis, San Antonio, Dayton, Des Moines, Long Beach, Jacksonville, Camden, Spokane, Wichita, Miami, Peoria, Canton (Ohio), El Paso, and Lowell (Massachusetts).

It is roughly estimated that about 2,000,000,000 gallons of groundwater are obtained daily through wells or similar structures for public water supplies in the United States. The largest groundwater development for public supplies is in the western part of Long Island, New York, in New York City, and the adjacent county of Nassau. The average pumpage from wells and infiltration galleries for all purposes in that area during the period from 1904 to 1937, has been somewhat over 158,000,000 gallons a day, of which about 113,000,000 gallons have been for the public supplies of New York City and suburbs. In the Houston–Galveston area, in Texas, the use of groundwater has increased progressively since the beginning of this century, and in 1936 it averaged about 98,000,000 gallons a day, of which about 25,000,000 gallons were for the public supply of Houston.

As might be expected, the many small waterworks supplied with groundwater vary greatly both as to quantity and quality of the water. However, with the constantly increasing vigilance of the State and local health departments, these supplies are with few exceptions of good sanitary quality. Moreover, in the recent droughts the public water supplies from wells have generally been adequate, and the difficulties on account of failing supplies have been chiefly in the smaller communities that depended on surface water without adequate storage.

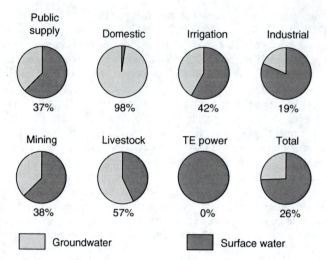

FIGURE 1.1 Usage of surface water and groundwater for different purposes in the United States, in percentage, for year 2000. (From USGS (United States Geological Survey), Web page available at: http://ga.water.usgs.gov/edu/waterdistribution.html, accessed on November 12, 2005.)

In the year 2000, groundwater withdrawals were 21% of the total freshwater withdrawal in the United States. Comparison of surface water and groundwater usage for different purposes is shown in Figure 1.1. Throughout the world, including the United States, groundwater is used for irrigation in greater proportion than for any other purpose (see Figure 1.2). It is this practice that has caused extensive depletion of some of the world's most important aquifers, such as Ogallala aquifer in the United States, Nubian sandstone aquifer in North Africa, and Great Artesian Basin aquifer in Australia.

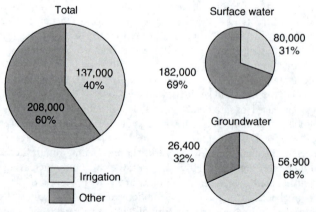

FIGURE 1.2 Usage of groundwater and surface water for irrigation in the United States, in million gallons per day, for year 2000. (From USGS (United States Geological Survey), Web page available at: http://ga.water.usgs.gov/edu/waterdistribution.html, accessed on November 12, 2005.)

1.2 HYDROLOGIC CYCLE AND GROUNDWATER

Groundwater is an important part of Earth's hydrologic cycle or movement of water between oceans, atmosphere, and land. Groundwater is derived mostly from percolation of precipitation and, to a lesser degree, from surface water streams and lakes that lose water to underlying aquifers. Only minute quantities of groundwater have their origin from processes in the deeper crust, associated with magmatism (this water is often called juvenile water). Groundwater from aquifers and aquitards discharges to fresh surface water bodies on land (streams, lakes, marshes) and to oceans. This discharge is either concentrated via springs and seeps, or directly into surface water bodies where it is normally not visible (Figure 1.3). The volume of groundwater stored and moving through aquifers and aquitards in the upper portion of Earth's crust is much larger than any other form of mobile freshwater on Earth, excluding glaciers and ice caps (see Figure 1.4).

1.2.1 TYPES OF WATER BELOW GROUND SURFACE

Water in the subsurface, from the practical hydrogeologic perspective, can be divided into two major zones: water stored in the unsaturated zone, also called vadose zone or zone of aeration, and water stored in the saturated zone (Figure 1.5).

Soil pore space in the vadose zone is filled with both air and water, in varying proportions, depending on the soil type, climatic and seasonal conditions. This zone may be divided, with respect to the occurrence and circulation of water, into the uppermost zone of soil water, the intermediate zone, and the capillary fringe immediately above water table. The zone of soil water is the part of the lithosphere from which water is discharged into the atmosphere in perceptible quantities by the action of plants or by direct evaporation. It differs greatly in thickness with different types of soil and vegetation, having only a few feet thick where the surface is covered with grass or ordinary field crops, but much thicker in forests and in tracts that support certain deep-rooting desert plants. The soil water is of primary interest in agriculture because it is near enough to the land surface to be available to the roots of plants. The depths to which the roots of plants reach for water vary greatly with different types of plants and with different soil and moisture conditions. Grasses and most field crops draw water from depths of up to 7 ft. However, alfalfa, once well established in fine sandy soils derived from loess, may obtain groundwater from as much as 20 to 30 ft below ground surface. Large trees and certain types of deep-rooted desert plants draw water from considerable depths. There is evidence that a certain type of mesquite obtains water as much as 50 ft below the surface and that other perennials may send their roots to depths of 50 ft or even 60 ft. With respect to its availability to plants, soil water is either available or not available for plant growth. The latter is so firmly held by adhesion or other forces that it cannot be taken up by plants rapidly enough to produce growth. More detail on various relationships between vegetative cover and water below land surface is given by Meinzer (1927b).

Hygroscopic water is the water in the soil that is in equilibrium with atmospheric water vapor. It is essentially the water with molecular attraction of soil grains can hold against evaporation. When a dry soil is in contact with the atmosphere, it absorbs water vapor from the atmosphere, and this water, after absorption, is called hygroscopic water. It can be removed from the soil only after heating to about $100°C$ to $110°C$ (i.e., after converting hygroscopic water back into vapor).

The capillary fringe is a zone directly above water table and contains capillary interstices some or all of which are filled with water that is continuous with the water in the zone of saturation but is held above that zone by capillarity acting against gravity (Figure 1.6). Capillarity is a term that describes joint action of two main molecular forces: adhesion

FIGURE 1.3 *Top*: Aerial thermal infrared scan of Town Cove, Nauset Marsh, Cape Cod, Massachusetts. Discharging fresh ground water is visible as dark (relatively cold) streams flowing outward from shore over light-colored (warm) but higher density estuarine water. Data were collected at low tide at 9.00 p.m. eastern daylight time on August 7, 1994. (From Barlow, P.M., Ground Water in Freshwater-Saltwater Environments of the Atlantic Coast *U.S. Geological Survey Circular* 1262, Reston, VA, 2003, 113 p. Photo courtesy of John Portnoy, Cape Cod National Seashore.) *Bottom*: One of numerous springs along the Tara River canyon, Montenegro.

(attraction between molecules of water and molecules of porous media) and cohesion (attraction between molecules of water). Experiments have shown that in cylindrical tubes the height at which water is held by capillarity is inversely proportional to the tube diameter (Figure 1.6a). The capillary fringe moves upward and downward together with the water table

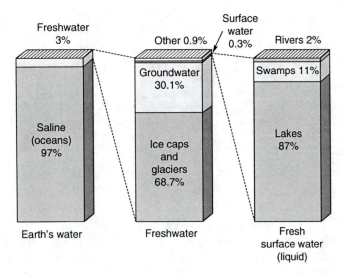

FIGURE 1.4 Distribution of water on Earth. (USGS (United States Geological Survey), Web page available at: http://ga.water.usgs.gov/edu/waterdistribution.html, accessed on November 12, 2005.)

due to seasonal patterns of aquifer recharge from the percolating precipitation. Figure 1.6b and c illustrates how different shapes and diameters of capillary tubes (i.e., pores in the subsurface) respond to a moving or stagnant water table. When the water table starts falling, the water in the capillary tubes will be held at the same level in all tubes that have the same diameter of the upper water–air contact, regardless of the shape and tube diameter below that contact (case B). As the water table stabilizes, the water in different capillary tubes will start rising slowly due to adhesion and cohesion (capillarity). The height of capillary rise will then depend on the tube diameter and will not be influenced by the shape of the tubes (case C). Tubes of different shape, but having the same diameter, will have the same height of capillary rise.

As can be seen, the thickness of the capillary fringe depends on the texture of the rock or soil. The fringe is relatively thin if it consists of materials in which all the capillary interstices

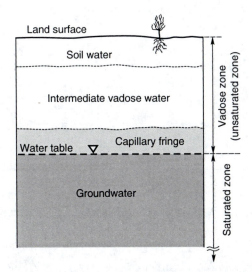

FIGURE 1.5 Hydrogeologic zones and types of water in the subsurface. (Modified from Meinzer, O.E., U.S. Geological Survey Water-Supply Paper 494, Washington, D.C., 1923b, 71 p.)

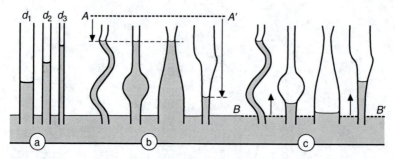

FIGURE 1.6 Principle of capillarity. (Modified from Meinzer, O.E., U.S. Geological Survey Water-Supply Paper 489, Washington, D.C., 1923a, 321 p.; and from Tolman, F.C., *Ground Water*, McGraw-Hill, New York, 1937, 593 p.)

are large. Materials that have only subcapillary interstices (such as fresh crystalline rocks) are not regarded as having any capillary fringe or as forming a functional part of such a fringe. In materials whose interstices are all supercapillary the capillary fringe is practically absent, such as in uniform coarse gravels. The thickness of the capillary fringe in silty materials has frequently been observed to be about 8 ft; in very fine-grained materials (clay) it is even thicker, and in coarse sand it is considerably thinner (Figure 1.7). Knowing the characteristics of the capillary fringe is important when designing an irrigation system, estimating potential evapotranspiration from water table, or when studying fate and transport of certain ground-water contaminants that may be accumulating at the water table or in the unsaturated zone (see Section 5.2.4).

Water that percolates beyond the soil water zone enters the intermediate zone. Here it is either drawn downward by gravity to the underlying zone of saturation or is drawn by molecular attraction into the capillary and subcapillary pores, where it may become station-ary (see Figure 1.8). Some of this trapped water may move again downward if the newly percolated water provides enough pressure height to overcome capillary forces. Whereas the zone of soil water and the capillary fringe are limited in thickness by definite local conditions, such as character of vegetation and texture of rock or soil, the intermediate zone is not limited in that respect. It is the residual part of the zone of aeration. It may be entirely absent or may

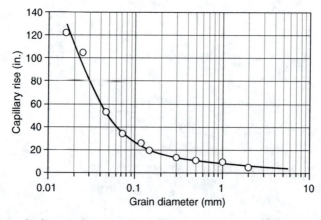

FIGURE 1.7 Capillary rise in porous materials of different grain sizes. (Data from Meinzer, O.E., U.S. Geological Survey Water-Supply Paper 489, Washington, D.C., 1923a, 321 p., after Hilgard, E.W., *Soils*, Macmillan Co., New York, 1906, 206 p.)

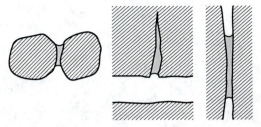

FIGURE 1.8 Nongravity water in the subsurface, held in small interstices by capillary and molecular forces; hatched areas represent rocks, shaded areas water, and blank spaces represent openings from which water has drained out. (Modified from Meinzer, O.E., U.S. Geological Survey Water-Supply Paper 557, Washington, D.C., 1927a, 94 p.)

attain a thickness of several hundred feet or more, depending on regional hydrogeologic and climatic conditions (Figure 1.9). In the arid and semiarid regions, such as in the southwest United States, the unsaturated zone is commonly thicker than 100–150 ft.

Groundwater is water below water table, filling entirely all rock interstices (void space) in the saturated zone. Groundwater may be divided, with respect to the force by which it is controlled, into water that can flow through the rock freely under the influence of gravity (gravity groundwater), and groundwater that is not under the control of gravity. The latter one is held against gravity by capillary forces and would be retained in the capillary and subcapillary pores (Figure 1.8).

Meinzer (1923a) provides this discussion on the nature of nongravity groundwater and its relationship to different porous materials:

> It can readily be shown, by applying the rule that the surface of a sphere is equal to 3.1416 times the square of its diameter, that the total interstitial surface of a cubic foot of sand composed of grains 1 millimeter in diameter is about 1,000 square feet, that of a cubic foot of sand composed of grains 0.02 millimeter in diameter is about 50,000 square feet, or more than 1 acre, and that of a cubic foot of material composed of grains only 0.001 millimeter in diameter is about 1,000,000 square feet, or more than 20 acres. From experiments made with the flow of air through various soils King (1900) calculated the aggregate surface of a cubic foot of ordinary loam soils to be about 1 acre and that of a cubic foot of fine clay soils to be about 4 acres. These figures give some conception of the vast areas of interstitial surface that are involved in fine-grained material and of the great influence that may be exerted upon water by the attraction of this surface, even though it acts through only a small range. The quantity of water held by a wet surface of a few square feet may be very small, but the quantity held by an entire acre is considerable even though the film of water adhering to this surface is very thin.

1.3 POROSITY AND EFFECTIVE POROSITY

Porosity (n) is the percentage of voids (empty space occupied by water or air) in the total volume of rock, which includes both solids and voids:

$$n = \frac{V_v}{V} \times 100\% \qquad (1.1)$$

where V_v is the volume of all rock voids and V is the total volume of rock (in geologic terms, *rock* refers to all of the following: soils, unconsolidated and consolidated sediments, and any type of rock in general). Assuming the specific gravity of water equals unity, total

FIGURE 1.9 Blue Lake, near Imotski in the Dinaric karst, is one of the deepest sinkholes in the world. The thickness of the unsaturated zone (depth to water table visible in the sinkhole) fluctuates between 150 and 200 m as seen from horizontal marks on the colluvium material.

porosity, expressed as a percentage, based on four common approaches, can be expressed as (Lohman, 1972):

$$n = \frac{V_i}{V} = \frac{V_w}{V} = \frac{V - V_m}{V} = 1 - \frac{V_m}{V} [\times 100\%] \qquad (1.2)$$

where n is porosity in percent per volume, V is total volume, V_i is volume of all interstices (voids), V_m is aggregate volume of mineral (solid) particles, and V_w is volume of water in a saturated sample.

Porosity can also be expressed as

$$n = \frac{\rho_m - \rho_d}{\rho_m} = 1 - \frac{\rho_d}{\rho_m} [\times 100\%] \qquad (1.3)$$

where ρ_m is average density of mineral particles (grain density) and ρ_d is density of dry sample (bulk density).

Porosity is the most important property of rocks that enables storage and movement of water in the subsurface. It directly influences the permeability and the hydraulic conductivity of rocks, and therefore the velocity of groundwater and other fluids that may be present. In general, rock permeability and groundwater velocity depend on the shape, amount, distribution, and interconnectivity of voids. Voids, on the other hand, depend on the depositional mechanisms of unconsolidated and consolidated sedimentary rocks, and on various other geologic processes that affect all rocks during and after their formation. Primary porosity is the porosity formed during the formation of rock itself, such as voids between the grains of sand, voids between minerals in hard (consolidated) rocks, or bedding planes of sedimentary rocks. Secondary porosity is created after the rock formation mainly due to tectonic forces (faulting and folding), which create micro- and macrofissures, fractures, faults, and fault zones in the solid rocks. Both the primary and secondary porosities can be successively altered multiple times, thus completely changing the original nature of the rock porosity. These changes may result in porosity decrease, increase, or altering of the degree of void interconnectivity without a significant change in the overall void volume.

The following discussion by Meinzer (1923a), and the figure that accompanies it (Figure 1.10), is probably the most cited explanation of rock porosity, and one can hardly add anything to it:

The porosity of a sedimentary deposit depends chiefly on (1) the shape and arrangement of its constituent particles, (2) the degree of assortment of its particles, (3) the cementation and compacting to which it has been subjected since its deposition, (4) the removal of mineral matter through solution by percolating waters, and (5) the fracturing of the rock, resulting in joints and other openings. Well-sorted deposits of uncemented gravel, sand, or silt have a high porosity, regardless of whether they consist of large or small grains. If, however, the material is poorly sorted small particles occupy the spaces between the larger ones, still smaller ones occupy the spaces between these small particles, and so on, with the result that the porosity is greatly reduced (Figure 1.10a and b). Boulder clay, which is an unassorted mixture of glacial drift containing particles of great variety in size, may have a very low porosity, whereas outwash gravel and sand,

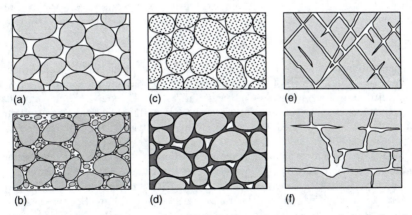

FIGURE 1.10 Diagram showing several types of rock interstices and the relation of rock texture to porosity. (a) Well-sorted sedimentary deposit having high porosity; (b) poorly sorted sedimentary deposit having low porosity; (c) well-sorted sedimentary deposit consisting of pebbles that are themselves porous, so that the deposit as a whole has a very high porosity; (d) well-sorted sedimentary deposit whose porosity has been diminished by the deposition of mineral matter in the interstices; (e) rock rendered porous by fracturing; and (f) rock rendered porous by solution. (Modified from Meinzer, O.E., U.S. Geological Survey Water-Supply Paper 489, Washington, D.C., 1923a, 321 p.)

derived from the same source but assorted by running water, may be highly porous. Well-sorted uncemented gravel may be composed of pebbles that are themselves porous, so that the deposit as a whole has a very high porosity (Figure 1.10c). Well-sorted porous gravel, sand, or silt may gradually have its interstices filled with mineral matter deposited out of solution from percolating waters, and under extreme conditions it may become a practically impervious conglomerate or quartzite of very low porosity (Figure 1.10d). On the other hand, relatively soluble rock, such as limestone, though originally dense, may become cavernous as a result of the removal of part of its substance through the solvent action of percolating water (Figure 1.10e). Furthermore hard, brittle rock, such as limestone, hard sandstone, or most igneous and metamorphic rocks, may acquire large interstices through fracturing that results from shrinkage or deformation of the rocks or through other agencies (Figure 1.10f). Solution channels and fractures may be large and of great practical importance, but they are rarely abundant enough to give an otherwise dense rock a high porosity.

Porosity of unconsolidated sediments (gravel, sand, silt, and clay) is often called *intergranular porosity* because the solids are loose detritic grains. When such rocks become consolidated, the former intergranular porosity is called *matrix porosity*. In general, the term matrix porosity is applied to primary porosity of all consolidated (hard) rocks, such as porosity between mineral grains (minerals) in granite, gneiss, slate, or basalt. Some unconsolidated or loosely consolidated (semiconsolidated) rocks may contain fissures and fractures, in which case the nonfracture portion of the overall porosity is also called matrix porosity. Good examples are fractured clays and glacial till sediments, or residuum deposits which have preserved fabric of the original bedrock in the form of fractures and bedding planes. Sometimes, microscopic fissures in rocks are also considered part of the matrix porosity, as opposed to larger fissures and fractures called *macro porosity*. Photographs in Figure 1.11 through Figure 1.17 help illustrate the very diverse nature of rock porosity at various scales.

When studying and observing porosity from the hydrogeologic perspective, it is very important to make a very clear distinction between the total porosity and the *effective porosity* of the rock. It is, however, unfortunate that some widely used hydrogeology textbooks do not make such distinction and even hypothesize that there is no such thing as effective porosity since "water molecules are shown to be able to move through any pore size." The following discussion by Meinzer (1932) may help in explaining why using the appropriate form of porosity and the appropriate corresponding number in quantitative hydrogeologic analyses does matter:

> To determine the flow of ground water, however, a third factor, which has been called the *effective porosity*, must be applied. Much of the cross section is occupied by rock and by water that is securely attached to the rock surfaces by molecular attraction. The area through which the water is flowing is therefore less than the area of the cross section of the water-bearing material and may be only a small fraction of that area. In a coarse, clean gravel, which has only large interstices, the effective porosity may be virtually the same as the actual porosity, or percentage of pore space; but in a fine-grained or poorly assorted material the effect of attached water may become very great, and the effective porosity may be much less than the actual porosity. Clay may have a high porosity but may be entirely impermeable and hence have an effective porosity of zero. The effective porosity of very fine grained materials is generally not of great consequence in determinations of total flow, because in these materials the velocity is so slow that the computed flow, with any assumed effective porosity, is likely to be relatively slight or entirely negligible. The problem of determining effective porosity, as distinguished from actual porosity, is, however, important in studying the general run of water-bearing materials, which are neither extremely fine nor extremely coarse and clean. Hitherto not much work has been done on this phase of the velocity methods of determining rate of flow. No distinction has generally been made between actual and effective porosity, and frequently a factor of 33 1/3 per cent has been used, apparently

FIGURE 1.11 *Top*: Photograph of coarse alluvial gravel. *Bottom*: ESM (electron scanning microscope) image of uniform, pure quartz sand. (ESM courtesy of Dr. Scott Chumbley, Iowa State University.)

without even making a test of the porosity. It is certain that the effective porosity of different water-bearing materials ranges between wide limits and that it must be at least roughly determined if reliable results as to rate of flow are to be obtained. It would seem that each field test of velocity should be supplemented by a laboratory test of effective porosity, for which the laboratory apparatus devised by Slichter (1905) could be used.

One of the reasons why some professionals today choose to ignore statements like this one by Meinzer is that determining the effective porosity is not straightforward since there is no one "magic" method of doing so, and different methods yield different results (e.g., see Stephens et al., 1998). On the other hand, determining the total porosity of a rock specimen has been a routine procedure for more than a century, as it involves simple volumetric–gravimetric techniques, i.e., measurement of simple quantities listed in Equation 1.2 and Equation 1.3.

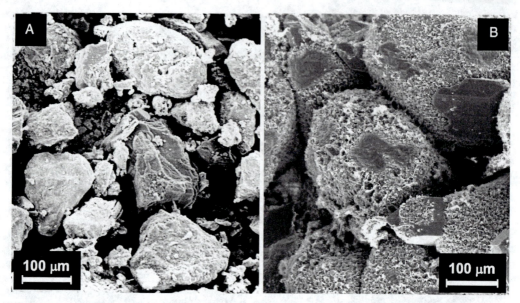

FIGURE 1.12 ESM images of two different types of sandstone. (a) Loosely cemented sandstone with high porosity. (Courtesy of Dr. Fred Longstaffe, The University of Western Ontario, Canada. With permission.) (b) Quartz grains and maybe feldspar grains coated with clay minerals; the clay minerals are likely illite or chlorite. (Courtesy of James Talbot. Copyright 2000, James Talbot, K/T GeoServices, Inc. With permission.)

Effective porosity is often equated to *specific yield* of the porous material, or that volume of water in the pore space that can be freely drained by gravity due to change in the hydraulic head. Effective porosity is also defined as the volume of interconnected pore space that allows free gravity flow of groundwater. The volume of water retained by the porous media, which cannot be easily drained by gravity, is called *specific retention*. Since drainage of pore space by gravity may take long periods of time, especially in fine-grained sediments, values of specific yield determined by various laboratory and field methods during necessarily limited times are probably somewhat lower than the "true" effective porosity. Specific yield determined by aquifer testing in the field is a lumped hydrodynamic response to pumping by all porosity types (porous media) present in the aquifer. This value cannot be easily related to values of total porosity, which are always determined in the laboratory for small samples. More detail on *aquifer specific yield* is given in Section 2.2.2. One important distinction between the specific yield and the effective porosity concepts is that the specific yield relates to volume of water that can be freely extracted from an aquifer, while the effective porosity relates to groundwater velocity and flow through the interconnected pore space. Unnecessary confusion is introduced by some professionals, trying to distinguish between the effective porosity for groundwater flow and the effective porosity for contaminant transport. If the contaminant is dissolved in groundwater, its advective transport will be governed by the same effective porosity since it moves with groundwater. Diffusive transport of the contaminant is its movement due to concentration gradient and is independent of the groundwater flow. Diffusion involves the entire (total) porosity: molecules of the contaminant (and water) can move through minute pores, which would otherwise not allow free gravity flow. In conclusion, there are no two different effective porosities and it is sufficient to determine two values for any quantitative analysis of groundwater flow, or contaminant fate and transport: one for the effective porosity and the other for the total porosity.

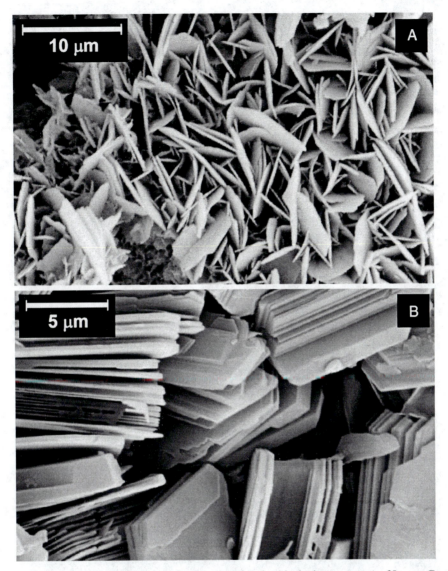

FIGURE 1.13 ESM images of clay minerals. *Top*: Authigenic chlorite in pore space of Lower Cretaceous Viking B sandstone, Alberta, Canada, showing house-of-cards texture. *Bottom*: Booklets and stacks of kaolinite filling pores space of the same sandstone. Chaotic arrangement of clay minerals like these result in high porosity and very low permeability of clay deposits due to prevalence of dead-end minute pores. (Courtesy of Dr. Fred Longstaffe, The University of Western Ontario, Canada; originally published in Schwartz and Longstaffe, 1988. Copyright The Geological Society of America, printed with permission.)

Most laboratory methods for determining effective porosity from the specific yield and specific retention involve a complete saturation of the undisturbed rock specimen with water (fluid), and then removal of the fluid by drainage. Alternatively, the fluid is first removed by forced drainage (e.g., by centrifuge), or by drying, and then completely saturated with the fluid. In either case, the volume of fluid used to fully saturate the sample at a given pressure and the volume of fluid that can be drained are easily measured. Testing of low-permeable materials and hard-rock cores may involve application of vacuum and pressures higher than

FIGURE 1.14 High porosity of residuum (saprolite) deposits comes from abundant clay minerals formed by weathering of the parent rocks. *Top*: residuum on the Piedmont crystalline rocks, North Carolina, showing preserved texture of the parent rock, such as vertical fractures and horizontal bedding planes. *Bottom*: ESM photograph of the Piedmont residuum soil, Rock Springs Farm, Rixeyville, Virginia. Macropores (black areas) in soils like these provide for moderate permeability, while the clay matrix provides for very high overall porosity. Soils in general have the highest total porosity among unconsolidated sediments, excluding organic peat formed in marshes. (ESM courtesy of Dr. Scott Chumbley, Iowa State University.)

the atmospheric pressure, and use of air or gasses such as helium. The pressures and procedures applied affect the degree or size of interconnections actually measured. Drying clay samples at temperatures close to 100°C and then fully saturating them may yield

FIGURE 1.15 Bedding planes, short vertical fractures connecting bedding planes, and fractures intersecting multiple bedding planes in limestone and other sedimentary rocks provide for high interconnectivity of secondary porosity. Limestone and marly limestone beds in the Djerdap National Park, Serbia.

erroneously high values of effective porosity as such temperatures are high enough to remove significant volumes otherwise immobile (hydrated) water stored between clay minerals that could not be removed by drainage alone.

American Society for Testing and Materials (ASTM) standard D2325 can be used to determine effective drainage porosity of medium to coarse grain sediments from water retention and full saturation water (Figure 1.18).

Figure 1.19 and Figure 1.20 are plots of average total porosity and porosity ranges for various rock types, processed from data given in Appendix C as compiled by Wolff (1982). A similar list of actually determined values of effective porosity, including a clear explanation of what was exactly tested, does not exist to the best of author's knowledge. On the other hand, values of specific yield are readily found in literature and, as expected, vary widely due to inevitable heterogeneity of natural aquifers and different field-testing methods. Unfortunately (again) some hydrogeology textbooks offer values for both the effective porosity and the specific yield for clays and some other low-permeable rocks as high as 35% or more, without providing an explanation of the unique method that discovered this amazing natural phenomenon.

Figure 1.21 shows values of average total porosity of uniform unconsolidated sediments (clay through coarse sand) processed from the data included in Appendix C versus consistent specific yield values listed by Johnson (1967). Total porosity values for gravel were not available, as it is usually not feasible to collect undisturbed gravel samples in the field. However, as first demonstrated by Slichter (1899), uniform sand and uniform gravel of the same grain packing (spatial arrangement of grains) have the same theoretical porosity regardless of the grain size. Caution should be exercised when using this graph for site-specific calculations since it shows that the specific yield (effective porosity) has a range of possible values, even for uniform materials. In general, presence of fine-grained sediments such as silt

FIGURE 1.16 Examples of secondary fracture porosity in granite (top) and slate (bottom).

and clay, even in relatively small quantities, can greatly reduce specific yield (effective porosity) of sands and gravels. This is illustrated in Figure 1.22, which can be used to find specific yield for various heterogeneous mixtures of sand, silt, and clay. A very detailed discussion of the specific yield concept, various methods of measurements, and case studies is given by Johnson (1967).

Total porosity of consolidated rocks generally decreases with depth. Matrix porosity decreases mainly due to compaction (Figure 1.23), while a decrease in the secondary fracture porosity is a combination of two main factors: (1) high fracture density near land surface and at shallow depths as the result of over-relaxation due to removal of the overlying solid matter

FIGURE 1.17 Voids of widely varying shapes and sizes, formed by dissolution of carbonate sediments such as limestone, dolomite, and gypsum, are the main characteristics of karst aquifers. Limestones in general have the highest variability of total porosity of all rocks: some old compacted and nonkartified limestones have porosity of less than 1%, while young oolitic karstified limestones may have dissolution-enhanced matrix porosity as high as 65%. (Top photo by George Sowers; printed with kind permission of Francis Sowers.)

by erosion and (2) lower fracture density and smaller fracture aperture at greater depths due to high pressures exerted by the overlying rock mass. In addition to its general decrease with depth, the fracture porosity, even when fractures appear to be frequent and of notable size, is almost always much lower than the surrounding matrix porosity of compact blocks which

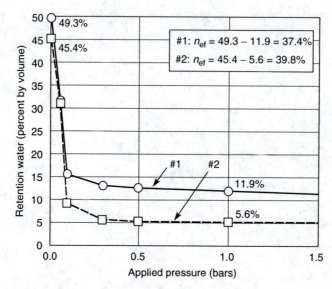

FIGURE 1.18 Typical graph of a water retention test by ASTM D2325, for two clean uniform beach sands. The effective drainage porosity is found for an applied pressure by subtracting the retained percent water (by volume) from the full saturation percent water. Applied pressure of 1 bar corresponds to free gravity drainage. (From Kresic, N., *Hydrologic and Hydrogeologic Study; Costa Serena Project, Loiza*, Law Environmental-Caribe, Santurce, Puerto Rico, 1998.)

make up the bulk of the overall rock volume. The following examples illustrate these two points:

1. Groundwater in the granites and gneisses of Connecticut occurs largely in vertical joints, which have an average spacing between 3 and 7 ft at the surface. At depths of more than 50 ft, the spacing is greater, owing to the dying out of subordinate joints. At still greater depths there appear to be very few water-bearing joints, 250 ft remaining the depth fixed as a limit beyond which is not advisable to drill for water. Of the horizontal joints, almost all are limited to the upper few feet of the rock, generally above the water table. While the joints may be half an inch or more in width at the surface, they rapidly narrow with depth, so that the common width in the upper 200 or 300 ft is 0.01 in. (Ellis, 1909).
2. In a rock cut by three sets of fractures, each set with fractures spaced 5 ft apart, if the average thickness of the void space in each fracture is 0.01 in., the total void space represented by the fractures is only 1/20th of 1% of the total volume of the rock (Meinzer, 1923a).

Continuing advancement of borehole geophysics provides various tools for quantitative analysis of different porosity types, including fracture frequency, orientation, and aperture measurements. Figure 1.24 shows an image of borehole walls obtained with optical televiewer (OTV). Application of logging tools like this greatly increases the utility and efficiency of investigative drilling, since continuous coring is not required, and digital information can be viewed and quantitatively processed at any time.

One feature unique to karst terrains, where secondary and primary porosities of carbonate sediments have been altered due to dissolution by flowing groundwater, is paleokarst. This term usually describes karst features, buried below noncarbonate rocks, which were developed during periods of karstification when the carbonate rocks were exposed at the surface. The same general

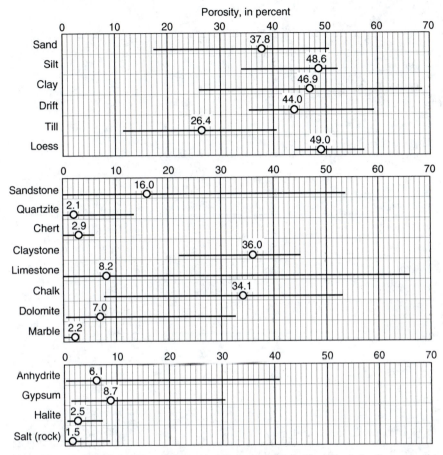

FIGURE 1.19 Porosity range (horizontal bars) and average porosities (circles) of unconsolidated and consolidated sedimentary rocks.

area with carbonate rocks may have been subjected to multiple periods of karstification depending on the depositional and tectonic history. In such cases, it is possible to find very porous zones, together with karst conduits, at greater and varying aquifer depths, including below overlying noncarbonates. These zones often mark position of a paleo water table where karstification intensity was the highest, as illustrated in Figure 1.25 and Figure 1.26.

In general, rocks that have both the matrix and the fracture porosity are referred to as dual-porosity media. This distinction is important in terms of groundwater flow, which has very different characteristics in fractures and conduits compared to the bulk of the rock. Carbonate and sandstone aquifers are the most important consolidated rock aquifers worldwide (see Chapter 2 for examples). However, the nature of various porosity types in these two rock types is notably different as shown in Table 1.1.

1.4 PRINCIPLES OF GROUNDWATER FLOW

Groundwater in the saturated zone is always in motion. It is often a matter of convention, or agreement by interested stakeholders in case of some practical question (project) like how will this motion be qualified or even named. As problems associated with groundwater contamination continue to multiply and, to a great extent, dominate in contemporary

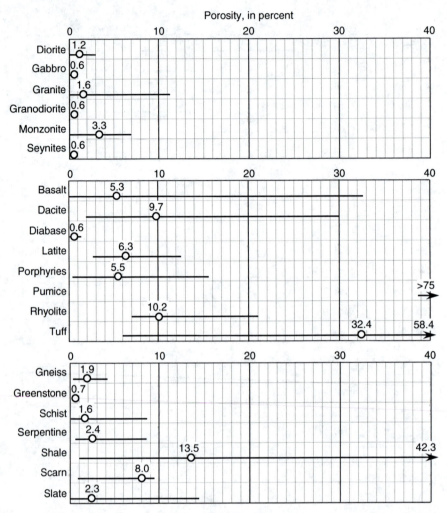

FIGURE 1.20 Porosity range (horizontal bars) and average porosities (circles) of magmatic and metamorphic rocks.

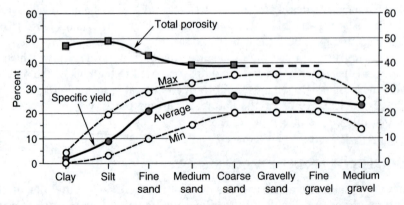

FIGURE 1.21 Total porosity versus specific yield (effective porosity) of unconsolidated sediments. (Porosity values processed from data in Appendix C; specific yield values from Johnson, A.I., U.S. Geological Survey Water-Supply Paper 1662-D, 1967, 74 p.)

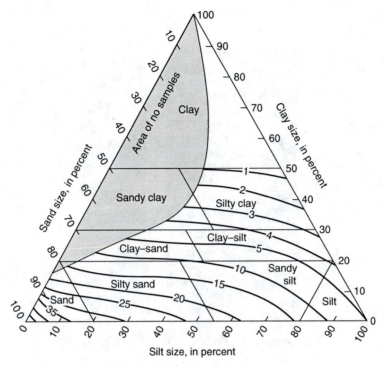

FIGURE 1.22 Soil classification triangle showing relationship between particle size and specific yield. Lines of equal specific yield are at 1% and 5% intervals. Particle size of sand is 2 to 0.0625 mm, of silt is 0.0625 to 0.004 mm, and of clay is <0.004 mm. (From Johnson, A.I., U.S. Geological Survey Water-Supply Paper 1662-D, 1967, 74 p.)

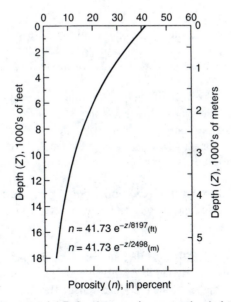

$$n = 41.73 \ e^{-z/8197} \text{(ft)}$$
$$n = 41.73 \ e^{-z/2498} \text{(m)}$$

FIGURE 1.23 Least-square exponential fit for the porosity versus depth data, south Florida carbonates. (Simplified from Schmoker and Halley, 1982. Copyright AAPG 1982; reprinted with permission of the AAPG whose permission is required for further use.)

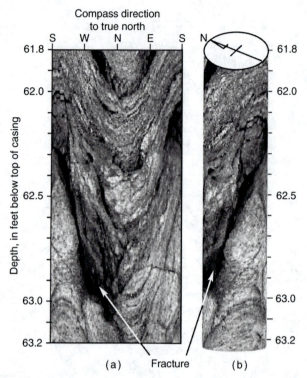

FIGURE 1.24 Optical televiewer (OTV) log of borehole MW109R in the University of Connecticut landfill study area, Storrs, Connecticut. (a) An unrolled 360° scan of the borehole wall. (b) OTV image rolled into a virtual core. (From Johnson, C.D. et al., U.S. Geological Survey Water-Resources Investigations Report 01-4033, Storrs, Connecticut, 2002, 42 p.)

groundwater studies and research, the issue of contaminant fate and transport versus groundwater flow becomes even more important. In short, groundwater flows in a particular three-dimensional direction in the saturated zone because of the two key factors:

1. The effective porosity of rock is of such magnitude that it allows gravity groundwater flow; in other words, the rock is permeable. If the flow of groundwater due to gravity forces is negligible compared to movement of solutes by diffusion, the rock is not permeable and the concepts of groundwater flow described further are not applicable (more on fate and transport of contaminants including diffusion is given in Chapter 6).
2. There is hydraulic gradient in the three-dimensional direction; in other words, there is difference in the hydraulic head of groundwater between various points in that three-dimensional direction.

A simple analogy would be flow of water through pipes or in open channels. The water will move if the hydraulic pressures at two ends of the pipe are different, or if the elevation at one end of the open channel is higher than the elevation at the other end. If there is no such difference in pressures or elevations, the water will not move. And, of course, everyone knows that water cannot flow uphill; in other words, it cannot flow from the point of lower pressure to the point of higher pressure. This pressure is called the hydraulic head in groundwater

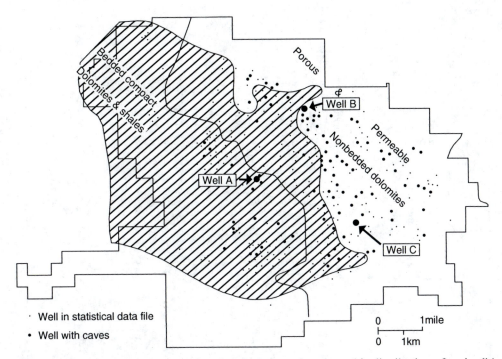

FIGURE 1.25 Wells with caves shown in relation to the paleogeographic distribution of major litho-facies exposed at the unconformity surface of the San Andres Dolomites. Cave is defined as a void detected in wells, which has minimum size (height) of 0.3 m. (From Craig, 1988, p. 346, Figure 16.4. Copyright Springer-Verlag, 1988. Reprinted with kind permission of Springer Science and Business Media.)

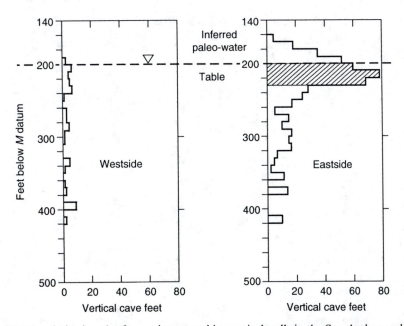

FIGURE 1.26 Cumulative length of caves intersected by vertical wells in the San Andres carbonates as a function of depth below *M* datum, a horizon in the Seven Rivers anhydrite. Cave is defined as a void detected in wells, which has minimum size (height) of 0.3 m. (From Craig, 1988, p. 354, Figure 16.10A. Copyright Springer-Verlag, 1988. Reprinted with kind permission of Springer Science and Business Media.)

TABLE 1.1
Comparison of Porosity in Sandstone and Carbonate Rocks

Aspect	Sandstone	Carbonate
Amount of primary porosity in sediments	Commonly 25%–40%	Commonly 40%–70%
Amount of ultimate porosity in rocks	Commonly half or more of initial porosity; 15%–30% common	Commonly none or only small fraction of initial porosity; 5%–15% common in reservoir facies
Types of primary porosity	Almost exclusively interparticle	Interparticle commonly predominates, but intraparticle and other types are important
Types of ultimate porosity	Almost exclusively primary interparticle	Widely varied because of postdepositional modifications
Size of pores	Diameter and throat sizes closely related to sedimentary particle size and sorting	Diameter and throat size commonly show little relation to sedimentary particle size or sorting
Shape of pores	Strong dependence on particle shape—a negative of particles	Greatly varied, ranges from strongly dependent positive or negative of particles to form completely independent of shapes of depositional or diagenetic components
Uniformity of size, shape, and distribution	Commonly fairly uniform within homogeneous body	Variable, ranging from fairly uniform to extremely heterogeneous, even within body made up of single rock type
Influence of diagenesis	Minor; usually minor reduction of primary porosity by compaction and cementation	Major; can create, obliterate, or completely modify porosity; cementation and solution important
Influence of fracturing	Generally not of major importance in reservoir properties	Of major importance in reservoir properties if present
Visual evaluation of porosity and permeability	Semiquantitative visual estimates commonly relatively easy	Variable; semiquantitative visual estimates range from easy to virtually impossible; instrument measurements of porosity, permeability, and capillary pressure commonly needed
Adequacy of core analysis for reservoir evaluation	Core plugs of 1 in. diameter commonly adequate for matrix porosity	Core plugs commonly inadequate; even whole cores (~3 in. diameter) may be inadequate for large pores
Permeability–porosity interrelations	Relatively consistent; commonly dependent on particle size and sorting	Greatly varied; commonly independent of particle size and sorting

Source: Choquette and Pray, 1970; Copyright AAPG, 1970, reprinted by permission of the AAPG whose permission is required for further use.

studies. However, the term "fluid pressure" has specific meaning in hydrogeology and should not be equated with the hydraulic head.

As groundwater always flows in a three-dimensional space, the hydraulic gradient is always three-dimensional as well. When one or two directions appear dominant, quantitative analyses may be performed assuming one- or two-dimensional flow for the purposes of simplification. When it is important to accurately analyze the entire flow field, which is often the case in contaminant fate and transport studies, a three-dimensional groundwater modeling may be the only feasible quantitative tool (three-dimensional analytical equations are rather complex and often cannot be solved in a closed form).

1.4.1 HYDRAULIC HEAD AND HYDRAULIC GRADIENT

The principle of the hydraulic head and the hydraulic gradient is illustrated in Figure 1.27. At the bottom of monitoring well #1, where the well screen is open to the saturated zone, the total energy (H) or the driving force for water at that point in the aquifer is

$$H = z + h_p + \frac{v^2}{2g} \tag{1.4}$$

where z is elevation above datum (datum is usually mean sea level, but it could be any reference level), h_p is pressure head due to pressure of fluid (groundwater) above that point, v is groundwater velocity, and g is acceleration of gravity.

Since the groundwater velocity in most cases is very low, the third member on the right-hand side may be ignored for practical purposes and Equation 1.4 becomes

$$H = h = z + h_p \tag{1.5}$$

where h is the hydraulic head, sometimes called piezometric head. Pressure head represents pressure of fluid (p) of constant density (ρ) at that point in aquifer:

$$h_p = \frac{p}{\rho g} \tag{1.6}$$

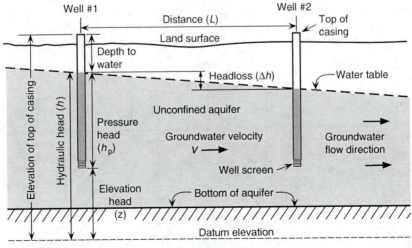

FIGURE 1.27 Schematic presentation of key elements for determining the hydraulic head and the hydraulic gradient in an unconfined aquifer.

In every day's practice, the hydraulic head is determined in monitoring wells or piezometers by subtracting measured depth to water level from the surveyed elevation of the top of casing:

$$h = \text{elevation of top of casing} - \text{depth to water in the well} \qquad (1.7)$$

As the groundwater flows from well #1 to #2 (Figure 1.27), it loses energy due to friction between groundwater particles and the porous media. This loss equates to decrease in the hydraulic head measured at the two wells:

$$\Delta h = h_1 - h_2 \qquad (1.8)$$

The hydraulic gradient (i) between the two wells is obtained when this decrease in the hydraulic head is divided by the distance (L) between the wells:

$$i = \frac{\Delta h}{L} [\text{dimensionless}] \qquad (1.9)$$

It is important to understand that the groundwater flow takes place from the higher hydraulic head towards the lower hydraulic head, and not necessarily from the higher pressure head to the lower pressure head. The latter is true only for strictly horizontal flow over strictly horizontal impermeable base of an unconfined aquifer; conditions that are seldom present. One possible example illustrating this point is shown in Figure 1.28. In conclusion, one only has to think in terms of differences in the hydraulic heads measured in the aquifer, not worrying about possible oddities such as an unusual geometric shape of the aquifer zone, the slope of the impermeable base, or the shape of confining layers. Some of these oddities are illustrated in Figure 1.29.

Except in case of some narrowly focused study of a very limited portion of the saturated (aquifer) zone, there is no such thing as strictly horizontal groundwater flow. Even in unconfined aquifers with a horizontal impermeable base, the flow of groundwater has a vertical component by default. This is illustrated in Figure 1.30 which shows groundwater flow from an area where aquifer recharge is dominant, towards an area where the groundwater discharges from the aquifer, such as to a surface water body or via a spring. In the recharge areas, the inflow of new water from percolating precipitation creates additional pressure head and displacement of the already stored water, which, in turn, displaces water in the discharge area. Although in an unconfined aquifer the recharge may take place everywhere, the discharge is

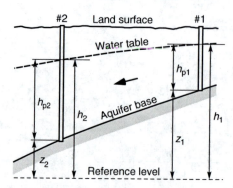

FIGURE 1.28 Groundwater flow is always from the higher hydraulic head towards the lower hydraulic head, not necessarily from the higher pressure head towards the lower pressure head. (Note that the pressure head at well #2 is higher than at well #1.)

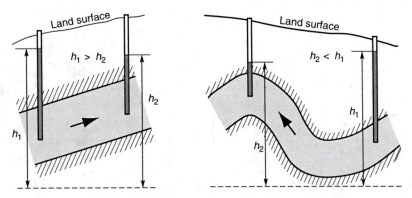

FIGURE 1.29 Some examples showing groundwater flow directions as they relate to hydraulic gradients.

localized by definition, and this displacement of water has to be accommodated with both vertical and horizontal gradients. In many cases, especially in contaminant fate and transport studies, it is critically important to correctly characterize and quantify hydraulic gradients and groundwater flow in all three dimensions. For example, having three or more monitoring wells screened at same depths will do nothing in determining if there is a vertical flow component.

Measuring hydraulic heads and subsequently determining hydraulic gradients and groundwater flow directions is by no means a straightforward task and requires good planning by an experienced hydrogeologist. Ultimately, the number of monitoring wells, their depths, screen lengths, and frequency of water level recordings will be based on the final goal of the study. One common mistake is to apply the same approach of hydraulic head measurements in different types of aquifers. For example, fractured rock and karst aquifers present great challenge even to more experienced professionals, and are easily explained by looking again at photographs in Figure 1.17 and reading the caption. Because one portion of the groundwater flow takes place in fractures or conduits, and the other part within the rock matrix porosity, measurements of the hydraulic heads do not provide a unique answer. Their interpretation

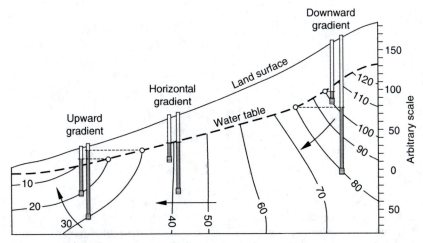

FIGURE 1.30 Movement of groundwater in an unconfined aquifer showing the importance of both vertical and horizontal hydraulic gradients. (Modified from Winter, C. et al., Ground water and surface water: a single resource. *U.S. Geological Survey Circular* 1139, Denver, CO, 1998, 79 p.).

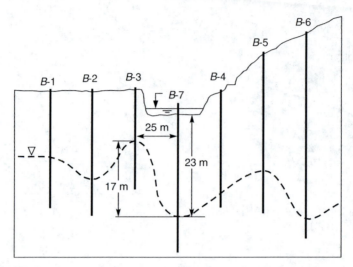

FIGURE 1.31 Hydraulic head measured at a transect of piezometers perpendicular to the Trebisnjica river in eastern Herzegovina, the largest sinking stream in Europe. (From Milanovic, P.T., *Karst Hydrogeology*, Water Resources Publications, Littleton, CO, 1981, 434 p. Copyright Water Resources Publications, printed with permission.)

should always be made in the overall hydrogeologic context. For example, a group of closely spaced wells (say, at a meter to decameter scale) may show a completely random distribution of the measured hydraulic head as illustrated in Figure 1.31 and Figure 1.32. In addition, one well may be completed in a homogeneous rock block, without any significant fractures and with low matrix porosity, and may even exhibit the so-called "glass effect" (no fluctuation of water table regardless of the precipitation–infiltration dynamics). On the other hand, a well 10 m or more away may show the hydraulic head fluctuations of several meters or more. Using only the hydraulic head information for the purposes of assessing representative groundwater flow directions (hydraulic gradients) in these two cases would obviously not be sufficient. For example, by looking at the hydraulic heads measured in piezometers P4, P3, and P2, one

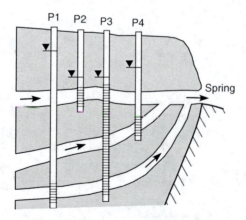

FIGURE 1.32 Example of how closely spaced monitoring wells in clusters may register very different hydraulic heads depending on the depth and length of well screens. Based on actual investigations near a large spring in Dinaric karst. (Modified from Kupusović, T., Measurements of piezometric pressures along deep boreholes in karst area and their assessment. Nas Krs, Vol. XV No. 26–27, 1989, pp. 21–30.)

could erroneously conclude that groundwater flows away from the spring (Figure 1.32). In general, the hydraulic head measurements should be combined with hydrogeologic mapping, possibly dye tracing, and certainly, a thorough understanding of various hydraulic factors such as flow through interconnected fractures and pipes (see Section 1.3.3).

When planning field measurements of the hydraulic head, the following facts should always be taken into consideration:

1. Hydraulic head changes in response to aquifer recharge, both seasonally and, especially in unconfined aquifers, after each recharge episode (rainfall). Measurements in multiple wells should therefore be performed within the shortest time interval feasible (so-called synoptic measurements). To accurately assess seasonal influences on the hydraulic head fluctuations, at least one round of synoptic measurements should be performed per season.
2. Hydraulic head in confined aquifers changes in response to barometric pressure fluctuations; this may also be true for unconfined aquifers in some cases. The only reasonable method to accurately determine the magnitude and importance of such changes is to measure the hydraulic head and the barometric pressure continuously using pressure transducers and data loggers.
3. Hydraulic head in coastal aquifers responds to harmonic tidal fluctuations. These changes can be accurately quantified only by performing continuous measurements.
4. Hydraulic head may change in response to some local hydraulic stress on the aquifer, such as cyclic operation of extraction wells in vicinity.

Any of the above influences must be properly accounted for when interpreting the hydraulic head measurements. This is especially important when performing an aquifer pumping test and subsequently interpreting the data. The data should be corrected by subtracting that portion of the hydraulic head change attributable to each applicable external factor.

1.4.2 PERMEABILITY AND HYDRAULIC CONDUCTIVITY

A permeable or pervious rock, with respect to groundwater, is one having a texture that permits water to move through it freely under the influence of gravity and hydraulic gradients commonly found in aquifers. Such a rock has communicating interstices (voids) of sufficient size to allow free groundwater flow. An impermeable or impervious rock, on the other hand, does not permit movement of water under common hydraulic heads. Such a rock may have subcapillary interstices or isolated interstices of larger size. It may possibly also have small communicating capillary interstices. A rock that is impermeable under common conditions may not be impermeable to water under a hydrostatic pressure in excess of those found in subsurface. It may also allow water to move through it under the influence of other forces, such as molecular attraction (Meinzer, 1923a).

Certain dense clays and shales are practically impermeable to water under the differential hydraulic heads usually found in the subsurface; they will apparently transmit no water and wells ending in them remain dry even though the clays or shales are saturated. An impermeable rock may be devoid of interstices, may contain only isolated interstices, or may have very minute communicating interstices. A clayey silt, with only minute pores, may transmit water very slowly, but a coarse clean gravel or a cavernous limestone with large openings that communicate freely with one another will transmit water very readily (Meinzer, 1923a).

To avoid confusion and controversy as to the definition of impermeability of rocks (still present decades after it was first recognized), Meinzer suggested using the term "hydraulic permeability" only with reference to a specified differential pressure or pressure gradient. In modern terms, and in light of contaminant migration in groundwater, the hydraulically permeable rocks would be those where the advective transport due to hydraulic gradient is dominant, whereas molecular attraction (term often used by Meinzer) would be applicable to low-permeable rocks where migration due to molecular diffusion is dominant (see Chapter 6).

French civil engineer Henry Darcy (1856) was the first to quantitatively analyze flow of water through sands as part of his design of water filters for the city of Dijon. His findings, published in 1856, are the foundation of all modern studies of fluid flow through porous media. At the time of this writing, a complete English translation of the Darcy's original work, as well as the French version, was available at http://biosystems.okstate.edu/darcy/. This Web site includes original experimental data and graphs used by Darcy for "determination of the laws of water flow through sand." Schematic presentation of experimental apparatus similar to the one used by Darcy is shown in Figure 1.33. Such apparatus is still, with various modifications, one of the main pieces of equipment in modern hydrogeologic laboratories. Based on numerous tests performed on siliceous sand, with varying hydraulic conditions, Darcy established the following quantitative relationship, now known as Darcy's law:

$$Q = KA \frac{\Delta h}{l} \ [m^3/s] \tag{1.10}$$

This linear law states that the rate of fluid flow (Q) through a sand sample is directly proportional to the cross-sectional area of flow (A) and the loss of the hydraulic head between

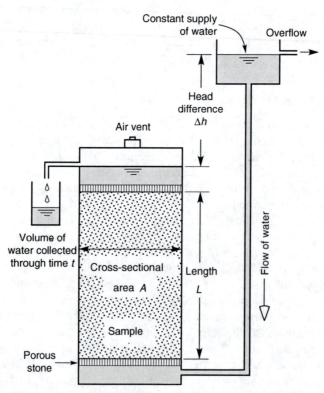

FIGURE 1.33 Schematic presentation of the constant head permeameter used to determine hydraulic conductivity of unconsolidated sediments.

two points of measurements (Δh), and it is inversely proportional to the length of the sample (l) as shown in Figure 1.33. K is the proportionality constant of the law called hydraulic conductivity and has units of velocity. This constant is arguably the most important quantitative parameter characterizing the flow of groundwater. It is not uncommon to find the minus sign in front of Equation 1.10 in the literature with the following explanation: it indicates that the flow is in the direction of decrease in the hydraulic head (i.e., from the higher toward the lower hydraulic heads). However, although the explanation is correct, the minus sign should not be placed in front of Equation 1.10, which is the solution of a differential form of Darcy's equation. The minus sign, which is correctly presented in the differential equation, disappears during its integration and it is mathematically incorrect to show it as a part of Equation 1.10.

The general hydraulic equation of the continuity of flow, which results from the principle of conservation of mass, is (for incompressible fluids)

$$Q = v_1 A_1 = v_2 A_2 = \text{constant} \tag{1.11}$$

which means that the volumetric flow rate (Q) through the cross-sectional areas A_1 and A_2 in a stream of fluid is the same (there is no loss or gain of water). Average flow velocities at cross sections 1 and 2 are v_1 and v_2, respectively. From Equation 1.11, the velocity of flow can be generally expressed as

$$v = \frac{Q}{A} \tag{1.12}$$

Relating Equation 1.10 and Equation 1.12 gives another form of Darcy's equation:

$$v = K \frac{\Delta h}{l} \tag{1.13}$$

Expressed in words, Equation 1.13 states that the velocity of fluid flow is proportional to the hydraulic gradient (i), which brings another common form of Darcy's equation:

$$v = K \times i \quad [\text{m/s}] \tag{1.14}$$

Equation 1.14 is used to experimentally determine the hydraulic conductivity of a sample by measuring the flow velocity through the apparatus for various hydraulic gradients and finding the linear relationship between different velocities and different hydraulic gradients (see Chapter 9 for examples).

Another quantitative parameter widely used in the studies of fluid flow though porous media is called intrinsic permeability, or simply permeability. It is defined as the ease with which a fluid can flow through a porous medium. In other words, permeability characterizes the ability of a porous medium to transmit a fluid (water, oil, gas, etc.). It is dependent only on the physical properties of the porous medium: grain size, grain shape and arrangement, or pore size and interconnections in general. On the other hand, the hydraulic conductivity is dependent on properties of both the porous medium and the fluid. The relationship between the permeability (K_i) and the hydraulic conductivity (K) is expressed through the following formula:

$$K_i = K \frac{\mu}{\rho g} \quad [\text{m}^2] \tag{1.15}$$

where μ is the absolute viscosity of the fluid (also called dynamic viscosity or simply viscosity), ρ is the density of the fluid, and g is the acceleration of gravity. The viscosity and the density of the fluid are related through the property called kinematic viscosity (v):

$$v = \frac{\mu}{\rho} \quad [\text{m}^2/\text{s}] \tag{1.16}$$

Inserting the kinematic viscosity into Equation 1.15 somewhat simplifies the calculation of the permeability since only one value (that of v) has to be obtained from tables or graphs such as the one shown in Figure 1.34 (note that for most practical purposes, the value of the acceleration of gravity (g) is 9.81 m/s^2, and is often rounded to 10 m/s^2):

$$K_i = K \frac{v}{g} \quad [\text{m}^2] \tag{1.17}$$

Although it is better to express the permeability in the units of area (m^2 or cm^2) for reasons of consistency and easier use in other formulas, it is more commonly given in darcys (which is a tribute to Darcy by the oil industry and Wyckoff et al. (1934), who proposed the name for the unit in 1933):

$$1 \text{darcy (D)} = 9.87 \times 10^{-9} \text{cm}^2 = 9.87 \times 10^{-13} \text{m}^2$$

When laboratory results of permeability measurements are reported in darcys (or square meters), the following two equations (derived from Equation 1.15 and Equation 1.17) can be used to find the hydraulic conductivity:

$$K = K_i \frac{g}{v} \quad \text{or} \quad K = K_i \frac{\rho g}{\mu} \quad [\text{m}/\text{s}] \tag{1.18}$$

Figure 1.34 shows kinematic viscosity (v), density (ρ), and dynamic viscosity (μ) of water for different temperatures. As can be seen, the temperature influences all three parameters and, consequently, the hydraulic conductivity is strongly dependent on groundwater temperature. Kinematic viscosity of water at temperature of 20°C is approximately 1×10^{-6} m^2/s, and gravity acceleration is often rounded to 10 m/s^2, which is the reason why the following approximate relation is used to convert permeability (given in m^2) to hydraulic conductivity (given in m/s):

$$K \quad [\text{m}/\text{s}] = K_i \quad [\text{m}^2] \times 10^7 \tag{1.19}$$

For example, permeability of 1×10^{-11} m^2 gives the following value of the hydraulic conductivity for water at 20°C:

$$K \quad [\text{m}/\text{s}] = 1 \times 10^{-11} \quad [\text{m}^2] \times 10^7 = 0.0001 \text{m}/\text{s} \quad \text{or} \quad 8.6 \text{ m}/\text{d}$$

Since the effective porosity, as the main factor influencing the permeability of a porous medium, varies widely for both different and same rock types, the hydraulic conductivity and permeability also have wide ranges as shown in Figure 1.35. Except in rare cases of uniform and nonstratified unconsolidated sediments, both parameters also vary in different directions within the same rock mass due to its anisotropy and heterogeneity (see Section 1.3.4). Most practitioners tend to simplify these inherent characteristics of the porous media by dividing

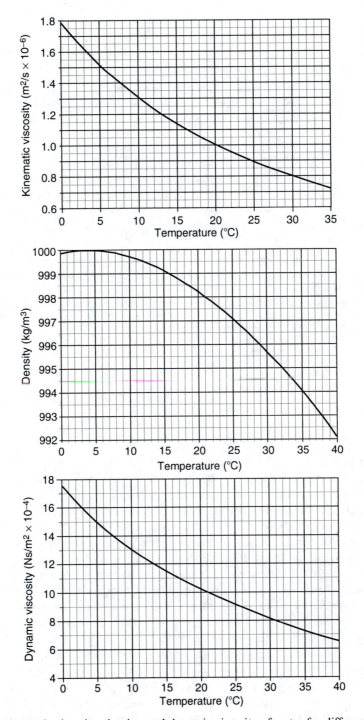

FIGURE 1.34 Kinematic viscosity, density, and dynamic viscosity of water for different temperatures. (Density and kinematic viscosity data from Maidment, D.R., Ed., *Handbook of Hydrology*, McGraw-Hill, New York (various paging), 1993; dynamic viscosity data from Giles, R.V., Evett, J.B., and Chiu, L., *Fluid Mechanics and Hydraulics*, 3rd ed., Schaum's Outline Series, McGraw-Hill, New York, 1994, 378 p.)

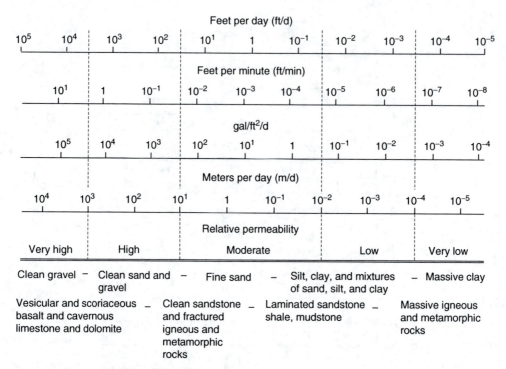

FIGURE 1.35 Range of hydraulic conductivites for different rock types. (From USBR, *Ground Water Manual*, U.S. Department of the Interior, Bureau of Reclamation, Washington, D.C., 1977, 480 p.)

the complex three-dimensional hydraulic conductivity tensor into just two main components: horizontal and vertical hydraulic conductivity. Furthermore, it seems common practice to apply some "rules of thumb" indiscriminately, such as vertical conductivity is ten times lower than the horizontal conductivity, without trying to better characterize the underlying hydrogeology. This difference in the two hydraulic conductivities can vary many orders of magnitude in highly anisotropic rocks and, in many cases, it may be completely inappropriate to apply the concept altogether: a highly transmissive fracture or a karst conduit may have any shape and spatial extent, at any depth.

As mentioned repeatedly, clear relationships exist between porosity and permeability, such that both generally decrease with depth in consolidated rock, or that fine-grained sediments (silt and clay) have lower permeability than coarse-grained sediments (sand and gravel). However, in nature the conditions are rarely met to permit an accurate quantitative prediction of permeability based on porosity data. Very extensive field and laboratory measurements of this relationship have been routinely performed in the oil industry for decades, with plenty of available related literature. Two primary targets have been carbonate and sandstone oil and gas reservoirs worldwide. For example, Nelson and Kibler (2003) have compiled porosity and permeability measurements on cored samples from siliciclastic formations (conglomerates, sandstones, siltstones, and shales) at 70 different locations throughout the world. Figure 1.36 and Figure 1.37 are examples from this source. The unconsolidated sand data are included for comparison. Three sands shown in general have much higher porosity and smaller range of permeability than sandstones in this example. At the same time, sands do not show any linear relationship between the two parameters. On the other hand, sandstones of all grain sizes clearly show a linear relationship between porosity and permeability.

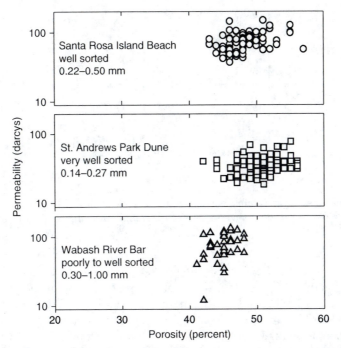

FIGURE 1.36 Porosity versus permeability for three freshly deposited unconsolidated sand deposits. (Plotted by Nelson, P.H. and Kibler, J.E., U.S. Geological Survey Open-file Report 03-420, Denver, Colorado, 2003; data from Pryor, W.A., *Am. Assoc. Petrol. Geol. Bull.*, 57 (1), 162, 1973.) Permeabilities of the undisturbed samples were measured in a commercially available constant head—falling head permeameter adapted to the dimensions of the thin-wall aluminum sample tubes. Water permeabilities, in Darcy units, were measured at relatively low head pressures (50 cm of water) to prevent disturbance of the unconsolidated sand samples by elutriation and repacking. Porosity was measured by a modification of Ludwick's volumeter and basing the calculations on the core volumes and grain volumes of the samples.

As is the case with porosity, limestones have the widest range of hydraulic conductivity of all rocks. The following examples illustrate the variability of hydraulic conductivity of carbonate rocks. Chalk and some limestones may have high porosity, but since the pores are small (usually less than 10 micrometers), primary permeability is low and specific retention is high (Cook, 2003). For example, the mean interconnected porosity of the Lincolnshire

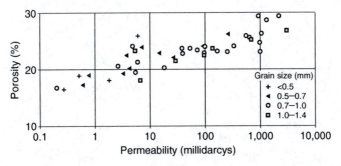

FIGURE 1.37 Porosity versus permeability of Oligocene and Miocene sandstones at Yacheng Field, South China Sea. (Plotted by Nelson, P.H. and Kibler, J.E., U.S. Geological Survey Open-file Report 03-420, Denver, Colorado, 2003; data from Bloch, S., *Am. Assoc. Petrol. Geol. Bull.*, 75 (7), 1145, 1991.)

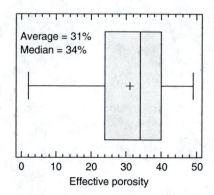

FIGURE 1.38 Box–Whisker plot of effective matrix porosity of the Upper Floridan carbonates in west-central Florida. The plot is based on analysis of 46 core samples from 10 different locations. (Raw data from Knochenmus, L.A. and Robinson, J.L., U.S. Geological Survey Water-Supply Paper 2475, Washington, D.C., 1996, 47 p.)

Limestone in England is 15%, while the mean matrix hydraulic conductivity is only 10^{-9} m/s (Cook, 2003, after Greswell et al., 1998). The groundwater flow is largely restricted to the fractures. The aquifer hydraulic conductivity determined from pumping tests ranges between approximately 20 and 100 m/d, which is more than five orders of magnitude greater than the matrix hydraulic conductivity (Cook, 2003, after Bishop and Lloyd, 1990). The San Antonio segment of the Edwards aquifer in Texas, the United States, consists of Cretaceous limestones and dolomites that have undergone multiple periods of karstification. The average aquifer hydraulic conductivity, based on over 900 well pumping tests, is approximately 7 m/d, while the mean matrix hydraulic conductivity is approximately 10^{-3} m/d (Cook, 2003, after Halihan et al., 2000). Based on the results of 191 aquifer pumping tests, the median horizontal hydraulic conductivity of the Upper Floridan aquifer in Georgia, the United States, is 140.3 ft/d. Such high average value is, in part, the result of generally high effective matrix porosity of the Upper Floridan aquifer Tertiary carbonates, as illustrated with the data from west-central Florida and

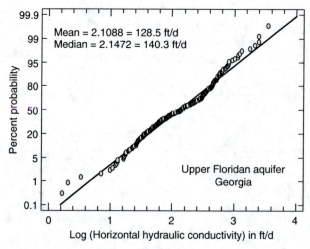

FIGURE 1.39 Distribution of horizontal hydraulic conductivity of the Upper Floridan aquifer in Georgia, the United States, based on results of 191 aquifer pumping tests. (Raw data from Clarke, J.S. et al., Georgia Geologic Survey Information Circular 109, Atlanta, Georgia, 2004, 50 p.)

Georgia (Figure 1.38 and Figure 1.39). Budd and Vacher (2004) provide comparison of matrix, fracture, and channel permeability for 12 different karst aquifers in northern America, Australia, England, and Germany, and discuss the importance of high matrix permeability for groundwater flow in the Floridan aquifer in the United States.

Vesicular basalts can also have very high hydraulic conductivity, but they are less permeable than limestone on average. Medium to coarse sand and gravel are rock types with the highest average hydraulic conductivities. Pure clays and fresh igneous rocks generally have the lowest permeability, although some field-scale bedded salt bodies were determined to have permeability of zero (Wolff, 1982). This is one of the reasons why salt domes are considered as potential depositories of high-radioactivity nuclear wastes in various countries.

It has been shown that individually determined hydraulic conductivities from various locations within the same aquifer, or within the same geologic unit, approximately follow an exponential probability distribution as shown with examples in Figure 1.39 and Figure 1.40. This observation is potentially useful when there is a need for an average value from multiple data. It is therefore advisable to use the geometric mean of individual measurements rather than the arithmetic mean.

Meinzer (1932) gives the following advice on determining permeability (or hydraulic conductivity in modern terms):

Efforts have been made by Hazen, King, Slichter, and others to compute the permeability of water-bearing material from its mechanical composition and porosity. For this purpose Hazen in his work on filter sands in 1889 to 1893, used the term 'effective size of grain' or size of grain that would give the actual permeability of a more or less heterogeneous material. He found that in the materials with which he was dealing the effective size was best shown by the '10 per cent size'–that is, the size that is not exceeded by the grains in 10 per cent of the material by weight. These indirect methods of computing permeability are useful for some purposes but have not always given consistent results and are not in general to be recommended. Direct tests of permeability require no more work and are generally more satisfactory.... One of the most promising methods of determining permeability is a field method proposed by G. Thiem, son of the German hydrologist A. Thiem, based on the performance of wells that enter the water-bearing formation ... This method has the advantage over the laboratory methods in that it deals with all the water-bearing materials in the vicinity of a well, undisturbed and in place.

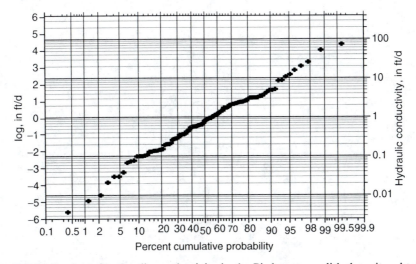

FIGURE 1.40 Distribution of hydraulic conductivity in the Piedmont regolith deposits, obtained from 105 slug tests at multiple sites in the greater Atlanta area and northern Georgia, the United States.

The determination of the hydraulic conductivity using laboratory tests and actual data are given in Chapter 9. For detailed explanations on the use of aquifer pumping test results in the determination of the hydraulic conductivity, as well as aquifer transmissivity and storage properties, see Section 2.2 for the theoretical basis, and Chapter 13 and Chapter 14 for solved problems.

1.4.3 GROUNDWATER VELOCITY AND FLOW

The velocity of groundwater flow (v), as defined by Darcy's law, is the product of the hydraulic conductivity (K) and the hydraulic gradient (i):

$$v = K \times i \tag{1.20}$$

However, this velocity, called Darcy's velocity, is not the real velocity at which water particles move through the porous medium. Darcy's law, first derived experimentally, assumes that the groundwater flow occurs through the entire cross-sectional area of a sample (porous medium) including both voids and grains (adequately, Darcy's velocity is called "smeared velocity" in Russian literature). As illustrated in Figure 1.41, the volumetric flow rate in Darcy's law (Q) is the product of Darcy's velocity and the total cross-sectional area of flow (A):

$$Q = v \times A \tag{1.21}$$

Since the flow takes place only through voids, the actual cross-sectional area of flow is the sum of the individual cross-sectional areas of voids:

$$A_v = A_1 + A_2 + A_3 \tag{1.22}$$

The total volumetric flow is the sum of the individual flows:

$$Q = Q_1 + Q_2 + Q_3 \tag{1.23}$$

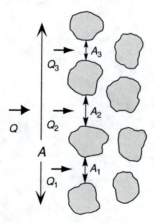

FIGURE 1.41 Schematic presentation of volumetric flow rates and the cross-sectional areas of groundwater flow through intergrannular porous media.

In order to preserve the principle of the continuity of flow (there is no gain or loss of water), a new equivalent velocity that describes the flow through the cross-sectional area of voids is introduced:

$$v \times A = v_L \times A_v \qquad (1.24)$$

where v_L is this new flow velocity often called the linear velocity. Since the actual cross-sectional area of flow is smaller than the total area ($A_v < A$), it follows that the linear velocity must be greater than Darcy's velocity: $v_L > v$. Equation 1.24 can be written as

$$v = v_L \times \frac{A_v}{A} \qquad (1.25)$$

The fraction at the right-hand side can be multiplied by the unit aquifer (sample) width to obtain volumes, which gives the following relationship:

$$v = v_L \times \frac{V_v}{V} \qquad (1.26)$$

where V_v is the volume of the interconnected voids that allow gravity groundwater flow and V is the total aquifer (sample) volume.

The ratio of these two volumes defines the effective porosity of a porous medium:

$$n_{ef} = \frac{V_v}{V} \qquad (1.27)$$

Combining Equation 1.26 and Equation 1.27 gives the following expression for the linear flow velocity:

$$v_L = \frac{v}{n_{ef}} \qquad (1.28)$$

or when inserting Equation 1.20 into Equation 1.28:

$$v_L = \frac{K \times i}{n_{ef}} \qquad (1.29)$$

The linear groundwater velocity is accurate when used to estimate the average travel time of groundwater, and Darcy's velocity is accurate for calculating flow rates. Neither, however, is the real groundwater velocity, which is the time of travel of a water particle along its actual path through the voids. It is obvious that, for practical purposes, the real velocity cannot be measured or calculated.

Two main forces act upon individual water particles that move through the porous medium: friction between the moving water particles and friction between the water particles and the solids surrounding the voids. This results in uneven velocities of individual water particles; some travel faster and some slower than the overall average velocity of a group of particles (Figure 1.42). This phenomenon is called mechanical dispersion and it is very important when quantifying transport of contaminants dissolved in groundwater (more on fate and transport of contaminants is presented in Chapter 6). Because of mechanical dispersion, spreading of individual water (or dissolved contaminant) particles is in all three

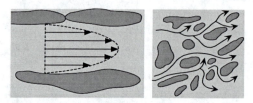

FIGURE 1.42 Schematic presentation of mechanical dispersion caused by varying velocity of water particles and tortuous flow paths between the porous medium grains. (From Franke, O.L. et al., U.S. Geological Survey Open-File Report 90-183, Reston, Virginia, 1990, 184 p.)

main directions with respect to the overall groundwater flow direction: longitudinal, transverse, and vertical. A very illustrative tool for demonstrating the effects varying hydraulic conductivity, porosity, and hydraulic gradients have on velocity, flow directions, and dispersion of fluid particles is available as a public domain computer program called *Particleflow* by Hsieh of the USGS (2001).

Calculation of groundwater flow rate (Q) in any aquifer setting, i.e., for both confined and unconfined aquifers of any cross-sectional area (A), and for the velocity vector in any direction (v), always starts with the basic equation of the continuity of groundwater flow:

$$Q = v \times A \qquad (1.30)$$

This simple principle is illustrated in Figure 1.43, which shows groundwater flow in a single flow tube of an aquifer. There is no gain or loss of water in the flow tube between two cross-sectional areas (A_1 and A_2), which can be of different sizes. Three-dimensional surfaces of the same hydraulic head within the flow tube are called equipotential surfaces. Their horizontal projections, often shown on hydrogeologic maps of aquifers, are called equipotential lines. Analytical equations describing groundwater flow in a truly three-dimensional setting, like the one shown in Figure 1.43, which would include all possible heterogeneities and anisotropies, are very complex and do not have closed solutions. It is for this reason that numeric models of groundwater flow are increasingly utilized given the complex nature of most groundwater studies today. However, approximate analytical solutions based on various assumptions (simplifications) are still widely used since they often provide satisfactory results for screening-level analyses, or final results in case of simple field conditions. The simplest and most common assumptions are (1) the hydraulic conductivity is uniform (porous medium is homogeneous and isotropic, and has one effective porosity), (2) the hydraulic gradient is uniform, (3) the saturated thickness of the aquifer is constant, (4) there is no gain or loss

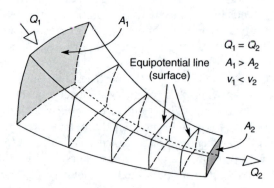

FIGURE 1.43 Groundwater flow tube in an aquifer, showing the principle of flow conservation.

of water, (5) the flow is horizontal, over a horizontal impermeable base, and (6) the flow is in steady state, i.e., it does not change in time.

1.4.3.1 Steady-State Flow in Confined Aquifer

Steady-state flow in a homogeneous, isotropic, confined aquifer of constant thickness, which is resting on a horizontal base, can be calculated for the unit width of the aquifer, or for some other width of interest, by directly applying Darcy's equation in its simplest form (same as Equation 1.10):

$$Q = KA\frac{h_1 - h_2}{l} \quad [\mathrm{m^3/s}] \tag{1.31}$$

This is illustrated in Figure 1.44 where the cross-sectional area of flow is either $A = 1 \times b$ for the unit flow (q), or $A = 800 \times b$, for the so-called volumetric flow (Q) through an aquifer segment that is 800 units wide; b is the aquifer thickness which stays constant between the two known hydraulic heads h_1 and h_2. The exact same equation can be obtained by applying simple differential and integral calculus, which is a more general mathematical approach to the same problem. For the flow between two known hydraulic heads, the mathematical reasoning is as follows:

1. The hydraulic gradient is an infinitesimally small change of hydraulic head ($\mathrm{d}h$) along an infinitesimally small length of the flow path ($\mathrm{d}l$):

$$i = \frac{\mathrm{d}h}{\mathrm{d}l} \tag{1.32}$$

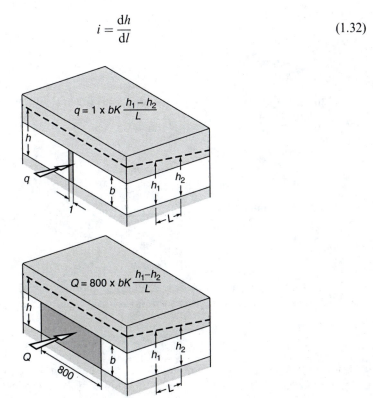

FIGURE 1.44 Schematic presentation of the unit flow rate (top) and the volumetric flow rate (bottom) in a confined aquifer.

2. In the case of one-dimensional groundwater flow in an isotropic, homogeneous, confined aquifer, the flow rate per unit width (q) is

$$q = -bK\frac{dh}{dl} \tag{1.33}$$

where the minus sign indicates that the flow is taking place from a higher hydraulic head towards a lower hydraulic head (the head is decreasing).

3. The linear differential Equation 1.33 can be written in the following form:

$$dh = -\frac{q}{bK}dl \tag{1.34}$$

4. This equation can be integrated with the following boundary conditions when the coordinate beginning for length in the horizontal direction is placed at h_1:

$$\begin{aligned}
\text{for} \quad l = 0, \quad h = h_1 \\
\text{for} \quad l = L, \quad h = h_2
\end{aligned} \tag{1.35}$$

which gives

$$\int_{h_1}^{h_2} dh = -\frac{q}{bK}\int_0^L dl \tag{1.36}$$

5. Solution of Equation 1.36 is (see Appendix B for basic differential and integral calculus if needed)

$$h_2 - h_1 = -\frac{q}{bK}L \tag{1.37}$$

or after rearrangement:

$$q = bK\frac{h_1 - h_2}{bK} \ [\text{m}^2/\text{s}] \tag{1.38}$$

Again, the result of this unit flow calculation can be multiplied by any desired width of the aquifer segment of interest to find the total (volumetric) flow through the aquifer:

$$Q = q \times \text{segment width} \tag{1.39}$$

Application of the general mathematical approach outlined above is illustrated with numeric examples in Chapter 10, including some more complicated situations where the aquifer thickness changes, the hydraulic conductivity is not uniform, and there are two sloping confined aquifers that merge into one.

1.4.3.2 Steady-State Flow in Unconfined Aquifer

Steady-state groundwater flow in an unconfined aquifer is in general somewhat more complex than in a confined aquifer for the following two reasons: (1) the aquifer is likely receiving water from areal recharge such as percolation of precipitation and (2) the saturated thickness changes by definition as the water table changes. Equation for these general conditions can be derived by analyzing an elementary aquifer volume (prism) shown in Figure 1.45. This volume has unit width, its length is dx, and its height is h. Any positive change in groundwater flow through this elementary volume equals a gain of water across the water table due to the infiltration of precipitation (for steady-state conditions). Similarly, any decrease of groundwater flow through the elementary prism is a consequence of water loss across the water table due to, for example, evapotranspiration. Assuming an increase of flow due to infiltration, denoted as w (infiltration rate has dimension of velocity), the following equation of continuity applies:

$$dq = wdx \qquad (1.40)$$

By applying Darcy's law and assuming that the velocity vector is horizontal (Dupuit's hypothesis), groundwater flow through an elementary aquifer prism of unit width is

$$q = -Kh\frac{dh}{dx} \qquad (1.41)$$

Differentiating Equation 1.41 and combining it with Equation 1.40 gives the following expression:

$$\frac{d^2(h^2)}{dx^2} = -\frac{2w}{K} \qquad (1.42)$$

Differential Equation 1.42 can be integrated to give the following general solution with the constants of integration C_1 and C_2:

$$h^2 = -\frac{w}{K}x^2 + C_1x + C_2 \qquad (1.43)$$

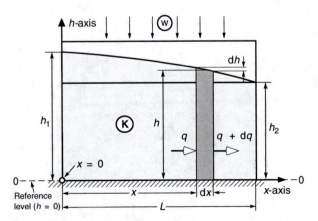

FIGURE 1.45 Calculation scheme for derivation of the equation for one-dimensional groundwater flow in an unconfined aquifer under the influence of infiltration.

The constants of integration, i.e., the particular solution, can be found if the boundary conditions are known. From Figure 1.45, the boundary conditions for the flow between the hydraulic heads h_1 and h_2 are

$$
\begin{aligned}
\text{for} \quad l &= 0, \quad h = h_1 \\
\text{for} \quad x &= L, \quad h = h_2
\end{aligned}
\tag{1.44}
$$

With these boundary conditions, Equation 1.43 becomes

$$
h^2 = h_1^2 - (h_1^2 - h_2^2)\frac{x}{L} + \frac{w}{K}(L - x)x
\tag{1.45}
$$

Equation 1.45 enables calculation of the hydraulic head (water table elevation) for any given position between two points with known saturated thickness. It should be noted that w denotes either gain or loss of water, and will have the corresponding sign: positive for gaining water (infiltration) and negative for losing water (evapotranspiration). If there is no gain or loss of water, Equation 1.45 reduces to

$$
h^2 = h_1^2 - (h_1^2 - h_2^2)\frac{x}{L}
\tag{1.46}
$$

Groundwater flow per unit width at any distance x from the coordinate beginning (i.e., from the origin, where $x = 0$) is

$$
q = -Kh\frac{dh}{dx}
\tag{1.47}
$$

while differentiation of Equation 1.45 leads to

$$
h\frac{dh}{dx} = -\frac{h_1^2 - h_2^2}{2L} - \frac{w}{K}(x - \frac{L}{2})
\tag{1.48}
$$

Combining Equation 1.47 and Equation 1.48, gives the general equation for the unit groundwater flow in an unconfined aquifer under the influence of precipitation, at any distance x from the coordinate beginning:

$$
q = K\frac{h_1^2 - h_2^2}{2L} + w\left(x - \frac{L}{2}\right)
\tag{1.49}
$$

At the location of h_1, i.e., at the coordinate beginning, where $x = 0$, the flow is

$$
q_0 = K\frac{h_1^2 - h_2^2}{2L} - w\frac{L}{2}
\tag{1.50}
$$

In the absence of any gain or loss of water (no infiltration and evapotranspiration), Equation 1.50 becomes

$$
q_0 = K\frac{h_1^2 - h_2^2}{2L}
\tag{1.51}
$$

Equation 1.51 can also be applied to a special case in which the slope of the aquifer impermeable base and the slope of the water table are the same, i.e., when the flow lines are parallel and the saturated thickness does not change (case a in Figure 1.46). This, however, is far less common than the other four cases shown in the same figure. Case a is called uniform flow, and the rest are called nonuniform flow. An important element to be considered in deriving the equation of flow for both uniform and nonuniform conditions is the normal aquifer depth, denoted h_o in Figure 1.46a. It is the depth perpendicular to both the hydraulic headline and the aquifer baseline, which are parallel in uniform flow conditions. Having in mind the principle of the conservation of mass, the flow rate in uniform conditions is the product of the velocity of flow and the cross-sectional area of flow:

$$Q = A \times v \qquad (1.52)$$

The unit flow, or the flow rate per unit width of the aquifer, is a product of the aquifer thickness (i.e., the normal depth h_o) and the flow velocity:

$$q = h_o \times v \qquad (1.53)$$

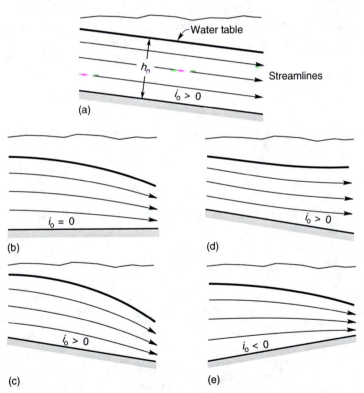

FIGURE 1.46 Possible cases of steady-state flow in an unconfined aquifer. (a) Uniform flow; flow lines are parallel and the hydraulic gradient equals the gradient of aquifer base. The normal depth is denoted with h_o. (b) Nonuniform flow; the aquifer thickness decreases downgradient, and the aquifer base is horizontal. (c) Nonuniform flow; the aquifer thickness decreases downgradient, and the gradient of aquifer base is greater than zero. (d) Nonuniform flow; the aquifer thickness increases downgradient, and the gradient of aquifer base is greater than zero. (e) Nonuniform flow; the aquifer thickness decreases downgradient, and the gradient of aquifer base is smaller than zero.

The flow velocity is in a discrete, nondifferential form given as

$$v = K \times i_o \tag{1.54}$$

where i_o is the hydraulic gradient equal to the aquifer base gradient. The unit flow is then

$$q = h_o \times K \times i_o \tag{1.55}$$

As already mentioned, the uniform flow is rare in natural conditions and a more general and complex equation has to be derived to describe nonuniform flow in which aquifer thickness changes. Dupuit's hypothesis allows derivation of equation that gives an accurate calculation of the flow rate, but is less accurate for calculating the position of the hydraulic head, especially near the points of aquifer discharge. Figure 1.47 illustrates the hypothesis which states:

- Equipotential lines are vertical.
- Velocity is constant along the vertical, i.e., along the equipotential line.
- The velocity vector has only the horizontal component.

The elements for deriving the general differential equation of nonuniform flow are shown in Figure 1.48. Note that the hydraulic head (h) is the sum of the pressure head (h_p) and the elevation head (z). The equation is derived for an infinitesimally small decrease of the hydraulic head (dh) along an infinitesimally small distance (dx). The difference in hydraulic heads between verticals 2 and 1 is

$$dh = dz + dh_p \tag{1.56}$$

The slope of the impermeable base is

$$i_o = -\frac{dz}{dx} \tag{1.57}$$

Note that the minus sign appears because i_o decreases in the direction of flow. From Equation 1.57, it follows that

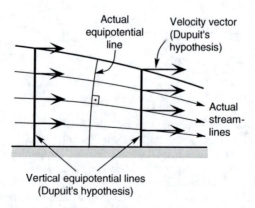

FIGURE 1.47 Schematic presentation of Dupuit's hypothesis.

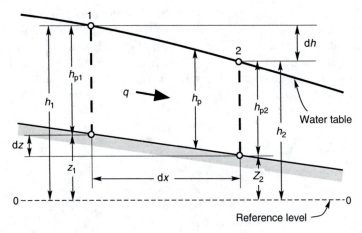

FIGURE 1.48 Elements for deriving general differential equation of nonuniform flow in an unconfined aquifer with a sloping base.

$$dz = -i_o \, dx \tag{1.58}$$

Inserting Equation 1.58 into Equation 1.56 gives

$$dh = -i_o \, dx + dh_p \tag{1.59}$$

Equation 1.59 can be divided by dx and multiplied by the hydraulic conductivity (K):

$$K\frac{dh}{dx} = K\left(\frac{dh_p}{dx} - i_o\right) \tag{1.60}$$

If Equation 1.60 is multiplied by (-1), the left-hand side is equal to Darcy's velocity, which gives

$$v = -K\left(\frac{dh_p}{dx} - i_o\right) \tag{1.61}$$

Finally, the general differential equation of unit groundwater flow in an unconfined aquifer is given as the product of the flow velocity and the saturated thickness:

$$q = -h_p \times K\left(\frac{dh_p}{dx} - i_o\right) \tag{1.62}$$

1.4.3.2.1 Solution for the Horizontal Aquifer Base

If the aquifer base is horizontal, then $i_o = 0$ and Equation 1.62 becomes

$$q = -h_p \times K\frac{dh_p}{dx} \tag{1.63}$$

To solve this equation, the variables x and h_p first have to be separated:

$$\frac{q}{K}dx = -h_p \times dh_p \tag{1.64}$$

and then integrated for known boundary conditions. Since the aquifer base is horizontal ($z =$ constant), the pressure head h_p can be substituted with the hydraulic head h for simplification when the reference level is placed at the aquifer base. Figure 1.49 shows a general case with the following boundary conditions: when $x = 0$ the hydraulic head is h_1, and when $x = L$ the hydraulic head is h_2. For these boundary condition, Equation 1.64 is

$$\frac{q}{K}\int_0^L dx = -\int_{h_1}^{h_2} h\,dh \tag{1.65}$$

and its solution is

$$q = K\frac{h_1^2 - h_2^2}{2L} \tag{1.66}$$

Equation 1.66 is valid for conditions where there are no gains of water due to aquifer recharge by precipitation or seepage through underlying aquitards, and where there are no losses such as evapotranspiration. The equation that includes the influence of external (positive or negative) fluxes is

$$q = K\frac{h_1^2 - h_2^2}{2L} \pm w\frac{L}{2} \tag{1.67}$$

where w is the vertical gain or loss of water with the velocity dimension. As can be seen, Equation 1.67 reduces to Equation 1.66 for $w = 0$. These two equations are identical to Equation 1.50 and Equation 1.51, respectively, which were derived in a slightly different way. If two hydraulic heads are known, the position of the hydraulic head at any distance x from the origin (see Figure 1.49) is found using the following equation:

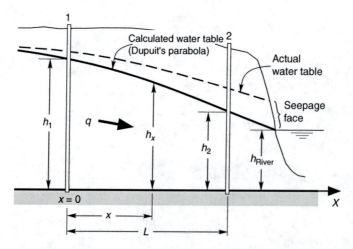

FIGURE 1.49 Boundary conditions for solving the general differential equation of nonuniform flow in an unconfined aquifer with the horizontal base. The position of the hydraulic head h_x can be found if the heads h_1 and h_2 are known. Note that the calculated water table is below the real water table because of Dupuit's assumption. The difference between the two is the biggest in the aquifer discharge zone where a seepage surface (face) commonly forms.

$$h_x = \sqrt{h_1^2 - (h_1^2 - h_2^2)\frac{x}{L}} \tag{1.68}$$

As mentioned earlier, the water table line calculated with Equation 1.68 is always below the actual water table because of Dupuit's assumption. This difference increases downgradient from the higher hydraulic head used in the calculation. In the zone of groundwater discharge, such as the one shown in Figure 1.49, this difference is the biggest. The position of the actual water table is difficult to formulate mathematically although various grapho-analytical methods for its determination have been suggested (e.g., see Kashef, 1987).

1.4.3.2.2   Solution for the Sloping Aquifer Base

The velocity of nonuniform flow in an unconfined aquifer with a gradient of its impermeable base greater than 0 ($i_o > 0$) is (see also Equation 1.61)

$$v = K\left(i_o - \frac{dh}{dx}\right) \tag{1.69}$$

The unit flow, in general, is

$$q = v \times h_p \tag{1.70}$$

where h_p is the saturated aquifer thickness (equal to pressure head). The equivalent unit flow in uniform conditions is

$$q = v \times h_o \tag{1.71}$$

or, as expressed with Equation 1.55:

$$q = K \times h_o \times i_o \tag{1.72}$$

Combining Equation 1.70 through Equation 1.72 gives another equation for the flow velocity in nonuniform conditions:

$$v = \frac{K \times h_o \times i_o}{h_p} \tag{1.73}$$

As can be seen, the velocity is expressed in terms of the aquifer base gradient (i_o), saturated thickness (h_p), and the equivalent normal aquifer depth (h_o) that would produce the same unit flow if the flow conditions were uniform. Equation 1.69 and Equation 1.73 give

$$K\frac{h_o \times i_o}{h_p} = K\left(i_o - \frac{dh_p}{dx}\right) \tag{1.74}$$

or, after simplifying and rearranging

$$\frac{dh_p}{dx} = i_o\left(1 - \frac{dh_p}{dx}\right) \tag{1.75}$$

To solve this basic differential equation of nonuniform flow when $i_o > 0$, a new variable, called the relative aquifer depth (η), is introduced so that

$$\eta = \frac{h_p}{h_o} \tag{1.76}$$

This transforms Equation 1.75 into

$$\frac{dh}{dx} = i_o \left(\frac{\eta - 1}{\eta}\right) \tag{1.77}$$

Differentiation of Equation 1.76 gives

$$dh = d\eta \times h_o \tag{1.78}$$

which transforms Equation 1.77 into

$$\frac{d\eta \times h_o}{dx} = i_o \left(\frac{\eta - 1}{\eta}\right) \tag{1.79}$$

To solve this equation, variables η and x have to be separated:

$$\frac{\eta}{\eta - 1} d\eta = \frac{i_o}{h_o} dx \tag{1.80}$$

Further preparation of Equation 1.80 is needed in order to use elementary integrals for solving it:

$$d\eta + \frac{d\eta}{\eta - 1} = \frac{i_o}{h_o} dx \tag{1.81}$$

Note that

$$\frac{\eta}{\eta - 1} = \frac{\eta - 1 + 1}{\eta - 1} = 1 + \frac{1}{\eta - 1}$$

Equation 1.81 is solved for the following general boundary conditions: when $x = x_1$ the relative aquifer depth is $\eta = \eta_1$, and when $x = x_2$ the relative depth is $\eta = \eta_2$:

$$\frac{i_o}{h_o} \int_{x_1}^{x_2} dx = \int_{\eta_1}^{\eta_2} d\eta + \int_{\eta_1}^{\eta_2} \frac{d\eta}{\eta - 1} \tag{1.82}$$

All three integrals in Equation 1.82 are elementary (see Appendix B) and the solution is easy:

$$\frac{i_o}{h_o}(x_2 - x_1) = \eta_2 - \eta_1 + \ln\frac{\eta_2 - 1}{\eta_1 - 1} \tag{1.83}$$

where

$$\eta_2 = \frac{h_{p2}}{h_o} \text{ and } \eta_1 = \frac{h_{p1}}{h_o}$$

and $(x_2 - x_1)$ is the distance between pressure heads h_{p2} and h_{p1}.

Equation 1.83 is the solution of the differential equation of nonuniform flow in an unconfined aquifer with the sloping base. Since h_o is not known, the equation is solved implicitly, i.e., the normal aquifer depth (h_o) is found by trial. All constants, known terms in Equation 1.83, can be grouped as follows:

$$i_o(x_2 - x_1) = h_o\left(\frac{h_{p2}}{h_o} - \frac{h_{p1}}{h_o}\right) + h_o \ln\frac{h_{p2} - h_o}{h_{p1} - h_o} \tag{1.84}$$

Note that

$$\ln\frac{\eta_2 - 1}{\eta_1 - 1} = \ln\frac{\dfrac{h_{p2}}{h_o} - 1}{\dfrac{h_{p1}}{h_o} - 1} = \ln\frac{\dfrac{h_{p2} - h_o}{h_o}}{\dfrac{h_{p1} - h_o}{h_o}} = \ln\frac{h_{p2} - h_o}{h_{p1} - h_o}$$

Equation 1.84 is further simplified:

$$i_o(x_2 - x_1) + h_{p1} - h_{p2} = h_o \ln\frac{h_{p2} - h_o}{h_{p1} - h_o} \tag{1.85}$$

or, since all the values on the left-hand side are known:

$$\text{Constant} = h_o \ln\frac{h_{p2} - h_o}{h_{p1} \quad h_o} \tag{1.85a}$$

The normal depth is found by trial so that the constant on the left-hand side equals the value on the right-hand side of Equation 1.85a. Again, note that h_p is the pressure head between the aquifer sloping aquifer base and the water table at that location. It is the saturated vertical thickness of the aquifer, and not the hydraulic head.

A solved problem of groundwater flow in unconfined aquifers, including cases with and without infiltration, is given in Chapter 10.

1.4.3.3 Transient Flow

General equation of two-dimensional transient groundwater flow (i.e., flow that changes in time) is derived by examining the water budget for a representative elementary volume (REV) of the aquifer. REV is a volume that has all the characteristics of the aquifer, including its homogeneity or heterogeneity and isotropy or anisotropy. In addition, its dimensions are, for the purposes of mathematical derivation of the differential equation, infinitesimally small. Figure 1.50 shows REV for a homogeneous, isotropic, unconfined aquifer resting on a horizontal, impermeable base. The aquifer is subjected to a vertical gain or loss of water due to infiltration and evapotranspiration. Elements of the water budget shown in Figure 1.50 are:

q_x is unit flow per elementary aquifer thickness in the X-axis direction
q_y is unit flow per elementary aquifer thickness in the Y-axis direction
$q_x\,dy$ is total flow entering REV in the X-axis direction (dy is the width of the flow)
$q_y\,dx$ is total flow entering REV in the Y-axis direction (dx is the width of the flow)
$\left(q_x + \frac{\partial q_x}{\partial x}dx\right)dy$ is total flow leaving REV in the X-axis direction

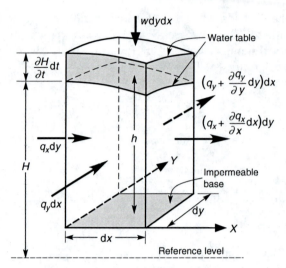

FIGURE 1.50 Representative elementary volume (REV) of an isotropic, homogeneous, unconfined aquifer resting on a horizontal, impermeable base, with elements for deriving Boussinesq's differential equation of groundwater flow. (Modified from Sestakov, V.M., *Dinamika podzemnih vod* (*Dynamics of Groundwater*, in Russian), Izdatelstvo Moskovskogo Universiteta, Moskva, 1979, 368 p.)

$\left(q_y + \dfrac{\partial q_y}{\partial y}\,dy \right)dx$ is total flow leaving REV in the Y-axis direction

$\dfrac{\partial q_x}{\partial x}\,dx$ is change of the unit flow in the X-axis direction

$\dfrac{\partial q_y}{\partial y}\,dy$ is change of the unit flow in the Y-axis direction

$w\,dx\,dy$ is vertical element of the water budget for REV (w is the intensity of the net gain or loss).

The specific flow in the X-axis direction can be expressed as the product of Darcy's velocity in the X-axis direction and the aquifer thickness (note that the hydraulic gradient in Darcy's velocity is the change of the hydraulic head (H) along the X-axis direction):

$$q_x = v_x h \tag{1.86}$$

$$v_x = -K\frac{\partial H}{\partial x} \tag{1.87}$$

$$q_x = -Kh\frac{\partial H}{\partial x} \tag{1.88}$$

Similarly, the specific flow in the Y-axis direction is

$$q_y = -Kh\frac{\partial H}{\partial y} \tag{1.89}$$

The volume of water stored in (or released from) REV as the water table changes its position in time is

$$V = \frac{\partial H}{\partial t}\,dt\,dx\,dy\,S \tag{1.90}$$

where S is the aquifer storativity which, for an unconfined aquifer, equals specific yield. The specific yield is roughly equal to the effective porosity, which is a percentage of voids within porous medium that store or release water without retention. Note that the partial differentials in all of the above equations are used to describe changes in time of the hydraulic head and the specific flow along an individual coordinate axis, while change in the other direction is not considered. This is because both the hydraulic head and the specific flow are functions of space (represented by two coordinate axes x and y) and time:

$$q = f(x, y, t)$$

$$H = f(x, y, t)$$

The water balance equation for the REV states that the difference between the flows entering REV (inputs) and leaving REV (outputs) in time must be equal to the volume of water stored in (or released from) REV:

$$q_x \, dy \, dt - \left(q_x + \frac{\partial q_x}{\partial x} dx \right) dy \, dt + q_y \, dx \, dt - \left(q_y + \frac{\partial q_y}{\partial y} dy \right) dx \, dt + w \, dx \, dy \, dt = \frac{\partial H}{\partial t} dt \, dx \, dy \, S \tag{1.91}$$

or, after simplifying

$$-\frac{\partial q_x}{\partial x} - \frac{\partial q_y}{\partial y} + w = S \frac{\partial H}{\partial t} \tag{1.92}$$

By expressing unit flows q_x and q_y in terms of Darcy's law, Equation 1.92 becomes (see Equation 1.86 and Equation 1.89):

$$\frac{\partial \left(hK \frac{\partial H}{\partial x} \right)}{\partial x} + \frac{\partial \left(hK \frac{\partial H}{\partial y} \right)}{\partial y} + w = S \frac{\partial H}{\partial t} \tag{1.93}$$

This nonlinear partial differential equation of two-dimensional groundwater flow in an unconfined aquifer is known in literature as the Boussinesq equation, after a French engineer, who first suggested its linearization. He introduced the following assumptions:

- The aquifer impermeable base is horizontal and the reference level can be placed at the base so that $h = H$ (see Figure 1.50).
- The change of the hydraulic head along flow lines is very slow so that the average thickness of the aquifer can be considered equal to the hydraulic head: $h_{av} = h = H$.
- The aquifer is homogeneous and isotropic so that $K = $ constant.

These assumptions give the following linear form of the Boussinesq equation:

$$\frac{\partial^2 h}{\partial x^2} + \frac{\partial^2 h}{\partial y^2} + \frac{w}{Kh_{av}} = \frac{S}{Kh_{av}} \frac{\partial h}{\partial t} \tag{1.94}$$

By introducing two new parameters called hydraulic diffusivity (a) and transimissivity ($T = hb$), which have the following relationship:

$$a = \frac{Kh_{av}}{S} = \frac{T}{S} \tag{1.95}$$

Equation 1.94 can be written in one of the following two forms:

$$\frac{\partial^2 h}{\partial x^2} + \frac{\partial^2 h}{\partial y^2} + \frac{w}{T} = \frac{1}{a}\frac{\partial h}{\partial t} \tag{1.96}$$

$$a\frac{\partial^2 h}{\partial x^2} + a\frac{\partial^2 h}{\partial y^2} + \frac{w}{S} = \frac{\partial h}{\partial t} \tag{1.97}$$

It has been proven that, even under the most unfavorable circumstances, the above-described linearization yields an error of less than 20% (Lebedev, 1968). For most practical purposes, the linear Boussinesq equation will give an acceptable result.

The linear Boussinesq equation of one-dimensional groundwater flow is

$$\frac{\partial h}{\partial t} = a\frac{\partial^2 h}{\partial x^2} + \frac{w}{S} \tag{1.98}$$

where $a = Kh_{av}/S$ is the hydraulic diffusivity, h_{av} is the average aquifer thickness (the thickness of the saturated zone), K is the hydraulic conductivity, S is the specific yield of the unconfined aquifer (its storativity), h is the hydraulic head (equal to H, see Figure 1.50), x is the distance along X-axis, and w is the intensity of aquifer recharge at the water table (e.g., recharge from precipitation).

If the unconfined aquifer of interest is receiving water from (or losing water to) an underlying aquifer, either directly or by leakage through an aquitard, Equation 1.98 becomes

$$\frac{\partial h}{\partial t} = a\frac{\partial^2 h}{\partial x^2} + \frac{w}{S} \pm \frac{\varepsilon}{S} \tag{1.99}$$

where ε is the intensity of leakage. Partial differential Equation 1.98 and Equation 1.99 can be solved for known boundary and initial conditions of the transient groundwater flow.

1.4.3.3.1 Solution for One-dimensional Flow with a Sudden Change at Boundary

The following assumptions are made in order to solve Equation 1.98 for an unconfined aquifer that has a strong equipotential boundary (e.g., river, channel, lake) on one side and has no boundary on the other side (note that, even though there must be a second boundary, for practical purposes it is placed far enough so that it does not influence the flow field):

- The change at the boundary (where $x = 0$) is sudden, and is given as (see Figure 1.51)

$$\Delta H(0,t) = \Delta H_0 \quad \text{(for } x = 0 \text{ and } t = t) \tag{1.100}$$

- The flow field is not disturbed on the other side (where there is no boundary) and the change in water table is 0:

$$\frac{\partial[\Delta H(x,t)]}{\partial x} = 0 \quad \text{(for } x \to \infty \text{ and } t = t) \tag{1.101}$$

- There is no aquifer recharge either from the top or from the bottom:

$$w = 0 \text{ and } \varepsilon = 0 \tag{1.102}$$

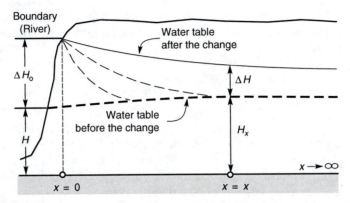

FIGURE 1.51 Boundary conditions for a transient one-dimensional flow with a sudden change at a boundary. The flow field has one strong equipotential boundary (river) while the boundary on the other side is placed in the infinity.

For these boundary and initial conditions, the solution is adapted from the theory of heat transfer and has the following form (modified from Lebedev, 1968):

$$\Delta H(x,t) = \Delta H_0 \times \text{erfc}(\lambda) \qquad (1.103)$$

where $\Delta H(x,t)$ is the head change as a function of space and time, ΔH_0 is the change at the boundary, and erfc(λ) is the complementary error function given as: erfc(λ) = 1 − erf(λ).

Error function (erf) is an integral of probability and is readily found in tables (see Appendix D). For our solution of transient one-dimensional flow with a sudden change at a boundary, it has the following form:

$$\text{erf}(\lambda) = \frac{2}{\sqrt{\pi}} \int_0^{\lambda} e^{-\lambda^2} \, d\lambda \qquad (1.104)$$

where λ is a parameter given as

$$\lambda = \frac{x}{2\sqrt{at}} \qquad (1.105)$$

In Equation 1.105, x is the distance from the boundary (river) at which the change in groundwater table (ΔH) is recorded, a is the hydraulic diffusivity ($a = Kh_{av}/S$), see also Equation 1.95, and t is the time of change. The average aquifer thickness needed to find the hydraulic diffusivity is calculated as follows (see Figure 1.51):

$$h_{av} = \frac{(H + \Delta H_0) + (H_x + \Delta H)}{2} \qquad (1.106)$$

Equation 1.103 can be used to predict a rise in water table at some distance x from the boundary after a sudden change of river stage such as during floods, or it can be used to estimate aquifer parameters if water levels in the river and the aquifer are recorded simultaneously. If two monitoring wells are placed along a streamline, the one closer to the boundary and having a bigger change can act as a boundary (ΔH_0), thus eliminating the need for measuring the river stage.

1.4.3.3.2 Solution for One-Dimensional Flow with a Gradual Change at Boundary

The general linear differential equation of a transient one-dimensional groundwater flow is (see Equation 1.98)

$$\frac{\partial h}{\partial t} = a\frac{\partial^2 h}{\partial x^2} + \frac{w}{S} \tag{1.107}$$

where $a = Kh_{av}/S$ is the hydraulic diffusivity, h_{av} is the average aquifer thickness (i.e., the thickness of the saturated zone), K is the hydraulic conductivity, S is the specific yield of the unconfined aquifer (i.e., its storativity), h is the hydraulic head (equal to H, see Figure 1.50), x is the distance along the X-axis, w is the intensity of aquifer recharge at the water table (e.g., recharge from precipitation), and t is the time.

If the unconfined aquifer of interest is receiving water from (or losing water to) an underlying aquifer, either directly or by leakage through an aquitard, Equation 1.107 becomes

$$\frac{\partial h}{\partial t} = a\frac{\partial^2 h}{\partial x^2} + \frac{w}{S} \pm \frac{\varepsilon}{S} \tag{1.108}$$

where ε is the intensity of leakage. Partial differential Equation 1.108 can be solved for known boundary and initial conditions of the transient groundwater flow. Figure 1.52 shows a situation in the vicinity of a strong equipotential boundary (river) developed after a gradual increase in the river stage. If data on the river stage changes are not readily available, measurements in two monitoring wells perpendicular to the boundary can be used to solve Equation 1.108. In such a case, monitoring well closer to the river becomes the boundary and the origin of the X-axis is placed at its location at the impermeable base of the aquifer. Boundary and initial conditions of the flow field are then:

1. $\Delta H(0,t) = bt$, where b is the velocity of change in the water table at $x = 0$ and t is the time of change
2. $\dfrac{\partial[\Delta H(x,t)]}{\partial x} = 0$ for $x \to \infty$

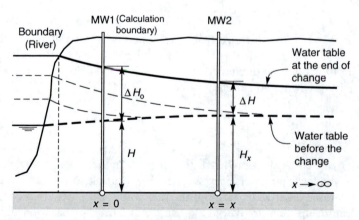

FIGURE 1.52 Boundary conditions for a transient one-dimensional flow with a gradual change at a boundary. The flow field has one strong equipotential boundary (river) while the boundary on the other side is placed in infinity.

3. There is a net infiltration across the water table: $w \neq 0$ (note that, in general, this net infiltration is the difference between the effective infiltration and the evaporation from the water table). The aquifer recharge by infiltration starts at time $t = 0$ and it has constant intensity during time t
4. There is no leakage into the aquifer from below: $\varepsilon = 0$

Solution of Equation 1.108 for these initial and boundary conditions is (Lebedev, 1968)

$$\Delta H(x,t) = \Delta H_0 \times R(\lambda) + \frac{wt}{S}[1 - R(\lambda)] \qquad (1.109)$$

where $\Delta H(x,t)$ is a function of space and time, ΔH_0 is the change at the boundary, $R(\lambda)$ is a special function, values of which are found in tables (see Appendix D), w is the intensity of infiltration, t is the time, and S is the aquifer storativity (specific yield).
Function $R(\lambda)$ is given as

$$R(\lambda) = 4i^2 \text{erfc}(\lambda) \qquad (1.110)$$

where the complementary error function is given as: $\text{ercf}(\lambda) = 1 - \text{erf}(\lambda)$. The error function (erf) is an integral of probability and is also readily found in tables (see Appendix D). Solution of transient one-dimensional flow with a gradual change at a boundary has the following form:

$$\text{erf}(\lambda) = \frac{?}{\sqrt{\pi}} \int_0^\lambda e^{-\lambda^2} \, d\lambda \qquad (1.111)$$

where

$$\lambda = \frac{x}{2\sqrt{at}} \qquad (1.112)$$

In Equation 1.112, x is the distance between the boundary MW1 and MW2 where the change in groundwater table (ΔH) is recorded (see Figure 1.52), a is the hydraulic diffusivity ($a = Kh_{av}/S$), and t is the time of change. Application of the transient one-dimensional equation of groundwater flow near boundaries, using field data, is given in Chapter 11.

1.4.3.4 Flow in Fractured Rock and Karst Aquifers

Except (arguably) for the application of Darcy's law in case of the matrix porosity dominated fractured rock and karst aquifers (which is rather uncommon), in all other situations there is a wide array of possible approaches to calculate groundwater flow rates. The simplest one is to assume that the porous media all behave similarly, at some representative scale, and then simply apply Darcy's equation. Although this equivalent porous medium (EPM) approach still seems to be the predominant one in hydrogeologic practice, it is hardly justified in a highly anisotropic fractured rock aquifer or a karst aquifer. Its inadequacy is emphasized when dealing with contaminant fate and transport analyses where all field scales are equally important, starting with contaminant diffusion into the rock matrix, and ending with predictions of most likely contaminant pathways in the subsurface. In the analytically most complicated, but at the same time the most realistic case, the groundwater flow rate is calculated by integrating the equations of flow through the rock matrix (Darcy's flow) with the hydraulic equations of flow through various sets of fractures, pipes, and channels. This integration, or interconnectivity between the four different flow components, can be deterministic, stochastic, or some combination of the two.

Deterministic connectivity is established by a direct translation of actual field measurements of the geometric fracture parameters such as dip and strike (orientation), aperture, and spacing between individual fractures in the same fracture set, and then doing the same for any other fracture set. Cavities are connected in the same way, by measuring the geometry of each individual cavity (cave). Finally, all of the discontinuities (fractures and cavities) are connected based on the field measurements and mapping. As can be easily concluded, such deterministic approach includes many uncertainties and assumptions by default ("You have walked and measured this cave, but what if there is a very similar one somewhere in the vicinity you don't know anything about?"). Stochastic interconnectivity is established by randomly generating fractures or pipes using some statistical and probabilistic approach based on field measures of the geometric fracture (pipe) parameters. An example of combining deterministic and stochastic approaches is when computer-generated fracture or pipe sets are intersected by a known major preferential flow path such as cave. Except for relatively simple analytical calculations using homogeneous, isotropic, EPM approach, most other quantitative methods for fractured rock and karst groundwater flow calculations include some type of modeling.

Figure 1.53 and Figure 1.54 show determination of fracture porosity for three highly simplified geometric models of fracture sets. Porosity determined in this way can be used to approximate cross-sectional area (A) of flow in the aquifer that takes place through fractures and, together with some equivalent hydraulic conductivity (K) and the hydraulic gradient (i), calculate the equivalent groundwater flow rate using the well-known simple equation based on Darcy's law: $Q = AKi$.

Fracture aperture and thickness are two parameters used most often in various single-fracture flow equations, while spacing between the fractures and fracture orientation is used when calculating flow through a set of fractures. However, these actual physical characteristics are not easily and meaningfully translated into equations attempting to describe flow at a realistic field scale for the following reasons:

1. As illustrated in Figure 1.55, fracture aperture is not constant and there are voids and very narrow or contact areas called asperities. Various experimental studies have shown that the actual flow in a fracture is channeled through narrow, conduit-like tortuous paths (Figure 1.56) and cannot be simply represented by the flow between two parallel

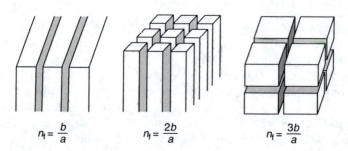

$$n_f = \frac{b}{a} \qquad\qquad n_f = \frac{2b}{a} \qquad\qquad n_f = \frac{3b}{a}$$

FIGURE 1.53 Fracture porosity (n_f) equations for the slides, matches, and cubes fracture models where a is the fracture spacing and b is the fracture aperture. (Modified from Cohen, R.M. and Mercer, J.W., *DNAPL Site Evaluation*, C.K. Smoley, CRC Press, Boca Raton, FL, 1993. Copyright C.K. Smoley, CRC Press, T&F.)

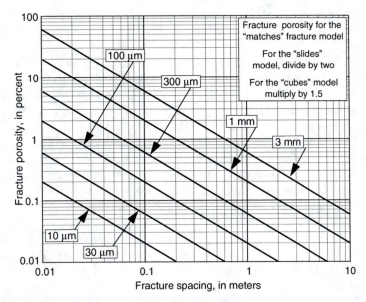

FIGURE 1.54 Fracture porosity is a function of fracture spacing and aperture. (From Cohen, R.M. and Mercer, J.W., *DNAPL Site Evaluation*, C.K. Smoley, CRC Press, Boca Raton, FL, 1993. Copyright C.K. Smoley, CRC Press, T&F.)

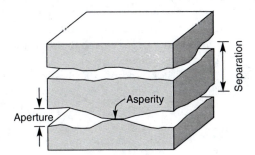

FIGURE 1.55 Fracture aperture, asperity, and separation between two parallel fractures in the same fracture set.

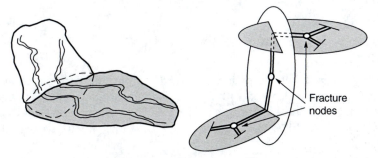

FIGURE 1.56 Channeling in a fracture plane and equivalent channel model. (After Cacas, M.C., Développement d'un modèle tridimensionel stochastique discret pour la simulation de l'écoulement et des transports de masse et de chaleur en milieu fracturé, Ph.D. thesis, Ecole des Mines de Paris, Fontainebleau, France, 1989, from Chilès, J.P. and de Marsily, G., *Flow and Contaminant Transport in Fractured Rock*, Bear, J., Tsang, C.F., and de Marsily, G., Eds., Academic Press, San Diego, 1993, pp. 169–236.)

plates separated by the mean aperture. In carbonate aquifers, this channeled flow may result in the formation of karst conduits as the rock mass is dissolved by the flowing groundwater.

2. Because of stress release, the aperture measure at outcrops or in accessible cave passages is not the same as an *in situ* aperture. Aperture measured on drill cores and in borings is also not a true one—the drilling process commonly causes bedrock adjacent to fractures to break out thereby increasing the apparent widths of fracture openings as viewed on borehole-wall images (Williams et al., 2001).

3. Fractures have limited length and width, which can also vary between individual fractures in the same fracture set. Spacing between individual fractures in the same set can also vary. Since all these variations take place in the three-dimensional space, they cannot be directly observed, except through continuous coring or logging of multiple closely spaced boreholes, which is the main cost-limiting factor.

Fractures by default have limited extent in the three-dimensional space of an aquifer, of which little is known. In order to simulate flow through a network of fractures, one has to decide on the spatial geometry of individual fractures and their interconnectivity within the entire aquifer volume of interest. This is done in many different ways, including simulation with two-dimensional fracture traces (orthogonal, or intersecting at an angle), three-dimensional orthogonal disks, three-dimensional disk clusters centered on seeds generated by a random process, or some two-dimensional or three-dimensional hierarchical model (Long, 1983; Long et al., 1985; Chilès, 1989a,b; Chilès and de Marsily, 1993). Whatever approach is selected, numeric models are arguably the only quantitative tools capable of solving groundwater flow through complex fracture networks.

Evolution of analytical equations and various approaches in quantifying fracture flow is given by Witherspoon (2000) and Faybishenko et al. (2000). In the simplest form, the hydraulic conductivity of a fracture with an aperture B (in Whiterspoon's notation $B = 2b$) represented with two parallel plates is

$$K = B^2 \frac{\rho g}{12\mu} \tag{1.113}$$

where ρ is fluid density, g acceleration of gravity, and μ is fluid viscosity. The groundwater flow rate through a cross-sectional area $A = Ba$ (where B is the fracture aperture and a is the fracture width perpendicular to the flow direction) is

$$Q = Av = -a\left(\frac{\rho g}{12\mu}\right)\frac{dh}{dx}B^3 \tag{1.114}$$

where dh/dx is the change in the hydraulic head (h) along the flow direction (x). In the definite form, this change is denoted with Δh and the minus sign disappears due to integration. The fracture flow approximation, called cubic law, assumes that the representative aquifer volume acts as an EPM (Darcian continuum). Witherspoon gives another form of the cubic law:

$$\frac{Q}{\Delta h} = \frac{CB^3}{f} \tag{1.115}$$

where C is a constant that depends on the geometry of the flow field and f is roughness that accounts for deviations from ideal conditions, which assume smooth fracture walls and laminar flow. The roughness f is related to the Reynolds number (Re; indicator if the flow is turbulent or laminar) and the friction factor (Ψ) through the following equation:

$$f = \frac{\Psi Re}{96} \qquad (1.116)$$

The Reynolds number is given as

$$Re = \frac{Dv\rho}{\mu} \qquad (1.117)$$

where D is the fracture hydraulic diameter assumed to be equal to four times the hydraulic radius. For a relatively smooth fracture, where the ratio between fracture asperity and aperture is less than 0.1, the transition to turbulent flow is at Reynolds number of about $Re = 2400$. As this ratio, which is the indicator of fracture roughness, increases, the Reynolds number for transition to turbulent flow decreases significantly. The friction factor is given by

$$\Psi = \frac{D}{v^2/2g}\Delta h \qquad (1.118)$$

where v is flow velocity. Equations of fluid flow in nonideal fractures with influences of various geometric irregularities and fracture network modeling approaches are discussed in detail by Bear et al. (1993), Zimmerman and Yeo (2000), and Faybishenko et al. (2000).

1.4.3.4.1 Flow through Pipes (Conduits) and Channels

Flow through a pipe of varying cross-sectional area is described by the Bernoulli equation for real viscous fluids as illustrated in Figure 1.57. Since there is no gain or loss of water in the pipe, the flow rate remains the same while other hydraulic elements change from one cross section to another. The total energy line (E) of the flow can only decrease from the upgradient cross section towards the downgradient cross section of the same flow tube (pipe or conduit) due to energy losses. On the other hand, the hydraulic head line (H) may go up and down along the same flow tube as the cross-sectional area increases or decreases, respectively. The total energy surface, which includes the flow velocity component ($\alpha v^2/2g$), can be directly measured only by the Pitot device whose installation is not feasible in most field conditions. Monitoring wells and piezometers, on the other hand, only record the hydraulic head, which

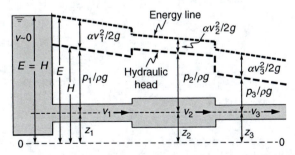

FIGURE 1.57 Illustration of the Bernoulli equation for the flow of real viscous fluids through a pipe with the varying cross-sectional area. Energy line (E) at any given cross section is the sum of the elevation head (z), the fluid pressure head ($p/\rho g$), and the velocity head ($\alpha v^2/2g$). The fluid pressure head and the elevation head give the hydraulic head (H).

does not include the flow velocity component. It is therefore conceivable that two piezometers in or near the same karst conduit with rapid flow may not provide useful information for calculation of the real flow velocity and flow rate between them, and may even falsely indicate the opposite flow direction. In fact, as discussed by Bögli (1980) it has been shown that water rising through a tube in an enlargement passage can flow backward over the main flow conduit into another tube, which begins at a narrow passage in the same main conduit.

There are additional complicating factors when attempting to calculate flow through natural karst conduits using the pipe approach:

1. Flow through the same conduit may be both under pressure and with free surface.
2. Since pipe and conduit walls are more or less irregular (rough), the related coefficient of roughness has to be estimated and inserted into the general flow equation.
3. Conduit cross section may vary significantly over short distances and in the same general area.
4. The flow may be both laminar and turbulent in the same conduit, depending on the flow velocity, cross-sectional area, and wall roughness. The irregularities that cause turbulent flow are mathematically described through the Reynolds number and the friction factor.

Bögli (1980), Ford and Williams (1989), and White (1988) provide a detailed discussion on various pipe and channel flow equations and their applicability in karst aquifers. Figure 1.58 is a good example of how complex or even impossible it would be to explicitly calculate

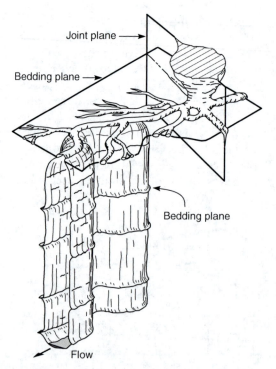

FIGURE 1.58 Drawing of a vertical shaft and cave passages developed along a bedding plane—a common pattern in Mammoth Cave, Kentucky, United States. (From Southeastern Friends of the Pleistocene, Hydrogeology and Geomorphology of the Mammoth Cave area, Kentucky. 1990 Field Excursion led by Quinlan, J.F., Ewers, R.O., and Palmer, A.N., Nashville, TN, 1990, 102 p.)

groundwater flow rates using the pipe approach, not to mention the existence of large vertical conduits (shafts) where water may be flowing as a thin film over the shaft walls. In addition, karst conduits and cave passages are often only partially filled with water and may have sediment accumulation in which cases it is more appropriate to apply hydraulic equations developed for surface channels.

Because of widely varying hydraulic conductivity and effective porosity of karstified carbonates, even within the same aquifer system, the groundwater velocity in karst can vary over many orders of magnitude. One should therefore be very careful when making a (surprisingly common) statement such as "groundwater velocity in karst is generally very high." Although this may be true for turbulent flow, taking place in karst conduits, a disproportionably larger volume of any karst aquifer has relatively low groundwater velocities (laminar flow) through small fissures and rock matrix. One common method for determining groundwater flow directions and apparent flow velocities in karst is dye tracing. However, most dye tracing tests in karst are designed to analyze possible connections between known (or suspect) locations of surface water sinking and locations of groundwater discharge (springs). Because such connections involve some kind of preferential flow paths (sink–spring type), the apparent velocities calculated from the dye-tracing data are usually biased towards the high end. Based on results of 43 tracing tests in karst regions of West Virginia in the United States (Jones, 1997), the median groundwater velocity is 716 m/d, while 50% of the tests show values between 429 and 2655 m/d (25th and 75th percentile of the experimental distribution, respectively). It is interesting that, based on 281 dye-tracing tests, the most frequent velocity (14% of all cases) in the classic Dinaric karst of Herzegovina (Europe), as reported by Milanović (1979), is quite similar: between 864 and 1728 m/d. Twenty-five percent of the results show groundwater velocity greater than 2655 m/d in West Virginia, and greater than 5184 m/d in Herzegovina. The West Virginia data do not show any obvious relationship between the apparent groundwater velocity and the hydraulic gradient.

Various approximate calculations of flow velocity have been made based on the geometry of hydraulic features, such as scallops and flutes visible on walls, floors, and ceilings of accessible cave passages (e.g., see Bögli, 1980; White, 1988). For example, the calculated flow velocity for a canyon passage in White Lady Cave, Little Neath Valley, U.K., is 1.21 m/s and the flow rate is 9.14 m^3/s, for the cross-sectional area of flow of 7.6 m^2 and scallop length of 4.1 cm (White, 1988). Confined karst aquifers, which do not have major concentrated discharge points in forms of large springs, generally have much lower groundwater flow velocities. This is regardless of the predominant porosity type because the whole system is under pressure and the actual displacement of "old" aquifer water with the newly recharged one is rather slow. Figure 1.59 shows the results of groundwater flow velocity estimates using ^{14}C isotope dating for the confined portion of the Floridan aquifer in central Florida (Hanshaw and Back, 1974). The average groundwater velocity based on 40 values is 6.9 m/y or 0.019 m/d.

1.4.4 ANISOTROPY AND HETEROGENEITY

The following discussion by Meinzer (1932) explains very eloquently what some researchers attempt to formulate with various graphs, usually of no practical use:

> The most serious difficulty…and one that up to the present time has not been effectively overcome is that of determining the true average permeability of the material that constitutes the water-bearing formation. Laboratory methods are available to determine accurately the permeability of the samples that are tested, but the difficulty lies in obtaining representative samples. Even apparently slight differences in texture may make great differences in permeability.

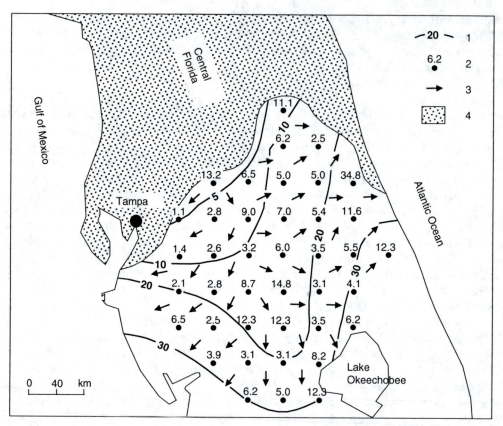

FIGURE 1.59 Velocity of groundwater in the Floridan aquifer determined from ^{14}C content of dissolved carbonate species and adjusted for solution of carbonate minerals. (a) ^{14}C hydroisochrons, in 1000 y; (b) Velocity of groundwater flow, in meters per year; (c) Approximate direction of groundwater flow; and (d) Approximate area of significant mixing. (From Hanshaw, B.B. and Back, W., Determination of Regional Hydraulic Conductivity through Use of ^{14}C Dating of Groundwater. Mémoires, Tome X, 1. Communications, 10th Congress of the International Association of Hydrogeologists, Montpellier, France, 1974, pp. 195–198.)

A rather inconspicuous admixture of colloidal clay to an otherwise permeable sand may cut down greatly its capacity to conduct water. In a sand formation a few thin strata of coarse clean sand may conduct more water than all the rest of the formation. These permeable strata may be overlooked in the sampling, or if samples from them are taken it may be impossible to give them the proper weight in comparison with samples from other parts of the formation. Consolidated rocks are likely to contain joints and crevices which conduct much of the water and which therefore render laboratory methods inapplicable. On the other hand, unconsolidated samples cannot easily be recovered and tested without disturbing the texture of the material and thus introducing errors of unknown but conceivably great amount. Moreover, samples taken at the outcrop of a formation may not be representative because of changes produced by weathering, and samples obtained from wells are generally nonvolumetric and greatly disturbed and may be either washed or mixed with clay of foreign origin. If the conditions of drilling can be controlled it may be possible to obtain an undisturbed or only moderately disturbed sample, especially if a core barrel is used, but such favorable conditions are rarely obtainable.

The points made by Meinzer are today lumped together into the following single term (to avoid somewhat lengthy explanations): REV. Simply stated, REV is a volume that has all the important characteristics of the aquifer, including any heterogeneity and anisotropy, so that a quantitative analysis performed on the REV can be applied elsewhere in the aquifer. However, this often does not mean the same thing to different people, it is highly problem-specific and, at the end, requires a lengthy explanation anyway. For example, academic researchers in the field of contaminant hydrogeology and groundwater remediation are usually occupied with testing new ideas on small laboratory samples of porous media, such as glass beads, clean sand, tight clay, or maybe a small plug from a limestone core. Extrapolating the results of such research to real field-scale problems always presents a major challenge. One approach is to assume that, although heterogeneous and anisotropic by default, the porous media at this particular *site* can be approximated by some average characteristics reasonably accurately. This approach is in many cases the only one feasible, given the limitations of available funds and time for investigations. For practicing hydrogeologists (or a project team), it is therefore probably the most appropriate to consider all of the factors shown in Figure 1.60 at the same time and draw their own conclusions as to what the problem-specific REV should be. In other words, every REV is both scale-dependent and problem-specific, and can vary even for the same porous medium at the same site depending on the final project goal.

In hydrogeology, heterogeneity refers to geologic fabric of the underlying rock volume subjected to investigations. This fabric is determined by the following two key elements: (1) rock types and their spatial arrangement and (2) discontinuities in the rock mass. Photographs in Figure 1.61 illustrate these two elements. Without elaborating much further on the geologic portion of hydrogeology, it is appropriate to state that groundwater professionals without a thorough geologic knowledge (education) would likely have various difficulties in understanding many important aspects of heterogeneity not shown in Figure 1.61. Geologic fabric directly determines the degree of anisotropy of the porous media and, consequently, the groundwater flow characteristics. In hydrogeology, anisotropy refers to groundwater velocity

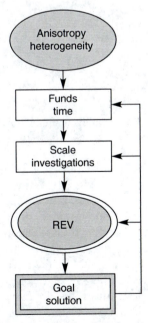

FIGURE 1.60 Schematic presentation of the representative elementary volume (REV) concept as it relates to real world problems.

FIGURE 1.61 *Top*: Alternating layers of coarse-grained (gravel) and fine-grained (silt and clay) uncon-solidated sediments are characteristics of most alluvial aquifers. *Bottom*: Fractures, faults, and beds of different lithology and competency are characteristics of most consolidated rock aquifers; complex geologic fabric is often partially preserved in the overlying regolith sediments and in the transition zone such as this one. (Photo taken in the Big Bend National Park, Texas.)

vectors. If the groundwater velocity is same in all spatial directions, the porous medium is isotropic. If the velocity varies in different directions, the porous medium is anisotropic. From Darcy's law (Equation 1.14) it follows that, if the groundwater velocity is anisotropic, the hydraulic conductivity is anisotropic as well. In fact, when talking about anisotropy in hydrogeology, one usually refers to anisotropy of the hydraulic conductivity, rather than the groundwater velocity. Figure 1.62 illustrates, in two dimensions, just some of many possible causes of anisotropy. It is important to understand that some degree of anisotropy can (and usually does) exist in all spatial directions. It is for reasons of simplification and computational feasibility that hydrogeologists consider only the three main perpendicular

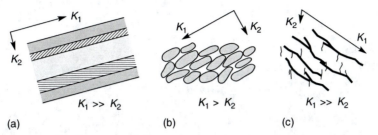

FIGURE 1.62 Some possible reasons for anisotropy of the hydraulic conductivity. (a) Sedimentary layers of varying permeability. (b) Orientation of gravel grains in alluvial deposit. (c) Two sets of fractures in massive bedrock.

directions of anisotropy: two in the horizontal plane and one in the vertical plane; in the Cartesian coordinate system, these three directions are represented with the X-, Y-, and Z-axes.

Again, for reasons of simplicity or feasibility, one may decide that the EPM approach is appropriate for solving a particular problem. EPM uses average values for porosity and hydraulic conductivity of all heterogeneous porous media present, including any discontinuities in the rock mass. A common example of this approach is when quantitative analyses based on Darcy's law are applied to a fractured rock aquifer, assuming that the fractures behave as an equivalent intergranular porous medium (as shown in the previous chapter, the flow through fractures is in reality of different nature). In conclusion, both the REV and the EPM concepts should be applied with care and, whenever possible, be accompanied with an analysis of the associated uncertainty.

1.5 CONTOUR MAPS AND FLOW NETS

Contour maps of the water table (unconfined aquifers) or the piezometric surface (confined aquifers) are made in the majority of hydrogeologic investigations and, when properly drawn, represent a very powerful tool in aquifer studies. Although commonly used for determination of groundwater flow directions only, contour maps, when accompanied with other data, allow for the analyses and calculations of hydraulic gradients, flow velocity and flow rate, particle travel time, hydraulic conductivity, and transmissivity. However, one should always remember that a contour map is a two-dimensional representation of a three-dimensional flow field, and as such it has limitations. If the area (aquifer) of interest is known to have significant vertical gradients, and enough field information is available, it is always recommended to construct at least two contour maps: one for the shallow and one for the deeper aquifer depth. As with geologic and hydrogeologic maps in general, a contour map should be accompanied with several cross sections showing locations and vertical points of the hydraulic head measurements with posted data, or ideally showing the entire cross-sectional flow net as illustrated in Figure 1.63. Probably the most incorrect and misleading case is when data from monitoring wells screened at different depths in an aquifer with vertical gradients are lumped together and contoured as one average data package. A perfect example would be a fractured rock or karst aquifer with thick residuum (regolith) deposits and monitoring wells screened in the residuum and at various depths in the bedrock. If data from all the wells were lumped together and contoured, it would be impossible to interpret where the groundwater is actually flowing for the following reasons: (1) the residuum is primarily an intergranular porous medium in unconfined conditions (it has water table) and horizontal flow directions may be influenced by local (small) surface drainage features and (2) the bedrock has discontinuous flow through fractures at different depths, which is often under pressure (confined

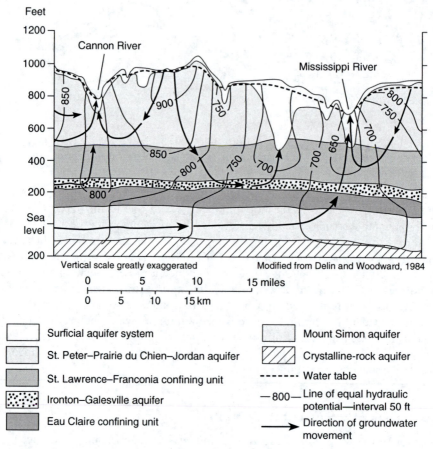

FIGURE 1.63 Regional hydrogeologic cross section showing groundwater flow towards the Mississippi River, the main discharge area for several aquifers (Miller, J.A., Introduction and national summary. *Ground-Water Atlas of the United Sates*, United States Geological Survey, A6, 1999.).

conditions), and may be influenced by regional features such as major rivers or springs. The flow in two distinct porous media (the residuum and the bedrock) may therefore be in two different general directions at a particular site, including strong vertical gradients from the residuum towards the underlying bedrock. Creating one average contour map for such a system does not make any hydrogeologic sense.

Contour map of the hydraulic head is one of the two parts of a flow net: flow net in a homogeneous isotropic aquifer is a set of streamlines and equipotential lines, which are perpendicular to each other (see Figure 1.64). Streamline (or flow line) is an imaginary line representing the path of a groundwater particle as it flows (or would flow) through an aquifer. Two streamlines bound a flow segment of the flow field and never intersect, i.e., they are roughly parallel when observed in a relatively small portion of the aquifer. Equipotential line is horizontal projection of the equipotential surface—everywhere at that surface the hydraulic head has constant value. Two adjacent equipotential lines (surfaces) never intersect and can also be considered parallel within a small aquifer portion. These characteristics are the main reasons why a flow net in a homogeneous, isotropic aquifer is sometimes called the net of small (curvilinear) squares. Practically every groundwater map that shows contours of the water table or the piezometric surface represents a flow net without streamlines, which can be

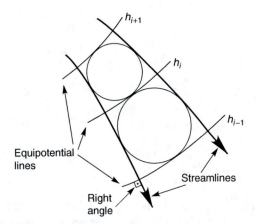

FIGURE 1.64 Flow net in a homogeneous and isotropic aquifer is a set of equipotential lines and streamlines which are perpendicular to each other. The equipotential line connects points with the same groundwater potential, i.e., hydraulic head h. The streamline is an imaginary line representing the path of a groundwater particle as it flows (or would flow) through an aquifer.

easily added following several simple rules. Groundwater contour maps and flow nets can be constructed:

1. Manually, from data recorded in the field (hydraulic head measurements at monitoring wells)
2. Using a computer program for contouring data recorded in the field, and adding the streamlines manually
3. Using a numerical or analytical groundwater flow model whose final result is a distribution of calculated hydraulic heads and flow lines within the aquifer, i.e., *XYZ* computer files that can be graphically contoured during model postprocessing

At least several data sets collected in different hydrologic seasons should be used to draw groundwater contour maps for the area of interest. In addition to recordings from piezometers, monitoring wells, and other water wells, every effort should be made to record elevations of water surfaces in the nearby surface streams, lakes, ponds, and other surface water bodies including cases when these bodies seem too far to influence groundwater flow pattern. In addition, one should gather information about hydrometeorologic conditions in the area for preceding months paying special attention to the presence of extended wet or dry periods. All of this information is essential for making a correct contour map. A flow net does not change over time for steady-state flow. In transient conditions, a flow net represents the instantaneous flow field at a particular time, and it changes for any other time. In practice, it can be viewed as a continuous succession of steady states, each of which is steady for short periods of time (Domenico and Schwartz, 1990).

When constructing a contour map (flow net), it is very important to remember the following hydrodynamic characteristics of groundwater flow:

- The velocity vector (of a moving groundwater particle) is a tangent to the streamline (see Figure 1.65). Therefore, a water particle cannot cross a streamline, i.e., there is no groundwater flow across a streamline. It follows that an impermeable boundary is also a streamline. This is the reason why streamlines are sometimes used as hydraulic

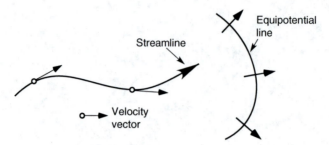

FIGURE 1.65 The groundwater flow velocity vector is a tangent to the streamline (left), and it is perpendicular to the equipotential line (right).

(artificial) no-flow boundaries in groundwater modeling. When a no-flow boundary, such as an impermeable contact, exists in the area to be contoured, a simple rule to remember is that equipotential lines have to end at that boundary, approximately at right angles—they cannot be parallel to the boundary.

- The velocity vector (of a moving groundwater particle) is perpendicular to an equipotential line and is oriented towards the decreasing hydraulic head. It follows that in homogeneous, isotropic aquifers the streamlines and equipotential lines are perpendicular to each other. Although not theoretically true, for practical purposes this can be applied to most inter-granular aquifers if they are not highly anisotropic. In the case of a pronounced anisotropy, flow nets cannot be constructed without complex analytical solutions, which defeat their purpose. Of course, a numeric groundwater flow model, which accounts for anisotropy, would do the trick but its development may be lengthy and costly.
- A drop in the hydraulic head between successive equipotential lines is the same (Δh = constant). The value of this decrease corresponds to the contour interval on groundwater contour maps.

In highly fractured and karst aquifers, where groundwater flow is discontinuous (i.e., it takes place mainly along preferential flow paths), the Darcy law does not apply and flow nets are not an appropriate method for the flow characterization. However, contours maps in highly heterogeneous aquifers are routinely made by many professionals who often find themselves excluding certain anomalous data points while trying to develop a "normal-looking" map. Contour maps showing regional (say, on a square-mile scale) flow pattern in a fractured rock and karst aquifer may be justified since groundwater flow generally is from recharge areas towards discharge areas and the regional hydraulic gradients will reflect this simple fact. The problems usually rise when interpreting local flow patterns as schematically shown in Figure 1.66.

The direction of groundwater flow in a localized area of an aquifer can be determined if at least three recordings of water table (or piezometric surface) elevations are available. Figure 1.67 illustrates the principle of finding the position of water table in three dimensions using data from three monitoring wells. The fastest way to construct contours is by linear triangulation. The direction of groundwater flow is indicated by the arrow drawn perpen-dicular to and down the contour lines. This is also the direction of the dip of the water table approximated by the flat shaded surface between the three wells. It is very important to understand that the flow direction determined in this way is representative only of the local area covered by the three monitoring wells. Depending on hydrogeologic conditions, this direction may change in a nearby aquifer portion. For that reason, more observation points are usually established during a hydrogeologic investigation in order to construct a reliable contour map.

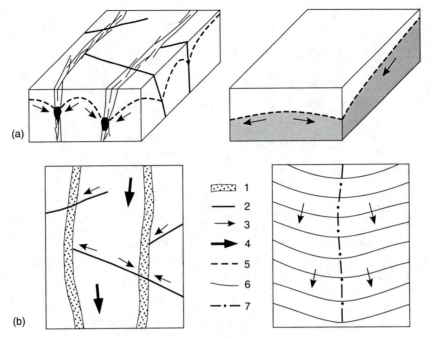

FIGURE 1.66 Groundwater flow and its map presentation for a fractured rock or karst (a) and an intergranular (b) aquifer. (1) Preferential flow path (e.g., fracture or fault zone or karst conduit and channel); (2) fracture or fault; (3) local flow direction; (4) general flow direction; (5) position of the hydraulic head (water table in the intergranular aquifer); (6) hydraulic head contour line; and (7) groundwater divide. (From Krešić, N., *Kvantitativna hidrogeologija karsta sa elementima zaštite podzemnih voda* (in Serbo-Croatian; *Quantitative Karst Hydrogeology with Elements of Groundwater Protection*), Naučna knjiga, Belgrade, 1991, 192 p.)

1.5.1 CONTOURING METHODS

1.5.1.1 Manual Contouring

Manual contouring is practically always utilized in groundwater studies, either as the only method or in conjunction with computer-based methods. A complete reliance on software contouring could lead to erroneous conclusions since computer programs are unable to recognize interpretations apparent to a groundwater professional such as presence of geologic boundaries, varying porous media, influence of surface water bodies, or principles of groundwater flow. Thus, manual contouring and manual reinterpretation of computer-generated maps are essential and integral parts of hydrogeologic studies. Manual contouring is based on the triangular linear interpolation (the principle of which is shown in Figure 1.67) combined with the hydrogeologic experience of the interpreter. The first draft map is not necessarily an exact linear interpolation between data points. Rather, it is an interpretation of the hydrogeologic and hydrologic conditions with contours that roughly follow numeric data on water table (or piezometric surface) elevations. Whenever possible, the contours should be drawn to satisfy principles described in the flow net analysis. This means that almost inevitable local irregularities in water table (hydraulic head) elevations should not blur the overall tendency of groundwater flow (remember that a groundwater contour map is also a flow net without shown streamlines and it represents a graphical solution of the two-dimensional flow field).

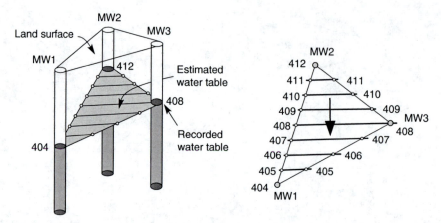

FIGURE 1.67 *Left*: Finding the position of the water table in three dimensions using data from three monitoring wells (numbers are water levels in meters or feet above seal level). *Right*: Construction of water table contour lines by triangulation with linear interpolation. The arrow, perpendicular to the contours, indicates the direction of groundwater flow.

A novice is often caught in drawing various depressions in water table from which there is no escape of groundwater. Unless there is a valid hydrogeologic explanation (e.g., presence of a pumping well, or vertical flow and gradient from the regolith into the underlying bedrock), these depressions are probably the result of erroneous contouring or erroneous data. Similarly, mysterious local mounds in water table should be carefully examined.

One of the most important aspects of constructing contour maps in alluvial aquifers is to determine the relationship between groundwater and surface waters (see Section 2.11, Figure 2.157, and Figure 2.158 in particular). In hydraulic terms, the contact between the aquifer and the surface stream is an equipotential boundary. This does not mean that everywhere along this contact the hydraulic head is the same. Since both the river water and the groundwater are flowing, there is a hydraulic gradient along the contact. If enough measurements of a river stage are available, it is relatively easy to draw the water table contours near the river and to finish them along the river–aquifer contact. However, often few or no precise data are available on river stage and, at the expense of precision, it has to be estimated from a topographic map or from the monitoring well data by extrapolating the hydraulic gradients until they intersect the river. Figure 1.68 is a manually drawn contour map using data listed in Section 12.2.

1.5.1.2 Contouring with Computer Programs

Even though an early-phase computer-generated contour map is likely to be inaccurate because the underlying hydrogeologic variability cannot be considered by any contouring program, it is desirable to have the final map as an *XYZ* computer file. For example, such a "digital" map could be used in developing a numeric groundwater flow model for the project, or within a geographic information system (GIS) of aquifer management. If the available computer program cannot produce a satisfactory contour map (for example, there is a complicated mixture of impermeable and equipotential boundaries), and it cannot be forced to do so by the interpreter, the solution is to digitize a manually drawn map. This, however, may be a lengthy process and it is better to acquire an appropriate software package for contouring. Quite a few computer programs today offer a wide range of contouring methods, allow the interpreters to interactively change generated grids, and display maps in

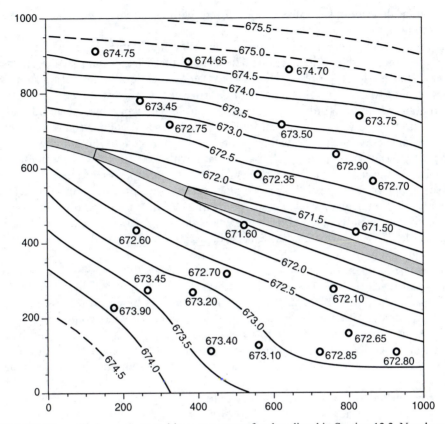

FIGURE 1.68 Manually drawn water table contour map for data listed in Section 12.2. Numbers next to well data points are recorded water table elevations in feet above sea level. Contour interval is 0.5 ft and the coordinate system is local, with the arbitrary origin placed at the lower left corner, with coordinates (0,0).

three dimensions (one of the most versatile and widely used such programs is SURFER by Golden Software).

Contouring programs require that the groundwater level data be organized as XYZ files where X and Y are plane coordinates of the measuring points (monitoring wells) and Z is the elevation of the water table or piezometric surface above a chosen reference level, usually above mean sea level (amsl). The most widely used plane coordinate system in geosciences and engineering is the universal transverse mercator (UTM) system, which is based on a transverse Mercator projection. Individual states in the United States also have state coordinate systems with unique origins and distances (coordinates) expressed in feet. Well and hydrologic databases maintained by state agencies are often based on such a system and can be easily converted into the UTM coordinate system. If these official systems are not, for various reasons, yet applied to the data collected in the field, a simple local (temporary) system may be required (e.g., the origin is arbitrarily set at (0,0)). Once the UTM coordinates of the origin (lower left corner of the local coordinate system used for contouring) are determined, the local coordinates can be easily converted. Although the latitude or longitude system is the most common way to specify geographic locations on the Earth, it is not used in hydrogeologic and engineering practice since it does not allow precise measurements of distances and areas. More examples of various contouring algorithms are given in Section 12.2 and Section 12.3.

Below are brief descriptions of several contouring methods widely used in geosciences and their applicability to groundwater studies.

Triangulation with linear interpolation is the exact linear interpolation between three neighboring points as shown in Figure 1.67. Because the original data points are used to define the triangles, they are preserved (honored) very closely on the map. In general, when there are few data points (e.g., less than 20), or data are not evenly distributed, triangulation is not effective since it creates triangular facets and holes in the map. However, trends indicated by triangulation may allow easy manual filling of the holes and completion of the contours. This also means that further adjustments are usually needed to obtain the final XYZ computer file. Triangulation is fast and accurate with 200 to 1000 data points evenly distributed over the map area. Note, however, that this number of original data is rarely available in groundwater investigations. To alleviate the problem with data availability, one can estimate additional auxiliary data and add them to the data set for contouring. If enough data are available, the advantage of triangulation is that it preserves breaks in contours such as faults, geologic boundaries, and streams when these are indicated (option available in SURFER for example).

Kriging is one of the most robust and widely used methods for interpolation and contouring in different fields. It is a geostatistical method that takes into consideration the spatial variance, location, and sample distribution in data, and it is based on the assumption that there is a spatial correlation between the data. It may be useful when contouring hydrogeologic parameters, which are thought to exhibit spatial correlation, such as the hydraulic conductivity or effective porosity. There are various interpolation models within the kriging method and the most appropriate one for the existing data is chosen after determining the semivariogram of the data, i.e., the geostatistical model that best describes the data. Although all major contouring programs on the market include kriging, few allow for a fully interactive modeling of the semivariogram. Without confirming that the data can be modeled with a particular semivariogram or, in other words, that the data show a degree of spatial correlation that can be described by the particular semivariogram model, blindly using available kriging options within a computer program would not make much sense. This is true, of course, for any other computer contouring method. Kriging is explained in detail in Section 12.3.

Inverse distance to a power is a method usually present in most contouring packages because it is very fast and can handle large data sets. However, it does not produce good results when used to contour groundwater levels often represented with small data sets. Inverse distance to a power weights data points during interpolation so that the influence of one data point relative to another decreases with the power of distance. The result is the generation of "bull's eyes" pattern around the positions of data points as shown in Figure 1.69. Software smoothing of contours can only slightly reduce this effect but does not eliminate unnatural depressions and mounds in water table.

Radial basis functions (splines) is a diverse group of functions analogous to variograms in kriging which produce similar results. They are all exact interpolators and closely preserve original data. Splines are frequently used in geosciences. More detail on contouring with radial basis functions can be found in the work of Carlson and Foley (1991).

1.5.2   CONTOUR MAPS AND REFRACTION OF STREAMLINES IN HETEROGENEOUS AQUIFERS

One clear advantage manual contouring has over computer programs is that the latter cannot fully account for the known heterogeneities in the aquifer. Although some advocates of computer-based contouring argue that it is the most objective method since it excludes possible bias by the interpreter, little can be added to the following statement: if something does not make hydrogeologic sense, it does not matter who or what created the senseless interpretation.

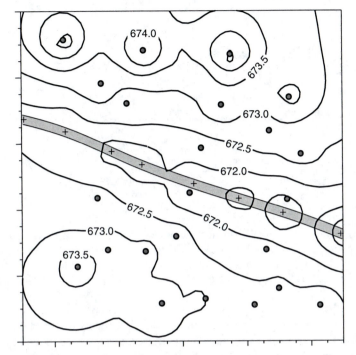

FIGURE 1.69 Water table map generated using data is listed in Section 12.2 and inverse distance to power method within SURFER computer program. The power is two and the smoothing factor is one. Note distinct unnatural (bull's eyes) pattern around the data points: there is no pumping from the aquifer and there are no sinks in the river.

A contour map of an aquifer, accompanied by data on aquifer thickness, is an excellent tool for the analysis of transmissivity and hydraulic conductivity distributions. In homogeneous, isotropic aquifers wider spacing of water table contours indicates higher transmissivity of porous media, while closely spaced contours show parts of the aquifer with lower transmissivity. Since the transmissivity is the product of the hydraulic conductivity and the aquifer thickness ($T = Kb$), the following is true:

- Widely spaced contours indicate larger aquifer thickness, while steeper hydraulic gradients (e.g., closely spaced contours) indicate smaller aquifer thickness.
- When an aquifer, or its portion of interest, has a uniform thickness, steeper hydraulic gradients indicate lower hydraulic conductivity of the porous media, while more widely spaced contours mean higher conductivity.

However, in aquifers that have varying hydraulic conductivity the above simple rules of thumb are not sufficient. Streamlines representing the direction of groundwater flow refract at the boundary between porous media with different hydraulic conductivities (an analogy would be refraction of light rays when they enter a medium with different density, e.g., from air to water). The refraction causes that the incoming angle, or angle of incidence and the outgoing angle of refraction are different (angle of incidence is the angle between the orthogonal line at the boundary and the incoming streamline; angle of refraction is the angle between the orthogonal at the boundary and the outgoing streamline). The only exception is when the streamline is perpendicular to the boundary in which case both angles

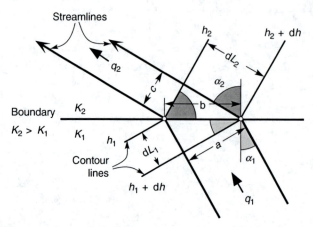

FIGURE 1.70 Refraction of two streamlines at the boundary between two porous media with different hydraulic conductivities (K_1 and K_2). The angle of incidence is α_1 and the angle of refraction is α_2. Other elements are explained in the text. Note that the same angles are shown elsewhere in the picture. This follows from the geometric rule that two angles with their sides perpendicular to each other are equal.

are the same, i.e., 90°. The mathematical relationship between the angle of incidence (α_1), angle of refraction (α_2), and the hydraulic conductivities of two porous media, K_1 and K_2, is explained by the elements shown in Figure 1.70. The figure applies to both map and cross-sectional views as long as there is a clearly defined boundary between two porous media. The unit flow bound by two streamlines remains constant through the two media with different hydraulic conductivities since there is no gain or loss of water across streamlines:

$$q_1 = q_2 \tag{1.119}$$

From the definition of the equipotential line, it follows that the value of the hydraulic head along the same equipotential line remains constant regardless of the changes in hydraulic conductivity of the porous media:

$$h_1 = h_2 \tag{1.120}$$

Also, the increment in the hydraulic head between two successive contour lines is the same:

$$h_1 + \mathrm{d}h = h_2 + \mathrm{d}h \tag{1.121}$$

By applying Darcy's law, the unit flows in two porous media are expressed as

$$q_1 = K_1 a \frac{\mathrm{d}h}{\mathrm{d}L_1} \tag{1.122}$$

$$q_2 = K_2 c \frac{\mathrm{d}h}{\mathrm{d}L_2} \tag{1.123}$$

As already mentioned, these two unit flows are equal:

$$K_1 a \frac{\mathrm{d}h}{\mathrm{d}L_1} = K_2 c \frac{\mathrm{d}h}{\mathrm{d}L_2} \tag{1.124}$$

From the rules of trigonometry (see Appendix B), it follows that

$$a = b \cos \alpha_1 \quad c = b \cos \alpha_2$$

$$\frac{dL_1}{b} = \sin \alpha_1 \quad \frac{b}{dL_1} = \frac{1}{\sin \alpha_1}$$

$$\frac{dL_2}{b} = \sin \alpha_2 \quad \frac{b}{dL_2} = \frac{1}{\sin \alpha_2}$$

When these transformations are included into Equation 1.124, it becomes

$$K_1 \frac{\cos \alpha_1}{\sin \alpha_1} = K_2 \frac{\cos \alpha_2}{\sin \alpha_2} \qquad (1.125)$$

Knowing that

$$\frac{\cos \alpha}{\sin \alpha} = \frac{1}{\tan \alpha}$$

Equation 1.125 can also be written as

$$\frac{K_1}{K_2} = \frac{\tan \alpha_1}{\tan \alpha_2} \qquad (1.126)$$

which is the equation describing the relationship between the angles of incidence and refraction, and the hydraulic conductivities of two porous media. Figure 1.71 shows the refraction of streamlines when they reach a boundary of higher hydraulic conductivity (top) and a boundary of lower hydraulic conductivity (bottom). The most noticeable effect is the change in spacing between both streamlines and contour lines. When interpreting or constructing a groundwater contour map, it is very important to understand the influence of geologic boundaries and a varying hydraulic conductivity on the overall picture. This knowledge is of particular importance during the calibration of groundwater models since the result of modeling groundwater flow is a distribution of the hydraulic heads in the flow field. This distribution, or the contour map of the piezometric surface, changes more or less obviously, whenever one or more hydrogeologic parameters such as the hydraulic conductivity and the position of boundaries change. Essentially, the modeler should know what to expect from the changes and how to adjust them to obtain a satisfactory contour map.

1.5.3 GROUNDWATER FLOW RATE

Once constructed, a contour map can be expanded with the streamlines (flow lines) to obtain a flow net. As mentioned earlier, flow net is a graphical solution for the two-dimensional flow field that can be used to calculate the volumetric flow rate in the aquifer. If the flow net represents a map view, the solution is extended through the overall aquifer thickness. If the flow net is constructed for a cross-sectional view, the solution is extended through the aquifer width perpendicular to the flow direction. Figure 1.72 shows elements for the flow calculations for both map and cross-sectional views.

The base for the flow determination is the unit flow rate per one flow net segment: Δq. The flow net segment is a portion of the aquifer bound by two streamlines and it extends throughout the aquifer. For a map view, when visualized in three dimensions, these bounding

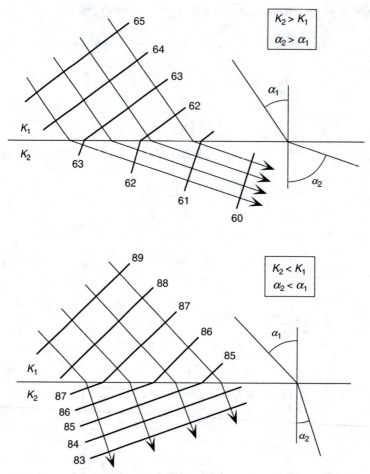

FIGURE 1.71 Refraction of streamlines at a boundary of higher hydraulic conductivity (top) and a boundary of lower hydraulic conductivity (bottom).

streamlines correspond to intersections of vertical stream surfaces and the aquifer top (see Figure 1.72a). The thickness of the segment equals the aquifer thickness (b), and the width of the segment is the distance between two bounding streamlines (ΔW). For a cross-sectional view (Figure 1.72b), the streamlines are intersections of the stream surfaces with the plane of the drawing. The thickness of the flow net segment is the spacing between the two bounding streamlines (ΔW), while the width of the segment equals the width of the aquifer. Note that, since the flow net method is a two-dimensional graphical solution of a three-dimensional flow field, it is very important to understand basic differences between the map view and the cross-sectional view approaches.

1.5.3.1 Map View Approach

The unit flow rate per flow net segment (Δq) is a product of the flow velocity (v) and the flow area (A):

$$\Delta q = v \times A \qquad (1.127)$$

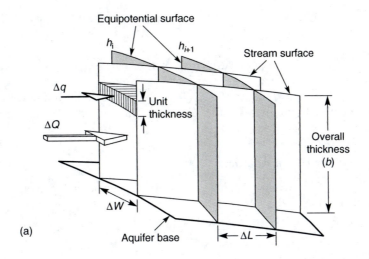

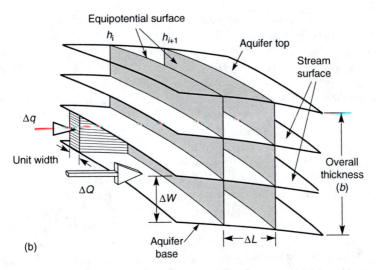

FIGURE 1.72 Elements for the calculations of flow rates using flow nets. (a) Map view approach and (b) cross-sectional view approach.

where the velocity is given as the product of the hydraulic conductivity and the hydraulic gradient:

$$v = K\frac{h_i - h_{i-1}}{\Delta L} = K\frac{\Delta h}{\Delta L} \tag{1.128}$$

and the flow area is the spacing between two bounding streamlines multiplied by one (unity):

$$A = \Delta W \times 1 \tag{1.129}$$

When combining Equation 1.128 and Equation 1.129, the unit flow rate per net segment becomes

$$\Delta q = K\Delta W \frac{\Delta h}{\Delta L} \tag{1.130}$$

The volumetric flow rate per net segment (ΔQ) is obtained by multiplying the unit flow rate with the aquifer thickness (b):

$$\Delta Q = \Delta q \times b \qquad (1.131)$$

Finally, the total flow rate in the aquifer (ΔQ), or in its part that is of interest, is calculated by summation of the flow rates in the corresponding number of N segments (ΔQ_1, $\Delta Q_2, \ldots, \Delta Q_N$).

1.5.3.2 Cross-Sectional View Approach

As seen from Figure 1.72b, the above equation of the unit flow per net segment derived for the map view approach remains the same:

$$\Delta q = K \Delta W \frac{\Delta h}{\Delta L} \qquad (1.132)$$

where K is the aquifer hydraulic conductivity, ΔW is the spacing between two adjacent streamlines in the cross-sectional view, Δh is the hydraulic head difference between two adjacent equipotential lines h_i and h_{i+1} (this corresponds to the contour interval on groundwater contour maps), and ΔL is the spacing between two adjacent equipotential lines. The volumetric flow rate per net segment is obtained by multiplying the unit flow with the aquifer width, and the total flow is the sum of flows in the corresponding number of segments.

The main difficulties in applying analytical solutions to groundwater flow occur when the aquifer has a varying hydraulic conductivity, and varying thickness and slope. A properly constructed flow net "takes care" of these irregularities and greatly simplifies calculations of groundwater flow. However, design of a flow net is often not easy and requires significant time and skill. In addition, this approach assumes that the flow is strictly two-dimensional. As illustrated earlier in Figure 1.43, the flow rate in one flow tube (aquifer segment between two streamlines) remains the same even though the aquifer thickness (and therefore the area of flow A) changes. This means that, in order to calculate the volumetric flow rate in one segment, it is enough to know the hydraulic conductivity and aquifer thickness at only one location within the segment. Also, because the flow net in a homogeneous, isotropic aquifer is a set of curvilinear squares, the unit flow rate in the adjacent flow net segments (flow tubes) is the same. This further simplifies the flow calculations (see Figure 1.73).

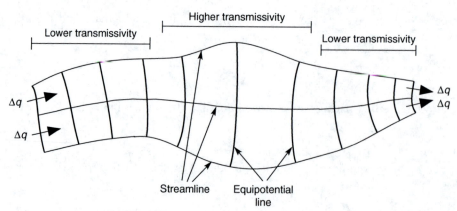

FIGURE 1.73 Flow rates through segments of the flow net are equal. The varying size of the curvilinear squares (i.e., spacing of the equipotential lines and streamlines) is a consequence of varying transmissivity.

1.6 FLOW IN UNSATURATED ZONE

The following discussion by Šimůnek et al. (1999) illustrates the importance of the unsaturated zone as an integral part of the hydrologic cycle:

> The zone plays an inextricable role in many aspects of hydrology, including infiltration, soil moisture storage, evaporation, plant water uptake, groundwater recharge, runoff, and erosion. Initial studies of the unsaturated (vadose) zone focused primarily on water supply studies, inspired in part by attempts to optimally manage the root zone of agricultural soils for maximum crop production. Interest in the unsaturated zone has dramatically increased in recent years because of growing concern that the quality of the subsurface environment is being adversely affected by agricultural, industrial and municipal activities. Federal, state and local action and planning agencies, as well as the public in large, are now scrutinizing the intentional or accidental release of surface-applied and soil-incorporated chemicals in the environment. Fertilizers and pesticides applied to agricultural lands inevitably move below the soil root zone and may contaminate underlying groundwater reservoirs. Chemicals migrating from municipal and industrial disposal sites also represent environmental hazards. The same is true for radionuclides emanating from energy waste disposal facilities.... The past several decades have seen considerable progress in the conceptual understanding and mathematical description of water flow and solute transport processes in the unsaturated zone. A variety of analytical and numerical models are now available to predict water and/or solute transfer processes between the soil surface and groundwater table. The most popular models remain the Richards' equation for variably saturated flow, and the Fickian-based convection-dispersion equation for solute transport. Deterministic solutions of these classical equations have been used, and likely will continue to be used in the near future, for (1) predicting water and solute movement in the vadose zone, (2) analyzing specific laboratory or field experiments involving unsaturated water flow and/or solute transport, and (3) extrapolating information from a limited number of field experiments to different soil, crop and climatic conditions, as well as to different soil and water management schemes.

The main hydraulic characteristic of the unsaturated zone is the negative fluid pressure head, which is lower than the atmospheric pressure. This negative pressure head is often called matric potential, and sometimes suction. The unsaturated soil will suck water or fluid from a porous cup of an instrument used to measure the pressure head above the water table. Conversely, sampling soil water requires application of enough pressure to overcome the soil matric potential: the water has to be extracted by suction. The matric potential, caused by capillary and adhesive forces, is a function of water content and sediment texture; it is stronger in less saturated and finer soils. As the soil saturation increases, the pressure head becomes less negative. At full saturation, at the water table, the pressure head is zero and equals the atmospheric pressure. Below the water table, the pressure head is positive and increases with depth. This is schematically illustrated in Figure 1.74 for a downward flux of infiltrating water. Small negative pressure heads (less than about 100 kPa) can be measured with tensiometers, which couple the measuring fluid in a manometer, vacuum gage, or pressure transducer to water in the surrounding partially saturated soil through a porous membrane. The pressure status of water held under large negative pressures (greater than 100 kPa) may be measured using thermocouple psychrometers, which measure the relative humidity of the gas phase within the medium, and with heat dissipation probes (HDPs) (Lappala et al., 1987; McMahon et al., 2003). The measuring instrumentation may be permanently installed in wells screened at different depths to measure soil moisture profile (Figure 1.75), or measurements may be performed on a temporary basis using direct push methods, such as cone penetrometers equipped with tension rings.

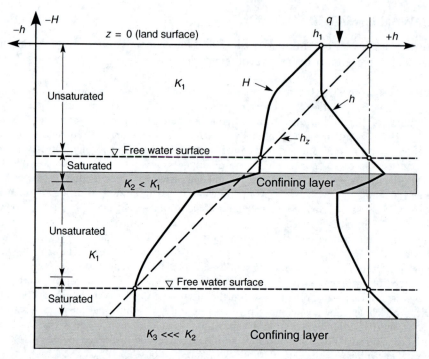

FIGURE 1.74 Relation among capillary (h), elevation (h_z), and total (H) potentials for downward flux (q) through layered media with a perched water table and a deep water table. (From Lappala, E.G., Healy, R.W., and Weeks, E.P., U.S. Geological Survey Water-Resources Investigations Report 83-4099, Denver, Colorado, 1987, 131 p.)

Similarly to the groundwater flow, the flow of water in the unsaturated zone is governed by two main parameters—the change in total potential (hydraulic head) along the flow path and the hydraulic conductivity. However, both parameters depend on the volumetric water content in the porous medium; they change in time and space as the soil becomes more or less saturated in response to water input and output (e.g., due to infiltration and evapotranspiration). The unsaturated hydraulic conductivity is always lower than the saturated hydraulic conductivity because of the presence of air in the voids, and it increases with the increasing saturation. Quantification of flow processes in the unsaturated zone is therefore more complex than in the saturated zone since it requires determination of two dynamic (time-dependent) parameters, which are constant in the saturated zone where saturation is always 100% by definition, and the hydraulic conductivity, although dependent on fluid characteristics, can be considered constant for most practical purposes.

1.6.1 Unsaturated Flow Equations

Water flow in variably saturated soils is traditionally described with the Richards equation (Richards, 1931) as follows (van Genuchten et al., 1991):

$$C\frac{\partial h}{\partial t} = \frac{\partial\left(K\dfrac{\partial h}{\partial z} - K\right)}{\partial z} \tag{1.133}$$

where h is the soil water pressure head, or matric potential (with dimension L), t is time (T), z is soil depth (L), K is the hydraulic conductivity (LT^{-1}), and C is the soil water capacity (L^{-1}) approximated by the slope ($d\theta/dh$) of the soil water retention curve, $\theta(h)$, in which θ is

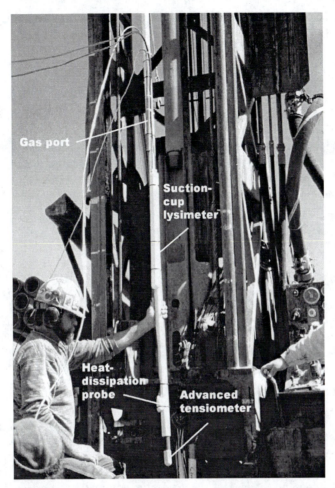

FIGURE 1.75 Unsaturated zone monitoring equipment prepared for installation. (From McMahon, P.B. et al., U.S. Geological Survey Water-Resources Investigations Report 03-4171, Reston, Virginia, 2003, 32 p.)

the volumetric water content ($L^3 L^{-1}$). Equation 1.133 may also be expressed in terms of the water content if the soil profile is homogeneous and unsaturated ($h \leq 0$):

$$\frac{\partial \theta}{\partial t} = \frac{\partial \left(D \frac{\partial \theta}{\partial z} - K \right)}{\partial z} \qquad (1.134)$$

where D is the soil diffusivity ($L^2 T^{-1}$) defined as

$$D = K \frac{\mathrm{d}h}{\mathrm{d}\theta} \qquad (1.135)$$

The unsaturated soil hydraulic functions in the above equations are the soil water retention curve, $\theta(h)$, the hydraulic conductivity function, $K(h)$ or $K(\theta)$, and the soil water diffusivity function $D(\theta)$. Several functions have been proposed to empirically describe the soil water retention curve. One of the most widely used is the equation of Brooks and Corey (1964; van Genuchten et al., 1991; Šimůnek et al., 1999):

$$\theta = \begin{cases} \theta_r + (\theta_s - \theta_r)(\alpha h)^{-\lambda} & (h < -1/\alpha) \\ \theta_s & (h \geq -1/\alpha) \end{cases} \tag{1.136}$$

where θ is the volumetric water content, θ_r is the residual water content, θ_s is the saturated water content, α is an empirical parameter (L^{-1}) whose inverse $(1/\alpha)$ is often referred to as the air entry value or bubbling pressure, α has negative value for unsaturated soils, λ is a pore-size distribution parameter affecting the slope of the retention function, and h is the soil water pressure head, which has negative value for unsaturated soil.

Equation 1.136 may be written in a dimensionless form as follows:

$$S_e = \begin{cases} (\alpha h)^{-\lambda} & (h < -1/\alpha) \\ 1 & (h \geq -1/\alpha) \end{cases} \tag{1.137}$$

where S_e, is the effective degree of saturation, also called the reduced water content $(0 < S_e < 1)$:

$$S_e = \frac{\theta - \theta_r}{\theta_s - \theta_r} \tag{1.138}$$

The residual water content, θ_r in Equation 1.138 specifies the maximum amount of water in a soil that will not contribute to liquid flow because of blockage from the flow paths or strong adsorption onto the solid phase (Luckner et al., 1989, from van Genuchten et al., 1991). Formally, θ_r may be defined as the water content at which both $d\theta/dh$ and K go to zero when h becomes large. The residual water content is an extrapolated parameter, and may not necessarily represent the smallest possible water content in a soil. This is especially true for arid regions where vapor phase transport may dry out soils to water contents well below θ_r. The saturated water content, θ_s, denotes the maximum volumetric water content of a soil. The saturated water content should not be equated to the porosity of soils; θ_s of field soils is generally about 5% to 10% smaller than the porosity because of entrapped or dissolved air (van Genuchten et al., 1991).

On a logarithmic plot, Equation 1.137 generates two straight lines that intersect at the air entry value, $h_\alpha = 1/\alpha$. Because of their simple form, Equation 1.136 and Equation 1.137 have been used in numerous unsaturated flow studies. The Brooks and Corey equation has been shown to produce relatively accurate results for many coarse-textured soils characterized by relatively narrow pore- or particle-size distributions. Results have generally been less accurate for many fine-textured and undisturbed field soils because of the absence of a well-defined air entry value for these soils. A continuously differentiable (smooth) equation proposed by van Genuchten (1980) significantly improves the description of soil water retention near saturation:

$$S_e = \frac{1}{[1 + (\alpha h)^n]^m} \tag{1.139}$$

where α, n, and m are empirical constants affecting the shape of the retention curve $(m = 1-1/n)$. By varying the three constants, it is possible to fit almost any measured field curve (see examples in Figure 1.76). It is this flexibility that made the van Genuchten equation arguably the most widely used in various computer models of unsaturated flow, and

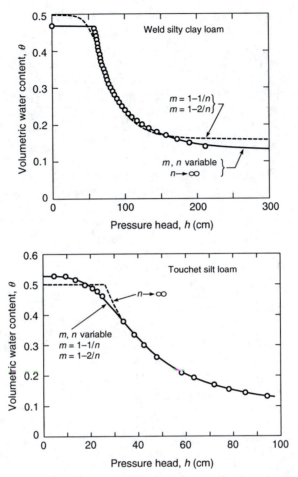

FIGURE 1.76 Examples of fitted and observed retention curves using van Genuchten equation. (From van Genuchten, M.Th., Leij, F.J., and Yates, S.R., The RETC Code for Quantifying the Hydraulic Functions of Unsaturated Soils, EPA/600/2-91/065, Ada, Oklahoma, 1991, 83 p.)

contaminant fate and transport. Combining Equation 1.138 and Equation 1.139 gives the following form of the van Genuchten equation:

$$\theta(h) = \theta_r + \frac{\theta_s - \theta_r}{[1 + (\alpha h)^n]^{1-1/n}} \tag{1.140}$$

The unsaturated hydraulic conductivity, which is a function of the soil water contents, can be estimated from the soil water retention curve. Applying restrictions to van Genuchten parameters m and n (such as $m = 1-1/n$ and $m = 1-2/n$) leads to relatively simple expressions for the hydraulic conductivity when Equation 1.139 is combined with the theoretical pore-size distribution model of Mualem (1976) for example. In contrast, the variable $m - n$ case leads to mathematical expressions for K and D, which may be too complicated for routine use in unsaturated flow studies. Imposing one of the three restrictions (including the restriction that $n \to \infty$) will, for a given value of $m - n$, fix the shape of the retention curve at the wet end when S_e approaches saturation. Of the three cases with restricted m, $m = 1 - 1/n$ seems to perform best for many but not all soils. Although the variable $m - n$ case usually produces superior results, its use is not necessarily recommended for all observed retention data sets. In many situations, especially when observed field data sets are involved, only a limited range of retention values (usually in the wet range) is available. Unless augmented with laboratory measurements at relatively low water contents, such data sets may not lead to an accurate definition of the retention curve in the dry range. Keeping m and n variables in those cases often leads to uniqueness problems in the parameter estimation process. Typically, m and n will then become strongly correlated, leading to poor convergence and ill-defined parameter values with large confidence intervals. More stable results are generally obtained when the restrictions $m = 1-1/n$ or $m = 1-2/n$ are implemented for these incomplete data sets (van Genuchten et al., 1991).

One of the widely used models for predicting the unsaturated hydraulic conductivity from the soil retention profile is the model of Mualem (1976), which may be written in the following form (van Genuchten et al., 1991):

$$K(S_e) = K_s S_e \left[\frac{f(S_e)}{f(l)} \right]^2 \tag{1.141}$$

$$f(S_e) = \int_0^{S_e} \frac{1}{h(x)} \, dx \tag{1.142}$$

where S_e is given by Equation 1.138, K_s is the hydraulic conductivity at saturation, and l is an empirical pore-connectivity (tortuosity) parameter estimated by Mualem to be about 0.5 as an average for many soils. Detailed solution of the Mualem's model by incorporating Equation 1.139 is given by van Genuchten et al. (1991). This solution, sometimes called van Genuchten–Mualem equation, has the following form:

$$K(S_e) = K_o S_e^l \left\{ 1 - \left[1 - S_e^{n/(n-1)} \right]^{1-1/n} \right\}^2 \tag{1.143}$$

where K_o is the matching point at saturation and similar, but not necessarily equal, to the saturated hydraulic conductivity (K_s). Fitting the van Genuchten–Mualem equation to field data gives good results in most cases (see Figure 1.77).

Rosetta and RETC are two very useful programs developed at the U.S. Salinity Laboratory for estimating unsaturated zone hydraulic parameters. Rosetta (Schaap, 1999) is Windows-based program for estimating unsaturated hydraulic properties from surrogate soil data such as soil texture data and bulk density. Models like this are often called pedotransfer functions (PTFs) because they translate basic soil data into hydraulic properties. Rosetta can be used to estimate the following properties:

1. Water retention parameters in the van Genuchten equation
2. Saturated hydraulic conductivity
3. Unsaturated hydraulic conductivity parameters in the van Genuchten–Mualem equation

Rosetta offers five PTFs that allow the prediction of the hydraulic properties with limited or more extended sets of input data. This hierarchical approach is of a great practical use because it permits the optimal use of available input data. The five models use the following input data:

1. Soil textural classes
2. Sand, silt, and clay percentages

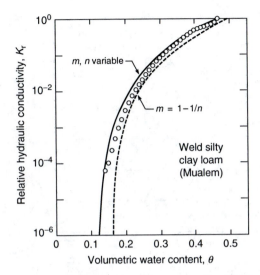

FIGURE 1.77 Observed and fitted relative conductivity curves for Weld silty clay loam; Mualem's model with $l = 0.5$. (From van Genuchten, M.Th., Leij, F.J., and Yates, S.R., The RETC Code for Quantifying the Hydraulic Functions of Unsaturated Soils, EPA/600/2-91/065, Ada, Oklahoma, 1991, 83 p.)

3. Sand, silt, and clay percentages and bulk density
4. Sand, silt, and clay percentages, bulk density, and a water retention point at 330 cm (33 kPa)
5. Sand, silt, and clay percentages, bulk density, and water retention point at 330 and 15,000 cm (33 and 1500 kPa)

The first model is based on a lookup table that provides class average hydraulic parameters for each USDA soil textural class (see Figure 1.78 and Table 1.2). The other four models are based on neural network analysis and provide more accurate predictions when more input data variables are used. In addition to the hierarchical approach, Rosetta includes a model that allows the prediction of unsaturated hydraulic conductivity parameters from fitted van Genuchten retention parameters. This model can be applied in the hierarchical approach where it automatically uses the predicted retention parameters as input instead of measured (fitted) retention parameters. All estimated hydraulic parameters are accompanied by uncertainty estimates that allow an assessment of the reliability of Rosetta's predictions.

The RETC computer program (van Genuchten et al., 1991) provides several options for describing or predicting the hydraulic properties of unsaturated soils. These properties define the soil water retention curve, $\theta(h)$, the hydraulic conductivity function, $K(h)$ or $K(\theta)$, and the soil water diffusivity function, $D(\theta)$. The soil water retention function is given by Equation 1.140, which contains five independent parameters, i.e., the residual water content θ_r, the saturated water content θ_s, and the shape factors α, n, and m. The predictive equations for K and D add two additional unknowns: the pore connectivity parameter, l, and the saturated hydraulic conductivity, K_s. Hence, the unsaturated soil hydraulic functions contain up to seven potentially unknown parameters. The restrictions $n \to \infty$ (i.e., the Brooks–Corey restriction), $m = 1-1/n$, and $m = 1-2/n$ will reduce the maximum number of independent parameters from seven to six. The RETC code may be used to fit any one, several, or all of the six or seven unknown parameters simultaneously to observed data. RETC can be applied to four broad classes of problems (van Genuchten et al., 1991):

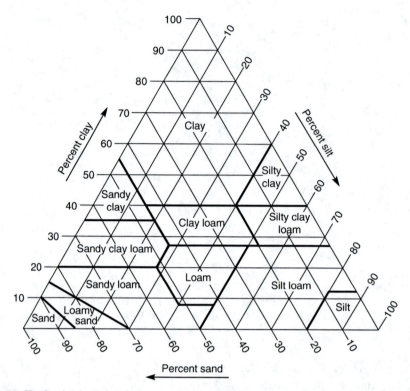

FIGURE 1.78 Twelve basic textural classes of soils with the percentages of sand, silt, and clay. (From USDA, Soil Survey Division Staff, *Soil Survey Manual*, Soil Conservation Service, U.S. Department of Agriculture Handbook 18, Available online at: http://soils.usda.gov/technical/manual/, accessed on December 15, 2005.)

1. Direct calculation of the unsaturated soil hydraulic characteristics from the user-specified parameters: α, n, m, θ_r, θ_s, l, and K_s. Values for l and K_s are not needed when only the retention function is calculated.
2. Fitting the unknown retention parameters (with or without restricted m, n values) to observed soil water retention data. The fitted retention parameters are subsequently used to predict the hydraulic conductivity and diffusivity functions by making use of the models of Mualem or Burdine.
3. Predicting $\theta(h)$ from observed K/D data. In some instances, experimental conductivity data may be available but no observed retention data. Such situations sometimes arise for certain coarse-textured or gravelly soils when tensiometers fail to operate correctly. RETC may then be used to fit the unknown hydraulic coefficients to observed conductivity data using one of the available predictive conductivity or diffusivity models.
4. Simultaneous fit of retention and K/D data. This option results in a simultaneous fit of the model parameters to observed water retention and hydraulic conductivity or diffusivity data.

Once the calculated unsaturated hydraulic conductivity and the related soil moisture profiles are fitted to the field data, or estimated based on general soil characteristics (e.g., grain size and density), the Richards equation can be used to calculate water flux moving through the unsaturated zone and eventually reaching the water table, thus becoming aquifer recharge. The solution of the flow equation requires boundary and initial conditions,

TABLE 1.2
Class-Average Values of the Seven Hydraulic Parameters for the 12 Basic USDA Textural Classes

Texture Class	N	θ_r cm³/cm³		θ_s cm³/cm³		$\log(\alpha)$ log(1/cm)		$\log(n)$ log10		K_s log(cm/d)		K_o log(cm/d)		l	
Clay	84	0.098	(0.107)	0.459	(0.079)	-1.825	(0.68)	0.098	(0.07)	1.169	(0.92)	0.472	(0.26)	-1.561	(1.39)
Clay loam	140	0.079	(0.076)	0.442	(0.079)	-1.801	(0.69)	0.151	(0.12)	0.913	(1.09)	0.699	(0.23)	-0.763	(0.90)
Loam	242	0.061	(0.073)	0.399	(0.098)	-1.954	(0.73)	0.168	(0.13)	1.081	(0.92)	0.568	(0.21)	-0.371	(0.84)
Loamy sand	201	0.049	(0.042)	0.390	(0.070)	-1.459	(0.47)	0.242	(0.16)	2.022	(0.64)	1.386	(0.24)	-0.874	(0.59)
Sand	308	0.053	(0.029)	0.375	(0.055)	-1.453	(0.25)	0.502	(0.18)	2.808	(0.59)	1.389	(0.24)	-0.930	(0.49)
Sandy clay	11	0.117	(0.114)	0.385	(0.046)	-1.476	(0.57)	0.082	(0.06)	1.055	(0.89)	0.637	(0.34)	-3.665	(1.80)
Sandy clay loam	87	0.063	(0.078)	0.384	(0.061)	-1.676	(0.71)	0.124	(0.12)	1.120	(0.85)	0.841	(0.24)	-1.280	(0.99)
Sandy loam	476	0.039	(0.054)	0.387	(0.085)	-1.574	(0.56)	0.161	(0.11)	1.583	(0.66)	1.190	(0.21)	-0.861	(0.73)
Silt	6	0.050	(0.041)	0.489	(0.078)	-2.182	(0.30)	0.225	(0.13)	1.641	(0.27)	0.524	(0.32)	0.624	(1.57)
Silty clay	28	0.111	(0.119)	0.481	(0.080)	-1.790	(0.64)	0.121	(0.10)	0.983	(0.57)	0.501	(0.27)	-1.287	(1.23)
Silty clay loam	172	0.090	(0.082)	0.482	(0.086)	-2.076	(0.59)	0.182	(0.13)	1.046	(0.76)	0.349	(0.26)	-0.156	(1.23)
Silty loam	330	0.065	(0.073)	0.439	(0.093)	-2.296	(0.57)	0.221	(0.14)	1.261	(0.74)	0.243	(0.26)	0.365	(1.42)

See Figure 1.78 for the percentages of sand, silt, and clay in the classes. The values are generated by computing the average values for each textural class. K_o and l are based on predicted parameters and may not be very reliable. The values in parenthesis give the one standard deviation uncertainties of the class average values.

Source: From Schaap, M.G., *Rosetta, Version 1.0,* U.S. Salinity Laboratory, U.S. Department of Agriculture, Riverside, California, 1999.

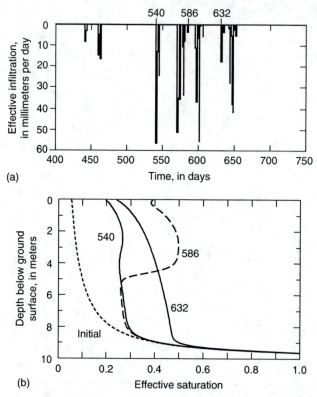

FIGURE 1.79 (a) Effective infiltration used as upper boundary condition for VS2DT model and (b) effective saturation profiles at indicated times, in days, simulated by using VS2DT for loamy sand with a 10-m-deep water table. Initial saturation for time is 0 d; shift to the right for 540 d is partly due to antecedent infiltration not shown on the graph. (Modified from O'Reilly, A.M., U.S. Geological Survey Scientific Investigations Report 2004-5195, Reston, Virginia, 2004, 49 p.)

such as soil moisture profile (initial condition) and one or more of the following: prescribed head and flux boundaries, boundaries controlled by atmospheric conditions (e.g., infiltration and evapotranspiration), and free drainage boundary (e.g., drainage to the underlying groundwater flow system). In most practical cases, application of computer programs for the variably saturated flow is preferable compared to complex and lengthy analytical solutions (see Figure 1.79). Two such programs, widely used for their versatility and robustness, and friendly graphical user interface, are VS2DT by the USGS, which is in public domain, and HYDRUS-2D initially developed at the U.S. Salinity Laboratory of the U.S. Department of Agriculture (this program, unlike HYDRUS-1D, is not in public domain). The U.S. Salinity Laboratory developed a number of other useful programs and has been the leader in unsaturated zone research under the guidance of van Genuchten.

A solved problem of contaminant fate and transport through unsaturated and saturated zones, utilizing VS2DT, is available for download at the followng URL: http://www.crcpress.com/e_products/downloads/download.asp?cat_no=3348

REFERENCES

Barlow, P.M., 2003. Ground Water in Freshwater-Saltwater Environments of the Atlantic Coast, *U.S. Geological Survey Circular* 1262, Reston, VA, 113 p.

Bear, J., C.F. Tsang, and G. deMarsily, Eds., 1993. *Flow and Contaminant Transport in Fractured Rock*, Academic Press, San Diego, 548 p.

Bishop, P.K. and J.W. Lloyd, 1990. Chemical and isotopic evidence for hydrogeological processes occurring in the Lincolnshire Limestone. *Journal of Hydrology*, 121, 293–320.

Bloch, S., 1991. Empirical prediction of porosity and permeability in sandstones. *American Association of Petroleum Geologists Bulletin*, 75 (7), 1145–1160.

Bögli, A., 1980. *Karst Hydrology and Physical Speleology*, Springer-Verlag, New York, pp. 85–93, 163–164.

Brooks, R.H. and A.T. Corey, 1964. Hydraulic properties of porous media. *Hydrology Paper*, No. 3, Colorado State University, Fort Collins, Colorado, 27 p.

Budd, D.A. and H.L. Vacher, 2004. Matrix permeability of the confined Floridan aquifer. *Hydrogeology Journal*, 12, 531–549.

Cacas, M.C., 1989. Développement d'un modèle tridimensionel stochastique discret pour la simulation de l'écoulement et des transports de masse et de chaleur en milieu fracturé, Ph.D. thesis, Ecole des Mines de Paris, Fontainebleau, France.

Carlson, R.E. and T.A. Foley, 1991. *Radial Basis Interpolation on Track Data*, Lawrence Livermore National Laboratory, UCRL-JC-1074238.

Chilès, J.P., 1989a. Three-dimensional geometric modeling of a fracture network. In: *Geostatistical Sensitivity, and Uncertainty Methods for Ground-Water Flow and Radionuclide Transport Modeling*, Buxton, B.E., Ed., Battelle Press, Columbus, Ohio, pp. 361–385.

Chilès, J.P., 1989b. Modélisation géostatistique de réseaux de fractures. In: *Geostatistics*, Armstrong, M., Ed., Kluwer Academic Publisher, Dordrecht, Vol. 1, pp. 57–76.

Chilès, J.P. and G. de Marsily, 1993. Stochastic models of fracture systems and their use in flow and transport modeling. In: *Flow and Contaminant Transport in Fractured Rock*, Bear, J., C.F. Tsang, and G. de Marsily, Eds., Academic Press, San Diego, pp. 169–236.

Clarke, J.S. et al., 2004. Hydraulic properties of the Floridan Aquifer System and equivalent clastic units in coastal Georgia and adjacent parts of South Carolina and Florida. Georgia Geologic Survey Information Circular 109, Atlanta, Georgia, 50 p.

Cohen, R.M. and J.W. Mercer, 1993. *DNAPL Site Evaluation*, C.K. Smoley, CRC Press, Boca Raton, FL.

Cook, P.G., 2003. *A Guide to Regional Groundwater Flow in Fractured Aquifers*, CSIRO Land and Water, Seaview Press, Henley Beach, South Australia, 108 p.

Craig, D.H., 1988. Caves and other features of Permian karst in San Andreas Dolomite, Yates Field reservoir, West Texas. In: *Paleokarst*, James, N.P. and P.W. Choquette, Eds., Springer-Verlag, New York, pp. 342–363.

Darcy, H., 1856. *Les fontaines publiques de la ville de Dijon; Exposition et application des principes a suivre et des formulas a employer dans les questions de distribution d'eau*, Appendice, Note D, Victor Dalmont, Ed., Libraire des Corps Impériaux des Ports et Chausses et des Mines, Paris.

Darton, N.H., 1902. Preliminary list of deep borings in United States. U.S. Geological Survey Water-Supply and Irrigation Paper, Washington, D.C., Part 1, 60 p. and Part 2, 67 p.

Darton, N.H., 1904. Geology and underground water resources of the Central Great Plains. U.S. Geological Survey Professional Paper, Washington, D.C.

Daubree, G.A., 1887. *Les eaux souterraines*, Ch. Dunod, 49 Quai des Augustins, Paris, 446 p.

Domenico, P.A. and F.W. Schwartz, 1990. *Physical and Chemical Hydrogeology*, John Wiley & Sons, New York, 824 p.

Ellis, E.E., 1909. Ground water in the crystalline rocks of Connecticut (Chapter IV). Underground water resources of Connecticut. U.S. Geological Survey Water-Supply Paper 232, Gregory, H.E., Ed., Washington, D.C., pp. 54–103.

Faybishenko, B., P.A. Witherspoon, and S.M. Benson, Eds., 2000. *Dynamics of Fluids in Fractured Rock*, Geophysical Monograph 122, American Geophysical Union, Washington, D.C., 400 p.

Ford, D.C. and P.W. Williams, 1989. *Karst Geomorphology and Hydrology*, Unwin Hyman, London, 601 p.

Franke, O.L. et al., 1990. Study guide for a beginning course in ground-water hydrology; Part 1, course participants. U.S. Geological Survey Open-File Report 90-183, Reston, Virginia, 184 p.

Fuller, M.L., 1904. Underground waters of eastern United States. U.S. Geological Survey Water-Supply Paper 118, Washington, D.C., 285 p., 18 pls.

Fuller, M.L., 1905. Contributions to the hydrology of eastern United States. U.S. Geological Survey Water-Supply and Irrigation Paper No. 145, Series 0, Underground Waters 46, Washington, D.C., 220 p.

Fuller, M.L., 1906. Total amount of free water in earth's crust. U.S. Geological Survey Water-Supply Paper 160, Washington, D.C., pp. 59–72.

Fuller, M.L., 1908. Summary of the controlling factors of artesian flows. U.S. Geological Survey Bulletin 319, Washington, D.C., 44 p.

Fuller, M.L., 1910. Underground waters for farm use. U.S. Geological Survey Water-Supply Paper 255, Washington, D.C.

Fuller, M.L. et al., 1911. Underground-water papers 1910. U.S. Geological Survey Water-Supply Paper 258, Washington, D.C., 128 p.

Ghilarov, A., 1998. Lamarck and the prehistory of ecology. *International Microbiology*, 1, 161–164.

Giles, R.V., J.B. Evett, and L. Chiu, 1994. *Fluid Mechanics and Hydraulics*, 3rd ed., Schaum's Outline Series, McGraw-Hill, New York, 378 p.

Greswell, R. et al., 1998. The micro-scale hydrogeological properties of the Lincolnshire Limestone, U.K. *Quarterly Journal of Engineering Geology and Hydrogeology*, 31, 181–197.

Halihan, T., R.E. Mace, and J.M. Sharp, Jr., 2000. Flow in the San Antonio segment of the Edwards aquifer: matrix, fractures, or conduits? In: *Groundwater Flow and Contaminant Transport in Carbonate Aquifers*, Wicks, C.M. and I.D. Sasowsky, Eds., A.A. Balkema, Rotterdam, pp. 129–146.

Hanshaw, B.B. and W. Back, 1974. Determination of Regional Hydraulic Conductivity through Use of ^{14}C Dating of Groundwater. Mémoires, Tome X, 1. Communications, 10th Congress of the International Association of Hydrogeologists, Montpellier, France, pp. 195–198.

Hazen, A., 1892. Experiments upon the purification of sewage and water at the Lawrence Experiment Station, November 1, 1889 to December 31, 1891. Massachusetts Board of Health Twenty-third Annual Report, Massachusetts, pp. 428–434.

Hazen, A., 1893. Some physical properties of sands and gravels with special reference to their use in filtration. Massachusetts Board of Health Twenty-fourth Annual Report, Massachusetts.

Hilgard, E.W., 1906. *Soils*, Macmillan Co., New York, p. 206.

Hsieh, P.A., 2001. Topodrive and Particleflow—two computer models for simulation and visualization of groundwater flow and transport of fluid particles in two dimensions. U.S. Geological Survey Open-file Report 01-286, Menlo Park, California, 30 p.

Johnson, A.I., 1967. Specific yield—compilation of specific yields for various materials. U.S. Geological Survey Water-Supply Paper 1662-D, 74 p.

Johnson, C.D. et al., 2002. Borehole-geophysical investigation of the University of Connecticut landfill. U.S. Geological Survey Water-Resources Investigations Report 01-4033, Storrs, Connecticut, 42 p.

Jones, W.K., 1997. Karst hydrology atlas of West Virginia. Special Publication 4, Karst Waters Institute, Charles Town, West Virginia, 111p.

Kashef, A.I., 1987. *Groundwater Engineering*, McGraw-Hill, New York, 512 p.

King, F.H., 1892. Observations and experiments on the fluctuations in the level and rate of movement of ground water on the Wisconsin Agricultural Experiment Station farm at Whitewater, Wisconsin. *U.S. Weather Bureau Bulletin*, 5, 67–69.

King, F.H., 1899a. Principles and conditions of the movements of ground water. U.S. Geological Survey Nineteenth Annual Report, Washington, D.C.

King, F.H., 1899b. *Irrigation and Drainage*, Macmillan Co., New York.

King, F.H., 1900. *A Textbook of the Physics of Agriculture*, Madison, Wisconsin, p. 124.

Knochenmus, L.A. and J.L. Robinson, 1996. Descriptions of anisotropy and heterogeneity and their effect on ground-water flow and areas of contribution to public supply wells in a karst carbonate aquifer system. U.S. Geological Survey Water-Supply Paper 2475, Washington, D.C., 47 p.

Konikow, L.F. and Bredehoeft, J.D., 1978. Computer model of two-dimensional solute transport and dispersion in ground water. U.S. Geological Survey Techniques of Water-Resources Investigations, Book 7, Chapter C2, 90 p.

Kresic, N., 1998. *Hydrologic and Hydrogeologic Study; Costa Serena Project, Loiza*, Law Environmental-Caribe, Santurce, Puerto Rico.

Krešić, N., 1991. *Kvantitativna hidrogeologija karsta sa elementima zaštite podzemnih voda* (in Serbo-Croatian; *Quantitative Karst Hydrogeology with Elements of Groundwater Protection*), Naučna knjiga, Belgrade, 192 p.

Kupusović, T., 1989. Measurements of piezometric pressures along deep boreholes in karst area and their assessment. Nas Krs, Vol. XV, No. 26–27, pp. 21–30.

Lamarck, J.-B., 1802. *Hydrogéologie*, L'Auter, au Muséum d'Histoire Naturelle (Jardin des Plantes), Paris, 268 p.

Lappala, E.G., R.W. Healy, and E.P. Weeks, 1987. Documentation of computer program VS2D to solve the equations of fluid flow in variably saturated porous media. U.S. Geological Survey Water-Resources Investigations Report 83-4099, Denver, Colorado, 131 p.

Lebedev, A.V., 1968. Determination of hydrogeological parameters by means of piezometric data. In: *Seminar on Groundwater Balance* (in Serbian), Boreli, M., Ed., Yugoslav Committee for the International Hydrologic Decade, Belgrade, 227 p.

Lee, C.H., 1912. An intensive study of the water resources of a part of Owens Valley California. U.S. Geological Survey Water-Supply Paper 294, 135 p.

Lee, C.H., 1915. The determination of safe yield of underground reservoirs of the closed basin type. *Transactions, American Society of Civil Engineers* 78, 148–251.

Lohman, S.W., 1972. Ground-water hydraulics. U.S. Geological Survey Professional Paper 708, 70 p.

Long, J.C.S., 1983. Investigation of Equivalent Porous Medium Permeability in Networks of Discontinuous Fractures, Ph.D. dissertation, University of California, Berkeley, 277 p.

Long, J.C.S., P. Gilmour, and P.A. Witherspoon, 1985. A model for steady fluid in random three-dimensional networks of disc-shaped fractures. *Water Resources Research*, 21 (8), 1105–1115.

Lovelock, J.E., 2000 (first published in 1979). *Gaia, a New Look at Life on Earth*, Oxford University Press, Oxford, 176 p

Maidment, D.R., Ed., 1993. *Handbook of Hydrology*, McGraw-Hill, New York (various paging).

McDonald, M.G. and A.W. Harbaugh, 1988. A modular three-dimensional finite-difference ground-water flow model. U.S. Geological Survey Techniques of Water-Resources Investigations, Book 6, Chapter A1, 586 p.

McMahon, P.B. et al., 2003. Water movement through thick unsaturated zones overlying the Central High Plains Aquifer, Southwestern Kansas, 2000–2001. U.S. Geological Survey Water-Resources Investigations Report 03-4171, Reston, Virginia, 32 p.

Meinzer, O.E., 1923a. The occurrence of ground water in the United States with a discussion of principles. U.S. Geological Survey Water-Supply Paper 489, Washington, D.C., 321 p.

Meinzer, O.E., 1923b. Outline of ground-water hydrology with definitions. U.S. Geological Survey Water-Supply Paper 494, Washington, D.C., 71 p.

Meinzer, O.E., 1927a. Large springs in the United States. U.S. Geological Survey Water-Supply Paper 557, Washington, D.C., 94 p.

Meinzer, O.E., 1927b. Plants as indicators of ground water. U.S. Geological Survey Water-Supply Paper 577, Washington, D.C., 95 p.

Meinzer, O.E., 1932 (reprint 1959). Outline of methods for estimating ground-water supplies. Contributions to the hydrology of the United States, 1931. U.S. Geological Survey Water-Supply Paper 638-C, Washington, D.C., pp. 99–144.

Meinzer, O.E., 1940. Ground water in the United States: a summary of ground-water conditions and resources, utilization of water from wells and springs, methods of scientific investigation, and literature relating to the subject. U.S. Geological Survey Water-Supply Paper 836-D, Washington, D.C., pp. 157–232.

Milanović, P., 1979. *Hidrogeologija karsta i metode istraživanja* (in Serbo-Croatian; *Karst Hydrogeology and Methods of Investigations*), HE Trebišnjica, Institut za korištenje i zaštitu voda na kršu, Trebinje, 302 p.

Milanovic, P.T., 1981. *Karst Hydrogeology*, Water Resources Publications, Littleton, CO, 434 p.

Miller, J.A., 1999. Introduction and national summary. *Ground-Water Atlas of the United States*, United States Geological Survey, A6.

Mualem, Y. 1976. A new model for predicting the hydraulic conductivity of unsaturated porous media. *Water Resources Research*, 12, 513–522.

Nelson, P.H. and J.E. Kibler, 2003. A catalog of porosity and permeability from core plugs in siliciclastic rocks. U.S. Geological Survey Open-file Report 03-420, Denver, Colorado.

Norton, W.H., 1897. Artesian wells of Iowa. *Iowa Geological Survey*, 6, 113–428.

O'Reilly, A.M., 2004. A method for simulating transient ground-water recharge in deep water-table settings in central Florida by using a simple water-balance/transfer-function model. U.S. Geological Survey Scientific Investigations Report 2004-5195, Reston, Virginia, 49 p.

Pryor, W.A., 1973. Permeability–porosity patterns and variations in some Holocene sand bodies. *American Association of Petroleum Geologists Bulletin*, 57 (1), 162–191.

Richards, L.A. 1931. Capillary conduction of liquids through porous mediums. *Physics*, 1, 318–333.

Schaap, M.G., 1999. *Rosetta, Version 1.0*, U.S. Salinity Laboratory, U.S. Department of Agriculture, Riverside, California.

Schmoker, J.W. and R.B. Halley, 1982. Carbonate porosity versus depth. A predictable relation for South Florida. *Am. Assoc. Petrol. Geol Bull.*, No. 65, p. 2561–2570.

Schwartz, F.W. and F.J. Longstaffe, 1988. Ground water and clastic diagenesis. In: *Hydrogeology, The Geology of North America*, Back, W. Rosenschein, J.R., and P.R. Seaber, Eds., Geological Society of America, Volume 0–2, pp. 413–434.

Sestakov, V.M., 1979. *Dinamika podzemnih vod (Dynamics of Groundwater*, in Russian). Izdatelstvo Moskovskogo Universiteta, Moskva, 368 p.

Slichter, C.S., 1899. Theoretical investigation of the motion of ground waters. U.S. Geological Survey Nineteenth Annual Report 1897-98, Washington, D.C., pp. 295–384.

Slichter, C.S., 1902. The motions of underground waters. U.S. Geological Survey Water-Supply and Irrigation Papers 67, Washington, D.C., 106 p.

Slichter, C.S., 1905. Field measurements of the rate of movement of underground waters. U.S. Geological Survey Water-Supply and Irrigation Paper 140, Series 0, Underground waters, 43, Washington, D.C., 122 p.

Southeastern Friends of the Pleistocene, 1990. Hydrogeology and Geomorphology of the Mammoth Cave area, Kentucky. 1990 Field Excursion led by Quinlan, J.F., R.O. Ewers, and A.N. Palmer, Nashville, TN, 102 p.

Stephens, D.B. et al., 1998. A comparison of estimated and calculated effective porosity. *Hydrogeology Journal*, 6, 156–165.

Šimůnek, J., M. Šejna, and M.Th. van Genuchten, 1999. The Hydrus-2D Software Package for Simulating the Two-Dimensional Movement of Water, Heat, and Multiple Solutes in Variably-Saturated Media, *Version 2.0*, U.S. Salinity Laboratory, U.S. Department of Agriculture, Riverside, California, 227 p.

Tolman, F.C., 1937. *Ground Water*, McGraw-Hill, New York, 593 p.

USDA, Soil Survey Division Staff, 2005. *Soil Survey Manual*, Soil Conservation Service, U.S. Department of Agriculture Handbook 18, Available online at: http://soils.usda.gov/technical/manual/, accessed on December 15, 2005.

USBR, 1977. *Ground Water Manual*, U.S. Department of the Interior, Bureau of Reclamation, Washington, D.C., 480 p.

USGS (United States Geological Survey), 2005. Web page available at: http://ga.water.usgs. gov/edu/waterdistribution.html, accessed on November 12, 2005.

van Genuchten, M.Th., 1980. A closed-form equation for predicting the hydraulic conductivity of unsaturated soils. *Soil Science Society of America Journal*, 44 (5), 892–898.

van Genuchten, M.Th., F.J. Leij, and S.R. Yates, 1991. The RETC Code for Quantifying the Hydraulic Functions of Unsaturated Soils. EPA/600/2-91/065, Ada, Oklahoma, 83 p.

Veatch, A.C., 1906. Fluctuations of the water level in wells, with special reference to Long Island, New York. U.S. Geological Survey Water-Supply and Irrigation Paper 155, Series 0, Underground Waters, 52, Washington, D.C., 83 p.

Vernadsky, V.I., 1926. *Biosphera (The Biosphere)*, Nauchnoe khimiko-technicheskoye izdatel'stvo (Scientific Chemico-Technical Publishing), Leningrad, 200 p.

Vernadsky, V.I., 1929. *La Biosphere*, Félix Alcan, Paris.

Vernadsky, V.I., 1997. *The Biosphere*, completed annotated edition. A Peter Nevraumont Book. Copernicus/Springer-Verlag, New York, 192 p.

White, B.W., 1988. *Geomorphology and Hydrology of Karst Terrains*, Oxford University Press, New York, 464 p.

Williams, J.H. et al., 2001. Application of advanced geophysical logging methods in the characterization of a fractured-sedimentary bedrock aquifer, Ventura County, California. U.S. Geological Survey Water-Resources Investigations Report 00-4083, 28 p.

Winter, C. et. al., 1998. Ground water and surface water: a single resource. *U.S. Geological Survey Circular* 1139, Denver, CO, 79 p.

Witherspoon, P.A., 2000. Investigations at Berkeley on fracture flow in rocks: from the parallel plate model to chaotic systems. In: *Dynamics of Fluids in Fractured Rock*, Faybishenko, B., P.A. Witherspoon, and S.M. Benson, Eds., Geophysical Monograph 122, American Geophysical Union, Washington, D.C., pp. 1–58.

Wolff, R.G., 1982. Physical properties of rocks—Porosity, permeability, distribution coefficients, and dispersivity. U.S. Geological Survey Open-File Report 82-166, 118 p.

Wyckoff, R.D. et al., 1934. Measurement of permeability of porous media. *American Association of Petroleum Geologists*, 18, 161–190.

Zimmerman, R.W. and I.W. Yeo, 2000. Fluid flow in rock fractures: from the Navier-Stokes equations to cubic law. In: *Dynamics of Fluids in Fractured Rock*, Faybishenko, B., P.A. Witherspoon, and S.M. Benson, Eds., Geophysical Monograph 122, American Geophysical Union, Washington, D.C., pp. 213–224.

2 Aquifers and Aquitards

2.1 DEFINITION, RECHARGE AND DISCHARGE AREAS

The word aquifer comes from two Latin words: *aqua* (water) and *affero* (to bring, to give). It was introduced in the United States by Norton in 1897, in reference to term *aquifère*, which was first used by French physicist and astronomer Arago in 1835 (Arago translated his paper to English the same year, Arago, 1835a,b). Aquifer is a geologic formation, or a group of hydraulically connected geologic formations, storing and transmitting significant quantities of potable groundwater. Although most dictionaries of geologic and hydrogeologic terms would have a very similar definition, it is surprising how many interpretations of the word exist in every day's practice, depending on the circumstances. The problem usually arises from the lack of common understanding of the following two terms that are not easily quantifiable: significant and potable. For example, a well yielding 2 gpm may be very significant for an individual household without any other available sources of water supply. However, if this quantity is at the limit of what the geologic formation could provide through individual wells, such "aquifer" would certainly not be considered as a potential source for any significant public water supply. Another issue is the question of groundwater quality. If the groundwater has naturally high total dissolved solids (TDS), say 5000 mg/L, it would disqualify it from being considered as a significant source of water supply, regardless of the groundwater quantity. On the other hand, water treatment technologies such as reverse osmosis (RO) make aquifers with slightly brackish groundwater potentially interesting for development. These are just some of the examples of difficulties that various shareholders face in the fields of water supply and contaminant hydrogeology. In any case, before attempting to solve a problem, all interested parties, including regulators, should have a common understanding of the two most important hydrogeologic terms that seem to be used somewhat arbitrarily and interchangeably: *groundwater* and *aquifer*. In other words, there should be at least some agreement of what the goal of the future effort is. For example, are we protecting (developing) the aquifer, or groundwater in general? Is it reasonable to assume that someone may use this particular groundwater for water supply? It is also important to set clear criteria how should the progress and the ultimate success of the effort be measured. For example, is the groundwater less contaminated then before (but maybe still contaminated), or how long would it take to restore the aquifer to its beneficial use?

Aquitard, which is closely related to aquifer, is also derived from two Latin words: *aqua* (water) and *tardus* (slow) or *tardo* (to slow down, hinder, delay). This means that aquitard does store water and is capable of transmitting it, but at a much slower rate than an aquifer so that it cannot provide significant quantities of potable groundwater to wells and springs. Use of the term aquitard is arguably even more contagious because of its role in protecting (or not protecting) adjacent aquifers from potential contamination. In cases where it is proven, or the available information suggests there is a high probability for water and contaminants to move through the aquitard within a reasonable time frame (say, less than one hundred years), such aquitard is called leaky. When the potential movement of groundwater and contaminants through the aquitard is estimated in hundreds or thousands of years, such aquitard is called competent. More details on the subject of competent and leaky aquitards is discussed in Section 2.10.

Aquiclude is another related term, generally much less used today in the United States but still in relatively wide use elsewhere (Latin word *claudo* means to confine, close, make inaccessible). Aquiclude is equivalent to an aquitard of very low permeability, which, for all practical purposes, acts as an impermeable barrier to groundwater and contaminant flow (note that there still is some groundwater stored in aquiclude, but it moves "very, very slowly"). A smaller number of professionals and some public agencies in the United States (such as the USGS, see Lohman et al., 1972) prefer to use term confining bed instead of aquitard and aquiclude. Accordingly, semiconfining bed would correspond to a leaky aquitard. USGS suggests additional qualifiers be specified to more closely explain nature of a confining layer (i.e., aquitard, aquiclude) of interest, such as slightly permeable or moderately permeable.

Figure 2.1 illustrates major aquifer types in terms of the character and position of the hydraulic head (pressure) in the aquifer, relative to the upper aquifer boundary. The top of the saturated zone of an unconfined aquifer is called water table. Water table is directly exposed to atmospheric pressure, so that the hydraulic head at the water table equals atmospheric pressure. The thickness of the saturated zone, and therefore the position of water table, may vary in time, but the hydraulic head at the water table is always equal to atmospheric pressure. Unconfined aquifers are sometimes called phreatic aquifers. Note that there may be a low-permeable layer, such as clay, at some distance above the water table, including all the way to the ground surface, but as long as there is an unsaturated zone above the water table, the aquifer is unconfined.

An impermeable or low-permeable bed of limited extent above the main (true) water table may cause accumulation of groundwater and formation of a relatively thin saturated zone

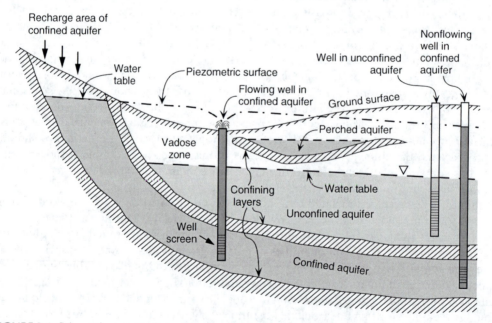

FIGURE 2.1 Schematic presentation of three main aquifer types based on position of the hydraulic head (pressure) in the aquifer. (Modified from USBR, *Ground Water Manual.* U.S. Department of the Interior, Bureau of Reclamation, Washington, D.C., 1977, 480pp.)

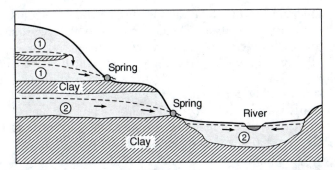

FIGURE 2.2 Example of several perched (1) and unconfined (2) aquifers, developed in terrace and alluvial sediments within a river valley. (From Milojević, N., *Hidrogeologija*. Univerzitet u Beogradu, Zavod za izdavanje udžbenika Socijalističke Republike Srbije, Beograd, 1967, 379pp.)

called perched aquifer. Groundwater in the perched aquifer may eventually flow over the edges of the impermeable bed due to recharge (water percolation) from the land surface, and continue to flow downward to the true water table, or it may discharge through a spring or seep if the confining bed intersects land surface (Figure 2.2).

Confined aquifer is bound above by a low-permeable confining bed, and the hydraulic head in the aquifer is above this contact. The top of the confined aquifer is at the same time the bottom of the overlying confining bed. Groundwater in a confined aquifer is under pressure, such that static water level in a well screened only within the confined aquifer would stand at some distance above the top of the aquifer. If the water level in such well rises above ground surface, the well is called flowing or artesian well, and the aquifer is sometimes called artesian aquifer. The imaginary surface of the hydraulic head (pressure) of the confined aquifer is called piezometric surface. Again, this imaginary surface is above the top of the confined aquifer and it can be located based on measurements of the hydraulic head at wells screened in the confined aquifer. Water table of unconfined aquifers, on the other hand, is not an imaginary surface—it is the top of the aquifer and, at the same time, the top of the saturated zone below which all voids are completely filled with water. The pressure at the water table is equal to the atmospheric pressure. Semiconfined aquifer receives water from, or loses water to, the adjacent aquifer from which it is separated by leaky aquitard.

Figure 2.3 shows key spatial features of an aquifer, which are also important for understanding its water balance. Recharge area is the actual land surface through which the aquifer

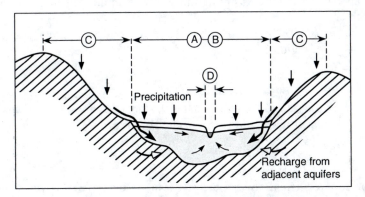

FIGURE 2.3 Key spatial elements of an unconfined aquifer. A, aquifer extent; B, recharge area (in this case recharge equals aquifer extent since there are no low-permeable layers at the land surface); C, drainage area (also called contributing area); D, discharge area.

receives water by percolation of precipitation and surface runoff, or directly from surface water bodies such as streams and lakes. Aquifer may also receive water from adjacent aquifers, but such contact between adjacent aquifers is not referred to as the recharge (or discharge) area. Discharge area is where the aquifer loses water to the land surface, such as through discharge to surface water bodies (streams, lakes, wetlands, and oceans) or discharge through springs. Note that in an unconfined aquifer with shallow water table, loss of water also happens by direct evaporation and plant root uptake, which may be significant if riparian vegetation is abundant. An area that gathers surface water runoff, which eventually ends up recharging the aquifer, is called aquifer drainage (or contributing) area. Aquifer extent is simply the surface projection of its overall limits. Except in cases of some simple alluvial unconfined aquifers (such as the one shown in Figure 2.3), the aquifer extent and recharge areas are usually not equal, and both can have different shapes depending on the geology and presence of confining layers. Some examples are shown in Figure 2.4.

In summary, defining the key geometric elements of an aquifer or aquifer system is the first and most important step in the majority of hydrogeologic studies. It is finding the answers to the following questions regarding the aquifer water: where is it coming from? (drainage area), where is it entering the aquifer? (recharge area), where is it flowing? (aquifer extent), and where is it discharging from the aquifer? (discharge area). Figure 2.5 shows these key elements for one of the largest, most important and most studied aquifers in the United States— Edwards Aquifer in Texas.

The flow of groundwater between the aquifer recharge and discharge areas may generally last anywhere from few months or years to millennia or more, depending on the aquifer geometry, characteristics of aquifer porous media, and interactions with adjacent aquitards and aquifers. Figure 2.6 illustrates this in case of several stratified aquifers, separated by aquitards, which all ultimately discharge into a major river.

2.2   AQUIFER TRANSMISSIVITY AND STORAGE

Once the geometry of the aquifer is determined, including areas where the water is coming into or leaving the aquifer, the next logical question is: how good is the aquifer? This can be quantified by answering the following:

1. How much water, from all sources, is actually entering the aquifer? Can we add more to it? (See Section 2.6 on natural and Section 2.7 on artificial aquifer recharge.)
2. How much water is being transmitted through the aquifer? (This chapter.)
3. Is the aquifer good storage of water? (This chapter.)
4. How much water can be reasonably extracted from the aquifer without adversely affecting it? (See Section 2.6.)

2.2.1   TRANSMISSIVITY

In quantitative terms, the transmissivity (T) is the product of hydraulic conductivity (K) of the aquifer material and the saturated thickness of the aquifer (b):

$$T = K \times b \tag{2.1}$$

It has units of squared length over time (e.g., m^2/d or ft^2/d). In practical terms, the transmissivity equals the horizontal groundwater flow rate through a vertical strip of aquifer one unit wide. The larger the transmissivity, the larger the hydraulic conductivity and the aquifer thickness. Figure 2.7 illustrates several possible cases of this relationship and the effect

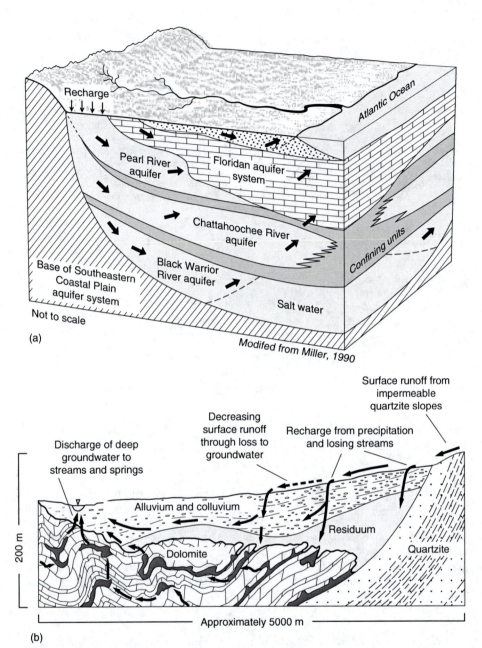

(a)

(b)

FIGURE 2.4 (a) In southeast Georgia, the United States, the general movement of water in the semiconsolidated sand aquifers of the Southeastern Coastal Plain aquifer system is from outcrop recharge areas down the hydraulic gradient of the aquifers until the water discharges upward to the overlying Floridan aquifer system. The downdip extent of flow is limited either by a marked decrease in permeability in the Chattahoochee River aquifer or by stagnant saline water in the Black Warrior River aquifer. (From Miller, J.A., 1999. Introduction and national summary. *Ground-Water Atlas of the United States*, United States Geological Survey, A6.) (b) Generalized hydrogeologic section of the western toe of the Blue Ridge karst region. (From Wolfe, W.J. et al., Preliminary conceptual models of the occurrence, fate, and transport of chlorinated solvents in karst regions of Tennessee. U.S. Geological Survey Water-Resources Investigations Report 97-4097, Nashville, TN, 1997, 80pp.)

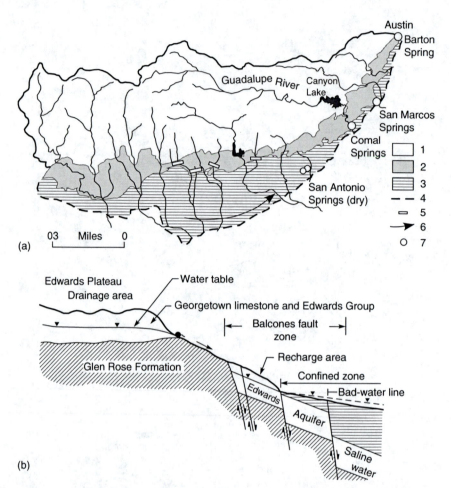

FIGURE 2.5 (a) Edwards aquifer in Texas (after Maclay and Small, 1986, and Edwards Underground Water District, simplified). 1, Drainage area; 2, recharge area (carbonate rock outcrops); 3, confined zone of the aquifer; 4, "bad water" line—east and south of this line groundwater has high salinity and is not potable; 5, recharge dam; 6, general direction of potable groundwater flow in the confined zone of the aquifer; 7, large karst spring. (b) Generalized cross-section of the Edwards aquifer in the San Antonio area. (From Maclay, R.W. and Small, T.A., Carbonate hydrology and hydrology of the Edwards aquifer in the San Antonio area, Texas. Texas Water Development Board Report 296, Austin, TX, 1986, 90pp.)

varying transmissivity has on distribution of the hydraulic head in the aquifer. Aquifer transmissivity can be determined from a variety of field tests as explained in Section 2.4, or it can be estimated by using Equation 2.1. It should be noted that an aquifer rarely has uniform thickness and uniform hydraulic conductivity so that application of Equation 2.1 always requires some degree of averaging using best professional judgment. On the other hand, transmissivity determined from aquifer pumping tests reflects this variability directly.

Higher aquifer transmissivity results in a lower drawdown at a pumping well as illustrated in Figure 2.8 and Figure 2.9. This relationship is inversely proportional (linear) for confined aquifers: decreasing the transmissivity by a factor would increase the drawdown approximately by the same factor. All other things being equal, such as the well pumping rate, the regional nonpumping hydraulic gradient, the initial saturated aquifer thickness, and the

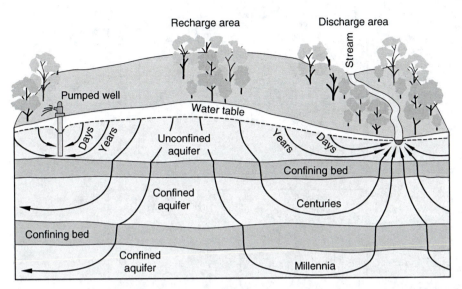

FIGURE 2.6 Groundwater flow paths vary greatly in length, depth, and travel time from points of recharge to points of discharge in the groundwater system. (From Winter, T.C. et al., Ground water and surface water. A single resource. U.S. Geological Survey Circular 1139, Denver, CO, 1998, 79pp; modified from Heath, R.C., Basic elements of ground-water hydrology with reference to conditions in North Carolina. U.S. Geological Survey Water-Resources Investigations Open-File Report 80-44, Raleigh, North Carolina, 1980, 87pp.)

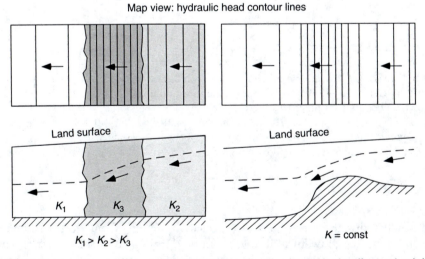

FIGURE 2.7 Examples of varying transmissivity due to changes in the hydraulic conductivity of the aquifer porous media (left) and aquifer thickness (right). The spacing between the hydraulic head contour lines reflects the aquifer transmissivity and the hydraulic gradient (map view on the top). When the hydraulic conductivity is lower, or aquifer thickness is smaller, the hydraulic gradient is steeper and the contour lines are more closely spaced.

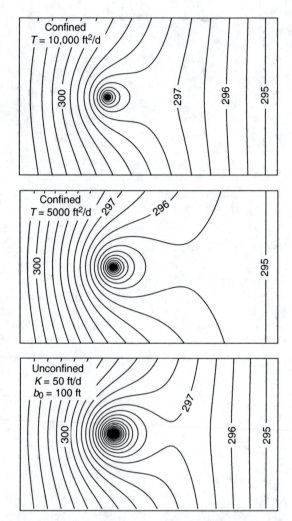

FIGURE 2.8 Maps of the hydraulic head contour lines (in feet above sea level) around a pumping well versus aquifer transmissivity (T) and aquifer type. Confined aquifer thickness, and the initial saturated thickness of the unconfined aquifer (b_0) are the same: 100 ft. The storage coefficient of the confined aquifer (S) is 0.0005, and the specific yield (S_y) of the unconfined aquifer is 0.2. The regional nonpumping hydraulic gradient is 0.0033 and the pumping rate is 300 gpm. No aquifer recharge is assumed for the duration of pumping. The maps, created by a numeric model, show the hydraulic head after 10 days of pumping.

hydraulic conductivity; the early drawdown at a well pumping from an unconfined aquifer would be similar to the drawdown at a well pumping from a confined aquifer (Figure 2.9). However, there are two distinct differences between the unconfined and confined aquifers in such case:

- The pumping hydraulic gradient is steeper near the unconfined aquifer well.
- The radius of well influence in an unconfined aquifer is smaller due to a larger volume of water actually withdrawn per aquifer unit area, i.e., due to differences in the nature of aquifer storage between unconfined and confined aquifers.

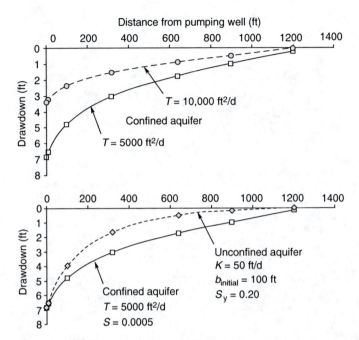

FIGURE 2.9 Cross sections of the drawdown caused by well pumpage shown in Figure 2.8.

As the pumping continues beyond several days, say for a year, the drawdown and the radius of influence of the well pumping from the confined aquifer would both be significantly greater than in the unconfined aquifer. Alley et al. (1999) show a modeling example illustrating this point: radius of influence of a well pumping 1000 gpm from an indefinite confined aquifer of moderate transmissivity (10,000 ft²/d) would be over 96 mi, but less than 2 mi in case of an unconfined aquifer with the same transmissivity.

2.2.2   STORAGE (STORATIVITY)

The storativity of an aquifer is the volume of water it releases from or takes into storage due to change in the component of hydraulic head normal to that surface. The storativity concept focuses attention on the volume of water that aquifer releases from or takes into storage. However, the physical mechanism that releases or stores water in the storage is not the same for unconfined and confined aquifers. When an unconfined aquifer is pumped, which results in the change of the hydraulic head (lowering of the water table), the water is released due to gravity drainage of the affected portion of the aquifer (Figure 2.10). The ratio between the volume of water the unconfined aquifer will yield due to gravity drainage, and the total affected volume of the aquifer is called specific yield (S_y). The volume of water that remains in the aquifer after this drainage has taken place is called specific retention. Consequently, the specific yield is equal to total porosity minus specific retention. This definition is very similar to effective porosity and it is not uncommon to use the two terms interchangeably. However, specific yield is a dynamic quantity, which also depends on rate of change of the water table, and can be relatively reliably determined only by aquifer testing in the field (see Section 2.4.2). In other words, lowering of the water table (release of water) and aquifer recharge (storage of water) occur over short periods of time so that gravity drainage is rarely or never complete (Lohman et al., 1972). Specific yield of aquifer porous media obtained from field-testing is

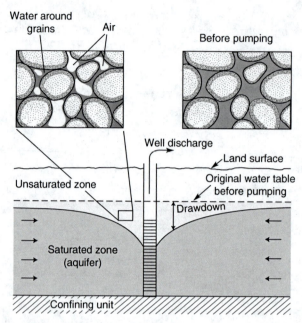

FIGURE 2.10 During pumping of an unconfined aquifer, water is released due to gravity drainage. Within the cone of depression (volume of aquifer affected by drawdown) not all water is released rapidly (the so-called delayed gravity drainage), and some may be retained permanently. (Modified from Alley, W.M., Reilly, T.E., and Lehn Franke, O., Sustainability of ground-water resources. U.S. Geological Survey Circular 1186, Denver, CO, 1999, 79pp.)

therefore usually somewhat lower than the equivalent effective porosity. It should be noted that field determination of specific yield using aquifer pumping tests often has an associated uncertainty due to nonuniqueness of test data analysis. In laboratory, the specific yield is usually determined in two ways:

1. From measurements of porosity (n) and specific retention (S_r) using the following equation

$$S_y = n - S_r$$

2. From column studies where column of undisturbed or repacked sediments is saturated from the bottom and then allowed to drain from the bottom of the column, which is maintained as a constant boundary of zero-pressure head (Johnson et al., 1963).

In two classic publications Meinzer (1923a,b), describes in much detail the concept of specific yield and early experimental determinations of it by Hazen (1892) and King (1899). As always with Meinzer, his observations and conclusions are timeless and it is worthwhile to repeat them here:

• The specific yield has frequently been called the "effective porosity" or "practical porosity," because it represents the pore space that will surrender water to wells and is therefore effective in furnishing water supplies.

- The distinction between gravity water and that which is retained by the rock or soil is not entirely definite, because the amount of water that will drain out depends on the length of time it is allowed to drain, on the temperature and mineral composition of the water, which affect its surface tension, viscosity, and specific gravity, and on various physical relations of the body of rock and soil under consideration. For example, a smaller proportion of water will drain out of a small sample than out of a large body of the same material. As the methods of determining specific yield have not been standardized, and as it may continue to be desirable to use different methods for different purposes and under different conditions, data as to specific yield should always be accompanied by a statement of the methods used in determining it.
- The importance of water that a saturated rock will furnish, and hence its value as a source of water supply, depends on its specific yield—not on its porosity. Clayey or silty formations may contain vast amounts of water and yet be unproductive and worthless for water supply, whereas a compact but fractured rock may contain much less water and yet yield abundantly.

Values of specific yield for unconfined aquifers generally range between 0.05 and 0.3, although lower or higher values are possible, especially in cases of finer grained and less uniform material (lower values), and uniform coarse sand or gravel or mature karst aquifers (higher values). Table 2.1 is an example of specific yield values for unconsolidated sediments determined using different techniques. The variability is attributed to natural heterogeneity of porous materials, different methods of analysis, and different times allotted to determination. In general, the coarser sediments drain more quickly so that S_y shows less time dependency and variability than for the finer sediments (Meinzer, 1923a,b; Healy and Cook, 2002).

One additional mechanism contributes to changes in storage of unconfined aquifers—compressibility of the water and aquifer material in the saturated zone. In most cases, however, the changes in water volume due to unconfined aquifer compressibility are minor and can be ignored for practical purposes. On the other hand, storage of confined aquifers is entirely dependent on compression and expansion of both aquifer water and solids. Namely, it is dependent on aquifer elastic properties. Figure 2.11 shows the forces acting at the interface between a confined aquifer and the confining material. These forces may be expressed as (Ferris et al., 1962)

TABLE 2.1
Statistics on Specific Yield from 17 Studies Compiled by Johnson (1967)

Texture	Average Specific Yield	Minimum Specific Yield	Maximum Specific Yield	Coefficient of Variation (%)	Number of Determinations
Clay	0.02	0.00	0.05	59	15
Silt	0.08	0.03	0.19	60	16
Sandy clay	0.07	0.03	0.12	44	12
Fine sand	0.21	0.10	0.28	32	17
Medium sand	0.26	0.15	0.32	18	17
Coarse sand	0.27	0.20	0.35	18	17
Gravelly sand	0.25	0.20	0.35	21	15
Fine gravel	0.25	0.21	0.35	18	17
Medium gravel	0.23	0.12	0.26	20	13

Source: Modified from Healy, R.W. and Cook, P.G., Using groundwater levels to estimate recharge, *Hydrogeol. J.*, 10(1), 91, 2002.

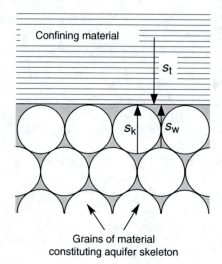

FIGURE 2.11 Microscopic view of forces acting at interface between confined aquifer and confining material. (Modified from Ferris, J.G., Knowles, D.B., Brown, R.H., and Stallman, R.W., Theory of aquifer tests. U.S. Geological Survey Water Supply Paper 1536-E, Washington, D.C., 1962, 173pp.)

$$s_t = s_w + s_k$$

where s_t is the total load exerted on a unit area of the aquifer, s_w is that part of the total load borne by the confined water, and s_k is that part borne by the structural skeleton of the aquifer. Assuming that the total load (s_t) exerted on the aquifer is constant, and if s_w is reduced because of pumping, the load borne by the skeleton of the aquifer will increase. This will result in a slight distortion of the component grains of material. At the same time, the water will expand to the extent permitted by its elasticity. Distortion of the grains of the aquifer skeleton means that they will encroach somewhat on pore space formerly occupied by water.

Conversely, if s_w increases, as in response to cessation of pumping, the piezometric head builds up again, gradually approaching its original value, and the water itself undergoes slight contraction. With an increase in s_w there is an accompanying decrease in s_k and the grains of material in the aquifer skeleton return to their former shape. This releases pore space that can now be reoccupied by water moving into the part of the formation that was influenced by the compression (Ferris et al., 1962). In summary, the water derived from storage in a confined aquifer because of a decline in hydraulic head comes from expansion of the water and compression of the aquifer solids. Water added to storage because of a rise in hydraulic head is accommodated partly by compression of the water and partly by expansion of the aquifer solids (Lohman et al., 1972). Storage properties (storativity) of confined aquifers are defined by the coefficient of storage. Although rigid limits cannot be established, the storage coefficients of confined aquifers may range from about 0.00001 to 0.001. In general, denser aquifer materials have smaller coefficient of storage. It is important to note that the value of coefficient of storage in confined aquifers may not be directly dependent on void content (porosity) of the aquifer material (USBR, 1977). Specific storage (S_s) of confined aquifers is the volume of water released (or stored) by the unit volume of porous medium, per unit surface of the aquifer, due to unit change in the component of hydraulic head normal to that surface. The unit of specific storage is inverse of length (e.g., m^{-1} or ft^{-1}) so that, when the specific storage is multiplied by aquifer thickness (b), it gives the coefficient of storage (S), which is a dimensionless number: $S = S_s b$. The specific storage is given as

$$S_s = \rho_w g(\alpha + n\beta) \qquad (2.2)$$

where ρ_w is density of water, g is acceleration of gravity, α is compressibility of the aquifer skeleton, n is total porosity, and β is compressibility of water.

Note that, when using transmissivity and storage coefficients in quantitative calculations and groundwater modeling, horizontal flow is assumed by default as the aquifer thickness is included within their values. It is more common, especially when studying fate and transport of contaminants (contaminant particles), that the analysis is performed using values of horizontal and vertical hydraulic conductivities, specific storage, and actual aquifer thickness. This enables calculations of three-dimensional velocity vectors and aquifer water balance or, in other words, changes in contaminant concentrations as it moves in three-dimensional space.

As mentioned in Section 2.2.1, all other things being equal, such as the well pumping rate, the regional nonpumping hydraulic gradient, the initial saturated aquifer thickness, and the hydraulic conductivity, the radius of well influence in a confined aquifer would be larger than in an unconfined aquifer. This is because less water is actually withdrawn from the same aquifer volume in case of confined aquifers due to their lower storativity. In other words, to provide the same well yield (volume of water), a larger aquifer area would be affected in a confined aquifer than in an unconfined aquifer, assuming that they initially have the same saturated thickness.

2.3 TYPES OF AQUIFERS

2.3.1 AQUIFERS IN UNCONSOLIDATED SEDIMENTS

Aquifers developed in unconsolidated sediments, which are composed of various mixtures of grains of varying size and shape, i.e., clay, silt, sand, and gravel, are called intergranular aquifers. Depending on the predominance of certain grain fraction, such aquifers may also be called sand or sand-and-gravel aquifers for example. It is common to call a particular intergranular aquifer by the depositional or geomorphologic process that created it such as alluvial aquifer or basin-fill aquifer. Due to fluvial depositional mechanisms, all alluvial aquifers show some degree of heterogeneity and stratification, with sand and gravel layers separated by silty and clay layers or lenses. In many cases, more than one process is responsible for creating unconsolidated deposits and every attempt should be made to at least understand the most important depositional mechanisms. This is important because characteristics of the intergranular porous media, such as anisotropy and heterogeneity, are a direct result of depositional processes. For example, an aquifer developed in thick aeolian sands (former sand dunes) should be very prolific, given enough historic or current natural recharge, because of the high storage capacity and effective porosity of uniform clean sands. On the other hand, alluvial deposits around a small or even medium-size stream in a drainage area where bedrock weathering creates mainly silts and clays may create a poor heterogeneous aquifer.

As a general rule, alluvial aquifers developed in floodplains of major rivers are among the most prolific and widely used for water supply throughout the world (see Figure 2.12). In addition to thick deposits of sand and gravel, they are in most part in direct hydraulic connection with the river, which provides for abundant aquifer recharge. Large well fields for public and industrial water supply are often designed to induce additional recharge from the river by creating increased hydraulic gradients from the river to the aquifer (Figure 2.13).

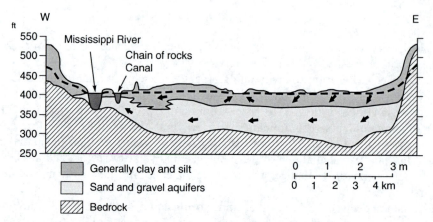

FIGURE 2.12 One of the thickest stream valley alluvial aquifers in the United States is present in the Mississippi River valley near East St. Louis. The aquifer is about 100 ft thick and it is hydraulically connected with the Chain of Rocks Canal and the Mississippi River. (From Lloyd, O.B. Jr. and Lyke, W.L., Illinois, Indiana, Kentucky, Ohio, Tennessee, *Ground Water Atlas of the United States*, United States Geological Survey, HA 730-K, 1995. Modified from Schicht, 1965).

Basins between mountains, filled with unconsolidated and semiconsolidated sediments, are another major group of aquifers present on all continents. The thickness of such deposits may sometimes exceed several thousand meters due to constant tectonic lowering of the basin by boundary faults, and supply of sediments brought in by fluvial and colluvial processes. Basin and range province in the western United States is an example of a large number of basin-fill aquifers utilized for water supply and irrigation in a major way. Groundwater extraction is usually from well-protected confined layers from deeper portions of basin, although there are examples of unwanted effects of such extraction due to induced upconing (vertical upward migration) of highly mineralized saline groundwater. This water is residing at greater depths where there is no flushing by fresh meteoric water. Another negative effect of groundwater extraction from basin-fill aquifers in arid climates is aquifer mining because of the lack of significant present-day natural aquifer recharge. Where precipitation is significant on the

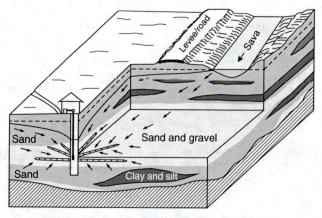

FIGURE 2.13 One of the Ranney (collector) wells in Makis, the Sava River valley, serving city of Belgrade, Serbia. (From Milojević, N., *Hidrogeologija*. Univerzitet u Beogradu, Zavod za izdavanje udžbenika Socijalističke Republike Srbije, Beograd, 1967, 379pp.)

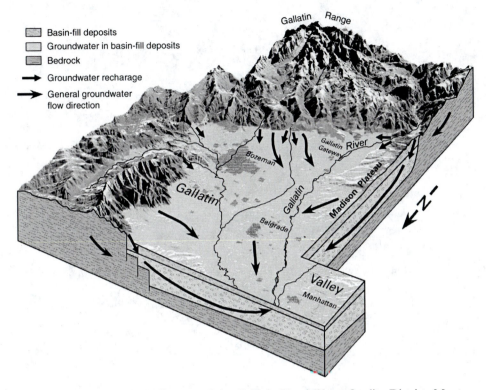

FIGURE 2.14 Perspective block diagram of the Gallatin Local Water Quality District, Montana, the United States. (Modified from Taylor, C.J. and Alley, W.M., Ground-water-level monitoring and the importance of long-term water-level data. U.S. Geological Survey Circular 1217, Denver, CO, 2001, 68pp; modified from Kendy, E., Magnitude, extent, and potential sources of nitrate in ground water in the Gallatin Local Water Quality District, southwestern Montana, 1997–98. U.S. Geological Survey Water-Resources Investigations Report 01-4037, 2001.)

surrounding mountains, concentrated recharge occurs when surface water flowing from the mountains infiltrates into the highly permeable coarse fill deposits along mountain front such as colluvial and alluvial fans (Figure 2.14).

Thick widespread sheet-like deposits that contain mostly sand and gravel form unconsolidated and semiconsolidated aquifers called blanket sand and gravel aquifers. They largely consist of alluvial deposits brought in from mountain ranges and deposited in lowlands. However, some of these aquifers, such as the High Plains aquifer in the United States (Ogallala aquifer), include large areas of windblown sand, whereas others, such as the surficial aquifer system of the southeastern United States, contain some alluvial deposits but are largely comprised of beach and shallow marine sands (Miller, 1999). The High Plains aquifer in the west-central United States extends over about 174,000 mi^2 in parts of eight states (Figure 2.15). The principal water-yielding geologic unit of the aquifer is the Ogallala Formation of Miocene age, a heterogeneous mixture of sand, gravel, silt, and clay that was deposited by a network of braided streams, which flowed eastward from the ancestral Rocky Mountains. Permeable dune sand is part of the aquifer in large areas of Nebraska and smaller areas in the other states. The Ogallala aquifer is principally unconfined and in direct hydraulic connection with the alluvial aquifers along major rivers which flow over it.

FIGURE 2.15 The High Plains aquifer (Ogallala aquifer) in the United States extends over about 174,000 mi^2 in parts of eight states. (Modified from Miller, J.A., Introduction and national summary. *Ground-Water Atlas of the United States*, United States Geological Survey, A6, 1999.)

The origin of water in the Ogallala aquifer is mainly from the last ice age, and the rate of present-day recharge is much lower. This has resulted in serious long-term water table decline in certain portions of the aquifer due to intensive groundwater extraction for water supply and irrigation. The High Plains aquifer is the most intensively pumped aquifer in the United States. During 1990, about 15 billion gallons per day (about 17 million acre-feet per year) of water was withdrawn from the aquifer. In 1992, the saturated thickness of the aquifer ranged from 0 ft where the sediments that comprise the aquifer are unsaturated to about 1000 ft in parts of Nebraska, and averaged about 190 ft (Miller, 1999).

Glacial deposit aquifers in general are good example of heterogeneous aquifers of varying scale and interconnectivity (Figure 2.16). The distribution of the numerous sand and gravel beds that make up the glacial deposit aquifers and the clay and silt confining units that are interbedded with them is extremely complex. The multiple advances of lobes of continental ice originating from different directions and different materials were eroded, transported, and deposited by the ice, depending on the predominant rock types in its path. When the ice is melted, coarse-grained sand and gravel outwash was deposited near the ice front, and the meltwater streams deposited successively finer material farther and farther downstream. During the next ice advance, heterogeneous deposits of poorly permeable till might be laid down atop the sand and gravel outwash. Small ice patches or terminal moraines dammed

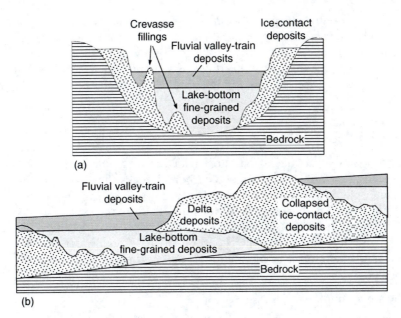

FIGURE 2.16 Coarse-grained glacial deposits are commonly found in bedrock valleys. Some of these deposits (a) formed at the ice-bedrock contact, and some filled cracks or crevasses in the ice. As the ice melted, outwash deposits of sand and gravel formed deltas at the ice front or in glacial lakes and fluvial valley-train deposits (b) downstream from the ice front. (From Trapp, H. Jr. and Horn, M.A., Delaware, Maryland, New Jersey, North Carolina, Pennsylvania, Virginia, West Virginia. *Ground Water Atlas of the United States*, United States Geological Survey, HA 730-L, 1997; Modified from Lyford, F.P. In: Sun, R.J. (Ed.), *Regional Aquifer-System Analysis Program of the U.S. Geological Survey—Summary of Projects, 1978–1984*, U.S. Geological Survey Circular 1002, 1986, pp. 162–167.)

some of the meltwater streams, causing large lakes to form. Thick deposits of clay, silt, and fine sand accumulated in some of the lakes and these deposits form confining units where they overlie sand and gravel beds (Miller, 1999). The glacial deposit aquifers either are localized in bedrock valleys or are in sheet-like deposits on outwash plains. The latter ones can be extensive, as in the coalescing gravel outwash plains of North America, the eastern Andes and the Himalayas–Pamir–Tienshan cordilleras, or quite narrow and sinuous, as in the glacial channels of the North German Plain and the Great Lakes in the United States (Morris et al., 2003). Yields of wells completed in the glacial deposit aquifers formed by continental glaciers in the United States are as much as 3000 gpm where the aquifers consist of thick sand and gravel. Locally, yields of 5000 gpm have been obtained from wells completed in glacial deposit aquifers that are located near rivers and can obtain recharge from the rivers. Aquifers that were formed by mountain glaciers yield as much as 3500 gpm in Idaho and Montana, and wells completed in mountain–glacier deposits in the Puget Sound area yield as much as 10,000 gpm (Miller, 1999).

Regolith is a layer of unconsolidated sediments derived from the underlying bedrock by various processes such as physical and chemical weathering and biological activities. It always varies in thickness, even in the same general area, due to heterogeneities of the parent rock and topography. In tropical climates thickness of regolith may reach several hundred feet, whereas in temperate climates it is usually not more than several tens of feet. Transition between regolith and the underlying bedrock is gradual and often appears "erratic" both laterally and vertically, with smaller or larger chunks of solid rock scattered throughout the unconsolidated regolith sediments. Regolith matrix usually contains a significant amount of clay minerals due to physical and chemical weathering processes. This limits the utility of regolith aquifers to local and individual water supply. Intensive weathering of granitic rocks may produce thick deposits of coarse sand regolith called grus, which, if other conditions are favorable (precipitation, topography) may have good potential for centralized water supply.

2.3.2 FRACTURED ROCK AND KARST AQUIFERS

Aquifers developed in consolidated rock are divided into two major groups: (1) fractured sedimentary deposits, mainly sandstone and carbonate rocks (limestone and dolomite) and (2) most metamorphic and magmatic rocks. Because of the soluble nature of sedimentary carbonate deposits such as limestone and dolomite, these rocks, when exposed to ground surface and direct infiltration of precipitation, usually form karst aquifers. These aquifers have enhanced secondary porosity where portion of the groundwater flow takes place in cavities and conduits developed by dissolution of the rock mass.

2.3.2.1 Sandstone Aquifers

Although generally less permeable, and usually with a lower natural recharge rate than surficial unconsolidated sand and gravel aquifers, sandstone aquifers in large sedimentary basins are one of the most important sources of water supply worldwide. Some loosely cemented sandstones retain a significant primary (intergranular) porosity, while secondary fracture porosity may be more important for well-cemented (usually older) sandstone. In either case, storage capacity of such deposits is high because of the thickness of major sandstone basins. Examples of continental-scale sandstone aquifers include the Guaraní aquifer system in South America, the Nubian Sandstone Aquifer in Africa, and the Great Artesian Basin in Australia.

The Guaraní aquifer system (also called Botucatu aquifer) includes areas of Brazil, Uruguay, Paraguay, and Argentina. Water of very good quality is exploited for urban supply, industry, irrigation, and for thermal, mineral, and tourist purposes. This aquifer is one of the most important fresh groundwater reservoirs in the world, due to its vast extension (about $1,200,000$ km^2) and volume (about $40,000$ km^3). The aquifer storage volume could supply a total population of 5500 million people (i.e., almost the entire present population of the world) for 200 years at a rate of 100 L/d per person (Puri et al., 2001). The gigantic aquifer is located in the Paraná and Chaco-Paraná Basins of southern South America (Figure 2.17). It is developed in consolidated aeolian and fluvial sands (now sandstones) from Triassic to Jurassic, usually covered by thick basalt flows (Serra Geral Formation) from the Cretaceous, which provide a high confinement degree. Its thickness ranges from a few meters to 800 m. The specific capacities of wells vary from 4 m^3/h/m of drawdown to more than 30 m^3/h/m. TDS contents are generally less than 200 mg/L. The production costs per cubic meter of water from wells of depths between 500 and 1000 m and yielding between 300 and 500 m^3/h vary from US\$0.01 to US\$0.08, representing only 10%–20% of the cost of storing and treating surface water sources (Rebouças and Mente, 2003).

The rocks of the Nubian aquifer system vary in thickness from zero in outcrop areas to more than 3000 m in the central part of the Kufra and Dakhla Basins and range in age from Cambrian to Neogene (Figure 2.18). The main productive aquifers, separated by regional confining units, are (from land surface down) Miocene sandstone, Mesozoic (Nubian) sandstone, Upper Paleozoic–Mesozoic sandstone, and Lower Paleozoic (Cambrian–Ordovician) sandstone. The groundwater of the Nubian Basin is generally characterized by its high quality. The TDS

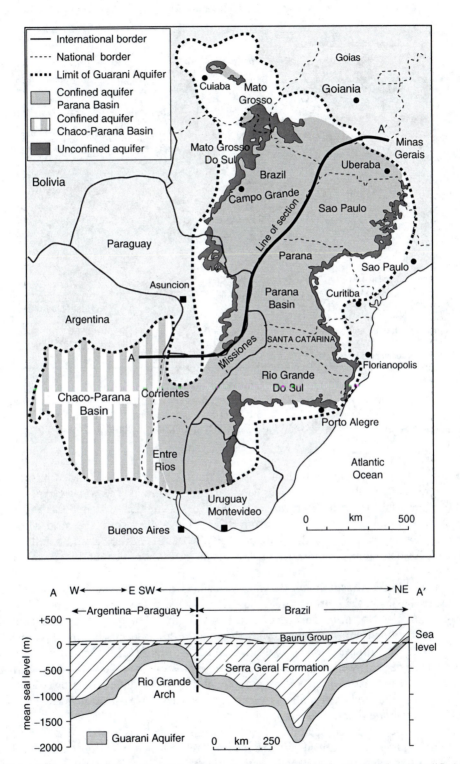

FIGURE 2.17 Map and cross section of the Guaraní aquifer system in South America. (Modified from Puri, S. et al., *Internationally Shared (Transboundary) Aquifer Resources Management: Their Significance and Sustainable Management. A Framework Document*, IHP-VI, Series on Groundwater 1, IHP Non-Serial Publications in Hydrology, UNESCO, Paris, 2001, 76pp.)

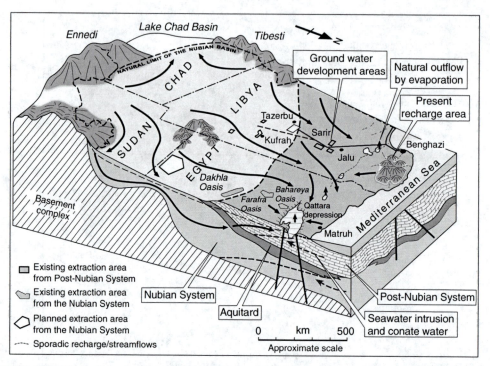

FIGURE 2.18 Block diagram of the Nubian Sandstone aquifer system. (Modified from Salem, O. and Pallas, P. In: Puri, et al. (Eds.), *Internationally Shared (Transboundary) Aquifer Resources Management: Their Significance and Sustainable Management. A Framework Document*, IHP-VI, Series on Groundwater 1, IHP Non-Serial Publications in Hydrology, UNESCO, Paris, 2001, pp. 41–44.)

range from 100 to 1000 ppm, with an increased salinity northward toward Mediterranean Sea where the freshwater–saltwater interface passes through Qattara depression in Egypt. In Libya, the TDS of the deep Nubian aquifers ranges from 160 to 480 mg/L and from 1000 to 4000 mg/L in the shallow aquifers (Khouri, 2004, from Salem, 1991). Because of the arid climate, the present-day natural recharge of the Nubian aquifer system is negligible and countries in the region have formed a joint commission for assessment and management of this crucial source of water supply.

The Great Artesian Basin covers 1.7 million square kilometers and is one of the largest groundwater basins in the world. It underlies parts of Queensland, New South Wales, South Australia, and the Northern Territory. The basin is up to 3000 m thick and contains a multilayered confined aquifer system, with the main aquifers occurring in Mesozoic sandstones interbedded with mudstone (Jacobson et al., 2004; Figure 2.19). As of 1980, more than 4000 flowing (artesian) wells have been drilled to depths of up to 2000 m, with the highest individual well yields exceeding 100 L/s or 1500 gpm. Many of these wells flowed uncontrollably wasting water and Australian government agencies have since engaged in a major project to rehabilitate them.

Because of the lower permeability at depths, and deep basin structure, consolidated sedimentary aquifers have not always been completely flushed with fresh water recharged from land surface and therefore may contain highly mineralized (connate) water at some depth. This sometimes poses challenge for more extensive groundwater withdrawal because of the potential upconing (upward migration) of the saline water due to drawdown of the overlying freshwater.

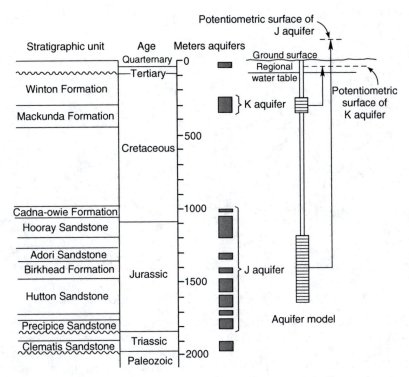

FIGURE 2.19 Multilayered confined aquifer system in the Bowen and Galilee Basins, Australia, with the main aquifers occurring in Mesozoic sandstone. (From Jacobson, G. et al. In: Zektser, I.S. and Everett, L.G. (Eds.), *Groundwater Resources of the World and Their Use*, IHP-VI, Series on Groundwater No. 6, UNESCO, Paris, France, 2004, pp. 237–276.)

2.3.2.2 Carbonate and Karst Aquifers

Large epicontinental carbonate shelf platforms, at the scale of hundreds to thousands of kilometers, and thousands of meters thick, together with isolated carbonate platforms in open-ocean basins, have developed through Paleozoic and Mesozoic, with the trend continuing into Cenozoic. At the same time, smaller (hundreds to tens of kilometers wide) carbonate platforms and buildups associated with intracratonic basins have also developed (James and Mountjoy, 1983). Both types have been redistributed and reshaped during various phases of plate tectonics and can presently be found throughout the world, both adjacent to oceans and seas and deep inside the continents. One of the largest carbonate aquifers in the world, the Floridan aquifer system in the southeast United States (North and South Carolinas, Georgia, and Florida) is developed on a thick epicontinental platform and consists of gently sloping thick sequences of carbonate sediments separated by less permeable clastic sediments (Figure 2.20). Regional flow directions are from the limestone outcrops toward the Atlantic Ocean and the Gulf of Mexico, with submerged discharge zones along the continental shelf. These discharge zones are either diffuse or in forms of submarine springs (Florida has over 20 well-documented large offshore springs and a number of undocumented ones). Where limestone is exposed at the surface, as in north-central Florida, karst features such as sinkholes, large springs, and caves are fully developed.

Classic Dinaric karst in the Balkans, Europe is an example of a tectonically disturbed large thick Mesozoic carbonate platform where Adriatic Sea is the regional erosional basis for

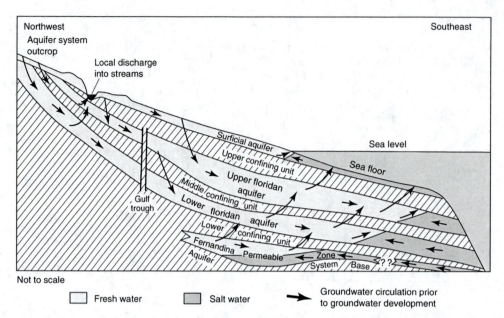

FIGURE 2.20 Schematic cross section of Floridan aquifer system from northwest to southeast Georgia. The Lower Floridan aquifer in this area includes a highly permeable unit called the Fernandina permeable zone. This zone is the source of a considerable volume of fresh to brackish water that leaks upward through the middle confining unit and ultimately reaches the Upper Floridan aquifer. (From Barlow, P.M., Ground water in freshwater–saltwater environments of the Atlantic coast. U.S. Geological Survey Circular 1262, Reston, VA, 2003, 113pp; modified from Krause, R.E. and Randolph, R.B., Hydrology of the Floridan aquifer system in southeast Georgia and adjacent parts of Florida and South Carolina. U.S. Geological Survey Professional Paper 1403-D, 1989, 65pp.)

groundwater discharge (word karst has its origin in the countries of former Yugoslavia, the Balkans). The regional groundwater flow is completely influenced by fully developed mature karst together with large closed depressions called "poljes" (Figure 2.21). Due to high precipitation and recharge rate, the Dinaric carbonate platform aquifer has some of the largest springs in the world (Figure 2.22). The presence of large springs is a common characteristics of karst aquifers. For example, it is estimated that there are nearly 700 springs in Florida of which 33 are the first magnitude springs with average discharge greater than 100 ft^3/s or 2.83 m^3/s (see Section 2.8 for detail discussion on springs).

The majority of karst areas in the United States constitute portions of carbonate platforms now located away from the coast lines: Edwards aquifer in Texas (see Figure 2.5) and karst of Alabama, Kentucky, Tennessee, West Virginia, Indiana, and Missouri for example. Regional groundwater flow in such cases is toward large springs located at the lowest contact between the carbonate (karst) and noncarbonate rocks, or toward the lowest large permanent surface stream intersecting the carbonates. Groundwater discharge along such streams is commonly through large springs, often submerged.

When of considerable thickness, recent coastal carbonate sediments may constitute important aquifers for both local and centralized water supply. Examples can be found in Jamaica, Cuba, Hispaniola, and numerous other islands in the Caribbean, the Yucatán peninsula of Mexico, the Cebu limestone of the Philippines, the Jaffna limestone in Sri Lanka, and some low-lying coral islands of the Indian oceans such as the Maldives (Morris et al., 2003).

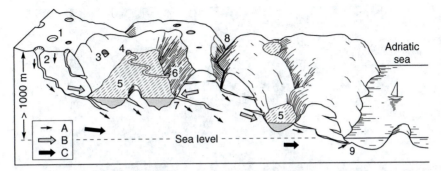

FIGURE 2.21 Karst and groundwater flow features in the thick Mesozoic carbonate platform of the Dinarides (modified from Krešić, 1988). 1, Sinkhole (doline); 2, pithole (jama); 3, dry cave—former spring; 4, active spring; 5, karst polje with unconsolidated fill sediments; 6, active sink; 7, estavelle (interchangeable spring and sink); 8, deep canyon with the loosing stream; 9, submarine spring in the Adriatic Sea. A, Predominant local groundwater flow direction and aquifer recharge; B, local groundwater flow direction during wet season and heavy rains; C, regional groundwater flow.

Permeability of recently deposited (Cenozoic) carbonates is high to very high thanks to initially high primary porosity and solutional enlargement of fractures (secondary or karst porosity). The high infiltration capacity of young carbonate sediments in coastal areas and islands means that there are few streams or rivers, and groundwater may be the only available source of water supply. This source is often very vulnerable to salt (sea) water encroachment due to overpumping of fresh groundwater.

Arguably the most problematic task in karst hydrogeology is determining drainage area that contributes water to a karst aquifer drained by a large spring. Using various techniques of dye tracing is still an irreplaceable method in doing so, together with the hydrogeologic judgment where to introduce the dye and where to monitor for its possible appearance. As illustrated with Figure 2.23 through Figure 2.25, surface streams in karst may have various stages of development and it is often difficult or even impossible, without detail hydrologic investigations, to determine if a stream actually loses water and at which sections. Portions of the same stream may be losing or gaining water or both, depending on the season, and may also be contributing water to more than just one subsurface drainage area.

2.3.2.3 Fractured (Bedrock) Aquifers

Groundwater occurs in bedrock (basement) aquifers over large parts of all continents. These aquifers are developed in crystalline magmatic and metamorphic rocks with little or no primary porosity, and groundwater is present only in fractures and near-surface weathered layers. In some cases, the bedrock has disintegrated into an extensive and relatively thick layer of unconsolidated highly weathered rock with a clayey residue of low permeability (regolith or residuum). Below this zone, the rock becomes progressively less weathered and more consolidated, transitioning into fresh fractured bedrock (Figure 2.26). Although the zone of weathering is generally only a few tens of meters, the storage of water in regolith is much larger than in the underlying bedrock as illustrated in Figure 2.27. Crystalline basement rocks are commonly used as a source of groundwater because of their wide extent but yields are typically small and the low storage makes wells prone to drying up during drought (Morris et al., 2003).

FIGURE 2.22 One of the largest springs in the world: the Buna spring near Mostar, Herzegovina, with the minimum and maximum discharge rate greater than 10 and 300 m³/s, respectively. The spring is of ascending type with submerged siphonal cave passages below the elevation of discharge. Note an oblique fault above the spring cave.

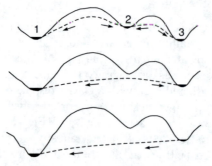

FIGURE 2.23 An example of stream piracy in karst. 1, Permanent stream with the largest flow rate or incision; 2, first gaining, then pirated (loosing), and finally permanently dry stream with the smallest drainage area and flow rate; 3, first gaining and then pirated (loosing) stream.

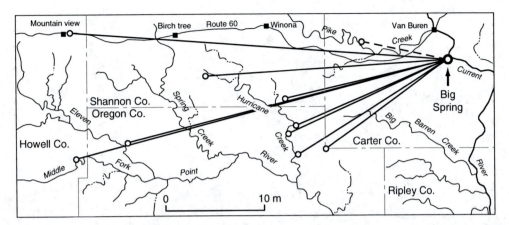

FIGURE 2.24 Water tracing experiments by Thomas J. Aley of Mark Twain National Forest, using fluorescein dyes and *Lycopodium* spores show that water flows through subterranean karst channels to Big Spring in Missouri, the United States from as far as 40 mi away. (From Vineyard, J.D. and Feder, G.L., Springs of Missouri. Water Resources Report No. 29, Missouri Department of Natural Resources, Division of Geology and Land Survey, 1982, 212pp.)

2.3.2.4 Aquifers in Basaltic and Other Volcanic Rocks

Aquifers in volcanic rocks are widely spread worldwide. However, as volcanic rocks have a wide range of chemical, mineralogic, structural, and hydraulic properties, these aquifers also range from some of the most prolific to those utilized only for limited individual supply. Pyroclastic rocks, such as tuff and ash deposits, might be emplaced by flowage of a turbulent mixture of gas and pyroclastic material, or might form as windblown deposits of fine-grained ash. Where they are unaltered, pyroclastic deposits have porosity and permeability characteristics like those of poorly sorted sediments and may act as regional aquitards. Where the rock fragments are very hot as they settle, however, the pyroclastic material might become welded and almost impermeable. Silicic lavas, such as rhyolite or dacite, tend to be extruded as thick, dense flows and have low permeability except where they are fractured (Miller, 1999). Basaltic lavas tend to be fluid and form thin flows that have a considerable amount of primary pore space at the tops and bottoms of the flows. Numerous basalt flows commonly overlap and the flows are often separated by topsoil or alluvial materials that form permeable zones.

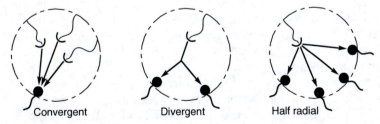

FIGURE 2.25 Sketch showing the three most common karst basin drainage patterns in West Virginia, the United States. (From Jones, W.K., Karst hydrology atlas of West Virginia. Special Publication 4, Karst Waters Institute, Charles Town, WV, 1977, 111pp.)

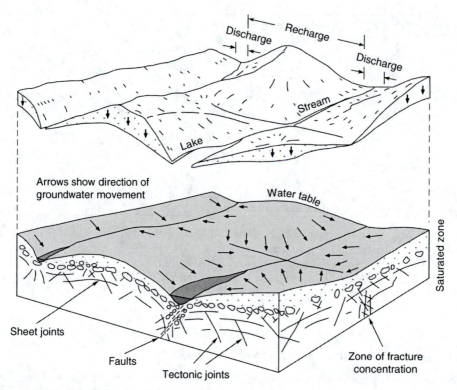

FIGURE 2.26 Conceptual view of the groundwater flow system of the North Carolina Piedmont showing the unsaturated zone (lifted up), the water table surface, the saturated zone, and directions of groundwater flow. (From Daniel, C.C. III, Evaluation of site-selection criteria, well design, monitoring techniques, and cost analysis for a ground-water supply in Piedmont crystalline rocks, North Carolina, U.S. Geological Survey Water-Supply Paper 2341-B, 1990, 35pp.)

Extensive lava flows occur in west-central India, where the Deccan basalts occupy an area of more than 500,000 km². Other extensive volcanic terrains occur in North and Central America, Central Africa, whereas many islands are entirely or predominantly of volcanic origin, such as Hawaii, Iceland, and the Canaries. Some of the older, more massive lavas can be practically impermeable (such as the Deccan) as are the dykes, sills, and plugs, which intrude them (Morris et al., 2003). Younger basic lavas provide some of the world's most prolific onshore and offshore springs such as Snake River Basalts in the United States and basalts of Hawaii. The Snake River Plain regional aquifer system in southern Idaho and southeastern Oregon is a large graben-like structure that is filled with basalt of Miocene and younger age. The basalt consists of a large number of flows, the youngest of which was extruded about 2000 years ago. The maximum thickness of the basalt, as estimated by using electrical resistivity surveys, is about 5500 ft (Miller, 1999). The permeability of basaltic rocks is highly variable and depends largely on the following factors: the cooling rate of the basaltic lava flow, the number and character of interflow zones, and the thickness of the flow. The cooling rate is most rapid when a basaltic lava flow enters water. The rapid cooling results in pillow basalt, in which ball-shaped masses of basalt form, with numerous interconnected open spaces at the tops and bottoms of the balls. Large springs that discharge thousands of gallons per minute issue from pillow basalt in the wall of the Snake River Canyon at Thousand Springs, Idaho (Figure 2.28 and Figure 2.29).

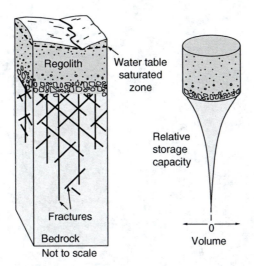

FIGURE 2.27 Differences in storage capacity of regolith and bedrock. (From Miller, J.A., Introduction and national summary. *Ground-Water Atlas of the United States*, United States Geological Survey, A6, 1999; modified from Heath, R.C., Basic elements of ground-water hydrology with reference to conditions in North Carolina. U.S. Geological Survey Water-Resources Investigations Open-File Report 80-44, Raleigh, North Carolina, 1980, 87pp and Daniel, C.C. III and Sharpless, N.B., Ground-water supply potential and procedures for well-site selection, upper Cape Fear River basin. North Carolina Department of Natural Resources and Community Development, 1983, 73pp.)

Highly permeable but relatively thin rubbly or fractured lavas act as excellent conduits but have themselves only limited storage. Leakage from overlying thick, porous but poorly permeable, volcanic ash may act as the storage medium for this dual system. The prolific aquifer systems of the Valle Central of Costa Rica and of Nicaragua and El Salvador are examples of such systems (Morris et al., 2003).

2.4   AQUIFER PARAMETERS AND METHODS OF DETERMINATION

As mentioned earlier, "true" aquifer parameters, in a strict, narrow sense, are aquifer transmissivity and aquifer storage (see Section 2.2). They are closely related to the hydraulic conductivity and effective porosity of the aquifer porous media so that these four quantities are often used interchangeably in the literature to describe aquifer characteristics. However, an aquifer is usually comprised of more or less heterogeneous sediments and rocks, which do have varying hydraulic conductivity and effective porosity. At the same time, an ideally homogeneous, and therefore unrealistic aquifer, the so-called "sand box," may have varying transmissivity because its thickness changes, even though the hydraulic conductivity is the same everywhere in the aquifer. Having these simple facts in mind explains why, at least theoretically speaking, the true hydraulic conductivity and effective porosity of a porous medium could only be determined in the laboratory from its sample, whereas the true transmissivity and storage could only be determined by testing (stressing) a portion of aquifer volume directly in the field. This, again, brings the question of best professional judgment needed for inevitable correlation and extrapolation between the data obtained at two very different scales: the laboratory and the field (see also Section 1.4.4 for discussion of representative elementary volume or REV).

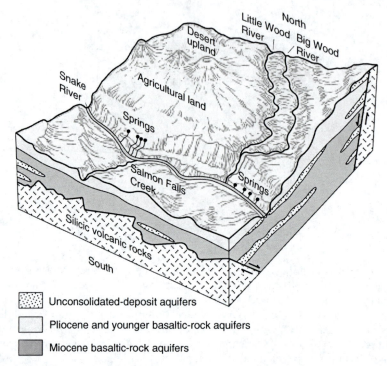

Unconsolidated-deposit aquifers

Pliocene and younger basaltic-rock aquifers

Miocene basaltic-rock aquifers

FIGURE 2.28 Basalt of Miocene and younger age fills the graben-like trough on which the Snake River Plain has formed. Low-permeability, silica-rich volcanic rocks bound the basalt, which is locally interbedded with unconsolidated deposits. (From Miller, J.A., Introduction and national summary. *Ground-Water Atlas of the United States*, United States Geological Survey, A6, 1999; modified from Whitehead, R.L., Ground Water Atlas of the United States—Segment 7: Idaho, Oregon, Washington. *U.S. Geological Survey Hydrologic Investigations Atlas* HA-730-H, 1994, 31pp.)

FIGURE 2.29 Thousand Springs in the Snake River valley, Idaho, the United States. (From Alley, W.M., Reilly, T.E., and Lehn Franke, O., Sustainability of ground-water resources. U.S. Geological Survey Circular 1186, Denver, CO, 1999, 79pp.)

2.4.1 LABORATORY TESTS

Samples (specimens) for laboratory testing of the hydraulic conductivity are collected in the field in an "undisturbed" form using various coring methods that extract cylinders (cores) of aquifer porous material during boring. The so-called split-barrel (split-spoon) samples are taken by driving (hammering) the sampler at the bottom of the drilled hole into the undisturbed soil. This method is applicable to granular unconsolidated sediments (American Society for Testing and Materials [ASTM] test method D 1586). Thin-walled tube sampling (often called "Shelby tube") is applicable to silts and clays. The sampler can be forced into the undisturbed sediment at the bottom of the boring either pneumatically or using a rotary technique (ASTM test method D 1587). Diamond core drilling is used to collect samples from consolidated sediments and rocks (ASTM test method D 2113).

Selection of the laboratory method depends mainly on the nature and relative permeability of the extracted material:

- Unconsolidated granular sediments of moderate to high permeability (sands, gravels) are tested with a constant-head permeameter, which does not have to be pressurized and can have rigid walls. This test is described in detail in ASTM test method D 2434.
- Unconsolidated sediments of low permeability (silts and clays) are usually tested with flexible wall permeameters that have the capability to apply backpressure to the specimen to facilitate saturation, and also simulate the actual field pressure at the depth where the sample is collected. Flexible wall (membrane) is used to encase the sample and prevent leakage. The test can be either with the constant or falling head (ASTM test method D 5084).
- Consolidated sediments and rocks are usually tested with gas (air) permeameters in which the fluid (e.g., helium gas or air) is forced under pressure through a small plug, up to couple of centimeters in diameter, taken from the rock core. This method, initially developed in the oil industry, usually gives the permeability of the sample in darcy units (darcys) and the obtained value must be converted into the units of hydraulic conductivity for migrating fluids (see Section 1.3.2). For more detail on the test application see American Petroleum Institute, 1998.

Wright et al. (2002) present an innovative and nondestructive method for measuring the hydraulic conductivity of drill core in horizontal and vertical directions within a triaxial cell at *in situ* effective stresses. It is important to note that the hydraulic conductivity value obtained at lower than effective field stresses is different than the probable *in situ* (real) value. The process of drilling and handling of samples, during which microfractures may develop, is one of several possible reasons. Such microfractures arguably close when the material is returned to its *in situ* effective stress during the laboratory test. Wright et al. (2002) discuss the importance of measuring the hydraulic conductivity at near *in situ* effective stresses, even for cores of confined aquifer materials from depths less than 160 m. More detail on the constant head and falling head permeameters, including analysis of test results using examples from practice, is given in Chapter 9.

Approximate values of the hydraulic conductivity of unconsolidated granular sediments can be determined in the laboratory from grain size analysis using sieves and, in case of a finer sediment fraction (silts and clays) using hydrometry. Detailed description of the grain size test and various analytical equations for determining the hydraulic conductivity are given in Chapter 9.

2.4.2 Field Tests

Aquifer testing in the field includes methods of recording changes of the hydraulic head in response to either natural or artificial stresses on the aquifer. These stresses may be as "simple" as the natural aquifer recharge by newly infiltrated water after a rainfall event, or as complex as a long-term pumping test for development of a major well field for water supply. In any case, the aquifer field tests are designed to collect sufficient information, on both the stresses and the hydraulic head changes, needed to establish some kind of a quantitative relationship between the two. In turn, this quantitative relationship, expressed by one or more analytical equations, enables an indirect estimation of the aquifer hydrogeologic parameters such as transmissivity (hydraulic conductivity), storage properties, and leakance from adjacent aquitards where applicable. It is important to understand that none of these parameters can be measured directly (like water temperature for example). Rather, their quantification is just an estimate, accuracy of which depends on both the field method applied and the interpretation of the collected information by a professional. In other words, interpretation of field test results is not unique and in most cases requires a solid hydrogeologic experience.

For many hydrogeologists the term "aquifer field test" implies pumping of a test well and recording of the related hydraulic head change (well drawdown) in the pumped well and one or more monitoring wells in its vicinity. The collected information on the time-dependent well pumping rates and well drawdown (in all monitored wells) is then analyzed to obtain estimates of the aquifer transmissivity and aquifer storage properties such as specific yield for unconfined aquifers and coefficient of storage for confined aquifers. This so-called aquifer (or well) pumping test is still the main and default type of field-testing in water supply hydrogeology. When designed properly, the test can also provide information on the test well efficiency, which is a very important factor for optimum groundwater extraction. Note that some professionals incorrectly use term "pump test" instead of pumping test; pump test should be what the term implies—a test of a well pump.

Pumping tests are much less frequent in contaminant hydrogeology primarily because of the two main concerns: (1) disposal issues related to large quantities of contaminated water extracted during the test and (2) possible alteration of contaminated groundwater (plume) flow directions. Consequently, various forms of slug tests are still considered default for contaminated aquifers because the contaminated groundwater is not extracted to the surface. When a detailed evaluation or implementation of groundwater remediation alternatives includes a pump and treat system, it would often be required that an aquifer pumping test is performed to collect information necessary for the remedial design. This is because slug tests characterize a limited volume of the aquifer porous media in the immediate vicinity of the test well, as opposed to the pumping tests, which evaluate response of a significantly larger aquifer volume.

Packer tests, borehole flowmeters, hydrophysical testing and advanced geophysical logging, often in conjunction with pumping tests, are increasingly utilized in the field to provide detail information on aquifer heterogeneity and anisotropy, particularly in fractured rock and karst aquifers.

Several simple field tests that only require recording of precipitation rates and continuous monitoring of the hydraulic head fluctuations in the aquifer and at a boundary may be very useful when estimating hydrogeologic parameters. For example, changes in surface stream levels and their impact on the hydraulic head in the adjacent aquifer can be used to estimate aquifer parameters such as hydraulic diffusivity and transmissivity, as well as aquifer recharge rate (see Section 2.6 and solved problems in Chapter 11). When two monitoring wells, located along the same known flow path, show response to major recharge events, a similar methodology can be applied to estimate aquifer hydraulic diffusivity and transmissivity between the two wells. Analyzing water table fluctuations in monitoring wells in response to precipitation is another simple way to estimate infiltration rates (see Section 2.6). Finally, influences of tidal fluctuations and barometric pressure changes on the hydraulic head have also been used to estimate aquifer parameters.

2.4.2.1 Aquifer Pumping Test

Aquifer pumping tests are among the most costly and labor-intensive field tests in hydrogeology. It is therefore crucial to approach their planning and execution by following well established and accepted practices and guidelines. The following discussion is based on recommendations and guidance documents by the United States Geological Survey (USGS: Ferris et al., 1962; Stallman, 1971; Heath, 1987), the United States Environmental Protection Agency (USEPA: Osborne, 1993), the United States Army Corps of Engineers (USACE, 1999), the United States Bureau of Reclamation (USBR, 1977), and the American Society for Testing and Materials (ASTM, 1999a,b). Other excellent reference books on the subject include *Groundwater and Wells* by Driscoll (1986), which covers all aspects of well design, drilling, and testing; *Analysis and Evaluation of Pumping Test Data* by Kruseman et al. (1991); and *Aquifer Testing: Design and Analysis of Pumping and Slug Tests* by Dawson and Istok (1992).

Aquifer pumping test consists of five main phases:

1. Definition of test objectives
2. Development of a focused site conceptual model
3. Test design
4. Data collection (recording)
5. Data analysis

The first four phases are described below; whereas the fifth (and arguably the most important) phase is described in more detail in Section 2.5.

2.4.2.1.1 Definition of Test Objectives

The question of defining the test objectives could simply be rephrased to "what exactly do you want to learn about the aquifer?" Is the test being conducted as part of a water budget study where the concern is defining transmissivity and storativity? Or is the test part of a water supply study where the main concern is long-term capacity of the well field? Is the test part of a groundwater contaminant transport study where the ultimate question is the velocity and direction of groundwater flow? Is there any concern about possible interconnection of two or more separated aquifers, such as an unconfined aquifer with shallow water table and a deeper artesian aquifer, with an aquitard between the two? A careful definition of the test objectives is therefore essential to ensure a successful test (USACE, 1999).

2.4.2.1.2 Development of a Focused Site Conceptual Model

Although aquifer pumping tests are also performed to provide data for development of a more detailed conceptual site model (CSM), at the same time it is very important that already available hydrogeologic information be fully utilized in the test design. Ideally, this focused CSM includes locations (or their estimates) of any hydraulic boundaries that may impact drawdown measured in the pumped well and the monitoring wells. Hydraulic boundaries are contacts with less permeable formations in both vertical and horizontal directions, and all natural and artificial surface water features near the test site including more "exotic" ones such as leaky sewers. Knowing locations and operational characteristics (e.g., screen depth, pumping rates, and times of pumping) of other extraction or injection wells located within a reasonable radius, is another very important component of the CSM because of the possible well interference (note that "reasonable radius" in case of confined aquifers and highly productive wells may extend several or more miles).

Preliminary estimates of the hydraulic conductivity and thickness of the aquifer to be tested, as well as of the underlying and overlying aquitards (where applicable), are used to predict possible pumping rates for the test, radius of influence, and drawdown at the test well and at various distances from it (i.e., at the existing or possible future monitoring wells). These estimates and other available information on the site hydrostratigraphy, integrated within the CSM, are then used in the next phase—the test design. In contaminant hydrogeology every attempt should be made to delineate horizontal and vertical extent of groundwater contamination before any pumping test. This delineation should be followed by an analysis of possible impact of the proposed test on contaminated groundwater flow directions, both horizontal and vertical.

2.4.2.1.3   Test Design
General design of an aquifer pumping test includes the following key elements:

- Design of test well
- Design of monitoring wells (observation wells, piezometers)
- Pumping design
- Duration of the test
- Water discharge control, measurements, and disposal

2.4.2.1.3.1   Design of Test Well   Whenever possible, the design of a test well should allow its easy conversion into a permanent extraction well in case the test results are favorable. Three main well design elements include well diameter, well screen, and well pump. More detail on these and other design elements such as selection of gravel pack, grouting, and well development methods can be found in Driscoll (1986) and Roscoe Moss Company (1990). In general, the well diameter should allow for 10% to 20% more pumping capacity than the maximum expected from the aquifer, and the pump should not be placed within the well screen but above or below it, in the cased portion of the well. This is to avoid rapid damage to the screen due to potential turbulent flow, clogging, or sanding.

Ideally, the pumping well screen should span the entire saturated thickness of the aquifer: the so-called fully penetrating well. However, due to budgetary and other constraints (e.g., thick aquifer, presence of contamination at different depths) this is often not feasible or desirable. If the aquifer is developed for water supply, and if certain aquifer portions are less permeable (e.g., there are thin layers or lenses of clay), the screen design may include perforated intervals separated by solid casing. In such cases, it would be difficult to estimate the overall aquifer transmissivity if the screens of monitoring wells are not designed in the same way. When single monitoring well screens are placed within different aquifer intervals, it would be equally difficult to estimate the transmissivity of discrete aquifer intervals without knowing their relative contribution to the overall pumping rate of the test well. Utilization of borehole flowmeters in such (and all other) cases provides invaluable information needed for the data analysis and interpretation of aquifer characteristics in general (see Section 2.4.2.4).

2.4.2.1.3.2   Design of Monitoring Wells (Observation Wells, Piezometers)   This component of the aquifer test design includes selection of the locations, screen depths, and the number of monitoring wells. It is always advisable to use existing wells for monitoring whenever possible, even when they appear not to be suitable for various "traditional" reasons such as: (1) the well is not completed in the same aquifer zone as the pumping well, (2) it is too far from the pumping well (inappropriately located), or (3) the well logs and well completion documents are not available or reliable. In any of these or similar cases the cost of equipping the existing wells with transducers, and collecting data on the hydraulic head changes during the test, would be minimal compared to the overall cost of the test. In general, having more data on aquifer response (or absence of response), even when it is not easily explainable, is always better than not having such a "headache." Finally, project budgets will usually provide a practical constraint for the number of new observation wells, so well locations must be optimized to fit the test objectives, and compromises often must be made (USACE, 1999).

The existing wells identified as potential observation wells should be field-tested to verify that they are suitable for monitoring aquifer response. The perforations of well screens of abandoned wells tend to become restricted by the buildup of iron compounds, carbonate compounds, sulfate compounds, or bacterial growth because of not pumping the well. Consequently, the response test is one of the most important prepumping examinations to be made if such wells are to be used for observation (Stallman, 1971; Osborne, 1993). The reaction of all wells to changing water levels should be tested by injecting or removing a known volume of water into each well and measuring the subsequent change of water level. Any wells which appear to have poor response should be either redeveloped, replaced, or dropped from consideration in favor of another available well selected (Osborne, 1993).

Many of the "rules of thumb" proposed in the past hydrogeologic practice regarding "appropriate" locations and screen depths of the observation wells are quickly losing importance because of the rapidly increasing utilization of numeric groundwater models in the analysis of aquifer test results. Carefully designed numeric model, based on a reasonably well-established CSM, is the best available tool for verification of any analytical method (i.e., "curve matching" method) used to interpret the test results. When applying the numeric modeling approach, virtually all collected data, including from "inappropriate wells" (such as those "too close" to, or "too far" from the pumping well, or with screen locations significantly offset from the center of the screen of the pumping well), becomes equally important in verifying assumptions inherent in any CSM.

Figure 2.30 illustrates an ideal number and placement of monitoring wells to determine aquifer parameters, including anisotropy, and the role of the overlying aquitard and unconfined aquifer:

- Monitoring wells with mutually isolated multiple screen intervals, or cluster wells, are used to assess possible vertical gradients in the system caused by pumpage, including within the fairly continuous aquitard. (Note that the "rule of thumb" for individual wells with one screen interval is to place their screen at approximately the same elevation in the aquifer as the center of the pumping well screen.)
- Placement of the monitoring wells at different radial angles from the pumping well enables determination of possible aquifer anisotropy. (Note that wells located along the same general direction, even when on the opposite sides of the pumping well, cannot be used for this purpose.)
- Wells placed at different distances from the pumping well, covering at least two log cycles (10, 80, and 600 m in this case) allow application of the distance–drawdown analytical method, which is useful for determining the radius of well influence and its development in time.

It is possible to estimate aquifer transmissivity using drawdown measurements directly from the pumping well when data from the monitoring wells is not available. This procedure is often referred to as the "well specific capacity method" (see Section 2.5.13). However,

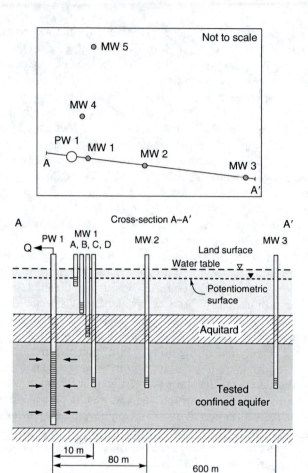

FIGURE 2.30 An example of a pumping test monitoring well network designed to determine characteristics and anisotropy of the tested confined aquifer, and nature of the aquitard including possible leakage from the aquitard and the unconfined aquifer into the underlying confined aquifer.

because of well loss in the pumping well, and assumptions that have to be made regarding storativity, the estimates are only approximate.

The number of monitoring wells to be installed depends on the purpose of the test, and especially on the funds available. The distance at which piezometers should be installed depends on the type of aquifer, its transmissivity, the duration of pumping, the discharge rate, the penetration rate of the screen, and whether the aquifer is stratified (Table 2.2).

The depth at which a monitoring well is installed depends on the aquifer type and the homogeneity of the aquifer. In an isotropic and homogeneous aquifer the monitoring well should be installed at a depth that is half the length of the well screen. For aquifers made up of sandy deposits with intercalations of less pervious layers, it is recommended to install a cluster of monitoring wells. If the aquifer is overlain by a partly saturated aquitard (leaky aquifer), monitoring wells should also be installed in the aquitard to check whether the water table is affected when the underlying aquifer is pumped (Griffioen and Kruseman, 2004).

2.4.2.1.3.3 Pumping Design It is usually desirable to pump at the maximum practical rate so as to stress the aquifer as much as possible for the duration of the test. This translates into more drawdown at the pumping well and the observation wells, and therefore more data

TABLE 2.2
Preferred Distance between Pumping Well and Monitoring Wells in Relation to Characteristics of the Pumping Test

Distance	Near	< ....>	Far
Type of aquifer	Unconfined		Confined
Transmissivity	Low		High
Pumping time	Short		Long
Discharge rate	Low		High
Penetration	Full		Partial[a]
Stratification	Little[b]		Strong

[a]The drawdown measured at a distance less than 1.5 times the thickness of the aquifer must be corrected for the influence of the vertical flow components close to the well.

[b]As a consequence of the differences in transmissivity at different depths, the drawdown observed close to the well may differ at different depths within the aquifer. With increasing pumping time and increasing distance to the well, the effect of stratification upon the drawdown diminishes.

Source: From Griffioen, J. and Kruseman, G.P. in *Groundwater Studies: An International Guide for Hydrogeological Investigations;* Kovalevsky, V.S., Kruseman, G.P., and Rushton, K.R. (Eds.), IHP-VI, Series on Groundwater No. 3, UNESCO, Paris, France, 2004, 217–238.

available for the final analysis. The maximum rate will be limited by the efficiency of the well construction and the specific capacity of the well, and should not cause dewatering below the pump intake during the duration of the test (USACE, 1999).

The maximum pumping rate is usually determined based on a three-step pumping test during which the well is pumped with three or more successively higher rates, each of which is kept constant for up to 8 h. The results of such test are used to determine the well efficiency and, consequently, the acceptability of inevitable well losses (see Section 3.1.2). In case the well loss is unacceptable, it will be necessary to redevelop the well before the aquifer test because of its inability to produce pumping rates needed to adequately stress the aquifer. In addition, the drawdown data from that well would be erroneous. Although pumping an inefficient well may not affect analysis of the test data collected at the monitoring wells (where there are no well losses), the overall validity of the test for any design purposes would be questionable.

Virtually all widely used methods of aquifer test analysis are based on the assumption of constant pumping rate for the duration of the drawdown phase. It is therefore very important to provide for a reliable energy source, and an adequate pump with enough power to maintain a constant pumping rate for the estimated maximum drawdown. In addition to the potentially large variation in discharge associated with the operation of pump motor or engine, the discharge rate is also related to the drop in water level near the pumping well during the aquifer test. As the pumping lift increases, the rate of discharge at a given level of power (such as engine rpm) will decrease. It will therefore be necessary to monitor and adjust the engine power during the test and keep the pumping rate constant. The pump should not be operated at its maximum rate. As a general rule, the pumping unit, including the engine, should be designed so that the maximum pumping rate is at least 20% more than the estimated long-term sustainable yield of the aquifer (Osborne, 1993).

If power is interrupted during the test, it may be necessary to terminate the test and allow for sufficient recovery so that prepumping water level trends can be extrapolated. At that point, a new test would be run. If, however, brief interruptions in power occur late in the test, the effect of the interruption can be eliminated by pumping at a calculated higher rate for some period so that the average rate remains unchanged. The increased rate must be calculated such that the final portion of the test compensates for the pumpage that would have occurred during the interruption of pumping (Osborne, 1993).

2.4.2.1.3.4 Duration of the Test Practical constraints usually limit the time available for the test, and at a maximum it is useless to run the test beyond the point at which a steady-state condition is reached (i.e., no more drawdown), or the point at which the pumping well intake screen begins to dewater. However, if delayed gravity response, dual porosity, or karst nature of the aquifer is suspected, then terminating the test soon after an apparent stabilization of the drawdown is reached may be erroneous (see Section 2.5.12).

Duration of aquifer pumping tests varies widely depending on the objectives and the aquifer characteristics. For example, in highly contaminated aquifers pumping test may last less than 6 hours in some cases to minimize the volume of extracted water. When developing a fractured or karst aquifer for major public water supply, some regulators require that the test may last more than 72 hours. Most tests, however, fall between 1 and 3 days for the pumping phase of the test, followed by a sufficient time to monitor the drawdown recovery to at least 90% of the prepumping level.

2.4.2.1.3.5 Water Discharge Control, Measurements, and Disposal Frequent and accurate control, and means of adjustment of the pumping rate during the test are essential for keeping it as nearly constant as possible. Common methods of measuring well discharge include the use of an orifice plate and manometer, an inline flowmeter, an inline calibrated pitot tube, a calibrated weir or flume, or, for low discharge rates, observing the length of time taken for the discharging water to fill a container of known volume (e.g., 5 gallon bucket; 55 gallon drum).

Discharging water immediately adjacent to the pumping well can cause problems with the aquifer test, especially in tests of permeable unconfined alluvial aquifers or karst aquifers. The water becomes a source of recharge, which will affect the results of the test. It is essential that the volumes of produced water, the storage needs, the disposal alternatives, and the treatment needs be assessed early in the planning process. The produced water from the test well must be transported away from all observation wells so it cannot return to the aquifer during the test. This may necessitate the laying of a temporary pipeline (sprinkler irrigation line is often used) to convey the discharge water a sufficient distance from the test site. In some cases, it may be necessary to have on-site storage, such as steel storage tanks or lined ponds. This is especially critical when testing contaminated zones where water treatment capacity is not available. In many cases it may be necessary to obtain permits for on-site storage and final disposal of the contaminated fluids. Final disposal could involve treatment and reinjection into the source aquifer or appropriate treatment and discharge (Osborne, 1993).

2.4.2.1.4 Design of Drawdown Measurements
The design of drawdown measurements should provide for collection of as many accurate measurements of the water level in the test and monitoring wells as possible. Prepumping water levels should be measured at least 24 h before start of the test pumping phase to detect any regional (background) trends and perform data adjustments as needed later during the data analysis phase. In addition to recording water levels in all the wells, every effort should be made to record barometric pressure for the same reason of possible data adjustment: piezometric pressure of confined and semiconfined aquifers fluctuates in response to changes in barometric pressure. Finally, if not feasible to record directly, provision should be made to secure data on precipitation and water levels at nearby hydraulic boundaries, such as streams

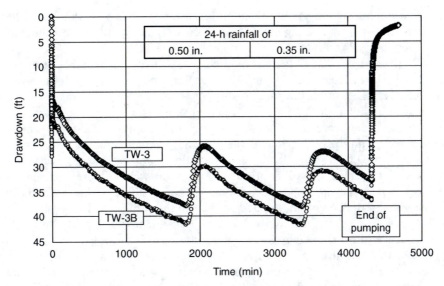

FIGURE 2.31 Drawdown observed during a 72-h aquifer test at well TW-3, City of Frederick, Maryland, the United States. Two drawdown recoveries during the test are due to two summer storm events, not because of cessation of pumping.

and lakes and reservoirs, which may be routinely collected by local, state, or federal agencies. Figure 2.31 shows time versus drawdown graph for a 72-h aquifer test at pumping well TW-3 and monitoring well TW-3B, City of Frederick, Maryland, the United States. Two recoveries during the test are not due to cessation of pumping—they are caused by aquifer recharge from two 24-h rainfalls of 0.50 and 0.35 in. recorded during 27 August and 28 August 2003 at the nearby gauging station. The rapid response of the hydraulic head in the aquifer is easily attributable to its karstic nature. "Unscheduled" events like these two summer storms limit the usefulness of the pumping test data for determining aquifer transmissivity. However, they may provide information for analyzing some other aquifer characteristics such as rates of aquifer recharge and effective porosity.

Before the era of automated data loggers and pressure transducers, various textbooks and manuals included recommendations as to the frequency of water level measurements during aquifer pumping tests. This was, to a large extent, influenced by the limited feasibility and high labor costs of manual recording at multiple well sites during a relatively short period of time. Data loggers and transducers eliminate these limitations as water levels are measured continuously and can be recorded at any time interval specified by the user. Manual recordings in such cases are made only sporadically to make sure the equipment is working properly. Whatever method of data measurements is selected, the main idea behind defining certain time intervals between data collection (economizing) is the fact that drawdown increases more rapidly at the beginning of the test then later into the test. Similarly, it would not make much sense to measure water level at a distant monitoring well every minute or so at the beginning of the test, when the drawdown may be measurable only after several hours or even days.

As most methods of aquifer test analyses include plotting time versus drawdown data on a semilog or log–log scale, the time interval between measurements should follow the logarithmic rule as well. Data loggers can be programmed to record drawdown on the logarithmic scale thus reducing the volume of stored information. In case of manual measurements, this means that the staff taking measurements can stop and catch some breath while running between various monitoring locations. What all this means in general is that there should be at least several measurements during first 10 min of the test, followed by 5- to 10-min intervals for the next 100 min or so, followed by hourly measurements during first 24 h. If the test is designed to last longer than 24 h, there should be at least five or so evenly spaced measurements every day for the reminder of the test. The frequency of water level measurements during the recovery phase (after the pump is shut off) should follow the same schedule as for the pumping phase until about 90%–95% of the drawdown is recovered.

In any case, continuous recording with data loggers and pressure transducers is much more reliable and precise, and provides invaluable information that otherwise may remain unknown when collecting data manually. This is especially important when drawdown is small such as in highly transmissive aquifers or at more distant wells. Figure 2.32 illustrates an example of a pumping test showing influence of one, and possibly two discrete and extremely transmissive zones in the Gray Limestone aquifer in southern Florida. The aquifer at the test location is unconfined, 25 ft thick, and underlain by a low-permeable confining unit. The major flow zone identified by geophysical logging and flowmeter testing in the test well is 5 ft thick, from 23 to 28 ft below ground surface. Another possible thin transmissive zone was identified at 17 ft (Reese, 2000, personal communication; see also Reese and Cunningham, 2000). The test well was pumped for 24 h at 297 gpm and the drawdown was observed at six monitoring wells screened at various depths. Drawdown observed at well C-1145, located 152 ft from the pumping well and screened in the Gray Limestone aquifer, is shown in Figure 2.32.

Very interesting disturbances in the recorded drawdown data, between approximately 50 and 300 min into the test, could be explained by a rapid pressure redistribution within the system as the upper transmissive zone at 17 ft bgs was being dewatered near the pumping well. Namely, the drawdown at the pumping well was lowered close to the 17-ft zone and then eventually below it, which first resulted in partial and then complete dewatering of the likely conduits near the extraction well. Thanks to the continuous data recording, the associated rapid changes in the hydraulic head are clearly visible. Note that, at the end of the test, the drawdown curve becomes even steeper, showing possible dewatering of the 23/28-ft zone as well.

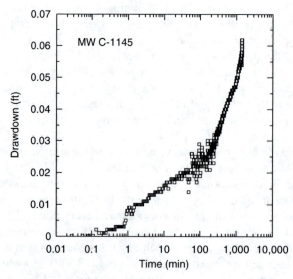

FIGURE 2.32 Time versus drawdown graph for data collected at monitoring well C-1145 during the Gray Limestone aquifer test at the FAA Radar site, southern Florida. (Data courtesy of Ronald R. Reese, U.S. Geological Survey.)

2.4.2.1.5 Data Analysis

Two main approaches to aquifer pumping test analysis are: (1) applying the so-called "type curve matching" methods and (2) inverse modeling. Each one has its advantages and disadvantages and an ideal solution would be to use both whenever there are no major budgetary constraints.

Curve matching methods commonly applied today are virtually all based, in one way or another, on the pioneering work of Theis (1935). Theis equation (see Section 2.5.3) can be used to determine aquifer transmissivity and storage if frequent measurements of drawdown versus time are performed in one or more observation wells. Since the equation has no explicit solution, Theis introduced a graphical method, which gives T and S if other terms are known.

Although the Theis method has quite a few assumptions (the aquifer is homogeneous, isotropic, and confined with an unlimited extent, horizontal impermeable base, horizontal flow lines, and no leakage), it has been shown that it often provides an approximate solution for true confined aquifers within acceptable error limits (e.g., 5%–10% error). With some minor changes and corrections, the Theis equation has also been applied to unconfined aquifers and partially penetrating wells (e.g., Hantush, 1961a,b; Jacob, 1963a,b; Moench, 1993), including when monitoring wells are placed closer to the pumping well where the flow is not horizontal (e.g., Stallman, 1961b, 1965).

However, heterogeneous hydrogeologic conditions in the field and presence of hydraulic boundaries commonly prevent direct application of the Theis and similar equations. Section 2.5 deals in detail with analytic solutions for various "non-Theisian" aquifers. Regardless of the aquifer type and hydraulic conditions in the field, it is important to understand that often more than just one type of curve or analytical method may be fitted to the observed time–drawdown curve, regardless of the underlying assumptions about aquifer type, and that different test analyzers may have different interpretations (see Figure 2.33 for just some of the common types of time–drawdown curves). Selection of the "correct" one therefore depends on the thorough overall hydrogeologic knowledge about the aquifer in question. This is particularly true in case of fractured and karst aquifers because of the absence of any prescribed, rigorous, and widely agreed-upon analytical method of aquifer pumping test analysis. As is the case with numeric groundwater modeling in fractured and karst aquifers, analysis of aquifer pumping tests is mostly performed by applying and combining methods developed for intergranular aquifers.

The results of aquifer tests are now routinely analyzed using computer programs and manual matching of the observed data to type curves is rather antiquated. Also commonly applied is inverse modeling of aquifer test results, using either analytical or numeric methods. AQTESOLV (HydroSOLVE, 2002) is one of the better-known and versatile computer programs for aquifer test analysis using match curves, and includes over 30 different methods for various types of aquifers. Particularly useful is a series of 10 public domain (free) Microsoft Excel-based programs for aquifer test analysis published by the U.S. Geological Survey (Halford and Kuniansky, 2002).

There are quite a few general textbooks and publications on analyzing aquifer pumping test results such as Ferris et al. (1962), Stallman (1971), Lohman (1972), USBR (1977), Driscoll (1986), Walton (1987), Kruseman et al. (1991), and Dawson and Istok (1992).

2.4.2.2 Slug Test

Slug tests are performed in exploratory boreholes and small-diameter monitoring wells to find an approximate hydraulic conductivity of the porous media in immediate vicinity of the well. Despite this limitation, and the fact that most methods in use cannot determine aquifer

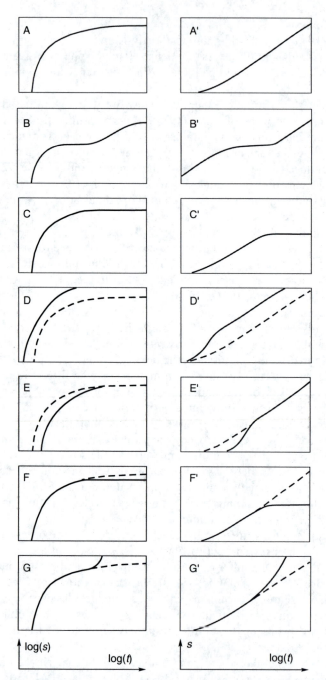

FIGURE 2.33 Log–log and semilog curves of drawdown versus time; A and A′, confined aquifer; B and B′, unconfined aquifer; C and C′, leaky (or semiconfined) aquifer; D and D′, effect of partial penetration; E and E′, effect of well-bore storage (large-diameter well); F and F′, effect of recharge boundary; G and G′, effects of an impervious boundary. (From Griffioen, J. and Kruseman, G.P. In: Kovalevsky, V.S., Kruseman, G.P., and Rushton, K.R. (Eds.), *Groundwater Studies: An International Guide for Hydrogeological Investigations*, IHP-VI, Series on Groundwater No. 3, UNESCO, Paris, France, 2004, pp. 217–238.)

storage properties, slug test are widely applied in groundwater studies because of the following advantages:

- The cost is incomparably smaller than the cost of an average pumping test (not including the cost of a pumping well itself).
- The time needed for its preparation and the duration of the test are short.
- There is no need for observation wells.
- In case of contaminated aquifers, there are no issues associated with groundwater extraction and disposal.
- The test can be performed in exploratory boreholes (including inexpensive soil borings) that are only temporarily equipped with a screen.
- The results are not influenced by short-term or long-term changes such as variations of barometric pressure, or interference from nearby pumping wells.

Since the hydraulic conductivity calculated from a slug test is an estimate for a very small part of the aquifer around the borehole, it is recommended to test as many boreholes in the area of interest as possible to find spatial distribution of the hydraulic conductivity.

In general, a slug test consists of quickly displacing a volume of water in the borehole (well) by adding or removing a "slug" (usually a solid cylinder of known volume) and then recording the change in the head at the well (Figure 2.34). As the displaced volume of water is relatively small, the head displacement is also small and the recovery is fast (usually not more than several minutes in moderately permeable aquifers, and less than 1 min in highly permeable aquifers). For this reason, it is necessary to quickly take as many measurements of the head in the well as possible in order to have enough data for the analysis. This is most accurately done with the help of a pressure transducer and a digital data logger, which can be programmed to take recordings every second, or with even smaller intervals during the initial portion of the test. Relatively quick recovery has been the main limitation of slug tests in highly permeable aquifers. However, recent research has lead to development of methodologies for slug test analysis in such cases (Butler et al., 2003). Slug tests with packers can be used to determine vertical distribution of the hydraulic conductivity around a test well (Zemansky and McElwee, 2005).

The modification of air-pressurized slug tests offers an efficient means of estimating the transmissivity (T) and storativity (S) of both low and highly permeable aquifers. Air-pressurized slug tests are conducted by pressurizing the air in the casing above the column of water in a well, monitoring the declining water level, and then releasing the air pressure and monitoring the rising water level. If the applied air pressure is maintained until a new equilibrium-water level is achieved and then the air pressure in the well is released instantaneously, the slug test solution of Cooper et al. (1967) can be used to estimate T and S from the rising water-level data (Shapiro and Greene, 1995; Greene and Shapiro, 1995, 1998). The total time to conduct the test can be reduced if the pressurized part of the test is terminated before achieving the new equilibrium-water level (Figure 2.34). This is referred to as a prematurely terminated air-pressurized slug test. Type curves generated from the solution of Shapiro and Greene (1995) are used to estimate T and S from the rising water-level data from prematurely terminated air-pressurized slug tests. The same authors also developed a public domain computer program called AIRSLUG for the analysis of air-pressurized slug test (Greene and Shapiro, 1995, 1998).

Three of the more commonly used methods of slug test analysis are Cooper et al. (1967) for confined aquifers, Hvorslev (1951) for confined aquifers, and Bouwer and Rice (1976) for unconfined and confined aquifers. There are several commercially available computer programs that include various methods of slug test analysis such as AQTESOLV by

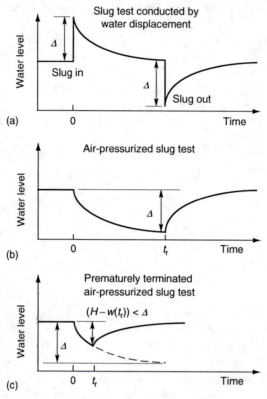

FIGURE 2.34 Schematic illustrating time-varying water level during (a) slug test conducted by water displacement, (b) air-pressurized slug test, and (c) prematurely terminated air-pressurized slug test, where Δ is the maximum change in water level due to water displacement or an applied air pressure, t_r is the time at which the pressurized part of the air-pressurized slug test is terminated, H is the initial water level at time $t = 0$, and $w(t_r)$ is the water level at time $t = t_r$. (From Green, E.A. and Shapiro, A.M., Methods of conducting air-pressurized slug tests and computation of type curves for estimating transmissivity and storativity. *U.S. Geological Survey Open-File Report* 95–424, Reston, VA, 43 p.)

HydroSOLVE. The U.S. Geological Survey series of 10 Microsoft Excel-based programs for aquifer test analyses including three slug test methods: Bouwer and Rice, Green and Shapiro, and van der Kamp (Halford and Kuniansky, 2002).

2.4.2.2.1   Bouwer and Rice Method

The well known Bouwer and Rice slug test can be used to determine the hydraulic conductivity in fully or partially penetrating and partially screened, perforated, or otherwise open wells (Bouwer and Rice, 1976; Bouwer, 1989). Although initially developed for unconfined aquifers, the test has been successfully used for confined aquifers given that the top of the screened section is below the overlying confining layer (Bouwer, 1989).

When water is instantaneously added to the well, the head is displaced upward and then starts to fall as water enters the aquifer (the hydraulic gradient is from the higher head in the well toward a lower head in the surrounding porous medium). This type of slug test is called the falling head test (see Figure 2.35). Water does not have to be added to the well in order to displace the head. A more common practice is to instantaneously lower into the well a heavy watertight cylinder on a rod, which produces the same effect of head displacement. When

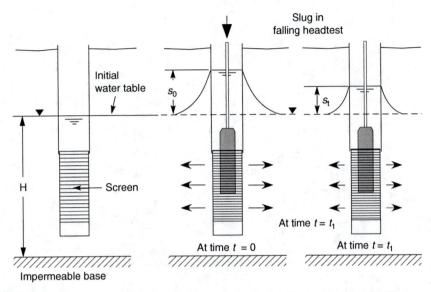

FIGURE 2.35 Elements of a falling head slug test. A watertight metal cylinder (slug) on a rod is instantaneously lowered in the well, which causes an upward displacement (rise) in the hydraulic head (s_0). The volume of displaced water is equal to the volume of the slug. Water from the well starts to enter the aquifer because of the created gradient, which causes the head in the well to fall.

water is quickly extracted from the well (equivalent to instantaneously pulling the cylinder out), the head is displaced downward and then starts to rise as water from the surrounding porous medium (aquifer) enters the well. This is called the rising head test (Figure 2.36). Common practice in the field is to first perform the falling head test (slug in), wait for the head stabilization, and then perform the rising head test by pulling the cylinder out (slug out). If performed correctly, the falling head and the rising head test results should give similar values of the hydraulic conductivity.

The elements needed for determining the hydraulic conductivity from the slug test data are shown in Figure 2.37. For reasons of consistency, the displacement of water level in the well is here denoted with s (equivalent to drawdown) rather than y as in the original work. The rate of water flow into the well, when the head in the well is lower than the initial water table in the surrounding aquifer, is given by the Thiem equation:

$$Q = 2\pi K L_e \frac{s}{\ln(R_e/r_w)} \tag{2.3}$$

where Q is the flow rate, K is the hydraulic conductivity, L_e is the length of screened (perforated, open) section of the well, s is the vertical difference between water level inside the well and water level outside the well (it is assumed that water table in the aquifer does not change for duration of the test), R_e is the effective radial distance over which s is dissipated, and r_w is the radial distance of undisturbed portion of aquifer from centerline, i.e., the radial distance from the center of the well to the undisturbed hydraulic conductivity of the aquifer.

The values of R_e were determined experimentally by the authors and expressed in terms of the dimensionless ratio $\ln(R_e/r_w)$ which is calculated with the aid of the curves shown in Figure 2.38. The value of r_w is the radius of the screen plus the thickness of the gravel pack

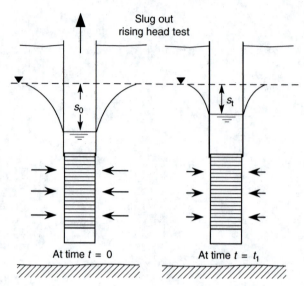

FIGURE 2.36 Elements of a rising head slug test. Cylinder is instantaneously withdrawn from the well, which causes downward displacement (drop) in the hydraulic head at the well (s_0). Water from the aquifer starts to enter the well because of the created gradient, which causes the head in the well to rise.

plus the thickness of the developed zone. In practice, the thickness of the developed zone is usually ignored because it is very difficult to define accurately.

The rate of water-level rise in the well is

$$\frac{ds}{dt} = -\frac{Q}{\pi r_c^2} \tag{2.4}$$

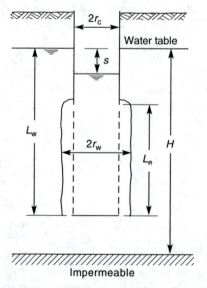

FIGURE 2.37 Geometry and symbols for the Bouwer and Rice slug test in an unconfined aquifer with gravel pack and/or developed zone around well screen. (From Bouwer, H., The Bouwer and Rice slug test-an update. *Ground Water*, 27(3), 304–309, 1989; contribution of the U.S. Department of Agriculture, Agricultural Research Service.)

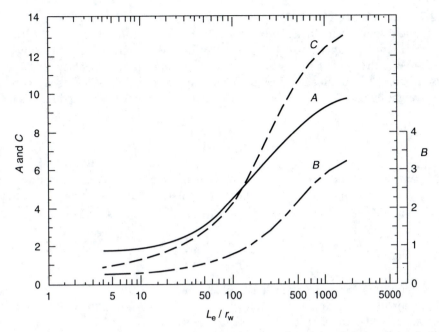

FIGURE 2.38 Dimensionless parameters A, B, and C as a function of L_e/r_w for calculation of $\ln(R_e/r_w)$ in the Bouwer and Rice slug test. (From Bouwer, H., The Bouwer and Rice slug test-an update. *Ground Water*, 27(3), 304–309, 1989; contribution of the U.S. Department of Agriculture, Agricultural Research Service.)

where r_c is the radius of the casing or other section of the well where the rise of water level is measured. If the water rises in the screened portion of the well and there is a gravel pack around the screen, r_c should be adjusted to account for the total free-water surface area in the well and the gravel pack:

$$r_c^* = [(1-n)r_c^2 + nr_w^2]^{1/2} \tag{2.5}$$

where n is the effective porosity of the gravel pack.

Solving Equation 2.4 for Q, equating the resulting expression to Equation 2.3, then integrating, and solving for K gives the Bouwer–Rice equation (Bouwer, 1989):

$$K = \frac{r_c^2 \ln(R_e/r_w)}{2L_e} \frac{1}{t} \ln\frac{s_0}{s_t} \tag{2.6}$$

where s_0 is the displacement at time zero and s_t is the displacement at time t.

When the well is partially penetrating ($L_w < H$), the dimensionless ratio $\ln(R_e/r_w)$ is

$$\ln\frac{R_e}{r_w} = \left[\frac{1.1}{\ln(L_w/r_w)} + \frac{A + B\ln[(H-L_w)/r_w]}{L_e/r_w}\right]^{-1} \tag{2.7}$$

For fully penetrating wells this ratio is

$$\ln\frac{R_e}{r_w} = \left[\frac{1.1}{\ln(L_w/r_w)} + \frac{C}{L_e/r_w}\right]^{-1} \tag{2.8}$$

where A, B, and C are dimensionless numbers plotted in Figure 2.38 as a function of L_e/r_w.

2.4.2.2.2 The Hydraulic Conductivity

As time (t) and displacement (s) are the only variables in logarithmic Equation 2.6, the plot of t versus s on a semilog paper must show a straight line. However, because the drawdown of the water table in the aquifer becomes more significant during the later part of the test, the basic assumption of Equation 2.6 does not hold any more and data points start to deviate from the straight line. The slope of the best-fitting straight line through field data is found as

$$\frac{\ln(s_1/s_2)}{(t_2 - t_1)}$$

and then inserted into Equation 2.6 with other known and calculated values to determine the hydraulic conductivity. s_1, t_1 and s_2, t_2 are the coordinates of any two points on the straight line. If the first point is chosen for time zero, the slope is then given as in Equation 2.6:

$$\frac{\ln(s_0/s_t)}{t - t_0} = \frac{1}{t} \ln \frac{s_0}{s_t}$$

Field data may often show the so-called double straight-line effect (Figure 2.39). The first straight-line portion is explained by a highly permeable zone around the well (gravel pack or developed zone), which quickly sends water into the well immediately after the water level in the well has been lowered (Bouwer, 1989). The second straight line is more indicative of the flow from the undisturbed aquifer into the well and its slope should be used in Equation 2.6.

A step-by-step example of using slug test data to find hydraulic conductivity with the Bouwer and Rice method is given in Section 14.6.

2.4.2.3 Packer Test

The origin of borehole packer tests is in geotechnical engineering for dam construction, particularly in fractured rock, where three-dimensional permeability of rock mass under pressure is tested to assess potential water loss around or below a dam. In general, the term packer test refers to any method designed to isolate portions of the borehole (well) and collect discrete vertical data on hydraulic conductivity, contaminant concentration, or other quantity

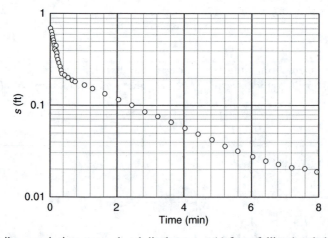

FIGURE 2.39 Semilog graph time versus head displacement (s) for a falling head slug test.

of interest. Borehole packers are pneumatic or mechanical devices that isolate sections of a borehole by sealing against the borehole wall.

Packer testing should test both the more permeable and less permeable portions of the borehole. In fractured and karst aquifers "more permeable" means borehole intervals with open fractures and cavities, while "less permeable" means nonfractured intervals that may be representative of the rock matrix porosity.

First widely applied method of pressure permeability testing was developed by Lugeon, primarily for designing geotechnical grouting during construction of dams. Original Lugeon method included grouting of the tested interval before drilling to deepen the borehole for the next test interval. Because of the high cost and time requirements, this method of pressure testing is now seldom used. However, the term "Lugeon" test is still the most widely used term for describing multiple pressure tests regardless of the drilling method. Figure 2.40 shows three most common methods of packer testing. Descending test is used mostly in unstable rock. The hole is drilled to the bottom of each test interval and an inflatable packer is set at the top of the interval to be tested. After the test, the boring is cased and then drilled to the bottom of the next test interval.

Ascending test is suited for stable rock where the boring can be drilled to the total depth without casing. Two inflatable packers, usually 5 to 10 ft (1.5 to 3 m) apart, are installed on the drill rod or pipe used for performing the test. The rod section between the packers is perforated. Tests start at the bottom of the hole. After each test, the packers are raised to the length of the test section, and another test is performed.

The length of the test section is governed by the character of the rock, but generally a length of 10 ft (3 m) is acceptable. Occasionally, a good packer seal cannot be obtained at the planned depth because of bridging, raveling, fractures and cavities, or a rough hole. If a good seal cannot be obtained, the test section length should be increased or decreased, or test sections overlapped to ensure that the test is made with well-seated packers. In case of highly permeable fractures and cavities, the 10 ft (3 m) section may take more water than the pump can deliver, and no stabilization of pressure (backpressure) can be developed. If this occurs, the length of the test section should be shortened until backpressure can be developed (USBR, 1977).

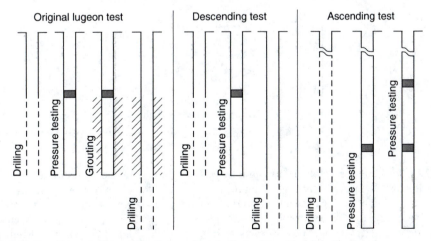

FIGURE 2.40 Possible configurations for packer testing with water injection under pressure. (From Cvetković-Mrkić, S., Metode geotehničkih melioracija, prva knjiga (in Serbian; *Methods of Geotechnical Meliorations*, Vol. 1). Rudarsko-geološki fakultet Univerziteta u Beogradu, Beograd, 1995, 177pp.)

Figure 2.41 shows test parameters used to determine the hydraulic conductivity of the tested borehole interval. The hydraulic conductivity is given as (USBR, 1977)

$$K = \frac{Q}{C_s r H} \tag{2.9}$$

where C_s is conductivity coefficient for semispherical flow in saturated material through partially penetrating cylindrical test wells. Values of C_s are found from the graph shown in Figure 2.42 for different values of l/r.

Multiple pressure permeability tests (Lugeon test) apply the pressure in three or more approximately equal steps. Each pressure is usually maintained for 15 min or so under the requirement of steady water intake. Flow readings are made at 5-min intervals. The pressure

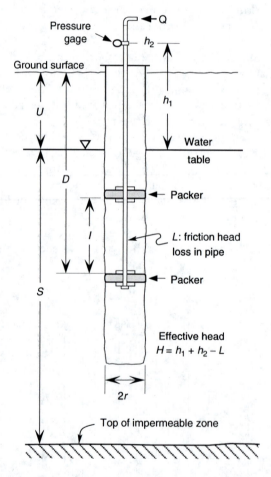

FIGURE 2.41 Pressure permeability test parameters for determining the hydraulic conductivity of consolidated rocks in saturated zone. l, Length of test section; D, distance from ground surface to bottom of test section; U, thickness of unsaturated material; S, thickness of saturated material; h_1, distance between gage and water table; h_2, applied pressure at gage; $H = h_1 + h_2 - L =$ effective head; L, head loss in pipe due to friction; ignore head loss for $Q < 4$ gallon/min in $\frac{1}{2}$ in. pipe; use length of pipe between gage and top of test section for computations; r, radius of test hole; Q, steady flow into well. (Modified from USBR, *Ground Water Manual*. U.S. Department of the Interior, Bureau of Reclamation, Washington, D.C., 1997, 480pp.)

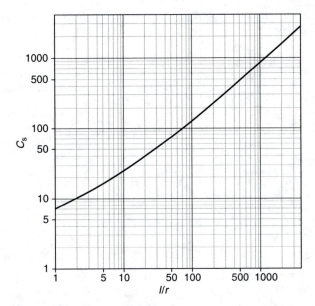

FIGURE 2.42 Conductivity coefficients for semispherical flow in saturated materials through partially penetrating cylindrical test well. (From USBR, *Ground Water Manual*. U.S. Department of the Interior, Bureau of Reclamation, Washington, D.C., 1997, 480pp.)

is then raised to the next step. After the highest step, the process is reversed and the pressure maintained for approximately the same time as during the first cycle. A plot of water taken against pressure for all test steps is then used to evaluate hydraulic conditions (Figure 2.43 and Figure 2.44).

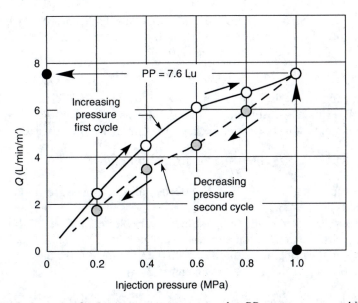

FIGURE 2.43 Typical graph of a Lugeon pressure test results. PP, pressure permeability; in the case shown it is 7.6 Lugeons for the default injection pressure of 1 Mpa. (From Cvetković-Mrkić, S., Metode geotehničkih melioracija, prva knjiga (in Serbian; *Methods of Geotechnical Meliorations*, Vol. 1). Rudarsko-geološki fakultet Univerziteta u Beogradu, Beograd, 1995, 177pp.)

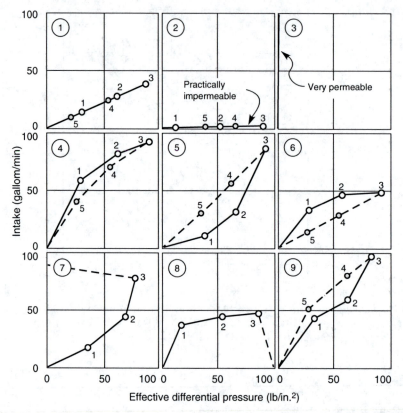

FIGURE 2.44 Plots of simulated, multiple pressure permeability tests. (From USBR, *Ground Water Manual*. U.S. Department of the Interior, Bureau of Reclamation, Washington, D.C., 1997, 480pp.)

Pressure permeability is often expressed in the unit called Lugeon (Lu), which is equal to the water intake of 1 L/1 m of the tested interval, during time interval of 1 min, under an injection pressure of 1 MPa or approximately 10 atm:

$$1Lu = 1L/min/m \text{ at } 1 \text{ MPa}$$

Specific permeability (q) is another measure of rock permeability obtained from the pressure and water intake measurements. It is defined as volume of water injected into the borehole per 1 m length, during 1 min, under a pressure of 1 m of water (0.1 atm):

$$q = \frac{Q}{Hl} \text{ L/min/m at } 0.1 \text{ atm} \qquad (2.10)$$

where Q is the water injection rate, H is the effective injection pressure, and l is the length of test interval (see Figure 2.41).

Hypothetical test results of multiple pressure tests are plotted in Figure 2.44 (USBR, 1977). The curves are typical of those often encountered. The test results should be analyzed using confined flow hydraulic principles combined with data obtained from the core or boring logs. Probable conditions represented by plots in Figure 2.44 are:

1. Very narrow, clean fractures. Flow is laminar, permeability is low, and discharge is directly proportional to head.
2. Practically impermeable material with tight fractures. Little or no intake regardless of pressure.

3. Highly permeable, relatively large, open fractures indicated by high rates of water intake and no backpressure. Pressure shown on gauge caused entirely by pipe resistance.
4. Permeability high with fractures that are relatively open and permeable but contain filling material which tends to expand on wetting or dislodges and tends to collect in traps that retard flow. Flow is turbulent.
5. Permeability high with fracture-filling material, which washes out, increasing permeability with time. Fractures probably are relatively large. Flow is turbulent.
6. Similar to 4, but fractures are tighter and flow is laminar.
7. Packer failed or fractures are large, and flow is turbulent. Fractures have been washed clean, and are highly permeable. Test takes capacity of pump with little or no backpressure.
8. Fractures are fairly wide but filled with clay gouge material that tends to pack and seal when under pressure. Takes full pressure with no water intake near end of test.
9. Open fractures with filling that tends to first block and then break under increased pressure. Probably permeable. Flow is turbulent.

The U.S. Geological Survey has developed a Multifunction Bedrock Aquifer Transportable Testing Tool (BAT[3]), which can be used to test discrete borehole intervals by either injecting or withdrawing water. The equipment is designed to perform the following operations by isolating a fluid-filled interval of a borehole using two inflatable packers (Shapiro, 2001):

- Collect water samples for chemical analysis
- Identify hydraulic head
- Conduct a single-hole hydraulic test by withdrawing water
- Conduct a single-hole hydraulic test by injecting water, and
- Conduct a single-hole tracer test by injecting and later withdrawing a tracer solution

The equipment can be configured to conduct these operations with only one of the borehole packers inflated, dividing the borehole into two intervals (above and below the inflated packer). The Multifunction BAT[3] is designed with two inflatable packers and three pressure transducers that monitor fluid pressure in the test interval (between the packers), as well as above and below the test interval; pressure transducers above and below the test interval are used to ensure that the borehole packers seal against the borehole wall during applications.

Sections of a borehole containing highly transmissive fractures are most easily tested by withdrawing water, whereas fractures with low transmissivity are tested by injecting small volumes of fluid. The multifunction BAT[3] is configured with both a submersible pump and a fluid-injection apparatus in the test interval to accommodate hydraulic tests that either withdraw or inject water. With this capability, the Multifunction BAT[3] can estimate transmissivity ranging over approximately eight orders of magnitude (Shapiro, 2001).

2.4.2.4 Borehole Flowmeter

The presence of varying permeability intervals in a borehole (well) may be indicated by various methods of geophysical logging, and their actual flow contribution may be measured and calculated using borehole flow meters (flowmeters). Flowmeters can be utilized in various ways, with or without pumping of the well. During well pumping packers can be used to isolate portions of the open borehole for a more precise characterization.

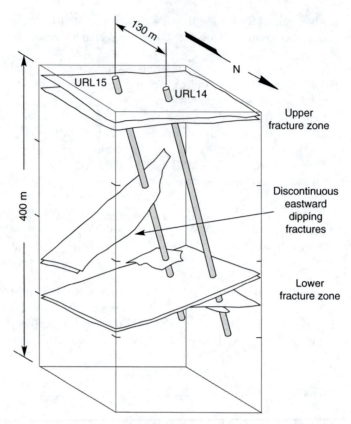

FIGURE 2.45 Diagram showing orientation of fractures intersecting boreholes URL14 and URL15 determined from acoustic televiewer logs and projections of fractures between the two boreholes. (From Paillet, F.L., Analysis of geophysical well logs and flowmeter measurements in boreholes penetrating subhorizontal fracture zones, Lac du Bonnet Batholith, Manitoba, Canada. U.S. Geological Survey Water-Resources Investigations Report 89-4211, Lakewood, CO, 1989, 30pp.)

Simultaneous use of geophysical logging tools and flowmeters is likely the best available method for characterization of aquifer heterogeneity (Molz et al., 1990; Paillet, 1994; Paillet and Reese, 2000). Integration of new geophysical methods with conventional logging techniques can be used to define flow zones, lithology, and structure, as well as their relations within the aquifer (Figure 2.45). Borehole flowmeters, whether vertical or horizontal, are used to both identify and quantify water-producing zones in a well. Flow logging tests between boreholes (cross-hole tests) can indicate the degree of connectivity of transmissive zones beyond individual borings (Figure 2.46).

Methods commonly used for quantitative hydraulic analysis of flow zones from flowmeter log data include proportion, analytical solution, and numerical modeling techniques. Proportion and analytical solution methods provide estimates of transmissivity for the flow zones, whereas numerical model method provides estimates of transmissivity and hydraulic head for the flow zones (Paillet, 1989, 1998, 2000, 2001; Molz and Melville, 1996; Wilson et al., 2001; USGS, 2004).

Differences in hydraulic head between two transmissive geologic units or fractures may produce vertical fluid flow in a borehole. Water enters the borehole at the unit with the higher head and flows toward and out of the unit with the lower head (see Figure 2.47). The vertical flow rate is limited by the geologic unit or fracture with the lower transmissivity. If the heads

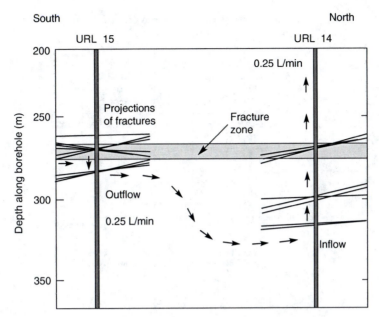

FIGURE 2.46 Diagram showing inferred hydraulic connection between boreholes URL14 and URL15 during cross-hole pumping tests. (From Paillet, F.L., Analysis of geophysical well logs and flowmeter measurements in boreholes penetrating subhorizontal fracture zones, Lac du Bonnet Batholith, Manitoba, Canada. U.S. Geological Survey Water-Resources Investigations Report 89-4211, Lakewood, CO, 1989, 30pp.)

of the different transmissive zones are the same, no vertical flow will occur in the well or borehole (USGS, 2004). However, it is recognized that the borehole may facilitate vertical flow between aquifers and fractures that would not normally be present, so any interpretation of vertical flows must be made with caution.

The nature of the aquifer heterogeneities can affect the flowmeter measurements, which are particularly sensitive to flowmeter positioning relative to the preferential flow zone (Molz et al., 1990; from Wilson et al., 2001). Steeply inclined fractures may also produce results that are difficult to interpret because of nonhorizontal flow across the borehole. The acoustic Doppler velocimeter is the only tool of the three-point measurement methods capable of measuring three-dimensional flow (Wilson et al., 2001).

2.4.2.5 Advanced Geophysical Logging

Hydrophysical logging, often combined with borehole flowmeter measurements, involves replacing the borehole fluid with deionized water, followed by a series of temperature and fluid-electrical-conductivity (FEC) logs. The logs profile the entire borehole to determine where formation water is entering and leaving (Wilson et al., 2001). A time series of FEC logs can identify the locations and rates of inflow and outflow. Hydrophysical logging can identify both vertical and horizontal flows since it surveys a length of the borehole, rather than providing point measurements. This makes hydrophysical logging a valuable method for obtaining profiles of flow characteristics and vertical permeability distribution in uncased wells or in wells with long screens (Tsang and Hale, 1989; Tsang et al., 1990).

The theory of hydrophysical logging is based on the law of mass balance and the linear relation between FEC and dissolved mass. By recording the changes in the electrical

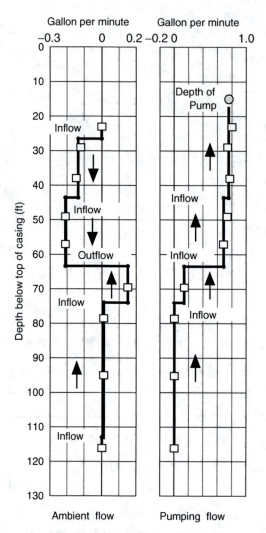

FIGURE 2.47 Sample heat-pulse flowmeter data from borehole in fractured-rock aquifer. Flow was measured under ambient (left) and pumping (right) conditions. Arrows indicate interpreted direction of flow, as fluid moves from zones of higher head to lower head. (From USGS (United States Geological Survey), Office of Ground Water, Branch of Geophysics, 2004. Vertical Flowmeter Logging, URL: http://water.usgs. gov/ogw/bgas/flowmeter/)

conductivity in the fluid column with depth, the locations of the water-bearing zones can be determined and the volumetric rate of inflow or outflow in discrete boring intervals can be calculated (Wilson et al., 2001; see also Figure 2.48). Hydrophysical logging can also be used to estimate hydraulic conductivity of discrete boring intervals.

Borehole radar provides a method to detect fracture zones at distances as far as 30 m or more from the borehole in electrically resistive rocks. Radar measurements can be made in a single borehole (transmitter and receiver in same borehole) or by cross-hole tomography (transmitter and receiver in separate boreholes). Single-hole, directional radar can be used to identify the location and orientation of fracture zones, and cross-hole tomography can be used to delineate fracture zones between boreholes. The movement of a saline tracer through fracture zones can be monitored by borehole radar (Williams and Lane, 1998).

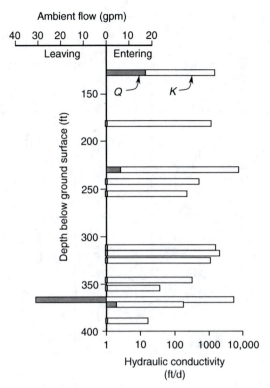

FIGURE 2.48 Results of hydrophysical logging in a deep boring in a fractured limestone aquifer. K, hydraulic conductivity; Q, ambient flow. (Data courtesy of Army Environmental Center.)

Acoustic televiewer (ATV) and optical televiewer (OTV) imaging result in continuous and oriented 360° views of the borehole wall from which the character, relation, and orientation of lithologic and structural planar features can be defined for studies of fractured-rock aquifers. Fractures are more clearly defined under a wider range of conditions on acoustic images than on optical images including dark-colored rocks, cloudy borehole water, and coated borehole walls. However, optical images allow for the direct viewing of the character of and relation between lithology, fractures, foliation, and bedding. The most powerful approach is the combined application of acoustic and optical imaging with integrated interpretation. Imaging of the borehole wall provides information useful for the collection and interpretation of flowmeter and other geophysical logs, core samples, and hydraulic and water-quality data from packer testing and monitoring (Williams and Johnson, 2004).

Lithology and structures such as fractures, fracture infillings, foliation, and bedding planes are viewed directly on the OTV images. OTV images can be collected in air or clear-water-filled intervals of boreholes. Unflushed drilling mud, chemical precipitation, bacterial growth, and other conditions that affect the clarity of the borehole water or produce coatings on the borehole wall impact the quality of OTV images.

ATV images can be collected in water or light mud-filled intervals of boreholes. Borehole enlargements related to structures such as fractures, foliation, and bedding planes scatter energy from the acoustic beam, reduce the signal amplitude, and produce recognizable features on the images (Paillet, 1994). Acoustic impedance contrast between the borehole fluid and the wall indicates the relative hardness of the borehole wall. Lithologic changes, foliation, bedding, and sealed fractures may be detected even when there is no change in borehole diameter if there

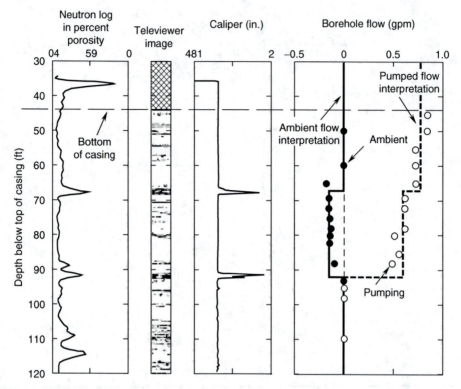

FIGURE 2.49 Background geophysical and vertical-flow logs with multiple geophysics for well FC-29 Fort Campbell, Kentucky, the United States. (Simplified from Wilson, J.T. et al., An evaluation of borehole flowmeters used to measure horizontal ground-water flow in limestones of Indiana, Kentucky, and Tennessee, 1999. U.S. Geological Survey Water-Resources Investigations Report 01-4139, Indianapolis, IN, 2001, 129pp.)

is sufficient acoustic contrast (Figure 2.49). Multiecho systems record the full wave train of the reflected acoustic signal and are capable of imaging behind plastic casing. Such systems are useful for imaging poorly competent intervals that will not stay open without being cased, and for inspecting annular grout seals. The newest digital televiewer system allows interactive determination of fracture orientation (Williams and Lane, 1998; Williams and Johnson, 2004). Figure 2.50 shows typical graphic presentation of the results of fracture analysis in a drilled borehole.

Television cameras are commonly used in groundwater studies to inspect the condition of well casing and screens, and they are also used to directly view (1) lithologic texture, grain size, and color, (2) water levels and cascading water, and (3) bedrock fractures. Television logs can be obtained in clear water and above the water level. The most sophisticated television systems are magnetically oriented and provide a 360° digital image of the borehole wall (Williams and Lane, 1998).

Nuclear magnetic resonance (NMR) is a promising new method, developed in the oil industry, for directly measuring the response from the fluids in the pore space of the rock. NMR logging results are presented as field log, integrated saturation analysis, and fluid identification. The field log generally includes total and effective porosity, bound fluid or free-fluid volumes and permeability. The integrated saturation analysis log uses resistivity data to determine whether the free-fluid volume contains hydrocarbons or water. NMR is

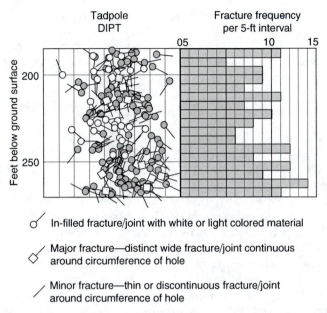

FIGURE 2.50 Results of fracture analysis in a section of a borehole completed in Proterozoic metavol-canic (mafic) bedrock near Phoenix, Arizona, the United States. (Courtesy of Malcolm Pirnie, Inc.)

arguably the first logging tool that allows direct measurement of permeability. Similarly, NMR is the only technology currently available that measures the pore size distribution of a sample, which determines permeability. NMR downhole measurements can also be directly correlated with core laboratory measurements thus greatly increasing its field applicability (Kenyon et al., 1995; Dunn et al., 2002).

2.5 ANALYSIS OF AQUIFER PUMPING TESTS

For a long time two basic methods of aquifer pumping test analysis were routinely applied by most groundwater practitioners to most field situations: steady-state solution developed by Thiem in 1906 and transient solution developed by Theis in 1936 (Thiem, 1906). Although both solutions are based on various simplified assumptions and are only applic-able to fully confined homogeneous aquifers, they are still often used, partially because of computational simplicity and partially because of "traditionalistic" inertia. However, hydrogeologic conditions in the field commonly prevent direct application of Thiem and Theis equations. Various analytical methods have been continuously developed to account for situations such as:

- Presence of leaky aquitards, with or without storage and above or below the pumped aquifer (Hantush and Jacob, 1955; Hantush, 1956, 1959, 1960; Cooper, 1963; Neuman and Witherspoon, 1969; Boulton, 1973; Streltsova, 1974; Moench, 1985)
- Delayed gravity drainage in unconfined aquifers (Boulton, 1954a and 1954b, 1963; Neuman, 1972, 1974; Moench, 1996)
- Other "irregularities" such as large-diameter wells and presence of bore skin on the well walls (Papadopulos and Cooper, 1967; Moench, 1985; Streltsova, 1988)
- Aquifer anisotropy (Papadopulos, 1965; Hantush, 1966a,b; Hantush and Thomas, 1966; Boulton, 1970; Boulton and Pontin, 1971; Neuman, 1975; Maslia and Randolph, 1986)

Various attempts have been made to develop analytical solutions for fractured aquifers, including dual-porosity approach and fractures with skin (e.g., Gringarten and Witherspoon, 1972; Gringarten and Ramey, 1974; Moench, 1984). However, due to the inevitable simplicity of analytical solutions, all such methods are limited to regular geometric fracture patterns such as orthogonal or spherical blocks, and single vertical or horizontal fractures.

2.5.1 Thiem Equation for Steady-State Flow toward a Well in a Confined Aquifer

In 1906, German engineer G. Thiem published as a dissertation for the degree of doctor-engineer from the Konigliche Technische Hochschule in Stuttgart the results of experiments and mathematical study relating to his field method for determining aquifer parameters. The experiments were made as a part of an investigation to find an additional water supply for the city of Prague, Czech Republic (then part of Austrian Empire). Ten sets of wells were installed for the purpose, each set including one well that was pumped and two observation wells. The observation wells were placed in line with the pumped well but in any direction from it regardless of the direction of the natural hydraulic gradient. The results from these tests were used to develop an equation for determining the hydraulic conductivity. The Thiem equation, as it is known since, describes time-independent (steady state) groundwater flow toward a well in a fully confined (nonleaky) aquifer. Detailed description of the Thiem equation application is given in the work by Wenzel (1936). The equation is based on the following assumptions:

- The well fully penetrates the confined aquifer and receives water from the entire thickness of the aquifer.
- The well is pumping water from the aquifer at a constant rate.
- The flow toward the well is radial, horizontal, and laminar, i.e., the flow lines are parallel along each radial cross section.
- The aquifer is homogeneous and isotropic, and has uniform thickness and a horizontal base.
- After a certain period of pumping, the drawdown does not increase anymore (it is stabilized), and the steady-state flow conditions are established.

Although these conditions are seldom completely satisfied in reality, there are several situations when a steady-state approach to the well pumping test analysis may be justified for a preliminary assessment, such as when the drawdown and the radius of well influence do not change in time. This includes pumping near a strong equipotential boundary (an island in a lake or river), or at a locality partly surrounded and completely hydraulically influenced by a large river. The boundary and the confined aquifer are in a direct hydraulic contact, the radius of well influence reaches the boundary relatively soon after the beginning of pumping, and the drawdown remains constant afterwards.

Figure 2.51 and Figure 2.52 show the elements for deriving the Thiem differential equation of groundwater flow toward a fully penetrating well in a confined aquifer. As is the case with most other equations in hydrogeology, the flow rate (Q) is given as the product of the cross-sectional area of flow (A) and the flow velocity (v):

$$Q = Av \tag{2.11}$$

The cross-sectional area of flow at distance r from the pumping well is the side of the cylinder with the radius r and the thickness b (which is the thickness of the confined aquifer):

$$A = 2\pi r b \tag{2.12}$$

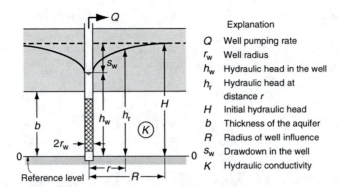

FIGURE 2.51 Elements of groundwater flow toward a fully penetrating well in a confined aquifer.

The velocity of flow at distance r is given as the product of the aquifer hydraulic conductivity (K) and the hydraulic gradient, which is an infinitesimally small drop in hydraulic head (dh) over an infinitesimally small distance (dr):

$$v = K(dh/dr) \tag{2.13}$$

Note that, unlike in the case of planar flow, there is no minus sign in the above equation. This is because the flow is toward the well, i.e., in the negative direction of the radial coordinate system, which cancels out the negative sign in front of the hydraulic gradient.

Inserting Equation 2.12 and Equation 2.13 into Equation 2.11 gives

$$Q = 2\pi r b K \frac{dh}{dr} \tag{2.14}$$

To solve this differential equation the variables h (hydraulic head) and r (radial distance from the well) first need to be separated:

$$dh = \frac{Q}{2\pi T} \frac{dr}{r} \tag{2.15}$$

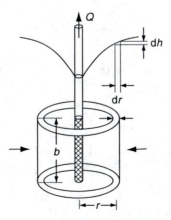

FIGURE 2.52 Scheme for deriving the differential equation of groundwater flow toward a fully penetrating well in a confined aquifer.

The above equation is then integrated with the following boundary conditions:

- At distance r_w (which is the well radius) the hydraulic head is h_w (head in the well).
- At distance r from the well the hydraulic head is h_r:

$$\int_{h_w}^{h_r} dh = \frac{Q}{2\pi T} \int_{r_w}^{r} \frac{dr}{r} \tag{2.16}$$

The integrals on both sides of the equation are readily solved because they are elementary integrals (see Appendix B):

$$h_r - h_w = \frac{Q}{2\pi T} \ln \frac{r}{r_w} \tag{2.17}$$

Finally, the hydraulic head at any distance from the pumping well is given as

$$h_r = h_w + \frac{Q}{2\pi T} \ln \frac{r}{r_w} \tag{2.18}$$

Equation 2.15 can also be integrated with the following boundary conditions:

- At distance r_w (which is the well radius) the hydraulic head is h_w (head in the well).
- At distance R from the well (which is the radius of well influence) the hydraulic head is H (which is the undisturbed head, equal to the initial head before the pumping started).

These boundary conditions enable introduction of the drawdown in the well (s_w) into Equation 2.18:

$$H - h_w = s_w = \frac{Q}{2\pi T} \ln \frac{R}{r_w} \tag{2.19}$$

The pumping rate of the well is

$$Q = \frac{2\pi T s_w}{\ln(R/r_w)} \tag{2.20}$$

Equation 2.19 can be rewritten to give the drawdown at any radial distance from the pumping well:

$$s_r = \frac{Q}{2\pi T} \ln \frac{R}{r} \tag{2.21}$$

Noting all constant terms in the above equation, the drawdown can be expressed as the function of distance only:

$$s_r = \frac{Q}{2\pi T} \ln R - \frac{Q}{2\pi T} \ln r \tag{2.22}$$

or, when the constant terms are replaced with general numbers:

$$s_r = a - b \ln r \tag{2.23}$$

Equation 2.23 indicates that the recorded data of drawdown versus the radial distance from the well would form a straight line when plotted on a semilogarithmic graph paper.

Figure 2.53 shows drawdown versus distance graph for three monitoring wells: MW 1, MW 2, and MW 3. Note that the drawdown recorded in the pumping well does not fall on the straight line connecting the monitoring well data; it is below the straight line, indicating that there is an additional drawdown in the well because of the well loss. The well loss, which is inevitable for any well, is explained in detail in Section 3.1.2. In short, it is a consequence of various factors such as disturbance of porous medium near the well during drilling, an improper (insufficient) well development, a poorly designed gravel pack or well screen, and turbulent flow through the screen.

Because of the well loss, at least two monitoring wells are needed to apply the Thiem equation properly. Using the pumping well drawdown and drawdown in one monitoring well would give erroneous results.

The steady-state radius of well influence (R) is the intercept of the straight line connecting the monitoring well drawdowns and the zero drawdown.

2.5.1.1 Transmissivity and Hydraulic Conductivity

The transmissivity of the aquifer can be determined from the graph drawdown versus distance such as the one shown in Figure 2.53. This is done by relating the coordinates of any two points on the straight line:

$$s_1 = \text{const} - \frac{Q}{2\pi T} \ln r_1 \tag{2.24}$$

$$s_2 = \text{const} - \frac{Q}{2\pi T} \ln r_2 \tag{2.25}$$

$$s_1 - s_2 = \Delta s = \frac{Q}{2\pi T} \ln \frac{r_2}{r_1} \tag{2.26}$$

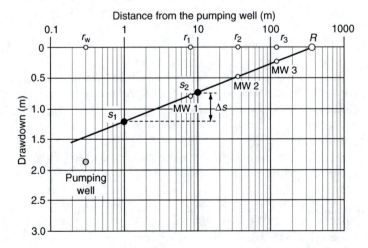

FIGURE 2.53 Graph drawdown versus distance for a pumping test with three monitoring wells.

The transmissivity is then found from this difference in drawdown for the chosen points:

$$T = \frac{Q}{2\pi\Delta s} \ln\frac{r_2}{r_1} \tag{2.27}$$

For practical purposes it is better to express transmissivity in terms of the logarithm to the base 10 (log) since the corresponding graph is easier to use. Knowing that

$$\log x = 0.4343 \ln x$$

and replacing "pi" (π) with number 3.14, Equation 2.27 becomes

$$T = \frac{0.366Q}{\Delta s} \log\frac{r_2}{r_1} \tag{2.28}$$

If two points on the straight line are chosen to be one log cycle apart (i.e., the ratio of the distance coordinates r_2 and r_1 is 10), Equation 2.28 reduces to the following simple form:

$$T = \frac{0.366Q}{\Delta s} \tag{2.29}$$

which is the final equation for determining transmissivity of a confined aquifer using the Thiem steady-state equation. The hydraulic conductivity of the aquifer porous media is found by dividing the transmissivity by the aquifer thickness. Note that aquifer storage cannot be found using the steady-state approach. Application of the Thiem equation to actual field data is given in Section 13.1.

2.5.2 Flow toward a Well in an Unconfined Aquifer

Analysis of wells pumping from unconfined aquifers is more complex than that of confined aquifers since the top of the aquifer corresponds to the water table and its position changes due to pumping. In other words, the thickness of the aquifer (saturated zone) changes within the radius of the well influence: it is the smallest at the well perimeter (see Figure 2.54). If the aquifer impermeable base is horizontal, and the reference level is set at the base, the hydraulic head equals the water table, which simplifies the derivation of the flow equation.

Applying Dupuit's hypothesis allows the exact calculation of the flow rate, whereas finding the accurate position of the water table is more complicated and is based on various experimental (approximate) equations. Figure 2.54 illustrates the concept of Dupuit's hypothesis together with the actual distribution of velocities around a pumping well. Dupuit's hypothesis states that

- Equipotential lines are vertical.
- Velocity is constant along any given vertical (i.e., equipotential line).
- The velocity vector has only the horizontal component (i.e., the streamlines are horizontal and parallel).

However, the velocity vector has a vertical component, which increases closer to the well and the water table. The actual position of the water table is above the calculated one (Dupuit's parabola) and this difference at the well perimeter corresponds to the seepage

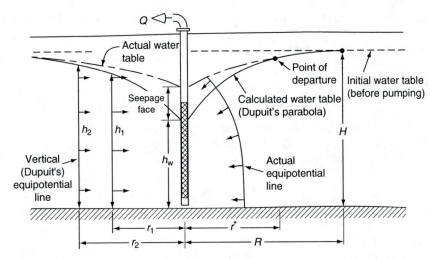

FIGURE 2.54 Illustration of the Dupuit's hypothesis (left) and the actual flow (right) toward a fully penetrating well in an unconfined aquifer. Note that the actual equipotential line is curvilinear, and that the velocity vector is not horizontal as assumed by Dupuit (the velocity vector has both the horizontal and the vertical components).

face. At a certain distance from the well (denoted r^* in Figure 2.54) the difference between the actual and the calculated water table becomes very small and can be ignored for practical purposes.

The flow rate (Q) is given as the product of the cross-sectional area of flow (A) and the flow velocity (v):

$$Q = Av \tag{2.30}$$

The cross-sectional area of flow at distance r from the pumping well is the side of the cylinder with radius r and thickness h (which is the thickness of the saturated zone; see Figure 2.55):

$$A = 2r\pi h \tag{2.31}$$

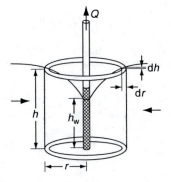

FIGURE 2.55 Scheme for deriving the differential equation of groundwater flow toward a fully penetrating well in an unconfined aquifer.

The velocity of flow at distance r is given as the product of the aquifer hydraulic conductivity (K) and the hydraulic gradient (i), which is an infinitesimally small drop in hydraulic head (dh) over an infinitesimally small distance (dr):

$$v = K \frac{dh}{dr} \tag{2.32}$$

Note that there is no minus sign in the above equation as is the case with Darcy's law in the Cartesian (rectangular) coordinate system. This is because the decrease in hydraulic head is in the negative direction of the r-axis (toward the well) and the two minus signs cancel each other:

$$v = -Ki = -K\left(-\frac{dh}{dr}\right) = K\frac{dh}{dr} \tag{2.32a}$$

The flow rate is then

$$Q = 2r\pi hK\frac{dh}{dr} \tag{2.33}$$

To solve this differential equation the variables h (hydraulic head) and r (radial distance from the well) have to be separated:

$$hdh = \frac{Q}{2\pi K}\frac{dr}{r} \tag{2.34}$$

The above equation is then integrated for known boundary conditions (see Figure 2.55):

- At distance r_w (which is the well radius) the hydraulic head is h_w (head in the well).
- At distance r from the well the hydraulic head is h:

$$\int_{h_w}^{h} hdh = \frac{Q}{2\pi K} \int_{r_w}^{r} \frac{dr}{r} \tag{2.35}$$

The integrals on both sides are elementary and are easily solved (see Appendix B):

$$\frac{1}{2}(h^2 - h_w^2) = \frac{Q}{2\pi K} \ln\frac{r}{r_w} \tag{2.36}$$

which gives the following common expression for the pumping rate of a fully penetrating well in an unconfined aquifer:

$$Q = \pi K \frac{(h^2 - h_w^2)}{\ln(r/r_w)} \tag{2.37}$$

Equation 2.34 can be integrated for any other pair of known hydraulic heads and the corresponding distances from the well. For example:

- At distance r_w (which is the well radius) the hydraulic head is h_w (head in the well).
- At distance R from the well (which is the radius of well influence) the hydraulic head is H (which is the initial, undisturbed head before pumping; see Figure 2.54):

$$Q = \pi K \frac{(H^2 - h_w^2)}{\ln(R/r_w)} \tag{2.38}$$

Equation 2.38 can be rewritten as

$$Q = \pi K \frac{(H - h_w)(H + h_w)}{\ln(R/r_w)} \tag{2.39}$$

Knowing that the drawdown in the well (s_w) is the difference between the initial hydraulic head (H) and the head registered in the well during pumping (h_w), i.e.,

$$s_w = H - h_w$$

and introducing the average aquifer thickness (thickness of the saturated zone) as

$$b^* = \frac{H + h_w}{2}$$

Equation 2.39 can be written in the following form:

$$Q = \frac{2\pi K b^* s_w}{\ln(R/r_w)} \tag{2.40}$$

which is similar to the equation describing steady-state flow toward a fully penetrating well in a confined aquifer:

$$Q = \frac{2\pi b K s_w}{\ln(R/r_w)}$$

where b is the thickness of the confined aquifer. For the analysis of steady-state groundwater flow toward wells in unconfined aquifers one should always apply Equation 2.39 rather than Equation 2.40 since the aquifer varying saturated thickness is represented more accurately.

2.5.2.1 Hydraulic Conductivity and Radius of Well Influence

As mentioned earlier, Equation 2.34 can be integrated for any pair of known hydraulic heads and their corresponding distances from the pumping well. If the heads (or drawdown) in the two monitoring wells are recorded, the boundary conditions are as follows:

- At distance r_1 (e.g., location of a monitoring well closer to the pumping well) the hydraulic head is h_1.
- At distance r_2 (e.g., location of a monitoring well farther away from the pumping well) the hydraulic head is h_2.

For these boundary conditions Equation 2.34 becomes

$$h_2^2 - h_1^2 = \frac{Q}{\pi K} \ln \frac{r_2}{r_1} \tag{2.41}$$

which, solved for the hydraulic conductivity, gives the following expression:

$$K = \frac{Q \ln(r_2/r_1)}{\pi(h_2^2 - h_1^2)} \tag{2.42}$$

The hydraulic conductivity and the radius of well influence can be found graphoanalytically, similar to the confined aquifer case described in Chapter 1. The general equation of groundwater flow toward a fully penetrating well in an unconfined aquifer can be written in the following form:

$$H^2 - h^2 = \frac{Q}{\pi K} \ln \frac{R}{r} \tag{2.43}$$

where h is the hydraulic head at distance r from the pumping well, H is the initial hydraulic head, and R is the radius of well influence. Knowing the basic rules of logarithmic calculus (see Appendix B), Equation 2.43 can also be written as

$$H^2 - h^2 = \frac{Q}{\pi K} \ln R - \frac{Q}{\pi K} \ln r \tag{2.44}$$

Since all the factors of the first term on the right-hand side are constant values (note that the radius of well influence does not change in time for steady-state conditions), Equation 2.44 becomes

$$H^2 - h^2 = \text{const} - \frac{Q}{\pi K} \ln r \tag{2.45}$$

or, when the natural logarithm is replaced with the logarithm to the base 10 (note that $\ln(x) = 2.3026 \log(x)$):

$$H^2 - h^2 = \text{const} - \frac{Q}{\pi K} 2.303 \log r \tag{2.46}$$

$$H^2 - h^2 = \text{const} - \frac{0.733Q}{K} \log r \tag{2.46a}$$

From Equation 2.46 it can be seen that the recorded cone of depression would appear as a straight line if plotted on a semilogarithmic graph since $H^2 - h^2 = f(\log r)$. Figure 2.56 shows data for a pumping well and two monitoring wells plotted on such graph. The coordinates of any two points on the straight line can be used to calculate the hydraulic conductivity with Equation 2.46a:

$$(H^2 - h^2)_1 = \text{const} - \frac{0.733Q}{K} \log r_1 \tag{2.47}$$

$$(H^2 - h^2)_2 = \text{const} - \frac{0.733Q}{K} \log r_1 \tag{2.47a}$$

$$(H^2 - h^2)_1 - (H^2 - h^2)_2 = \Delta(H^2 - h^2) = \frac{0.733Q}{K} \log \frac{r_2}{r_1} \tag{2.48}$$

$$K = \frac{0.733Q}{\Delta(H^2 - h^2)} \log \frac{r_2}{r_1} \tag{2.49}$$

If two points on the straight line are chosen to be one log cycle apart (i.e., the ratio of the distance coordinates r_2 and r_1 is 10), Equation 2.49 reduces to

$$K = \frac{0.733Q}{\Delta(H^2 - h^2)} \tag{2.50}$$

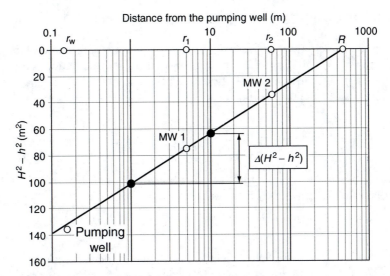

FIGURE 2.56 Semilogarithmic graph $H^2 - h^2$ versus radial distance from a pumping well (r) for determining the hydraulic conductivity and the radius of well influence.

The radius of well influence is determined by the interception of the straight line drawn through the monitoring well points with the line of zero drawdown. Note that the straight line should be drawn only through the monitoring well points, as the actual drawdown recorded in the pumping well includes an additional drawdown component. This additional drawdown is a consequence of well losses, as well as formation of a seepage face along the well screen perimeter.

2.5.2.2 Position of Water Table around Pumping Well

An equation, based on Dupuit's assumption, which describes the position of the water table (h) at a distance r from the pumping well, is derived from Equation 2.37:

$$h = \sqrt{h_w^2 + \frac{Q}{\pi K} \ln \frac{r}{r_w}} \tag{2.51}$$

where h_w is the hydraulic head recorded in the well and r_w is the well radius. This equation is also known as Dupuit's parabola. The well pumping rate is given as (see Equation 2.38)

$$Q = \pi K \frac{(H^2 - h_w^2)}{\ln(R/r_w)} \tag{2.52}$$

By inserting Equation 2.52 into Equation 2.51, the hydraulic head is expressed only in terms of other known heads and distances at which they are recorded. This is a more convenient method for calculating Dupuit's parabola since one does not need to know the hydraulic conductivity and the well pumping rate:

$$h = \sqrt{h_w^2 + (H^2 - h_w^2) \frac{\ln(r/r_w)}{\ln(R/r_w)}} \tag{2.53}$$

where H is the initial hydraulic head (before pumping) and R is the radius of well influence.

The position of the actual water table is above the Dupuit's parabola and several experimental (approximation) equations have been proposed for its calculation. One of the more common is the Babbitt–Caldwell equation (adapted from Kashef, 1987), which estimates the actual hydraulic head at the well perimeter as

$$h_0 = H - \frac{0.6}{H} \frac{H^2 - h_w^2}{\ln(R/r_w)} \ln\frac{R}{0.1H} \tag{2.54}$$

The difference between the actual hydraulic head outside the well (h_0) and the head directly measured in the well (h_w) gives the length of the seepage face (Δh):

$$\Delta h = h_0 - h_w$$

Another equation, proposed by Schneebeli, is also used for determining the length of the seepage face (Δh) at the well perimeter (adapted from Vukovic and Soro, 1984):

$$\Delta h = \sqrt{h_w^2 + \frac{Q}{\pi K}\left[0.4343 \ln \frac{Q}{\pi K r_w^2} - 0.4\right]} - h_w \tag{2.55}$$

or

$$\Delta h = \sqrt{h_w^2 + \frac{H^2 - h_w^2}{\ln(R/r_w)}\left[0.4343 \ln \frac{H^2 - h_w^2}{r_w^2 \ln(R/r_w)} - 0.4\right]} - h_w \tag{2.56}$$

Equation 2.55 and Equation 2.56 should be used when the following condition is satisfied:

$$\frac{Q}{K r_w^2} > 8 \tag{2.57}$$

When this fraction is less than 8, the seepage face is very small and can be ignored for practical purposes.

The position of the water table near the pumping well is estimated based on the calculated seepage face and a distance from the well at which Dupuit's assumption is valid, i.e., where Dupuit's parabola and the actual water table can be considered close enough for practical purposes. This distance (r^*) is found from Equation 2.51 for the condition that $dh/dr = 0.2$, which, for most practical purposes, provides that the actual water table and the Dupuit's parabola are close enough to each other (Vukovic and Soro, 1984):

$$\frac{2.5(H^2 - h_w^2)}{\ln(R/r_w)} = r^* \sqrt{h_w^2 + (H^2 - h_w^2)\frac{\ln(r^*/r_w)}{\ln(R/r_0)}} \tag{2.58}$$

or

$$\text{Const} = r^* \sqrt{h_w^2 + (H^2 - h_w^2)\frac{\ln(r^*/r_w)}{\ln(R/r_w)}} \tag{2.58a}$$

As can be seen from Equation 2.58, distance r^* is expressed implicitly and its value is found by trial. As practice shows, distance r^* is usually within the radius of 10 m from the pumping well. In cases of extremely high well losses, r^* can be placed at greater distances.

Detailed step-by-step determination of aquifer hydraulic properties and water table position is given in Section 13.2. Steady-state flow solution for a well in an unconfined aquifer under influence of infiltration is given in Section 13.3. Transition from confined to unconfined aquifer conditions due to pumping is given in Section 13.4.

2.5.3 Theis Equation for Transient Flow toward a Well in a Confined Aquifer

The Theis equation, which describes transient (nonequilibrium) groundwater flow toward a fully penetrating well in a confined aquifer, is the basis for most methods of transient pumping test analysis. Using the equation, transmissivity and storage can be determined from the drawdown measurements without requirement for drawdown stabilization. In addition, only one observation well is enough to estimate aquifer hydrogeologic parameters, as opposed to steady-state calculations where at least two observation wells are needed.

The Theis equation gives drawdown (s) at any time after the beginning of pumping:

$$s = \frac{Q}{4\pi T} W(u) \tag{2.59}$$

where Q is the pumping rate and is kept constant during the test, T is transmissivity, and $W(u)$ is called well function of u, also known as the Theis function, or simply well function. Dimensionless parameter u is given as

$$u = \frac{r^2 S}{4Tt} \tag{2.60}$$

where r is the distance from the pumping well where the drawdown is recorded, S is the storage coefficient, and t is the time since the beginning of pumping. The well function of u is the exponential integral

$$W(u) = -E(-i) = -\int_u^\infty \frac{e^{-u}}{u} du \tag{2.61}$$

which can also be expressed with the following series:

$$W(u) = -0.5772 - \ln u + u - \frac{u^2}{2 \cdot 2!} + \frac{u^3}{3 \cdot 3!} - \frac{u^4}{4 \cdot 4!} + \cdots \tag{2.62}$$

or as

$$W(u) = \ln \frac{0.5615}{u} + \sum_{n=1}^{n=\infty} (-1)^{n+1} \frac{u^n}{n \cdot n!} \tag{2.63}$$

Values of $W(u)$ for various values of the parameter u are given in Appendix D and can be readily found in groundwater literature. The graph in Figure 2.57 shows this relationship in the form of $W(u)$ versus $1/u$.

Theis derived his equation based on quite a few assumptions and it is very important to understand its limitations. If the aquifer tested and the test conditions significantly deviate

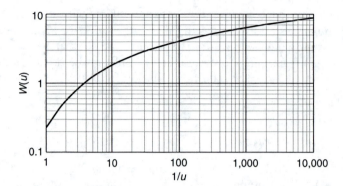

FIGURE 2.57 Part of theoretical curve (type curve) of $W(u)$ versus $1/u$ used for the Theis curve matching method of well pumping test analysis.

from these assumptions (which, in fact, is very often the case in reality), other methods of analysis using appropriate analytical equations should be used. Equation 2.59 is valid for the following conditions (modified from Driscoll, 1986):

- The aquifer is homogeneous and isotropic.
- The aquifer is uniform in thickness and the pumping never affects its exterior boundary (the aquifer can be considered infinite in areal extent).
- The aquifer is confined and it does not receive any recharge.
- Well discharge is derived entirely from aquifer storage.
- The water removed from storage is discharged instantaneously when the head is lowered.
- The pumping rate is constant.
- The radius of the well is infinitesimally small, i.e., the storage in the well can be ignored.
- The pumping well is fully penetrating: it receives water from the entire thickness of the aquifer and it is 100% efficient (there are no well losses).
- The initial potentiometric surface (before pumping) is horizontal.

Theis equation can be used to determine aquifer transmissivity and storage if frequent measurements of drawdown versus time are performed in one or more observation wells. Equation 2.59 has no explicit solution and Theis introduced a graphical method, which gives T and S if other terms are known. Theoretical curve $W(u)$ versus $1/u$ is plotted on a graph paper with logarithmic scales as shown in Figure 2.57. Field data of drawdown (s) versus time (t) for an observation well is plotted separately on a graph with the same scale. This field curve for the observation well is then superimposed on the theoretical curve, also called type curve (Figure 2.58). It is essential that both graphs/curves have identical logarithmic scales and cycles as shown in Figure 2.58. Keeping coordinate axes of the curves parallel, the field data is matched to the type curve.

Once a satisfactory match is found, a match point on the overlapping graphs is selected. The match point is defined by four coordinates, the values of which are read on two graphs: $W(u)$ and $1/u$ on the type curve graph, s and t on the field graph. The match point can be any point on the overlapping graphs, i.e., it does not have to be on the matching curves. Figure 2.58 shows two possibilities of choosing the match point: one is on the theoretical curve/measured data (which looks somewhat more logical), and one is chosen outside to obtain convenient values of $W(u)$ and $1/u$: 1 and 100, respectively. Both matching points should lead to the same values of T and S if the coordinates are read carefully.

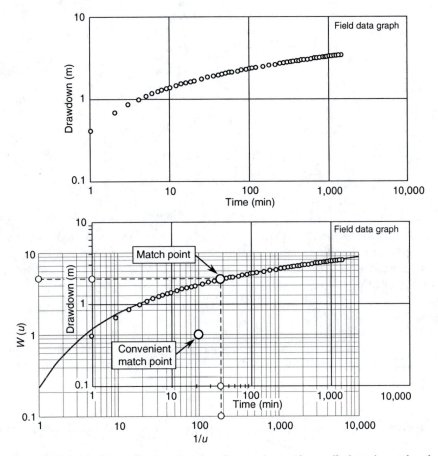

FIGURE 2.58 Field data of drawdown versus time for an observation well plotted on a log–log graph paper at the same scale as the theoretical type curve, and superimposed on it.

The transmissivity is calculated using Equation 2.59 and the match point coordinates s and $W(u)$:

$$T = \frac{Q}{4\pi s} W(u) \tag{2.64}$$

The storage coefficient is calculated using Equation 2.60, the match point coordinates $1/u$ and t, and the transmissivity value already determined:

$$S = \frac{4Ttu}{r^2} \tag{2.65}$$

For the example shown in Figure 2.58 the match point coordinates are:

$$W(u) = 4.75 \qquad 1/u = 180 \quad u = 0.0055$$
$$s = 2.0 \text{ m} \qquad t = 39 \text{ min} = 2440 \text{ s}$$

The transmissivity is calculated using Equation 2.64

$$T = \frac{0.008 \text{m}^3/\text{s}}{4\pi \times 2.0 \text{m}} 4.75 = 1.51 \times 10^{-3} \text{ m}^2/\text{s}$$

The storage coefficient is calculated using Equation 2.65:

$$S = \frac{4 \cdot 1.51 \times 10^{-3} \text{m}^2/\text{s} \cdot 2340\text{s} \cdot 0.0055}{(40.5\text{m})^2} = 4.7 \times 10^{-5}$$

The hydraulic conductivity (K) is calculated from the transmissivity (T) and the aquifer thickness (b):

$$K = \frac{T}{b} = \frac{1.51 \times 10^{-3} \text{m}^2/\text{s}}{18\text{m}} = 8.39 \times 10^{-5} \text{m/s}$$

2.5.3.1 General Note on Application of the Theis Method

If the focused CSM developed before the test suggests that the aquifer should "behave" like a homogeneous confined aquifer, but the test field data cannot be matched to the theoretical Theis curve because of an "odd" shape, the reason may be that the CSM was not appropriate because of one or more of the following:

- The aquifer is not confined.
- The aquifer is heterogeneous and anisotropic.
- There is a source of recharge and discharge such as leaky aquifer and aquitard conditions.
- The aquifer behaves as a dual-porosity medium (fractured and karst aquifers).
- There is one or more hydraulic boundaries.

In such cases, a hydrogeologic assessment of the possible causes should be made and the pumping test data should be analyzed with an appropriate method as described further in this chapter.

2.5.4 COOPER–JACOB STRAIGHT LINE SOLUTION OF THEIS EQUATION

Theis equation can be modified, as pointed out by Cooper and Jacob (1946), to simplify calculations and determination of the aquifer parameters using well pumping data. For small values of parameter u, i.e., sufficiently large values of pumping time t, or small values of r (i.e., when a monitoring well is close to the pumping well), the well function $W(u)$ can be approximated by the first two members of the series in Equation 2.62:

$$W(u) = -0.5772 - \ln u = \ln 0.5615 - \ln u \tag{2.66}$$

$$W(u) = \ln \frac{0.5615}{u} \tag{2.67}$$

For values of u equal to 0.05, the above equation approximates the well function with an error of 2% (errors for $u = 0.03$ and $u = 0.07$ are 1% and 3.2%, respectively). When u is equal to or less than 0.05, the well function expressed in the following form is considered acceptable for practical purposes:

$$W(u) = \ln \frac{2.25 T t}{r^2 S} \tag{2.68}$$

and the Theis Equation 2.59 becomes

$$s = \frac{Q}{4\pi T} \ln \frac{2.25Tt}{r^2 S}$$

(2.69)

or

$$s = \frac{0.183Q}{T} \log \frac{2.25Tt}{r^2 S}$$

(2.69a)

when the logarithm to the base e or "natural logarithm" (ln) is replaced by the logarithm to the base 10 (log). The logarithmic form of Theis equation is much easier to work with and enables usage of semilogarithmic graphs for well pumping test analyses.

If well discharge (Q) is kept constant, and it is assumed that transmissivity (T) and storage coefficient (S) do not change in time, Equation 2.69 can be rewritten to separate these constant terms from the logarithm of time (t) which is the only variable on the right-hand side:

$$s = \frac{Q}{4\pi T} \ln \frac{2.25T}{r^2 S} + \frac{Q}{4\pi T} \ln t$$

(2.70)

Simplification of Equation 2.70, i.e., replacement of all constants with new terms a and b, gives the following function:

$$s = a + b \ln t$$

(2.71)

Equation 2.71 shows that, if the condition that $u < 0.05$ is satisfied, recorded data of drawdown versus time would form a straight line when plotted on semilogarithmic graph paper (Figure 2.59). Time needed for the condition regarding parameter u to be satisfied can be determined using the following equation:

$$t = \frac{r^2 S}{4Tu}$$

(2.72)

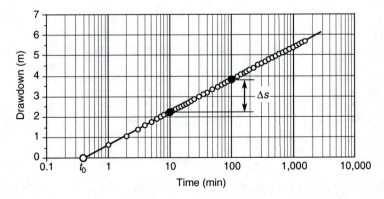

FIGURE 2.59 Semilog graph drawdown versus time used for the calculation of aquifer transmissivity and storage coefficient by the Cooper–Jacob method (straight-line method).

Any other deviations from the straight line, provided the pumping rate during the test was kept constant, indicate that the aquifer tested is not a perfect confined aquifer, in the Theis sense.

Drawdown and time coordinates of any two points on a straight line drawn through the recorded (field) data are used to determine aquifer transmissivity as shown with the following equations:

$$s_1 = \frac{Q}{4\pi T} \ln \frac{2.25T}{r^2 S} + \frac{Q}{4\pi T} \ln t_1 \tag{2.73}$$

$$s_2 = \frac{Q}{4\pi T} \ln \frac{2.25T}{r^2 S} + \frac{Q}{4\pi T} \ln t_2 \tag{2.74}$$

$$s_2 - s_1 = \Delta s = \frac{Q}{4\pi T} (\ln t_2 - \ln t_1) \tag{2.75}$$

Finally, the transmissivity is given as

$$T = \frac{Q}{4\pi \Delta s} \ln \frac{t_2}{t_1} \tag{2.76}$$

For practical purposes it is better to express transmissivity in terms of logarithm to the base 10 since the corresponding graph papers are easier to use:

$$T = \frac{0.183Q}{\Delta s} \log \frac{t_2}{t_1} \tag{2.76a}$$

If two points on the straight line are chosen to be one log cycle apart (i.e., ratio of time coordinates t_2 and t_1 is 10), Equation 2.76a reduces to

$$T = \frac{0.183Q}{\Delta s} \tag{2.77}$$

In addition to transmissivity, the Cooper–Jacob method is used to determine the storage coefficient of the aquifer. This is done graphically using the same graph shown in Figure 2.59. The straight line is extended to intercept the zero drawdown. The time of the intercept (t_0) is related to the zero drawdown with the following equation:

$$s = 0 = \frac{0.183Q}{T} \log \frac{2.25T t_0}{r^2 S} \tag{2.78}$$

For the above equation 2.78 to be valid, the term under the logarithm must be equal to 1:

$$\frac{2.25T t_0}{r^2 S} = 1 \tag{2.79}$$

Knowing that $\log 1 = 0$, the storage coefficient is then expressed as

$$S = \frac{2.25T t_0}{r^2} \tag{2.80}$$

Only data recorded at monitoring wells (piezometers) can be used to determine the storage coefficient because the actual drawdown measured in the pumping well includes a well loss. This additional drawdown is not known beforehand and neither is the theoretical position of the pumping well data (without the well loss). Therefore the zero-drawdown intercept of the straight line for the pumping well data is not the theoretical value of t_0 from Equation 2.78 through Equation 2.80 and would lead to an erroneous result.

2.5.5 Drawdown versus Ratio of Time and Squared Distance

This method is an obligatory part of every well pumping test analysis because, in addition to aquifer transmissivity and storage coefficient, it allows the determination of well loss and efficiency. It is also based on the application of the modified Theis Equation 2.69 introduced by Cooper and Jacob:

$$s = \frac{Q}{4\pi T} \ln \frac{2.25Tt}{r^2 S}$$

This equation can be rewritten in the following form:

$$s = \frac{Q}{4\pi T} \ln \frac{2.25T}{S} + \frac{Q}{4\pi T} \ln \frac{t}{r^2} \tag{2.81}$$

which, similarly to Equation 2.70, separates the constant terms (i.e., transmissivity and storage coefficient) from the t/r^2 ratio which now becomes a variable related to drawdown (s). By replacing all the constant terms with new parameters a and b, Equation 2.81 is further simplified:

$$s = a + b \ln \frac{t}{r^2} \tag{2.82}$$

The above equation shows that, if the condition that $u < 0.05$ is satisfied, the recorded data of drawdown (s) versus t/r^2 ratio would form a straight line if plotted on a semilogarithmic graph paper. Figure 2.60 shows a plot of this function for a monitoring well drawdown data. Coordinates of any two points on the straight line can be used to determine aquifer transmissivity:

$$s_2 - s_1 = \Delta s = \frac{Q}{4\pi T} \left[\ln\left(\frac{t}{r^2}\right)_2 - \ln\left(\frac{t}{r^2}\right)_1 \right] \tag{2.83}$$

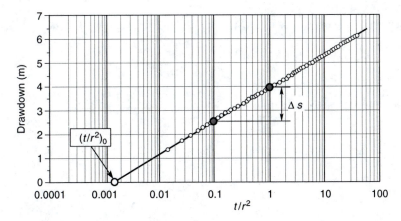

FIGURE 2.60 Semilog graph drawdown versus t/r^2 for a monitoring well showing two values needed for calculating the transmissivity and the storage coefficient: Δs and $(t/r^2)_0$.

Solving the above equation for transmissivity and introducing logarithm to the base 10 gives

$$T = \frac{0.183Q}{\Delta s}\left[\log\left(\frac{t}{r^2}\right)_2 - \log\left(\frac{t}{r^2}\right)_1\right] \tag{2.84}$$

If two points on the straight line are chosen to be one log cycle apart (i.e., the ratio of coordinates $(t/r^2)_2$ and $(t/r^2)_1$ is 10), Equation 2.84 reduces to a simple-to-use form:

$$T = \frac{0.183Q}{\Delta s} \tag{2.85}$$

The storage coefficient is, similarly to the Cooper–Jacob method, determined by substituting the previously calculated transmissivity and the zero-drawdown intercept of the straight line into the following equation:

$$S = 2.25T\left(\frac{t}{r^2}\right)_0 \tag{2.86}$$

Determination of well loss using this method and a detailed explanation of the concept of well loss and efficiency are given in Section 14.1.2 and Section 3.1.2, respectively.

2.5.6 DRAWDOWN VERSUS DISTANCE

The drawdown–distance analysis can be performed when there are two or more piezometers (monitoring wells) available. The method allows calculation of both the aquifer transmissivity and storativity. Assuming that the aquifer is homogeneous and that parameter $u < 0.05$ (i.e., the Cooper–Jacob equation is valid), the drawdown at distance r_1 and at time t after the start of pumping is given as

$$s_1 = \frac{Q}{4\pi T}\ln\frac{2.25Tt}{r_1^2 S} \tag{2.87}$$

or

$$s_1 = \frac{0.183Q}{T}\log\frac{2.25Tt}{r_1^2 S} \tag{2.88}$$

At the same time t, the drawdown at distance r_2 from the pumping well is

$$s_2 = \frac{0.183Q}{T}\log\frac{2.25Tt}{r_2^2 S} \tag{2.89}$$

The difference between two drawdowns is

$$\Delta s = s_1 - s_2 = \frac{0.183Q}{T}\log\frac{r_2^2}{r_1^2} \tag{2.90}$$

knowing that

$$\log\frac{r_2^2}{r_1^2} = \log\frac{1}{r_1^2} - \log\frac{1}{r_2^2}$$

and that

$$\log\frac{1}{r^2} = 2\log\frac{1}{r}$$

Equation 2.90 can also be written as

$$\Delta s = \frac{0.366Q}{T}\log\frac{r_2}{r_1} \tag{2.91}$$

Equation 2.91 shows that drawdown data recorded at two or more monitoring wells, which are at various distances from the pumping well, can be plotted on a semilog graph drawdown versus distance forming a straight line in case of homogeneous isotropic aquifers. Figure 2.61 shows such a graph with the drawdown data recorded at three monitoring wells at two different times. As can be seen, all the lines drawn through the monitoring well data are indeed straight and parallel to each other thus confirming the assumption that the aquifer is homogeneous and isotropic.

The drawdown registered in the pumping well would be above the straight lines shown in Figure 2.61 as result of a well loss.

From Equation 2.91 the aquifer transmissivity is

$$T = \frac{0.366Q}{\Delta s}\log\frac{r_2}{r_1} \tag{2.92}$$

The drawdown difference (Δs) can be found for any two points on the straight line drawn through the piezometers data for one time period. Considering this difference for one log cycle simplifies Equation 2.92 to

$$T = \frac{0.366Q}{\Delta s} \tag{2.93}$$

The drawdown–distance method is also used to calculate the storage coefficient of the aquifer: any of the straight lines is extended to intersect the zero-drawdown axis and find the corresponding value of r_0. By definition, this distance for which the drawdown equals zero is the radius of well influence. As can be seen in Figure 2.61, the radius of well influence (R) increases in time, and from the condition that the drawdown equals zero, i.e.:

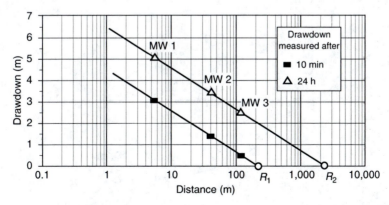

FIGURE 2.61 Semilog graph drawdown versus distance for determining aquifer transmissivity and the radius of well influence as it changes with time.

$$s = 0 = \frac{0.183Q}{T} \log \frac{2.25Tt}{r_0^2 S} \tag{2.94}$$

it follows that the term under the logarithm must be equal to 1 since $\log 1 = 0$:

$$\frac{2.25Tt}{r_0^2 S} = 1 \tag{2.95}$$

The storage coefficient is then

$$S = \frac{2.25Tt}{r_0^2} \tag{2.96}$$

2.5.7 DRAWDOWN RECOVERY METHOD

Recording and analysis of the drawdown recovery is an obligatory part of every aquifer pumping test. Pumping at a test well lowers the potentiometric surface around the well and creates the drawdown. After pumping stops, there is an inflow of water from the aquifer into the previously created cone of depression in the potentiometric surface. This inflow causes a rise in the depleted hydraulic head, i.e., the drawdown recovery (see Figure 2.62).

The inflow of water from the aquifer during the recovery can be simulated with imaginary well injecting water into the aquifer. This well has the same pumping rate as the real test well, only with the negative sign. Accordingly, the entire aquifer test can be simulated as a superposition of flows at two wells. This principle is illustrated in Figure 2.63: at the time when actual pumping stops, the imaginary well starts adding water to the aquifer and the real well continues extracting it at the same rate. Regarding the aquifer water balance, the effect of the two wells is the same as in case of the one real well. The resulting drawdown at any time after the actual pumping stops is the algebraic sum of the drawdown from the extraction well (that continues to operate) and the buildup (negative drawdown) from the introduced imaginary injection well (see Figure 2.63):

$$s' = s + s_{\text{rcv}} \tag{2.97}$$

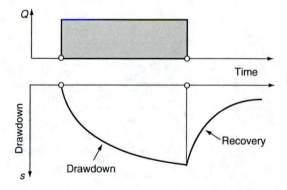

FIGURE 2.62 Drawdown and recovery versus time, and the corresponding pumping rate hydrograph for a well pumping test.

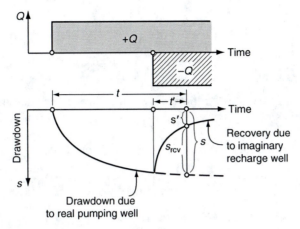

FIGURE 2.63 The principle of superposition of flows used to simulate drawdown recovery during a well pumping test.

where s' is the resulting drawdown measured at the well (residual drawdown), s is the drawdown from the extraction well, and s_{rcv} is the recovery (negative drawdown) from the recharge well. When these drawdown components are expressed with the Theis equation, it follows that

$$s' = \frac{+Q}{4\pi T} W(u) + \frac{-Q}{4\pi T} W(u_{rcv}) \tag{2.98}$$

The parameter u for the extraction well is

$$u = \frac{r^2 S}{4Tt} \tag{2.99}$$

where r is the radial distance from the pumping well where the drawdown is recorded, S is the storage coefficient, T is the transmissivity, and t is the time since the actual pumping started.

The parameter u_{rcv} for the recharge (imaginary) well is

$$u_{rcv} = \frac{r^2 S}{4Tt'} \tag{2.100}$$

where t' is the time since the actual pumping stopped. When parameter u is < 0.05, the Theis equation can be simplified as suggested by Jacob and Cooper (see Equation 2.69), and Equation 2.98 becomes

$$s' = \frac{Q}{4\pi T} \left(\ln \frac{2.25Tt}{r^2 S} - \ln \frac{2.25Tt'}{r^2 S} \right) \tag{2.101}$$

or

$$s' = \frac{Q}{4\pi T} \ln \frac{t}{t'} \tag{2.102}$$

Knowing that $\ln(x) = 2.3 \log(x)$, the above equation can be written as

$$s' = \frac{0.183Q}{T} \log \frac{t}{t'} \qquad (2.103)$$

2.5.7.1 Residual Drawdown

As seen from Equation 2.103, the t/t' ratio versus residual drawdown data recorded during the recovery would form a straight line on a semilog graph paper. This is the case, for the most part, for the data shown in Figure 2.64.

The residual drawdown coordinates of any two points on the straight line are

$$s'_1 = \frac{0.183Q}{T} \log\left(\frac{t}{t'}\right)_2$$

$$s'_2 = \frac{0.183Q}{T} \log\left(\frac{t}{t'}\right)_2$$

and their difference is

$$\Delta s' = \frac{0.183Q}{T} \log \frac{(t/t')_2}{(t/t')_1} \qquad (2.104)$$

If the two points are chosen over one log cycle, and Equation 2.104 is solved for the aquifer transmissivity, it follows that

$$T = \frac{0.183Q}{\Delta s'} \qquad (2.105)$$

Note that the residual drawdown method of the recovery analysis does not allow calculation of the storage coefficient, either in pumping wells or monitoring wells. This is obvious from the absence of the storage coefficient in the basic equation describing the method (Equation 2.102).

For an infinite, homogeneous, and isotropic aquifer, and the groundwater flow that satisfies the Theis assumptions, the straight line on the semilog graph s' versus t/t' intersects

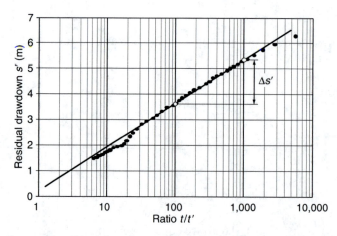

FIGURE 2.64 Semilog graph residual drawdown versus t/t' ratio for determining aquifer transmissivity.

the zero-drawdown axis at $t/t' = 1$. This theoretical value is obtained from the condition that the residual drawdown equals 0 in Equation 2.103:

$$s' = 0 = \frac{0.183Q}{T} \log \frac{t}{t'}$$

For this equation to be valid, the logarithm of t/t' must equal 0 and the ratio t/t' is then 1. Deviation of the experimental straight line from the theoretical position, i.e., when the t/t' intercept is different than 1, indicates that the aquifer is not "ideal" and may suggest a possible cause. For example, the t/t' intercept of 2 or more suggests that there is a source of recharge resulting in a full recovery in a shorter time than the corresponding time of drawdown (the imaginary well adds more water to the aquifer than what is extracted by the real well). The t/t' intercept noticeably smaller than 1 means that the recovery is slow and incomplete thus indicating an aquifer of limited extent or poor storage characteristics.

2.5.7.2 Calculated Recovery

The residual drawdown recorded in the well after the end of pumping is the result of the combined influence of one extraction and one recharge (imaginary) well. Although the individual influence of each well cannot be found explicitly, it is possible to estimate it using a graphoanalytical approach as shown in Figure 2.65. Once the influence of the imaginary well is separated, it can be considered a simple case of the time-drawdown method of pumping test analysis (the drawdown here corresponds to the negative drawdown, or the recovery from the imaginary well). The method requires that the drawdown curve from the actual well be extended through the monitored recovery period. The recovery is then found as

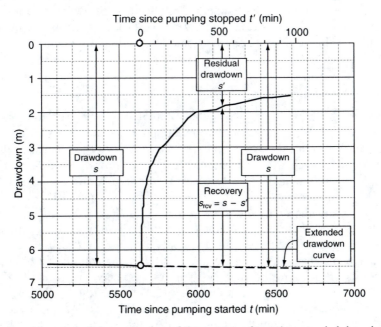

FIGURE 2.65 Graphoanalytical determination of the recovery from the extended drawdown curve.

the difference between the extended (estimated) drawdown and the residual drawdown recorded at the well:

$$s_{rcv} = s - s' \tag{2.106}$$

Although relatively simple, this procedure is often lengthy (the analyst must manually define "enough" points and read their drawdown values) and there is subjectivity in extending the drawdown line.

The aquifer transmissivity and the storage coefficient are found from a semilog graph of the calculated recovery (s_{rcv}) versus t' (time relevant for the operation of the imaginary well) in the same way as explained earlier for the Cooper–Jacob straight-line method:

$$T = \frac{0.183Q}{\Delta s_{rcv}} \tag{2.107}$$

$$S = \frac{2.25Tt'_0}{r^2} \tag{2.108}$$

One very important advantage of both recovery methods described is that, during the "operation" of the imaginary well, the well loss is practically minimal since the pump is turned off. Although there will still be some linear resistance to the flow in the near-well zone due to aquifer disturbance during the well completion, the turbulent well loss is absent. In addition, the water level recorded in the pumping well is much more accurate when the pump is not working. This all means that the recovery data from the pumping well itself can be used to determine the aquifer parameters with reasonable accuracy. (Note, however, that for results that are more reliable the recovery analyses should always be performed with the monitoring well data when available.)

When the recovery is correctly calculated, values of the transmissivity and the storage coefficient should be close to those obtained using the actual drawdown data from the first part of the test (Cooper–Jacob straight-line method). Detailed step-by-step calculation of aquifer parameters using drawdown recovery data after a three-step pumping test is given in Section 14.3.

2.5.8 VARYING PUMPING RATE

One of the advantages of the recovery analysis is that it can be used to calculate aquifer parameters with reasonable accuracy when the pumping rate was not kept constant during the test. The pumping rate can vary for various reasons, intentional (such as during a step test) or unintentional, which is more common. A frequent problem during pumping tests in newly constructed wells is an improper (insufficient) well development, which results in erratic surges and drops of the pumping rate. Another example is a constant decrease of the pumping rate due to lowering of the hydraulic head in the well and the consequent decrease of the pump capacity. In both cases, an attempt to sporadically adjust the pumping rate may sometimes further aggravate the problem. Figure 2.66 shows the hydrograph of a pumping test initially designed to have the pumping rate of 8 L/s and duration of 8 h. Due to various problems, the pumping rate (recorded every 30 min) varied between 7.8 and 6.6 L/s. Practice shows that when the pumping rate varies less than 10% to 15% during the test, the recovery data can be used for the transmissivity determination after necessary adjustments. In our case this variation is 15%, just "enough" to consider the result acceptable.

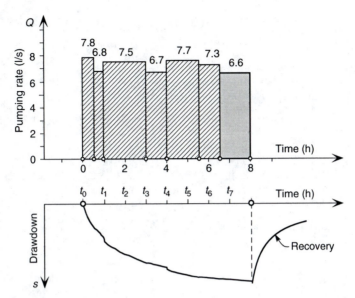

FIGURE 2.66 Hydrograph of a varying pumping rate and the corresponding drawdown–recovery curve at an aquifer test well.

The actual varying pumping rate is replaced with an equivalent constant pumping rate, which will be used in calculations. The new pumping rate is not the average of all the recorded rates. Rather, it is the last pumping rate of a significant duration—6.6 L/s in our case (see Figure 2.66). This last recorded pumping rate has the most influence on the future development of the cone of depression (drawdown) around the pumping well.

The introduction of the new pumping rate requires an adjustment to the hydrograph time base so that the aquifer water balance does not change. In other words, the volume of water actually withdrawn must equal the volume of water that would have been withdrawn with the newly adjusted pumping rate. The duration of pumping with the adjusted pumping rate, i.e., the new time base of the hydrograph (t^*), is found using the following general formula:

$$t^* = \frac{Q_1 t_1 + Q_2(t_2 - t_1) + \cdots + Q_n(t_n - t_{n-1})}{Q_n} \tag{2.109}$$

where t_1, t_2, t_{n-1}, t_n are the times of the pumping rate change, Q_1, Q_2, Q_{n-1}, Q_n are the recorded pumping rates, and Q_n is the last recorded pumping rate.

In this case the time is (see Figure 2.66 and Figure 2.67) calculated as follows:

$$t^* = \frac{7.8\,l/s \cdot 0.5\,h + 6.8\,l/s \cdot 0.5\,h + 7.5\,l/s \cdot 2\,h + 6.7\,l/s \cdot 1\,h + 7.7\,l/s \cdot 1.5\,h + 7.3\,l/s \cdot 1\,h + 6.6\,l/s \cdot 1.5\,h}{6.6\,l/s}$$

$$t^* = 8.75\,h$$

As can be seen, the new time base is longer than the time of actual pumping. The adjusted time at the beginning of pumping is set as 8.75 h backward from the time at which the actual pumping stopped. Note that the point in time at which the actual pumping stopped remains the same, i.e., it is "fixed." The new time-zero (t_0), which is to the left of the beginning of the actual pumping, is now the reference point for the recovery analysis (see Figure 2.67).

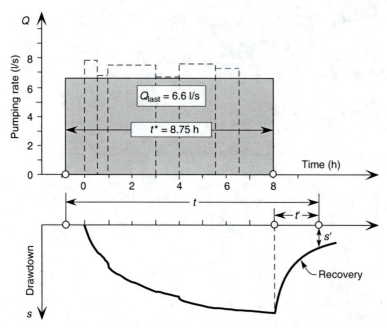

FIGURE 2.67 New adjusted hydrograph with the constant pumping rate of 6.6 L/s and the new time base of 8.75 h. Note that the new beginning time of the test is shifted to the left in this case.

When the pumping rate varies during the test, the drawdown data cannot be used for the transmissivity calculation using the graphoanalytical approaches already described. However, the residual drawdown recorded at the well during recovery can be used since it is assumed that the aquifer supplies water to the well at a constant rate (i.e., the imaginary recharge well injects water to the aquifer at a constant rate). Again, this constant rate equals the last recorded actual pumping rate of a significant duration (6.6 L/s in our case).

The transmissivity is found using a semilog graph residual drawdown (s') versus t/t' ratio and choosing two points on the straight line over one log cycle as explained in Section 2.5.7. Note that time t is now changed—it is the time since the new adjusted beginning of the test as shown in Figure 2.67. This means that the actual time data have to be corrected accordingly. Time since the pumping stopped (t') remains the same and does not require correction. The transmissivity is then calculated using Equation 2.105 where Q_{last} is the last recorded pumping rate of significant duration during the test:

$$T = \frac{0.183 Q_{last}}{\Delta s'}$$

2.5.9 TRANSIENT FLOW TOWARD A WELL IN A LEAKY CONFINED AQUIFER

The Theis nonequilibrium equation describes flow toward a fully penetrating well in a confined aquifer and is based on numerous assumptions, one of which is that the aquifer receives no recharge for the duration of pumping. However, this condition is seldom entirely satisfied as most confined aquifers receive recharge, either continuously or intermittently. One common source of recharge is leakage through an overlying or underlying aquitard, which separates the pumped aquifer from another aquifer. The rate of leakage can significantly increase during pumping due to an increased hydraulic gradient between the pumped aquifer and the adjacent aquitard or aquifer. Eventually, if the rate of leakage balances the pumping rate, the drawdown will stabilize at a certain level and the radius of well influence will cease to expand. Figure 2.68 is a log–log plot of the drawdown versus time, showing this effect of leakage. Approximately 100 min after pumping started, the leakage balances the withdrawal of water from the confined aquifer and the drawdown remains practically unchanged for the duration of the test. (Note that the influence of an equipotential boundary would look very similar and the analyst should verify its absence before assuming leaky conditions.)

The equation describing flow toward a well in a leaky confined aquifer, which is separated from an unconfined aquifer by an aquitard, was derived in 1955 by Hantush and Jacob in the following dimensionless form (Lohman, 1972):

$$\frac{s}{Q/4\pi T} = 2K_0(2v) - \int_{v^2/u}^{\infty} \frac{1}{y} \exp\left(\frac{-y - v^2}{y}\right) dy \tag{2.110}$$

where s is the drawdown in the confined aquifer, Q is the pumping rate of the well, T is the transmissivity of the confined aquifer, K_0 is the modified Bessel function of the second time of the zero order, y is the variable of integration, and u is the same parameter (dimensionless) as in the Theis equation:

$$u = \frac{r^2 S}{4Tt} \tag{2.111}$$

where r is the distance from the pumping well to the monitoring well, S is the storage coefficient of the confined aquifer, and t is the time since pumping started,

and

$$v = \frac{r}{2}\sqrt{\frac{K'}{b'T}} \tag{2.112}$$

where K' is the vertical hydraulic conductivity of the confining bed, b' is the thickness of the confining bed, and T is the transmissivity of the aquifer.

The authors gave two sets of equations for the solution of Equation 2.110, both for large and small values of time. The method was later made more friendly by various authors who

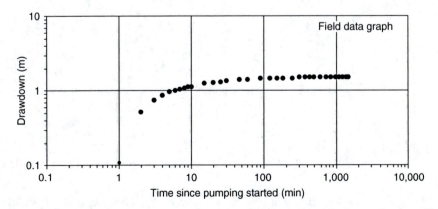

FIGURE 2.68 Time versus drawdown recorded at a monitoring well in a leaky confined aquifer.

computed corresponding well function and derived graphoanalytical methods for solving Equation 2.110 with the aid of type curves (see Hantush, 1956; Walton, 1960, 1962; Cooper, 1963). Presented here is the Walton graphoanalytical method, based on type curves, for solving the Hantush–Jacob equation in the following form:

$$s = \frac{Q}{4\pi T} W(u, r/B) \tag{2.113}$$

where $W(u, r/B)$ is the well function for the leaky confined aquifer and B is the leakage factor given as:

$$B = \sqrt{\frac{Tb'}{K'}} \tag{2.114}$$

The values of the well function are given in Appendix D, and type curves for various values of $1/u$ and r/B are shown in Figure 2.69.

The procedure for finding aquifer hydrogeologic parameters is as follows:

- Plot log–log graph time versus recorded drawdown at the same log-cycle scale as the graph with type curves (see Figure 2.68).
- Superimpose the field graph on the graph with type curves and, keeping the axes of both graphs parallel, match the field data with a type curve r/B (Figure 2.70).
- Choose a match point anywhere where the two graphs overlap and, noting the r/B value, read the four coordinates as follows: drawdown (s) and time (t) from the field graph, and $W(u, r/B)$ and $1/u$ from the type curve graph. It is recommended to choose rounded values for $1/u$ and $W(u, r/B)$ for easier calculations.
- Calculate transmissivity by substituting known values into equation

$$T = \frac{Q}{4\pi s} W(u, r/B) \tag{2.115}$$

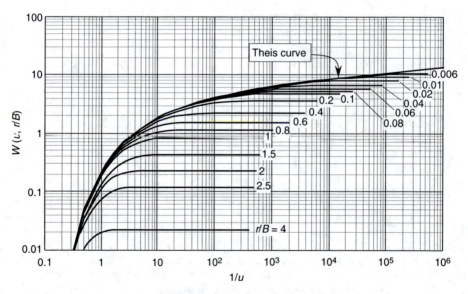

FIGURE 2.69 Type curves for leaky confined aquifer without the release of water from storage in confining layer. The curves are plotted using values computed by Hantush (1956).

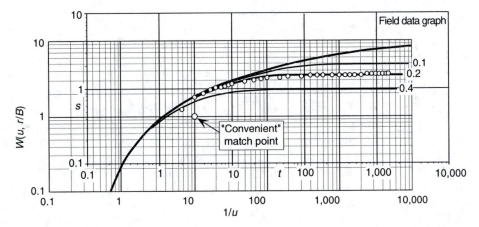

FIGURE 2.70 Matching field data to a type curve. The best fit is for curve $r/B = 0.2$. The match point is chosen to have convenient coordinates $W(u, r/B) = 1$, and $1u = 10$.

- Find the coefficient of storage using equation

$$S = \frac{4Tut}{r^2} \tag{2.116}$$

- Find the hydraulic conductivity of the confining layer (aquitard) from

$$K' = \frac{Tb'(r/B)^2}{r^2} \tag{2.117}$$

It is important to remember that Equation 2.110 has limited application because of the following assumptions introduced as part of its solution (in addition to the Theis equation assumptions):

- The water table in the unconfined aquifer does not change during pumping.
- There is no water released from the aquitard.
- The piezometric surface of the confined aquifer does not fall below top of the overlying aquitard.

The first assumption is valid if the following condition is satisfied (after Neuman and Witherspoon, 1969):

$$T_{\text{UA}} = 100 \times T_{\text{CA}} \tag{2.118}$$

where T_{UA} is the transmissivity of the unconfined aquifer and T_{CA} is the transmissivity of the confined (pumped) aquifer.

The rate of flow attributed to the release of water from storage in the aquitard can be ignored if the following condition is satisfied (Hantush, 1960):

$$t > \frac{0.036b'S'}{K'} \tag{2.119}$$

where S' is the storage coefficient of the aquitard.

Detailed step-by-step calculation of aquifer parameters using the leaky aquifer solution is given in Section 14.4.

2.5.10 Transient Flow toward a Well in an Unconfined Aquifer

The Theis equation, with all its assumptions, is derived for confined aquifers. In general, if the drawdown in monitoring wells does not exceed 25% of the saturated thickness, the Theis equation can be applied to unconfined aquifers with certain adjustments. For drawdowns that are less than 10% of the aquifer's prepumping thickness, it is not necessary to adjust the recorded data since the error introduced by using the Theis equation is small. When the drawdown is between 10% and 25%, it is recommended to correct the measured values using the following equation derived by Jacob (1963a):

$$s' = s - \frac{s^2}{2h} \tag{2.120}$$

where s' is the corrected drawdown, s is the measured drawdown in a monitoring well, and h is the saturated thickness of the unconfined aquifer before pumping started.

This correction is needed since the transmissivity of the aquifer changes during the test as the saturated thickness decreases (remember that, for unconfined aquifers, $T = Kb$ where b is the saturated thickness). If the drawdown in a monitoring well is more than 25%, the Theis equation should not be used in the aquifer analysis.

The response of unconfined aquifers to well pumping tests is usually considerably different than the response of confined aquifers. One of the Theis assumptions in deriving the well equation is that water is withdrawn from storage instantaneously with a decrease in hydraulic head. Although this assumption, for practical purposes, may be considered "true enough" for most confined aquifers, it is not acceptable for most unconfined aquifers because of a lag in releasing water from storage. This lag is caused by slow gravity drainage of the porous media within the cone of depression and above the hydraulic head surface, especially at the beginning of pumping. As a result, the storage coefficient determined from the early drawdown data and using the (unmodified) Theis equation, will be underestimated. A more realistic value of the storage coefficient is obtained from the late drawdown data when the cone of depression spreads at a lower rate and the gravity drainage comes to an equilibrium with other pumping influences. Figure 2.71 shows a typical delayed response of unconfined aquifers to pumping tests. This effect is particularly pronounced in stratified unconfined aquifers with alternating fine and coarse sediments. When the porous medium has a high effective porosity (specific yield), the aquifer response will be less delayed and may look similar to confined aquifers.

A log–log graph of the recorded drawdown versus time in Figure 2.71 indicates three distinct segments as the result of a delayed aquifer response:

1. The early segment, which represents the first 3 min of pumping, shows a rapid drawdown similar to confined conditions. Almost all water supplied to the well comes from the aquifer storage in the saturated zone. Gravity water above the hydraulic head within the cone of depression still has not reached the saturated zone. The coefficient of storage during this stage is close to a confined aquifer of the same porous material as that of the unconfined aquifer.
2. The intermediate segment, between 5 and approximately 100 min, is a subhorizontal (flat) curve, which indicates that the gravity water is reaching the saturated zone, but is still not in equilibrium with the saturated flow.

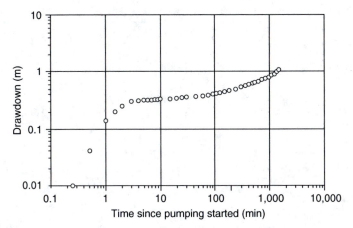

FIGURE 2.71 Log–log graph time versus drawdown recorded at a monitoring well in an unconfined aquifer with delayed gravity response.

3. The third (late) segment represents equilibrium between the gravity drainage and the saturated flow when the delayed response ceases. During this third period of the pumping test, the storage properties are those of a truly unconfined aquifer and are termed specific yield. This is the segment that should be used for determination of true aquifer parameters.

Transient flow toward a well in an unconfined aquifer with a delayed response to pumping was first analyzed by Boulton (1954a,b). Based on his theory, commonly called the Boulton method, an approach of analyzing water wells with delayed gravity response using type curves can be found in most textbooks on groundwater. Slight differences between circulating versions originate from modifications, simplifications, and graphical solutions that have been introduced to the method by various authors, including Boulton (e.g., Stallman, 1961a,b, 1963, 1965; Boulton, 1963, 1970; Prickett, 1965; Boulton and Pontin, 1971; Neuman, 1972, 1975). Presented here is the Neuman method, which allows determination of the hydrogeologic parameters of anisotropic unconfined aquifers with delayed response to pumping: the transmissivity, the horizontal and vertical hydraulic conductivities, the elastic storage coefficient, and the specific yield.

2.5.10.1 Neuman Method

During pumping from a fully penetrating well, the drawdown at a fully penetrating observation well at distance r and at time t is (Neuman, 1975):

$$s_{r,t} = \frac{Q}{4\pi T} \int_0^\infty 4y J_0(y\beta^{1/2}) \left[u_0(y) + \sum_{n=1}^\infty u_n(y) \right] dy \qquad (2.121)$$

where

$$u_0(y) = \frac{\{1 - \exp[-t_s\beta(y^2 - \gamma_0^2)]\} \tan h(\gamma_0)}{\{y^2 + (1 + \sigma)\gamma_0^2 - [(y^2 - \gamma_0^2)^2/\sigma]\}\gamma_0} \qquad (2.122)$$

$$u_n(y) = \frac{\{1 - \exp[-t_s\beta(y^2 + \gamma_n^2)]\}\tan h(\gamma_n)}{\{y^2 - (1 + \sigma)\gamma_n^2 - [(y^2 + \gamma_n^2)^2/\sigma]\}\gamma_n} \qquad (2.123)$$

and the terms γ_0 and γ_n are the roots of the equations

$$\sigma\gamma_0 \sin h(\gamma_0) - (y^2 - \gamma_0^2)\cos h(\gamma_0) = 0 \qquad (2.124)$$

for $\gamma_0^2 < y^2$

$$\sigma\gamma_n \sin(\gamma_n) + (y^2 + \gamma_n^2)\cos(\gamma_n) = 0 \qquad (2.125)$$

for $(2n - 1)(\pi/2) < \gamma_n < n\pi$ and $n \geq 1$.

The expression used in Equation 2.121 through Equation 2.125 is as follows: s, drawdown at observation well; r, radial distance from pumping well; t, time since pumping started; Q, pumping rate; T, transmissivity; $J_0(x)$, zero-order Bessell function of the first kind; y, a variable given as $y = x(K_D)^{1/2}$; K_D, degree of anisotropy given as $K_D = K_z/K_r$; K_z, vertical hydraulic conductivity; K_r, horizontal hydraulic conductivity; β, dimensionless parameter given as $\beta = K_D r^2/b^2$; b, initial saturated thickness of the aquifer; σ, dimensionless parameter given as $\sigma = S/S_y$; S, storage coefficient given as $S = S_s b$; S_s, specific (elastic) storage; and S_y, specific yield.

Equation 2.121 is expressed in terms of independent dimensionless parameters σ and β, and dimensionless time given either as time with respect to S_s (early data when the elastic storage is dominant):

$$t_s = \frac{Tt}{Sr^2} \qquad (2.126)$$

or as time with respect to S_y (specific yield is dominant):

$$t_y = \frac{Tt}{S_y r^2} \qquad (2.127)$$

For field drawdown data, Equation 2.121 is solved with the aid of type curves shown in Figure 2.72. The corresponding numerical values computed by Neuman are given in Appendix D. The graph consists of two merged families of curves: type A curves lying to the left of the central (flat) graph area, and type B curves lying to the right. The curves are plotted for various values of dimensionless parameter β, dimensionless times t_s and t_y, and dimensionless drawdown s_D, which is given as

$$s_D = \frac{4\pi T}{Q}s \qquad (2.128)$$

Note that type A curves are used with early drawdown data and correspond to the top scale expressed in terms of t_s (elastic storage is dominant). Type B curves are intended for use with late drawdown data when specific yield is dominant (true unconfined conditions without delayed response). They correspond to the bottom scale expressed in terms of s_y.

The procedure for calculating hydrogeologic parameters is as follows:

- Plot log–log graph time versus recorded drawdown at the same log-cycle scale as the graph with type curves (see Figure 2.71).
- Superimpose the field curve (data) on the type B curves (curves on the right-hand side) and, keeping the axes of both graphs parallel, match as much of the late drawdown data as possible to a particular β curve.

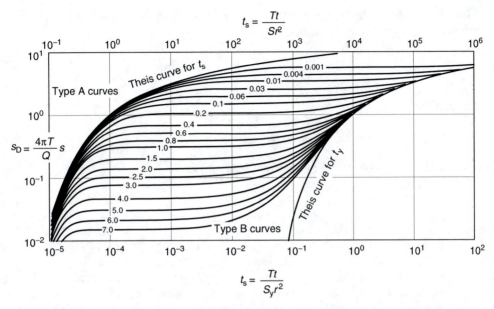

FIGURE 2.72 Type curves for fully penetrating wells in unconfined aquifers with delayed gravity response. (From Neuman, S., *Water Resources Research*, 11(2), 329, 1975. Copyright by the American Geophysical Union. With permission.)

- Choose a match point anywhere where the two graphs overlap and read the four coordinates as follows: s_D and t_y from the type graph, and s and t from the field graph. It is recommended to choose rounded values for s_D and t_y for easier calculations.
- Calculate the transmissivity using the following equation:

$$T = \frac{1}{4\pi}\frac{Qs_D}{s} \qquad (2.129)$$

- Calculate the specific yield using equation:

$$S_y = \frac{Tt}{r^2 t_y} \qquad (2.130)$$

- Overlap the early field data (curve) and the type A curve, which has the same value of β as the type B curve. While doing this try to match as much data as possible to the type A curve keeping the axes on both graphs parallel. (Note that since data have to be matched to the same β curve, the graphs should move only in the horizontal direction.)
- Choose a match point anywhere where the two graphs overlap and read the four coordinates as follows: s_D and t_s from the type graph, and s and t from the field graph. Again, it is recommended to choose rounded values for s_D and t_s for easier calculations.
- Calculate the transmissivity again using Equation 2.129. The value should be close to the one calculated for the late data.
- Calculate the elastic storage coefficient using equation:

$$S = \frac{Tt}{r^2 t_s} \qquad (2.131)$$

- Find the horizontal hydraulic conductivity knowing the average aquifer thickness and the previously calculated transmissivity:

$$K_r = \frac{T}{b} \qquad (2.132)$$

- Find the degree of anisotropy (K_D), which is needed to calculate the vertical hydraulic conductivity, using equation:

$$K_D = \frac{\beta b^2}{r^2} \qquad (2.133)$$

- And finally, find the vertical hydraulic conductivity from

$$K_z = K_D K_r \qquad (2.134)$$

If the saturated thickness of the aquifer decreases more than 10% during the test, Neuman (1975) recommends that the recorded drawdown should be corrected using Jacob's formula (Equation 2.120) for the late data only. Correcting early drawdown would lead to erroneous results as the response of the aquifer to pumping is primarily due to the elastic properties of the porous media and water (i.e., the elastic storage is predominant).

Calculation of aquifer parameters using the Neuman solution is given in Chapter 14, Problem 14.5.

2.5.11   EFFECT OF BOUNDARIES ON FLOW TOWARD A WELL

Although, for various reasons (the most important being simplicity), it is not uncommon to analyze aquifers as if they were of infinite extent, in reality all aquifers have more or less clearly defined boundaries. When a well is completely surrounded by an equipotential boundary (e.g., when it is located on an island or in a meander—see Section 13.1), application of the steady-state formulas described in Section 2.5.1 is justified. In all other cases one either has to assume "quasi-steady-state" conditions without boundaries (if their influence is not apparent), or account for the boundaries by introducing appropriate hydraulic conditions.

2.5.11.1   Equipotential Boundary

The method of images is a convenient and effective way to account for physical aquifer boundaries by replacing them with hydraulic entities that allow application of the equations describing flow toward wells. These new hydraulic entities are also wells, either discharging or recharging depending on the nature of the boundaries. Figure 2.73 illustrates the method of images in case of a discharging (extraction) well near a large perennial river. If the river stage is not affected by the well, the river is considered a strong equipotential boundary equivalent to a line source having a constant head. This is the case with both fully and partially (as in our case) penetrating streams.

The equipotential boundary is simulated with an imaginary recharging well which is the exact image of the real extraction well, only with the opposite (negative) sign for the flow rate. This image well, which operates simultaneously with the real well, is placed on the other side of the boundary at the same distance from the boundary as the real well. The line connecting the two wells is perpendicular to the boundary (see Figure 2.74). The recharge from the image well simulates the flow from the stream toward the real well since it builds up the water table. This additional head along the boundary cancels the drawdown caused by the extraction well so that there is no change in the hydraulic head along the boundary. Elsewhere in the aquifer

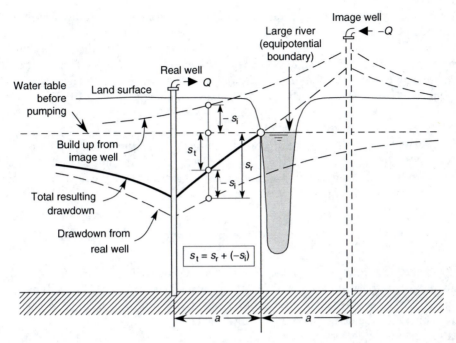

FIGURE 2.73 Schematic cross-sectional presentation of the resulting drawdown when an equipotential boundary near the extraction well is simulated with the imaginary recharge well of the same flow rate. Note that the drawdown at the boundary is (correctly) zero as the result of this substitution of the real physical boundary (river) with an imaginary hydraulic boundary (recharge well).

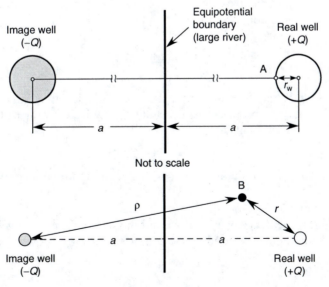

FIGURE 2.74 Scheme for calculating the drawdown at the extraction well, at point A on its perimeter, and at any point (B) in the aquifer influenced by the equipotential boundary.

the resulting drawdown is the algebraic sum of the drawdown from the real well and the hydraulic head buildup (negative drawdown) from the image well.

It is important to remember an important rule of the image theory: the image well on the other side of the equipotential boundary changes the character—it has the same flow rate as the real well but with the opposite sign. If the real well is extracting water, its image is adding water to the aquifer. If the real well is recharging the aquifer, the image well is extracting water.

2.5.11.1.1 Drawdown at the Real Well

The drawdown at the real well is the algebraic sum of the drawdown caused by groundwater withdrawal, and the buildup (negative drawdown) caused by the image well. This can be expressed with the following equation when the resulting drawdown is calculated for point A at the well perimeter (see Figure 2.74):

$$s_A = \frac{+Q}{2\pi T} \ln \frac{R}{r_w} + \frac{-Q}{2\pi T} \ln \frac{R}{2a - r_w} \tag{2.135}$$

where Q is the flow rate at the wells, R is the radius of well influence, T is the aquifer transmissivity, r_w is the well radius, and $2a - r_w$ is the distance between point A and the image well. (Note that the actual recorded drawdown in the real well will be larger because of the well loss which must be determined by testing—see Section 15.2.)

Equation 2.135 can be written as

$$s_A = \frac{Q}{2\pi T} \ln \frac{2a - r_w}{r_w} \tag{2.136}$$

Since the well radius (r_w) is much smaller than twice the distance between the wells ($2a$), it can be ignored and Equation 2.136 becomes

$$s_A = \frac{Q}{2\pi T} \ln \frac{2a}{r_w} \tag{2.137}$$

When the above equation is compared with the general equation of groundwater flow toward a fully penetrating well in a confined aquifer of infinite extent:

$$s = \frac{Q}{2\pi T} \ln \frac{R_D}{r_w}$$

where R_D is the "Dupuit's radius of well influence" (i.e., the calculated, but not the real radius of influence), it becomes apparent that the hydraulically equivalent radius of influence of the well near the equipotential boundary is

$$R_D = 2a \tag{2.138}$$

where a is the distance between the real well and the equipotential boundary.

2.5.11.1.2 Drawdown in the Aquifer

As already mentioned, the drawdown at any point in the aquifer is the algebraic sum of the drawdown from the pumping well and the buildup (negative drawdown) from the injection well (see Figure 2.74):

$$s_P = \frac{+Q}{2\pi T} \ln \frac{R}{r} + \frac{-Q}{2\pi T} \ln \frac{R}{\rho} \tag{2.139}$$

where r is the distance between the point and the real well and ρ is the distance between the point and the image well.

Equation 2.139 can also be written as

$$S_P = \frac{Q}{2\pi T} \ln \frac{\rho}{r} \tag{2.140}$$

As can be seen, the drawdown at any point in the aquifer is dependent on the distances between that point and the two wells (one real and one image). It also means that, to find the drawdown (i.e., the hydraulic head), one does not need to know the real radius of well influence.

2.5.11.2 Impermeable Boundary

An impermeable boundary near a discharging well is simulated by another discharging well placed on the other side of the boundary. This new well is an exact image of the real well and has the same pumping rate. It is placed at the same distance from the boundary as the real well (see Figure 2.75). The two wells are on the same line perpendicular to the boundary and operate simultaneously. The resulting drawdown in the aquifer is the algebraic sum of their individual drawdowns. The formation of a groundwater divides along the boundary, due to the operation of two extracting wells, satisfies the hydraulic condition that there is no flow across the impermeable boundary. Note that the actual drawdown along the boundary line is two times larger than it would be if the real well was operating in an infinite aquifer without the boundary.

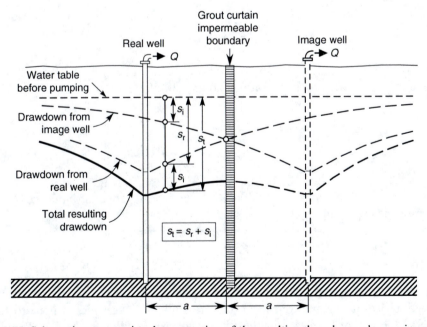

FIGURE 2.75 Schematic cross-sectional presentation of the resulting drawdown when an impermeable boundary near an extraction well is simulated with an imaginary extraction well of the same flow rate. Note that the pumping water table between the real well and the boundary is much flatter than on the other side of the real well.

2.5.11.2.1 Drawdown at the Real Well

The drawdown at the real well is calculated for a point placed at the well perimeter as in the case of the equipotential boundary. It is the algebraic sum of the drawdown caused by groundwater withdrawal at two extraction wells—the real and the image wells. This is expressed by the following equation:

$$s_A = \frac{+Q}{2\pi T} \ln \frac{R}{r_w} + \frac{+Q}{2\pi T} \ln \frac{R}{2a - r_w} \tag{2.141}$$

where Q is the flow rate at the wells, R is the radius of well influence, T is the aquifer transmissivity, r_w is the well radius, and $2a - r_w$ is the distance between point A and the image well. (Again, note that the actual recorded drawdown in the real well will be larger because of the well loss, which must be determined by testing.)

Equation 2.141 can be written as

$$s_A = \frac{Q}{2\pi T} \ln \frac{R^2}{2ar_w - r_w^2} \tag{2.142}$$

Since the squared well radius (r_w^2) is much smaller than the product $2ar_w$, it can be ignored and Equation (2.142) becomes

$$s_A = \frac{Q}{2\pi T} \ln \frac{R^2}{2ar_w} \tag{2.143}$$

or

$$s_A = \frac{Q}{2\pi T} \ln \frac{(R^2/2a)}{r_w} \tag{2.144}$$

When the above equation 2.144 is compared with the general equation of groundwater flow toward a fully penetrating well in a confined aquifer of infinite extent:

$$s = \frac{Q}{2\pi T} \ln \frac{R_D}{r_w}$$

where R_D is "Dupuit's radius of well influence" (i.e., the calculated but not the real radius of influence), it becomes apparent that hydraulically equivalent radius of influence of the well near the impermeable boundary is

$$R_D = \frac{R^2}{2a} \tag{2.145}$$

where a is the distance between the real well and the impermeable boundary and R is the real radius of influence of a single well in an infinite aquifer. It also means that, to find the drawdown at the well one must know the real radius of well influence (unlike in case of the equipotential boundary where $R_D = 2a$). Since the real radius of well influence in an infinite aquifer depends on time, it follows that flow toward a well near an impermeable boundary has to be analyzed as transient (time-dependent).

2.5.11.2.2 Drawdown in the Aquifer

As mentioned earlier, the drawdown at any point in the aquifer is the algebraic sum of the individual drawdowns from two pumping wells—the real and the image wells:

$$s_P = \frac{+Q}{2\pi T} \ln \frac{R}{r} + \frac{+Q}{2\pi T} \ln \frac{R}{\rho} \qquad (2.146)$$

where r is the distance between the point and the real well and ρ is the distance between the point and the image well.

Equation 2.146 can also be written as

$$S_P = \frac{Q}{2\pi T} \ln \frac{R^2}{r\rho} \qquad (2.147)$$

As can be seen, the drawdown at any point in the aquifer depends mainly on the distances between that point and the two wells (one real and one image), as well as on the real, time-dependent radius of well influence.

Application of the image theory for finding drawdown at a location in an aquifer influenced by both equipotential and impermeable boundaries using steady-state solution is given in Section 13.5.

2.5.11.3 Multiple Boundaries

The principle of hydraulic simulation of two or more boundaries that intersect, or are parallel, remains the same as in the case of a single boundary: the boundaries are replaced with image wells whose character depends mainly on the boundary type. An image well over an impermeable boundary has the flow rate with the same sign as the real well (e.g., the image of an extraction well is also an extraction well, and the image of a recharge well is also a recharge well). When the imaging is done for an equipotential boundary, the well flow rate changes the sign. For both types of boundaries the image well and the real well are on the same line perpendicular to the boundary and at the same distance from the boundary.

Figure 2.76 illustrates, with map views, some of the possible relationships between aquifer boundaries. It is important to remember that, when boundaries intersect, they have to be extended beyond their actual physical limits. The image wells from the real well are then imaged over these extended boundaries as well as over the real boundaries creating "images of the images." This process should continue until the loop is closed, i.e., until there is no room for another image (see Figure 2.76e and Figure 2.76f).

In the case of parallel boundaries, every new image well is imaged back over the next "available" boundary (i.e., the boundary that was not used for its imaging—see Figure 2.76b and Figure 2.76d), thus creating an infinite number of image wells. In practice, it will be enough to create several images since the solution quickly reaches an asymptotic value. If a more accurate analysis is needed, it is recommended to draw a graph drawdown versus number of wells included in the calculation. This will provide a convenient way to choose the right number of wells for a given level of accuracy. The graph is particularly useful in the case of one equipotential and one impermeable parallel boundary since the solution oscillates along a straight line.

For multiple boundaries, the drawdown at any point in the aquifer is calculated by adding drawdown components from each real and image well that are considered simply as a group of wells with superimposed flows. Again, it is very important to remember that the individual drawdowns can be positive (from extraction wells) or negative (from recharge wells).

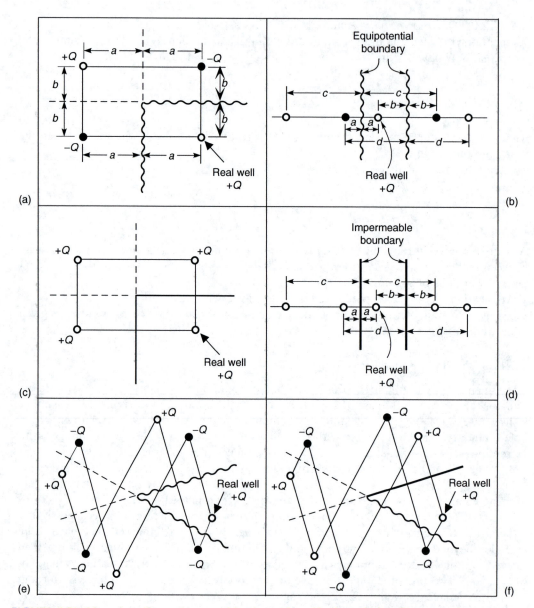

FIGURE 2.76 Map views for various combinations of aquifer boundaries. The extraction wells are shown with empty circles and $(+Q)$, and the recharge wells are shown with filled circles and $(-Q)$. (a) Two equipotential boundaries intersecting at a right angle. (b) Two parallel equipotential boundaries (note that the number of image wells is theoretically infinite). (c) Two impermeable boundaries intersecting at the right angle. (d) Two parallel impermeable boundaries. (e) Two equipotential boundaries intersecting at a sharp angle. (f) One equipotential and one impermeable boundary intersecting at a sharp angle.

2.5.12 Special Considerations for Fractured Rock and Karst Aquifers

The nature of the unique porosity of fractured rock and karst aquifers is best illustrated with a typical time–drawdown curve in response to groundwater withdrawal from a well (Figure 2.77). Given enough pumping time and presence of all porosity types, the time–drawdown

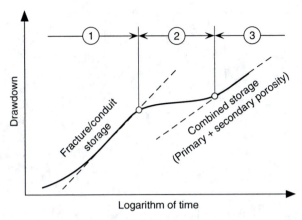

FIGURE 2.77 Theoretical response of the time–drawdown curve caused by effects of dual porosity in a fractured rock or karst aquifer.

curve would show three distinct segments. The first portion of the curve, with a uniform slope, indicates a quick response from a well-connected network of secondary porosity, which includes fractures of varying size and may include large dissolutional openings (cavities) in case of fully developed karst aquifers. Drainage of this type of porosity, in the early stages of the test, is characterized by storage properties generally similar to that of confined aquifers. Unconfined intergranular aquifers often exhibit similar early response to pumping, due to sudden change in hydraulic pressure, with the storage coefficient significantly less than 1% (1×10^{-3} or less).

The flattening of the curve (curve portion 2) indicates that the initial source of water in large solutional openings (channels) or fractures is being supplemented by water coming from another set of porosity. This additional inflow of water starts when the fluid pressure in the large fractures (or karst conduits) decreases enough, resulting in the hydraulic gradients from the smaller fractures and fissures toward the larger fractures or karst conduits. Again, a similar response often happens in unconfined intergranular aquifers when additional water, due to gravity drainage, starts reaching the lowered water table (the so-called delayed gravity response to pumping).

As the source of water from different sets of secondary porosity features, and water from the primary (matrix) porosity attain similar level of influence, the drawdown curve exhibits another relatively uniform slope (curve portion 3). As explained in Section 1.2, such rock formation is often referred to as a "dual-porosity" formation because of the distinct hydraulic characteristics of different types of porosity present (Barenblatt et al., 1960). It is also not uncommon to see the term "triple porosity" in case of well-developed karst aquifers with conduit flow. An exact determination of individual storage properties of different porosity types is beyond capability of common aquifer pumping tests. Aquifer storage parameters related to the effective matrix porosity may be approximately assessed by laboratory tests, which would ideally include core samples from multiple locations and depths within the aquifer.

Figure 2.78 illustrates some of the possible responses of karst and fractured aquifers to groundwater withdrawal. In addition to the discussion presented within the figure caption, it is important to emphasize the significance of step-drawdown tests in karst aquifers. Such tests, routinely performed to determine well efficiency and optimum pumping rate for the main aquifer test, may reveal the nature of aquifer porosity surrounding the well.

A schematic of a formation having a network of dissolutional openings intersecting a network of "diffuse" fractures (in which the groundwater flow is slow), and being pumped

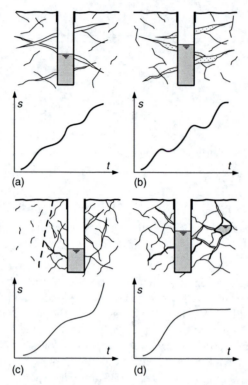

FIGURE 2.78 Characteristic time–drawdown curves for some examples of well pumping tests in fractured rock and karst aquifers. (a) Pumping from a limited number of large fracture or conduit sets which are being consecutively dewatered, (b) pumping from large fractures and conduits filled with clastic sediments which are being washed out, (c) cone of depression reaches a less permeable (less fractured) portion of the aquifer, (d) cone of depression reaches an equipotential boundary such as a large karst conduit with flowing water or a surface stream. (Modified from Larsson, I., Chairman of the Project Panel, *Ground water in Hard Rocks*, Project 8.6 of The International Hydrological Programme, UNESCO, Paris, 1982, 228pp.).

by a fully penetrating well, is shown in Figure 2.79. The equations and boundary conditions governing the pumping of water from a well in a dual-porosity karst formation are given by Greene et al. (1999) and are described below.

Flow in dissolutional openings:

$$S\frac{\partial h}{\partial t} - T\frac{1}{r}\frac{\partial}{\partial r}\left(r\frac{\partial h}{\partial r}\right) = \beta(h_f - h) \tag{2.148}$$

Initial condition in dissolutional openings:

$$h(r, t = 0) = H \tag{2.149}$$

Boundary conditions in dissolutional openings:

$$2\pi r T\frac{\partial h}{\partial r}\bigg|_{r\to 0} = Q, \;\; h(r \to \infty, t) = H \tag{2.150}$$

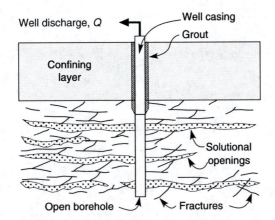

FIGURE 2.79 Conceptual model of groundwater flow to a well pumping in a karst formation consisting of solutional openings and a network of diffuse fractures. (From Greene, E.A., Shapiro, A.M., and Carter, J.M., Hydrogeologic characterization of the Minnelusa and Madison Aquifers near Spearfish, South Dakota. U.S. Geological Survey Water-Resources Investigations Report 98-4156, Rapid City, South Dakota, 1999, 64pp.)

Flow in fractures:

$$S_f \frac{\partial h_f}{\partial t} = -\beta(h_f - h) \tag{2.151}$$

Initial condition in fractures:

$$h_f(r, t = 0) = H \tag{2.152}$$

where S is the storativity of dissolutional openings, S_f is the storativity of the network of diffuse fractures, t is the time, T is the transmissivity of dissolutional openings, h is the hydraulic head in the dissolutional openings, h_f is the hydraulic head in the network of diffuse fractures, H is the initial head, r is the radial distance from the pumping well, and β is the rate of fluid exchange between the network of fractures and the dissolutional openings.

Assuming that the aquifer is homogeneous and isotropic, its properties are estimated by solving the above equations for various choices of T, S, S_f, and β, and comparing the calculated type curves with the observed data until an acceptable fit is obtained. Figure 2.80 shows the results of such curve fitting to the data observed during testing of the Dickey well in the confined Madison limestone karst aquifer near Spearfish, South Dakota, the United States (Greene et al., 1999). The observed data are for the Kyte well, located 1800 ft from the Dickey well, which was pumped for six days at 680 gpm. Note that, if the test had lasted only one day for example, the effects of dual porosity would have not been apparent.

As mentioned earlier, unconfined aquifers often exhibit similar time–drawdown curves due to delayed gravity response. For illustration purposes, Figure 2.81 shows the results of fitting the Kyte well data to a type curve obtained using Neuman solution for unconfined aquifers with delayed gravity response. Arguably, the fitted type curve looks even "better" than the one on Figure 2.80. Boulton (1973) showed analytically that the same set of equations can be applied to: (1) the delayed flow to a well in an unconfined aquifer and (2) leakance from the upper aquifer through the aquitard overlying the pumped confined aquifer. Two possible

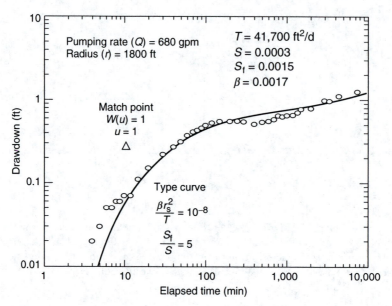

FIGURE 2.80 Best-fit type curve match to the drawdown data observed at the Kyte well during the 6-d aquifer pumping test at the Dickey well, Madison limestone aquifer. (From Greene, E.A., Shapiro, A.M., and Carter, J.M., Hydrogeologic characterization of the Minnelusa and Madison Aquifers near Spearfish, South Dakota. U.S. Geological Survey Water-Resources Investigations Report 98-4156, Rapid City, South Dakota, 1999, 64pp.)

interpretations of the aquifer test results shown in Figure 2.80 and Figure 2.81 thus become more ambiguous knowing that the Madison limestone aquifer at the test location is overlain by the Minnelusa aquifer, which in its lower part acts as a confining layer consisting of interbedded sandstones and dolomitic limestone.

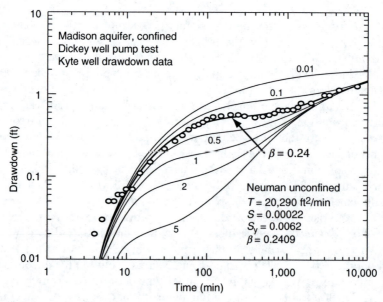

FIGURE 2.81 Neuman solution type curve match to the drawdown data observed at the Kyte well, Madison limestone aquifer, during the 6-day aquifer pumping test at the Dickey well (time–drawdown data from Greene et al., 1999).

Whatever the reason for the apparent dual porosity-like response to the pumping test is (e.g., true dual porosity nature of the karst aquifer, or leakance from the overlying aquitard, or both), the selection of aquifer parameters for some future quantitative analysis should always consider the late drawdown information. Interestingly enough, as illustrated by Figure 2.82, a simple Cooper–Jacob solution for both the early and the late drawdown data gives values of T and S very similar to the dual-porosity solution of Greene et al. (1999). The Cooper–Jacob early storativity is almost identical with the fracture and conduit storativity of Greene et al., whereas the Cooper–Jacob late storativity is very similar to the "diffuse fracture" storativity of Greene et al.

2.5.12.1 Aquifer Anisotropy

Fractured and karst aquifers are by definition heterogeneous and anisotropic when analyzed at a realistic field scale. Determining quantitative parameters of these two very important characteristics is virtually impossible using aquifer pumping test results alone. At least some information on tectonic (fracture) fabric, geometry and nature of aquifer top and bottom, and depositional environment should be available to successfully interpret results of a pumping test in fractured rock and karst aquifers.

Various methods of determining anisotropy of aquifer transmissivity, developed for homogeneous intergranular aquifers, have been more or less successfully applied using the equivalent porous medium approach. Greene (1993) explains in detail a method developed by Hantush (1966b) using an example from a pumping test conducted in the Madison limestone aquifer in the western Rapid City area, South Dakota, the United States (see Figure 2.83 and Figure 2.84). The method assumes homogeneous, leaky confined aquifer conditions where anisotropy is in the horizontal plane. Figure 2.83 shows field-data plots for five observation wells located at different distances (r) and angles from the production well. The five plots would theoretically fall on the same curve if transmissivity was not anisotropic or leakage did not occur. The displacement of the plots from one another is the result of directional

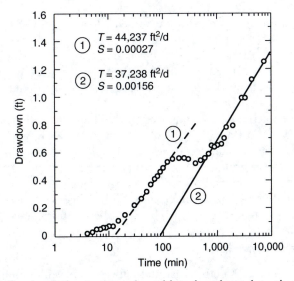

FIGURE 2.82 Cooper–Jacob solution to the early and late drawdown data observed at the Kyte well, Madison limestone aquifer, during the 6-d aquifer pumping test at the Dickey well (time–drawdown data from Greene et al., 1999).

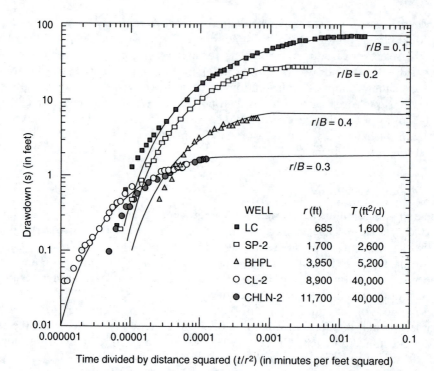

FIGURE 2.83 Drawdown data for five observation wells superimposed on the best-fit leaky confined aquifer type curves for RC-5 aquifer test, Madison limestone aquifer, South Dakota, the United States. (From Greene, E.A., Hydraulic properties of the Madison aquifer system in the western Rapid City area, South Dakota. U.S. Geological Survey Water-Resources Investigations Report 93-4008, Rapid City, South Dakota, 1993, 56pp.)

transmissivities in the Madison aquifer and because leakage through the confining bed does not equally affect drawdown in the five wells (Greene, 1993).

The results of the anisotropy analysis are presented in Figure 2.84. The major axis of transmissivity is 56,000 ft^2/d at an angle of 42° clockwise from north. The minor axis of transmissivity is 1300 ft^2/d at an angle of 48° counterclockwise from north. It should be noted that any such analysis, based on the equivalent porous medium approach, is just an indicator of a possible anisotropy due to, for example, some preferential flow paths (conduits) in the karst aquifer.

Warner (1997) applies methodology developed by Papadopulos (1965) and the related computer program by Maslia and Randolph (1986) to estimate anisotropy of the Upper Floridan aquifer transmissivity using pumping test data from the southwestern Albany area, Georgia in the United States. A degree of professional judgment is exercised by excluding from the analysis, as an outlier, a well with the highest transmissivity. Although debatable, such practice of a selective interpretation of data is often present in karst hydrogeology given its complexity and nonconformity with many methods developed by the "classic intergranular" hydrogeology.

2.5.13 WELL SPECIFIC CAPACITY METHOD

Specific capacity of a well (S_c) is defined as ratio between the well pumping rate (Q) and the stabilized drawdown in the pumping well (s):

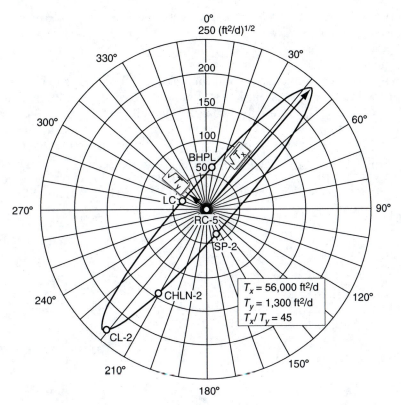

FIGURE 2.84 Theoretical ellipse showing the angle and magnitude of the major and minor axes of transmissivity from the anisotropic analysis of the RC-5 aquifer test, Madison limestone aquifer, South Dakota, the United States. (From Greene, E.A., Hydraulic properties of the Madison aquifer system in the western Rapid City area, South Dakota. U.S. Geological Survey Water-Resources Investigations Report 93-4008, Rapid City, South Dakota, 1993, 56pp.)

$$S_c = \frac{Q}{s} \qquad (2.153)$$

In the United States the specific capacity is expressed in gallons per minute per one foot of drawdown (gpm/ft). For example, a well pumped at 300 gpm with 30 ft of drawdown would have reported specific capacity of 10 gpm/ft. For most individual and small-scale public water supply wells the specific capacity is determined by well drillers upon well completion, by pumping the well for several hours at a constant rate and measuring the drawdown before the pump is turned off. This means that, in most cases, the well specific capacity is likely overestimated as the pumping duration was not long enough for drawdown to stabilize (become quasistable). It should be noted that drawdown might never truly stabilize, especially in case of confined aquifers, regardless of the duration of pumping (unless there is an equipotential boundary influencing the drawdown). However, in the absence of a long-term pumping test, which is rather expensive and requires expertise in interpreting the test results, well specific capacity is often used to estimate aquifer transmissivity. This practice is useful in regional aquifer studies by providing many more data points for transmissivity interpolation and better understanding of groundwater flow characteristics. For example, in the Edwards aquifer of central Texas, over 1000 specific-capacity tests were found in over 1000 wells compared to only 71 time–drawdown tests in 21 wells (Mace, 2001). Figure 2.85 shows different empirical relationships between aquifer transmissivity and well specific capacity for various aquifers.

In addition to aquifer characteristics and pumping duration, the specific capacity is a function of the well characteristics, such as well radius, degree of penetration, and well loss. Larger aquifer transmissivity, storativity, and well radius all result in greater specific capacity (smaller drawdown). In general, a well that only penetrates part of the aquifer (a partially penetrating well) has a lower specific capacity than if the well penetrated the entire aquifer thickness. Furthermore, a well that partially penetrates an aquifer may overestimate the specific capacity for the penetrated portion of the aquifer owing to vertical flow components caused by the partial penetration. Laminar and turbulent well loss cause specific capacity to

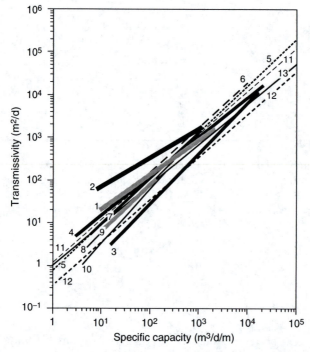

FIGURE 2.85 Comparison among the different empirical relationships and between the empirical relationships and the analytical approaches including: 1, Eagon and Johe (1972) for a carbonate aquifer; 2, Razack and Huntley (1991) for a heterogeneous alluvial aquifer; 3, Huntley et al. (1992) for a fractured hard rock aquifer; 4, El-Naqa (1994) for a fractured carbonate aquifer; 5, Mace (1997) for a karstic aquifer in Texas; 6, Mace (1997) for a karstic aquifer in Florida; 7, Fabbri (1997) for a fractured carbonate aquifer; 8, Mace et al. (2000) for the sandstone Trinity aquifer in north-central Texas; 9, Mace (2000) for the sandstone Paluxy aquifer in north-central Texas; 10, Mace et al. (2000) for the Woodbine sandstone aquifer in north-central Texas; 11, the Thomasson et al. (1960) approach for alluvium; 12, the Thomasson et al. (1960) approach for fractured hard rocks; 13, the Theis et al. (1963) approach for an assumed value of C' of 10^5 d/m^2. For the empirical relationships (lines 1–7), the length of the line corresponds to the applicable range of the relationship and the prediction intervals are about an order of magnitude. Note that line 7 is coincident with line 5 (From Mace, R.E., Estimating transmissivity using specific-capacity data. Geological Circular 01-2, Bureau of Economic Geology, The University of Texas at Austin, Austin, TX, 2001, 44pp.)

be underestimated because the measured drawdown at the well is greater than the actual water-level decline in the aquifer (Mace, 2001).

The relationship between the well specific capacity and aquifer transmissivity is presented by the following equation based on the classic Theis nonsteady-state equation and introduced by Theis et al. (1963; modified after Mace, 2001):

$$S_c = \frac{4\pi T}{[\ln(2.25Tt_p/r_w^2 S)]} \quad (2.154)$$

where S is the storativity of the aquifer and t_p is the pumping time. This equation assumes (1) a fully penetrating well, (2) homogeneous, isotropic porous media, (3) negligible well loss, and (4) effective radius equal to the radius of the production well (Walton, 1970). Application of Equation 2.154 is probably the most frequently used technique for estimating transmissivity from specific capacity. Walton (1970), Lohman (1972), and Bradbury and Rothschild (1985) offer examples of using this approach.

Since aquifer transmissivity (T) is in both the numerator and denominator, Equation 2.154 must be solved iteratively, or graphically as presented by Meyer (1963) and Theis et al. (1963). Iteratively, the equation is solved as follows:

1. Estimate initial transmissivity value, e.g., based on known values from some other location in the same aquifer, or by multiplying probable aquifer thickness with the hydraulic conductivity representative of the aquifer porous media; assume storativity based on known aquifer characteristics; for unconfined aquifers use value of storativity between 0.05 and 0.3, and for confined aquifers between 0.00001 and 0.001.
2. Calculate the right-hand side of Equation 2.154 and compare the resulting number to the measured specific capacity.

If the calculated and measured values of specific capacity are close, then the transmissivity value used in step 1 is acceptable. If the two numbers do not agree, then the transmissivity value must be adjusted up or down and step 2 is repeated. Transmissivity value used in Equation 2.154 is much more sensitive than the value of assumed storativity, and it will have more effect on the final result than any other parameter in the equation. For detail discussion on the sensitivity of different parameters in Equation 2.154 see Mace (2001).

Bradbury and Rothschild (1985) derived an equation based on Theis et al. (1963) that considers partial penetration and well loss:

$$T = \frac{Q}{4\pi(s_w - s_L)\left[\ln\left(\frac{2.25Tt}{r_w^2 S} + 2s_p\right)\right]} \quad (2.155)$$

where s_L is the well loss (see Section 3.1.2), s_w is the drawdown measured in the well, and s_p is the partial penetration factor defined by Brons and Marting (1961) as

$$s_p = \frac{1 - (L_w/b_a)}{L_w/b_a}\left[\ln\left(\frac{b_a}{r_w}\right) - G\left(\frac{L_w}{b_a}\right)\right] \quad (2.156)$$

where L_w is the length of the well screened in the aquifer, b_a is the aquifer thickness, and G is a function of the ratio of L_w to b_a.

The same authors evaluated G for different values of the ratio of L_w to b_a, and Bradbury and Rothschild (1985) fit a polynomial (correlation coefficient = 0.992) to the Brons and Marting (1961) data to get the following expression (after Mace, 2001):

$$G\left(\frac{L_w}{b_a}\right) = 2.948 - 7.363\left(\frac{L_w}{b_a}\right) + 11.447\left(\frac{L_w}{b_a}\right)^2 - 4.675\left(\frac{L_w}{b_a}\right)^3 \qquad (2.157)$$

Another equation considering partial well penetration factor (s_p) is given by Sternberg (1973; after Mace, 2001):

$$S_c = \frac{4\pi T}{\left[\ln(2.252Tt/r_w^2 S) + 2s_p\right]} \qquad (2.158)$$

When a partially penetrating well is pumped, the component of vertical flow to the well causes drawdowns to be less than expected, thus overestimating specific capacity for the penetrated section. However, because the well does not completely penetrate the aquifer, the specific capacity for the entire thickness of the aquifer is underestimated. Muskat (1946) and Turcan (1963) present another possible solution (after Mace, 2001):

$$S_c' = \frac{S_c}{\left[\dfrac{L_w}{b_a}\right]\left\{1 + 7\left(\dfrac{r_w}{2L_w}\right)^{1/2}\cos\left(\dfrac{\pi L_w}{2b_a}\right)\right\}} \qquad (2.159)$$

where S_c' is the specific capacity corrected for partial penetration, L_w is the length of the well screened in the aquifer, b_a is the aquifer thickness, and r_w is the well radius. This equation accounts for vertical flow to the well and estimates the specific capacity for entire aquifer thickness assuming that the unpenetrated interval of the aquifer has the same characteristics as the penetrated interval of the aquifer. S_c determined in this way is then used in Equation 2.154 to estimate the aquifer transmissivity iteratively as described earlier.

Well loss is more of a concern at higher pumping rates: the higher the pumping rate the more likely there is turbulent well loss. Furthermore, turbulent well loss increases as the square of the pumping rate. In many cases, higher specific capacities are positively correlated with higher pumping rates because wells that can support greater yields tend to be pumped at higher rates. Greater turbulent well losses for greater specific capacities should result in specific capacities that are more in error for large values than for small values. Well loss corrections can be incorporated into approaches for estimating transmissivity from specific capacity using the following equation (Mace, 2001):

$$S_c' = S_c \frac{s_L}{s_w - s_L} = \frac{Q}{s_w - s_L} \qquad (2.160)$$

where S_c' is the corrected specific capacity, s_w is the drawdown measured in the well, s_L is the well loss (see Section 3.1.2), and Q is the well pumping rate.

The corrected specific capacity is then used in Equation 2.154 for iterative calculation of the transmissivity.

2.6   AQUIFER RECHARGE AND SUSTAINABILITY

The determination of aquifer recharge is one of the most important and, at the same time, the most difficult tasks in hydrogeology. This is because recharge commonly takes place over large areas (aquifer recharge area) with varying land cover, soil, geomorphologic and geologic characteristics and, in addition to direct infiltration of precipitation, may involve infiltration of sheet and concentrated surface runoff. Except in rare cases such as a sinking stream in karst (see Figure 2.86), the individual components of aquifer recharge cannot be directly measured in the field. Rather, they are estimated using indirect water balance methods and measurements of related "simple" physical quantities such as precipitation rate, air temperature, water table fluctuations, surface streamflows, and soil-moisture content or infiltration capacity for example. Recharge is also often estimated (calibrated) as part of groundwater modeling studies where other model parameters and boundary conditions are considered to be more certain (better known). In such cases, however, the model developer should clearly discuss uncertainties related to "calibrated" recharge rates, and the sensitivity of all key model parameters. For example, a difference of 5% or 10% in aquifer recharge rate may not be all that sensitive compared to aquifer transmissivity (hydraulic conductivity) when matching field water table measurements; however, this difference is very significant in terms of aquifer water budget and analyses of aquifer "safe" yield (sustainability). Recharge rate also has implications on the shape and transport characteristics of groundwater contaminant plumes. For example, more or less direct recharge from land surface may result in a more or less diving plume, respectively, even without simulated (by a model, for example) vertical hydraulic gradients (Figure 2.87). Different recharge rates also result in different overall concentrations as illustrated in the same figure.

Spatial variability in recharge at local and intermediate scales may not be critical for water-resource assessment, but it is crucial for analysis of contaminant transport, because focused

FIGURE 2.86 Permanent sinking stream Rak at the entrance to Tkalca cave in classic karst of Slovenia.

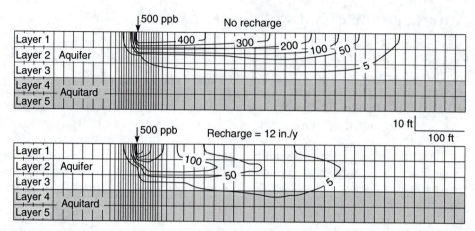

FIGURE 2.87 Influence of varying recharge rate applied to water table on a model-simulated plume, 3 y after release of contaminant; cross-sectional view. Constant concentration at the source in Layer 1 is 500 μg/L (micrograms per liter, or parts per billion, or ppb). Regional hydraulic gradient in the shallow aquifer (i.e., in all of the first three layers) from left to right is 0.0043; there is no vertical hydraulic gradient within the first three layers. The hydraulic conductivity of the first three layers is 10 ft/d, the longitudinal transverse and vertical dispersivities are 5, 0.5, and 0.05 ft, respectively. No retardation or degradation of contaminant.

recharge and preferential flow allow contaminants to migrate rapidly through the unsaturated zone to underlying aquifers; therefore, delineation of zones of high recharge is critical to defining zones that are vulnerable to contamination (Scanlon et al., 2002). The timescales of recharge estimates are also important. Information on recharge at decadal timescales is generally required for water-resource planning, whereas timescales required for contaminant transport range from days to thousands of years, depending on the particular contaminants being considered (for example, an incidental gasoline spill in a shallow unconfined aquifer, and a long-term radioactive waste disposal site study, respectively).

Various terms related to groundwater recharge are used interchangeably, often causing confusion. In general, infiltration refers to any water movement from the surface into the subsurface. This water is sometimes called potential recharge, indicating that only a portion of it may eventually reach water table (saturated zone). Term actual recharge is increasingly used to avoid any possible confusion: it is the portion of infiltrated water that reaches the aquifer and it is estimated (confirmed) based on groundwater studies. Effective (net) infiltration, or deep percolation describe water movement below the root zone, and these are often equated to actual recharge. In hydrologic studies the term effective rainfall describes portion of precipitation that reaches surface streams through direct overland flow or near-surface flow, and does not have any relationship with infiltrated water. Rainfall excess describes part of rainfall that does not infiltrate and it generates surface runoff. *Interception* is part of rainfall intercepted by vegetative cover before it reaches ground surface and it is not available for either infiltration or surface runoff. The term net recharge is used to differentiate between the following two water fluxes: recharge reaching the water table due to vertical downward flux from the unsaturated zone, and evapotranspiration from the water table, which is an upward flux (negative recharge). Areal (or diffuse) recharge refers to recharge derived from precipitation or irrigation that occurs fairly uniformly over large areas, whereas concentrated recharge refers to loss of ponded or flowing surface water (playas, lakes, streams) to the

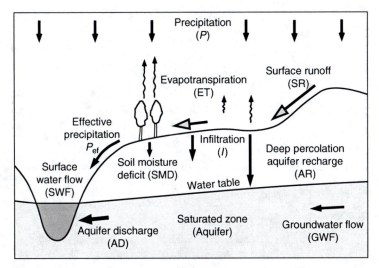

FIGURE 2.88 Key elements of the rainfall–runoff–recharge process.

subsurface. Figure 2.88 shows several key elements of the complex rainfall–runoff–recharge process. Some of the relationships between different elements in Figure 2.88 include:

$$I = P - SR - ET$$
$$AR = I - SMD$$
$$SWF = P_{ef} + AD$$
$$AD = AR + GWF \qquad (2.161)$$

2.6.1 FACTORS INFLUENCING AQUIFER RECHARGE

2.6.1.1 Climate

Humid and arid systems represent end members for different climates and generally require different approaches to quantify recharge. Humid regions are usually characterized by shallow water tables and gaining streams where areal (diffuse) recharge from precipitation is dominant. Groundwater is discharged mainly through evapotranspiration and baseflow to streams. In contrast, deep water tables and losing streams are common in alluvial valleys and basins in arid regions where concentrated recharge dominates. Recharge rates are limited in large part by the availability of water at the land surface, which is controlled by climatic factors, such as precipitation and evapotranspiration, and by surface geomorphic features (Scanlon et al., 2002).

The simplest classifications of climate are based on annual precipitation. Strahler and Strahler (1978, p. 129, from Bedinger, 1987) refer to arid climates as having 0 to 250 mm/y precipitation, semiarid as having 250 to 500 mm/y precipitation, and subhumid as having 500 to 1000 mm/y precipitation. However, a climate classification based on the relationship of general factors that affect the availability of water to plants is more useful than a classification based simply on annual precipitation. The soil-water balance of Thornthwaite (1948) provides such a classification of climate by calculating the annual cycle of soil-moisture availability or deficiency, thus providing measures of soil moisture available for plant growth. Based on the

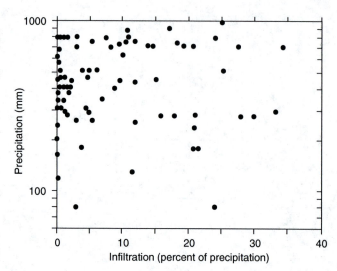

FIGURE 2.89 Relationships between infiltration and average precipitation based on 29 studies of arid and semiarid regions of the world. (From Bedinger, M.S., Summary of infiltration rates in arid and semiarid regions of the world, with an annotated bibliography. U.S. Geological Survey Open-File Report 87-43, Denver, CO, 1987, 48pp.)

soil-water balance, Strahler and Strahler (1978) discuss the worldwide distribution of 13 climatic types based on the average annual variations of precipitation, potential evapotranspiration, and the consequent soil-moisture deficit or surplus. Soil-water balance models, which are very similar to the soil-moisture model for classifying worldwide climate, have been developed to estimate recharge. These models utilize more specific data, such as soil type and moisture-holding capacity, vegetation type and density, surface-runoff characteristics, and spatial and temporal variations in precipitation. Various soil-water balance models used in estimating recharge in arid and semiarid regions of the world are discussed by Bedinger (1987) who includes an annotated bibliography of 29 references. Figure 2.89 shows relationship between infiltration and average precipitation for the 29 studies, some of which have multiple values. As can be seen, data points have wide scatter, which is attributed to differences in applied methods, real differences in rates of infiltration to various depths and net recharge, and varying characteristics of soil types, vegetation, precipitation, and climatic regime.

Based on a detailed study of a wide area in the midcontinent United States spanning six states, Dugan and Peckenpaugh (1985) concluded the following: both the magnitude and proportion of potential recharge from precipitation decline as the total precipitation declines (see Figure 2.90), although other factors, including other climatic conditions, vegetation, and soils also affect potential recharge. The limited scatter among the points in Figure 2.90 indicates a close relationship between precipitation and recharge. Furthermore, the relationship becomes approximately linear where mean annual precipitation exceeds 30 in. and recharge exceeds 3 in. Presumably, when the precipitation and recharge are less than these values, disproportionately more infiltrating water is spent on satisfying the moisture deficit in dry soils. The extremely low recharge in the western part of the study area, particularly Colorado and New Mexico, appears to be closely related to the high potential evapotranspiration, solar radiation, percent of possible sunshine, and lower relative humidity. Seasonal distribution of precipitation also shows a strong relationship to recharge. Areas of high cool-season precipitation tend to receive higher amounts of recharge. Where potential evapotranspiration is low and long winters prevail, particularly in the Nebraska and South Dakota

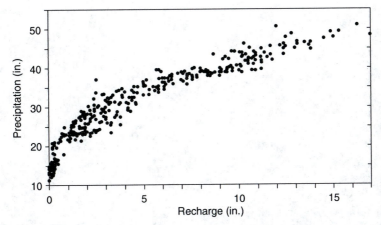

FIGURE 2.90 Computed mean annual recharge using soil-moisture program versus mean annual precipitation, by model grid element, mid-central United States. (Modified from Dugan, J.T. and Peckenpaugh, J.M., Effects of climate, vegetation, and soils on consumptive water use and ground-water recharge to the central Midwest regional aquifer system, mid-continent United States. U.S. Geological Survey Water-Resources Investigations Report 85-4236, Lincoln, NE, 1985, 78pp.)

parts of the study area, effectiveness of cool-season precipitation as a source of recharge increases. Overall, however, when cool-season precipitation is less than 5 in., recharge is minimal. Dugan and Peckenpaugh (1985) conclude that generalized patterns of potential recharge are determined mainly by climatic conditions. Smaller variations within local areas, however, are related to differences in land cover, soil types, and topography.

Three physical characteristics of the soil (permeability, available water capacity [AWC], and slope) affect the availability of water for consumptive use by regulating both infiltration and the ability of the soil profile to store water. Infiltration is largely a function of permeability and slope, whereas the water storage capacity is determined by the product of the AWC and the root zone depth. Finer-textured soils on steeper slopes generally have less deep percolation than those with coarser textures and lesser slopes. They also have lower permeabilities and higher AWC, which slows infiltration but increases the overall soil-moisture storage capacity. Steeper slopes, with the same vegetative cover and soil permeability, generally have more runoff and less infiltration. Sandy soils limit runoff, thus increasing infiltration, which results in greater deep percolation or potential recharge. The higher actual evapotranspiration associated with finer-textured soils indicates greater AWC; therefore, more water is available for consumptive water use by vegetative cover (Dugan and Peckenpaugh, 1985). Rainfall intensity also plays an important role in recharge generation. Generally, higher rainfall intensities are indicative of less infiltration and more runoff; the converse occurs with lower rainfall intensities.

Air temperature, humidity, wind pattern, and insolation (number of hours with sunshine) all influence direct evaporation, as well as development of vegetative cover and thus plant transpiration. These two factors of overall water balance are usually estimated together as evapotranspiration, using a variety of proposed empirical equations (for example, see Singh, 1993; Shuttleworth, 1993; Dingman, 1994). However, there should be a clear distinction between potential evapotranspiration, determined by some empirical equations, and actual evapotranspiration. Quite often the potential evapotranspiration, determined from direct measurements of evaporation with pan meters, and a few other measurements of weather elements (e.g., air temperature, hours of sunshine, wind speed; see Figure 2.91) exceed the amount of total precipitation on an annual basis, which is not reasonable except in most

FIGURE 2.91 Hydrometeorologic station equipped with instruments for measurements of direct evaporation (pan evapometer in the front), wind speed, precipitation, air temperature, and insolation (hours of sunshine).

extreme cases. Unfortunately, the actual evapotranspiration cannot be measured directly and has to be estimated from other elements of the drainage water balance or from potential evapotranspiration. In general, aquifer recharge is higher in climatic regions with moderate air temperatures, less sunshine, less wind, and higher humidity. Some arid areas with extreme conditions experience phenomenon of falling rain that never reaches ground surface because of almost instantaneous evaporation from the air below clouds. Little rain that reaches the land also evaporates quickly because of the high surface temperatures.

2.6.1.2 Land Cover and Land Use

The soil surface can become encrusted with, or sealed by, the accumulation of fines or other arrangements of particles that prevent or retard the entry of water into the soil. As rainfall starts, the fines accumulated on bare soil may coagulate and strengthen the crust or enter the soil pores and seal them off. A soil may have excellent subsurface drainage characteristics but still have a low infiltration rate because of the retardant effect of surface crusting or sealing (King, 1992). This low-permeable crust is the main reason why bare soil has much lower infiltration capacity than soil with vegetative cover. Consequently, bare soil is much more easily eroded by surface runoff. Presence of vegetation protects the soil surface from the impact of rainfall. Root systems of vegetation tend to enhance soil porosity and permeability. Organic matter greatly increases pore sizes and pore size distribution (King, 1992). The influence of different land cover on aquifer recharge is illustrated in Table 2.3 and Table 2.4.

Urban areas and development generally decrease infiltration rates and increase surface runoff because of the increasing presence of various impervious surfaces (rooftops, asphalt, concrete). However, as illustrated in Table 2.4, infiltration rate varies significantly within an

TABLE 2.3
Mean Annual Values (in.) of Selected Output from the Soil-Moisture Program in the CMRASA Study Area for Soil Group 1 at Paris, Arkansas, the United States

Vegetation Type	Infiltration	Surface Runoff	CWR	STD	AET	CIR	DP	DP (%)
Row crop	36.45	8.25	32.14	7.36	24.78	8.98	11.67	26.11
Alfalfa	39.51	5.19	40.59	6.82	33.77	9.13	5.74	12.84
Small grain	39.51	5.19	32.35	2.55	29.80	3.26	9.71	21.72
Grassland	39.51	5.19	33.29	6.02	27.27	7.61	12.24	27.38
Woodland	39.51	5.19	34.76	4.41	30.35	7.92	9.16	20.49
Fallow	36.45	8.25	22.26	1.19	21.07	2.10	15.38	34.41

Mean annual precipitation is 44.70 in., mean annual potential evapotranspiration is 51.53 in. CWR, consumptive water requirements; STD, soil-moisture deficit under dryland conditions; AET, actual evapotranspiration (consumptive water use under dryland conditions); DP, deep percolation (aquifer recharge); CIR, consumptive irrigation requirements.

Source: From Dugan, J.T. and Peckenpaugh, J.M. Effects of climate, vegetation, and soils on consumptive water use and ground-water recharge to the central Midwest regional aquifer system, mid-continent United States. U.S. Geological Survey Water-Resources Investigations Report 85-4236, Lincoln, NE, 1985, 78pp.

urban area based on actual land use. This is particularly important when evaluating fate and transport of contaminant plumes, including development of groundwater models for such diverse areas. For example, a contaminant plume may originate at an industrial facility with high percentage of impervious surfaces resulting in hardly any actual infiltration, and then migrate toward a residential area where infiltration rates may be rather high because of the open space (yards) and various vegetative cover (not to mention residents that prefer to see their lawns always green and use extensive watering).

Agricultural activities have had direct and indirect effects on the rates and compositions of groundwater recharge and aquifer biogeochemistry. Direct effects include dissolution and transport of excess quantities of fertilizers and associated materials and hydrologic alterations related to irrigation and drainage. Some indirect effects include changes in water–rock

TABLE 2.4
Estimates of Mean Annual Recharge on the Basis of Mean Annual Precipitation, Generalized Surficial Geology, and Land-Use and Land-Cover Categories from the Willamette Lowland Regional Aquifer System Analysis

Land Use and Land Cover	Area (mi²)	Precipitation (in./y)	Recharge (in./y)	Recharge (%)
Undeveloped and nonbuilt-up	641	44.2	24.1	54.5
Residential	13	43.3	12.7	29.3
Built-up	35	45.0	13.3	29.6
Urban	99	43.7	8.1	18.5
All categories	788	44.2	21.4	48.4

Source: Modified from Lee, K.K. and Risley, J.C. Estimates of ground-water recharge, base flow, and stream reach gains and losses in the Willamette River Basin, Oregon. U.S. Geological Survey Water-Resources Investigations Report 01-4215, Portland, OR, 2002, 52pp.

reactions in soils and aquifers caused by increased concentrations of dissolved oxidants, protons, and major ions. Agricultural activities have directly or indirectly affected the concentrations of a large number of inorganic chemicals in groundwater, for example NO_3^-, N_2, Cl, SO_4^{2-}, H^+, P, C, K, Mg, Ca, Sr, Ba, Ra, and As, as well as a wide variety of pesticides and other organic compounds (Böhlke, 2002).

2.6.1.3 Porosity and Permeability of Soil Cover

Arguably, the most significant factors affecting infiltration are the physical characteristics and properties of soil layers. The rate at which water enters soil cannot exceed the rate at which water is transmitted downward through the soil. Thus, soil-surface conditions alone cannot increase infiltration unless the transmission capacity of the soil profile is adequate. Under conditions where the surface entry rate (rainfall intensity) is slower than the transmission rate of the soil profile, the infiltration rate will be limited by the rainfall intensity (water supply). Until the topsoil horizon is saturated (i.e., the soil-moisture deficit is satisfied), the infiltration rate will be constant as shown in Figure 2.92. For higher rainfall intensities, all the rain will infiltrate into the soil initially until the soil surface becomes saturated ($\theta = \theta_s$, $h \geq 0$, $z = 0$), i.e., until the so-called ponding time is reached. At that point, the infiltration is less than the rainfall intensity and surface runoff begins. These two conditions are expressed as (Rawls et al., 1993):

$$-K(h)\frac{\partial h}{\partial z} + 1 = R \qquad \theta(0, t) \leq \theta_s \qquad \leq t_p \tag{2.162}$$

$$h = h_0 \qquad \theta(0, t) = \theta_s \qquad t \geq t_p \tag{2.163}$$

where $K(h)$ is the hydraulic conductivity for given soil-water potential (degree of saturation), h is the soil-water potential, h_0 is a small positive ponding depth on the soil surface, θ is the volumetric water content, θ_s is the volumetric water content at saturation, z is the depth from land surface, t_p is the ponding time, and R is the rainfall intensity.

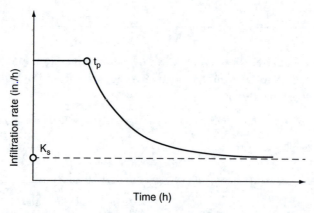

FIGURE 2.92 Typical infiltration curve for a homogeneous uniform soil showing constant infiltration rate while it is higher than the rainfall intensity, and the topsoil horizon is still not completely saturated (horizontal portion of the curve). Once the topsoil horizon is saturated, the ponding at the land surface occurs and the surface runoff starts (at the ponding time t_p). The infiltration capacity decreases with time and eventually reaches the soil-saturated hydraulic conductivity (K_s).

The transmission rates can vary at different horizons in the unsaturated soil profile. After saturation of the uppermost horizon, the infiltration rate is limited to the lowest transmission rate encountered by the infiltrating water as it travels downward through the soil profile.

As water infiltrates through successive soil horizons and fills in the pore space, the available storage capacity of the soil will decrease. The storage capacity available in any horizon is a function of the porosity, horizon thickness, and the amount of moisture already present. Total porosity and the size and arrangement of the pores have significant effect on the availability of storage. During the early stage of a storm, the infiltration process will be largely affected by the continuity, size, and volume of the larger-than-capillary (noncapillary) pores, because such pores provide relatively little resistance to the infiltrating water. If the infiltration rate is controlled by the transmission rate through a retardant layer of the soil profile, then the infiltration rate, as the storm progresses, will decrease as a function of the decreasing storage availability above the restrictive layer. The infiltration rate will then equal the transmission rate through this restrictive layer until another, more restrictive layer is encountered by the water (King, 1992). The soil infiltration capacity decreases in time and eventually asymptotically reaches the value of overall saturated hydraulic conductivity (K_s) of the affected soil column as illustrated in Figure 2.92.

In general, the infiltration rate decreases with increasing clay content in the soil and increases with increasing noncapillary porosity through which water can flow freely under the influence of gravity. Presence of certain clays such as montmorillonite, even in relatively small quantities, may dramatically reduce infiltration rate as they become wet and swell. Runoff conditions on low-permeable soils develop much sooner and more often than on uniform, clean sands, and gravels, which have infiltration rates higher than most rainfall intensities. Although clay layers generally impede flow due to their lower saturated hydraulic conductivity, when these layers are near or at the surface and initially very dry and with drying cracks, the initial infiltration rate may be much higher and then drop off rapidly.

The initial soil-water content and saturated hydraulic conductivity of the soil media are the primary factors affecting the soil-water infiltration process. The wetter the soil initially, the lower will be the initial infiltration rate (due to a smaller suction gradient), and the constant infiltration rate will be attained more quickly. In general, the higher the saturated hydraulic conductivity of the soil, the higher the infiltration rate. More on the unsaturated flow characteristics and permeability of different soil types is given in Section 1.6.

2.6.1.4 Geologic and Geomorphologic Characteristics

When soil cover is thin or absent, the lithologic and tectonic characteristics of the bedrock play a dominant role in aquifer recharge. Fractured bedrock surface and steep or vertical bedding greatly increase infiltration rates, whereas layers of unfractured bedrock sloping at the same angle as the land surface may eliminate infiltration altogether (Figure 2.93 through Figure 2.95).

Mature karst areas, where rock porosity is greatly increased by dissolution, generally have the highest infiltration capacity of all geologic media (Figure 2.96). For example, actual aquifer recharge rates of over 80% of total precipitation, even for high-intensity rainfall events of more than 250 mm/d, have been recorded in classic karst areas of Montenegro. These rates were determined by measuring flow of large temporary karst springs, which would become active within only a few hours after the start of rainfall. Several temporary springs in the area have recorded maximum discharge rate of over 300 m^3/s and are among the largest such springs in the world (Figure 2.97).

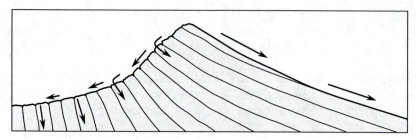

FIGURE 2.93 Influence of bedding dip and angle on infiltration.

As might be expected, the slope of the land can also affect the infiltration rate. Steep slopes will result in runoff, which will affect the amount of time the water will be available for infiltration. In contrast, gentle slopes will have less of an impact on the infiltration process

FIGURE 2.94 McCardy Shale in Saltville, Virginia, the United States, sloping at the same angle as the land surface, resulting in minimal infiltration.

FIGURE 2.95 Vertical limestone layers in Durmitor National Park, Montenegro.

due to decreased runoff. Depressions in land surface will accumulate surface runoff and increase both infiltration and evaporation rates.

2.6.1.5 Depth to Water Table

Infiltration can lead directly to recharge when the wetting front reaches the regional water table while infiltration is still occurring. In the arid and semiarid regions, however, deep water tables and short infiltration times lead to infiltration ending long before the wetting front crosses the water table. In these instances, water that has infiltrated into the unsaturated zone can be removed by evapotranspiration or, if the water has moved past the root zone, remain in storage until it is displaced by water from a subsequent infiltration event. Unsaturated zone water in storage below the root zone can continue to redistribute over time. This long-term movement of water, or percolation flux, occurs both vertically, by drainage, and horizontally in response to heterogeneity or anisotropy of the sediments. Long-term percolation can eventually result in aquifer recharge (Coes and Pool, 2005).

2.6.2 Methods for Estimating Aquifer Recharge

Methods based on basin water balance and unsaturated zone data provide estimates of potential recharge, whereas those based on groundwater data generally provide estimates of actual recharge. Uncertainties in each approach to estimating recharge underscore the need for application of multiple techniques to increase reliability of recharge estimates. Recommendations on choosing appropriate techniques for quantifying groundwater recharge are given by Scanlon et al. (2002).

FIGURE 2.96 Limestone surface in northwest Ireland with a dissolution-enlarged fracture network. (Courtesy of George Sowers; reproduced with kind permission of Frances Sowers).

FIGURE 2.97 Temporary karst spring Sopot in Montenegro, discharging over 200 m^3/s on September 23, 2005, within 24 h of a 220 mm/d rainfall in its drainage area.

2.6.2.1 Methods Based on Unsaturated Zone Data

2.6.2.1.1 Soil-Moisture Content

The degree of soil saturation with water directly influences infiltration rates as illustrated with Equation 2.162 and Equation 2.163. Measurements of soil moisture are therefore a widely used method for estimating potential infiltration. Lysimeters are permanent field installations usually extending deeper in the soil column to measure actual vertical infiltration from below the root intake zone. Measurements of this flux are therefore the closest to the actual recharge of unconfined aquifers with relatively shallow water table in unconsolidated sediments. Lysimeters may be equipped with a variety of instruments, which measure soil-water potential (matrix potential) at different depths, such as tensiometers, and instruments that measure actual flux (flow rate) of infiltrating water at different depths by collecting the infiltrate. Some lysimeters may also include piezometers for recording water table fluctuations. More detail on lysimeter construction can be found in Pruitt and Lourence (1985). Data collected from lysimeters is often used to calibrate empirical equations or numeric models for determining other water balance elements such as evapotranspiration.

The major difficulty when extrapolating lysimeter data to a much wider aquifer recharge area is the inevitably high variability in soil and vegetative characteristics. Figure 2.98 illustrates this with the range of cumulative percolation rates at seven lysimeters all installed on a 100-ft^2 plot located within a 2.8 mi^2 watershed in Pennsylvania in the United States. This watershed was subject to a comparative study of several different methods for estimating aquifer recharge (Risser et al., 2005b). Even at the 100-ft^2 scale, the coefficient of variation between monthly rates at individual lysimeters is greater than 20% for 6 months, with the June, July, and August values of about 50%, 100%, and 60%, respectively. Figure 2.99 shows mean monthly recharge rates for seven lysimeters during the 1994–2001 period of record. The mean annual recharge rate was 12.2 in. or 29% of precipitation. In comparison, this rate was 24% as calculated with the water table fluctuations method, 25% as calculated with the PART hydrograph baseflow separation method, and 29% using a daily water balance HELP3 model.

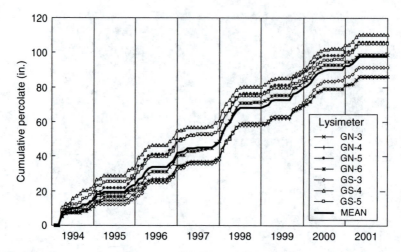

FIGURE 2.98 Range of cumulative percolation rates at seven lysimeters installed on a 100 ft^2 plot located within a 2.8 mi^2 watershed in Pennsylvania, the United States. (From Risser, D.W., Gburek, W.J., and Folmar, G.J., Comparison of methods for estimating ground-water recharge and base flow at a small watershed underlain by fractured bedrock in the eastern United States. U.S. Geological Survey Scientific Investigations Report 2005-5038, Reston, VA, 2005b, 31pp.)

2.6.2.1.2 Environmental Tracers

The methods of computation of infiltration employing environmental tracers rely on (1) the principle of conservation of mass using naturally introduced radioactive or other environmental tracers and (2) the use of intentionally introduced radioactive or other tracers to calculate the rate of soil-moisture movement (tracer dating). Suitable natural environmental tracers are those that are not products of pedogenesis, do not react chemically in the soil system, are not removed substantially by exfiltration, and are highly soluble (Bedinger, 1987). Environmental tracers produced in significant amounts by thermonuclear explosions (e.g., tritium and chlorine-36), and chloride have been most widely used in the measurement of infiltration rates. In addition to tritium, other natural isotopes (principally, deuterium,

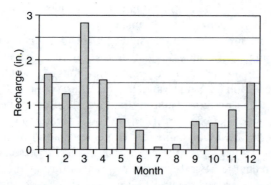

FIGURE 2.99 Mean monthly recharge rates for seven closely spaced lysimeters during the 1994–2001 period of record. (Data from Risser, D.W., Gburek, W.J., and Folmar, G.J., Comparison of methods for estimating ground-water recharge and base flow at a small watershed underlain by fractured bedrock in the eastern United States. U.S. Geological Survey Scientific Investigations Report 2005-5038, Reston, VA, 2005b, 31pp.)

carbon-13, carbon-14, and oxygen-18) have been used in the qualitative and quantitative understanding of hydrologic regimes in general. More on environmental tracers and ground-water dating in general is presented in Section 4.2.5.

Mass balance of a natural environmental tracer in the soil pore water may be used to measure infiltration where the sole source of the tracer is atmospheric precipitation and where runoff is known or negligible. The relationship for a steady-state mass balance is (Bedinger, 1987)

$$R_q = (P - R_s)\frac{C_p}{C_z} \qquad (2.164)$$

where R_q (infiltration) is a function of P (precipitation), R_s (surface runoff), C_p (tracer concentration in precipitation), and C_z (tracer concentration in the soil moisture).

In the ideal model, the tracer content of soil water increases with depth by the loss of soil water to evapotranspiration and conservation of tracer. The tracer content of soil water attains a maximum at the maximum depth of evapotranspiration. The model postulates a constant content of tracer in the soil moisture from this point to the water table. Departures from the ideal model of tracer variation with depth are common and have been attributed to changes in land use, such as clearing of native vegetation, replacement of native vegetation by cropped agriculture, bypass mechanisms for infiltrating water through the soil profile, and changes in climate (Bedinger, 1987).

Mass balance equation 2.164 assumes that recharge occurs by piston flow. However, it has been shown that diffusion and dispersion are important components of chloride flux in the unsaturated zone. Johnston (1983, from Bedinger, 1987) proposed the following equation for mass balance where runoff is negligible:

$$PC_p = -D_s - \frac{\partial C_z}{\partial Z} + C_z R_q \qquad (2.165)$$

where D_s is the diffusion–dispersion coefficient and $\partial C/\partial C_z$ is the rate of change in soil-moisture concentration with depth. P, C_z, and R_q are as defined in Equation 2.164.

Chloride (Cl) is continuously deposited on the land surface by precipitation and dry fallout. The high solubility of Cl enables its transport into the subsurface by infiltrating water. Because Cl is essentially nonvolatile and its uptake by plants is minimal, it is retained in the sediment when water is removed by evaporation and transpiration. An increase in Cl within the root zone of the shallow subsurface, therefore, is proportional to the amount of water lost by evapotranspiration (Allison and Hughes, 1978, from Coes and Pool, 2005). In areas where active infiltration is occurring, an increase in Cl in the shallow subsurface will generally be absent, and concentrations will be very low through the unsaturated zone. In areas where little to no active infiltration is occurring, an increase in Cl in the shallow subsurface will be present. After reaching maximum Cl concentrations will stay relatively constant down to the water table. However, there are published studies showing that Cl concentrations in arid areas with little infiltration may decrease below the peak concentration in the root zone due to varying factors such as paleoclimatic variations and nonpiston flow.

If the Cl deposition rate on the land surface is known, the average travel time of Cl (t_{Cl}) to a depth in the unsaturated zone (z) can be calculated by (Coes and Pool, 2005):

$$t_{Cl} = \frac{\int_0^z Cl_{soil} dz}{Cl_{dep}} \qquad (2.166)$$

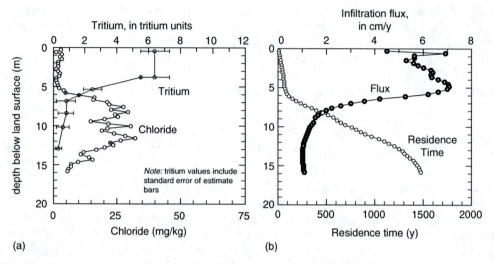

FIGURE 2.100 Determination of basin-floor infiltration at borehole BF1, Sierra Vista subwatershed, Arizona, the United States. (a) Sediment-chloride and pore-water tritium data, (b) residence time and infiltration flux, calculated from chloride data. (From Coes, A.L. and Pool, D.R., Ephemeral-stream channel and basin-floor infiltration and recharge in the Sierra Vista subwatershed of the upper San Pedro basin, Southeastern Arizona. U.S. Geological Survey Open-File Report 2005-1023, Reston, VA, 2005, 67pp.)

where Cl_{soil} is the chloride mass in the sample interval $[M/L^3]$ and Cl_{dep} is the chloride deposition rate $[M/L^2/t]$.

The above calculation entails several assumptions: (1) flow in the unsaturated zone is downward-vertical and piston type, (2) bulk precipitation (precipitation plus dry fallout) is the only source of Cl and there are no mineral sources of Cl, (3) the Cl deposition rate has stayed constant over time, and (4) there is no recycling of Cl within the unsaturated zone. Figure 2.100 shows an example of chloride concentration in a thick unsaturated zone in semiarid climate of Arizona in the United States.

Tritium (^3H), a naturally occurring radioactive isotope of hydrogen, can be used to estimate the age of subsurface water and infiltration fluxes into the subsurface. Large quantities of tritium were released to the atmosphere during thermonuclear weapons testing from 1952 until the late 1960s; maximum releases occurred in the early 1960s. As a result, the amount of tritium in precipitation sharply increased during testing as tritium was introduced into the water cycle, and decreased after the testing ended. The amount of tritium in subsurface water at a given time is a function of the amount of tritium in the atmosphere when infiltration occurred and the radioactive decay rate of tritium. If flow in the unsaturated zone is assumed to be downward-vertical and piston type, the average infiltration flux (q_i) can be estimated by (Coes and Pool, 2005):

$$q_i = \frac{\Delta z}{\Delta t} \theta_v \tag{2.167}$$

where Δz is the depth to maximum tritium activity $[L]$, Δt is the elapsed time between sampling and maximum historic atmospheric tritium activity, and θ_v is the volumetric soil-water content $[L^3/L^3]$.

2.6.2.2   Methods Based on Groundwater Data

2.6.2.2.1   Environmental Tracers

Similarly to the unsaturated zone pore water, environmental tracers dissolved in groundwater can be used to measure infiltration by determining their concentration at different depths within the saturated zone and calculating residence times (age of water). However, caution should be exercised when interpreting the results of such analysis because movement of groundwater in the saturated zone is generally faster, including mixing with the water flux coming to the sample location laterally or vertically. Figure 2.101 shows chloride concentrations at different depths within a shallow alluvial aquifer in Norman, Oklahoma, for two sampling events. Low concentrations indicate new recharged water, and there is a clear increase associated with older water. More on environmental tracers and groundwater dating in general is presented in Section 4.2.5.

2.6.2.2.2   Water Level Fluctuations

A rise in water table after rainfall events is arguably the most accurate indicator of actual aquifer recharge. Figure 2.102 shows a simple principle of expressing this direct aquifer recharge through the rise in water table and the specific yield. The main uncertainty in applying this approach is the value of specific yield, which, in most cases, would have to be assumed. Even in cases where the specific yield is estimated from aquifer test data, it may not be representative of the subsurface interval that became saturated after the water table rise.

The influence of heterogeneity of the aquifer porous media on the representative specific yield is especially important in fractured and karst aquifers as illustrated in Figure 2.103. The effective porosity and specific yield in highly fractured portions of the aquifer, which in addition may have interconnected conduit networks, would be very high and would result

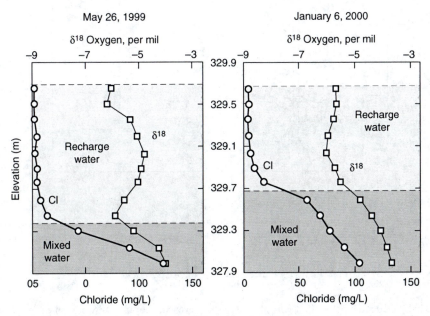

FIGURE 2.101 Changes of recharge water thickness at the control site, Norman Landfill, Norman, Oklahoma, the United States. (Modified from Scholl, M. et al., Recharge processes in an alluvial aquifer riparian zone, Norman landfill, Norman, Oklahoma, 1998–2000. U.S. Geological Survey Scientific Investigations Report 2004-5238, Oklahoma City, OK, 2004, 53pp.)

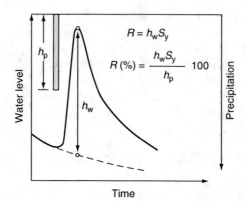

FIGURE 2.102 Estimation of recharge (R) from rise in water table (h_w), and known or estimated specific yield (S_y). The equation shows recharge estimated as percentage of precipitation height (h_p).

in a rapid rise of the hydraulic head. In contrast, portions of the aquifer with low matrix porosity and no significant fracture network would react slowly, showing a much smaller rise in the hydraulic head. Selecting both the representative rise in the hydraulic head and the value of specific yield in such cases would rely heavily on best professional judgment.

The location of water level measurement relative to aquifer recharge zone and preferential flow path also plays an important role when interpreting the hydraulic head fluctuations. All

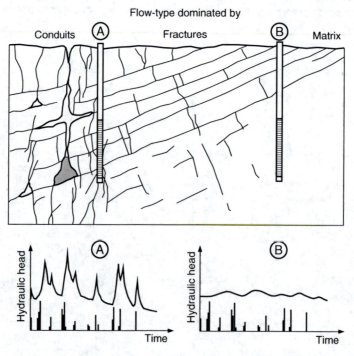

FIGURE 2.103 Dependence of the hydraulic head measured in monitoring wells on different types of effective porosity (specific yield) in karst aquifers. A, Rapid rise of the hydraulic head after major recharge events in portions of the aquifer with large conduits and no significant storage in the matrix. B, Delayed and dampened response of aquifer matrix. Flow dominated by fractures may include any combination of these two extremes.

other things being equal (hydraulic conductivity, specific yield, uniform areal rate of recharge), a well closer to the discharge zone, such as a surface stream, would have a smaller
hydraulic head rise than a well located farther upgradient (Figure 2.104). To illustrate this
point, Risser et al. (2005b) constructed a cross-sectional MODFLOW model having a length
of 1000 ft, transmissivity of 1000 ft^2/d, specific yield of 0.01, and steady-state recharge rate of
1ft/y. Recharge of 0.1 ft was added to the model (in addition to the 1 ft/y steady rate) for a
period of 1 day and the resulting water table rise was plotted for headwaters, midslope, and
near-stream well locations. A water table rise of 10 ft is predicted by the water table rise
method for a recharge event of 0.1 ft in an aquifer with specific yield of 0.01 (see equation in
Figure 2.102). As expected, a water-level rise of about 10 ft was simulated for the upland well
location, but water levels at the midslope and valley locations rose less—only 8 and 1.8 ft,
respectively. This result is caused by the movement of water away from the water table during
the 1-day period of recharge, which is most rapid near the stream boundary. Such conditions
are most pronounced for aquifers with high hydraulic diffusivity (transmissivity divided by
storage coefficient) and high stream density (short distance from streams to divides). Risser
et al. (2005b) conclude that, other factors being equal, wells in upland settings would be the
best candidates for estimating aquifer recharge using water table fluctuations method.

2.6.2.2.3   *Spring Flow Hydrograph*
Increases in spring flow after rainfall events are another direct indicator of actual aquifer
recharge. Knowing the exact area where this direct recharge from rainfall takes place, and the
representative (average) amount of rainfall, enables a very precise determination of the
aquifer recharge. Although it is often difficult to accurately determine a spring drainage
area, especially in karst, the spring hydrograph reflects the response to rainfall of all porosity
types in the entire volume of the aquifer, as well as the response to all water inputs into the
aquifer. In addition to direct infiltration of rainfall, these inputs may include percolation of
sheet surface runoff from less permeable areas beyond the aquifer extent, or direct percolation
from surface streams. For this reason, the spring hydrograph method of determining overall
aquifer recharge, from all sources, is arguably the most objective one.

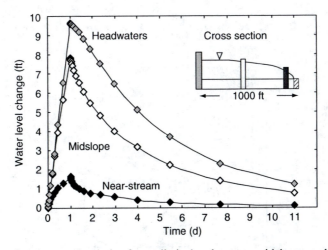

FIGURE 2.104 Simulated water-level rise for wells in headwaters, midslope, and near-stream locations. (From Risser, D.W., Gburek, W.J., and Folmar, G.J., Comparison of methods for estimating
ground-water recharge and base flow at a small watershed underlain by fractured bedrock in the
eastern United States. U.S. Geological Survey Scientific Investigations Report 2005-5038, Reston,
VA, 2005b.)

Although the processes that generate hydrographs of springs and surface streams are quite different, there is much that is analogous between them, and the hydrograph terminology is the same. Figure 2.105 shows main elements of a discharge hydrograph. The beginning of discharge after a rainfall episode is marked with point A, and the time between the beginning of rainfall and the beginning of discharge, called the starting time, with t_s. The time in which the hydrograph raises to its maximum (point C), is the concentration time—t_c. The time from maximum discharge until the end of the hydrograph, when the discharge theoretically equals zero (point E), is the falling time—t_f. Together, the concentration time and the falling time are called the base time of the hydrograph—t_b. The time between the centroid of the precipitation episode (C_P) and the centroid of the hydrograph (C_H) is called the retardation time (t_r). The time interval for recording the amount of precipitation and the flow rate at the spring is Δt.

The shape of the hydrograph is defined by its base (AE), the rising limb (AB), the crest (BCD), and the falling limb (DE). The falling limb also corresponds to the recession period. B and C are inflexion points where the hydrograph curve changes its shape from convex to concave and vice versa. For surface streams, point D is the end of direct runoff after the rain. In the case of a spring hydrograph, point D is the beginning of recession, i.e., discharge without direct influence of precipitation.

The shape of a discharge hydrograph depends upon the size and shape of the drainage area, as well as the precipitation intensity. When a rainfall episode lasts longer, and the intensity is lower, the hydrograph has a bigger time base and vice versa: intensive short storm events cause sharp hydrographs with small time bases. The area under the hydrograph is the volume of discharged water for the recording period. In reality, unless a spring or a surface stream is intermittent, the recorded hydrograph has a more complex shape, which reflects the influence of antecedent precipitation and other possible water inputs. Such hydrographs are formed by the superposition of single hydrographs corresponding to separate precipitation events (see Figure 2.106).

The separation of a surface stream hydrograph is the determination of the individual elements that participate in flow formation. Theoretically, they are divided into: flow formed by direct precipitation over the surface stream, surface runoff collected by the stream, near-surface flow of the newly infiltrated water (also called underflow), and groundwater inflow. However, it is practically impossible to separate all these components of discharge from a real physical drainage area. In practice, the problem of component separation is therefore reduced

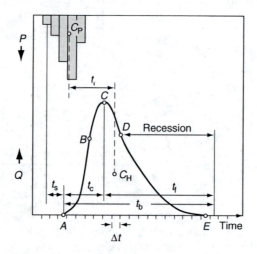

FIGURE 2.105 Components of a discharge hydrograph (explanation in text).

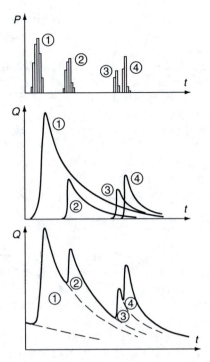

FIGURE 2.106 A complex discharge hydrograph (bottom) composed of single hydrographs (middle) as the result of several rainfall events (top) (modified from Jevdjevic, 1956).

to an estimation of the baseflow, formed by groundwater, and surface runoff, which is the integration of all the other components. In the case of a spring hydrograph, the equivalents are the discharge formed by the draining of previously accumulated groundwater, and the discharge that is the result of newly infiltrated water from rainfall.

The first method of hydrograph component separation shown in Figure 2.107 (line *ABC*) is commonly applied to surface streams with significant groundwater inflow. Assuming that point *C* represents the end of all surface runoff, and the beginning of flow generated solely by groundwater discharge, the late near-straight line section of the hydrograph is

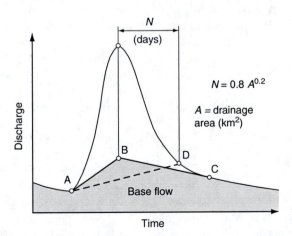

FIGURE 2.107 Two common methods of graphical separation of flow hydrograph components.

extrapolated backward until it intersects the ordinate of the maximum discharge (point B). Point A, representing the beginning of surface runoff after rainfall, and point B are then connected with the straight line. The area under the line ABC is the baseflow, i.e., the groundwater component of the surface streamflow. Obviously, for spring hydrographs this area does not have an equivalent explanation since all discharged water is by definition groundwater. However, the portion of a spring hydrograph that corresponds to the base-flow component of a surface stream hydrograph could partially be attributed to a release of water from smaller voids (small fissures, matrix porosity) under the pressure from newly infiltrated water.

The second graphical method of baseflow separation is used for surface streams in low-permeable terrain without significant groundwater flow. It is conditional since point D (i.e., the hydrograph falling time) is found by the following empirical formula (Linsley et al., 1975):

$$t_f = 0.8A^{0.2} \text{ (d)} \tag{2.168}$$

where A is the drainage area (km^2). In general, this method gives short falling times: for an area of 100 km^2 t_f is 2 days, and for 10,000 km^2 t_f is 5 days. Thus, the method should be applied cautiously after analyzing a sufficient number of single hydrographs and establishing an adequate area–time relationship.

The impact of new rainwater on spring discharge varies with respect to the predominant type of porosity and the stage of groundwater level. In any case, the first reaction of karst or fissured aquifers to recharge in form of the rapid initial increase in discharge rate is in most cases the consequence of pressure propagation through karst conduits and large fractures, and not the outflow of newly infiltrated water. The new water arrives at the spring with certain delay and its contribution is just a fraction of the overall flow rate. After a recharge episode and the initial response of the system are over, two cases may occur (see Figure 2.108) (a) an increase in volume of groundwater accumulated in storage, which is reflected in the shifting of the baseflow to a higher hydrograph level and (b) new water is mostly transmitted through a well-developed network of fissures or karst channels and conduits, and is dis-charged without significant accumulation of groundwater in the matrix porosity. The base-flow continues along the same extrapolated recession line present before the rainfall event.

The first case is characteristic for periods of main aquifer recharge, i.e., March through June in moderate climates: both groundwater level and moisture content in the unsaturated zone are high, while evapotranspiration loss compared to other periods of the year is small. Newly infiltrated water raises an already high hydraulic head (pressure) and the groundwater is more easily injected into the aquifer matrix porosity and narrow fissures. During the

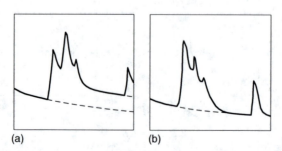

(a)   (b)

FIGURE 2.108 Possible reaction of aquifers to infiltration of precipitation as seen from spring discharge hydrographs. (a) Lifting of the baseflow to a higher hydrograph level because of storage water increase, and (b) discharge without increase in storage.

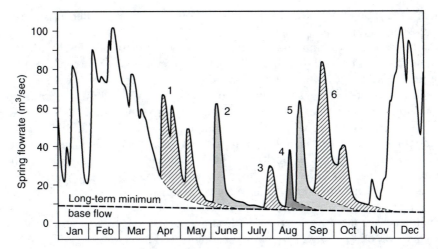

FIGURE 2.109 Determination of aquifer recharge using individual spring flow hydrographs (Ombla spring near Dubrovnik in Croatia). See Table 2.5 for the results. (From Krešić, N., *Kvantitativna hidrogeologija karsta sa elementima zaštite podzemnih voda* (in Serbo-Croatian; Quantitative karst hydrogeology with elements of groundwater protection), Naučna knjiga, Belgrade, 1991, 192pp.)

summer and autumn period, the hydraulic heads and the hydraulic gradients are low, and new water is transmitted mainly through large fissures and conduits. This second case may also indicate lack of significant matrix and fissure porosity in the aquifer.

Figure 2.109 and Figure 2.110 show the principle of aquifer recharge determination from individual spring flow hydrographs and the corresponding rainfall amount. The volume of discharged water above the long-term baseflow (shaded hydrograph areas) is equal to aquifer recharge from individual rainfall events. The volume of total rainfall is obtained by multiplying the spring drainage area (600 km^2) and the height of rainfall (expressed here in

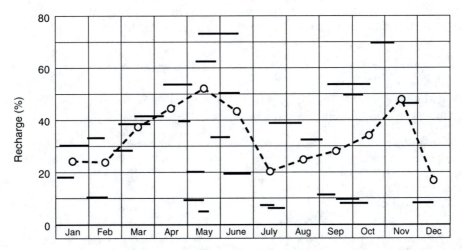

FIGURE 2.110 Average monthly recharge of the Grza spring limestone aquifer, eastern Serbia, determined using hydrograph method shown in Figure 2.109. (From Krešić, N., *Kvantitativna hidrogeologija karsta sa elementima zaštite podzemnih voda* (in Serbo-Croatian; Quantitative karst hydrogeology with elements of groundwater protection), Naučna knjiga, Belgrade, 1991, 192pp.)

millimeters). The ratio of two volumes, multiplied by 100, gives the aquifer recharge expressed as percentage of total rainfall (Table 2.5).

When spring hydrograph and precipitation data is available for multiple years, the above-described method can be used to estimate average seasonal or monthly recharge rates for the drainage area. These rates would vary from year to year depending on various seasonal factors such as antecedent saturation of the vadose zone, snow pack, and evapotranspiration. In general, however, such analysis should provide a reasonable assessment of a long-term recharge pattern and characteristics in the spring drainage area. Figure 2.110 is an example of average monthly recharge rates in a limestone aquifer drained by a permanent springs with a well-defined drainage area. The horizontal lines represent base lengths of 29 individual hydrographs and the corresponding recharge rates determined as shown in Figure 2.109. Average monthly rates (circles) are time-weighted averages of the individual recharge rates within that month. As can be seen, these rates show a wide scatter, particularly in May and June, which have the largest number of hydrographs analyzed. However, the characteristic seasonal character of aquifer recharge in temperate humid climates is clearly visible.

More on using spring flow hydrographs in quantitative aquifer analysis is given in Chapter 16.

2.6.2.3 Methods Based on Analytical Equations and Numeric Modeling

Numerous equations and mathematical models have been developed for estimating soil-water movement and infiltration rates for various purposes such as irrigation and drainage, groundwater development, soil and groundwater contamination studies, managed aquifer recharge (MAR) and wastewater management, to name just a few. Ravi and Williams (1998) and Williams et al. (1998) have prepared a two-volume publication for the U.S. Environmental Protection Agency, in which they present a number of easy-to-use, widely applied methods. The compiled methods are divided into three types: (a) empirical models, (b) Green–Ampt models, and (c) Richards equation models. These methods (except the empirical models) are based on widely accepted concepts of soil physics, and soil hydraulic and climatic parameters representative of the prevailing site conditions. The two volumes (1) categorize infiltration models presented based on their intended use, (2) provide a conceptualized scenario for each infiltration model that includes assumptions, limitations, mathematical boundary conditions, and application, (3) provide guidance for model selection for site-specific

---

TABLE 2.5

Determination of Recharge from Precipitation in the Ombla Spring Drainage Area Using Spring Flow Hydrograph Method Shown in Figure 2.109

Number of Hydrograph	Discharged Water (m³)	Rainfall (mm)	Recharge (%)
1	8.704×10^7	201.5	72
2	3.451×10^7	69.3	83
3	1.614×10^7	35.4	76
4	1.067×10^7	27.8	64
5	5.046×10^7	164.9	51
6	1.011×10^8	271.8	62

Source: From Krešić, N. *Kvantitativna hidrogeologija karsta sa elementima zaštite podzemnih voda* (in Serbo-Croatian; Quantitative karst hydrogeology with elements of groundwater protection), Naučna knjiga, Belgrade, 1991, 192pp. Drainage area is estimated at 600 km²; data for year 1969.

scenarios; (4) provide a discussion of input parameter estimation, (5) provide example application scenarios, for each model, and (6) provide a demonstration of sensitivity analysis for selected input parameters.

Since a universal mathematical model is not available to address water infiltration for all field conditions, assumptions are commonly made, and limitations are identified during the model development and application process. Assumptions used in developing water infiltration models commonly include the following (Williams et al. 1998):

1. *Initial soil-water content profile*: Most infiltration models assume a constant and uniform initial soil-water content profile. However, under field conditions, the soil-water content profile is seldom constant and uniformly distributed.

2. *Soil profile*: There are two types of soil profiles that exist under field conditions: (a) homogeneous and (b) heterogeneous. Infiltration models developed for the homogeneous soil profile cannot be used for heterogeneous soil profile without simplifying assumptions concerning the heterogeneity.

3. *Constant and near saturated soil-water content at the soil surface*: This is a common assumption for most infiltration models, which will allow for the ignoring of the initially high hydraulic gradient across the soil's surface. The length of time the surface is not near saturation is very small when compared to the length of time associated with the infiltration event.

4. *Duration of infiltration process*: Some infiltration models are only valid for a short-term period of infiltration, which can limit their usefulness to field applications where infiltration may last for longer time periods. A short-term period might be representative of a rainfall or irrigation event.

5. *Surface crust and sealing*: None of the selected models can be used for the surface crusting boundary condition. This boundary condition is complex and dynamic.

6. *Flat and smooth surface*: These surface boundary conditions can be easily incorporated into mathematical models, whereas an irregular surface or sloping surface can result in an added dimension to the mathematical modeling.

Boundary conditions are those conditions defined in the modeling scenario to account for observed conditions at the boundaries of the model domain. These conditions must be defined mathematically, the most common of which are given as follows:

1. *Constant or specified flux at the surface*: The rate of water applied to the soil surface can be constant or time-varying. Because of its simplicity as compared to other boundary conditions, a constant flux condition has been used in most infiltration models.

2. *Surface ponding condition*: The surface ponding or nonponding conditions are dependent on the rate of water application as well as on soil infiltrability. Whenever the rate of water application to the soil exceeds the soil infiltration capacity, the surface ponding occurs. The opposite case will result in surface nonponding conditions.

3. *Finite column length at the lower boundary*: Some infiltration models are limited to a finite column length and others may allow for an infinite column length. Infiltration models with the finite column length condition may limit their applications for deeper water infiltration.

4. *Based on Richards equation*: Several infiltration models (e.g., Philip's model and Eagleson's model) were developed based on Richards equation. These models commonly have well-defined physically based theories and give considerable insight into the processes governing infiltration. These would require information related to the water retention characteristics of the soil media.

A summary of assumptions, limitations, and boundary conditions for six commonly used models is given in Table 2.6. For a detailed explanation of Richards equation see Section 1.6.

There are several versatile public domain unsaturated flow, and fate and transport numeric models that can be used to estimate aquifer recharge rates. Examples with friendly graphic user interface (GUI) include VS2DT developed by the U.S. Geological Survey (see Figure 2.111), and Hydrus-1D and UNSATCHEM developed by the U.S. Salinity Laboratory, U.S. Department of Agriculture (1957). Combined, these programs include a variety of options such as heterogeneous porous media, and changes in hydraulic conductivity based on volumetric moisture contact and different chemical characteristics of both the infiltrating fluid and the porous media. Hydrus-1D and UNSATCHEM also include various options specifically developed for agricultural applications such as root zone uptake and specific drainage conditions.

O'Reilly (2004) proposes the use of a simple water-balance transfer function (WBTF) model for estimating transient recharge in deep water table settings. The model represents a one-dimensional column from the top of the vegetative canopy to the water table and consists of two components: (1) a water-balance module that simulates the water storage capacity of the vegetative canopy and root zone and (2) a transfer-function module that simulates the travel-time of water as it percolates from the bottom of the root zone to the water table. Data requirements include two time series for the period of interest: (1) precipitation (or precipitation minus surface runoff, if surface runoff is not negligible) and (2) evapotranspiration. The model requires values for five parameters that represent water storage capacity or soil-drainage characteristics. A limiting assumption of the WBTF model is that the percolation of water below the root zone is a linear process. That is, percolating water is assumed to have the same travel-time characteristics, experiencing the same delay and attenuation, as it moves through the unsaturated zone. This assumption is more accurate if the moisture content, and consequently the unsaturated hydraulic conductivity, below the root zone does not vary substantially with time. The model is available for free download at the USGS Web site.

2.6.2.4 Hydrograph Baseflow Separation

As explained earlier in Section 2.6.2.2, groundwater contribution to surface streamflow is most often estimated using technique called graphical separation of flow hydrographs. Although some professionals view this method as a "convenient fiction" because of its subjectivity and lack of rigorous theoretical basis, it does provide useful information in the absence of detail (and expensive) data on many surface water runoff processes and drainage basin characteristics that all contribute to streamflow generation.

There are at least several widely applied graphical techniques that separate flow hydrograph into two major components: flow generated by surface and near-surface runoff and flow generated by discharge of groundwater (see Figure 2.107). The groundwater flow component is called stream baseflow. In natural long-term conditions (steady state), i.e., in the absence of artificial groundwater withdrawal, the rate of groundwater recharge in a drainage basin of a permanent gaining stream is equal to the rate of groundwater discharge. Assuming that all groundwater discharges into the surface stream, either directly or through springs, it follows that the stream baseflow equals the groundwater recharge in the basin. This simple concept is illustrated in Figure 2.112. However, its application is not always straightforward and it should be based on a thorough understanding of geologic and hydrogeologic characteristics of the basin. The following examples illustrate some situations where baseflow alone should not be used to estimate actual groundwater recharge:

TABLE 2.6
Summary of Assumptions, Limitations, and Mathematical Boundary Conditions for Various Infiltration Models, Where θ Is the Volumetric Water Content

Model Name	Assumptions and Limitations				Model Features		Boundary Conditions			
	Constant and Uniform Initial θ	Homogeneous Soil Profile	Valid for Only Short Term	Surface Crust and Sealing	Constant Water Content at Top Boundary	Constant Flux at the Top Boundary	Surface Ponding Condition	Finite Column Length	Based on Richards Equation	Based on Empirical Equation
SCS runoff curve	N	N	N	N	N	N	N	N	N	Y
Philip's two term	Y	Y	Y	N	Y	N	N	N	Y	N
Green–Ampt model for layered systems	Y	Y	N	N	Y	N	Y	Y	N	N
Green–Ampt explicit	Y	Y	N	N	Y	N	Y	Y	N	N
Constant flux Green–Ampt	Y	Y	N	N	Y	Y	N	Y	N	N
Infiltration–exfiltration	Y	Y	N	N	Y	N	N	N	Y	N

Source: Williams, J.R., Ouyang, Y., and J-S Chen, 1998. Estimation of infiltration rate in the vadose zone: application of selected mathematical models, Volume II. EPA/600/R-97/128b, U.S. Environmental Protection Agency, National Risk Management Research Laboratory, Ada, Oklahoma, 44p. Features included in the particular model are denoted with Y, and those not available in the model are denoted with N.

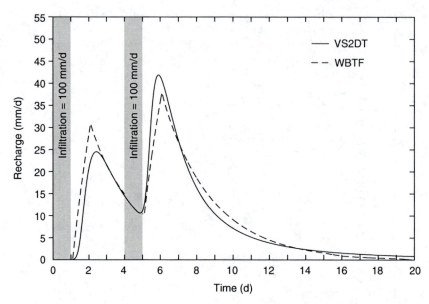

FIGURE 2.111 Groundwater recharge simulated by using the WBTF and VS2DT models for a sandy loam soil, 2.5-m-deep water table, and two infiltration events. (From O'Reilly, A.M., A method for simulating transient ground-water recharge in deep water-table settings in central Florida by using a simple water-balance/transfer-function model. U.S. Geological Survey Scientific Investigations Report 2004-5195, Reston, VA, 2004, 49pp.)

1. Surface streamflows through a karst terrain where topographic divide and ground-water divide are not the same. The groundwater recharge based on baseflow may be grossly overestimated or underestimated depending on the circumstances (see Figure 2.23 and Figure 2.24).
2. The stream is not permanent, or some river segments are losing water (either always or seasonally), and locations and timing of the flow measurements are not adequate to assess such conditions.
3. There is abundant riparian vegetation in the stream floodplain, which extracts a significant portion of groundwater through evapotranspiration.

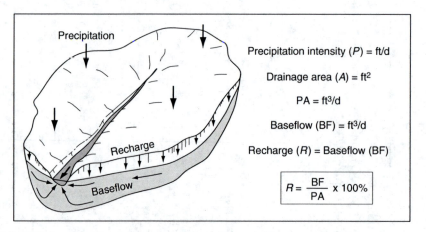

FIGURE 2.112 Estimation of groundwater recharge from surface stream baseflow.

4. There is discharge from deeper aquifers, which have remote recharge areas in other drainage basins (for example see Figure 1.63 and note discharge of the deep Mount Simon aquifer into the Mississippi River).

5. A dam regulates the flow in the stream.

Risser et al. (2005a) present detail application and comparison of two automated methods of hydrograph separation for estimating groundwater recharge based on data from 197 stream-flow gaging stations in Pennsylvania. The two computer programs—PART and RORA (Rutledge, 1993, 1998) developed by the USGS are in public domain and available for free download from the USGS Web site. The PART computer program uses a hydrograph separation technique to estimate baseflow from the streamflow record. The RORA computer program uses the recession-curve displacement technique of Rorabaugh (1964) to estimate groundwater recharge from each storm period. The RORA program is not a hydrograph-separation method; rather, recharge is determined from displacement of the streamflow-recession curve according to the theory of groundwater drainage.

The PART program computes baseflow from the streamflow hydrograph by first identi-fying days of negligible surface runoff and assigning baseflow equal to streamflow on those days; the program then interpolates between those days. PART locates periods of negligible surface runoff after a storm by identifying the days meeting a requirement of antecedent-recession length and rate of recession. It uses linear interpolation between the log values of baseflow to connect across periods that do not meet those tests. A detailed description of the algorithm used by PART is provided in Rutledge (1998, p. 33–38). An example illustrating the separation of the baseflow component from a streamflow hydrograph is shown in Figure 2.113.

Rorabaugh's method utilized by RORA is a one-dimensional analytical model of ground-water discharge to a fully penetrating stream in an idealized, homogenous aquifer with uniform spatial recharge. Because of the simplifying assumptions inherent in the equations, Halford and

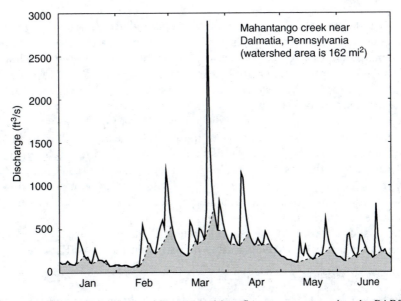

FIGURE 2.113 Streamflow hydrograph with separated baseflow component using the PART program. (From Risser, D.W. et al., Estimates of ground-water recharge based on streamflow-hydrograph methods: Pennsylvania. U.S. Geological Survey Open File Report 2005-1333, Reston, VA, 2005a, 30pp.)

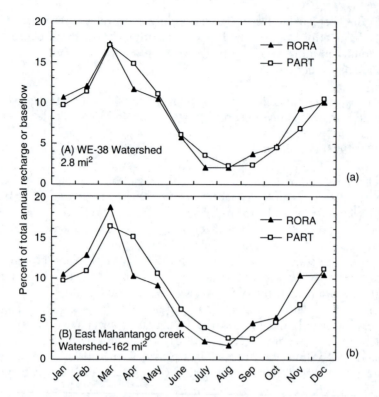

FIGURE 2.114 Mean monthly recharge from RORA and baseflow from PART for watershed scales of (a) 2.8 and (b) 162 mi^2. (From Risser, D.W., Gburek, W.J., and Folmar, G.J., Comparison of methods for estimating groundwater recharge and base flow at a small watershed underlain by fractured bedrock in the eastern United States. U.S. Geological Survey Scientific Investigations Report 2005-5038, Reston, VA, 2005b, various pages.)

Mayer (2000) caution that RORA may not provide reasonable estimates of recharge for some watersheds. In fact, in some extreme cases, RORA may estimate recharge rates that are higher than the precipitation rates. Figure 2.114 shows mean monthly recharge estimated with RORA and PART programs as a percentage of mean annual precipitation for two watersheds of different size. The pattern for both watersheds is very similar and in agreement with the average for all 197 watersheds studied: the monthly values indicate the major groundwater recharge period in Pennsylvania typically occurs in November through May, during which about 80% of the annual recharge occurs. About 18% of the annual recharge typically occurs in March, the month with the greatest average recharge. Estimates of mean monthly recharge from RORA are probably less reliable than estimates for longer periods. Rutledge (2000, p. 31) recommends that results from RORA not be reported at timescales smaller than seasonal (3 months), because results differ most greatly from manual application of the recession-curve displacement method at small timescales.

Figure 2.115 shows variabilty of the calculated monthly recharge rates for the same experimental watershed in Pennsylvania using five different methods (Risser et al., 2005b).

2.6.3 CONCEPT OF AQUIFER SUSTAINABILITY

In an essay prepared for UNESCO as part of the *International Hydrological Programme*, entitled *Water and Ethics. Use of Groundwater*, Llamas (2004) provides the following enlightening assessment:

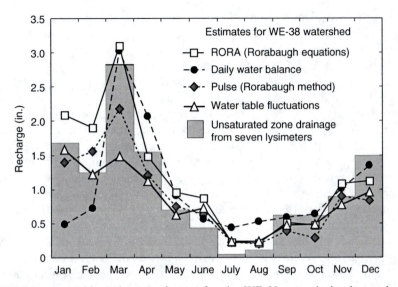

FIGURE 2.115 Mean monthly recharge estimates for the WE-38 watershed using various methods versus lysimeter percolate at the Masser Recharge Site. (Modified from Risser, D.W., Gburek, W.J., and Folmar, G.J., Comparison of methods for estimating ground-water recharge and base flow at a small watershed underlain by fractured bedrock in the eastern United States. U.S. Geological Survey Scientific Investigations Report 2005-5038, Reston, VA, 2005b, various pages.)

Groundwater development significantly increased during the second half of the last century in most semiarid or arid countries. This development has been mainly undertaken by a large number of small (private or public) developers and often the scientific or technological control of this development by the responsible Water Administration has been scarce. In contrast, the surface water projects developed during the same period are usually of larger dimension and have been designed, financed and constructed by Government Agencies which normally manage or control the operation of such irrigation or urban public water supply systems. This historical situation has often produced two effects: 1) most Water Administrations have limited understanding and poor data on the groundwater situation and value; 2) in some cases the lack of control on groundwater development has caused problems such as depletion of the water level in wells, decrease of well yields, degradation of water quality, land subsidence or collapse, interference with streams or surface water bodies, ecological impact on wetlands or gallery forests. These problems have been sometimes magnified or exaggerated by groups with lack of hydrogeological know-how, professional bias or vested interests. Because of this in recent decades groundwater overexploitation has become a kind of "hydromyth" that has pervaded water resources literature. A usual axiom derived from this pervasive "hydromyth" is that groundwater is an unreliable and fragile resource that should only be developed if it is not possible to implement the conventional large surface water projects. The aim of this essay is to present a summary of: 1) the many and confusing meanings of the term overexploitation and the main factors of the possible adverse effects of groundwater development; 2) the criteria to diagnose aquifers prone to situations of overuse; 3) the strategies to prevent or correct the unwanted effects of groundwater development in stressed or intensively used aquifers.

Two terms that relate to the issues brought up by Llamas have been widely used in groundwater literature: the older one is aquifer safe yield and the more recent one is aquifer sustainability. The safe yield concept comes from water supply engineering studies. Originally, it was focused on the relation between the size (capacity) of a surface water reservoir and its safe yield, defined as the maximum quantity of water that could be supplied from the reservoir during a critical period (Alley and Leake, 2004). With respect to groundwater resources, Lee (1915) first defined safe yield as the quantity of water that can be pumped "regularly and permanently without dangerous depletion of the storage reserve" (from Alley and Leake, 2004). Meinzer (1923a, p. 55) defined safe yield as "the rate at which water can be withdrawn from an aquifer for human use without depleting the supply to such extent that withdrawal at this rate is no longer economically feasible."

These first concepts show the human consumption and related economic issues as priorities. Arguably, they also show a certain lack of consideration for other environmental users (beneficiaries) of aquifer water. Even today, some state regulators in the United States define aquifer safe yield simply as the rate of annual groundwater withdrawal that does not exceed the rate of annual aquifer recharge. One immediate problem rising from such a simplistic approach is the question of long-term climatic fluctuations and the availability of historic groundwater monitoring information to assess the long-term aquifer recharge and depletion of the hydraulic heads due to historic withdrawals (see Figure 2.116).

The introduction of aquifer sustainability, as viewed by many, is an attempt to address both the environmental and the socioeconomic aspects of the overall issue of aquifer management (see also Section 2.7). Periods of prolonged droughts, such as several years of annual precipitation less than a long-term average, usually bring focus to the overall issue of available groundwater reserves within an aquifer. This focus often results in confrontations between various stakeholders such as regulatory agencies, current users (agriculture, public water supply, industry), and environmental groups trying to protect the less powerful ones (flora and fauna in surface streams, wetlands, riparian and spring habitats). Mega groundwater supply projects may also bring issues of transboundary cooperation in management of major world aquifers by "host" countries.

Aquifer mining is a term often used to describe withdrawal of groundwater from storage that would not be replenished by natural recharge within a reasonable timeframe. This approach

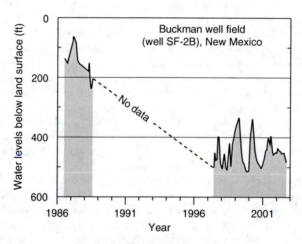

FIGURE 2.116 A hydrograph showing groundwater level declines in the Buckman well field, which supplies water for Santa Fe, New Mexico, the United States. No measurements were made between August 1988 and June 1997, during which time water levels declined nearly 300 ft (100 m), emphasizing the importance of continual monitoring. Long-term data that document the evolving response of aquifers to groundwater development are particularly important for calibrating groundwater flow models used to forecast future conditions. (From USGS (United States Geological Survey), Groundwater depletion across the Nation. USGS Fact Sheet 103–03, U.S. Department of the Interior, U.S. Geological Survey, 2003, 4pp.)

to "aquifer management" did not require much justification in the past since the agricultural, industrial, and urban development needs had an unquestionable priority. In rare cases when most stakeholders were aware of aquifer mining, decisions to continue with it were based on vague future expectations. For example, Los Angeles, California, the United States, relied on groundwater in storage even though the supply was being depleted because of the expectation that imported water eventually would take place of water used from storage. Thus, when talking about sustainability, it may be necessary to stipulate the period over which the use is planned and any assumptions about future sources of water supply (Hiscock et al., 2002, from Alley and Leake, 2004). In that respect, the following quote from Slichter (1902) shows a great foresight:

> The fundamental disadvantage in the utilization of underground sources of water is the danger of overdrawing the natural supply. In regions in which the rainfall is light and catchment areas are small, as in parts of southern California, it is easy to extend development of underground sources so as to greatly exceed the natural rate of annual replenishment. In this way underground reservoirs are depleted which have been ages in filling; principal as well as interest is drawn upon, and much disappointment must inevitably follow.

As already mentioned, safe yield is usually defined in terms of annual water withdrawal, which is balanced by the annual recharge. The temporal patterns (short-term or long-term) are more open-ended in definitions of aquifer sustainability (Alley and Leake, 2004). Whatever the case-specific definition of aquifer safe yield and sustainability may be, it is safe to say that both terms carry a great amount of ambiguity and potential conflict between various stakeholders. The role of hydrogeologists is therefore a crucial one in presenting all consequences of aquifer development for any number of possible scenarios. Use of properly developed and transparent groundwater models, including clear presentation of all model assumptions and limitations, should be an integral part of such hydrogeologic studies.

2.7 MANAGED AQUIFER RECHARGE (MAR) AND AQUIFER STORAGE AND RECOVERY (ASR)

The following general discussion on sustainability serves as a good introduction to the issue of aquifer management. The WCED (1987) introduced the sustainability concept as the "development which meets the needs of the present without compromising the ability of future generations to meet their own needs." A more specific definition has been adopted by the WCED (1987): "Sustainable development is a process of change in which the exploitation of resources, the direction of investments, the orientation of technological development, and institutional change are all in harmony and enhance both current and future potential to meet human needs and aspirations." Food and Agricultural Organization (FAO), an agency of the United Nations, revised concepts proposed by many authors and formulated its own definition focusing on agriculture, forestry, and fisheries:

> Sustainable development is the management and conservation of the natural resource base and the orientation of technological and institutional change in such a manner as to ensure the attainment and continued satisfaction of human needs for the present and future generations. Such sustainable development (in the agriculture, forestry and fisheries sectors) conserves land, water, plant and animal genetic resources, is environmentally non-degrading, technically appropriate, economically viable and socially acceptable.

This definition fully agrees with that of WCED (1987), uses the same conceptual components, and introduces the biodiversity implications (Pereira et al., 2002).

This discussion by Meinzer (1932) shows that early considerations and implementations of artificial aquifer recharge in the United States were made as early as at the beginning of the twentieth century:

Artificial recharge can be accomplished in some places by draining surface water into wells, spreading it over tracts underlain by permeable material, temporarily storing it in leaky reservoirs from which it may percolate to the water table, or storing it in relatively tight reservoirs from which it is released as fast as it can seep into the stream bed below the reservoir. Artificial recharge by some of these methods has been practiced in the United States and other countries. It was suggested by Hilgard in 1902 for southern California, where it has since received considerable investigation and has been adopted as a conservation measure. Drainage into wells has been practiced in many parts of the United States, chiefly to reclaim swampy land or to dispose of sewage and other wastes. The drainage of sewage or other wastes into wells cannot be approved because it may produce dangerous pollution of water supplies. Drainage of surface water into wells to increase the ground-water supply for rice irrigation in Arkansas is now under consideration. Water spreading has been practiced to a considerable extent in southern California partly to decrease the effects of flood but largely to increase the supply of ground water. Storage in ordinary reservoirs and subsequent release has frequently been considered and the unavoidable leakage of some reservoirs has been used to increase the ground-water supply. Artificial recharge by damming stream channels in the permeable lava rocks of the Hawaiian Islands has been considered. In ground-water investigations that involve the question of safe yield attention should as a matter of course be given to the possibilities of artificial recharge.

Artificial aquifer recharge is now a rapidly expanding hydrogeologic and engineering solution to the question of sustainability of groundwater resources. Two terms have gained popularity in describing the concept: MAR as a more general description for a variety of engineering solutions, and aquifer and storage recovery (ASR), which describes injection and extraction of potable water with dual-purpose wells. Artificial aquifer recharge should not be confused with induced aquifer recharge, which is a response of the surface water system to groundwater withdrawal as illustrated in Figure 2.117. Sometimes the induced recharge is referred to as indirect artificial aquifer recharge. Many groundwater supply systems in alluvial aquifers near large streams are intentionally designed to induce aquifer recharge for two main reasons: (1) increased capacity, and (2) filtration of the river water en-route to the supply well. Such systems, however, are vulnerable to accidental or existing contamination of the surface water and in many cases require additional treatment of the extracted groundwater. In addition, excessive lowering of the water table in the river floodplain may adversely impact riparian vegetation.

The important factors that have to be considered for any scheme of artificial aquifer recharge are:

- Regulatory requirements.
- The availability of an adequate source of recharge water of suitable chemical and physical quality.
- Geochemical compatibility between the recharge water and the existing groundwater (e.g., possible carbonate precipitation, iron hydroxide formation, mobilization of trace elements).
- The hydrogeologic properties of the porous media (soil and aquifer) must facilitate desirable infiltration rates and allow direct aquifer recharge. For example, existence of extensive low-permeable clays in the unsaturated (vadose) zone may exclude a potential recharge site from future consideration.

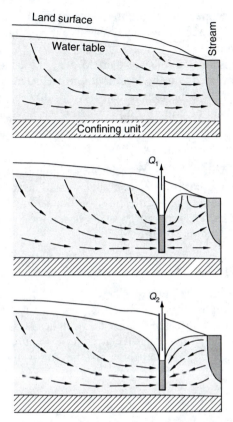

FIGURE 2.117 Induced aquifer recharge due to groundwater withdrawal near a surface water body. (Modified from Alley, W.M., Reilly, T.E., and Lehn Franke, O., Sustainability of ground-water resources. U.S. Geological Survey Circular 1186, Denver, CO, 1999, 79pp.)

- The water-bearing deposits must be able to store the recharged water in a reasonable amount of time, and allow its lateral movement toward extraction locations at acceptable rates. In other words, the specific yield (storage) and the hydraulic conductivity of the aquifer porous media must be adequate.
- Presence of fine-grained sediments may have an advantage of improving the quality of recharged water due to their high filtration and sorption capacities. Other geochemical reactions in the vadose zone below recharge facilities may also influence water quality.
- Engineering solution should be designed to facilitate efficient recharge when there is available surplus of water, and efficient recovery when the water is most needed.
- The proposed solution must be cost-efficient, environmentally sound, and competitive to other water-resource development options.

Aquifers that can store large quantities of water but do not transmit them away quickly are best suited for artificial recharge. For example, karst aquifers may accept large quantities of recharged water but tend to transmit them very quickly away from the system, significantly reducing its efficiency. Alluvial aquifers are usually the most suitable, because of the generally shallow water table and vicinity to source water (surface stream). Sandstone aquifers are in many cases very good candidates due to their high storage capacity and moderate hydraulic conductivity. Aquifers developed in carbonate rocks of high matrix porosity, even when karstified, can be successfully used for artificial recharge if hydrogeologic conditions regarding retention of groundwater are favorable.

Many artificial recharge projects in the world are located in coastal settings, mainly because these areas are heavily populated and the local aquifers are extensively exploited which may cause saltwater intrusion. These projects often serve double purpose of augmenting the reserves of groundwater and protecting the aquifer from seawater intrusion.

2.7.1 METHODS OF ARTIFICIAL AQUIFER RECHARGE

Three general methods of artificial aquifer recharge include: (1) spreading water over land surface, (2) delivering it to the unsaturated zone below land surface, and (3) injecting water directly into the aquifer. A variety of engineering solutions and combinations of these three general methods have been used to accomplish a simple common goal—to deliver more water to the aquifer (see Figure 2.118 and Figure 2.119).

2.7.1.1 Spreading Structures

Artificial recharge by spreading is applied when the aquifer extends close to the ground surface. Recharge is accomplished by spreading water over the ground surface or by conveying the raw water to infiltration basins and ditches. Use of spreading grounds is the most common method for artificial recharge (see Figure 2.120). The operational efficiency of the spreading grounds depends on the following factors (modified from Pereira et al., 2002):

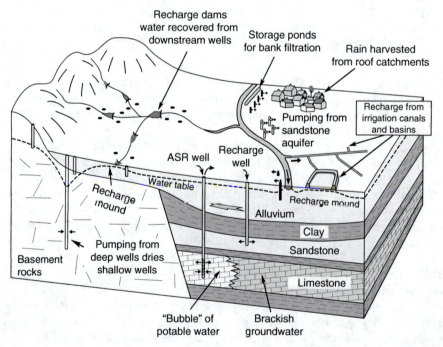

FIGURE 2.118 Possible options for managed aquifer recharge or MAR. (Modified from IAH (International Association of Hydrogeologists), Managing aquifer recharge. Commission on Management of Aquifer Recharge, IAH-MAR, 2002, 12pp.)

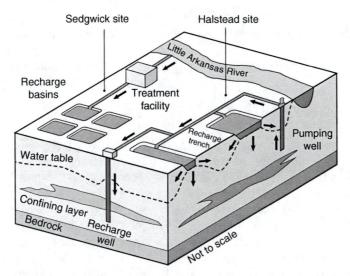

FIGURE 2.119 At the Halstead recharge demonstration site, the aquifer is artificially recharged by water pumped from a diversion well immediately adjacent to the Little Arkansas River at Wichita, Kansas, the United States. The diverted water is piped about 3 mi to the recharge site and recharged to the aquifer through direct-well injection, surface-spreading basins, and a recharge trench. At the Sedgwick recharge demonstration site water comes from a surface water intake in the Little Arkansas River. The surface water is treated with powdered activated carbon and polymers to remove organic contaminants and sediments. The treated water is then piped about 3 mi to the Sedgwick site and recharged through surface-spreading basins. (Modified from Galloway, D.L. et al., Evolving issues and practices in managing ground-water resources: case studies on the role of science. U.S. Geological Survey Circular 1247, Reston, VA, 2003, 73pp.)

• Presence of sufficiently pervious layers between the ground surface and the water table (aquifer)
• Enough thickness and storage capacity of the unsaturated layers above the water table

FIGURE 2.120 Photograph of the Rio Hondo Coastal Basin Spreading Grounds in the Montebello Forebay, looking south toward the Pacific Ocean. (From Anders, R. and Schroeder, R.A., Use of water-quality indicators and environmental tracers to determine the fate and transport of recycled water in Los Angeles County, California. U.S. Geological Survey Water-Resources Investigations, Report 03-4279, Sacramento, CA, 2003, 104pp.)

- Appropriate transmissivity of the aquifer horizons
- Surface water without excessive particulate matter (low turbidity) to avoid clogging

The quantity of water that can enter the aquifer from spreading grounds depends on three basic factors:

- The infiltration rate at which the water enters the subsurface
- The percolation rate, i.e., the rate at which water can move downward through the unsaturated zone until it reaches the water table (saturated zone)
- The capacity for horizontal movement of water in the aquifer, which depends on the hydraulic conductivity and thickness of the saturated zone

The infiltration rate tends to reduce over time due to the clogging of soil pores by sediments carried in the raw water, growth of algae, colloidal swelling, soil dispersion, and microbial activity. The infiltration rate may recover when adopting alternate wet and dry periods of spreading and by scraping away the clogged surface layer, among other techniques. A spreading basin is constructed with a flat bottom that is covered evenly by shallow water. This requires the availability of large surfaces of land for meaningful size recharge works. Several basins may be arranged in line so that excess water runs between the basins. Retaining basins may be used for settlement of suspended sediments before water enters the spreading basins. The settling of sediments may also be assisted with the addition of coagulation agents. Variations of the spreading ground technique consist of the use of ditches that are often easier to handle and for which clogging is a lesser problem since a major part of the sediment is carried out of the ditches by the slow-flowing water (Pereira et al., 2002).

2.7.1.2   Dams

Widely applied artificial recharge structures are floodwater retention dams, the purpose of which is to delay the runoff of water in surface streams and provide the time needed for recharge into the local aquifer. The structures usually consist of low dams, including earth walls and gabions built to be toppled by floods. For example, a number of such dams have been built in the Edwards aquifer recharge zone (see Figure 2.5). More permanent structures (real recharge dams) maintain water reservoirs located at sites overlying aquifers. Permanent dams are equipped with proper spillways and could be used for artificial groundwater recharge at the site and to release water through outlets toward spreading grounds or other types of recharge schemes. In these cases, the recharge dams act to facilitate both direct recharge and sediment retention, providing clear water for recharge purposes (Pereira et al., 2002).

2.7.1.3   Wells and Trenches in Vadose Zone

Vadose-zone wells (also called recharge shafts or dry wells) are normally installed with a bucket auger, and they are about 1 m in diameter and as much as 60 m deep. The wells are also backfilled with coarse sand or fine gravel. Water is normally applied through a perforated or screened pipe in the center. Free-falling water in this pipe should be avoided to prevent air entrainment in the water and formation of entrapped air in the backfill and the soil around the vadose-zone well. To do this, water is supplied through a smaller pipe inside the screened or perforated pipe that extends to a safe distance below the water level in the well. Pipes with various diameters also can be installed. Water is then applied through the pipe that gives sufficient head loss to avoid free-falling water. Also, a special orifice type of valve can be placed at the bottom of the supply pipe that can be adjusted to restrict the flow enough to avoid free-falling water (Bouwer, 2002).

The main advantage of recharge trenches or wells in the vadose zone is that they are relatively inexpensive. The disadvantage is that eventually they clog up at their infiltrating surface because of accumulation of suspended soils or biomass. Because they are in the vadose zone, they cannot be pumped for backwashing the clogging layer, or redeveloped or cleaned to restore infiltration rates. To minimize clogging, the water should be pretreated to remove suspended solids. For recharge trenches, pretreatment is accomplished in the trench itself by placing a sand filter with possibly a geotextile filter fabric on top of the backfill. Where this would reduce the flow into the backfill too much, the recharge trench could be widened at the top to create a T-trench with a larger filter area than the surface area of the trench itself. Economically, the choice is between pretreatment to extend the useful life of the trench or vadose-zone well, and constructing new ones (Bouwer, 2002).

2.7.1.4 Injection Wells

Clear advantage of injection wells is that they can be used to recharge any type of aquifer, at any depth, thus eliminating problems associated with low-permeable surficial soils and low-permeable layers in general. Since the water is directly injected into the aquifer, it should be of high quality to minimize any potential groundwater contamination. Even if such high quality water is used, trouble-free operations cannot be guaranteed for longer periods of well operation due to various geochemical and mechanical processes, which tend to clog well screens and reduce permeability of well gravel pack and adjacent aquifer material. A common practice is to inject water through dual-purpose wells, which are used to occasionally pump water back from the aquifer thus removing screen-clogging materials (backwashing).

2.7.1.5 Aquifer Storage and Recovery

ASR has been widely applied in coastal areas of the eastern United States, particularly in Florida, where population growth and demand for water resources are of great concern. Figure 2.121 illustrates the basic principle of ASR: potable water is injected (stored) into portion of aquifer with brackish, nonpotable water during periods of low water demand or

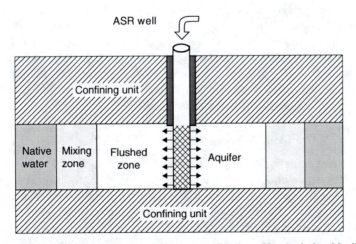

FIGURE 2.121 Aquifer storage and recovery well in a confined aquifer depicting idealized flushed and mixing (transition) zones created by recharge. Flushed zone contains mostly recharged water. (From Reese, R.S., Inventory and review of aquifer storage and recovery in southern Florida. U.S. Geological Survey Water-Resources Investigations Report 02-4036, Tallahassee, FL, 2002, 56pp.)

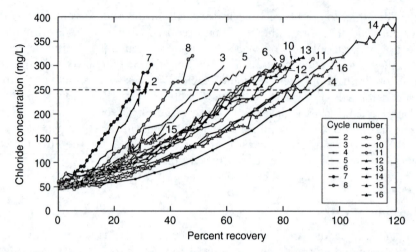

FIGURE 2.122 Percent recovery of recharged water during operational cycles in relation to chloride concentration of recovered water at the Boynton Beach East Water Treatment Plant site in Palm Beach County. Recovery for cycle 14 was continued until reaching a chloride concentration of about 1000 mg/L. (From Reese, R.S., Inventory and review of aquifer storage and recovery in southern Florida. U.S. Geological Survey Water-Resources Investigations Report 02-4036, Tallahassee, FL, 2002, 56pp.)

high water availability, and then extracted (recovered) during periods of increased demand. The success of such systems is measured by recovery efficiency, expressed as percentage of the potable water meeting a preset criterion (e.g., chloride concentration less than 250 mg/L) that can be recovered relative to the quantity injected, over one full cycle that includes injection, storage, and recovery. Recovery efficiency in most cases increases with the number of full cycles as more of the injected potable water remains in the aquifer after each cycle (Figure 2.122). Recovery of the stored water is dependent on the effective emplacement of a relatively stable, thick lens of low-density recharge water during the injection phase. To form this lens, enough water must be injected to displace a large volume of saline water, the mixing of the injected and native waters must not be significant, and confinement must be sufficiently tight to prevent rapid vertical migration of the less dense recharged water (Rosenshein and Hickey, 1977, from Yobbi, 1997). Among various problems associated with ASR operation is formation of a thin lens of fresh water floating on the denser native (brackish/saline) water indicating that the displacement was not successful. Figure 2.123 shows formation of such lens in the first screened interval of a monitoring well. Alternatively, the logs may indicate presence of a more transmissive aquifer zone in that interval, which would be another problem for a more effective operation of the system.

2.7.1.6   Hydraulic Influence of Recharge

The main quantitative analysis for all recharge structures consists of two basic questions: (1) what is the infiltration rate and its possible reduction in time, and (2) what are the geometric characteristics of the hydraulic head mound created by recharge (vertical and horizontal extent of the mound, altered directions of groundwater flow, travel times of recharged water). These calculations are most efficiently performed with a combination of unsaturated and saturated zone models (Figure 2.124 through Figure 2.126) due to the complexity of various flow components and hydrogeologic parameters involved (e.g., unsaturated and saturated vertical and horizontal hydraulic conductivities, soil-moisture distribution, changing aquifer transmissivity, and varying boundary conditions during cycling recharge

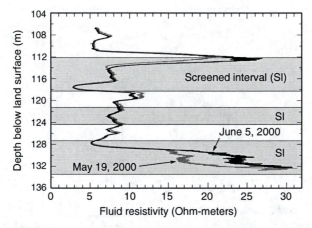

FIGURE 2.123 Formation resistivity profiles collected in observation well CHN-809 during the injection phase of cycle 2 of the aquifer storage and recovery study in Charleston, South Carolina. (Modified from Petkewich, M.D. et al., Hydrologic and geochemical evaluation of aquifer storage recovery in the Santee Limestone/Black Mingo aquifer, Charleston, South Carolina, 1998–2002. U.S. Geological Survey Scientific Investigations Report 2004-5046, Reston, VA, 2004, 81pp.)

operations). Alternatively, and usually for a screening level analysis, various analytical equations can be used to estimate the influence of recharge on groundwater levels. For injection wells, the methodology is exactly the same as when determining the drawdown and the radius of well influence from extraction wells (Section 2.5), except that the pumping rate is negative and the drawdown is "negative" (rise in the hydraulic head).

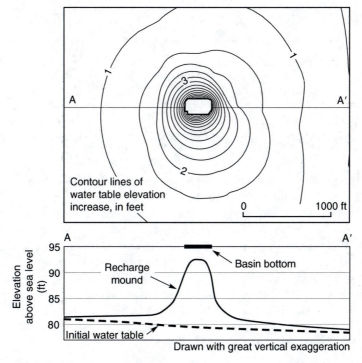

FIGURE 2.124 Modeled recharge mound below a groundwater recharge basin (drawn with great vertical exaggeration).

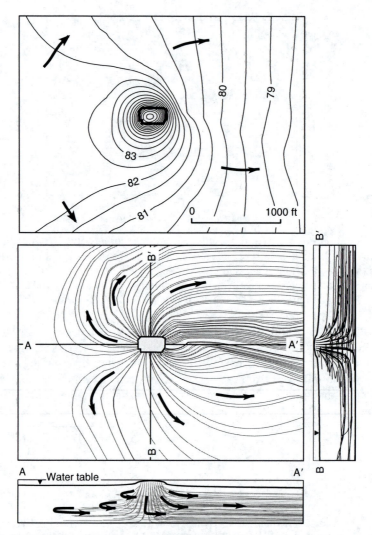

FIGURE 2.125 Hydraulic head contour map (top) and flow lines (bottom) for the recharge basin in Figure 2.124. The flow lines are also shown in two cross sections through the basin: AA' and BB'.

Bouwer (2002) presents a set of simple equations, including default values for key parameters of different soil types, that can be used to estimate infiltration rates at recharge basins (including in case of soil clogging), recharge at vadose-zone wells, and effects of recharge on groundwater levels. Other papers have been published on the rise of a groundwater mound in the aquifer in response to infiltration from a recharge system, and some also on the decline of the mound after infiltration has stopped (Glover, 1964; Hantush 1967; Marino 1975a,b; Warner et al., 1989). The usual limiting assumption for the application of such equations is a uniform isotropic aquifer of infinite extent with no other recharge or discharge.

2.7.1.7 Source Water Treatment before Recharge

In the United States, the water used for well injection is usually treated to meet drinking-water quality standards for two reasons. One is to minimize clogging of the well–aquifer

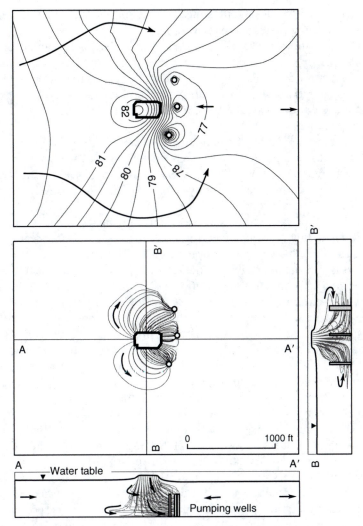

FIGURE 2.126 Hydraulic head contour map (top) and flow lines (bottom) for a scenario when three extraction wells are operating downgradient from the recharge basin.

interface, and the other is to protect the quality of the water in the aquifer, especially where it is pumped by other wells in the aquifer for potable uses. Direct injection into the saturated zone does not have benefits of fine-textured unsaturated soils below surface and vadose-zone infiltration systems, which often improve quality of the recharged water. Thus, whereas secondary sewage effluent can readily be used in surface infiltration systems for soil-aquifer treatment and eventual potable reuse, effluent for well injection should at least receive tertiary treatment (sand filtration and chlorination). This treatment removes remaining suspended solids and some microorganisms like *Giardia* (protozoa), and *Cryptosporidium* and parasites, like helminth eggs, by filtration. Bacteria and viruses are removed by chlorination, ultraviolet irradiation, or other disinfection (Bouwer, 2002). Even with an extensive treatment, recharged water may cause well clogging due to growth of native bacteria (injected water usually have higher oxygen content than the native water and acts as stimulant for bacterial growth), as well as other interactions between the injected and native water and porous media. Injection wells therefore require continuous maintenance and rehabilitation, including frequent pumping for removal of clogging materials (backwashing).

Where groundwater is not used for drinking, water of lower quality can be injected into the aquifer. For example, in Australia stormwater runoff and treated municipal wastewater effluent are injected into brackish aquifers to produce water for irrigation by pumping from the same wells. Clogging is alleviated by a combination of low-cost water treatment and well redevelopment, and groundwater quality is protected for its declared beneficial uses (Dillon and Pavelic 1996; Dillon et al., 1997).

Turbidity of source water is often the main quality problem for surface infiltration facilities such as basins, trenches, and drains. Fine suspended solids tend to clog contact surfaces and have to be regularly removed after infiltration basins and trenches are drained (infiltration drains with this problem cannot be efficiently rehabilitated). For this reason, large recharge facilities often include separate settling basins.

Although the degree of raw water (surface water and wastewater) treatment before recharge may vary depending on the engineering solutions and applicable regulations, various stakeholders would likely have opposing opinions since the body of scientific evidence and accepted practices is still not definitive, particularly regarding the disinfection requirements. The issue of emerging contaminants (see Section 5.4), which are not regulated, and are often not removed by conventional drinking water or wastewater treatment technologies, is another major point of disagreement between advocates and opponents of artificial aquifer recharge. These contaminants include recalcitrant organic chemicals such as 1,4-dioxane, various pharmaceuticals, endocrine disruptors (hormones) and disinfection byproducts formed during drinking water and injection water treatment such as trihalomethanes (THMs), chloroform, and total organic halide (TOX). Some of the disinfection byproduct are carcinogenic and may be persistent for long periods in certain aquifer environments (Rostad, 2002). Anders and Schroeder (2003) describe in detail a methodology used to investigate: (1) the fate and transport of wastewater constituents as they travel from the point of recharge to points of withdrawal and (2) the long-term effects that artificial recharge using tertiary-treated municipal wastewater has on the quality of the groundwater in the Central Basin in Los Angeles County (see Figure 2.120).

Several issues regarding quality of recovered water and geochemical reactions between the injected and native aquifer water and porous media have emerged during the last several years after more monitoring data became available at various ASR sites in Florida (Arthur et al., 2002). The current focus is primarily on mobilization of arsenic, uranium, and other trace elements at some locations within the Floridan aquifer due to introduction of imported water with higher dissolved oxygen content than native water. Once these waters are introduced into a reduced aquifer, selective leaching or mineral dissolution may release metals into the injected water (Figure 2.127). This example amplifies the requirement that feasibility studies, design, construction, and operation of ASR facilities, including monitoring well placement and monitoring schedules, should take into account the possibility of water–rock interaction and mobilization of trace elements into recovered waters.

Large-scale MAR and ASR projects are currently implemented or considered by major water users and government agencies throughout the United States and the world (see Figure 2.128). Although still challenging in many ways, the benefits they offer for sustainable management of water resources clearly makes MAR and ASR projects the trend of the future. More detail information on artificial aquifer recharge can be found in National Research Council (1994), Pyne (1995), ASCE (2001), and Bouwer (2002). International Association of Hydrogeologists (IAH) maintains a MAR-dedicated page on its Web site, including various useful links to related information.

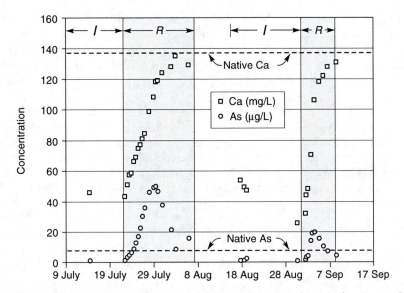

FIGURE 2.127 Changes in arsenic (As) and calcium (Ca) concentrations during Punta Gorda ASR project cycle tests. During recovery, mixing of low-Ca injected waters and higher-Ca native groundwater is observed. Arsenic concentrations in recharge and native groundwaters are less than 10 μg/L. An As peak (up to 50 μg/L) is observed indicating water–rock interaction. (From Arthur, J.D., Dabous, A.A., and Cowart, J.B. In: Aiken, G.R. and Kuniansky, E.L. (Eds.), *U.S. Geological Survey Artificial Recharge Workshop Proceedings*, April 2–4, 2002, Sacramento, CA, U.S. Geological Survey Open-File Report 02-89, 2002, pp. 47–50.)

2.8   SPRINGS AND SPRING DEVELOPMENT

Spring is a location at the land surface where groundwater discharges from the aquifer, creating a visible flow. When the flow is not visible, but the land surface is wet compared to the surrounding area, such discharge of groundwater to the surface is called seep. Seepage

FIGURE 2.128 This typical artificial recharge well in Las Vegas, Nevada, has the dual function of pumping water from the aquifer for public supply and injecting water into the aquifer for storage. The tall object on the far right is the electric motor for the pump. (Courtesy of Devin Galloway, USGS.)

spring is a term often used to indicate discharge of water through numerous small intergranular openings of unconsolidated sediments (e.g., sand and gravel). They are usually marked by abundant vegetation and commonly occur where valleys are cut downward into the zone of saturation of a uniform water-bearing deposit. Fracture (or fissure) spring refers to discharge of water along bedding planes, joints, cleavage, faults, and other breaks in the consolidated (hard) rock (Figure 2.129).

Geysers are those springs in which at more or less regular intervals the water is ejected with some force. The waters are always warm or hot and, therefore, come from considerable depth. Geyser springs generally emerge from tubular conduits that are lined with silica, deposited by the water, and end at the surface in a cone of similar material. The ejection of the water is probably due to the expelling force of steam generated deep below the surface under certain peculiar conditions. (Fuller, 1905).

Meinzer's description of springs along Snake River in Idaho, the United States, is a textbook example of all key elements of spring characterization (Meinzer, 1927. Large springs in the United States, pp. 42–50). The following short excerpts from the Meinzer's work illustrate these elements (italicized):

Spring size

Many large springs issue on the north side of Snake River between Milner and King Hill, Idaho, nearly all of them in the canyon below Shoshone Falls or in short tributary canyons (fig. 12). According to the measurements that have been made the total discharge of these springs was 3,885 second-feet in 1902, before any irrigation developments had been made on the north side, and averaged 5,085 second-feet in 1918, after the north-side irrigation project had been developed. The great volume of water discharged by these springs can perhaps be better appreciated by recalling that in 1916 the aggregate consumption of New York, Chicago, Philadelphia, Cleveland, Boston, and St. Louis, with more than 12 million inhabitants, averaged only 1,769 million gallons a day (2,737 second-feet), or only slightly more than one-half of the yield of these Snake River springs in 1918. In fact, these springs yield enough water to supply all the cities in the United States of more than 100,000 inhabitants with 120 gallons a day for each inhabitant. As shown in Figure 12 and in the following table, there are 11 springs or groups of springs that yield more than 100 second-feet, of which 1 yields more than 1,000 second-feet, 3 yield between 500 and 1,000 second-feet, and 7 yield between 100 and 500 second-feet. Moreover here are 5 springs that yield between 50 and 100 second-feet and numerous so-called "small" springs which would be considered huge in most localities.

Fluctuation in discharge

The flow of these springs is relatively constant, and in this respect they differ notably from most of the large limestone springs.

Role of geology

These springs issue chiefly from volcanic rocks or closely related deposits. The water-bearing volcanic rocks are largely basalt, but they also include jointed obsidian and rhyolite. A large part of the basin of Snake River above King Hill, Idaho, was inundated with basaltic lava during the Tertiary and Quaternary periods, and the lava rock is in many parts so broken or vesicular that it absorbs and transmits water very freely. " ... " At the Thousand Springs the water can be seen gushing from innumerable openings in the exposed edge of a scoriaceous zone below a more compact sheet of lava rock. At Sand, Box Canyon, and Blind Canyon Springs the water, according to Russell, comes from a stratum consisting largely of white sand, which is overlain by a thick sheet of lava. At most of the springs there is so much talus that the true source of the water can not be observed, but it probably issues chiefly from the large openings in scoriaceous or

FIGURE 2.129 Frozen discharge of water from fractures in sandstone near Sedona, Arizona, the United States.

shattered basalt where the basalt overlies more dense rocks. The fact that most of the springs are confined to rather definite localities and issue at points far above the river indicates that the flow of the ground water to the springs is governed by definite rock structure. The great body of groundwater is obviously held up in the very permeable water-bearing rocks by underlying impermeable formations. It may be that the underlying surface which holds up the water is a former land surface and that the principal subterranean streams which supply the springs follow down the valleys of this ancient surface. ... At most of the springs the water issues at considerable heights above river level.

Drainage area, recharge, source of water

The lava plain lying north and northeast of these large springs extends over a few thousands of square miles and receives the drainage of a few thousand square miles of bordering mountainous country. The great capacity of the broken lava rock to take in surface water is well established and is, moreover, shown by the fact that in the entire stretch of more than 250 miles from the head of Henrys Fork of Snake River to the mouth of Malade River no surface stream of any consequence enters Henrys Fork or the main river from the north. The greater part of this vast lava plain discharges no surface water into the Snake, and a number of rather large streams that drain the mountain area to the north lose themselves on this lava plain. A part of the water that falls on the plain and adjoining mountains is lost by evaporation and transpiration, but a large part percolates into the lava rock and thence to the large springs.

Water quality

The water of these springs does not contain much mineral matter. It is generally very clear, although the water of some of the springs, such as the Blue Lakes, has a beautiful blue color and a slight opalescence due to minute particles in suspension. So far as is known, all the springs have about normal temperatures.

Conflict between utilization and preservation:

On account of the notable height above the river at which most of these springs issue, together with the great volume of water which they discharge, they are capable of developing a great amount of water power. Large power plants have already been installed at Malade Springs and at Thousand and Sand Springs, and other large plants could be installed at other springs, especially at Clear Lakes, Box Canyon, and Bickel Springs. ... At the Crystal Springs the clear, cold water is utilized in a fish hatchery, where great quantities of trout are raised. ... The most spectacular feature of these springs is the cataracts which they form, or which they formed before they were harnessed to develop electric energy (pls. 9 and 10). Thousand Springs formerly gave rise to a strikingly beautiful waterfall 2000 feet long and 195 feet high. Snowball Spring, which discharges 150 to 160 second-feet, is at the east end of the Thousand Springs and is included with them in Figure 12. Formerly its water dashed over the rough talus slope, forming a cataract of great beauty that suggested a snowbank. The Niagara Springs, which issue from the canyon wall 125 feet above the river level, also form a spectacular cataract.

2.8.1 SPRING CLASSIFICATION

There have been various proposed classifications of springs, based on different character-istics, of which the most common are:

- Discharge rate and uniformity
- Character of the hydraulic head (pressure) creating the discharge
- Geologic structure controlling the discharge
- Water quality and temperature

Meinzer's classification of springs based on average discharge expressed in the U.S. units is still widely used in the United States (Table 2.7). The same table also includes Meinzer's classification based on the metric system. However, the classification based solely on average spring discharge, without specifying other discharge parameters, is not very useful when evaluating the potential for spring utilization. For example, a spring may have a very high average discharge but it may be dry or just trickling most of the year (see Figure 2.97 and Figure 2.130). Practice in most countries is that the springs are evaluated based on the minimum discharge recorded over a long period, typically longer than several hydrologic years (hydrologic year is defined as spanning all wet and dry seasons within a full annual cycle). When evaluating availability of spring water, it is important to include a measure of spring discharge variability, which should also be based on periods of record longer than one hydrologic year. The simplest measure of variability is the ratio of the maximum and minimum discharge:

TABLE 2.7
Classification of Springs Based on Average Discharge Rate

(a) Metric System

Magnitude	Discharge in Metric Units	Discharge in English Units (Approximate)
First	10 m^3/s or more	353 second-feet
Second	1 to 10 m^3/s	35 to 353 second-feet
Third	0.1 to 1 m^3/s	3.5 to 35 second feet
Fourth	10 to 100 L/s	158 gpm to 3.5 second-feet
Fifth	1 to 10 L/s	16 to 158 gpm
Sixth	0.1 to 1 L/s	1.6 to 16 gpm
Seventh	10 to 100 cm^3/s	1.25 pints to 1.6 gpm
Eight	Less than 10 cm^3/s	Less than 1.25 pints per minute

(b) U.S. Units

Magnitude	Discharge
First	100 second-feet (ft^3/s) or more
Second	10 to 100 second-feet
Third	1 to 10 second-feet
Fourth	100 gpm to 1 second-foot
Fifth	10 to 100 gpm
Sixth	1 to 10 gpm
Seventh	1 pint to 1 gpm
Eight	less than 1 pint per minute

Source: From Meinzer, O.E. The occurrence of ground water in the United States with a discussion of principles. U.S. Geological Survey Water-Supply Paper 489, Washington, D.C., 1923a, 321pp.

FIGURE 2.130 Large karstic spring in Montenegro discharging more than 250 m^3/s in September 2005, close to its maximum recorded discharge rate. The minimum discharge rate of this spring is less than 10 L/s, or 25 thousand times less than the maximum.

$$I_v = \frac{Q_{max}}{Q_{min}} \qquad (2.169)$$

Springs with the index of variability (I_v) greater than 10 are considered highly variable, and those with $I_v < 2$ are sometimes called constant or steady springs. Meinzer proposed the following measure of variability expressed in percentage:

$$V = \frac{Q_{max} - Q_{min}}{Q_{av}} \times 100 \ (\%) \qquad (2.170)$$

where Q_{max}, Q_{min}, and Q_{av} are maximum, minimum, and average discharge, respectively. Based on this equation, a constant spring would have variability less than 25%, and a variable spring would have variability greater than 100%. Intermittent spring discharges only for a period of time, whereas at other times it is dry, reflecting directly the aquifer recharge pattern. Ebb-and-flow springs, or periodic springs, are usually found in limestone (karst) terrain and are explained by existence of a siphon that fills up and empties out with certain regularity, irrespective of the recharge (rainfall) pattern (Figure 2.131). Periodic springs can be permanent or intermittent. Estavelle has dual function: it acts as a spring during high hydraulic heads in the aquifer, and as a surface water sink during periods when the hydraulic head in the aquifer is lower than in the body of surface water (estavelles are located within or adjacent to surface water features).

When a spring naturally discharges below surface of a water body (lake, river, and sea), it is called subaqueous (or submerged) spring (Figure 2.132 and Figure 2.133). If this water body is sea or ocean, it is also called submarine spring. This type of springs is most common in limestone (karst) terrains.

Meinzer (1940) gives this account of large springs in the United States:

> According to a study completed about 10 years ago, there are in the United States 65 springs of the first magnitude. Of these springs, 38 rise in volcanic rock or in gravel associated with volcanic rock, 24 in limestone, and 3 in sandstone. Of the springs in volcanic rock or associated gravel 16 are in Oregon, 15 in Idaho, and 7 in California. Of the springs in limestone, 9 rise in limestone of Paleozoic age, 8 of them in the Ozark area of Missouri and Arkansas; 4 are in Lower Cretaceous limestone in the Balcones fault belt in Texas; and 11 are in Tertiary limestone in Florida. The 3 springs that issue from sandstone are in Montana. The great discharge of these springs is believed

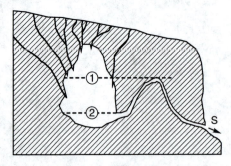

FIGURE 2.131 Schematic of an ebb-and-flow (periodic) spring. When water in the cavity reaches level 1, the siphon becomes active and the spring (S) starts flowing. When the level drops to position 2, the spring stops flowing. (From Radovanovic, S., *Podzemne vode; izdani, izvori, bunari, terme i mineralne vode (Ground Waters; Aquifers, Springs, Wells, Thermal and Mineral Waters; in Serbian)*, Srpska kniževna zadruga, 42, 1897, 152pp.)

FIGURE 2.132 One of the submerged springs discharging into the Adriatic Sea near Brela, Croatia.

to be due to faults or to other special features. With the additional data now available, some revision of these figures could be made but it would be of minor character.

Since this account by Meinzer and the USGS, the numbers have changed due to more precise flow measurements and contributions of other agencies and investigators across the United States. In Florida alone there are 33 documented first-magnitude springs

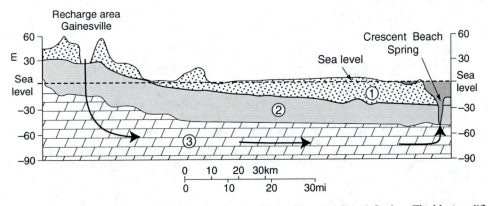

FIGURE 2.133 Idealized cross section of groundwater flow to Crescent Beach Spring, Florida (modified from Barlow, 2003). 1, Post-Miocene deposits (Green clay, sand, and shell); 2, confining unit (Hawthorn Formation); 3, upper Floridan aquifer (Eocene Ocala Limestone). The morphology of the spring vent and discharge characteristics were investigated in detail.

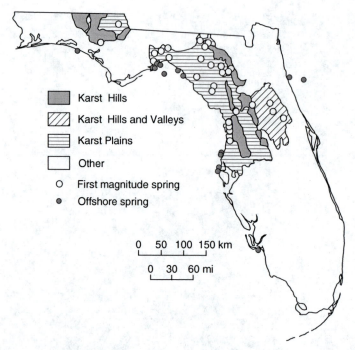

FIGURE 2.134 First magnitude springs and large offshore springs in Florida, the United States, and their relation to main geomorphologic karst types. (Modified from Scott, T.M. et al., *Springs of Florida*. Florida Geological Survey, Bulletin No. 66, Tallahassee, FL, 2004, 658pp.)

(see Figure 2.134) and nearly 700 other significant springs (Scott et al., 2004). Florida represents perhaps the largest concentration of freshwater springs on Earth. Other regions of the world with large springs are also located in karst areas such as the Dinarides (the Balkans), France, Turkey, and China. Arguably, the largest spring in the world is Dumanli spring in Turkey with the average flow of about 50 m³/s, although several springs in the Dinarides, some of which have been artificially submerged by man-made reservoirs, are of similar magnitude or larger.

Analysis of spring discharge hydrographs may reveal useful information regarding the nature of the aquifer system drained by the spring. In many cases, it is also the only available direct quantitative information about the aquifer, which is the main reason why various methods of spring hydrograph analyses have been continuously introduced.

2.8.2 HYDRAULIC HEAD AND GEOLOGIC CONTROL OF SPRING DISCHARGE

Although all springs, except when associated with young volcanism and hydrothermal activity, ultimately discharge at the land surface because of the force of gravity, they are usually divided into two main groups based on the hydraulic head in the underlying aquifer at the point of discharge:

- Gravity springs emerge under unconfined conditions where water table intersects land surface. They are also called descending springs.
- Artesian springs discharge under pressure due to confined conditions in the underlying aquifer, and are also called ascending or rising springs.

FIGURE 2.135 Williford Spring, Florida "has a circular spring pool in a conical depression whose sand bottom is rippled by issuing spring currents. The pool measures 57 ft (17.4 m) in diameter. The vent is under a limestone ledge roughly in the center of the pool, and the depth measured over the vent is 10.1 ft (3.1 m). There is a sizeable boil over the vent, and the color of the water is light blue-green." (Photo courtesy of R. Means, from Scott, T.M. et al., *Springs of Florida*. Florida Geological Survey, Bulletin No. 66, Tallahassee, FL, 2004, 658pp.)

Geomorphology and geologic fabric (rock type and tectonic features such as folds and faults) play the key role in the emergence of springs. When site-specific conditions are rather complicated, springs of formally different type based on some classifications may actually appear next to each other causing confusion. For example, a lateral impermeable barrier in fractured rock, caused by faulting, may force groundwater from greater depth to ascend and discharge at the surface. This water may have high temperature due to normal geothermal gradient in the Earth crust—such springs are called thermal springs. At the same time, groundwater of normal temperature may issue at a spring located very close to the thermal spring. Yet, a third spring may be present with water temperature varying between "hot" and "cold." All three springs are caused by the same lateral contact between the aquifer and the impermeable barrier, and can all be called barrier springs, although the hydraulic mechanism of groundwater discharge is quite different.

Figure 2.136 shows several common spring types. In general, when the contact between the water-bearing porous medium and the impermeable medium is sloping toward the spring, in the direction of groundwater flow, and the aquifer is above the impermeable contact, the spring is called contact spring of descending type (Figure 2.136A). When the impermeable contact slopes are away from the spring, in the direction opposite of the groundwater flow, the spring is called overflowing (Figure 2.136B). When river cuts through alternating layers of permeable and impermeable rocks, contact springs may emerge at various elevations above stream channel (Figure 2.137).

Depression springs are formed in unconfined aquifers when topography intersects the water table, usually due to surface stream incision (Figure 2.136C). Possible contact between the aquifer and the underlying low-permeable formation is not the reason for spring emergence (this contact may or may not be known).

Figure 2.136D through Figure 2.136F show some examples of barrier springs, the term generally referring to springs at steep (vertical) or hanging lateral contacts between the aquifer and the impermeable rock. When such contact forces groundwater to ascend under hydrostatic pressure, i.e., because the hydraulic head in the aquifer is higher than the land surface elevation at the spring location, the spring is called ascending or artesian. Artesian springs are usually caused by tectonic structures (faults, fractures, and folds) and often have steady

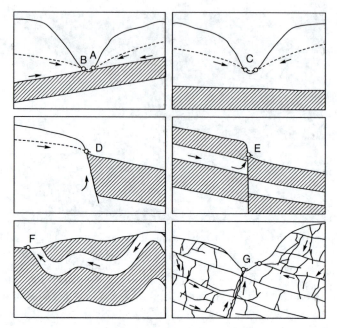

FIGURE 2.136 Different spring types based on the hydraulic head and geologic controls (explanation given in text).

temperatures and discharge, not directly exposed to the atmosphere and recharge from precipitation. Thermal springs are almost always ascending. Figure 2.136G shows springs in fractured-rock aquifers, which can be both descending and ascending.

Secondary springs issue from locations located away from the primary location of spring discharge, which is covered by colluvium or other debris and therefore not visible.

Faults play major role in emergence of springs, especially in fractured rock and karst aquifers. They are also not uncommon in unconsolidated and semiconsolidated sediments.

FIGURE 2.137 View of the Colorado River canyon from Deadhorse Point State Park near Moab, Utah, showing alternating layers of permeable (sandstone, limestone) and impermeable beds. Native Americans in the Southwest made their settlements near contact springs in canyons like this one.

In any case, faults themselves may have one of the following three roles: (1) conduits for groundwater flow, (2) storage of groundwater due to increased porosity within the fault (fault zone), and (3) barriers to groundwater flow due to decrease in porosity within the fault.

Meinzer (1923a): Faults differ greatly in their lateral extent, in the depth to which they reach, and in the amount of displacement. Minute faults do not have much significance with respect to ground water except, as they may, like other fractures, serve as containers of water. But the large faults that can be traced over the surface for many miles, that extend down to great depths below the surface, and that have displacements of hundreds or thousands of feet are very important in their influence on the occurrence and circulation of ground water. Not only do they affect the distribution and position of aquifers, but they may also act as subterranean dams, impounding the ground water, or as conduits that reach into the bowels of the earth and allow the escape to the surface of deep-seated waters, often in large quantities. In some places, instead of a single sharply defined fault, there is a fault zone in which there are numerous small parallel faults or masses of broken rock called fault breccia. Such fault zones may represent a large aggregate displacement and may afford good water passages. (see Figure 2.136G).

The impounding effect of faults is caused by three main mechanisms:

- The displacement of alternating permeable and impermeable beds in such a manner that the impermeable beds are made to abut against the permeable beds, as shown in Figure 2.136D and Figure 2.136E
- Due to clayey gouge along the fault plane produced by the rubbing and mashing during displacement of the rocks, this gouge is smeared over the edges of the permeable beds. The impounding effect of faults is most common in unconsolidated formations that contain considerable clayey material (Figure 2.138)
- Cementation of the pore space by precipitation of material, such as calcium carbonate, from the groundwater circulating through the fault zone (Figure 2.139)
- Rotation of elongated flat clasts parallel to the fault plane so that their new arrangement reduces permeability perpendicular to the fault

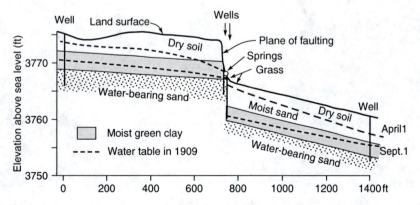

FIGURE 2.138 Section of Owens Valley, California, showing a spring produced by impounding effects of a fault. The "moist sand" on the downthrown side is equivalent to some of the "dry soil" on the upthrown side and apparently has an impounding effect. (After C.H. Lee, from Meinzer, O.E., The occurrence of ground water in the United States with a discussion of principles. U.S. Geological Survey Water-Supply Paper 489, Washington, D.C., 1923a, 321pp.)

FIGURE 2.139 Rock core (6 cm in diameter) from a fault in semiconsolidated Quaternary sediments near Phoenix, Arizona. The impounding effect of the fault is caused by deposition of calcium carbonate from circulating groundwater and rotation of flat elongated clasts. The hydraulic head difference between the two sides of the fault is 35 ft. (Photo courtesy of Jeff Manuszak.)

Mozley et al. (1996) discuss reduction in hydraulic conductivity associated with high-angle normal faults that cut poorly consolidated sediments in the Albuquerque Basin, New Mexico. Such fault zones are commonly cemented by calcite, and their cemented thickness ranges from a few centimeters to several meters, as a function of the sediment grain size on either side of the fault. Cement is typically thickest where the host sediment is coarse grained and thinnest where it is fine grained. In addition, the fault zone is widest where it cuts coarser-grained sediments. Such sediments comprise most of the boring core shown in Figure 2.139. Extensive discussion on deformation mechanisms and hydraulic properties of fault zones in unconsolidated sediments is given in Bense et al. (2003).

Advances in cave diving in last couple of decades have made some important revelations regarding major ascending springs in karst terrains (Touloumdjian, 2005). In most cases, such springs are issuing from deep vertical or subvertical conduits formed at the lateral contact of karstified carbonate (limestone) rocks with noncarbonates or nonkarstified rock (Figure 2.140). They are also called vauclusian springs after the Fontaine de Vaucluse, the source of river Sorgue in Provance, France:

> At the moment the shaft is explored to a depth of 315 m. "..." This exploration was done using a small submarine robot called MODEXA 350. The camera of the robot showed a sandy floor at this dept, leads were not visible. The water table is most time of the year below the rim of the shaft. The Fontaine appears as a very deep and blue lake. Small caves below lead to several springs in the dry bed of the river, just 10 m below the lake. The source is fed by the rainfall on the Plateau de Vaucluse. In spring or sometimes, after enormous rainfall, the water table rises higher than the

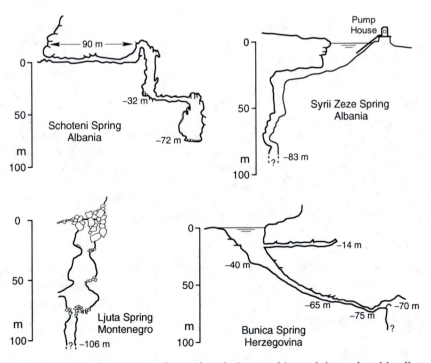

FIGURE 2.140 Examples of large ascending springs in karst, with conduits explored by divers. (From Touloumdjian, C. In: Stevanovic, Z. and Milanovic, P. (Eds.), *Water Resources and Environmental Problems in Karst—Cvijić 2005*, Proceedings of the International Symposium, Institute of Hydrogeology, University of Belgrade, Belgrade, 2005, pp. 443–450.)

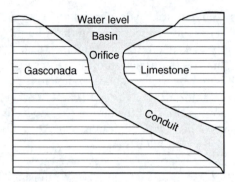

FIGURE 2.141 Diagrammatic section showing the outlet of Bennett Spring, Dallas County, Missouri. (After Shepard, from Meinzer, O.E., Large springs in the United States. U.S. Geological Survey Water-Supply Paper 557, Washington, D.C., 1927, 94pp.)

rim. In these periods the Fontaine de Vaucluse really is a spring, producing more than 200 m^3/sec of water. (Showcaves, 2005).

In many cases, regardless of the size, ascending springs in limestone may issue from pools, which have bottoms covered by sand, gravel, or rock colluvium (fragments), and without visible vertical conduits. However, the conduits or fractures may be masked and bridged over by clastic sediment—a possibility that cannot be excluded. For example, Meinzer (1927) gives the following account of the Bennett Spring in Dallas County, Missouri (see Figure 2.141):

> It issues from a circular basin in gravel about 30 feet in diameter and gives rise to a stream that is about 1$\frac{1}{2}$ miles long and that empties into Niangua River at a level about 22 feet below the spring. The spring furnishes power for a mill in the village of Brice. The temperature of the water is about 58°F. The spring was visited in 1903 by Shepard, who states that at that time it boiled up with great force from a vertical cavelike opening through the limestone into a large oval basin.

The location, discharge rate, and other spring characteristics are very useful indicators of aquifer conditions and its behavior as one system. For example, a study of carbonate-rock aquifers in southern Nevada by Dettinger (1989) shows that within 10 miles of regional springs, aquifers are an average of 25 times more transmissive than they are farther away. These are areas where flow is converging, flow rates are locally high and the conduit-type of flow likely plays a significant, if not predominant, role.

As a conclusion, springs should always be analyzed in as much detail as possible, together with other techniques of aquifer investigations such as drilling and geophysics.

2.8.3 THERMAL AND MINERAL SPRINGS

Thermal springs can be divided into warm springs and hot springs depending on their temperature relative to the human body temperature of 98°F or 37°C: hot springs have a higher and warm springs a lower temperature. Warm spring has temperature higher than the average annual air temperature at the location of discharge.

Meinzer (1940) gives the following illustrative discussion regarding the occurrence and nature of thermal springs:

> An exact statement of the number of thermal springs in the United States is, of course, arbitrary, depending upon the classification of springs that are only slightly warmer than the normal for their localities and upon the groupings of those recognized as thermal springs.

A recently published report lists 1,059 thermal springs or spring localities. Of these 52 are in the East-Central region (46 in the Appalachian Highlands and 6 in the Ouachita area in Arkansas), 3 are in the Great Plains region (in the Black Hills of South Dakota), and all the rest are in the Western Mountain region. The States having the largest number of thermal springs, according to the listing in the report, are Idaho 203, California 184, Nevada 174, Wyoming 116, and Oregon 105. The geyser area of Yellowstone National Park, however, exceeds all others in the abundance of springs of high temperature (29). Indeed, the number of thermal springs in this area might be given as several thousand if the springs were counted individually instead of being grouped.... Nearly two-thirds of the recognized thermal springs issue from igneous rocks—chiefly from the large intrusive masses, such as the great Idaho batholith, which still retain some of their original heat. Few, if any, derive their heat from the extrusive lavas, which were widely spread out in relatively thin sheets that cooled quickly. Many of the thermal springs issue along faults, and some of these may be artesian in character, but most of them probably derive their heat from hot gases or liquids that rise from underlying bodies of intrusive rock. The available data indicate that the thermal springs of the Western Mountain region derive their water chiefly from surface sources, but their heat largely from magmatic sources.

One of such springs is shown in Figure 2.142. The spring is located in the Rio Grande fault zone, at the contact between bedrock and alluvium, in the general area of both young and paleomagmatic activity.

Meinzer (1940): The thermal springs in the Appalachian Highlands owe their heat to the artesian structure, the water entering the aquifer at a relatively high altitude, passing to considerable depth through a syncline or other inverted siphon and reappearing at a lower altitude; in the deep part of its course the water is warmed by the normal heat of the deep-lying rocks. Stearns et al. (1937) give detail description of thermal springs in the United States.

FIGURE 2.142 Ascending thermal spring in the channel of Rio Grande, Big Bend National Park, Texas. The spring hydraulic head is higher than the river, which maintains higher water level in the little pool (water in the pool is always clear as opposed to often muddy water of the river).

The term mineral spring (or mineral water for that matter) has very different meaning in different countries, and could be very loosely defined as a spring with water having one or more chemical characteristics different from normal potable water used for public supply. For example, water can have elevated content of free gaseous carbon dioxide (naturally carbonated water), or high radon content (radioactive water—still consumed in some parts of the world as "medicinal" water of "miraculous" effects), or high hydrogen sulfide content ("good for skin diseases" and "soft skin"), or high-dissolved magnesium, or simply have the TDS higher than 1000 mg/L. Some water bottlers, exploiting a worldwide boom in the use of bottled spring water, label water derived from a spring as "mineral" even when it does not have any unusual chemical or physical characteristics. In the United States, public use and bottling of spring and mineral water is under control of the Food and Drug Administration and such water must conform to strict standards including source protection.

2.8.4 Spring Development and Regulation

Springs were a preferred source of water supply for millennia. Three basic types of spring utilization are:

1. Use of the spring "as is," without any artificial intervention
2. Spring capture with varying degrees of engineering, aimed at securing the source for reliable use and protection from surface contamination
3. Engineering aimed at increasing spring discharge rate, possibly combined with an artificial extraction of the aquifer water in the general location of the spring (spring regulation)

Figure 2.143 shows typical capture of a contact gravity spring using a watertight basin constructed in place with reinforced concrete. The basin has one side open to the aquifer, allowing inflow of water. In case of ascending springs issuing from a horizontal surface without a clearly defined impermeable contact, the basin is open at the bottom. In either case, the side of the basin open to the inflow of groundwater should be stabilized with a gravel-sand pack or rock fragments (for fracture springs). If the water carries suspended solids and is turbid after heavy rains, this pack may be supplemented by a fine replaceable screen. The basin should be vented to the surface and have an easy access for maintenance.

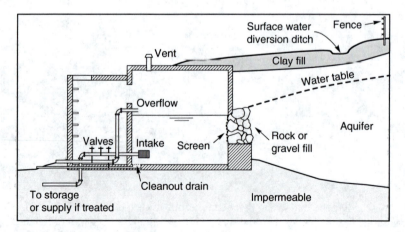

FIGURE 2.143 Typical capture of a contact gravity spring.

Three pipes equipped with valves should allow for water overflow, a complete basin drainage for the cleanout and maintenance, and transfer of water to supply or storage. All pipes should have screens on either end. If the water is of such quality that disinfection is the only treatment required, a chlorination tank or UV equipment may be housed in the maintenance room adjacent to the basin. Sanitary protection of the spring at the surface is achieved by fencing, placement of an impermeable clay fill, and surface drainage ditches located uphill from the spring to intercept surface water runoff and carry it away from the source. Spring capture should not be built at the location of secondary springs since they may move in time.

Depending on the site-specific conditions, the basic configuration of spring capture shown in Figure 2.143 may include additional features such as drainage pipes (or galleries in case of fractured-rock aquifers) extending into the saturated zone for intercepting more flow.

The rate of natural discharge has often been a limiting factor when considering the use of a particular spring for public water supply. Springs with discharge hydrographs similar to the one shown in Figure 2.144 may have a potential for regulation, i.e., may be amenable to artificial increasing of their minimum and average annual flows. The basic idea is to take advantage of the fact that the spring is capable of discharging large quantities of water during periods of nonpeak demand such as in spring or late fall when natural aquifer recharge is the highest. This volume of "surplus" water may be regulated in two basic ways:

1. By using it to naturally recharge the volume of aquifer drained by overpumping below spring elevation during periods of peak demand (summer-early fall) as shown in Figure 2.145 and Figure 2.146
2. By storing it in the aquifer above the elevation of natural spring discharge, i.e., by creating a surface or underground dam and groundwater impoundment behind it (Figure 2.147)

In either case, the main prerequisite is that the aquifer has an adequate storage capacity, below or above the spring elevation, respectively. Key additional requirement for case 2 is that there would not be an uncontrollable water loss around or below the dam. This means that the spring must issue from a V-shaped land-surface contact between the aquifer and the

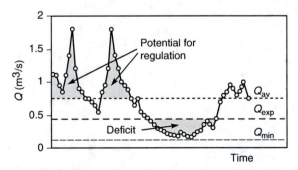

FIGURE 2.144 Hydrograph of a spring with potentially exploitable reserves higher than the minimum spring discharge. Q_{av}, average spring discharge; Q_{min}, minimum spring discharge; Q_{expl}, potentially secure exploitable reserves. (From Stevanovic, Z. et al. In: Stevanovic, Z. and Milanovic, P. (Eds.), *Water Resources and Environmental Problems in Karst—Cvijić 2005*, Proceedings of the International Conference, Institute of Hydrogeology, University of Belgrade, Belgrade, 2005, pp. 283–290.)

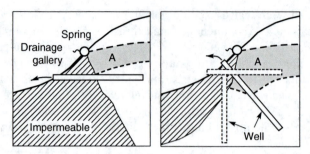

FIGURE 2.145 Some of the potentially favorable conditions for spring regulation using drainage galleries or wells for overpumping. A, Additional volume of water that may be extracted from aquifer storage during peak demand, assuming recovery during main periods of natural aquifer recharge.

impermeable barrier. The dam is keyed into the impermeable barrier and, in combination with a drainage gallery, is used to control the hydraulic head in the aquifer and water flow. Spring regulation in this way often enables generation of hydroelectric power because of the high hydraulic head in the aquifer behind the dam.

Large barrier karst springs are the best candidates for the flow regulation as the nature of karst porosity allows rapid recharge and transmittal of groundwater. For example, over-pumping of the Krupac spring (Figure 2.146) with one extraction well drilled into the main ascending channel 64 m below spring elevation, currently secures additional 80 L/s for water supply of Nis during two summer months of peak demand. Extensive aquifer tests and

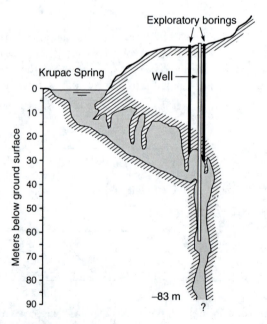

FIGURE 2.146 Regulation, by overpumping, of Krupac Spring used for water supply of Nis, Serbia. (From Milanovic, S. In: Stevanovic, Z. and Milanovic, P. (Eds.), *Water Resources and Environmental Problems in Karst—Cvijić 2005*, Proceedings of the International Conference, Institute of Hydrogeo-logy, University of Belgrade, Belgrade, 2005, pp. 451–458.)

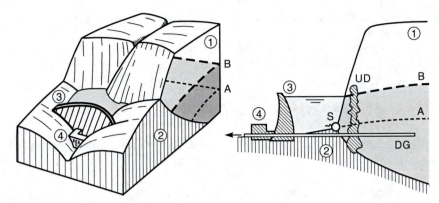

FIGURE 2.147 Possible spring regulation using surface or underground dam in case of favorable geologic and geomorphologic conditions. 1, Aquifer; 2, impermeable base; 3, surface dam with impoundment; 4, hydropower plant and water treatment. S, Original spring; A, original natural water table before impoundment; B, water table after impoundment; DG, drainage gallery; UD, underground dam, such as grout curtain, instead of surface dam.

monitoring show no long-term depletion of aquifer storage due to the current overpumping and, based on modeling studies two more wells are installed down to elevation -80 m in the main channel. Once completed, the system will provide additional 200 L/s during critical months (Jevtic et al., 2005). The natural minimum and maximum spring discharge is 35 L/s and over 1 m^3/s, respectively.

Feasibility studies and design of spring regulation require extensive hydrogeologic, hydraulic, and modeling studies, but may pay big dividends due to many advantages of storing water underground as explained in Section 2.9.

2.9 UNDERGROUND DAMS

Slichter (1902) gives the first account of an underground dam built in the United States:

Another method of recovering the underflow of a stream is by means of a subsurface dam. Such a dam is constructed by excavating a trench at right angles to the direction of the underflow and extending in depth to the impervious stratum, and then filling the trench with impervious material. If the underflow is confined within an impervious trough or canyon, it is obvious that such a construction must result in bringing it to the surface. An example of this is found on Pacoima Creek, Los Angeles County, Cal., where a subsurface dam was constructed in 1887–1890. It is claimed that by means of this dam the owners have been enabled to use the bedrock flow of water for the three dry years, 1898–1900, and thereby to successfully carry through the orange, lemon, and olive growing in Fernando Valley. This dam is described in the Eighteenth Annual Report of the United States Geological Survey, Part IV, pages 693 to 695; also in Reservoirs for Irrigation, Water Power, etc., by James D. Schuyler, 1901, page 205.

Compared with surface water reservoirs formed by conventional dams, use of groundwater impoundments behind underground dams has the following major advantages:

1. Very limited or negligible evaporation loss.
2. Land use above the groundwater reservoir can continue without change (there is no submergence of houses, infrastructure, and property in general).

3. Life span of the underground reservoir may be permanent since there is no accumulation of sediment.
4. There is no danger of dam failure and catastrophic loss of life and property.
5. Overall impact on the environment and natural habitat of plants and animals is of much lower magnitude.

However, the following are the main disadvantages:

1. Storage capacity of the underground reservoir cannot be accurately determined and has to be estimated based on more or less limited field data on, by default, heterogeneous porous media.
2. Virtually none of the feasible (cost-effective) construction methods can guarantee a complete dam impermeability (although practical impermeability, i.e., a tolerable level of dam leakage is often possible to achieve).
3. Operations and maintenance cost of the underground dam and groundwater extraction facilities (e.g., wells or drains) may sometimes be higher than for surface dams.

Similarly to surface dams and reservoirs, underground dams and groundwater storage basins cannot be constructed "everywhere" and require certain favorable hydrogeologic conditions such as high effective porosity of the aquifer materials, sufficient thickness of the unsaturated zone, large areal extent of the aquifer, and natural lateral and vertical containment of groundwater flow. Figure 2.148 illustrates the general layout of several underground dams successfully built on the Miyakojima Island in Japan where such favorable conditions exist.

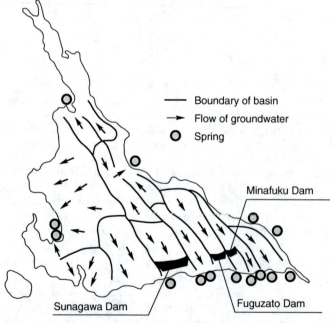

FIGURE 2.148 Location of the groundwater basins and underground dams on the Miyakojima Islands, Japan. (From Osuga, K. In: Uitto, J.I. and Schneider, J. (Eds.), *Freshwater Resources in Arid Lands*. UNU Global Environmental Forum V, United Nations University Press, Tokyo, 1997. http://www.unu.edu/unupress/unupbooks/uu02fe/uu02fe02.htm (accessed on October 27, 2005). Copyright United Nations University.)

As explained in detail by Osuga (1997), the Shimajiri mudstone layer of bedrock has low average hydraulic conductivity of 2×10^{-6} cm/s. The highly permeable Lyukyu limestone layer with the hydraulic conductivity of 3.5×10^{-1} cm/s forms the overlying aquifer with thickness ranging between 10 and 70 m. The effective porosity of the Lyukyu limestone is between 10% and 15%. Tectonic movements along several major faults have formed underground valleys, with groundwater flow parallel to the "ridges" between them. About 40% of the abundant rainfall on Miyakojima recharges the limestone aquifers but quickly flows out into the ocean unused. In addition, the subtropical climate results in evaporation of as much as 50% of the rainfall. The result is that only 10% of the precipitation result in surface flows on the island. Figure 2.149 shows the simple concept of regulating the groundwater flow at the island using underground dams. Figure 2.150 shows changes in groundwater levels after the Sunagawa dam cutoff wall was completed in November 1993. The full underground reservoir capacity was reached in September 1995, or less than 2 y after completion of the dam.

Sunagawa dam has capacity of about 10 million m³ of stored groundwater. This megadam is 1677 m long and was built with a mixed-in-place slurry wall method (Ishida et al., 2003). Since the beginning of groundwater withdrawal with vertical wells in 2001, the groundwater levels were somewhat higher than the designed full underground reservoir water level, indicating that the rates of groundwater withdrawal were slightly lower than the rate of natural aquifer recharge. The Sunagawa dam provided 2.7, 4.8, and 5 million m³ of water a year for irrigation in 2001, 2002, and 2003, respectively (Ishida, 2005). An analysis of nitrate concentrations in groundwater before and after the dam completion and operation for 3 years showed that the dam resulted in their overall decrease. After 3 years, the nitrates were also more uniformly distributed throughout the reservoir due to longer retention of nitrate-free natural recharge, mixing, and circulation of water caused by pumping (Ishida, 2005).

Hydrogeologic investigations and engineering design for underground dams include virtually all quantitative methods covered in this book and will not be listed again. In short,

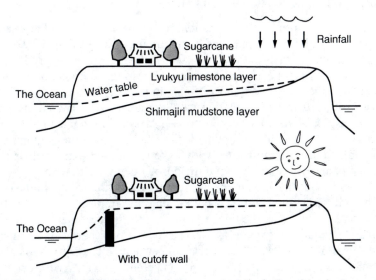

FIGURE 2.149 The concept of the underground dams on the Miyakojima Islands, Japan. (From Osuga, K. In: Uitto, J.I. and Schneider, J. (Eds.), *Freshwater Resources in Arid Lands*. UNU Global Environmental Forum V, United Nations University Press, Tokyo, 1997. http://www.unu.edu/unupress/unup-books/uu02fe/uu02fe02.htm (accessed on October 27, 2005). Copyright United Nations University.)

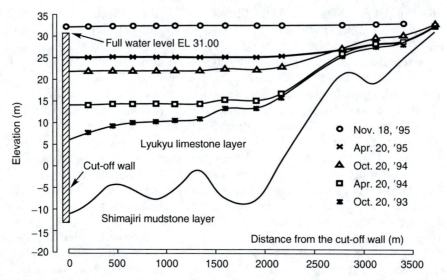

FIGURE 2.150 Changes in groundwater levels at the Sunagawa underground dam, Miyakojima Islands, Japan. (From Osuga, K. In: Uitto, J.I. and Schneider, J. (Eds.), *Freshwater Resources in Arid Lands.* UNU Global Environmental Forum V, United Nations University Press, Tokyo, 1997. http://www .unu.edu/unupress/unupbooks/uu02fe/uu02fe02.htm (accessed on October 27, 2005). Copyright United Nations University.)

because of the significant investment and benefits, but also possible failure due to uncertainties associated with the "invisible" subsurface, large underground dam projects are perfect examples of necessary interdisciplinary approach to aquifer management and protection.

2.10 COMPETENT AND LEAKY AQUITARDS

As mentioned briefly in Section 2.1, the question of hydrogeologic role of aquitards is one of the most important in both water supply and contaminant hydrogeology. One can think of an aquitard, when continuous and thick, and when overlying a highly productive confined aquifer, as a perfect "protector" of the valuable groundwater resource. Some professionals, however, would state that "every aquitard leaks" and it is only a matter of time when the existing shallow groundwater contamination would enter the confined aquifer and threaten the source. These professionals often rely on "best professional judgment" and are much less specific in terms of the "reasonable amount of time" after which the contamination would break through the aquitard. If confronted with some field-based data, such as the thickness and the hydraulic conductivity of the aquitard porous material, they may have the "best" answer ready in hand: "But the measurements did not include flow through the fractures, and we all know that all rocks and sediments comprising an aquitard, including clay, do have some fractures, somewhere." And the final argument is the hardest one to address: "But how do we know that the aquitard is continuous? There must be a pathway through it, such as interconnected lenses of some "sandy" material somewhere." The truth is, as always, somewhere in between. There are perfectly protective competent aquitards, which would not allow migration of shallow contamination to the underlying aquifer for thousands of years or more, regardless of the site-specific conditions in the underlying aquifer. And, there are leaky aquitards, which do not prevent such migration for more than several decades or so. Of course, if an aquitard is not continuous, and is only a few feet thick in places, all bets are off.

In such cases, the site-specific conditions in the underlying aquifer would play the key role. These conditions, in a "worst" case, may include large regional drawdowns caused by pumping in the confined aquifer, and the resulting steep hydraulic gradients between the two aquifers (the shallow and the confined) separated by the aquitard. However, it is surprising how many investigations in contaminant hydrogeology fail to collect more (or any) field information on the aquitard, even though determining its role may be crucial for a groundwater remediation project.

Possible migration of contaminants is not the only pressing question as to the role of aquitards. Where contamination is not an issue, the groundwater quantity available for water supply is the key question. Aquitard may allow leakance from the underlying or overlying aquifers into the pumped aquifer, or may release water stored in the aquitard itself.

Whatever the preliminary conceptual site model may be, the only correct way to confirm its applicability is to conduct field investigations and collect information about the key parameters of the aquifer–aquitard system. In some cases, it may be sufficient to collect the following data to reasonably accurately determine if there is an actual hydraulic connection between the two aquifers, through the intervening aquitard:

- Discrete measurements of the hydraulic head along the same vertical in the overlying aquifer; such measurements should be made at at least three locations to determine both horizontal and vertical vectors of groundwater flow in the overlying aquifer
- Discrete measurements of the hydraulic head in the aquitard at same locations
- Discrete measurements of the hydraulic head in the underlying aquifer at same locations

Figure 2.151 and Figure 2.152 show possible relationship between the hydraulic heads measured in the aquifer–aquitard–aquifer system. If the hydraulic heads in the three-unit stratified system are successively increasing or decreasing, this is an indicator of the possible hydraulic connections, and groundwater flow downward (Figure 2.151a) or upward (Figure 2.151b) through the aquitard. The existence of the actual flow can be indirectly confirmed only by hydraulically stressing (pumping) one of the aquifers and confirming the obvious related hydraulic head change in the other two units. When interpreting the hydraulic

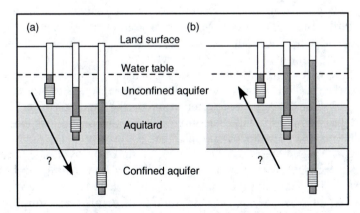

FIGURE 2.151 Some of the possible relationships between the hydraulic heads measured in a stratified aquifer–aquitard–aquifer system. The arrows show only the possible flow directions, not the actual flow. If the aquitard is competent, there would not be any significant free gravity (advective) flow between the two aquifers—the flow would be diffusion-dominated.

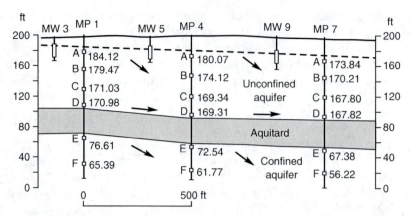

FIGURE 2.152 Measurements of the hydraulic head at multiport monitoring wells screened above and below aquitard for assessment of its hydraulic role. The confined aquifer is being pumped for water supply with an extraction well located approximately 4600 ft from MP-7.

head changes (fluctuations), all possible natural causes such as barometric pressure or tidal influences should be excluded. Note that the only direct method of confirming the actual flow through the aquitard is dye tracing, but it is of no practical use due to normally very long travel times through the aquitard.

Figure 2.152 is a good example of possibly misleading conclusions based on measuring the hydraulic heads at only one depth in the surficial aquifer (say, at MP-4A, where the head is 180.07 ft), and only one depth in the confined aquifer (MP-4F, the head is 61.77 ft). The vertical difference between these two hydraulic heads is 118.3 ft, which may lead one to believe that there must be vertical flow downward through the aquitard caused by such a strong vertical hydraulic gradient (incidentally, the confined aquifer is pumped for water supply). However, when considering all the measurements in the multiport wells, it becomes obvious that the 35-ft thick aquitard is competent and that groundwater flow in the unconfined aquifer just above the aquitard is horizontal. The head difference between the last two ports above the aquitard, at all multiport wells, is absent for all practical purposes: it is within one hundredth of one foot, upward or downward. The situation like this indicates absence of advective flow (free gravity flow) of groundwater from the unconfined aquifer into the underlying aquitard. The higher downward vertical gradients at shallow depths in the unconfined aquifer may be the result of recharge, possibly combined with the influence of some lateral flow boundary in the unconfined aquifer.

Hydraulic conductivity of the aquitard porous media is another key field information required to assess the rates (velocities) of groundwater flow through an aquitard. In addition to collecting samples for laboratory determination of the matrix hydraulic conductivity, conducting field tests would provide measurements of the overall response of different porosity types that may be present (for example, matrix porosity, micro- and macrofractures in clay). Vargas and Ortega-Guerrero (2004) present results of the hydraulic conductivity tests conducted in 225 piezometers installed in a regional lacustrine clay aquitard in the metropolitan area of Mexico City. The aquitard (split into first and second subaquitards) has thickness between 50 and 300 m, and covers the main aquifer used for water supply of 25 million people. The results of the study show notable differences between the matrix hydraulic conductivity determined in the laboratory, which is on the order of 1×10^{-10} to 1×10^{-11} m/s, and the field-determined hydraulic conductivity at various depths. In general, the aquitard is more heterogeneous and contains more microfractures at shallow depths of

25–40 m. This is reflected in the hydraulic conductivity values spanning as much as five orders of magnitude at some locations: between 1×10^{-11} and 1×10^{-7} m/s. The range of variation generally narrows down with depth, so that field values for the second regional aquitard are between 1×10^{-11} and 1×10^{-9} m/s. Figure 2.153 illustrates this trend of decreasing hydraulic conductivity with depth, evident in the shallow aquitard as well. For example, in this general area, all 14 values determined in the field at depths greater than 15 m are less than 1×10^{-9} m/s, which would label the aquitard as competent for all practical purposes.

Chemical composition of groundwater present in an aquitard is yet another key element for assessing its hydraulic role relative to the adjacent aquifers. Farvolden and Cherry (1988) present results of hydrogeologic investigations of thick clayey aquitards in Ontario and Quebec, Canada, conducted as part of studying possible locations for waste disposal sites. These deposits are primarily glacial till and glaciolacustrine in southwestern Ontario, and glaciomarine in eastern Ontario and Quebec. The clay deposits exhibit a surficial-weathered zone with vertical and near-vertical ubiquitous fractures and relic rootlets common to depths of about 3–4 m. Beneath the weathered zone, the clayey material is relatively unfractured, varying in thickness between 25 and 45 m. The hydraulic conductivity of the unweathered clayey deposits has been determined at many locations by means of piezometer-response monitoring and by consolidation and triaxial cell tests on undistorted core samples. Values typically range from 4×10^{-5} m/d to 9×10^{-6} m/d. The surficial-weathered zone commonly has a much higher bulk hydraulic conductivity caused by the fractures, root holes, and other secondary features. The average linear groundwater velocity determined from the vertical hydraulic head profiles (see Figure 2.154) is less than 1 cm/y.

Vertical profiles of major ions and environmental isotopes, together with the hydraulic head and conductivity profiles, were used to interpret the mechanisms of groundwater movement in the aquitards. The relatively high concentrations of major ions in and near the weathered zone are attributed to chemical weathering that took place primarily during Altithermal time when a warmer, drier climate caused the average water table to be 2 or 3 m deeper than the present-day water table. Desiccation caused the fractures to form during this drier period. The vertical changes in concentrations of the analyzed constituents in the unweathered clay are primarily caused by molecular diffusion, which causes their migration due to concentration gradients. The vertical flow of groundwater (advection) has negligible effect in this respect. Cl^-, Na^+, and CH_4 are diffusing upward from the bedrock where the high concentrations of these constituents originate. Upward diffusion dominates over the

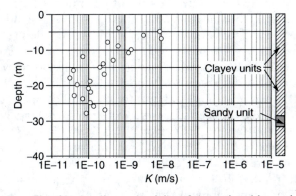

FIGURE 2.153 Vertical profile of hydraulic conductivity of the regional lacustrine aquitard, measured in the field at the Medical Center in Mexico City. (From Vargas, C. and Ortega-Guerrero, A., *Hydrogeol. J.*, 12, 336, 2004. Copyright Springer-Verlag. With permission.)

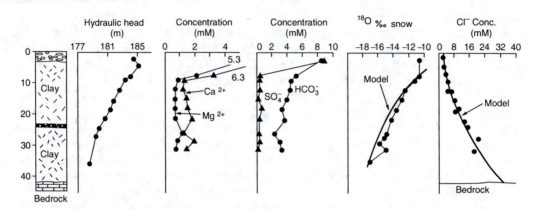

FIGURE 2.154 Hydraulic head, major ions, and oxygen-18 in thick clayey deposits at a site near Sarnia, Ontario, Canada. (Modified from Farvolden and Cherry, 1988. Copyright The Geological Society of America, Inc., with permission.)

downward advective flow; therefore, the net movement is upward. Ca^{2+}, Mg^{2+}, HCO_3^-, and SO_4^{2-} are diffusing downward from the weathered zone where they originate. ^{18}O and ^2H (deuterium) are also diffusing downward from the bottom of the weathered zone into the unweathered clayey till. This interpretation was confirmed with mathematical modeling based on Fick's Laws of diffusion (examples in Figure 2.154 are modeled concentrations of ^{18}O and Cl^-). As a conclusion, the clayey material beneath the weathered zone contains groundwater that is thousands of years old and that exhibits diffusion-controlled distribution of major ions and isotopes (Farvolden and Cherry, 1988, based on studies by Desaulniers, 1986, and Desaulniers et al., 1986). Carbon-14 dating in the Sarnia district is additional evidence of the age of deep groundwater: it is between 10,000 and 14,000 years old. Based on all the information presented above, one would easily conclude that the clayey aquitard in question is competent.

Juana–Diaz Formation in southern Puerto Rico is an example of a thick aquitard of marine origin that still contains highly mineralized water entrapped in the sediment during its original deposition. This formation overlies reef limestone, which contains groundwater with sharply different chemical characteristics (Figure 2.155). Most of the monitoring wells drilled in the Juana–Diaz Formation at the waste disposal site were initially dry or had negligible yields. Little water that did collect in the exposed horizon of a boring was likely through exudation of the trapped connate water with movement initiated by the pressure differences between the corresponding horizon of the formation and the drill-hole. Age dating of this water using environmental isotopes of oxygen and carbon shows groundwater residence time in the Juana–Diaz Formation below the site is on the order of thousands of years. However, the most compelling evidence of the absence of any significant flow in the formation, due to possible present-day flushing with fresh water, is the chemical composition of groundwater: chloride concentrations across the site range from 18,500 to 25,700 mg/L, sulfate concentrations range from 1780 to 4010 mg/L, TDS range from 31,100 to 45,300 mg/L, and sodium concentrations range from 4160 to 8270 mg/L. For comparison, average concentrations of these species in seawater are 19,000 mg/L for chloride, 2,700 mg/L for sulfate, and 10,500 mg/L for sodium (Hem, 1989). If the saline groundwater within the formation were moving downward due to free advective (gravity) flow, the concentrations of various constituents in the underlying limestone would have to be much higher, especially considering density differences between the saline water in the Juana–Diaz Formation and the fresh water in

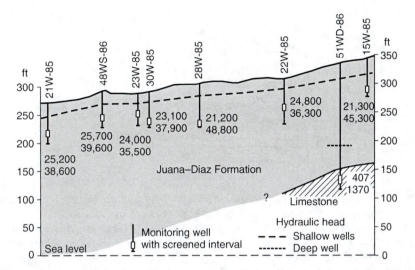

FIGURE 2.155 Chloride concentrations (top number) and total dissolved solids (bottom number), in milligrams per liter, in groundwater at a waste disposal site, Puerto Rico. (From Krešić, N., Relevance of ASTM Standard D 5717-95 to hydrogeologic conditions at PROTECO landfill facility, Peñuelas, Puerto Rico. LAW Engineering and Environmental Services, Kennesaw, GA, 1999. With kind permission of Dr. Jorge Fernandez.)

the limestone. On the other hand, downward migration of any dissolved constituents that may be entering the Juana–Diaz sediments from the land surface through percolating rainfall, would not be possible due to the same density differences (newly infiltrated low-mineralized water cannot flow through the denser saline water). In conclusion, solely based on the collected hydrochemical information, the Juana–Diaz Formation at the site is a competent aquitard that should prevent any migration of potential surface contamination through the dense, saline water down to the reef limestone. This includes any possible faults or fractures in the formation, since those would also contain saline water by definition.

2.11 SURFACE WATER–GROUNDWATER INTERACTIONS

Surface water and groundwater are inseparable parts of the same hydrologic cycle. However, they are often studied separately, even in cases of closely related problems. This division still holds firm with many professionals, but the artificial boundaries between the surface water and groundwater projects are quickly fading away as more and more people realize that management and protection of surface water resources cannot be accomplished without the management and protection of groundwater resources, and vice versa. As already mentioned in Section 2.6.2.4, most perennial surface streams, as well as natural lakes, would not have permanent flow without groundwater contribution (stream baseflow). Excessive withdrawal of groundwater may cause depletion or complete cessation of flow in a surface stream, or disappearance of wetlands. Conversely, changes in land use, such as urban development, may completely alter patterns of surface water runoff and severely reduce aquifer recharge. Contamination of surface streams adjacent to well fields used for water supply may, in some cases, threaten groundwater quality in the underlying aquifer (see Figure 2.156), while discharge of contaminated groundwater into a surface water body may have negative impact on human health and the environment.

Figure 2.157 illustrates basic hydraulic relationships between surface streams and underlying unconfined aquifers. However, depending on local hydrogeologic and climatic conditions,

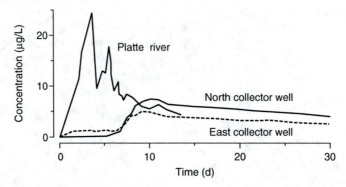

FIGURE 2.156 Atrazine concentration, measured in late May and early June, in water from the collector wells increases after a time delay of 5 to 7 d in response to the increased concentration in the Platte River, Nebraska. The peak concentration in groundwater is substantially lower due to attenuation processes. (From Galloway, D.L. et al., Evolving issues and practices in managing ground-water resources: case studies on the role of science. U.S. Geological Survey Circular 1247, Reston, VA, 2003, 73pp; modified from Verstraeten, I.M., Minimizing the risk of herbicide transport into public water supplies—A Nebraska case study. U.S. Geological Survey Fact Sheet 078-00, 2000, 4pp.)

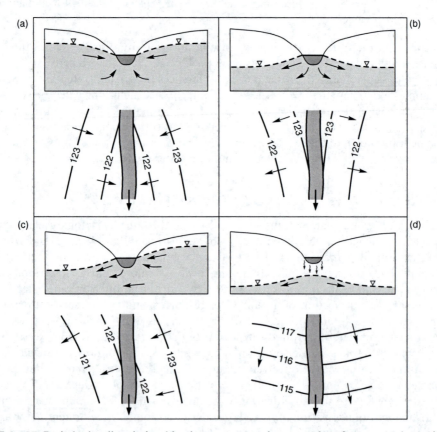

FIGURE 2.157 Basic hydraulic relationships between groundwater and surface water shown in cross-sectional and map views. A, Perennial gaining stream; B. Perennial losing stream; C. Perennial stream gaining water on one side and losing water on the other side; D. Losing stream disconnected from the underlying water table, also called ephemeral stream.

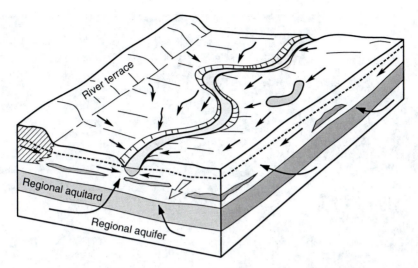

FIGURE 2.158 Groundwater flow directions in a floodplain of a meandering stream, which acts as a base for regional groundwater discharge (aquifer discharge zone).

and human impacts (e.g., dams and locks), the same stream may be losing or gaining water in different sections, and this pattern may change in time. Major unregulated meandering streams with large floodplains that are seasonally flooded may have quite complicated surface water–groundwater relationships, especially if the stream is meandering and there are old inactive (oxbow) or buried meanders. In addition, such streams are often regional bases for groundwater discharge from deeper confined aquifers, which complicates things even further (Figure 2.158).

The simplest way to confirm if a stream segment is gaining water from groundwater discharge is also the most obvious: one has to only observe stream banks and channel bottom for signs of discharge in the form of springs, seepage springs, or bubbles of gas freed from the groundwater due to pressure drop (Figure 2.159 and Figure 2.160).

Current hydraulic status of a stream segment can be determined by using a number of different methods, of which at least two methods should be combined for the verification of results. Detail description of various instruments and methodologies is given in Winter et al. (1998),

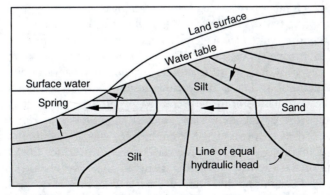

FIGURE 2.159 Submerged seepage spring discharging into a gaining stream. (From Winter, T.C. et al., Ground water and surface water. A single resource. U.S. Geological Survey Circular 1139, Denver, CO, 1998, 79pp.)

FIGURE 2.160 Bubbles of gas freed from the discharging groundwater due to pressure drop; streambed of Honey Creek in Ann Arbor, Michigan.

Galloway et al. (2003), Stonestrom and Constantz (2003), and Greswell (2005). The most widely used approach, and arguably still the most accurate, is the traditional continuous recording of the hydraulic heads in piezometers placed perpendicular to the stream (or a surface water body of interest), including in the aquifer material below the stream channel. Piezometer differs from a monitoring well in that it has only its bottom open to the aquifer, as opposed to a well, which has a screen of certain length. Piezometer therefore accurately records the hydraulic head at the elevation of its bottom. Monitoring wells with longer screens may be used if in conjunction with three-dimensional borehole flowmeters. An ideal placement of piezometers, on either side of the stream, is shown in Figure 2.161. Note that, if the line of piezometers is just on one side of the stream, it would not be possible to accurately state that the stream is truly gaining or losing water since the other side may behave differently. Care should be taken to place piezometers both outside and inside the so-called hyporheic zone, if such zone is present.

Hyporheic zone (*hypo*—Greek word for under or beneath, *rheos*—stream, *rheo*—to flow) has a number of definitions based on various scientific fields that have interest in it. The majority of the current literature is presented by biologists and ecologists, who have concentrated on the role of the hyporheic zone as a habitat and refuge for freshwater invertebrate fauna and as a location for salmonid egg development. Probably second in scale, the hydrology literature considers processes that control water flow within the hyporheic zone in terms of the exchange of water between the river channel and the adjacent hyporheic sediments. A smaller pool of literature exists within the hydrogeological, geochemical, and geomorphological fields (Smith, 2005). Some researchers use the term riparian zone either interchangeably or in conjunction with hyporheic zone, to emphasize presence and influence of organics-rich sediments and vegetation.

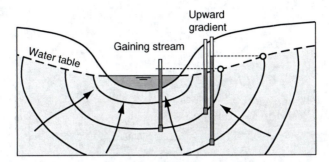

FIGURE 2.161 Placement of piezometers for determining the hydraulic role of a stream segment. In the case shown, the stream segment is gaining water (groundwater discharges into the stream). Contour lines are of equal hydraulic head and arrows show direction of groundwater flow.

Common themes in the definitions of the hyporheic zone in the literature are:

1. It is the zone below and adjacent to a streambed in which water from the open channel exchanges with interstitial water in the bed sediments.
2. It is the zone around a stream which has characteristic (specific) fauna.
3. It is the mixing zone between groundwater and surface water (Figure 2.162).

In any case, the hydraulic head measured in the hyporheic zone may not be the true head in the underlying aquifer; it may be influenced by the "microflow" pattern within the streambed sediments where surface water is entering and leaving in short intervals, depending on the channel bottom morphology. The same is true for small-scale seepage measurements in the streambed sediments, which may not be registering real groundwater inflow.

Hyporheic zone is receiving an increasing attention in groundwater contamination studies because of its unique biological, chemical, and hydraulic characteristics. It has been argued that, in some cases, certain agricultural contaminants such as phosphorus and nitrates, and organic contaminants such as BTEX (common components of gasoline) and chlorinated solvents may be attenuated in the hyporheic zone (Burt et al., 1999; Reddy et al., 1999; Conant et al., 2004; Puckett, 2004; Westbrook et al., 2005).

A zone of enhanced biogeochemical activity usually develops in shallow groundwater as a result of the flow of oxygen-rich surface water into the subsurface environment, where bacteria and geochemically active sediment coatings are abundant (Figure 2.163). This input of oxygen to the streambed stimulates a high level of activity by aerobic (oxygen-using) microorganisms if dissolved oxygen is readily available. It is not uncommon for dissolved oxygen to be completely used up in hyporheic flow paths at some distance into the streambed, where anaerobic microorganisms dominate microbial activity. Anaerobic

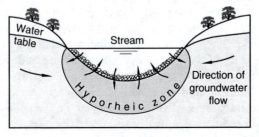

FIGURE 2.162 Schematic of a hyporheic zone. (Modified from Winter, T.C. et al., Ground water and surface water. A single resource. U.S. Geological Survey Circular 1139, Denver, CO, 1998, 79pp.)

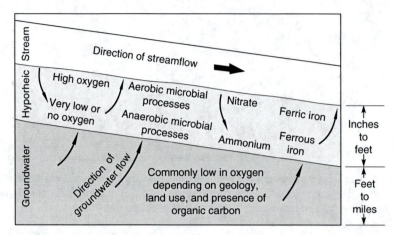

FIGURE 2.163 Some of the processes and chemical transformations that may take place in the hyporheic zone. Actual chemical interactions depend on numerous factors including aquifer mineralogy, shape of the aquifer, types of organic matter in surface water and groundwater, and nearby land use. (From Winter, T.C. et al., Ground water and surface water. A single resource. U.S. Geological Survey Circular 1139, Denver, CO, 1998, 79pp.)

bacteria can use nitrate, sulfate, or other solutes in place of oxygen for metabolism. The result of these processes is that many solutes are highly reactive in shallow groundwater near the streambeds (Winter et al., 1998).

The movement of nutrients and other chemical constituents, including contaminants, between groundwater and surface water is affected by biogeochemical processes in the hyporheic zone. For example, the rate at which organic contaminants biodegrade in the hyporheic zone can exceed rates in stream water or in groundwater away from the stream. Another example is the removal of dissolved metals in the hyporheic zone. As water passes through the hyporheic zone, dissolved metals are removed by precipitation of metal oxide coatings on the sediments (Winter et al., 1998).

Measuring temperature of water, and temperature of streambed sediment at various depths including below the hyporheic zone in the aquifer, is a relatively inexpensive method for an immediate indication of the flow regime character. Typically, heat movement caused by temperature differences is traced by continuous monitoring of temperature pattern in the stream and streambed followed by interpretation using numerical models. For example, reaches of the channel in which sediment-temperature fluctuations are highly damped relative to in-stream fluctuations indicate high rates of groundwater discharge to the stream (that is, groundwater discharge to a strongly gaining reach). Conversely, segments of the stream channel where fluctuations in streambed temperatures closely follow in-stream fluctuations indicate high rates of water loss through streambed sediments (that is, groundwater recharge from a losing reach). To quantify the rates, location, and timing of streamflow gains and losses, an array of temperature sensors is deployed in the stream and adjacent sediments (Stonestrom and Constantz, 2003). Figure 2.164 illustrates thermal and hydraulic responses to four possible streambed conditions. A and B show gaining and losing perennial streams that are connected to the local groundwater system. C and D show dry and flowing ephemeral streams that are separated from the local groundwater system by an intervening unsaturated zone (ephemeral streams lose water when flowing). The flowing stream has large diurnal variations in temperature, which are generally dampened beneath the channel bottom. For a gaining stream (A) diurnal variation in the streambed is only slight because water is flowing up from depths where temperatures are constant on diurnal timescales. For

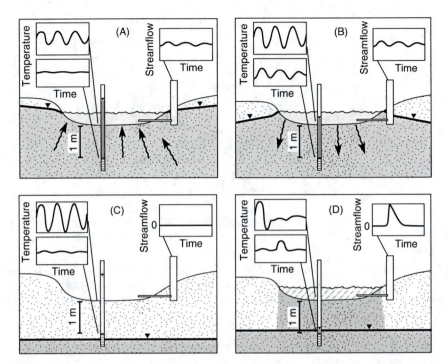

FIGURE 2.164 Four basic types of hydraulic and thermal responses in surface streams and streambed sediments. A, Perennial stream, gaining reach; B, perennial stream, losing reach; C, ephemeral stream without flow; D, ephemeral stream with flow. (Modified from Stonestrom, D.A. and Constantz, J. (Eds.), *Heat as a Tool for Studying the Movement of Ground Water Near Streams*, U.S. Geological Survey Circular 1260, Reston, VA, 2003, pp. 1–5.)

a losing stream (B) the diurnal variations, although dampened, are obvious as the downward flow of water transports heat from the stream into the sediments. In either case, larger or smaller flow of water (advective heat transfer) causes larger or smaller contrasts. Consequently, enough temperature sensors should be placed at varying depths to cover all possible variations. In general, surface water temperature in the losing stream has greater diurnal variations than in the gaining stream as regional groundwater is not flowing into the stream.

Figure 2.165 is an example of a gaining perennial stream (case A in Figure 2.164) where streambed temperatures at shallow depths are similar to the stream temperature, whereas at greater depths they are generally colder and slowly increase with time during the summer, reflecting the groundwater temperature with no effect from fluctuating surface temperatures (Conlon et al., 2003). Measured temperatures can be used for calibration of numeric models capable of simulating both the heat transfer and flow through porous media, such as VS2DH/T by USGS, which can then be used to estimate the flow exchanges between the surface stream and the underlying aquifer (Figure 2.166).

Flow measurement at successive stream segments is a common method of determining if the stream is losing or gaining water between the segments. However, because of the large variability of flow conditions in the same stream and the associated potential measurement errors, this method should be applied with great care in order to avoid false conclusions. For example, if only one set of measurements is made at a few locations, with several or more hours separating them, and the streamflow is under the influence of recent precipitation, the results would almost certainly be misleading as the flow wave is moving fairly rapidly. It is

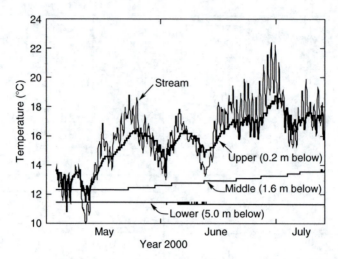

FIGURE 2.165 Observed temperatures of Zollner Creek, Central Willamette Basin, Oregon. (Modified from Conlon, T., Lee, K., and Risley, J. In: Stonestrom, D.A. and Constantz, J. (Eds.), *Heat as a Tool for Studying the Movement of Ground Water near Streams*, U.S. Geological Survey Circular 1260, Reston, VA, 2003, pp. 29–34.)

therefore best if the flow measurement method is based on continuous recording of the stream stage (flow) at successive stream segments for at least several days. In addition, the derived flow hydrographs provide information on the actual change of volume of water between the

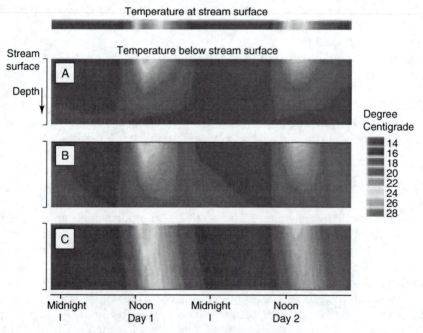

FIGURE 2.166 Computer simulation of the streambed temperature contours for cases in which a stream with the given temperature (thin band at top) is gaining groundwater (A), is neutral (B), and is losing water (C). (Modified from Constantz, J. et al. In: Stonestrom, D.A. and Constantz, J. (Eds.), *Heat as a Tool for Studying the Movement of Ground Water near Streams*, U.S. Geological Survey Circular 1260, Reston, VA, 2003, pp. 21–27.)

segments, which is the only real measure of gain or loss. Rantz et al. (1982) provide detail explanation of various methods for measurement and computation of streamflow.

2.11.1   GROUNDWATER UNDER THE DIRECT INFLUENCE OF SURFACE WATER

Criteria have been established under the U.S. primary drinking-water regulations for conditions in which a public water system is required to filter a surface water source or a groundwater source that is under the direct influence of surface water. Groundwater under the direct influence of surface water is defined as "any water beneath the surface of the ground with significant occurrence of insects or other macroorganisms, algae, or large-diameter pathogens such as *Giardia lamblia* or *Cryptosporidium*, or significant and relatively rapid shifts in water characteristics such as turbidity, temperature, conductivity, or pH which closely correlate to climatological or surface water conditions" (section 141.2, Title 40, U.S. Code of Federal Regulations, 2000). A groundwater source that is determined to be under the influence of surface water must be filtered by a conventional filtration technique or by an alternative filtration technique in combination with disinfection to inactivate pathogenic organisms. The alternative filtration technique must be approved for use by the State in which the public water system resides and must have been demonstrated to consistently achieve 99.9% removal and (or) inactivation of *Giardia lamblia* cysts, 99.99% removal and (or) inactivation of viruses, and 99% removal of *Cryptosporidium oocysts* (section 141.173, Title 40, U.S. Code of Federal Regulations, 2000). Natural bank filtration is one of the alternative filtration systems that have been used to meet these requirements for drinking-water filtration (Galloway et al., 2003). This type of filtration refers to passage of surface water through the streambed and alluvial sediments to the groundwater extraction system, which may be a collector well, or a vertical well or an underdrain near the stream (see Chapter 3).

Before bank filtration can be considered as an alternative for removal of certain surface water influences, travel time between the stream and the groundwater intake must be determined among other parameters. USGS field-based research (Sheets et al., 2002) at a public-supply well field near the Great Miami River in Ohio illustrates the potential use of an easily measured parameter, such as specific conductance, to estimate groundwater travel times from a river to nearby wells. The network of monitoring wells installed at various distances between the river and the extraction well (Figure 2.167) serves to monitor water

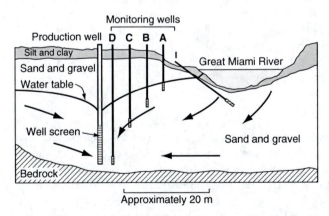

FIGURE 2.167 Monitoring wells placed between the river and production well used to estimate groundwater travel times. (Modified from Galloway, D.L. et al., Evolving issues and practices in managing ground-water resources: case studies on the role of science. U.S. Geological Survey Circular 1247, Reston, VA, 2003, 73pp.)

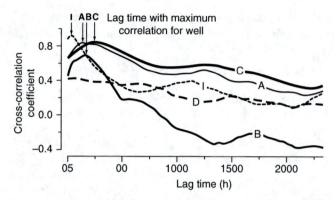

FIGURE 2.168 Cross-correlograms of specific conductance in monitoring wells and specific conductance in the Great Miami River. The maximum cross-correlation coefficient identified for each well
indicates the lag time for which the concentration in the well is most correlated with the peak
concentration that had previously occurred in the river. (Modified from Galloway, D.L. et al., Evolving
issues and practices in managing ground-water resources: case studies on the role of science. U.S.
Geological Survey Circular 1247, Reston, VA, 2003, 73pp and Sheets, R.A., Darner, R.A., and
Whitteberry, B.L., *J. Hydrology*, 266, 162, 2002.)

levels and various geochemical parameters, and to ultimately assess the processes of pathogen
transport from the river to the well field. The basic approach was to determine lag times
between a marked change in the specific conductance of the Great Miami River and a
subsequent change in specific conductance in the monitoring wells. Specific conductance of
the river water was highly correlated with chloride, which is chemically conservative and used
as environmental tracer in groundwater studies.

The results of the time series analysis are shown in Figure 2.168. The cross-correlation
coefficients for most of the monitoring wells indicate that specific conductance of the well
water is highly correlated with specific conductance of the river water at particular lag times.
Lag times ranged from 29 h for the slanted monitoring well I below the river channel to 235 h
(10 d) for monitoring well C. Data from the deepest monitoring well (well D) are not highly
correlated and a lag time for which the correlation coefficient is the highest was not determined. The lack of correlation between the specific conductance of water from the deepest
monitoring well (well D) with that of the river, even though the monitoring well completed
above it responds readily, suggests that pumping from the production well captures most of the
water near the top of the screen from the water table (ultimately the Great Miami River) and
much of the water near the bottom of the screen is captured from more regional groundwater
flow (Galloway et al., 2003).

2.11.2 COASTAL AQUIFERS

As groundwater use has increased in coastal areas due to rapid population growth, so has the
recognition that groundwater supplies are vulnerable to overuse contamination. Groundwater
development depletes the amount of groundwater in storage, causes reductions in groundwater
discharge to streams, wetlands, and coastal estuaries, and lowers water levels in ponds and
lakes. Contamination of groundwater resources has resulted in degradation of some drinking-
water supplies and coastal waters. Although, overuse and contamination of groundwater are
not uncommon throughout the United States, the proximity of coastal aquifers to saltwater
creates unique issues with respect to groundwater sustainability in coastal regions. These issues
are primarily saltwater intrusion into freshwater aquifers and changes in the amount and

quality of fresh groundwater discharge to coastal saltwater ecosystems. Saltwater intrusion is the movement of saline water into freshwater aquifers and most often is caused by groundwater pumping from coastal wells. Because saltwater has high concentrations of TDS and certain inorganic constituents, it is unfit for human consumption and many other anthropogenic uses. Saltwater intrusion reduces fresh groundwater storage and, in extreme cases, leads to the abandonment of supply wells when concentrations of dissolved ions exceed drinking-water standards. The problem of saltwater intrusion was recognized as early as 1854 on Long Island, New York (Back and Freeze, 1983), thus predating many other types of drinking-water contamination issues in the news (Barlow, 2003).

In natural conditions, lighter (less dense) fresh groundwater overlies more dense saline water and the thickness of the fresh water above the interface with saline water can be estimated based on the ratio of their respective densities. This relationship was first recognized by Ghyben and Herzberg, two European scientists who derived it independently in the late 1800s:

$$z = \frac{\rho_f}{\rho_s - \rho_f} h \qquad (2.171)$$

where z is the thickness of fresh water between the interface and the sea level, ρ_f is the density of fresh water, ρ_s is the density of saltwater, and h is the thickness of fresh water between the sea level and the water table.

Freshwater has a density of about 1.000 g/cm^3 at 20°C, whereas that of seawater is about 1.025 g/cm^3. Although this difference is small, Equation 2.172 indicates that it results in 40 ft of freshwater below sea level for every 1 ft of freshwater above sea level as illustrated with the example in Figure 2.169:

$$z = 40h \qquad (2.172)$$

Although in most applications this equation is sufficiently accurate, it does not describe the true nature of freshwater–saltwater interface since it assumes hydrostatic conditions (no movement of either water). In reality, fresh groundwater discharges into the saltwater body (sea, ocean) with certain velocity and through a seepage surface of certain thickness, thus creating a transition zone in which two waters of different density mix by the processes of dispersion and molecular diffusion. Mixing by dispersion is caused by spatial variations (heterogeneities) in the geologic structure, the hydraulic properties of an aquifer, and by dynamic forces that operate over a range of timescales, including daily fluctuations in tide

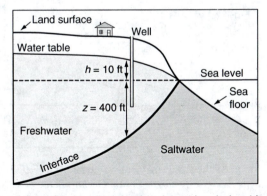

FIGURE 2.169 Illustration of the Ghyben–Herzberg hydrostatic relationship between fresh water and saltwater. (Modified from Barlow, P.M., Ground water in freshwater–saltwater environments of the Atlantic coast. U.S. Geological Survey Circular 1262, Reston, VA, 2003, 113pp.)

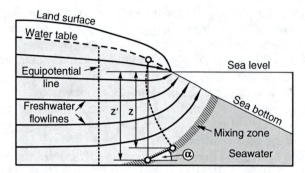

FIGURE 2.170 Hydrodynamic relationship between freshwater and saltwater in an unconfined coastal aquifer. α is angle of the interface slope. True depth to salt water (z') is greater than the one assumed based on the Ghyben–Herzberg relationship (z).

stages, seasonal and annual variations in groundwater recharge rates, and long-term changes in sea level position. These dynamic forces cause the freshwater and saltwater zones to move seaward at times and landward at times. Because of the mixing of freshwater and saltwater within the transition zone, a circulation of saltwater is established in which some of the saltwater is entrained within the overlying freshwater and returned to the sea, which in turn causes additional saltwater to move landward toward the transition zone (Barlow, 2003). By convention, freshwater is defined as water having TDS less than 1000 mg/L and chloride concentration less than 250 mg/L. For seawater, these values are 35,000 and 19,000 mg/L, respectively. Everything in between would correspond to a mixing zone. The thickness of a mixing zone depends on local conditions in the aquifer but, in general, it is much smaller than the general vertical field scale of interest. In most cases, quantitative analyses and groundwater modeling codes are based on the assumption of sharp interface between fresh water and saltwater.

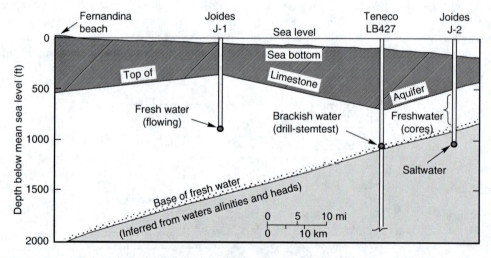

FIGURE 2.171 Inferred position of the freshwater–saltwater interface based on hydraulic testing and water analyses at offshore exploratory oil wells, Atlantic Ocean off the coast of Georgia and Florida. (From Johnston, R.H. et al., Summary of hydrologic testing in tertiary limestone aquifer, Tenneco offshore exploratory well—Atlantic OCS, lease-block 427 (Jacksonville NH 17-5). U.S. Geological Survey Water-Supply Paper 2180, Washington, D.C., 1982, 15pp.)

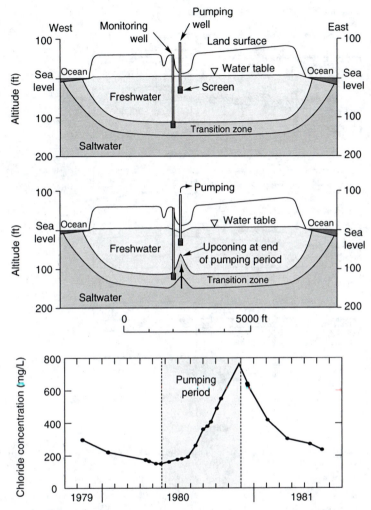

FIGURE 2.172 Groundwater pumping from a well in the town of Truro on Cape Cod, Massachusetts, caused upcoming of the transition zone beneath the well, which in turn caused increased chloride concentrations in a nearby monitoring well. After pumping at the well stopped in November 1980, the transition zone moved downward, and chloride concentrations at the monitoring well slowly decreased with time. (Modified from LeBlanc, D.R. et al., Ground-water resources of Cape Cod, Massachusetts. *U.S. Geological Survey Hydrologic Investigations Atlas* 692, 1986, 4 sheets and Barlow, P.M., Ground water in freshwater–saltwater environments of the Atlantic coast. U.S. Geological Survey Circular 1262, Reston, VA, 2003, 113pp.)

Discharge of freshwater causes flow lines in the aquifer to deviate from vertical as illustrated in Figure 2.170. Because the Dupuit's hypothesis does not apply, the true vertical thickness of freshwater is somewhat greater than the one estimated using the Ghyben–Herzberg equation, as first recognized by Hubbert (1940). The slope of the interface (α) can be calculated using the following equation (Davis and DeWiest, 1991):

$$\sin \alpha = \frac{\partial z}{\partial s} = -\left[\frac{1}{K_f} \frac{\rho_f}{\rho_f - \rho_s} V_{f,s} - \frac{1}{K_s} \frac{\rho_f}{\rho_f - \rho_s} V_{s,s} \right] \qquad (2.173)$$

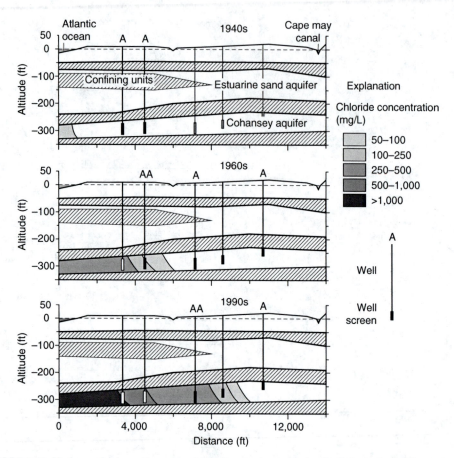

FIGURE 2.173 Hydrogeologic sections showing saltwater intrusion in the Cohansey aquifer at the Cape May City, New Jersey well field, 1940s to 1990s; "A" above well indicates groundwater was pumped from the well during part or all of the decade. (Modified from Lacombe, P.J. and Carleton, G.B, Hydrogeologic framework, availability of water supplies, and saltwater intrusion, Cape May County, New Jersey. U.S. Geological Survey Water-Resources Investigations Report 01-4246, 2002, 151pp and Barlow, P.M., Ground water in freshwater–saltwater environments of the Atlantic coast. U.S. Geological Survey Circular 1262, Reston, VA, 2003, 113pp.)

where s is the trace of the interface in a vertical plane, K_f and K_s are hydraulic conductivities for freshwater and saltwater, respectively, and $V_{f,s}$ and $V_{s,s}$ are velocities of fresh water and saltwater along the interface.

Equation 2.173 can be simplified if it is assumed that saltwater is stagnant compared to freshwater which flows over it, so that the second term in the brackets becomes zero.

Regional coastal aquifers of continental scale overlain by thick competent aquitards may extend well off the coastline as illustrated in Figure 2.171. Multiple stratified confined aquifers along the Atlantic coast of the United States contain large quantities of freshwater that extend to various distances off the coastline. These aquifers have enabled continuing development along the coast, including numerous barrier islands.

When the natural conditions in a coastal aquifer are altered by groundwater withdrawal, the position and shape of the fresh water–saltwater interface, as well as the mixing zone thickness, may change in all three dimensions and result in saltwater encroachment, as illustrated in Figure 2.172 and Figure 2.173. The presence of leaky and discontinuous aquitards, and pumping from different aquifers or different depths in the same aquifer may create rather complex spatial relationship between freshwater and saltwater.

REFERENCES

Alley, W.M. and Leake, S.A., 2004. The journey from safe yield to sustainability. *Ground Water*, 42(1), 12–16.

Alley, W.M., Reilly, T.E., and Lehn Franke, O., 1999. Sustainability of ground-water resources. U.S. Geological Survey Circular 1186, Denver, CO, 79pp.

Allison, G.B. and Hughes, M.W., 1978. The use of environmental chloride and tritium to estimate total recharge to an unconfined aquifer. *Australian Journal of Soil Resources*, 16, 181–195.

American Petroleum Institute (API), 1998. *Recommended Practices for Core Analysis*, RP40, 2nd ed., API Publishing Services, Washington, D.C.

American Society of Civil Engineers (ASCE), 2001. *Standard Guidelines for Artificial Recharge of Groundwater*, ASCE Standard No. 34–01, 120pp.

American Society for Testing and Materials (ASTM), 1999a. *ASTM Standards on Determining Subsurface Hydraulic Properties and Ground Water Modeling*, 2nd ed., West Conshohocken, PA, 320pp.

ASTM, 1999b. *ASTM Standards on Ground Water and Vadose Zone Investigations; Drilling, Sampling, Geophysical Logging, Well Installation and Decommissioning*, 2nd ed., West Conshohocken, PA, 561pp.

Anders, R. and Schroeder, R.A., 2003. Use of water-quality indicators and environmental tracers to determine the fate and transport of recycled water in Los Angeles County, California. U.S. Geological Survey Water-Resources Investigations Report 03-4279, Sacramento, CA, 104pp.

Arago, D.F.J., 1835a. Sur les puits forés, connus sous le nom de Puits Artésiens, des fontaines artésiennes, ou de fontaines jaillissantes. *Paris, Bureau des Longitudes Annuaire*, pp. 181–258.

Arago, D.F.J., 1835b. On springs, artesian wells, and spouting fountains. *New Philosophical Journal*, 18, 205–246.

Arthur, J.D., Dabous, A.A., and Cowart, J.B., 2002. Mobilization of arsenic and other trace elements during aquifer storage and recovery, southwest Florida. In: Aiken, G.R. and Kuniansky, E.L. (Eds.), U.S. Geological Survey Artificial Recharge Workshop Proceedings, April 2–4, 2002, Sacramento, CA, U.S. Geological Survey Open-File Report 02–89, pp. 47–50.

Back, W. and Freeze, R.A. (Eds.), 1983. *Chemical Hydrogeology: Benchmark Papers in Geology*, Hutchinson Ross Publication Company, Stroudsburg, PA, 416pp.

Barenblatt, G.E., Zheltov, I.P., and Kochina, I.N., 1960. Basic concepts in the theory of seepage of homogeneous liquids in fissured rocks: *Jour. Appl. Math. and Mechanics*, V. 24, pp. 1286–1303.

Barlow, P.M., 2003. Ground water in freshwater–saltwater environments of the Atlantic coast. U.S. Geological Survey Circular 1262, Reston, VA, 113pp.

Bedinger, M.S., 1987. Summary of infiltration rates in arid and semiarid regions of the world, with an annotated bibliography. U.S. Geological Survey Open-File Report 87–43, Denver, CO, 48pp.

Bense, V.F., Van den Berg, E.H., and Van Balen, R.T., 2003. Deformation mechanisms and hydraulic properties of fault zones in unconsolidated sediments; the Roer Valley Rift System, The Netherlands. *Hydrogeology Journal*, 11, 319–332.

Böhlke, J.-K., 2002. Groundwater recharge and agricultural contamination. *Hydrogeology Journal*, 10(1), 153–179.

Boulton, N.S., 1954a. Unsteady radial flow to a pumped well allowing for delayed yield from storage. *International Association for Scientific Hydrology*, 37, 472–477.

Boulton, N.S., 1954b. The drawdown of the water table under nonsteady conditions near a pumped well in an unconfined formation. *Institute for Civil Engineers Proceedings (London)*, pt. 3, 564–579.

Boulton, N.S., 1963. Analysis of data from non-equilibrium pumping tests allowing for delayed yield from storage. *Institute for Civil Engineers Proceedings (London)*, 26, 469–482.

Boulton, N.S., 1970. Analysis of data from pumping tests in unconfined anisotropic aquifers. *Journal of Hydrology*, 10, 369.

Boulton, N.S., 1973. The influence of delayed drainage on data from pumping tests in unconfined aquifers. *Journal of Hydrology*, 19(2), 157–169.

Boulton, N.S. and Pontin, J.M.A., 1971. An extended theory of delayed yield from storage applied to pumping tests in unconfined anisotropic aquifers. *Journal of Hydrology*, 19, 157–169.

Bouwer, H., 1989. The Bouwer and Rice slug test—an update. Contribution of the U.S. Department of Agriculture, Agricultural Research Service. *Ground Water*, 27(3), 304–309.

Bouwer, H., 2002. Artificial recharge of groundwater: hydrogeology and engineering. *Hydrogeology Journal*, 10, 121–142.

Bouwer, H. and Rice, R.C., 1976. A slug test method for determining hydraulic conductivity of unconfined aquifers with completely or partially penetrating wells. *Water Resources Research*, 12(3), 423–428.

Bradbury, K.R. and Rothschild, E.R., 1985. A computerized technique for estimating the hydraulic conductivity of aquifers from specific capacity data. *Ground Water*, 23(2), 240–246.

Brons, F. and Marting, V.E., 1961. The effect of restricted fluid entry on well productivity. *Journal of Petroleum Technology*, 13(2), 172–174.

Burt, T.P. et al., 1999. Denitrification in riparian buffer zones: the role of floodplain hydrology. *Hydrological Processes*, 13, 1451–1463.

Butler, J.J. Jr., Garnett, E.J., and Healey, J.M., 2003. Analysis of slug tests in formations of high hydraulic conductivity. *Ground Water*, 41(5), 620–630.

Coes, A.L. and Pool, D.R., 2005. Ephemeral-stream channel and basin-floor infiltration and recharge in the Sierra Vista subwatershed of the upper San Pedro basin, Southeastern Arizona. U.S. Geological Survey Open-File Report 2005-1023, Reston, VA, 67pp.

Conant, B. Jr., Cherry, J.A., and Gillham, R.W., 2004. A PCE groundwater plume discharging into a river: influence of the streambed and near-river zone on contaminant distributions. *Journal of Contaminant Hydrology*, 73, 249–279.

Conlon, T., Lee, K., and Risley, J., 2003. Heat tracing in streams in the central Willamette Basin, Oregon. In: Stonestrom, D.A. and Constantz, J. (Eds.), *Heat as a Tool for Studying the Movement of Ground Water near Streams*, U.S. Geological Survey Circular 1260, Reston, VA, pp. 29–34.

Constantz, J. et al., 2003. The Santa Clara River—the last natural river of Los Angeles. In: Stonestrom, D.A. and Constantz, J. (Eds.), *Heat as a Tool for Studying the Movement of Ground Water near Streams*, U.S. Geological Survey Circular 1260, Reston, VA, pp. 21–27.

Cooper, H.H. Jr., 1963. Type curves for nonsteady radial flow in an infinite leaky artesian aquifer. In: Bentall, R. (Compiler), *Shortcuts and Special Problems in Aquifer Tests*. U.S. Geological Survey Water-Supply Paper 1545-C, pp. C48–C55.

Cooper, H.H. Jr. and Jacob, C.E., 1946. A generalized graphical method for evaluating formation constants and summarizing well-field history. *Transactions, American Geophysical Union*, 27, 526–534.

Cooper, H.H., Bredehoeft, J.D., and Papadopulos, S.S., 1967. Response of a finite-diameter well to an instantaneous charge of water. *Water Resources Research*, 3(1), 263–269.

Cvetković-Mrkić, S., 1995. Metode geotehničkih melioracija, prva knjiga (in Serbian; *Methods of Geotechnical Meliorations*, Vol. 1). Rudarsko-geološki fakultet Univerziteta u Beogradu, Beograd, 177pp.

Daniel, C.C. III, 1990. Evaluation of site-selection criteria, well design, monitoring techniques, and cost analysis for a ground-water supply in Piedmont crystalline rocks, North Carolina. U.S. Geological Survey Water-Supply Paper 2341-B, 35pp.

Daniel, C.C. III and Sharpless, N.B., 1983. Ground-water supply potential and procedures for well-site selection, upper Cape Fear River basin. North Carolina Department of Natural Resources and Community Development, 73pp.

Davis, S.N. and DeWiest, R.J.M., 1991. *Hydrogeology*, Krieger Publishing Company, Malabar, FL, 463pp.

Dawson, K. and Istok, J., 1992. *Aquifer Testing: Design and Analysis*, Lewis Publishers, Boca Raton, FL, 280pp.

Desaulniers, D.E., 1986. Groundwater origin, geochemistry, and solute transport in three major clay plains of east-central North America, Ph.D. thesis, Department of Earth Sciences, University of Waterloo, Waterloo, 450pp.

Deasulniers, D.E. et al., 1986. ^{37}Cl–^{35}Cl variations in a diffusion-controlled groundwater system. *Geochimica et Cosmochimica Acta*, 50, 1757–1764.

Dettinger, M.D., 1989. Distribution of carbonate-rock aquifers in southern Nevada and the potential for their development, summary of findings, 1985–1988. Program for the Study and Testing of Carbonate-Rock Aquifers in Eastern and Southern Nevada, Summary Report No. 1, Carson City, NV, 37pp.

Dillon, P. and Pavelic, P., 1996. Guidelines on the quality of stormwater and treated wastewater for injection into aquifers for storage and reuse. Research Report No. 109, Urban Water Research Association of Australia, Water Services Association of Australia, Melbourne.

Dillon, P. et al., 1997. Aquifer storage and recovery of stormwater runoff. *Australian Water and Wastewater Association Journal of Water*, 24(4), 7–11.

Dingman, S.L., 1994. *Physical Hydrology*, Macmillan, New York, 575pp.

Driscoll, F.G., 1986. *Groundwater and Wells*, 3rd ed., Johnson Filtration Systems Inc., St. Paul, MN, 1089pp.

Dugan, J.T. and Peckenpaugh, J.M., 1985. Effects of climate, vegetation, and soils on consumptive water use and ground-water recharge to the central Midwest regional aquifer system, mid-continent United States. U.S. Geological Survey Water-Resources Investigations Report 85–4236, Lincoln, NE, 78pp.

Dunn, D.J., Bergman, K.J., and LaTorraca, G.A., 2002. *Nuclear Magnetic Resonance—Petrophysical and Logging Applications*. Seismic Exploration No. 32, Pergamon Press, Amsterdam, The Netherlands, 312pp.

Eagon, H.B. Jr. and Johe, D.E., 1972. Practical solutions for pumping tests in carbonate-rock aquifers. *Ground Water*, 10(4), 6–13.

Edwards Underground Water District, Water resources of the Edwards Aquifer region (poster).

El-Naqa, A., 1994. Estimation of transmissivity from specific capacity data in fractured carbonate rock aquifer, central Jordan. *Environmental Geology*, 23(1), 73–80.

Fabbri, P., 1997. Transmissivity in the geothermal Euganean Basin; a geostatistical analysis. *Ground Water*, 35(5), 881–887.

Farvolden, R.N. and Cherry, J.A., 1988. Chapter 18, Region 15, St. Lawrence Lowland. In: *The Geology of North America, Vol. O-2, Hydrogeology*, Back, W., Rosenshein, J.S., and Seaber, P.R. (Eds.), The Geological Society of American, Boulder, Colorado, pp. 133–140.

Ferris, J.G., Knowles, D.B., Brown, R.H., and Stallman, R.W., 1962. Theory of aquifer tests. U.S. Geological Survey Water Supply Paper 1536-E, Washington, D.C., 173pp.

Galloway, D.L. et al., 2003. Evolving issues and practices in managing ground-water resources: case studies on the role of science. U.S. Geological Survey Circular 1247, Reston, VA, 73pp.

Glover, R.E., 1964. Ground water movement. U.S. Bureau of Reclamation, Engineering Monograph 31, 67pp.

Greene, E.A., 1993. Hydraulic properties of the Madison aquifer system in the western Rapid City area, South Dakota. U.S. Geological Survey Water-Resources Investigations Report 93–4008, Rapid City, South Dakota, 56pp.

Greene, E.A. and Shapiro, A.M., 1995. Methods of conducting air-pressurized slug tests and computation of type curves for estimating transmissivity and storativity. U.S. Geological Survey Open-File Report 95–424.

Greene, E.A. and Shapiro, A.M., 1998. AIRSLUG: A FORTRAN program for the computation of type curves to estimate transmissivity and storativity of prematurely terminated air-pressurized slug tests. *Ground Water*, 36(2), 373–376.

Greene, E.A., Shapiro, A.M., and Carter, J.M., 1999. Hydrogeologic characterization of the Minnelusa and Madison Aquifers near Spearfish, South Dakota. U.S. Geological Survey Water-Resources Investigations Report 98–4156, Rapid City, South Dakota, 64pp.

Greswell, R.B., 2005. *High-Resolution In Situ Monitoring of Flow between Aquifers and Surface Waters*. Environment Agency, Science Report SC030155/SR4, Bristol, United Kingdom, 32pp.

Griffioen, J. and Kruseman, G.P., 2004. Determining hydrodynamic and contaminant transfer parameters of groundwater flow. In: Kovalevsky, V.S., Kruseman, G.P., and Rushton, K.R. (Eds.), *Groundwater Studies: An International Guide for Hydrogeological Investigations*, IHP-VI, Series on Groundwater No. 3, UNESCO, Paris, France, pp. 217–238.

Gringarten, A.C. and Ramey, H.J., 1974. Unsteady state pressure distributions created by a well with a single horizontal fracture, partial penetration or restricted entry. *Society for Petroleum Engineers Journal*, SPE 3819, 413–426.

Gringarten, A.C. and Whiterspoon, P.A., 1972. A method of analyzing pump test data from fractured aquifers. *International. Society for Rock Mechanics and International Association for Engineering Geology Proceedings Symposium Rock Mechanics*, Stuttgart, 3-B, pp. 1–9.

Halford, K.J. and Kuniansky, E.L., 2002. Documentation of spreadsheets for the analysis of aquifer-test and slug-test data. U.S. Geological Survey Open-File Report 02-197, Carson City, NV, 51pp.

Halford, K.J. and Mayer, G.C., 2000. Problems associated with estimating ground-water discharge and recharge from stream-discharge records. *Ground Water*, 38(3), 331–342.

Hantush, M.S., 1956. Analysis of data from pumping tests in leaky aquifers. *Transactions, American Geophysical Union*, 37(6), 702–714.

Hantush, M.S., 1959. Nonsteady flow to flowing wells in leaky aquifers. *Journal of Geophysical Research*, 64(8), 1043–1052.

Hantush, M.S., 1960. Modification of the theory of leaky aquifers. *Journal of Geophysical Research*, 65, 3713–3725.

Hantush, M.S., 1961a. Drawdown around a partially penetrating well. *Journal of Hydrology Division, Proceedings of the American Society for Civil Engineers*, 87(HY4), 83–98.

Hantush, M.S., 1961b. Aquifer tests on partially penetrating wells. *Journal of Hydrology Division, Proceedings of the American Society for Civil Engineers*, 87(HY5), 171–194.

Hantush, M.S., 1966a. Wells in homogeneous anisotropic aquifers. *Water Resources Research*, 2(2), 273–279.

Hantush, M.S., 1966b. Analysis of data from pumping tests in anisotropic aquifers. *Journal of Geophysical Research*, 71(2), 421–426.

Hantush, M.S., 1967. Growth and decay of ground water mounds in response to uniform percolation. *Water Resources Research*, 3, 227–234.

Hantush, M.S. and Jacob, C.E., 1955. Nonsteady radial flow in an infinite leaky aquifer. *Transactions, American Geophysical Union*, 36(1), 95–100.

Hantush, M.S. and Thomas, R.G., 1966. A method for analyzing a drawdown test in anisotropic aquifers. *Water Resources Research*, 2(2), 281–285.

Hazen, A., 1892. Experiments upon the purification of sewage and water at the Lawrence Experiment Station, Nov. 1, 1889 to Dec. 31, 1891. *Massachusetts Board of Health Twenty-third Annual Report*, pp. 428–434.

Healy, R.W. and Cook, P.G., 2002. Using groundwater levels to estimate recharge. *Hydrogeology Journal*, 10(1), 91–109.

Heath, R.C., 1980. Basic elements of ground-water hydrology with reference to conditions in North Carolina. U.S. Geological Survey Water-Resources Investigations Open-File Report 80-44, Raleigh, North Carolina, 87pp.

Heath, R.C., 1987. Basic ground-water hydrology. U.S. Geological Survey Water-Supply Paper 2220, 4th ed., Denver, CO, 84pp.

Hem, J.D., 1989. Study and interpretation of the chemical characteristics of natural water, 3rd ed. U.S. Geological Survey Water-Supply Paper 2254, Washington, D.C., 263pp.

Hiscock, K.M., Rivett, M.O., and Davison, R.M. (Eds.), 2002. *Sustainable Groundwater Development*, Special Publication No. 193, Geological Society, London.

Hubbert, M.K., 1940. The theory of ground-water motion. *Journal of Geology*, 48(8), 785–944.

Huntley, D., Nommensen, R., and Steffey, D., 1992. The use of specific capacity to assess transmissivity in fractured-rock aquifers. *Ground Water*, 30(3), 396–402.

HydroSOLVE, Inc., 2002. *AQTESOLV for Windows, User's Guide*. HydroSOLVE, Inc., Reston, VA, 185pp.

Hvorslev, M.J., 1951. Time lag and soil permeability in ground-water observations. *Bull. No. 36, Waterways Experiment Station, Corps of Engineers*, U.S. Army, Vicksburg, MS, 50pp.

IAH (International Association of Hydrogeologists), 2002. Managing aquifer recharge. Commission on Management of Aquifer Recharge, IAH-MAR, 12pp.

Ishida, S. et al., 2003. Construction of subsurface dams and their impact on the environment. *RMZ Materials and Geoenvironment*, 50(1), 149–152.

Ishida, S. et al., 2005. Evaluation of impact of an irrigation project with a mega-subsurface dam on nitrate concentration in groundwater from the Ryukyu limestone aquifer, Miyako Island, Okinawa, Japan. In: Stevanovic, Z. and Milanovic, P. (Eds.), *Water Resources and Environmental Problems in Karst—Cvijić* 2005, Proceedings of the International Conference, Institute of Hydrogeology, University of Belgrade, Belgrade, pp. 121–126.

Jacob, C.E., 1963a. Determining the permeability of water-table aquifers. In: Bentall, R. (Compiler), *Methods of Determining Permeability, Transmissibility, and Drawdown*. U.S. Geological Survey Water-Supply Paper 1536-I, pp. 245–271.

Jacob, C.E., 1963b. Corrections of drawdown caused by a pumped well tapping less than the full thickness of an aquifer. In: Bentall, R. (Compiler), *Methods of Determining Permeability, Transmissibility, and Drawdown*, U.S. Geological Survey Water-Supply Paper 1536-I, pp. 272–292.

Jacobson, G. et al., 2004. Groundwater resources and their use in Australia, New Zealand and Papua New Guinea. In: Zektser, I.S. and Everett, L.G. (Eds.), *Groundwater Resources of the World and Their Use*, IHP-VI, Series on Groundwater No. 6, UNESCO, Paris, France, pp. 237–276.

James, N.P. and Mountjoy, E.W., 1983. Shelf-slope break in fossil carbonate platforms: an overview. In: Stanley, D.J. and Moore, G.T. (Eds.), *The Shelfbreak: Critical Interface on Continental Margins*, SEPM Special Publication No. 33, pp. 189–206.

Jevdjevic, V., 1956. Hidrologija, I deo (Hydrology, Part 1). Hidrotehnicki Institut Jaroslav Cerni, Beograd, 404 p.

Jevtic, G. et al., 2005. Regulation of the Krupac spring outflow regime. In: Stevanovic, Z. and Milanovic, P. (Eds.), *Water Resources and Environmental Problems in Karst—Cvijić 2005*, Proceedings of the International Conference, Institute of Hydrogeology, University of Belgrade, Belgrade, pp. 321–326.

Johnson, A.I., 1967. Specific yield—compilation of specific yields for various materials. U.S. Geological Survey Water-Supply Paper 1662-D, 74pp.

Johnson, A.I., Prill, R.C., and Morris, D.A., 1963. Specific yield—column drainage and centrifuge moisture content. U.S. Geological Survey Water-Supply Paper 1662-A, 60pp.

Johnston, C.D., 1983. Estimation of groundwater recharge from the distribution of chloride in deeply weathered profiles from south-west Western Australia. In: *Papers of the International Conference on Groundwater and Man, Vol. 1, Investigation and Assessment of Groundwater Resources*, Sydney, 1983, Canberra, Australian Water Resources Council, Conference Series 8, pp. 143–152.

Johnston, R.H. et al., 1982. Summary of hydrologic testing in tertiary limestone aquifer, Tenneco offshore exploratory well—Atlantic OCS, lease-block 427 (Jacksonville NH 17-5). U.S. Geological Survey Water-Supply Paper 2180, Washington, D.C., 15pp.

Jones, W.K., 1977. Karst hydrology atlas of West Virginia. Special Publication 4, Karst Waters Institute, Charles Town, WV, 111pp.

Kashef, A.A.I., 1987. *Groundwater Engineering*, McGraw Hill, Singapore, 512pp.

Kendy, E., 2001. Magnitude, extent, and potential sources of nitrate in ground water in the Gallatin Local Water Quality District, southwestern Montana, 1997–98. U.S. Geological Survey Water-Resources Investigations Report 01-4037.

Kenyon, B. et al., 1995. Nuclear magnetic resonance imaging—technology for the 21st century. *Oil Field Review*, Autumn 1995, pp. 19–33.

Khouri, J., 2004. Groundwater resources and their use in Africa. In: Zektser, I.S. and Everett, L.G. (Eds.), *Groundwater Resources of the World and their Use*, IHP-VI, Series on Groundwater No. 6, UNESCO, Paris, France, pp. 209–237.

King, F.H., 1899. Principles and conditions of the movements of ground water. *U.S. Geological Survey Nineteenth Annual Report.*, Washington, D.C.

King, R.B., 1992. Overview and bibliography of methods for evaluating the surface-water-infiltration component of the rainfall-runoff process. U.S. Geological Survey Water-Resources Investigations Report 92-4095, Urbana, IL, 169pp.

Kovalevsky, V.S., Kruseman, G.P., and Rushton, K.R. (Eds.), 2004. *Groundwater Studies: An International Guide for Hydrogeological Investigations*, IHP-VI, Series on Groundwater No. 3, UNESCO, Paris, France, 430pp.

Krause, R.E. and Randolph, R.B., 1989. Hydrology of the Floridan aquifer system in southeast Georgia and adjacent parts of Florida and South Carolina. U.S. Geological Survey Professional Paper 1403-D, 65pp.

Krešić, N., 1988. *Karst i pećine Jugoslavije* (in Serbo-Croatian; Karst and caves of Yugoslavia), Naučna knjiga, Belgrade, 149pp.

Krešić, N., 1991. *Kvantitativna hidrogeologija karsta sa elementima zaštite podzemnih voda* (in Serbo-Croatian; Quantitative karst hydrogeology with elements of groundwater protection), Naučna knjiga, Belgrade, 192pp.

Krešić, N., 1999. Relevance of ASTM Standard D 5717-95 to hydrogeologic conditions at PROTECO landfill facility, Peñuelas, Puerto Rico. LAW Engineering and Environmental Services, Kennesaw, GA.

Kruseman, G.P., de Ridder, N.A., and Verweij, J.M., 1991. *Analysis and Evaluation of Pumping Test Data* (completely revised 2nd ed.), International Institute for Land Reclamation and Improvement (ILRI) Publication 47, Wageningen, The Netherlands, 377pp.

Lacombe, P.J. and Carleton, G.B, 2002. Hydrogeologic framework, availability of water supplies, and saltwater intrusion, Cape May County, New Jersey. U.S. Geological Survey Water-Resources Investigations Report 01-4246, 151pp.

Larsson, I., Chairman of the Project Panel, 1982, *Ground water in Hard Rocks*, Project 8.6 of The International Hydrological Programme, UNESCO, Paris, 228pp.

LeBlanc, D.R. et al., 1986. Ground-water resources of Cape Cod, Massachusetts. U.S. Geological Survey Hydrologic Investigations Atlas 692, 4 sheets.

Lee, C.H., 1915. The determination of safe yield of underground reservoirs of the closed basin type. *Transactions, American Society of Civil Engineers*, 78, 148–251.

Lee, K.K. and Risley, J.C., 2002. Estimates of ground-water recharge, base flow, and stream reach gains and losses in the Willamette River Basin, Oregon. U.S. Geological Survey Water-Resources Investigations Report 01-4215, Portland, OR, 52pp.

Linsley, R.K., Kohler, M.A., and Paulhus, J.L.H. 1975. *Hydrology for Engineers*. McGraw-Hill, New York, 482 p.

Llamas, R., 2004. *Water and Ethics. Use of Groundwater*. UNESCO International Hydrological Programme, Series on Water and Ethics, Essay 7, UNESCO, Paris, 33pp.

Lloyd, O.B. Jr. and Lyke, W.L., 1995. Illinois, Indiana, Kentucky, Ohio, Tennessee, *Ground Water Atlas of the United States*, United States Geological Survey, HA 730-K.

Lohman, S.W., 1972. Ground-water hydraulics. U.S. Geological Survey Professional Paper 708, Washington, D.C., 70pp.

Lohman, S.W. et al., 1972. Definitions of selected ground-water terms—revisions and conceptual refinements. U.S. Geological Survey Water Supply Paper 1988 (Fifth Printing 1983), Washington, D.C., 21pp.

Lyford, F.P., 1986. Northeast glacial regional aquifer-system study. In: Sun, R.J. (Ed.), *Regional Aquifer-System Analysis Program of the U.S. Geological Survey—Summary of Projects, 1978–1984*, U.S. Geological Survey Circular 1002, pp. 162–167.

Mace, R.E., 1997. Determination of transmissivity from specific capacity tests in a karst aquifer, *Ground Water*, 35(5), 738–742.

Mace et al., 2000. Hydraulic conductivity and storativity of the Carrizo–Wilcox acquifer in Texas. The University of Texas at Austin, Bureau of Economic Geology, contract report prepared for the Texas Water Development Board, 76 p.

Mace, R.E., 2001. Estimating transmissivity using specific-capacity data. Geological Circular 01-2, Bureau of Economic Geology, The University of Texas at Austin, Austin, TX, 44pp.

Maclay, R.W. and Small, T.A., 1986. Carbonate hydrology and hydrology of the Edwards aquifer in the San Antonio area, Texas. Texas Water Development Board Report 296, Austin, TX, 90pp.

Marino, M.A., 1975a. Artificial ground water recharge. I. Circular recharging area. *Journal of Hydrology*, 25, 201–208.

Marino, M.A., 1975b. Artificial ground water recharge. II. Rectangular recharging area. *Journal of Hydrology*, 26, 29–37.

Maslia, M.L. and Randolph, R.B., 1986. Methods and computer program documentation for determining anisotropic transmissivity tensor components of two-dimensional ground-water flow. U.S. Geological Survey Open-File Report 86-227, 64pp.

Meinzer, O.E., 1923a. The occurrence of ground water in the United States with a discussion of principles. U.S. Geological Survey Water-Supply Paper 489, Washington, D.C., 321pp.

Meinzer, O.E., 1923b. Outline of ground-water hydrology with definitions. U.S. Geological Survey Water-Supply Paper 494, Washington, D.C., 71pp.

Meinzer, O.E., 1927. Large springs in the United States. U.S. Geological Survey Water-Supply Paper 557, Washington, D.C., 94pp.

Meinzer, O.E., 1932 (reprint 1959). Outline of methods for estimating ground-water supplies. Contributions to the hydrology of the United States, 1931. U.S. Geological Survey Water-Supply Paper 638-C, Washington, D.C., pp. 99–144.

Meinzer, O.E., 1940. Ground water in the United States; a summary of ground-water conditions and resources, utilization of water from wells and springs, methods of scientific investigation, and literature relating to the subject. U.S. Geological Survey Water-Supply Paper 836-D. Washington, D.C., pp. 157–232.

Meyer, R.R., 1963. A chart relating well diameter, specific capacity, and the coefficient of transmissibility and storage. In: Bentall, R. (Ed.), *Methods of Determining Permeability, Transmissibility and Drawdown*, U.S. Geological Survey Water-Supply Paper 1536-I, pp. 338–340.

Milanovic, S., 2005. Hydrogeological characteristics of some deep siphonal springs in Serbia and Montenegro karst. In: Stevanovic, Z. and Milanovic, P. (Eds.), *Water Resources and Environmental Problems in Karst—Cvijić 2005*, Proceedings of the International Conference, Institute of Hydrogeology, University of Belgrade, Belgrade, pp. 451–458.

Miller, J.A., 1999. Introduction and national summary. *Ground-Water Atlas of the United States*, United States Geological Survey, A6.

Milojević, N., 1967. *Hidrogeologija*. Univerzitet u Beogradu, Zavod za izdavanje udžbenika Socialističke Republike Srbije, Beograd, 379pp.

Moench, A.F., 1984. Double-porosity models for a fissured groundwater reservoir with fracture skin. *Water Resources Research*, 21(8), 1121–1131.

Moench, A.F., 1985. Transient flow to a large-diameter well in an aquifer with storative semiconfining layers. *Water Resources Research*, 8(4), 1031–1045.

Moench, A.F., 1993. Computation of type curves for flow to partially penetrating wells in water-table aquifers. *Ground Water*, 31(6), 996–971.

Moench, A.F., 1996. Flow to a well in a water-table aquifer: an improved Laplace transform solution. *Ground Water*, 34(4), 593–596.

Molz, F.J. and Melville, J.G., 1996, Discussion of combined use of flowmeter and time–drawdown data to estimate hydraulic conductivities in layered aquifer systems. *Ground Water*, 34(5), 770.

Molz, F.J., Güven, O., and Melville, J.G., (with contributions by Javandel, I., Hess, A.E., and Paillet, F.L.), 1990. A new approach and methodologies for characterizing the hydrogeologic properties of aquifers, EPA/600/2-90/002, U.S. Environmental Protection Agency, Ada, OK.

Morris, B.L. et al., 2003. Groundwater and its susceptibility to degradation: a global assessment of the problem and options for management. *Early Warning and Assessment Report Series*, RS. 03-3. United Nations Environment Programme, Nairobi, Kenya, 126pp.

Mozley, P.S. et al., 1996. Using the spatial distribution of calcite cements to infer paleoflow in fault zones: examples from the Albuquerque Basin, New Mexico [abstract]. American Association of Petroleum Geologists 1996 Annual Meeting.

Muskat, M., 1946. *The Flow of Homogeneous Fluids through Porous Media*, J.W. Edwards, Ann Arbor, MI.

National Research Council, 1994. *Ground Water Recharge using Waters of Impaired Quality*, National Academy Press, Washington, D.C., 382pp.

Neuman, S.P., 1972. Theory of flow in unconfined aquifers considering delayed response to the water table. *Water Resources Research*, 8(4), 1031–1045.

Neuman, S.P., 1974. Effects of partial penetration on flow in unconfined aquifers considering delayed gravity response. *Water Resources Research*, 10(2), 303–312.

Neuman, S.P., 1975. Analysis of pumping test data from anisotropic unconfined aquifers considering delayed gravity response. *Water Resources Research*, 11(2), 329–342.

Neuman, S.P. and Witherspoon, P.A., 1969. Applicability of current theories of flow in leaky aquifers. *Water Resources Research*, 5, 817–829.

Norton, W.H., 1897. Artesian wells of Iowa. Iowa Geological Survey, 6, 113–428.

Osborne, P.S., 1993. Suggested operating procedures for aquifer pumping tests. *Ground Water Issue*, United States Environmental Protection Agency, EPA/540/S-93/503, 23pp.

O'Reilly, A.M., 2004. A method for simulating transient ground-water recharge in deep water-table settings in central Florida by using a simple water-balance/transfer-function model. U.S. Geological Survey Scientific Investigations Report 2004-5195, Reston, VA, 49pp.

Osuga, K., 1997. The development of groundwater resources on the Miyakojima Islands. In: Uitto, J.I. and Schneider, J. (Eds.), *Freshwater Resources in Arid Lands*. UNU Global Environmental Forum V, United Nations University Press, Tokyo. http://www.unu.edu/unupress/unup-books/uu02fe/uu02fe02.htm (accessed on October 27, 2005).

Paillet, F.L., 1989. Analysis of geophysical well logs and flowmeter measurements in boreholes penetrating subhorizontal fracture zones, Lac du Bonnet Batholith, Manitoba, Canada. U.S. Geological Survey Water-Resources Investigations Report 89-4211, Lakewood, CO, 30pp.

Paillet, F.L., 1994. Application of borehole geophysics in the characterization of flow in fractured rocks. U.S. Geological Survey Water-Resources Investigations Report 93-4214, Denver, CO, 36pp.

Paillet, F.L., 1998. Flow modeling and permeability estimation using borehole flow logs in heterogeneous fractured formations. *Water Resources Research*, 34(5), 997–1010.

Paillet, F.L., 2000. A field technique for estimating aquifer parameters using flow log data. *Ground Water*, 38(4), 510–521.

Paillet, F.L., 2001. Hydraulic head applications of flow logs in the study of heterogeneous aquifers. *Ground Water*, 39(5), 667–675.

Paillet, F.L. and Reese, R.S., 2000. Integrating borehole logs and aquifer tests in aquifer characterization. *Ground Water*, 38(5), 713–725.

Papadopulos, I.S., 1965. Nonsteady flow to a well in an infinite anisotropic aquifer. *Proceedings of the Dubrovnik Symposium on the Hydrology of Fractured Rocks*, International Association of Scientific Hydrology, pp. 21–31.

Papadopulos, I.S. and Cooper, H.H., 1967. Drawdown in a well of large diameter. *Water Resources Research*, 3, 241–244.

Pereira, L.S., Cordery, I., and Iacovides, I., 2002. *Coping with Water Scarcity*, International Hydrological Programme VI, Technical Documents in Hydrology No. 58, Paris, 269pp.

Petkewich, M.D. et al., 2004. Hydrologic and geochemical evaluation of aquifer storage recovery in the Santee Limestone/Black Mingo aquifer, Charleston, South Carolina, 1998–2002. U.S. Geological Survey Scientific Investigations Report 2004-5046, Reston, VA, 81pp.

Prickett, T.A., 1965. Type-curve solution to aquifer tests under water-table conditions. *Ground Water*, 3(3), 5–14.

Pruitt, W.O. and Lourence, F.J., 1985. Experience in lysimetry for ET and surface drag measurements. In: *Advances in Evaporation, ASAE Publication* 14-85, American Society of Agricultural Engineers, St. Joseph, MI, pp. 51–69.

Puckett, L.J., 2004. Hydrogeologic controls on the transport and fate of nitrate in ground water beneath riparian buffer zones: results from thirteen studies across the United States. *Water Science and Technology*, 49(3), 47–53.

Puri, S. et al., 2001. Internationally Shared (transboundary) Aquifer Resources Management: their Significance and Sustainable Management. *A Framework Document*, IHP-VI, Series on Groundwater 1, IHP Non Serial Publications in Hydrology, UNESCO, Paris, 76pp.

Pyne, R.D.G., 1995. *Groundwater Recharge and Wells: A Guide to Aquifer Storage Recovery*, Lewis Publishers, Boca Raton, FL, 375pp.

Radovanovic, S., 1897. *Podzemne vode; izdani, izvori, bunari, terme i mineralne vode (Ground Waters; Aquifers, Springs, Wells, Thermal and Mineral Waters; in Serbian)*, Srpska kniževna zadruga, 42, 152pp.

Rantz, S.E. et al., 1982. *Measurements and Computation of Streamflow; Vol. 1: Measurements of Stage and Discharge; Vol. 2: Computation of Discharge*, U.S. Geological Survey Water-Supply Paper 2175, 631pp.

Ravi, V. and Williams, J.R. 1988. Estimation of infiltration rate in the vadose zone: compilation of simple mathematical models, Volume 1. EPA 600/R-97/128a. U.S. Environmental Protection Agency, National Risk Management Research Laboratory, Ada, Oklahoma, 26 p.

Rawls, W.J., et al., 1993. Infiltration and soil water movement. In: Maidment, D.R. (Ed.), *Handbook of Hydrology*, McGraw-Hill, New York, pp. 5.1–5.51.

Razack, M. and Huntley, D., 1991. Assessing transmissivity from specific capacity in a large and heterogeneous alluvial aquifer. *Ground Water*, 29(6), 856–861.

Rebouças, A. and Mente, A., 2003. Groundwater resources and their use in South America, Central America and the Caribbean. In: Zektser, I.S. and Everett, L.G. (Eds.), *Groundwater Resources of the World and their Use*, IHP-VI, Series on Groundwater No. 6, UNESCO, Paris, France, pp. 189–208.

Reddy, K.R. et al., 1999. Phosphorus retention in streams and wetlands: a review. *Critical Reviews in Environmental Science and Technology*, 29, 83–146.

Reese, R.S., 2002. Inventory and review of aquifer storage and recovery in southern Florida. U.S. Geological Survey Water-Resources Investigations Report 02-4036, Tallahassee, FL, 56pp.

Reese, R.S. and Cunningham, K.J., 2000. Hydrogeology of the Gray Limestone aquifer in Southern Florida. U.S. Geological Survey Water-Resources Investigations Report 99-4213, Tallahassee, Florida, 244pp.

Risser, D.W. et al., 2005a. Estimates of ground-water recharge based on streamflow-hydrograph methods: Pennsylvania. U.S. Geological Survey Open File Report 2005-1333, Reston, VA, 30pp.

Risser, D.W., Gburek, W.J., and Folmar, G.J., 2005b. Comparison of methods for estimating ground-water recharge and base flow at a small watershed underlain by fractured bedrock in the eastern United States. U.S. Geological Survey Scientific Investigations Report 2005-5038, Reston, VA, 31pp.

Rorabaugh, M.I., 1964. Estimating changes in bank storage and ground-water contribution to streamflow. *Extract of Publication No. 63 of the International Association of Scientific Hydrology Symposium Surface Waters*, pp. 432–441.

Roscoe Moss Company, 1990. *Handbook of Ground Water Development*, John Wiley & Sons, New York, 493pp.

Rosenshein, J.S. and Hickey, J.J., 1977. Storage of treated sewage effluent and storm water in a saline aquifer, Pinellas Peninsula, Florida. *Ground Water*, 15(4), 289–293.

Rostad, K., 2002. Fate of disinfection by-products in the subsurface. In: Aiken, G.R. and Kuniansky, E.L. (Eds.) *U.S. Geological Survey Artificial Recharge Workshop Proceedings*, April 2–4, 2002, Sacramento, California, U.S. Geological Survey Open-File Report 02-89, pp. 27–30.

Rutledge, A.T., 1993. Computer programs for describing the recession of ground-water discharge and for estimating mean ground-water recharge and discharge from streamflow records. U.S. Geological Survey Water-Resources Investigations Report 93-4121, 45pp.

Rutledge, A.T., 1998. Computer programs for describing the recession of ground-water discharge and for estimating mean ground-water recharge and discharge from streamflow records—update. U.S. Geological Survey Water-Resources Investigations Report 98-4148, 43pp.

Rutledge, A.T., 2000. Considerations for use of the RORA program to estimate ground-water recharge from streamflow records. U.S. Geological Survey Open-File Report 00-156, Reston, VA, 44pp.

Salem, O. and Pallas, P., 2001. The Nubian Sandstone Aquifer System (NSAS). In: Puri, et al. (Eds.), *Internationally Shared (transboundary) Aquifer Resources Management: Their Significance and Sustainable Management; A Framework Document*, IHP-VI, Series on Groundwater 1, IHP Non-Serial Publications in Hydrology, UNESCO, Paris, pp. 41–44.

Scanlon, B.R., Healy, R.W., and Cook, P.G., 2002. Choosing appropriate techniques for quantifying groundwater recharge. *Hydrogeology Journal*, 10(1), 18–39.

Schicht, R.J., 1965. Ground-water development in East St. Louis area, Illinois. *Illinois State Water Survey Report of Investigations* 51, 70p.

Scholl, M. et al., 2004. Recharge processes in an alluvial aquifer riparian zone, Norman landfill, Norman, Oklahoma, 1998–2000. U.S. Geological Survey Scientific Investigations Report 2004-5238, Oklahoma City, OK, 53pp.

Scott, T.M. et al., 2004. *Springs of Florida*. Florida Geological Survey, Bulletin No. 66, Tallahassee, FL, 658pp.

Shapiro, A.M., 2001. Characterizing ground-water chemistry and hydraulic properties of fractured rock aquifers using the multifunction bedrock-aquifer transportable testing tool (BAT^3). U.S. Geological Survey Fact Sheet FS-075-01, 4pp.

Shapiro, A.M. and Greene, E.A., 1995. Interpretation of prematurely terminated air-pressurized slug tests. *Ground Water*, 33(5), 539–546.

Sheets, R.A., Darner, R.A., and Whitteberry, B.L., 2002. Lag times of bank filtration at a well field, Cincinnati, Ohio, USA. *Journal of Hydrology*, 266, 162–174.

Showcaves, 2005. http://www.showcaves.com/english/fr/springs/Vaucluse.html (accessed on November 15, 2005).

Shuttleworth, W.J., 1993. Evaporation. In: Maidment, D.R. (Ed.), *Handbook of Hydrology*, Mc-Graw Hill,, New York, pp. 4.1–4.53.

Singh, V.P., 1993. *Elementary Hydrology*, Prentice Hall, Englewood Cliffs, NJ, 973pp.

Slichter, C.S., 1902. The motions of underground waters. U.S. Geological Survey Water-Supply and Irrigation Papers 67, Washington, D.C., 106pp.

Smith, J.W.N., 2005. *Groundwater–Surface Water Interactions in the Hyporheic Zone*, Environment Agency, Science Report SC030155/SR1, Bristol, United Kingdom, 65pp.

Stallman, R.W., 1961a. Boulton's integral for pumping-test analysis. In: *Short Papers in the Geologic and Hydrologic Sciences*. U.S. Geological Survey Professional Paper 424-C, Washington, D.C., pp. C24–C29.

Stallman, R.W., 1961b. The significance of vertical flow components in the vicinity of pumping wells in unconfined aquifers. In: *Short Papers in the Geologic and Hydrologic Sciences*. U.S. Geological Survey Professional Paper 424-B, Washington, D.C., pp. B41–B43.

Stallman, R.W., 1963. Type curves for the solution of single boundary problems. In: *Shortcuts and Special Problems in Aquifer Tests*, U.S. Geological Survey Water Supply Paper 1545-C, Washington, D.C., pp. C45–C47.

Stallman, R.W., 1965. Effects of water-table conditions on water-level changes near pumping wells. *Water Resources Research*, 1(2), 295–312.

Stallman, R.W., 1971. Aquifer-test, design, observation and data-analysis. *U.S. Geological Survey Techniques of Water-Resources Investigations*, Book 3, Chap. B1, 26pp.

Stearns, N.D., Stearns, H.T., and Waring, G.A., 1937. Thermal springs in the United States. U.S. Geological Survey Water-Supply Paper 679-B.

Sternberg, Y.M., 1973. Efficiency of partially penetrating wells. *Ground Water*, 11(3), pp. 5–7.

Stevanovic, Z. et al., 2005. Management of karst aquifers in Serbia for water supply—achievements and perspectives. In: Stevanovic, Z. and Milanovic, P. (Eds.), *Water Resources and Environmental Problems in Karst—Cvijić 2005*, Proceedings of the International Conference, Institute of Hydrogeology, University of Belgrade, Belgrade, pp. 283–290.

Stonestrom, D.A. and Constantz, J. (Eds.), 2003. *Heat as a Tool for Studying the Movement of Ground Water Near Streams*. U.S. Geological Survey Circular 1260, Reston, VA, pp. 1–5.

Strahler, A.N. and Strahler, A.H., 1978. *Modern Physical Geography*, John Wiley & Sons, New York, 502pp.

Streltsova, T.D., 1974. Drawdown in compressible unconfined aquifer. *Journal of Hydrology Division, Proceedings of the American Society for Civil Engineers*, 100(HY11), 1601–1616.

Streltsova, T.D., 1988. *Well Testing in Heterogeneous Formations*, John Wiley & Sons, New York, 413pp.

Taylor, C.J. and Alley, W.M., 2001. Ground-water-level monitoring and the importance of long-term water-level data. U.S. Geological Survey Circular 1217, Denver, CO, 68pp.

Theis, C.V., 1935. The lowering of the piezometric surface and the rate and discharge of a well using ground-water storage. *Transactions, American Geophysical Union*, 16, 519–524.

Theis, C.V., Brown, R.H., and Meyer, R.R., 1963. Estimating the transmissibility of aquifers from the specific capacity of wells. In: Bentall, R. (Compiler), *Methods of Determining Permeability, Transmissibility, and Drawdown*. U.S. Geological Survey Water-Supply Paper 1536-I, pp. 331–341.

Thiem, G., 1906. *Hydrologische Methoden*, Leipzig, Gebhardt, 56pp.

Thomasson, H.J., Olmstead, F.H., and LeRoux, E.R., 1960. Geology, water resources, and usable ground water storage capacity of part of Solano County, CA. U.S. Geological Survey Water Supply Paper 1464, 693pp.

Thornthwaite, C.W., 1948. An approach toward a rational classification of climate. *The Geological Review*, January, pp. 55–94.

Touloumdjian, C., 2005. The springs of Montenegro and Dinaric karst. In: Stevanovic, Z. and Milanovic, P. (Eds.), *Water Resources and Environmental Problems in Karst—Cvijić 2005*, Proceedings of the International Symposium, Institute of Hydrogeology, University of Belgrade, Belgrade, pp. 443–450.

Trapp, H. Jr. and Horn, M.A., 1997. Delaware, Maryland, New Jersey, North Carolina, Pennsylvania, Virginia, West Virginia. *Ground Water Atlas of the United States*, United States Geological Survey, HA 730-L.

Tsang, C.-F. and Hale, F.V., 1989. A direct integral method for the analysis of borehole fluid conductivity logs to determine fracture inflow parameters. In: *Proceedings of the National Water Well Conference on New Field Techniques for Quantifying the Physical and Chemical Properties of Heterogeneous Aquifers*, Dallas, Texas, March 20–23, 1989, Rep. LBL-27930, Lawrence Berkeley Laboratory, Berkeley, CA, 1989.

Tsang, C.-F., Hufschmeid, P., and Hale, F.V., 1990. Determination of fracture inflow parameters with a borehole fluid conductivity logging method, *Water Resources Research*, 26(4), 561–578.

Turcan, A.N. Jr., 1963. Estimating the specific capacity of a well. U.S. Geological Survey Professional Paper 450-E.

United States Army Corps of Engineers (USACE), 1999. *Groundwater hydrology. Engineer Manual* 1110-2-1421, Washington, D.C. pp. 1–1 to D-8.

USBR, 1977. *Ground Water Manual*. U.S. Department of the Interior, Bureau of Reclamation, Washington, D.C., 480pp.

U.S. Department of Agriculture, 1957. Monthly precipitation and runoff for small watersheds in the United States. *Agricultural Research Service, Soil and Water Conservation Research Branch*, p. 691.

USGS (United States Geological Survey), 2003. Ground-water depletion across the Nation. *USGS Fact Sheet* 103-03, U.S. Department of the Interior, U.S. Geological Survey, 4pp.

USGS (United States Geological Survey), Office of Ground Water, Branch of Geophysics, 2004. Vertical Flowmeter Logging, URL: http://water.usgs.gov/ogw/bgas/flowmeter/

Vargas, C. and Ortega-Guerrero, A., 2004. Fracture hydraulic conductivity in the Mexico City clayey aquitard: field piezometer rising-head tests. *Hydrogeology Journal*, 12, 336–344.

Verstraeten, I.M., 2000. Minimizing the risk of herbicide transport into public water supplies—a Nebraska case study. U.S. Geological Survey Fact Sheet 078-00, 4pp.

Vineyard, J.D. and Feder, G.L., 1982. Springs of Missouri. Water Resources Report No. 29, Missouri Department of Natural Resources, Division of Geology and Land Survey, 212pp.

Vukovic, M. and Soro, A., 1984. *Dinamika Podzemnih Voda* (*Dynamics of Groundwater*, in Serbo-Croatian). Posebna Izdanja, Knjiga 25, Institut Jaroslav Cerni, Beograd, 500pp.

Walton, W.C., 1960. Leaky artesian aquifer conditions in Illinois. Illinois State Water Survey Report of Investigation, 39.

Walton, W.C., 1962. Selected analytical methods for well and aquifer evaluation. *Illinois State Water Survey Bulletin*, 49, 81pp.

Walton, W.C., 1970. *Groundwater Resource Evaluation*, McGraw-Hill, New York.

Walton, W.C., 1987. *Groundwater Pumping Tests, Design & Analysis*, Lewis Publishers, Chelsea, MI, 201pp.

Warner, D., 1997. Hydrogeologic evaluation of the Upper Floridan aquifer in the southwestern Albany area. Georgia: U.S. Geological Survey Water-Resources Investigations Report 97-4129, Atlanta, GA, 27 p.

Warner, J.W. et al., 1989. Mathematical analysis of artificial recharge from basins. *Water Resources Bulletin*, 25, 401–411.

World Commission for Environment and Development (WCED), 1987. *Our Common Future*, Oxford University Press, New York.

Wenzel, L.K., 1936. The Thiem method for determining permeability of water-bearing materials and its application to the determination of specific yield; results of investigations in the Platte River valley, Nebraska. U.S. Geological Survey Water Supply Paper 679-A, Washington, D.C., 57pp.

Westbrook, S.J. et al., 2005. Interaction between shallow groundwater, saline surface water and contaminant discharge at a seasonally and tidally forced estuarine boundary. *Journal of Hydrology*, 302, 255–269.

Whitehead, R.L., 1994. Ground Water Atlas of the United States—Segment 7: Idaho, Oregon, Washington. U.S. Geological Survey Hydrologic Investigations Atlas HA-730-H, 31pp.

Williams, J.H. and Johnson, C.D., 2004. Acoustic and optical borehole-wall imaging for fractured-rock aquifer studies. *Journal of Applied Geophysics*, 55(1–2), 151–159.

Williams, J.H. and Lane, J.W., 1998. Advances in borehole geophysics for ground-water investigations. U.S. Geological Survey Fact Sheet 002-98, 4pp.

Williams, J.H. et al., 2002. Application of advanced geophysical logging methods in the characterization of a fractured-sedimentary bedrock aquifer, Ventura County, California. U.S. Geological Survey Water-Resources Investigations Report 00-4083, 28pp.

Williams, J.R., Ouyang, Y., and Chen, J.-S. 1998. Estimation of infiltration rate in the vadose zone: application of selected mathematical models, Volume II. EPA/600/R-97/128b, U.S. Environmental Protection Agency, National Risk Management Research Laboratory, Ada, Oklahoma, 44p.

Wilson, J.T. et al., 2001. An evaluation of borehole flowmeters used to measure horizontal ground-water flow in limestones of Indiana, Kentucky, and Tennessee, 1999. U.S. Geological Survey Water-Resources Investigations Report 01-4139, Indianapolis, IN, 129pp.

Winter, T.C. et al., 1998. Ground water and surface water. A single resource. U.S. Geological Survey Circular 1139, Denver, CO, 79pp.

Wolfe, W.J. et al., 1997. Preliminary conceptual models of the occurrence, fate, and transport of chlorinated solvents in karst regions of Tennessee. U.S. Geological Survey Water-Resources Investigations Report 97-4097, Nashville, TN, 80pp.

Wright, M. et al., 2002. Measurement of 3-D hydraulic conductivity in aquifer cores at *in situ* effective stresses. *Ground Water*, 40(5), 509–517.

Yobbi, D.K., 1997. Simulation of subsurface storage and recovery of effluent using multiple wells, St. Petersburg, Florida. U.S. Geological Survey Water-Resources Investigations Report 97-4024, Tallahassee, FL, 30pp.Zemansky, G.M. and McElwee, C.D., 2005. High-resolution slug testing. *Ground Water*, 43(2), 222–230.

Zemansky, G.M. and McElwee, C.D., 2005. High-resolution slug testing. *Ground Water*, 43(2), 222–230.

3   Groundwater Extraction

The extraction of groundwater by artificial means for various purposes is arguably the most important final stage for the majority of hydrogeologic projects. When surface water, or treated contaminated groundwater, is injected into the aquifer as part of a project, such as aquifer recharge or groundwater remediation, in many cases this injection can also be considered as extraction, but with a negative sign (water is added to the aquifer), because the basic hydraulic principles and engineering design are quite similar. Probably the first image that comes to mind to many people when mentioning groundwater extraction and use in general is that of a *well*. For nonhydrogeologists and those who are not in a related water supply profession, well usually means a nondescript hole in the ground that somehow produces water; this may include an image of a "mysterious" fenced well house, or a picturesque countryside image of a dug well with a rotating wooden wheel and a bucket. In any case, a relatively small number of people fully understand the complexity, importance, and cost of a properly constructed well used for public water supply. The same is true in many developed countries where modern drilling technologies are routinely used to construct wells for domestic supply: the end users usually leave this "well business" to well drillers and do not care to learn much about their own "hole" in the ground. Hydrogeologists and groundwater professionals, however, think of wells in many different contexts, and some of them spend lifetimes trying to better understand wells and their various interactions with aquifers and groundwater. The main difficulty in all this is, and will continue to be, a very simple fact: every well is a hole in the ground (with more or less equipment in it) that cannot be visually examined so that everything about it has to be determined indirectly.

Groundwater extraction, in general, is performed either because of water supply needs, or because groundwater causes some undesirable effects, such as when contaminated, or when impacting agricultural fields and human-made structures (excavations, mine works, buildings, or transportation infrastructure). In addition to classical vertical wells, groundwater extraction is performed in a number of other ways, including with horizontal and slanted wells, collector wells with drains, infiltration galleries, drainage galleries, trenches, and drains. Regardless of the type of groundwater extraction design, they all are based on the following general design criteria and requirements:

1. Structural and operational safety
2. Capacity (yield or groundwater discharge rate), both short-term and long-term
3. Aquifer material (porous media) must stay in place, i.e., it cannot be extracted together with groundwater
4. Radius of influence
5. Impact on quality and quantity of groundwater in the targeted aquifer and the adjacent aquifers
6. Protection from contamination (from both the land surface and the subsurface)

7. Impact on the environment
8. Capital cost
9. Operations and maintenance requirements and cost

3.1 VERTICAL WELLS

3.1.1 WELL DESIGN

Vertical wells have been used for centuries for domestic and public water supply throughout the world. Their depth, diameter, and construction methods vary widely and there is no such thing as "one size fits all" approach to well design. This author's favorite source of answers to just about any question regarding well design is the classic 1000-page book *Groundwater and Wells* by Driscoll (1986, published by Johnson Filtration Systems, now available from Johnson Screens, a Weatherford Company). Useful publications by the U.S. government agencies include those by the U.S. Environmental Protection Agency (USEPA, 1975, 1991) and the U.S. Bureau of Reclamation (USBR, 1977).

Three main factors influence well design parameters, as illustrated in Figure 3.1: (i) expected yield, (ii) depth to aquifer productive zone, and (iii) physical and chemical characteristics of porous media and groundwater. An optimum well design should be based on information obtained by a pilot boring, drilled prior to the main well bore. Geophysical logging and coring (sample collection) of the pilot boring provide the following key information: depth to and thickness of the water-bearing intervals in the aquifer, grain size, and permeability of the targeted intervals, and physical and chemical characteristics of the porous media and groundwater. If pilot boring is not feasible, some design parameters would have to be estimated, including on the conservative side (for example, selecting smaller screen openings or gravel pack grain size to prevent entrance of fines). This may significantly reduce well efficiency. The drilling method selection depends on the geologic formation and well depth as illustrated in Table 3.1. Deep wells or thick stratification of permeable and low-permeable porous media

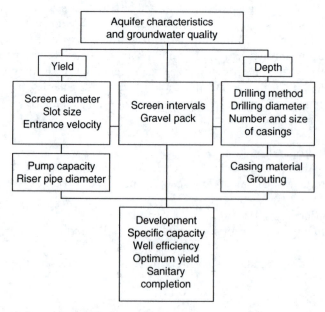

FIGURE 3.1 Factors influencing well design parameters.

TABLE 3.1
Applicable Drilling Method for Different Types of Geologic Formations

Characteristics	Dug	Bored	Driven	Drilled			
					Rotary		
				Percussion	Hydraulic	Air	Jetted
General range of common depths	0–50 ft	0–100 ft	0–50 ft	0–1000 ft	0–1000 ft	0–750 ft	0–100 ft
Diameter	3–20 ft	2–30 in.	$1\frac{1}{4}$–2 in.	4–18 in.	4–24 in.	4–10 in.	2–12 in.
Type of geologic formation							
Clay	Yes	Yes	Yes	Yes	Yes	No	Yes
Silt	Yes	Yes	Yes	Yes	Yes	No	Yes
Sand	Yes	Yes	Yes	Yes	Yes	No	Yes
Gravel	Yes	Yes	Finer size	Yes	Yes	No	¼ in. pea gravel
Cemented gravel	Yes	No	No	Yes	Yes	No	No
Boulders	Yes	Yes, if less than well diameter	No	Yes, when in firm bedding	Difficult	No	No
Sandstone	Yes, if soft or fractured	Yes, if soft or fractured	Thin layers	Yes	Yes	Yes	No
Limestone	No	No	No	Yes	Yes	Yes	No
Dense igneous rock	No	No	No	Yes	Yes	Yes	No

Source: From USEPA, *Manual of Small Public Water Supply Systems*, Office of Water, EPA 570/9-91-003, 1991, 211pp.

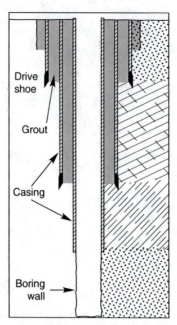

FIGURE 3.2 Drilling with reduced diameters and installation of multiple casing may be required in deep wells, unstable formations, and to isolate undesirable intervals. (Modified from Driscoll, F.G., *Groundwater and Wells*, Johnson Filtration Systems Inc, St. Paul, MN, 1986, 1089pp; reprinted by permission of Johnson Screens, a Weatherford Company.)

may require drilling with several diameters, and installation of several casings of progressively smaller diameter, called telescoping casing (Figure 3.2). This is done to provide for stable and plumb borehole in deep wells, or to bridge difficult or undesirable intervals (e.g., flowing sands, highly fractured and unstable walls prone to caving, thick sequences of swelling clay). The cost of drilling increases progressively with the drilling diameter and it is important to balance this cost with other design requirements, some of which may be desirable but not always necessary. For perspective, a public supply, high-capacity well that is several thousands feet deep (say, 1000 m) may easily cost well over US$ 1 million. Such wells are drilled with large drill rigs and may utilize special bits as shown in Figure 3.3.

Ultimately, the expected well capacity is the parameter that will define the last drilling diameter sufficient to accommodate the required screen diameter, including thickness of any gravel pack, for that capacity. The relationship between the two diameters is not linear—doubling the screen diameter will not result in doubling the well yield as illustrated in Figure 3.4. For example, for the same drawdown and radius of influence, an increase in diameter from a 6-in. well to a 12-in. well will yield only 10% more water. In some cases, however, it may be worthwhile to increase well diameter to obtain 15% to 25% more water, depending on the cost factors involved. In addition to the screen diameter, the *riser pipe* (*inner casing*) diameter also plays a limiting role in determining the effective well yield. The riser pipe may have the same diameter as the screen, or it may be larger in which case the screen and the casing are connected with a diameter reducer. In either case, the riser pipe diameter must satisfy two requirements: (i) the casing must be large enough to accommodate the pump of required capacity, and to provide for easy maintenance access, and (ii) the diameter of the casing must be sufficient to assure that the uphole velocity is 5 ft/s (1.5 m/s) or less to avoid an excessive pipe loss (Driscoll, 1986). Table 3.2 lists optimum and minimum size of well inner

FIGURE 3.3 Roller or cone-type bits are preferred when drilling consolidated rock and very deep wells. They are often constructed in configurations such as this one to enlarge the borehole in stages and maintain plumbness.

casing (riser pipe) for various pumping rates, and Table 3.3 lists the maximum pumping rates for which the uphole velocity is 5 ft/s and the friction losses are still acceptable.

All casings in a completed well have to be grouted and cannot be left loose. Grouting prevents possible short-circuiting of groundwater along the boring walls and between various aquifer intervals or aquifers, and possible contamination from the land surface. Casing material has to be compatible with the groundwater chemistry to prevent corrosion or other failures. Casing has to be strong enough to provide for structural stability, especially in deep wells where high formation pressures may cause it to collapse.

Wells in stable bedrock are in most cases completed as an open borehole intersecting as many fractures as possible to maximize well yield. Such wells have the casing in the upper portion of the well, appropriately grouted to prevent possible contamination of the aquifer and the well from land surface (Figure 3.5). Although the final boring diameter still has to accommodate the pump assembly and easy maintenance access, it is not limited by various well screen design parameters.

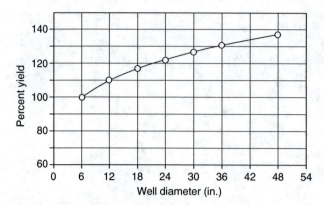

FIGURE 3.4 Graph showing well diameter vs. percent yield increase from the basic 6-in. screen diameter well in an unconfined aquifer, pumping at 100 gallons/min and having a 400-ft radius of influence. (Data from Driscoll, F.G., *Groundwater and Wells*, Johnson Filtration Systems Inc, St. Paul, MN, 1986, 1089pp.)

Well screen is the most important part of a well since this is where groundwater enters the well and where the entire well may be lost due to an inadequate design. It is understood that the screen should be placed within a thick aquifer interval with the highest hydraulic conductivity. However, when the aquifer is relatively thin or stratified, it is sometimes preferable to use multiple screen intervals separated by solid casing, including screens with varying slot size. Screen intervals will also depend on the aquifer transmissivity and well specific capacity, which would determine acceptable drawdown; in other words, any screen portion of a water supply well should not be dewatered during pumping of the well. One additional requirement is that well pump intake should not be placed within a screen interval but within a solid riser pump section to avoid hydraulic stresses on the screen. Recommendations as to the screen intervals and their lengths for four typical hydrogeological situations are as follows (modified from Driscoll, 1986):

TABLE 3.2
Recommended Well Diameters for Various Pumping Rates

Anticipated Well Yield		Optimum Casing Size		Smallest Casing Size	
gpm	l/s	In.	mm	In.	mm
<100	<5	6 ID	152 ID	5 ID	127 ID
75 to 175	5 to 10	8 ID	203 ID	6 ID	152 ID
150 to 350	10 to 20	10 ID	254 ID	8 ID	203 ID
300 to 700	20 to 45	12 ID	305 ID	10 ID	254 ID
500 to 1000	30 to 60	14 OD	356 OD	12 ID	305 ID
800 to 1800	50 to 110	16 OD	406 OD	14 OD	356 OD
1200 to 3000	75 to 190	20 OD	508 OD	16 OD	406 OD
2000 to 3800	125 to 240	24 OD	610 OD	20 OD	508 OD
3000 to 6000	190 to 380	30 OD	762 OD	24 OD	610 OD

Source: From Driscoll, F.G., *Groundwater and Wells*, Johnson Filtration Systems Inc, St. Paul, MN, 1986, 1089pp; reprinted by permission of Johnson Screens, a Weatherford Company.

ID, inside diameter; OD, outside diameter. gpm, gallons/min.

TABLE 3.3
Maximum Discharge Rates for Certain Diameters of Standard-Weight Casing, Based on an Uphole Velocity of 5 ft/s (1.5 m/s)

Casing Size		Maximum Discharge	
In.	mm[a]	gpm	l/s
4	102	200	13
5	127	310	20
6	152	450	28
8	203	780	49
10	254	1230	78
12	305	1760	111
14	337	2150	136
16	387	2850	180
18	438	3640	230
20	489	4540	286
24	591	6620	418

[a]Actual inside diameter.

Source: From Driscoll, F.G., *Groundwater and Wells*, Johnson Filtration Systems Inc, St. Paul, MN, 1986, 1089pp; reprinted by permission of Johnson Screens, a Weatherford Company.
gpm, gallons/min.

1. *Homogeneous unconfined aquifer*: Screening of the bottom one third to one half of an aquifer less than 150 ft (45 m) thick provides the optimum design for homogeneous unconfined aquifers. In some cases, however, particularly in thick, deep aquifers, as much as 80% of the aquifer may be screened to obtain higher specific capacity and greater efficiency, even though the total yield is less. A well in an unconfined aquifer is

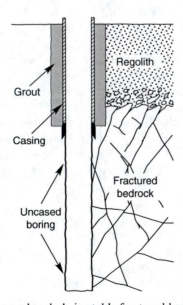

FIGURE 3.5 Well completed as open borehole in stable fractured bedrock.

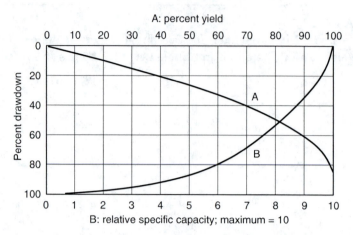

FIGURE 3.6 Relationship between well yield and drawdown in an unconfined aquifer with 100% penetration. (From USBR, *Ground Water Manual*, U.S. Department of the Interior, Bureau of Reclamation, Washington, D.C., 1977, 480pp.)

usually pumped so that, at maximum capacity, the pumping water level is maintained slightly above the top of the pump intake or screen. The well screen is positioned in the lower portion of the aquifer because the upper part is dewatered during pumping. Maximum drawdown should not exceed two thirds of the saturated thickness because larger drawdown does not provide significant additional yield (see Figure 3.6) but increases well loss and energy cost for pumping.

2. *Heterogeneous unconfined aquifer*: The basic principles of well design for homogeneous unconfined aquifers also apply to this type of aquifer. The only variation is that the screen or screen sections are positioned in the most permeable layers of the lower portion of the aquifer so that maximum drawdown of two thirds of the aquifer saturated thickness is available.

3. *Homogeneous confined aquifer*: In this type of aquifer, 80% to 90% of the thickness of the water-bearing sediment should be screened, assuming that the pumping water level is not expected to be below the top of the aquifer. Maximum available drawdown for wells in confined conditions should be the distance from the hydraulic head (potentiometric) surface to the top of the aquifer. If the available drawdown is limited, it may be necessary to lower the hydraulic head below the aquifer top in which case the aquifer will respond like an unconfined aquifer during pumping.

4. *Heterogeneous confined aquifer*: Most relatively thick confined aquifers are heterogeneous and screen sections should be placed in 80% to 90% of the permeable layers, interspaced with blank casing in the less permeable (silt and clay) zones of the formation. Continuous screens of varying slot size (multiple-slot screens) can be successfully utilized in generally permeable, water-bearing aquifer sections, consisting of alternating layers of finer and coarser sediments as shown in Figure 3.7. Two recommendations should be followed when selecting slot openings for such screens to avoid entrance of the finer material into the well (Driscoll, 1986):
   i. If the fine material overlies the coarse material, extend at least 3 ft (0.9 m) of the screen designed for the fine material into the coarse material below.
   ii. The slot size for the screen section installed in the coarse layer 3 ft beneath the formation contact should not be more than double the slot size for the overlying finer material. Doubling of the slot size should be done over screen increments of 2 ft (0.6 m) or more.

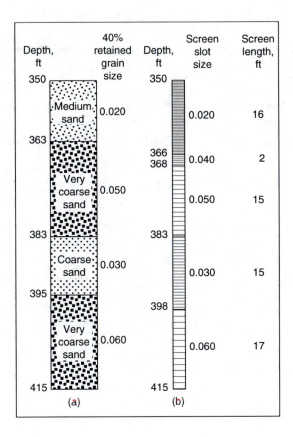

FIGURE 3.7 (a) Stratigraphic section that will be screened with slot size corresponding to various layers. (b) Sketch of screen showing the slot sizes selected based on rules 1 and 2. (From Driscoll, F.G., *Groundwater and Wells*, Johnson Filtration Systems Inc, St. Paul, MN, 1986, 1089pp; reprinted by permission of Johnson Screens, a Weatherford Company.)

The size of screen openings (*screen slot size*) should be large enough to permit efficient entrance of groundwater but, at the same time, it must prevent aquifer material from entering the well. Selecting the optimum balance between these two requirements is arguably the most critical phase of screen (and well) design. Two parameters dictate this selection: entrance velocity of groundwater and the size of various fractions of the aquifer porous media. The screen entrance velocity should be less than 0.1 ft/s (0.03 m/s) since it has been shown that higher velocities cause turbulent well loss and may accelerate various screen problems such as corrosion and incrustation. The average entrance velocity is calculated by dividing the well yield by the total area of screen openings. It follows that three possible changes in the screen design parameters will decrease the entrance velocity: larger screen diameter, longer screen section, and larger slot size. While larger screen diameter is limited only by cost, the other two screen elements are directly influenced by the aquifer porous material, i.e., its grain sizes. In a naturally developed well, the screen slot size is selected so that most of the finer formation materials near the borehole are brought into the screen and pumped from the well during development. The typical approach is to select a slot through which 60% of the material will pass and 40% will be retained. However, only coarse-grained heterogeneous material can be developed in this way and in most cases it will be necessary to use a gravel (filter) pack around the screen. Screen slot openings for either method are selected based on the grain size analysis of aquifer materials performed in the laboratory on representative aquifer samples. This is

TABLE 3.4
Slot Size and Corresponding Open Areas of Screens for Several Common Slot Configurations

Screen Diameter	Slot Size[a]	Continuous Slot		Louvered (Maximum Open Area)		Bridge Slot		Mill Slotted (Vertical)		Plastic Continuous Slot		Slotted Plastic	
		in.²/ft	%	in.²/ft	%	in.²/ft	%	in.²/ft	%	in.²/ft	%	in.²/ft	%
4(in) ID	20	44	25	—	—	—	—	—	—	22	13	—	—
	60	90	52	—	—	19	12	8	5	52	30	18	11
	30	80	25	—	—	—	—	—	—	57	18	26	8
8(in) ID	60	135	41	10	3	17	6	15	5	93	29	47	14
	95	165	51	15	5	—	—	23	7	—	—	—	—
	30	77	16	—	—	12	3	—	—	—	—	—	—
12(in) ID	60	135	28	20	4	33	7	21	5	—	—	52	11
	95	182	38	30	7	—	—	32	7	—	—	—	—
	125	214	45	39	9	68	14	43	9	—	—	—	—
	30	97	16	—	—	16	3	—	—	—	—	52	9
16(in) OD	60	169	28	24	4	35	6	27	5	—	—	—	—
	95	228	38	35	6	—	—	41	7	—	—	—	—
	125	268	45	47	8	78	13	55	9	—	—	—	—

[a]Slot size 30 is 0.03 in.; slot size 125 is 0.125 in.

Source: From Driscoll, F.G., *Groundwater and Wells*, Johnson Filtration Systems Inc, St. Paul, MN, 1986, 1089pp; reprinted by permission of Johnson Screens, a Weatherford Company.

explained in detail in Sections 9.3 and 15.1. Slot openings have been designated by numbers, which correspond to the width of the openings in thousands of inch. A number 30 slot, for example, is an opening of 0.030 in. Table 3.4 lists slot sizes for several common screen types and the corresponding open area per 1 ft of the screen. Continuous-slot screens, such as those pioneered by Johnson Screens, have much larger open areas than any other screen type and assure maximum specific capacity. The same screen can also be made to have varied individual slot sizes matching variations of geologic conditions in the aquifer productive zone.

Gravel pack has the following main purposes:

• Stabilize the formation
• Prevent or reduce pumping of fines and sand
• Enable larger screen openings
• Establish transitional velocity and pressure fields between the formation and the well screen

The placement of a gravel pack makes the zone around the well screen more permeable and increases the effective hydraulic diameter of the well. The gravel pack allows the removal of some formation material during well development, and it retains most of the aquifer fine material during the well exploitation. Gravel pack is particularly useful in fine-grained, uniformly graded formations, and in extensively laminated aquifers where there are alternating layers of silt, sand, and gravel. In addition to sand pumping, which can destroy the pump and the screen, another very important problem in well maintenance can be avoided or deferred by a properly installed gravel pack—chemical and biological incrustation of the

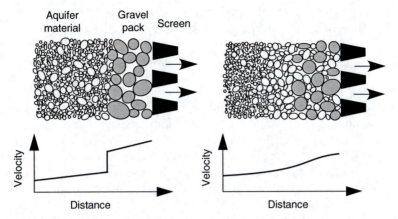

FIGURE 3.8 Distribution of the grains and the groundwater velocity in the well screen zone before (left) and after (right) development of the well.

well screen and the material adjacent to the well screen. Carbonate, iron, and manganese incrustation are the most common problems. They are related to the velocity-induced pressure changes that disturb the chemical equilibrium of the groundwater in the well screen zone. Figure 3.8 shows the distribution of the grains and the groundwater velocity in the screen section of a well with the gravel pack, before and after the well development. Due to the

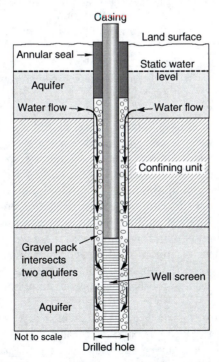

FIGURE 3.9 Long gravel pack can cause aquifer contamination and can lead to uncertainty as to the source of water to a well. (From Lapham. W.W., Franceska, W.D., and M.T. Koterba, 1997. Guidelines and standard procedures for studies of ground-water quality: selection and installation of wells, and supporting documentation. U.S. Geological Survey Water-Resources Investigations Report 96-4233, Reston, VA, 110pp.)

permeability change, there is an abrupt increase of groundwater velocity at the contact between the aquifer material and the gravel pack before the well development. This increase corresponds to an equally abrupt decrease of pressure, which causes the precipitation of calcium carbonate, iron, and manganese. This precipitation may occur either in an improperly developed well with a gravel pack, or in a well without a gravel pack (here the pressure drop occurs at the contact between the aquifer material and the screen). Well development results in a significant change in velocity and pressure fields, and permeability in the screen zone. Most of the fine grains are removed from the aquifer material resulting in an increased permeability. At the same time, the permeability of the gravel pack portion is somewhat smaller due to the filling of the pore space between the pack grains with the aquifer material. The result is a gradually increasing groundwater velocity and a gradually decreasing pressure in the well screen zone, which allows groundwater to carry its chemical load into the well rather than to precipitate it onto the screen and the aquifer or pack material.

In order to successfully retain the formation particles, the thickness of the gravel (filter) pack in ideal conditions does not have to be more than 0.5 in. according to laboratory tests made by Johnson Screens: "Filter-pack thickness does little to reduce the possibility of sand pumping, because the controlling factor is the ratio of the grain size of the pack material in relation to the formation material" (Driscoll, 1986). However, for practical purposes, the thickness of the gravel pack should be at least 3 in. to ensure its accurate placement and complete surrounding of the screen. On the other hand, a filter pack that is more than 8 in. (203 mm) thick can make the final development of the well more difficult "... because the energy created by the development procedure must be able to penetrate the pack to repair the damage done by drilling, break down any residual drilling fluid on the borehole wall, and remove fine particles (from the formation) near the borehole" (Driscoll, 1986). The selection of gravel pack characteristics for uniform and nonuniform aquifer materials is explained in detail in Chapter 15.

Figure 3.10 shows key design elements for a typical well completed in an unconfined aquifer. After the well is completed and properly developed to remove drilling cuttings and fines from the near-well zone, gravel pack and screen, it has to be tested for specific yield and well efficiency. These parameters are required to determine the optimum well capacity, including the need for additional development, to select and size the pump, and design the pumping schedule. Well completion includes installation of the permanent well pump and construction of the well housing (wellhead) with all its sanitary requirements. It is also highly recommended that a dedicated small-diameter (less than 1 in.) perforated line be installed in the well for future water level measurements and depth-discrete sampling. Probably the only design and construction element that cannot be reasonably guaranteed by any well driller or consultant, even though it is the most important one, is the exact well yield and its long-term sustainability. It is not uncommon that, for various reasons, a well that costs several hundred thousands dollars to complete disappoints all stakeholders by actually producing just a fraction of the designed capacity; the opposite is also true. However, many such surprises can be avoided by following well-established hydrogeologic principles of aquifer evaluation and testing, and, of course, well design itself. Arguably, a little bit of luck could never hurt when installing a well as illustrated in Figure 3.11, which is a photograph of the highest-capacity drilled well in the world, located near San Antonio, Texas and known as *Ron Pucek's Catfish Farm well*. This well is not active any more as part of the efforts to better manage Edwards aquifer for the overall beneficial use. At its heyday, the 30-in. well was capable of producing an incredible 40,000 gallons/min (2500 l/s or 2.5 m^3/s), thus creating a considerable-size surface stream as shown in Figure 3.11. More information on the "world's biggest well" and the prolific Edwards aquifer in general can be found on a web site, maintained by Gregg Eckhardt (http://www.edwardsaquifer.net).

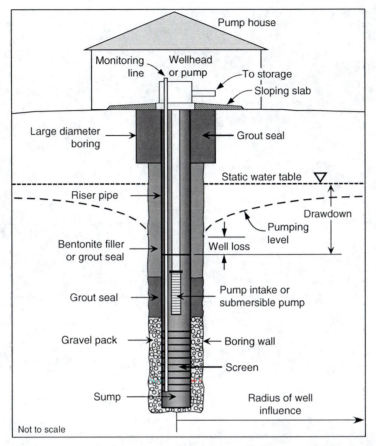

FIGURE 3.10 Typical shallow well completed in an unconfined aquifer, with key design elements.

Monitoring wells are an integral part of the overall extraction well design, construction, and testing for either water supply or groundwater contamination studies and remediation purposes. Since they are not covered in this book, the reader is directed to the following useful publications with emphasis on environmental applications: *Handbook of Suggested Practices for the Design and Installation of Ground-Water Monitoring Wells* by Aller et al. (USEPA, 1991) and *Selection and Installation of Wells, and Supporting Documentation* by Lapham et al., 1997 of USGS (1997).

3.1.2 WELL LOSS, EFFICIENCY, AND SPECIFIC YIELD

Well loss is the difference between the actual measured drawdown in the pumping well and the theoretical drawdown expressed by an equation describing the flow of groundwater toward the particular well in the particular aquifer. This theoretical drawdown is also called the formation loss, and the related equations should be applicable to the actual aquifer (formation) conditions, such as confined, unconfined, leaky, with delayed gravity response, quasisteady state or transient, as explained in detail in Section 2.5. Well loss is the result of various factors, such as an inevitable disturbance of the porous medium near the well during drilling, an improper well development (e.g., drilling fluid is left in the formation, mud cake along the borehole is not removed), a poorly designed gravel pack or well screen, and turbulent flow through the pack or well screen. Well loss is always present in pumping

FIGURE 3.11 Top: the highest-capacity drilled well in the world, capable of producing 40,000 gallons/min (2.5 m^3/s), located on the Ron Pucek's Catfish Farm near San Antonio, Texas. The well is completed in karstic Edwards aquifer. Bottom: water from the well conveyed toward the catfish basins. (Photo by Gregg Eckhardt; printed with kind permission.)

wells and its evaluation is an important part in deciding if the well performance is satisfactory or not. All wells will also experience a decrease in well efficiency (well aging) sooner or later, as indicated by an increased well loss. Three-step pumping test is the only reliable means of quantifying the well loss, and it should be performed not only after well completion, but also periodically during well exploitation to evaluate the well performance and needs for possible well rehabilitation.

The total measured drawdown at a well is a combination of the linear losses and turbulent losses:

$$s_w = AQ + BQ^2 \tag{3.1}$$

where A is the coefficient of the linear losses, B is the coefficient of turbulent losses, and Q is the pumping rate. The turbulent losses are usually assumed to be quadratic, but other powers may be used to describe it. The linear losses include both formation loss and linear loss in the near-screen zone. Their respective coefficients are A, A_0, and A_1:

$$A = A_0 + A_1 \tag{3.2}$$

Again, the formation loss, or the theoretical drawdown in the well (s_0), is determined by using the appropriate equation for the specific flow condition (Section 2.5). For example, in case of a quasi steady-state flow in a confined aquifer, the equation is

$$s_0 = \frac{Q}{2\pi T} \ln \frac{R}{r_w} \tag{3.3}$$

The coefficient of linear formation loss (A_0) can be calculated as

$$A_0 = \frac{1}{2\pi T} \ln \frac{R}{r_w} \tag{3.4}$$

or determined graphically as explained in detail in Section 15.2. The coefficients of the total linear loss (A) and the quadratic loss (B) can be determined from a graph pumping rate (Q) vs. drawdown–pumping rate ratio (s/Q) as shown in Figure 3.12. The graph is a straight line of the following form:

$$\frac{s_w}{Q} = A + BQ \tag{3.5}$$

where A is the intercept and B is the slope of the best-fit straight line drawn through the experimental data from the multiple-step pumping test (such a test may include more than

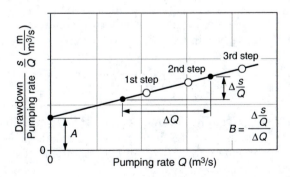

FIGURE 3.12 Graph pumping rate (Q) vs. drawdown–pumping rate ratio (s/Q) used to determine components of drawdown in a pumping well for steady-state conditions. A: Coefficient of the linear losses; B: coefficient of the turbulent losses.

just three steps). After substituting values of A and B determined from the graph into Equation 3.1, it is possible to calculate the total (i.e., expected to be actually recorded) drawdown in the pumping well for any pumping rate.

In case of transient conditions in a confined aquifer, the coefficient of formation loss is easily found by applying the Theis equation:

$$A_0 = \frac{1}{2\pi T} W(u) \tag{3.6}$$

Parameter u for the test well is given as

$$u = \frac{r_w^2 S}{4Tt} \tag{3.7}$$

where r_w is the well radius, S is the storage coefficient, T is the aquifer transmissivity, and t is the time since the pumping started.

As shown by Cooper and Jacob, for small values of parameter u ($u < 0.05$), i.e., sufficiently long pumping time, the well function $W(u)$ is:

$$W(u) = \frac{2.25Tt}{r_w^2 S} \tag{3.8}$$

and the formation loss (i.e., theoretical drawdown s) can be written as

$$s = \frac{Q}{2\pi T} \frac{1}{2} \ln \frac{2.25Tt}{r_w^2 S} \tag{3.9}$$

$$s = \frac{Q}{2\pi T} \ln \sqrt{\frac{2.25Tt}{r_w^2 S}} \tag{3.10}$$

$$s = \frac{Q}{2\pi T} \ln \frac{\sqrt{\frac{2.25Tt}{S}}}{r_w} \tag{3.11}$$

$$s = \frac{Q}{2\pi T} \ln \frac{1.5 \cdot \sqrt{Tt/S}}{r_w} \tag{3.12}$$

Notice that Equation 3.12 looks similar to the steady-state equation describing groundwater flow toward a fully penetrating well in a confined homogeneous aquifer:

$$s = \frac{Q}{2\pi T} \ln \frac{R_D}{r_w} \tag{3.13}$$

where R_D is the radius of the well influence that does not change with time (steady-state flow) and is also called Dupuit's radius of well influence. From the analogy between Equation 3.12 and Equation 3.13 it is apparent that, in transient conditions, Dupuit's radius of well influence is time dependent and is expressed as

$$R_D = 1.5\sqrt{Tt/S} \tag{3.14}$$

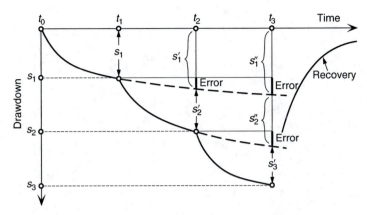

FIGURE 3.13 Components of drawdown recorded at the end of each pumping step. Note the error made if drawdowns s_1, s_2, and s_3 were used directly to draw graph s/Q vs. Q.

Theoretically, for an infinite confined aquifer, the groundwater flow forms in infinity and reaches the well pumping rate at the well perimeter (r_w). The corresponding radius of well influence also approaches infinity for a long pumping period ($t \to \infty$), which means that Dupuit's radius of well influence does not have a real physical meaning. For most practical purposes, however, Dupuit's radius of well influence given with Equation 3.14 will yield satisfactory results in various analytical calculations involving the Theis equation. Again, it should be noted that a definite real radius of well influence could not be formed in a homogeneous confined aquifer unless there is a source of recharge, such as from a boundary or from leakage. Using the expression for Dupuit's radius of well influence, the coefficient of the linear formation loss is

$$A_0 = \frac{1}{2\pi T} \ln \frac{1.5\sqrt{Tt/S}}{r_w} \tag{3.15}$$

Similarly to steady-state conditions, the coefficients of linear and turbulent well losses are found graphoanalytically from a graph s/Q vs. Q. However, since the radius of well influence in transient conditions is not constant (it increases with time), the drawdown data recorded during the three-step test must be corrected in order to plot the graph. Figure 3.13 shows the components of drawdown recorded at the end of each step and the error made if the three drawdowns (s_1, s_2, and s_3) were used to draw a graph s/Q vs. Q without necessary corrections. Section 15.2 explains in detail how to correct the data and determine well loss in transient conditions.

The coefficient of the turbulent well loss (B) is the slope of the straight line drawn through the corrected points (Figure 3.14). Theoretically, B is not time dependent and should remain the same for different pumping rates. A common exception is pumping from karst and fractured rock aquifers where turbulent well loss may increase with an increasing pumping rate. In such a case, the points on the graph s/Q vs. Q would form a parabola rather than a straight line.

Well efficiency is the ratio between the theoretical drawdown and the actual drawdown measured in the well. It is expressed in percent:

$$\text{Well efficiency} = \frac{\text{Theoretical drawdown } (s_0)}{\text{Measured drawdown } (s_w)} \times 100\% \tag{3.16}$$

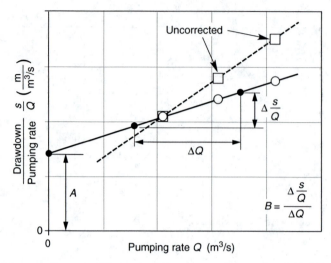

FIGURE 3.14 Graph pumping rate (Q) vs. drawdown–pumping rate ratio (s/Q) used to determine components of drawdown in a pumping well for transient conditions, after correcting the data. A: Coefficient of the linear losses; B: coefficient of the turbulent losses.

The theoretical drawdown is determined by applying an appropriate equation of groundwater flow toward a well (theoretical drawdown equals the formation loss). It can also be found graphoanalytically as explained in Section 15.2. In general, the difference between the theoretical drawdown and the measured drawdown increases with the increasing pumping rate. Consequently, the well efficiency decreases with the increasing pumping rate. Although some professionals argue that quantifying well efficiency does not have much practical use (as long as everyone is satisfied with the well), this practice is highly recommended because it provides valuable information about the well performance and can be used to make an informed decision regarding the well operations, maintenance, and rehabilitation. A well efficiency of 70% or more is usually considered acceptable. If a newly developed well has less than 65% efficiency, it should not be approved without a thorough analysis of the possible underlying reasons.

The specific capacity of the well is given as

$$\text{Well specific capacity} = \frac{\text{Pumping rate } (Q)}{\text{Measured drawdown } (s_w)} \qquad (3.17)$$

and is expressed as the pumping rate per unit drawdown (e.g., liters per second per 1 m of drawdown). Similarly to the well loss, the well specific capacity also decreases with the increasing pumping rate. Choosing an optimum rate at which a well will be pumped is a decision based on numerous factors. For example, if the well will be used for a short-term construction dewatering, maintaining a desired drawdown may be the only relevant criterion. In some cases, where there are no alternatives, a certain pumping rate is all that matters. On the other hand, if the well is designed for a long-term exploitation, in addition to the energy cost of pumping, the hydraulic criteria are the most important in deciding which pumping rate is the optimum one. This includes a comparative analysis of well losses and well efficiency. Graphs showing drawdown vs. pumping rate for several pumping steps, together with the well efficiency, such as the ones shown in Figure 3.15, are a simple tool for examining various options for well management.

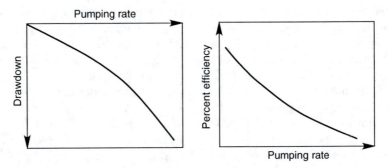

FIGURE 3.15 Graph pumping rate vs. drawdown and well efficiency used for decision making.

3.2 COLLECTOR WELLS AND HORIZONTAL WELLS

Collector wells have been used primarily for the development of water supply from alluvial aquifers formed around large permanent streams with two main goals: (i) to take advantage of induced aquifer recharge from the inexhaustible (for all practical purposes) recharge boundary and (ii) to use natural filtration of river water through the underlying aquifer sediments thus improving the quality of water; this is sometimes referred to as riverbank filtration. The basic design of a collector well is shown in Figure 3.16. A large-diameter

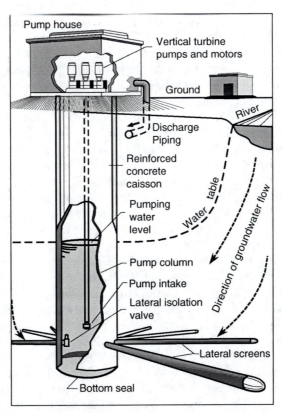

FIGURE 3.16 Schematic of a typical radial collector well adjacent to riverbank. (Courtesy of Ranney Division.)

reinforced concrete caisson is installed adjacent to the riverbank, extending sometimes tens of feet below the bottom of the channel. Well screens are jacked out horizontally in sections from the caisson to various distances, depending on the locations of most productive aquifer zones. The screens may be injected in all radial directions, not only beneath the river, to maximize the yield and take advantage of favorable hydrogeologic conditions where present. The length of the individual screens can be substantial, sometimes more than 2000 ft. One common problem with the screens (laterals) in collector wells is their relatively rapid clogging and deterioration when aquifer sediments contain fines. This is mainly due to less efficient means of placing gravel pack around the screens, but also because of oxygen-rich river water, which sometimes accelerates growth of iron bacteria and screen incrustation in general. Collector well technology, however, is constantly improved, including placement of screens with prefabricated filter packs and developing more efficient methods of screen rehabilitation. Radial collector wells are particularly popular in Europe where they supply water for major cities located in floodplains of large rivers. For example, Belgrade has over 90 Ranney-type collector wells, located in floodplains of the Danube and Sava rivers, with some individual wells yielding over 400 l/s, or 9 million gallons/d.

Horizontal wells have been used mainly in environmental applications for fluid recovery as part of groundwater remediation projects. Their popularity experienced a peak during mid- and late 1990s, as the regulatory demands for a more active cleanup of complex contaminated sites increased. In certain situations, horizontal wells may offer an advantage over conven- tionally drilled vertical wells. Examples include general recovery of elongated plumes of dissolved contaminants, recovery of light nonaqueous phase liquids floating on groundwater, recovery or containment of small-thickness, relatively wide, or long dissolved contaminant plumes, and recovery of accumulation of dense nonaqueous phase liquids over a shallow horizontal impermeable layer (Figure 3.17). In many cases, however, accessibility issues, such as contamination beneath structures, have played the primary role in deciding to install horizontal wells vs. vertical wells. For example, one of the longest, deepest, and most expensive horizontal wells in the world was installed because of the community opposition to vertical wells, and regulatory requirements for cleanup of a dissolved contaminant plume at any cost (Figure 3.18). An overview of various issues related to installation and perform- ance of horizontal wells is given in USEPA (1994a).

Unrelated to any regulatory or political requirements are many hydraulic, hydrogeologic, and engineering challenges facing horizontal well projects. In general, horizontal wells are much more prone to collapse due to various forces during both the installation and the operation, clogging of screens due to limitations with gravel pack installation, inadequate development due to specific well hydraulics, and an uneven influx into the well (Figure 3.19).

Two key hydrogeologic (noncontaminant) reasons for the installation of horizontal wells are the aquifer anisotropy and heterogeneity, which are also the main reason for the failure of many horizontal wells. Figure 3.20 shows possible orientation of a horizontal well relative to aquifer anisotropy. In general, a horizontal well drains the largest volume if placed normal to the plane that contains the axes of maximum and intermediate permeability. This only can be possible if the direction of greatest permeability is vertical (Figure 3.20a). Consequently, a horizontal well will have limited utility in hydrogeologic settings where the vertical hydraulic conductivity is significantly lower than the horizontal. The exception would be a slow-moving thin plume in a thin horizontal layer where a horizontal well may be used to either intercept the plume or recover it. Compared to vertical wells, a proper design of horizontal wells requires a much more intensive site characterization and exploratory vertical drilling along its future trace. This is, in part, because the cost of horizontal wells per linear foot installed is significantly higher than for vertical wells so that a failure of a horizontal well, by default, is more serious.

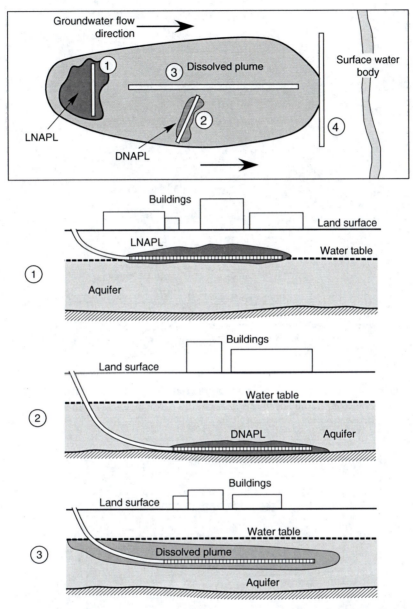

FIGURE 3.17 Some applications of horizontal wells in groundwater remediation. (1) Recovery of light nonaqueous phase liquids (LNAPLs) floating on water table; (2) recovery of dense nonaqueous phase liquids (DNAPLs) resting on impermeable aquifer base; (3) pump and treat of dissolved contaminant plumes; (4) interception of contaminant plumes.

As with any relatively new approach and the accompanying technology development, utilization of horizontal wells should eventually become a matter of routine, as long as site-specific hydrogeologic and hydraulic parameters of their installation are fully characterized. A useful resource for public exchange of information regarding horizontal wells is Horizontal Well Interest Group of the National Ground Water Association (www.ngwa.org). The association also facilitates work of other professional interest groups, such as for groundwater modeling.

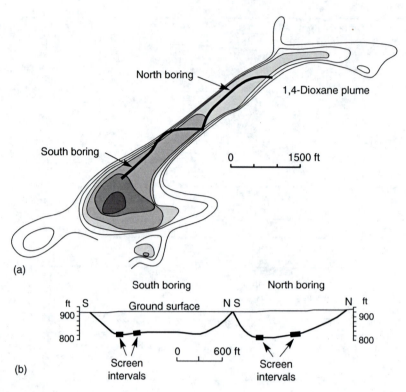

(a)

(b)

FIGURE 3.18 One of the longest and deepest horizontal wells in the world installed for pump-and-treat remediation of a dissolved contaminant plume in Ann Arbor, Michigan: (a) map view; (b) cross section. (Courtesy of Pall Corporation.)

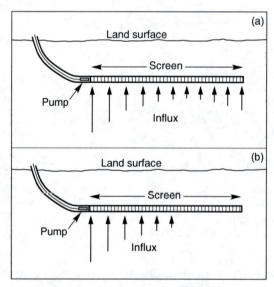

FIGURE 3.19 Cross section schematic of two possible variations of influx into a horizontal well. (a) Groundwater flow in the aquifer converges toward the well from beyond the end of the well; (b) limited pump capacity creates critical distance beyond which the effect of the well is negligible. (Modified from USEPA, Manual; alternative methods for fluid delivery and recovery. EPA/625/R-94/003, U.S. Environmental Protection Agency, Office of Research and Development, Cincinnati, OH, 1994a, 87pp.)

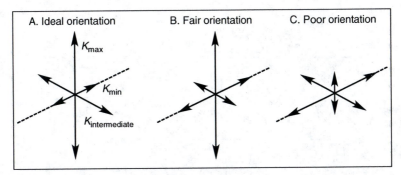

FIGURE 3.20 Orientation of horizontal well (dashed line) relative to principal directions of permeability. (From USEPA, Manual; alternative methods for fluid delivery and recovery. EPA/625/R-94/003, U.S. Environmental Protection Agency, Office of Research and Development, Cincinnati, OH, 1994a, 87pp.)

3.3 INFILTRATION GALLERIES AND DRAINS

An infiltration gallery may be considered a horizontal well or subsurface drain that intercepts shallow groundwater flow or underflow of surface water streams when conditions are such that use of normal vertical wells is not economically or technically feasible. This includes thin aquifers, thin deposits of sand and gravel underlying ephemeral or intermittent streams, and thin freshwater lenses that are underlain by brackish or saline water. Infiltration galleries are often used in place of direct intake of surface water with high loads of suspended materials, carrying inert and physiologically harmless contaminants, or with rapid and unpredictable changes in water level (USBR, 1977). Infiltration galleries are usually constructed to discharge into a sump whose bottom is some distance below the invert of the drain and casing. The sump may be of almost any dimensions but commonly is a circular or square structure 4 to 8 ft in diameter or on a side. The top of the sump is sealed to prevent any possible leakage from the surface. One or more perforated drains extend from the sump into the permeable (aquifer) material and should always be set in gravel pack, which should have a minimum of 6 in. thickness below the drain bottom. Gravel pack design is similar to that for a vertical well, but with a slightly more liberal multiplier of 6 to 7 times the 70%-retained size (Driscoll, 1986). The excavation trench is usually backfilled above the gravel pack to the land surface (Figure 3.21). The drains can have any configuration based on local orientation of permeable sediments and the stream character (permanent, ephemeral, or intermittent flow); they can be placed parallel to the riverbank at some distance to maximize effects of natural river water filtration through porous material, or beneath the river channel and perpendicular to the underflow if the riverbank conditions are not favorable. In any case, the design should provide for an average entrance velocity of 0.1 ft/s (0.03 m/s) or less. It is also advisable to design for possibility of backwashing the drains since they tend to clog by the silt load carried by the stream. For the same reason, it is good practice to overdesign the drain parameters by increasing its open area in estimation of the capacity decrease. This is done by increasing the drain length, diameter, or slot size.

Assuming a relatively thin saturated thickness of the sediments supplying water to the drain, and placing the reference level at the channel bottom, the flow rate entering the drain per unit length is estimated as the sum of flows from two directions (see notification in Figure 3.21). The unit flow from the river direction is

$$q_1 = K \frac{h_1^2 - h_s^2}{2L_1} \tag{3.18}$$

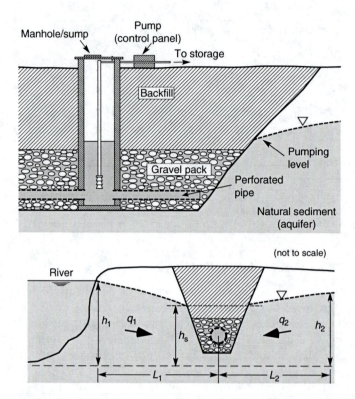

FIGURE 3.21 Schematic of an infiltration gallery placed parallel to the riverbank. Top: cross section along the gallery. Bottom: cross section perpendicular to the gallery and the river.

and the unit flow from the opposite direction is

$$q_2 = K \frac{h_2^2 - h_s^2}{2L_2} \tag{3.19}$$

where K is the coefficient of filtration of native sediments, h_1 is the river stage measured from the reference level, h_2 is the hydraulic head in a piezometer placed at a distance L_2 from the gallery, h_s is the pumping level at the sump, measured from the reference level, L_1 is the distance between the river and the drain, and L_2 is the distance between the drain and a piezometer placed on a section perpendicular to the drain.

The total flow (Q) for a drain of length L is calculated as

$$Q = L \cdot (q_1 + q_2) \tag{3.20}$$

If there is a good hydraulic connection between the river and the drain, q_1 is in most cases much higher than q_2.

When riverbank conditions are not favorable for an on-land infiltration gallery, and in cases where the streambed has relatively low hydraulic conductivity, the infiltration gallery can be placed directly beneath the channel as shown in Figure 3.22. The drain with the gravel pack is placed in the excavation, which is sized to accommodate the desired flow rate. The excavation is backfilled with the permeable clean sand to provide for direct hydraulic connection with the river. Such a design should anticipate partial clogging of the gravel

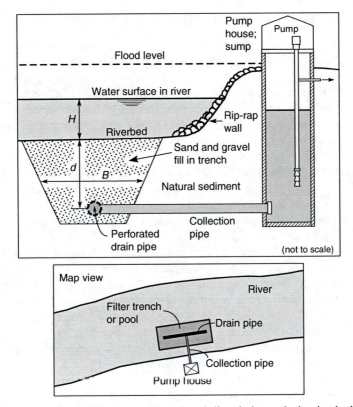

FIGURE 3.22 Schematic of an infiltration gallery placed directly beneath the riverbed.

pack and the drain (screen) by silt from the river and compensate for the anticipated decrease in capacity. In case of low-permeable riverbed sediment, the flow occurs directly downward from the water body into the sand-filled excavation, the gravel pack and the perforated pipe. The flow rate per unit length of the gallery is (see notification in Figure 3.22)

$$q = BK\frac{(H+d)}{d} \tag{3.21}$$

where B is the average width of trench backfilled with sand and gravel, K is the hydraulic conductivity of the backfill, H is the depth of water in the river above the riverbed, and d is the depth to drain from the riverbed (channel bottom).

Note that $(H+d)/d$ is the hydraulic gradient for this vertical flow when the reference level is placed at the center of the drain.

When the gallery is placed in relatively permeable riverbed sediment, the inflow of water occurs from the lateral directions as well. The flow rate per unit length of the gallery is (modified from USBR, 1977; see notification in Figure 3.22)

$$q = \frac{2\pi KH}{\ln(2d/r)} \tag{3.22}$$

where r is the radius of the perforated pipe, K is the hydraulic conductivity of the native sediment, and the other notification is the same as in Equation 3.21.

3.4   SOURCE WATER PROTECTION ZONES

The following discussion by Kraemer et al. (2005) of the USEPA summarizes the concept of source water protection zones and the role of the USEPA and state agencies in its implementation: "Ground water is a valued source of public drinking water in many parts of the country. Not only is the groundwater resource renewable if properly managed, its quality can be excellent, due to the natural cleansing capabilities of biologically active soil cover and aquifer media. It has come to the public attention in the past few decades that groundwater wells can be threatened by over-drafting and pollution. Source water protection is based on the idea of protecting public water wells from contamination up front, and working to maintain the quality of the resource. Source water protection means taking positive steps to manage potential sources of contaminants and doing contingency planning for the future by determining alternate sources of drinking water. The Safe Drinking Water Act has required the USEPA to develop a number of programs that involve the source water protection concept for implementation at the State level, such as the Wellhead Protection Program and the Source Water Assessment Program. The main steps of the source water protection process involve the assessment of the area contributing water to the well or well field, a survey of potential contaminant sources within this area, and an evaluation of the susceptibility of the well to these contaminants. This includes the possibility of contaminant release and the likelihood of transport through the soil and aquifer to the well screen. The designation of the wellhead protection area is then a commitment by the community to source area management. Delineation of the wellhead protection area is often a compromise between scientific and technical understanding of geohydrology and contaminant transport, and practical implementation for public safety. The EPA Office of Ground Water and Drinking Water established guidance on the criteria and methods for delineating protection areas (USEPA, 1993a; USEPA, 1994b)." The Agency maintains a web page dedicated to source water protection, with downloadable publications and various useful links at http://www.epa.gov/safewater/protect.html.

Preceding the USEPA by almost 400 years is this very precise wellhead protection guidance by Governor Gates of Virginia, issued in the form of Proclamation for Jamestown in 1610: "There shall be no man or woman dare to wash any unclean linen, wash clothes . . . nor rinse or make clean any kettle, pot, or pan, or any suchlike vessel within twenty feet of the old well or new pump. Nor shall anyone aforesaid, within less than a quarter mile of the forte, dare to do the necessities of nature, since by these unmanly, slothful, and loathsome immodesties, the whole forte may be choked and poisoned."

USEPA has recently released a significantly updated and enhanced version of WhAEM2000, a public domain and open source general-purpose groundwater flow modeling system, capable of representing regional flow systems, and groundwater–surface water interactions. This computer program was initially designed to facilitate capture zone delineation and protection area mapping in support of the State's Wellhead Protection Programs (WHPP) and Source Water Assessment Planning (SWAP) for public water supply wells in the United States. WhAEM2000 provides an interactive computer environment for design of protection areas based on radius methods, well in uniform flow solutions, and groundwater modeling methods. It can import a variety of electronic base maps, including several USGS formats, so that the modeled protection zones can be determined directly in the applicable geographic coordinate system. WhAEM2000 is an analytic element model limited to steady-state and two-dimensional flow conditions, and includes options for simulating the influence of flow boundaries, such as rivers, recharge, and no-flow contacts (Kraemer et al., 2005). Figure 3.23 illustrates just some of the complexities associated with wellhead protection zone delineation using several approaches outlined by the USEPA. As can be seen, the shapes of

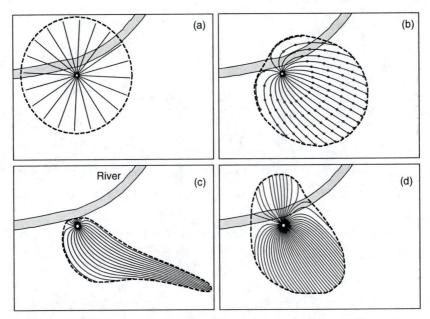

FIGURE 3.23 Five-year capture zones of a single well near a permanent river, delineated using different options in W*h*AEM2000 computer program. General groundwater flow direction is from the southeast toward the river. (a) Fixed radius volumetric method. (b) Uniform flow field method; small circles correspond to 1-y intervals. (c) Zone delineated taking into consideration hydrologic boundaries (not shown). (d) Zone delineated taking into consideration possible resistance of the riverbed sediments. (Modified from Kraemer et al., 2005.)

the four zones are quite different reflecting critically different assumptions, all of which would be valid based on the USEPA guidance. In all four cases, the same (single) well near a relatively large permanent river operates at a constant (steady-state) pumping rate, pumping water from an unconfined aquifer in which the groundwater flow is assumed to be horizontal and with a "regional" hydraulic gradient from the southeast to the northwest (toward the river). It is obvious that the "correct" wellhead protection zone would have to be selected in a constructive discussion between professional hydrogeologists and all other stakeholders (nonhydrogeologists), including inevitable considerations of time and budgetary constraints. Unfortunately, most states in the United States were confronted with the task of developing wellhead protection programs for thousands of public drinking water supply systems in only a few years' time. According to Kraemer et al. (2005), conducting an "extensive" groundwater modeling campaign for each individual drinking water well (or well field) was out of question, both in view of the time involved and the cost. "The USEPA recognized this reality from the start and proposed a series of simplified capture zone delineation methods to facilitate a timely implementation of the States wellhead protection programs."

Consequently, it is almost certain that many public drinking water supply systems have delineated wellhead capture zones that are not based on any hydrogeologic reality, and may be overprotecting their water supply by unnecessarily restricting various land uses, or may not be protecting them at all when having a false sense of security. Figure 3.24 shows that the majority of community water supply systems serving fewer than 3300 people use groundwater, and many larger systems also depend on groundwater. Small systems use groundwater as a source because groundwater usually requires less treatment than surface water and is therefore more affordable. This is an important consideration since many small systems

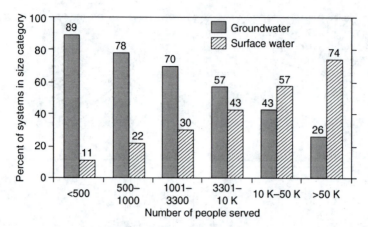

FIGURE 3.24 Community water supply systems by source. (From USEPA and DWA (Drinking Water Academy), 2003. Introduction to the public water system supervision program, January 2003. http://www.epa.gov/safewater/protect.html (accessed on November 12, 2005).)

without a large, rate-paying base cannot afford extensive hydrogeologic studies for wellhead (source water) protection purposes. At the same time, wellhead protection efforts are often among the most cost-effective ways to ensure safe drinking water. These efforts prevent contamination from occurring rather than treating contamination after it has occurred.

In general, wellhead protection zone delineation methods can be divided into the following three categories:

1. *Nonhydrogeologic*
   • Arbitrary fixed radius

2. *Quasihydrogeologic*
   • Calculated fixed radius
   • Well in a uniform flow field
   • Groundwater modeling not based on hydrogeologic mapping

3. *Hydrogeologic*
   • Hydrogeologic mapping (aquifer vulnerability mapping)
   • Groundwater modeling based on hydrogeologic mapping

Except for the first category, which is simply an arbitrarily set wellhead protection zone, all other methods can be used to delineate wellhead protection zones by addressing the residence time criterion, which is based on the assumption that (i) nonconservative contaminants, subject to various fate and transport processes (e.g., sorption, diffusion, degradation), may be attenuated after a given time in the subsurface; (ii) detection of conservative contaminants (not subject to attenuation) entering the wellhead protection area will give enough lead time for the public water supply entity to take necessary action, including groundwater remediation or development of a new (alternative) water supply; and (iii) detection of any contaminants already in the wellhead protection area would require an immediate remedial action. The most critical decision regarding an appropriate residence time is left to the stakeholders in each individual case, although it appears that 5-, 10-, and 20-year wellhead capture zones have been most widely used to delineate certain subzones of various land-use restrictions within the main wellhead capture zone.

Quasihydrogeologic methods use very simple assumptions, which, in many cases, do not have much in common with the site-specific hydrogeologic conditions. Since they include application of certain equations, it may appear to nonhydrogeologists that such a method must have some credibility. When involved in application of quasihydrogeologic methods, hydrogeologists should clearly explain the limitations of various unrealistic assumptions and their implications on the final wellhead zone delineation. For example, these methods do not consider aquifer stratification and presence of confining layers, vertical flow components, aquifer heterogeneity and anisotropy, or interference between multiple wells screened at different depths in the same well field. None of these methods should be applied in aquifers known to be heterogeneous and isotropic by default, such as fractured rock and karst aquifers, and many glacial and glaciofluvial aquifers. For information purposes, a brief discussion of two quasihydrogeologic methods is presented below.

3.4.1 Calculated Fixed Radius

The fixed radius is calculated based on a simple two-dimensional static water balance analysis, assuming negligible ambient flow in the aquifer (the initial hydraulic head is horizontal). Assuming radial flow toward a well in an aquifer with a horizontal impermeable base and constant saturated thickness, the cylindrical boundary of capture is delineated by an isochrone of residence time t, which means that any water particle that enters the cylinder or is present in the cylinder will travel no longer than t days before being pumped up by the well (Kraemer et al., 2005; see Figure 3.25). A water balance for the period t is

$$N\pi R^2 t + n\pi R^2 H = Qt \tag{3.23}$$

where H is the initial constant saturated thickness of the aquifer, R is the radius of capture for the time base of t, N is the areal recharge, n is the effective aquifer porosity (specific yield), and Q is the well pumping rate.

The first term in Equation 3.23 represents the inflow due to aquifer recharge, the second term represents the amount of water contained inside the cylindrical portion of the aquifer, and the term of the right-hand side is the total amount of water removed by the well for the pumping period t. The radius R can be expressed as

$$R = \sqrt{\frac{Qt}{N\pi t + n\pi H}} \tag{3.24}$$

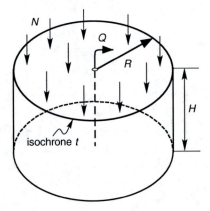

FIGURE 3.25 Water balance for radial flow to a well in a domain bounded by an isochrone of residence time t. (From Kraemer et al., 2005.)

When t becomes infinitely large, the radius R represents the complete capture zone (not time dependent):

$$R \approx \sqrt{\frac{Q}{\pi N}} \tag{3.25}$$

The use of this equation is called the "recharge method" (USEPA, 1993a). If the term $N\pi t$ becomes small, due to a small value of t or N or both, Equation 3.24 reduces to

$$R \approx \sqrt{\frac{Qt}{n\pi N}} \tag{3.26}$$

The use of this equation is called the "volumetric method" (USEPA, 1993a). According to Kraemer et al. (2005), since the saturated thickness of an unconfined aquifer subject to pumping will not be constant, even assuming a horizontal base, a conservative (protective) capture zone will be obtained by using the smallest saturated thickness H in Equation 3.26. The major limitation of this and similar methods is the assumption of a sandbox-type groundwater flow.

The radially symmetric isochrone in Figure 3.25 will only occur in the absence of a significant regional hydraulic gradient. When the regional gradient cannot be ignored (which is a much more realistic assumption), the isochrone will be elongated in the upgradient direction, beyond the circular representation of that isochrone calculated with Equation 3.26. Under such conditions, the direction and magnitude of the ambient flow should be determined and isochrones should be estimated using a solution for a well in a uniform flow field.

3.4.2 WELL IN UNIFORM FIELD

This method assumes that the regional groundwater flow, generated from far field aquifer recharge, may be approximated by a uniform flow field (straight streamlines) in the immediate vicinity of the well. The capture zone for the well in a uniform flow field will no longer be circular and centered about the well, but will be an elongated domain upgradient from the well, parallel to the direction of uniform flow. Several analytical equations have been used to delineate infinite capture zones in steady-state conditions using simplifying sandbox assumptions. This approach is shown in Figure 3.26. The equation describing the edge of the capture zone (groundwater divide) for a confined aquifer is (Grubb, 1993, from USACE, 1999)

$$x = \frac{-y}{\tan\left(\dfrac{2\pi Kbiy}{Q}\right)} \tag{3.27}$$

where Q is the pumping rate, K is the hydraulic conductivity, b is the aquifer thickness, i is the hydraulic gradient of the flow field in the absence of the pumping well, and tan is tangens function in radians.

The distance from the pumping well downgradient to the stagnation point that marks the end of the capture zone is

$$x_0 = -\frac{Q}{2\pi Kbi} \tag{3.28}$$

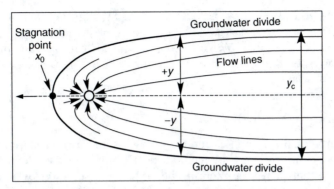

FIGURE 3.26 Capture zone of a pumping well in a uniform flow field, in plan view. The well is located at the origin (0,0) of the x,y-plane. (From USACE (U.S. Army Corps of Engineers), *Groundwater Hydrology*, Engineer Manual 1110-2-1421, Washington, D.C., 1999.)

The maximum width of the capture zone as the distance (x) upgradient from the pumping well approaches infinity is

$$y_c = \frac{Q}{Kbi} \tag{3.29}$$

The equation of the groundwater divide for unconfined steady-state conditions is (Grubb, 1993, from USACE, 1999)

$$x = \frac{-y}{\tan\left[\dfrac{\pi K(h_1^2 - h_2^2)y}{QL}\right]} \tag{3.30}$$

where Q is the pumping rate, K is the hydraulic conductivity, h_1 is the upgradient hydraulic head above horizontal aquifer base before pumping, h_2 is the downgradient hydraulic head above horizontal aquifer base before pumping, L is the distance between h_1 and h_2, and tan is tangens function in radians.

The distance from the pumping well downgradient to the stagnation point that marks the end of the capture zone is

$$x_0 = -\frac{QL}{\pi K(h_1^2 - h_2^2)} \tag{3.31}$$

The maximum width of the capture zone as the distance (x) upgradient from the pumping well approaches infinity is

$$y_c = \frac{2(QL)}{K(h_1^2 - h_2^2)} \tag{3.32}$$

Inclusion of the time factor (isochrone approach) would change the shape of the complete (infinite) capture zone shown in Figure 3.26. Because the flowpaths (streamlines) are not all of the same length, the zone would have an oval shape, like the one shown in Figure 3.23b; this shape would be progressively longer as the calculation time increases. Numeric and analytic groundwater flow models, such as WhAEM2000, can quickly calculate and draw flowlines of various lengths corresponding to the same time of travel. Capture zones are lines connecting end points of such flowlines.

3.4.3 HYDROGEOLOGIC MAPPING

Hydrogeologic mapping first identifies and then presents, in forms of maps and accompanying graphics and documentation, geologic, hydrologic, and hydraulic features that control groundwater flow within an area of interest. The capture zone of a well, well field, or a spring used for water supply can be quite complex and may include multiple interconnected aquifers and surface water features. Mapping three-dimensional physical and hydraulic boundaries of such a flow system is therefore the key for a successful hydrogeologic map of the wellhead capture zone. It is important to understand that hydrogeologic map is not simply a geologic map with different colors; a geologic or lithostratigraphic unit (or formation) is not necessarily directly translatable into an aquifer or aquitard. Several geologic units may act as one aquifer, and there may be several aquifers separated by aquitards within the same geologic formation. In hydrogeologic studies, it is therefore common to use the term hydrostratigraphic unit, which describes one or more geologic units that have the same porous media characteristics and act as one hydrodynamic entity (aquifer or aquitard). Tectonic fabric of the mapped area, including faults and fault systems, may also play an important role in directing groundwater flow within its boundaries. In short, the hydrogeologic map must show where the water is coming from into the flow system captured by the wells, and how it is flowing toward the wells, which act as the local discharge area within the aquifer. This is possible only if there is sufficient three-dimensional field information on the actual hydraulic heads within the flow system, including their seasonal variations. Once the geometry of the flow system is understood, it is mapped by showing three-dimensional equipotential lines and flowlines. The final step is to estimate the velocities of groundwater flow within the capture zone and to delineate aquifer volumes with the same residence time, i.e., show them in the three-dimensional space. It is obvious that a thorough hydrogeologic mapping may require substantial resources for more complex hydrogeologic conditions, and therefore may not be feasible for some water supply systems. Whatever the case may be, hydrogeologic mapping for delineation of wellhead capture zones is most resourceful when used for a concurrent mapping of the aquifer vulnerability in its recharge area. Figure 3.27 illustrates the key difference between the hydrogeologic mapping and the sandbox-type analytical methods of wellhead capture zone delineation: the latter type methods cannot distinguish between the "theoretical" two-dimensional capture zone and the actual contributing zone in the aquifer recharge area based on groundwater flow balance (hatched in Figure 3.27); they also cannot account for the vertical flow component. As a result, such methods may underestimate or overestimate the land surface expression (two-dimensional map) of the real capture zone.

Aquifer vulnerability (or sensitivity) mapping is a type of hydrogeologic mapping that focuses on areas at the land surface vulnerable to groundwater (aquifer) contamination due to both hydrogeologic and anthropogenic factors. Some examples of hydrogeologically vulnerable areas include thin unsaturated zones with permeable sediments (e.g., sandy and gravely soils), shallow or exposed fractured bedrock, and karst topography with solution cavities and sinkholes. Aquifer vulnerability maps are used for land-use planning purposes and are often created as part of a systematic, large mapping program by government agencies. They can also be created specifically for groundwater protection and management purposes including for delineation of specific protection zones within one general wellhead protection area.

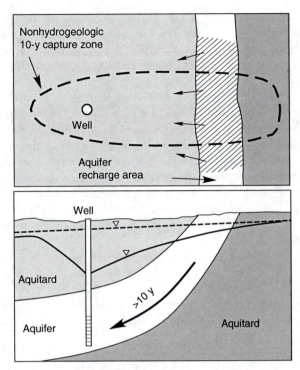

FIGURE 3.27 Example of hydrogeologic mapping of the wellhead protection zone for one well in a confined aquifer. Top: map view. Bottom: cross section. The aquifer is overlain and underlain by competent aquitards and has a clearly defined recharge area (outcrop). The nonhydrogeologic 10-year capture zone determined using analytic Equation 3.27 for well in uniform flow field, and the actual contributing area (vulnerable zone) of the well are shown in the map view. The travel time from the vulnerable zone to the deep well screen may be significantly longer than the one determined based on the assumption of uniform horizontal flow toward the well.

Commonly, there are three wellhead protection zones, although the number may vary based on local hydrogeologic conditions and regulatory requirements:

- Zone 1 (sanitary zone or zone of strict protection) is essentially an administrative zone of physical protection of the water source, such as fencing and restricted access measures. Its purpose is to prevent accidental or deliberate damage or contamination of the source itself including contamination of the aquifer through the source. The size of Zone 1 may vary between several tens and several hundreds of feet (meters) for simple systems consisting of one well for example, or it may be larger in case of several closely spaced springs or multiple wells.

- Zone 2, sometimes called the inner protection zone, is based on time-of-travel analysis. It reflects the site-specific understanding of the minimum time of potential contaminant travel between its introduction into the subsurface and the groundwater withdrawal locations during which it would be possible to initiate remediation activities and execute contingency plans for water supply if necessary. In some states, Zone 2 is considered an attenuation zone in which pathogens entering the zone from septic systems or surface water bodies would be attenuated before reaching the well (e.g., see Wyoming DEQ, 2005). It is obvious that Zone 2 in general may vary in shape and size widely, depending on the aquifer type and hydraulic conductivity of the most permeable zones. Zone 2 is entirely inside the groundwater divide and corresponds to the aquifer volume from which all groundwater would discharge through the wells or springs within the given time. Glacial gravel deposits and karst aquifers are examples of porous media where contaminant travel times may be extremely short (e.g., several days) over distances of miles. Consequently, Zone 2 may cover a rather large area and pose a serious challenge for the development of an adequate source water management and protection plan. For example, an extreme but possible scenario is that a contaminant may be introduced into a karst aquifer 10 miles away from the well and it may show up at the well within 24 h. Prescribing a Zone 2 that corresponds to the 50-d or even 10-d time of travel in such a case obviously does not make much sense, although it may seem quite "conservative" when compared to 5- or 10-y capture zones commonly considered as protective in intergranular aquifers. On the other hand, delineation of Zone 2 may not be necessary or even meaningful in case of deep confined aquifers protected by competent aquitards, distant aquifer recharge zones, and groundwater resident times measured by hundreds or thousands of years.

- Zone 3, also called the contributing area, includes the entire aquifer volume within the groundwater divide from which all groundwater will eventually discharge through the wells or springs regardless of the time of travel. This zone may be subdivided based on various criteria, including time of travel. Its main importance is for long-term planning of groundwater resources management, and the related land-use and aquifer protection regulations.

Hydrogeologic maps and aquifer vulnerability maps should be developed in a geographic information system (GIS) environment to combine various data layers. In case of comprehensive hydrogeologic maps, data layers include geology, hydrostratigraphy, tectonics, topography, hydrologic features (hydrography), climate factors influencing aquifer recharge, land cover, land use, soil types, depth to saturated zone, aquifer parameters (transmissivity and hydraulic conductivities and porosity and storage properties), hydraulic head contour lines, aquifer recharge and discharge areas and locations (both natural and artificial), and known or potential sources of soil and groundwater contamination. On the opposite side of the mapping spectrum are simple and pragmatic approaches to vulnerability mapping of unconfined aquifers, such as various index methods. These methods do not require hydrodynamic information and combine just a few data layers (Figure 3.28). Probably the most widely used index method is DRASTIC, named for the seven factors considered: Depth to water, Recharge, Aquifer media, Soil media, Topography, Impact of the vadose zone, and hydraulic Conductivity of the aquifer (Aller et al., 1985). The seven factors are incorporated into a relative ranking scheme that uses a combination of ratings and weights to produce a numerical value, called the DRASTIC index. Each of the factors is ranked between 1 and 10 and the rank is multiplied by an assigned weighting, which ranges between 1 and 5. The weighted ranks are summed to give a score for the particular hydrogeologic unit with the higher scores indicating greater vulnerability to contamination. The DRASTIC method has been used to develop groundwater vulnerability maps in many parts of the United States; however, the effectiveness of the method has met with mixed success due to its subjectivity since the maps are not calibrated to any measured contaminant concentrations or specific contaminants (USEPA, 1993b; USGS, 1999). Basic DRASTIC maps may be improved by calibration of the point ratings based on the results of statistical correlations between groundwater quality and hydrogeologic and anthropogenic factors. For example, one of the significant weaknesses of the relative vulnerability maps developed for agricultural nitrate contamination in an area of Idaho is that soil permeability was not the primary soil factor; there was no correlation between the nitrite, nitrate, and nitrogen concentrations in

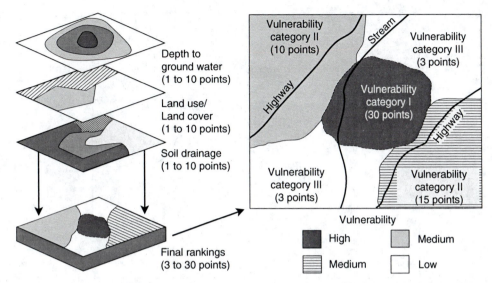

FIGURE 3.28 Schematic of index mapping of groundwater vulnerability. Some indices are based on quantitative data, but the individual and the final rankings are subjective. (Modified from Focazio, M.J., et al., Assessing ground-water vulnerability to contamination: providing scientifically defensible information for decision makers. U.S. Geological Survey Circular 1224, Reston, VA, 2002, 33pp.)

groundwater and the soil permeability, but there was a strong correlation with soil drainage types, presumably because soil drainage is a better indicator of nitrate leaching conditions (USGS, 1999). Calibration of the aquifer vulnerability maps with groundwater quality information is the most effective way to determine which hydrogeologic and anthropogenic factors are related to the chemical compound of interest.

3.4.4 GROUNDWATER MODELING BASED ON HYDROGEOLOGIC MAPPING

Numeric models based on the results of hydrogeologic mapping are the best tools for well-head capture zone delineation and groundwater resources management in general. Regardless of the effort and resources invested into hydrogeologic mapping for any purpose, there will inevitably remain a certain level of uncertainty as to the true representative hydrodynamic characteristics of the groundwater flow system analyzed. Numeric models provide for quantitative analysis of this uncertainty and enable decision makers to analyze different "what if" scenarios of their source management and protection. Numeric models can take into account all important aspects of the three-dimensional hydrogeologic mapping and convert them into quantitative description of the flow system. Consequently, they can delineate complex shapes and aquifer volumes contributing water to well fields, and perform sensitivity analysis of the delineation results by changing various model input parameters, such as hydraulic conductivity or aquifer recharge rate (more detail on groundwater modeling is given in Chapter 7). Figure 3.29 through Figure 3.31 illustrate the importance of groundwater modeling as a tool for decision making in groundwater resources management. Some of the possible utilizations of the modeling results shown are to restrict certain land uses in the aquifer recharge areas directly contributing water to the well fields, and to plan installation of monitoring wells in appropriate locations for early warning on contaminant migration.

Figure 3.32 shows some of the results of contributing area delineation for a well field in the Central Swamp region near Tampa, Florida. The method used was a numeric three-dimensional flow and particle-tracking model. Time-related areas of contribution were

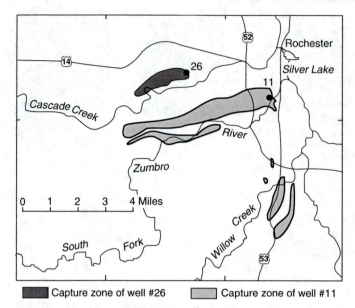

FIGURE 3.29 Long-term (steady-state) model-calculated contributing recharge areas for wells 11 and 26 near Rochester, Minnesota, which are screened in the St. Peter–Prairie du Chien–Jordan aquifer. Contributing areas shown for the case when no other high-capacity wells are pumping. (Modified from Franke O.L., et al., Estimating areas contributing recharge to wells; lessons from previous studies. U.S. Geological Survey Circular 1174, 1998, 14pp; Delin, G.N. and Almendinger, J.E., Delineation of recharge areas for selected wells in the St. Peter–Prairie du Chien–Jordan aquifer, Rochester, Minnesota: U.S. Geological Survey Water-Supply Paper 2397, 1993, 39pp.)

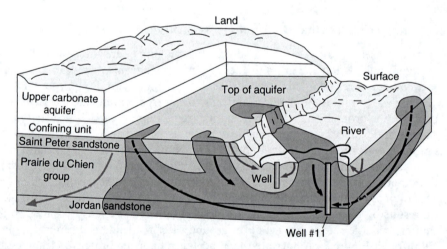

FIGURE 3.30 Irregularly shaped volumes in the subsurface that contain flowpaths originating at the water table and discharging at well 11, shown on the map in Figure 3.29. Any additional discharging wells (one shown) would capture their own subsurface flowpaths and related contributing recharge areas at the water table, thereby changing local flow patterns in the surrounding groundwater flow system. (Modified from Franke, O.L., et al., Estimating areas contributing recharge to wells; lessons from previous studies. U.S. Geological Survey Circular 1174, 1998, 14pp; Delin, G.N. and Almendinger, J.E., Delineation of recharge areas for selected wells in the St. Peter–Prairie du Chien–Jordan aquifer, Rochester, Minnesota: U.S. Geological Survey Water-Supply Paper 2397, 1993, 39pp.)

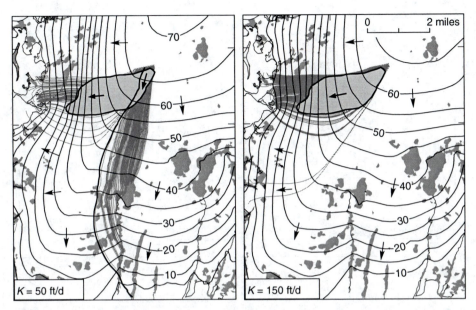

FIGURE 3.31 Simulated effects of increasing horizontal hydraulic conductivity of moraine sediments on the water table configuration and groundwater flowpaths near the Landfill-1 contaminant plume, western Cape Cod, Massachusetts. Hydraulic conductivity of the moraine sediments is 50 ft/d (left) and 150 ft/d (right). In the 50 ft/d simulation, flowpaths split in two directions, west and south, but predominantly to the south. In the 150 ft/d simulation, although the configuration of the water table changed very little at this scale, virtually all of the flowpaths moved to the west and followed the known configuration of the contaminant plume. (Modified from Franke, O.L., et al., Estimating areas contributing recharge to wells; lessons from previous studies. U.S. Geological Survey Circular 1174, 1998, 14pp; Masterson, J.P., et al., Use of particle tracking to improve numerical model calibration and to analyze ground-water flow and contaminant migration, Massachusetts Military Reservation, Western Cape Cod, MA: U.S. Geological Survey Water-Supply Paper 2482, 1997, 50pp.)

analyzed for six carbonate aquifer system types: an isotropic and homogeneous single-layer system; an anisotropic in a horizontal plane single-layer system; a discrete vertically fractured single-layer system; a multilayered system; a double porosity single-layer system; and a vertically and horizontally interconnected heterogeneous system. The simulated aquifer anisotropy was 5:1 and was determined from TENSOR2D results. The simulated vertical discrete fracture network represents locations of photolineaments. The simulated enhanced flow zones were determined from borehole video and geophysical logs. This study indicates that the distribution and nature of aquifer heterogeneities will affect the size, shape, and orientation of areas of contribution in a karst carbonate aquifer system. The size of the 50-y time-related areas of contribution ranged from 8.2 to 39.1 square miles. Simulations showed that the size of areas of contribution is primarily affected by simulated withdrawal rates, effective porosity of the carbonate rock, and transmissivity. The shape and orientation of the simulated areas of contribution primarily result from aquifer anisotropy, well distribution, flow along solution-enhanced zones, and short-circuiting of flow through fracture networks. Flow velocities and particle path length respond to withdrawal rates, simulated effective porosity, short-circuiting of flow by fractures, and transmissivity. The results of simulations incorporating aquifer heterogeneity indicate that oversimplification of the flow system may result in erroneous definition of flow fields. For example, aquifer anisotropy, typical of many carbonate aquifers, creates elliptical flow fields. Circular protection zones do not adequately characterize the

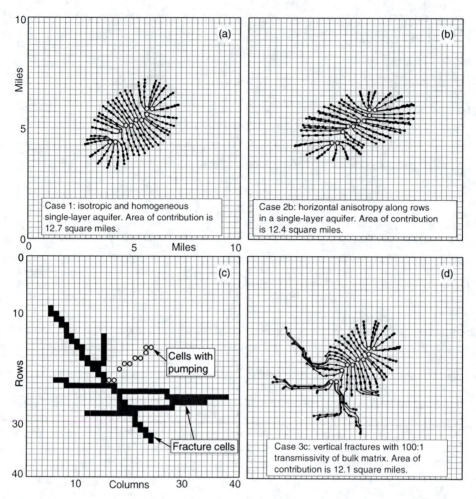

FIGURE 3.32 Influence of aquifer anisotropy and heterogeneity on the modeled capture zone for the Central Swamp region, Cypress Creek well field near Tampa, Florida. (a) Isotropic and homogeneous single-layer aquifer; (b) anisotropic hydraulic conductivity with 5 times greater value along rows; (c) and (d): simulation of vertical fractures with "fracture" cells where transmissivity is 100 times greater than in the surrounding "matrix" cells. (Modified from Knochenmus, L.A., and Robinson, J.L., Descriptions of anisotropy and heterogeneity and their effect on ground-water flow and areas of contribution to public supply wells in a karst carbonate aquifer system. U.S. Geological Survey Water-Supply Paper 2475, Washington, D.C., 1996, 47pp.)

elliptical areas of contribution. Also, areas of contribution were larger when the carbonate aquifer system was simulated as a multilayered system with discrete flow zones (Knochenmus and Robinson, 1996).

REFERENCES

Aller, L., et al., 1985. DRASTIC—A standardized system for evaluating ground water pollution potential using hydrogeologic settings. U.S. Environmental Protection Agency, Robert S. Kerr Environmental Research Laboratory, EPA/600/2-85/018, 163pp.

Aller, L.T., et al., 1991. *Handbook of Suggested Practices for the Design and Installation of Ground-Water Monitoring Wells*. EPA160014-891034, Environmental Monitoring Systems Laboratory, Office of Research and Development, Las Vegas, NV, 221pp.

Delin, G.N. and J.E. Almendinger, 1993. Delineation of recharge areas for selected wells in the St. Peter–Prairie du Chien–Jordan aquifer, Rochester, Minnesota: U.S. Geological Survey Water-Supply Paper 2397, 39pp.

Driscoll, F.G., 1986. *Groundwater and Wells*. Johnson Filtration Systems Inc, St. Paul, MN, 1089pp.

Focazio, M.J., et al., 2002. Assessing ground-water vulnerability to contamination: providing scientifically defensible information for decision makers. U.S. Geological Survey Circular 1224, Reston, VA, 33pp.

Franke, O.L., et al., 1998. Estimating areas contributing recharge to wells; lessons from previous studies. U.S. Geological Survey Circular 1174, 14pp.

Grubb, S., 1993. Analytical model for estimation of steady-state capture zones of pumping wells in confined and unconfined aquifers. *Ground Water*, 31(1), 27–32.

Knochenmus, L.A., and J.L. Robinson, 1996. Descriptions of anisotropy and heterogeneity and their effect on ground-water flow and areas of contribution to public supply wells in a karst carbonate aquifer system. U.S. Geological Survey Water-Supply Paper 2475, Washington, D.C., 47pp.

Kraemer, S.R., Haitjema, H.M., and V.A. Kelson, 2005. Working with WhAEM2000; Capture zone delineation for a city wellfield in a valley fill glacial outwash aquifer supporting wellhead protection. EPA/600/R-00/022. U.S. Environmental Protection Agency, Office of Research and Development, Washington, D.C., 77pp.

Lapham, W.W., Franceska, W.D., and M.T. Koterba, 1997. Guidelines and standard procedures for studies of ground-water quality: selection and installation of wells, and supporting documentation. U.S. Geological Survey Water-Resources Investigations Report 96-4233, Reston, VA, 110pp.

Masterson, J.P., Walter, D.A., and J. Savoie, 1997. Use of particle tracking to improve numerical model calibration and to analyze ground-water flow and contaminant migration, Massachusetts Military Reservation, Western Cape Cod, MA: U.S. Geological Survey Water-Supply Paper 2482, 50pp.

USACE (U.S. Army Corps of Engineers), 1999. *Groundwater Hydrology*. Engineer Manual 1110-2-1421, Washington, D.C.

USBR, 1977. *Ground Water Manual*. U.S. Department of the Interior, Bureau of Reclamation, Washington, D.C., 480pp.

USEPA, 1975. *Manual of Water Well Construction Practices*. Office of Water Supply, EPA-570/9-75-001, 156pp.

USEPA, 1991. *Manual of Small Public Water Supply Systems*. Office of Water, EPA 570/9-91-003, 211pp.

USEPA, 1993a. Guidelines for delineation of wellhead protection areas. EPA/440/5-93-001, U.S. Environmental Protection Agency, Office of Water Office of Ground Water Protection, Washington, D.C.

USEPA, 1993b. A review of methods for assessing aquifer sensitivity and ground water vulnerability to pesticide contamination: U.S. Environmental Protection Agency, EPA/813/R–93/002, 147pp.

USEPA, 1994a. Manual; alternative methods for fluid delivery and recovery. EPA/625/R-94/003, U.S. Environmental Protection Agency, Office of Research and Development, Cincinnati, OH, 87pp.

USEPA, 1994b. Handbook: ground water and wellhead protection. EPA/625/R-94/001, U.S. Environmental Protection Agency, Office of Research and Development, Cincinnati, OH.

USEPA and DWA (Drinking Water Academy), 2003. Introduction to the public water system supervision program, January 2003. http://www.epa.gov/safewater/protect.html (accessed on November 12, 2005).

USGS, 1999. Improvements to the DRASTIC ground-water vulnerability mapping method. U.S. Geological Survey Fact Sheet FS-066-99, 6pp.

Wyoming DEQ (Department of Environmental Quality), 2005. http://deq.state.wy.us/wqd/groundwater/pollution.asp (accessed on December 21, 2005).

4 Groundwater Chemistry

Water is the most powerful solvent of geochemical materials and plays the main role in their continuous redistribution in the environment, below, at, and above the land surface. This capability of water is the result of a unique structure of the water molecule, which has an asymmetric distribution of the hydrogen nuclei with respect to the oxygen nucleus and the two pairs of unshared electrons as illustrated in Figure 4.1. The water molecule has a shape of a distorted tetrahedron, with the oxygen atom located at its center. The covalent bonds between the hydrogen atoms and the oxygen atom form an angle of $104.5°$, while this angle in the molecule of ice is $109°$, resulting in a regular tetrahedron, which is one of the reasons for the firm (solid) structure of ice. The centers of gravity and electric charges in the water molecule are asymmetric, which makes water molecule a dipole. Two corners of the tetrahedron with the hydrogen atoms are strongly positive because of the lack of electrons (hydrogen nuclei share electrons with the oxygen nucleus), while two corners with the unshared electron pairs that belong to the oxygen atom are strongly negative. Water molecules are connected between themselves with strong hydrogen bonds: hydrogen atom (proton) of one water molecule is bonded to the negatively charged side of another water molecule. This hydrogen bonding is one additional reason for the unique behavior of water. For example, breaking hydrogen bonds to boil and evaporate water takes considerable energy—water has higher specific heat than any common substance (1 cal/g per 1°C, or 4.186 J/g °C).

The polarity of molecules, in general, is quantitatively expressed with the dipole moment, which is the product of the electric charge and the distance between the electric centers. Dipole moment for water is 6.17×10^{-30} C m (coloumb-meters), which is higher than for any other substance and explains why water can dissolve more solids and liquids and in greater

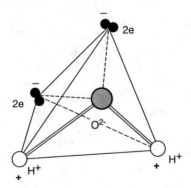

FIGURE 4.1 Molecule of water is a distorted tetrahedron, with the oxygen nucleus in its center. Two positively charged hydrogen atoms are connected to the oxygen atom with covalent bonds. Together with two pairs of unshared electrons (negatively charged) that belong to oxygen, they form a strong dipole which is the main reason why water is the most powerful solvent.

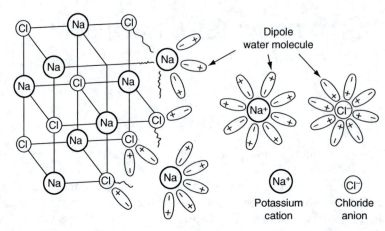

FIGURE 4.2 Schematic presentation of sodium chloride dissolution by water. Electrostatic forces holding together the crystal ionic lattice of halite are broken by strongly polar water molecules. Sodium cation (positively charged) and chloride anion (negatively charged) are finally separated in the solution by shells of dipole water molecules. (From Dimitrijevic, N.S., *Hidrohemija* (*Hydrochemistry*, in Serbian), Univerzitet u Beogradu, Beograd, 1988, 313pp.).

concentrations than any other liquid. Figure 4.2 illustrates schematically the process of dissolution of ionized substances, such as sodium chloride, by water. Salt ions are easily and quickly separated by shells of dipole water molecules, which explains their high solubilities in water. Organic substances with polarized molecules, such as methanol, are also highly soluble in water: hydrogen bonds between water and methanol molecules can readily replace the very similar hydrogen bonds between different methanol molecules and different water molecules. Methanol is therefore said to be miscible in water (its solubility in water is infinite for practical purposes). On the other hand, many nonpolar organic molecules, such as benzene and carbon tetrachloride for example, have very low water solubility.

In addition to the common water molecule with mass 18 (common oxygen-16 plus two common hydrogen-1: $^{16}O + {}^1H + {}^1H$), there are water molecules with masses of 19 through 24 because of the existence of natural and artificial isotopes of oxygen and hydrogen. Three natural and stable isotopes of oxygen are ^{16}O, ^{17}O, and ^{18}O (with ^{16}O constituting 99.76%). The unstable isotopes (^{14}O, ^{15}O, and ^{19}O) have very short half-lives (77, 118, and 30 s, respectively) and therefore no importance in natural waters (Matthess, 1982). The proportion of deuterium (^2H or D), the stable natural isotope of hydrogen, is 0.000149 (compared to 99.9844% of common ^1H), while the proportion of the radioactive isotope tritium (^3H or T) is 1.3×10^{-18}. All the isotopes of oxygen and hydrogen combine to form 18 water molecules and 12 ion species. In this mixture, the proportion of heavy water ($^1H^2H^{16}O$ and $^2H^2H^{16}O$) is very low, about 0.03%.

4.1 SOLUBILITY, WATER CHEMISTRY UNITS, REACTIONS

4.1.1 SOLUBILITY OF SOLIDS AND LIQUIDS

Overall, water is the most effective solvent of environmental substances in all three forms— solid, liquid, and gas. Substances subject to dissolution by water (or any other liquid) are called solutes. True solutes are in the state of separated molecules and ions, which all have very small dimensions (commonly between 10^{-6} and 10^{-8} cm), thus making water solution transparent to light. Colloidal solutions have solid particles and groups of molecules that are larger than ions and molecules of the solvent. When colloidal particles are present in large enough quantities, they give water an opalescent appearance by scattering light. Although there is no one-agreed-to definition of what exactly colloidal sizes are, a common range cited is between 10^{-6} and 10^{-4} cm (Matthess, 1982). The amount of a solute in water is expressed in terms of its concentration, usually in milligrams per liter (mg/L or parts per million, ppm) and micrograms per liter (parts per billion, ppb). It is sometimes difficult to distinguish between certain true solutes and colloidal solutions that may carry particles of the same source substance. Filtering or precipitating colloidal particles before determining the true dissolved concentration of a solute may be necessary in some cases. This especially because drinking water standards for most substances are based on dissolved concentrations. Laboratory analytical procedures are commonly designed to determine the total concentrations of a substance and do not necessarily provide indication of all the individual species (chemical forms) of it present. If needed, however, such speciation can be requested. For example, the determination of individual chromium(Cr) species, rather than the total chromium concentration, may be important in groundwater contamination studies since hexavalent chromium or Cr(VI) is more toxic and has different mobility than trivalent chromium, Cr(III), in groundwater.

Another common way to express the concentration of a substance in solution is using moles per liter. A mole of a substance is its atomic or molecular weight in grams. A solution having a concentration of one mole per liter is a molar solution; thus, the molarity of a solution is its concentration in a weight per volume unit. The SI unit for molarity is mole per cubic meter (mol/m^3), which is the same as millimole per liter (mmol/L). A molal solution is one that contains one mole of solute per 1000 g of a solvent. For dilute solutions up to about 0.01 molar, these two units (molar and molal) are equal, within ordinary experimental error (Hem, 1989). Concentrations in milligrams per liter are readily converted to millimoles per liter by dividing them by the atomic or formula weight of the constituent. For example, to convert 1.8 mg/L of aluminum to millimoles, this concentration has to be divided by the atomic weight of aluminum:

$$\text{millimoles/L} = \frac{\text{concentration of aluminum, in mg/L}}{\text{atomic weight of aluminum}} = \frac{1.8}{26.98} = 0.07 \text{ mM/L}$$

Table 4.1 shows the conversion factors based on the atomic and formula weights for common constituents dissolved in water. Concentrations in millimoles per liter are used in mass-law calculations, which describe various chemical processes including dissolution of substances in water and chemical reactions between different species in solution.

Ionic charge of a substance can be taken into account to compare different ions present in the solution, i.e., to determine their chemical equivalence. Equivalent weight of an ion is obtained when the formula weight of that ion is divided by its charge (valence). When the concentration value in milligrams per liter of a species is divided by this equivalent weight, the result is an equivalent concentration called milligram-equivalents per liter, or milliequivalents, or "meq" when abbreviated. For example, to find milliequivalents of 1.8 mg/l of aluminum, which has valence (charge) of 3, the first step is to find its equivalent weight:

$$\text{Equivalent weight} = \frac{\text{atomic weight of aluminum}}{\text{valence of aluminum ion}} = \frac{26.98}{3} = 8.99$$

The second step is to divide the aluminum concentration, in milligrams per liter, by the equivalent weight:

$$\text{Milliequivalents per liter} = \frac{\text{concentration of aluminum}}{\text{equivalent weight}} = \frac{1.8}{8.99} = 0.20 \text{ meq}$$

TABLE 4.1
Conversion Factors

Element/Species	F_1	F_2	Element/Species	F_1	F_2
Aluminum (Al^{3+})	0.11119	0.03715	Magnesium (Mg^{2+})	0.08229	0.04114
Ammonium (NH_4^+)	0.05544	0.05544	Manganese (Mn^{2+})	0.03640	0.01820
Antimony (Sb)	—	0.00821	Mercury (Hg)	—	0.00499
Arsenic (As)	—	0.01334	Molybdenum (Mo)	—	0.01042
Barium (Ba^{2+})	0.01456	0.00728	Nickel (Ni)	—	0.01704
Beryllium (Be^{2+})	0.22192	0.11096	Nitrate (NO_3^-)	0.01613	0.01613
Bicarbonate (HCO_3^-)	0.01639	0.01639	Nitrite (NO_2^-)	0.02174	0.02174
Boron (B)	—	0.09250	Phosphate (PO_4^{3-})	0.03159	0.01053
Bromide (Br^-)	0.01252	0.01252	Phosphate (HPO_4^{2-})	0.02084	0.01042
Cadmium (Cd^{2+})	0.01779	0.00890	Phosphate ($H_2PO_4^-$)	0.01031	0.01031
Calcium (Ca^{2+})	0.04990	0.02495	Potassium (K^+)	0.02558	0.02558
Carbonate (CO_3^{2-})	0.03333	0.01666	Rubidium (Rb^+)	0.01170	0.01170
Cesium (Cs^+)	0.00752	0.00752	Selenium (Se)	—	0.01266
Chloride (Cl^-)	0.02821	0.02821	Silica (SiO_2)	—	0.01664
Chromium (Cr)	—	0.01923	Silver (Ag^+)	0.00927	0.00927
Cobalt (Co^{2+})	0.03394	0.01697	Sodium (Na^+)	0.04350	0.04350
Copper (Cu^{2+})	0.03147	0.01574	Strontium (Sr^{2+})	0.02283	0.01141
Fluoride (F^-)	0.05264	0.05264	Sulfate (SO_4^{2-})	0.02082	0.01041
Hydrogen (H^+)	0.99216	0.99216	Sulfide (S^{2-})	0.06238	0.03119
Hydroxide (OH^-)	0.05880	0.05880	Thorium (Th)	—	0.00431
Iodide (I^-)	0.00788	0.00788	Titanium (Ti)	—	0.02088
Iron (Fe^{2+})	0.03581	0.01791	Uranium (U)	—	0.00420
Iron (Fe^{3+})	0.05372	0.01791	Vanadium (V)	—	0.01963
Lead (Pb^{2+})	0.00965	0.00483	Zinc (Zn^{2+})	0.03059	0.01530
Lithium (Li^+)	0.14407	0.14407			

Milligrams per liter $\times F_1$ = milliequivalents per liter; milligrams per liter $\times F_2$ = millimoles per liter; based on 1975 atomic weights, referred to as carbon-12.

Source: Modified from Hem, J.D., Study and interpretation of the chemical characteristics of natural water; 3rd edition. U.S. Geological Survey Water-Supply Paper 2254. Washington, D.C., 1989, 263pp.

Table 4.1 shows direct concentration conversion factors from milligrams per liter to milliequivalents per liter for common ions present in natural water solutions.

In an analysis expressed in milliequivalents per liter, unit concentrations of all ions are chemically equivalent. This means that if all ions have been correctly determined, the total milliequivalents per liter of cations is equal to the total milliequivalents per liter of anions. When this is not the case, in addition to possible analytical and computational errors, it may be that some of the ionic species usually considered minor and therefore not analyzed, are present in higher concentrations. Laboratory analyses do not always provide information on the exact form and charge of ionic species detected, which is one disadvantage of using milliequivalents. In addition, an equivalent weight cannot be computed for species whose charge is zero but may be present in significant concentrations, as is often the case with silica. Table 4.2 shows a single water analysis for major dissolved constituents expressed in milligrams per liter, milliequivalents per liter, percentage of total cation and anion equivalents, and millimoles per liter. All the numbers in Table 4.2 were derived from the same original analytical data. It is assumed that dissolved bicarbonate would be converted to carbonate in the dry residue, with loss of an equivalent amount of carbon dioxide and water

TABLE 4.2
Chemical Analysis of a Water Sample Expressed in Four Ways

Constituent	mg/L or ppm	meq/L or epm	percent of meq/L	mM/L
Calcium (Ca)	37	1.85	6.1	0.925
Magnesium (Mg)	24	1.97	6.5	0.985
Sodium (Na) and Potassium (K)	611	26.58	87.4	26.58
Bicarbonate (HCO_3)	429	7.03	23.1	7.03
Sulfate (SO_4)	1010	21.03	69.2	10.52
Chloride (Cl)	82	2.31	7.6	2.31
Fluoride (F)	0.6	0.03	0.1	0.03
Nitrate (NO_3)	0.0	0.00	0.0	0.000
Boron (B)	0.2	—	—	0.019
Silica (SiO_2)	7.9	—	—	0.131
Iron (Fe)	0.17	—	—	0.003
Calculated dissolved solids	1980	60.80	200	48.535
Hardness as $CaCO_3$, Total	191			1.91
Specific conductance (micromhos at 25°C)			2880	
pH			7.3	

Source: Modified from Hem, J.D., Study and interpretation of the chemical characteristics of natural water; 3rd edition. U.S. Geological Survey Water-Supply Paper 2254. Washington, D.C., 1989, 263pp.

(Hem, 1989). This and additional conversion factors for the computations in Table 4.2 are given in Table 4.3.

The water solubility of a given substance is the maximum amount of that substance water can dissolve and maintain in solution that is in equilibrium with the solid or liquid source of the substance. Solubilities of various inorganic and organic substances are extremely variable, from those that are infinite (e.g., for liquid substances miscible with water such as methanol) to those that are quite low, such as for nonaqueous phase liquids, which are immiscible with water. The terms hydrophilic (water-loving) and hydrophobic (water-hating) are sometimes used in reference to water solubility and water insolubility, respectively. The water solubility

TABLE 4.3
Conversion Factors for Quality-of-Water Data

To Convert	To	Multiply by
Ca^{2+}	$CaCO_3$	2.497
$CaCl_2$	$CaCO_3$	0.9018
HCO_3^-	$CaCO_3$	0.8202
*HCO_3^-	CO_3^{2+}	0.4917
Mg^{2+}	$CaCO_3$	4.116
Na_2CO_3	$CaCO_3$	0.9442
NO_3^-	N	0.2259
N	NO_3^-	4.4266

*In the reaction $2HCO_3^- = CO_3^{2-} + H_2O + CO_2$ for computing total dissolved solids.

Source: Modified from Hem, J.D., Study and interpretation of the chemical characteristics of natural water; 3rd edition. U.S. Geological Survey Water-Supply Paper 2254. Washington, D.C., 1989, 263pp.

of a substance is controlled by quite a few factors, including water temperature, pressure, concentrations of hydrogen (H^+) and hydroxyl (OH^-) ions, i.e., pH of water, redox potential (Eh), and relative concentrations of other substances in the solution. The relationships between these variables in actual field conditions are complex and constantly changing, so that exact solubilities of various substances of interest cannot be determined. However, principles of analytic laboratory chemistry combined with some general assumptions and geochemical modeling, can be used to establish reasonable limits of natural solubilities of common substances. Various general texts in chemistry list aqueous solubilities for inorganic and organic compounds, which can be used for initial analyses (e.g., Lide, 2005).

The solubility product, which is a concept of physical chemistry, is an equilibrium constant for the solution of a compound that dissociates into ions. For a saturated solution of a compound, the product of the molar concentrations of the ions is a constant at any fixed temperature. For example, the chemical equation representing the dissolution of gypsum is

$$CaSO_4 + 2H_2O = Ca^{2+} + SO_4^{2-} + 2H_2O \qquad (4.1)$$

and the corresponding solubility product (K_{sp}) expression is (with water having activity of 1):

$$[Ca^{2+}][SO_4^{2-}] = K_{sp} \qquad (4.2)$$

For pure water, i.e., an ideal solution conditions, the solubility product for gypsum is $10^{-4.6}$ at a temperature of 25°C. The solubility of gypsum in moles per liter is equal to the activity of calcium and sulfate ions since 1 mole of gypsum dissolves in pure water to give 1 mole of each in solution. Thus, the saturation solubility of gypsum in pure water at 25°C is

$$Solubility = [Ca^{2+}][SO_4^{2-}] = \sqrt{10^{-4.6}} = 10^{-2.3} mol/L$$

However, the dissolution of gypsum is not this simple, as sulfate ion would always react with some hydrogen ion (or with some water molecules) to form some HSO_4^- ions, and this activity will be highly dependent on pH and other variables. Published values of individual solubility products often are in disagreement with each other because of experimental difficulties. In addition, applying mineral solubility products for pure water (ideal solution) when estimating solubility in the field conditions would be erroneous since, to be precise in saturation calculations, it is necessary to know the chemical form (single ions, complex ions, or neutral molecules) and activity of all ionic species in a given solution, including possible chemical reactions between various ions. Therefore, simple geochemical solubility calculations based on published solubility products are only rough approximations. An additional complicating factor is that the so-called activity of an ionic species is not constant and changes with concentration of both the individual ion considered and the other ions present in the solution. In other words, for dilute solutions, it is often assumed that the activity of an ion equals its mole fraction (concentration) in the solution, but this assumption may lead to significant errors for more concentrated solutions. In either case, it would be more correct to determine the actual ionic activities of particular ions for which the solubility calculations are made, as these activities are influenced by all the constituents in the solution. The solubility product, and therefore solubility, generally increases with temperature, but not in all cases. For substances that exhibit phase changes at a certain temperature, such as inversion of gypsum to anhydrite, the solubility product decreases after the inversion temperature (e.g., 60°C for the gypsum anhydrite inversion). The solubility of solid inorganic substances is independent of pressure in common groundwater systems. Figure 4.3 shows the dependence of solubility on temperature

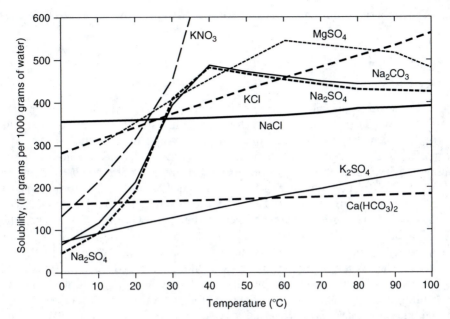

FIGURE 4.3 Solubilities of some solid inorganic substances in water at different temperatures. (Data from Matthess, G., *The Properties of Groundwater*, John Willey & Sons, New York, 1982, 406pp.)

for some common solid inorganic substances. The solubility product also increases with the rising concentration of the solution. This effect is especially noticeable in highly concentrated mineralized water and NaCl brines, in which the concentration can reach very high values (Matthess, 1982).

Ionic activity or simply activity is a concept introduced to account for real-life (nonideal) solutions in thermodynamic relationships involving determined molar concentrations of individual solutes. The activity of an ionic species (α_i) is related to its concentration (C_i) on the molal or molar base as follows (Hem, 1989):

$$\alpha_i = \gamma_i C_i \tag{4.3}$$

where γ_i is known as the activity coefficient of component i. For an ideal solution, $\gamma_i = 1$, and the activity of a component is equal to the mole fraction of that component in the solution (in dilute solutions molar and molal scales are nearly the same). For nonideal solutions, α_i can be considered as the effective concentration of a component; the solution behaves as though the amount of i present is α_i, even though the actual amount present is C_i (Brownlow, 1996). For dilute electrolyte solutions, a common method of calculating activity coefficients is the use of the Debye–Hückel equation (Matthess, 1982; Hem, 1989):

$$\log \gamma_i = -\frac{A z_i^2 \sqrt{I}}{1 + B a_i \sqrt{I}} \tag{4.4}$$

where A is a pressure- and temperature-dependent constant specific to the solvent and related to its unit weight (for water at 25°C, it is 0.5085); z_i is the ionic charge (valence) of the ion i; B is another pressure- and temperature-dependent constant specific to the solvent and related to its unit weight (for water at 25°C, it is 0.3281); a_i is a constant related to the effective diameter of the ion in solution; values of a_i for various ions are given in Hem (1989; Table 4.5, p. 16) and Matthess (1982; Table 4.14, p. 33), and I is the ionic strength of the solution being studied.

The ionic strength of a solution is a measure of the strength of the electrostatic field caused by the ions and is computed as

$$I = \frac{1}{2} \sum m_i z_i^2 \tag{4.5}$$

where m_i is the concentration of component i in moles per liter, and z_i is charge on that ion. The terms in the summation include one for each ionic species present in the solution. A nomograph, which simplifies calculations of ionic strength from analytical data in milligrams per liter, is given in Hem (1989; plate 1). The same reference includes a graph for determining activity coefficients for various ions when the ionic strength I is known (Hem, 1989; Figure 4.1, p. 17). For ionic strengths greater than 10^{-3} molal, Equation 4.4 has to be modified, while other equations have to be used for ionic strengths above 0.1 molal to obtain activity coefficients that agree with experimental data (Brownlow, 1996). Generally, the solubility of a solid increases with increasing ionic strength, because other ions in solution reduce the activity of the ions involved in the dissolution–precipitation reaction (Domenico and Schwartz, 1990). When other ions are present in a solution, the solubility of a solid is different from the one in pure water: it would generally increase if other ions are different (not common) and decrease in the presence of the common (same) ion.

When an ionic substance (AB), such as halite (HCl; A = H, B = Cl), dissociates in an ideal solution to form positively and negatively charged ions, the following relationship is true at equilibrium:

$$K = \frac{[A^+][B^-]}{[AB]} \tag{4.6}$$

where AB is the solid substance, A^+ is the cation part in dissolved phase, B^- is the anion part in the dissolved phase, and K is the general constant of the reaction. At saturation, this general constant is exactly equal to the solubility product (K_{sp}) and the solution is in equilibrium between the solid and dissolved phases of the substance (no additional dissolution or precipitation of the substance from solution should occur). When $K < K_{sp}$, the solution is undersaturated with respect to AB and can dissolve (hold) more of it. When $K > K_{sp}$, the solution is supersaturated and precipitation of AB should occur.

As can be seen from the previous and foregoing discussions about natural aqueous systems, many of the simple analytical laboratory and theoretical considerations cannot fully describe all the complexities of site-specific geochemical conditions that influence chemical characteristics of groundwater. Arguably, the use of geochemical models is the only satisfactory approach to a more comprehensive understanding of various chemical reactions simultaneously taking place in a groundwater system, including evolution of groundwater chemistry along a flowpath. The most widely used geochemical model for natural and contaminated waters is PHREEQC developed at the United States Geological Survey (USGS) and in public domain (Parkhurst and Appelo, 1999). The user-friendly version with a graphical user interface (PHREEQCI; USGS, 2002) includes the following modeling capabilities:

- Aqueous, mineral, gas, surface, ion-exchange, and solid-solution equilibria
- Kinetic reactions
- One-dimensional diffusion or advection and dispersion with dual-porosity medium
- A powerful inverse modeling capability that allows identification of reactions that account for the chemical evolution in observed water compositions
- Extensive geochemical databases

Speciation modeling available in PHREEQC uses a chemical analysis of water to calculate the distribution of aqueous species based on an ion-association aqueous model. The most important results of speciation calculations are saturation indices for minerals, which indicate whether a mineral should dissolve or precipitate. Speciation modeling is useful in situations where the possibility of mineral dissolution or precipitation needs to be known, as in water treatment, aquifer storage and recovery, artificial recharge, and well injection. Inverse modeling capability of PHREEQC can be used to deduce geochemical reactions and mixing in local and regional aquifer systems, and in aquifer storage and recovery studies. It calculates geochemical reactions that account for the change in chemical composition of water along a flowpath. For inverse modeling, at least two chemical analyses of water are needed at different points along the flow path, as well as a set of minerals and gases that are potentially reactive. Mole transfers of phases are calculated that account for the change in water composition along the flowpath. The numerical method applied accounts for uncertainties in analytical data.

4.1.2 SOLUBILITY OF ORGANIC SUBSTANCES

It is common to refer to organic liquids and solids as generally insoluble in water when compared to inorganic salts, bases, and acids. However, this is only partially true because there are organic liquids, such as some alcohols and solvent stabilizers (1,4-dioxane, for example), that are highly soluble or even completely miscible with water due to their polar nature and hydrogen bonds. Organic liquids that are hydrophobic (immiscible) because of their nonpolar nature will also always dissolve in water to some extent, even though the bulk of their volume will remain in a separate (free) liquid phase for long periods of time. Such liquids are called nonaqueous phase liquids (NAPLs). Free NAPL phase may be distributed in a number of ways in the subsurface, such as relatively extensive contiguous volumes of the liquid called NAPL pools, or small globules and ganglia partially or completely surrounded by water (such globules and ganglia are often called residual phase). Dissolution of different NAPL phases in the actual field conditions will depend on a number of factors such as effective and total porosities of the porous media, groundwater flow rates, surface contact area between the NAPL and groundwater (i.e., geometry of different NAPL phases), and fate and transport characteristics including adsorption and diffusion. Assuming that an NAPL would dissolve in the flowing groundwater following some published solubility value for pure water would therefore be erroneous. The actual dissolution of NAPL in the subsurface is often referred to as rate-limited because of the various limiting factors, which all act to decrease the pure-water solubility. The true dissolution rate for an NAPL will be highly site-specific and will change in time. Published solubilities for different hydrophobic organic substances should therefore be used with care and only as a starting point for the related analyses. Cohen and Mercer (1993) provide a detailed discussion on different approaches to determining aqueous and field solubilities of NAPLs, including tables with published literature values. Literature values for aqueous solubility of common NAPL organic contaminants are given in Section 5.2.4.

Another important consideration when estimating real-life solubilities of organic contaminants in the subsurface is that they are often commingled (there are more than one in a mixture), which decreases their individual aqueous solubilities. The exception is the presence of cosolvents, such as alcohols, which would act to increase solubilities of some NAPLs. Cosolvents and surfactants are liquids often used in groundwater remediation projects to increase NAPL solubility and mobility in the porous media, and enhance their recovery. Surfactant stands for surface active agents, which are active ingredients in soaps and detergents and are common commercial chemicals. Two properties of surfactants are central to remediation technologies: the ability to lower interfacial tension and the ability to increase solubility of hydrophobic organic compounds. Both properties arise from the fact that surfactant molecules have a hydrophobic portion and a hydrophilic portion. As a result, when water containing surfactant and NAPL come into contact, surfactant molecules will concentrate along the interface, with their polar ends in water and their nonpolar ends in the NAPL; this lowers the interfacial tension between the two immiscible fluids. When present in sufficient concentration (the critical micellar concentration), surfactant molecules form oriented aggregates, termed micelles. In water, the molecules in a micelle are arranged with their polar ends outwards and their nonpolar ends inwards, forming a nonpolar interior to the micelle. Micelles can incorporate hydrophobic molecules in their interior, producing an apparent increase in solubility. The process of dissolving by incorporation into micelles is termed solubilization. Once solubilized, a compound is transported as if it were a typical dissolved phase (Fountain, 1998).

As a general rule, the most soluble organic species are those with polar molecular structure, or those containing oxygen or nitrogen in a simple, short molecular structure. Examples include alcohols or carboxylic acids, which form hydrogen bonds with water molecules and fit easily into the structure of water. Without hydrogen bonding, the solubility of organic substances diminishes since forcing a nonpolar organic molecule into the tetrahedral structure of water requires considerable energy. The importance of hydrogen bonding decreases with an increasing molecular length and size. The larger the organic molecule is, the larger the space that is required in the water structure, and the less soluble the compound is. For example, smaller alcohol molecules such as methanol and ethanol are infinitely soluble whereas octanol is only slightly soluble. This effect is also clear when examining the solubility of aromatic compounds such as benzene, toluene, naphthalene, and biphenyl. The solubility is inversely proportional to the molecular mass or size of the molecules: it is highest for benzene, which has the smallest molecular mass (78.0 g/mol) of the four, and the lowest for biphenyl, which has the largest molecular mass (154.0 g/mol). The respective reported aqueous solubilities of benzene and biphenyl are 1780 and 7.48 g/m^3 (Domenico and Schwartz, 1990). Because of the double bonds, aromatic organic compounds in general are more soluble than their nonaromatic analogs.

Henry's law is applicable to aqueous dissolution of individual organic components and their theoretical solubility where the component's chemical potentials in the solute and in the pure chemical phase are in equilibrium. In the mixture of NAPLs, however, an individual component will not dissolve according to its theoretical aqueous solubility because its chemical potential in the NAPL mixture is different from that in a pure NAPL. Effective aqueous solubility (S_i^e) of an individual liquid compound i, in mg/L, is given as

$$S_i^e = X_i S_i \tag{4.7}$$

where X_i is the mole fraction of the individual compound, and S_i is the pure-phase solubility of liquid phase of the individual compound. This relationship is based on Raoult's law initially developed for gaseous compounds (see Chapter 5). Laboratory analyses suggest that Equation 4.7 is a reasonable approximation for mixtures of sparingly soluble hydrophobic organic liquids that are structurally similar, and that effective solubilities calculated for complex mixtures (e.g., gasoline and other petroleum products) are unlikely to be in error by more than a factor of 2 (Leinonen and Mackay, 1973; Banerjee, 1984; Cohen and Mercer, 1993).

4.1.3 SOLUBILITY OF GASSES AND VOLATILIZATION

Two types of gasses have to be considered when estimating gaseous solubility in water: those that do not chemically react with water, such as hydrogen, oxygen, nitrogen, and helium for example, and those that do react with water, such as hydrogen sulfide (H_2S), hydrogen chloride (HCl), or carbon dioxide (CO_2). In the latter case, a fraction of the gas would not follow Henry's law, which states that the amount of the dissolved gas in water is directly proportional to its pressure above the interface with water. However, because the reactive gasses are transformed in water usually in small amounts (e.g., CO_2 transforms to carbonic acid to the extent of about 1%), the deviation from Henry's law is not large. Henry's law can be expanded to state that the dissolved concentration of a gas is also inversely proportional to its temperature: at a constant pressure, the solubility of gasses increases with the decreasing temperature and vice versa. At the water boiling temperature, all gasses escape the solution. When there is more than one gas in the mixture (as is the case with air), Henry's law applies to the partial pressure of each individual gas. Henry's law relationship is expressed as

$$C_g = \frac{P_g}{K_H} \tag{4.7}$$

where C_g is the concentration of the gas in water (solution), in moles per volume (m^3 or liter); P_g is the (partial) pressure of the gas in the gaseous phase above the interface with water (the unit is in atmospheres, atm, or Pascals, Pa); and K_H is Henry's law constant, or simply Henry's constant, in atm/(mol/dm^3), or Pa/(mol/m^3).

Published values of Henry's constant for gasses, both common inorganic and various organic, vary greatly and should be used with caution when attempting to calculate gas concentrations in groundwater. A rather extensive compilation of Henry's constant values for inorganic and organic substances is given by Sander (1999).

In general, reactive gasses have much higher water solubilities than the nonreactive gasses (see Table 4.4). Note that most reactive gasses, such as H_2 and H_2S, cannot persist in the atmosphere because their partial pressure is close to zero.

The tendency of a dissolved organic chemical to volatilize from the aqueous solution increases with an increase in Henry's law constant. The same is true for the free liquid phase of a chemical that has a high Henry's constant. Again, it is emphasized that Henry's constant has been determined for many organic liquids and is readily available in reference books (see also Section 6.2); however, the published values vary from source to source and should be used with caution. Organic chemicals with a high Henry's constant are called

---

TABLE 4.4
Solubility of Common Gasses, at Three Different Temperatures, When in Equilibrium with the Atmosphere (Mixture of Gasses), and with the Pure Gas

Gas	In Equilibrium with Pure Gas			In Equlibrium with Atmosphere		
	20°C	10°C	0°C	20°C	10°C	0°C
O_2	43.39	53.70	69.48	9.6	11.9	15.4
N_2	19.01	23.12	29.42	15.1	18.3	23.4
CO_2	1689	2319	3347	0.5	0.7	1
H_2S	3929	5112	7027	N/A	N/A	N/A

Sources: From Alekin, O.A., *Osnovi gidrohemii* (*Principles of Hydrochemistry*, in Russian). Gidrometeoizdat, Leningrad, 1953; Alekin, O.A., *Grundlages der wasserchemie. Eine einführung in die chemie natürlicher wasser*. VEB Deutsch. Verl., Leipzig, 1962, 260 pp. (originally published in Russian in 1953), 1962; Milojevic, N., *Hidrogeologija* (*Hydrogeology*, in Serbian). Univerzitet u Beogradu, Zavod za izdavanje udzbenika Socijalisticke republike Srbije, Beograd, 1967, 379pp; Matthess, G., *The Properties of Groundwater*, John Willey & Sons, New York, 1982, 406pp.

volatile, and include a very important group of common organic groundwater contaminants such as aromatic and halogenated hydrocarbons (e.g., benzene and trichloroethylene). Transport of these chemicals in the subsurface may occur in both the dissolved and gaseous phases, and may involve multiple exchanges between the two phases as the site-specific conditions change in time and space. According to Henry's law, a decrease in vapor (gas) pressure above the solution would cause volatilization (escape) of the dissolved gas into the gaseous phase above the interface. This phenomenon has been widely exploited in the remediation of groundwater contaminated with volatile organic compounds (VOCs) through use of various techniques that increase pressure gradients between the dissolved and gaseous phases. For example, application of vacuum in the unsaturated zone above water table would volatilize an aromatic organic compound from both the dissolved phase and the free phase (if present), which is the principle of a remediation technology called soil vapor extraction.

Raoult's law has been used to quantify the volatilization of individual constituents from a mixture of NAPLs. This law relates the ideal vapor pressure and relative concentration of a chemical in solution to its vapor pressure over the NAPL solution (Cohen and Mercer, 1993):

$$P_A = X_A P_A^o \tag{4.8}$$

where P_A is the vapor pressure of chemical A over the NAPL solution, X_A is the mole fraction of chemical A in the NAPL solution, and P_A^o is the vapor pressure of the pure chemical A.

In natural systems, dissolution and volatilization of gasses from groundwater are rather slow processes. In cases where there are significant natural changes in pressure and temperature within the porous media along the groundwater flow path, dissolution and volatilization of gasses may be fast or even abrupt. A good example is escape of carbon dioxide or hydrogen sulfide into the atmosphere at mineral springs fed by deeply circulating groundwater. Groundwater at great depths is exposed to high subsurface pressures, which may result in a high content of dissolved gasses if there is a source (e.g., a magmatic or a metamorphic process). When such water discharges at the land surface, the gasses escape because of the pressure difference, creating bubbles or odor (e.g., smell of rotten eggs in case of H_2S). Deep artesian wells often show the same mechanism.

4.1.4 SPECIFIC CONDUCTANCE, TOTAL DISSOLVED SOLIDS, AND SALINITY

Solids and liquids that dissolve in water can be divided into electrolytes and nonelectrolytes. Electrolytes, such as salts, bases, and acids, dissociate into ionic forms (positively and negatively charged ions) and conduct electrical current. Nonelectrolytes, such as sugar, alcohols, and many organic substances, occur in aqueous solution as uncharged molecules and do not conduct electrical current. The ability of one cubic centimeter of water to conduct electrical current is called specific conductance (or sometimes simply conductance, although the units are different). Conductance is the reciprocal of resistance and is measured in units called siemens (international system) or mho (one siemen equals one mho; the name mho is derived from the unit for resistance ohm, by spelling it in reverse). Specific conductance is expressed as Siemen/cm or mho/cm. Since the mho is usually too large for most groundwater types, the specific conductance is reported in micromhos/cm or microsiemens/cm (μS/cm), with instrument readings adjusted to 25°C, so that variations in conductance are only a function of the concentration and type of dissolved constituents present (water temperature also has a significant influence on conductance). The measurements of specific conductance can be made rapidly in the field with a potable instrument, which provides for a convenient method to quickly estimate total dissolved solids and compare general types of water quality. For a preliminary (rough) estimate of total dissolved solids, in milligrams per liter, in fresh potable water, the specific conductance in micromhos/cm can be multiplied by 0.7. Pure water has a conductance of 0.055 micromhos at 25°C, laboratory distilled water between 0.5 and 5 micromhos, rainwater usually between 5 and 30 micromhos, potable groundwater ranges from 30 to 2000 micromhos, sea water from 45,000 to 55,000 micromhos, and oil-field brines have commonly more than 100,000 micromhos (Davis and DeWiest, 1991).

The total concentration of dissolved material in groundwater is commonly determined by weighing the dry residue after heating the sample usually to 103°C or 180°C (the higher temperature is used to eliminate more of the crystallization water). This concentration, also called total dissolved solids, may be calculated if the concentrations of major ions are known. However, for some water types, a rather extensive list of analytes may be needed to accurately obtain the total. During evaporation, approximately one half of the hydrogen carbonate ions are precipitated as carbonates and the other half escapes as water and carbon dioxide. This loss is taken into account by adding half of the HCO_3^- content to the evaporation (dry) residue. Some other losses, such as precipitation of sulfate as gypsum, and partial volatilization of acids, nitrogen, boron, and organic substances, may contribute to a discrepancy between the calculated and the measured total dissolved solids.

The term salinity is often used for total dissolved salts (ionic species) in groundwater, in the context of water quality for agricultural uses or human and livestock consumption. Various salinity classifications based on certain salts and their ratios have been proposed (see Matthess, 1982). One problem with the term salinity is that a salty taste may be already noticeable at somewhat higher concentrations of NaCl (e.g., 300–400 mg/L), even though the overall concentration of all dissolved salts may not "qualify" a particular groundwater to be called "saline." In practice, it is common to call water with less than 1000 mg/L dissolved solids potable or fresh, and water with more than 10,000 mg/L saline.

4.1.5 HYDROGEN-ION ACTIVITY (pH)

Hydrogen-ion activity, or pH, is probably the best-known chemical characteristics of water. It is also the one that either directly affects or is closely related to most geochemical and biochemical reactions in groundwater. Whenever possible, pH should be measured directly in the field since groundwater, once outside its natural environment (aquifer), quickly undergoes several changes which directly impact pH, the most important being temperature and the CO_2-carbonate system. Incorrect values of pH may be a substantial source of error in geochemical equilibrium and solubility calculations.

By convention, the content of hydrogen ion in water is now expressed in terms of its activity "pH" rather than its concentration in milligrams or millimoles per liter. Although hydrogen ion (H^+) is being continuously formed by dissociation of water, its concentration in natural waters that are not strongly acidic is very low. For example, at pH 7 (neutral conditions), there are only 1×10^{-7} moles of hydrogen ion in 1 L of water, compared to 10^{-4} mol/L and up for major constituents in groundwater. The dissociation of water into H^+ and OH^- (hydroxyl) ions is a reversible reaction and this equilibrium for a dilute solution is expressed with the following equation:

$$K_w = \frac{[H^+][OH^-]}{[H_2O]} \tag{4.8}$$

where K_w is the constant equal to the product of activities of H^+ and OH^- when the activity of the pure liquid water is assumed to be 1 at 25°C by convention. This ion-activity product of H^+ and OH^- at 25°C is $10^{-14.000}$ in exponential terms. This can be expressed in terms of the negative logarithm as $-\log(K_w) = 14.00$. In other words, when omitting the negative sign by convention, the logarithmic product of the H^+ and OH^- activities is 14. By definition, when the number of hydrogen ions equals the number of hydroxyl ions, the solution is neutral and the hydrogen ion activity is pH $= 7$ (note that log[7] times log[7] is log[14]). Theoretically, when there are no hydrogen ions, pH $= 14$ and the solution is purely alkaline (base); when there are no hydroxyl ions, the solution is a pure acid with pH $= 1$. Accordingly, when the activity (concentration) of hydrogen ions decreases, the activity of hydroxyl ions must increase because the product of the two activities is always the same, i.e., 14. For example, when pH $= 5$ (slightly acidic solution), pOH $= 9$. In terms of actual concentrations of the two ions, this would translate into H^+ concentration of 10^{-5}, or 1 mole of H^+ in 100,000 m^3 of water (which is equal to 1 millimole in 100,000 liters); the concentration of OH^- would be 10^{-9} or 1 mole in 1,000,000,000 m^3 of water. The reaction of dissolved carbon dioxide with water is one of the most important in establishing pH in natural water systems. This reaction is represented by the following three steps (Hem, 1989):

$$CO_2(g) + H_2O(l) = H_2CO_3(aq) \qquad (4.9)$$

$$H_2CO_3(aq) = H^+ + HCO_3^- \qquad (4.10)$$

$$HCO_3^- = H^+ + CO_3^{2-} \qquad (4.11)$$

where g, l, and aq denote gaseous, liquid, and aqueous phases, respectively. The second and third steps produce hydrogen ions which influence the acidity of solution. Other common reactions that create hydrogen ions involve dissociation of acidic solutes such as:

$$H_2PO_4^- = HPO_4^{2-} + H^+ \qquad (4.12)$$

$$H_2S(aq) = HS^- + H^+ \qquad (4.13)$$

$$HSO_4^- = SO_4^{2-} + H^+ \qquad (4.14)$$

Many of the reactions between water and solid species consume H^+, resulting in the creation of OH^- and alkaline conditions. One of the most common is hydrolysis of solid calcium carbonate (calcite):

$$CaCO_3 + H_2O = Ca^{2+} + HCO_3^- + OH^- \qquad (4.15)$$

Note that Equation 4.15 explains why lime dust (calcium carbonate) is often added to acidic soils in agricultural applications to stimulate growth of crops that do not tolerate such soils. The reaction described by Equation 4.15 is accelerated when dissolved CO_2 creates carbonic acid in reaction with water (Equation 4.9), which then also attacks calcite.

The pH of water has a profound effect on the mobility and solubility of many substances. Only a few ions such as sodium, potassium, nitrate, and chloride remain in solution through the entire range of pH found in normal groundwater. Most metallic elements are soluble as cations in acid groundwater but will precipitate as hydroxides or basic salts with an increase in pH. For example, all but traces of ferric ions will be absent above a pH of 3, and ferrous ions diminish rapidly as the pH increases above 6 (Davis and DeWiest, 1991).

4.1.6 ALKALINITY

The alkalinity of a solution may be defined as the capacity for solutes it contains to react with and neutralize acid. An example illustrating the practical meaning of this definition in earth-related sciences is the ability of some soils rich with calcium carbonate to neutralize acid rains. This is often referred to as the soil-buffering capacity. In almost all natural waters, the alkalinity is produced by the dissolved carbon dioxide species, bicarbonate, and carbonate. In other words, except for waters having high pH (greater than about 9.50) and some others having unusual chemical compositions, the alkalinity of groundwater can be assigned entirely to dissolved bicarbonate and carbonate without serious error (Hem, 1989). The most common practice is to report alkalinity in terms of an equivalent amount of calcium carbonate. Table 4.3 provides factors for converting common calcium and carbonate ionic species into equivalent calcium carbonate.

4.1.7   REDUCTION–OXIDATION (REDOX) POTENTIAL (E_h)

Reduction and oxidation can be broadly defined as gain of electrons and loss of electrons, respectively. For a particular chemical reaction, an oxidizing agent is any material that gains electrons, and a reducing agent is any material that loses electrons. The reduction process is illustrated with the following expression (Hem, 1989):

$$Fe^{3+} + e^- = Fe^{2+} \tag{4.16}$$

where ferric iron (Fe^{3+}) is reduced to the ferrous state by gaining one electron. The symbol e^- represents the electron, or unit negative charge. This expression is a "half-reaction" for the iron reduction–oxidation couple; for the reduction to take place there has to be a source of electrons, i.e., another element has to be simultaneously oxidized (lose electrons). Together with the hydrogen ion activity (pH), reduction and oxidation reactions play a key role in the solubility of various ionic substances. Microorganisms are involved in many of the reduction–oxidation reactions and this relationship is especially important when studying the fate and transport of contaminants subject to biodegradation.

The electric potential of a natural electrolytic solution with respect to the standard hydrogen half-cell measuring instrument is expressed (usually) in millivolts. This measured potential is known as reduction–oxidation potential or redox and is denoted with E_h (h stands for hydrogen). Observed E_h range for groundwater is between +700 and −400 mV. Positive sign indicates that the system is oxidizing, and negative that the system is reducing. The magnitude of the value is a measure of the oxidizing or reducing tendency of the system. E_h, just like pH, should be measured directly in the field. Concentration of certain elements is a good indicator of the range of possible E_h values. For example, notable presence of H_2S (>0.1 mg/L) always causes negative E_h. If oxygen is present in concentrations greater than 1 mg/L, E_h is commonly between 300 and 450 mV. Generally, an increase in content of salts decreases E_h of the solution.

The redox state is determined by the presence or absence of free oxygen in groundwater. Newly percolated (recharge) water often supplies oxygen to groundwater in the range from 6 to 12 mg/L. At the water table, oxygen is also dissolved directly from the soil air and can be carried into deeper groundwater levels by diffusion and flow dispersion. The depth of the oxidizing zone in an aquifer varies greatly based on local hydrogeologic and geochemical conditions. Oxidation processes are faster in warm and humid climates than in cold and arid climates. Oxygen can be consumed in a number of different geochemical reactions, the most direct being oxidation of iron and manganese compounds. Microbial activity also consumes oxygen and may rapidly create reducing environment in a saturated zone with an excess in dissolved organic carbon (DOC, which is a nutrient for microbes) such as in cases of groundwater contamination with organic liquids. Determining redox potential in an aquifer is particularly important in contaminant fate and transport and remediation studies. For example, oxidizing (aerobic) conditions favor biodegradation of petroleum hydrocarbons such as gasoline, while reducing (anaerobic) conditions favor biodegradation of chlorinated compounds such as tetrachloroethene (PCE). Another example of the importance of redox conditions in an aquifer is the state of metal ions, such as iron and chromium. In slightly acidic to alkaline environments, Fe(III) precipitates as a highly adsorptive solid phase (ferric hydroxide), whereas Fe(II) is very soluble and does not retain other metals. The reduction of Fe(III) to Fe(II), therefore, releases not only Fe^{2+} to the water, but also any contaminants that were adsorbed to the ferric hydroxide surfaces. Hexavalent Cr(VI) is a toxic, relatively mobile ion whereas trivalent Cr(III) is inert, relatively insoluble, and strongly adsorbs to solid surfaces (Johnson et al., 1989).

Based on oxygen demand of the various bacterial species, an oxygen content between 0.7 and 0.01 mg O_2/L at 8°C water temperature has been commonly defined as threshold oxygen concentration for the boundary between oxidizing and reducing conditions. However, field observations suggest that reducing conditions may appear at considerably higher oxygen contents (Matthess, 1982). Bacteria present in groundwater catalyze almost all important redox reactions. This catalytic capability is produced by the activity of enzymes that normally occur within the bacteria. Enzymes, which are protein substances formed by living organisms, have the power to increase the rate of redox reactions by decreasing the activation energies of the reactions. They accomplish this by strongly interacting with complex molecules representing molecular structures halfway between the reactant and the product (Pauling and Pauling, 1975, from Freeze and Cherry, 1979). Bacteria and their enzymes are involved in redox processes in order to acquire energy for synthesis of new cells and maintenance of old cells. If the redox reactions that require bacterial catalysis are not occurring at significant rates, a lack of one or more of the essential nutrients for bacterial growth is likely the cause. There are various types of nutrients, some of which are required for incorporation into the cellular mass of the bacteria. Carbon, nitrogen, sulfur, and phosphorous compounds and many metals are in this category. Other nutrients are substances that function as electron donors or energy sources, such as water, ammonia, glucose, and hydrogen sulfide (H_2S). Substances that function as electron acceptors in redox reactions are oxygen, nitrate, and sulfate (Freeze and Cherry, 1979). The key principle of bioremediation of contaminated groundwater is the stimulation of bacterial activity that produces certain redox reactions. This stimulation is accomplished by delivering nutrients or electron acceptors into the saturated zone targeted for remediation.

The redox potential generally decreases with rising temperature and pH, and this decrease results in an increasing reducing power of the aqueous system. Reducing systems, in addition to the absence or very much reduced oxygen content, have noticeable content of iron and manganese, occurrence of hydrogen sulfide, nitrite and methane, absence of nitrate, and often a reduction or absence of sulfate (Matthess, 1982). Stability diagrams, which relate pH and E_h to various chemical forms of a substance, are often used to draw conclusions about its occurrence and chemical behavior in a particular system. An example for the various forms of iron is given in Figure 4.4.

4.2   NATURAL GROUNDWATER CONSTITUENTS AND THEIR RELATION TO GEOLOGIC MEDIA

As the most effective solvent for inorganic geologic materials, groundwater contains a large number of dissolved natural elements. Complete chemical analyses of groundwater (those looking for "all" possible naturally occurring elements) would commonly turn out to be more than 50 at the lowest levels detectable in commercial laboratories. However, less than

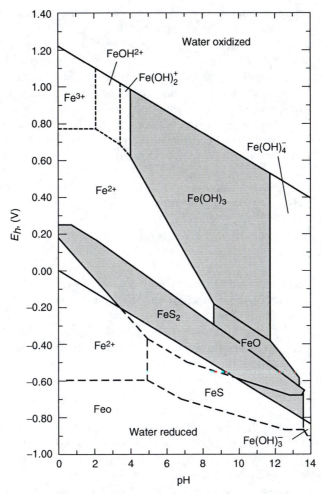

FIGURE 4.4 Fields of stability for solid and dissolved forms of iron as a function of E_h and pH at 25°C and 1 atm pressure. Activity of sulfur species is 96 mg/L as SO_4^{2-}, carbon dioxide species 61 mg/L as HCO_3^-, and dissolved iron 56 μg/L. (From Hem, J.D., Study and interpretation of the chemical characteristics of natural water; 3rd edition. U.S. Geological Survey Water-Supply Paper 2254. Washington, D.C., 1989, 263pp.)

that are usually present in normal potable groundwater at notable concentrations, which vary based on local geologic conditions and mineral contents of the porous media (rock minerals). Constituents that are commonly found to be present at concentrations greater than 1 mg/L are sometimes called major or macro constituents of groundwater. Such components are analyzed by default because they most obviously reflect the type of rocks present in the aquifer and are therefore used to compare general genetic types of groundwater. Some of the elements that are commonly present in groundwater at concentrations between 0.01 and 10 mg/L, and are significant for understanding of its genesis, are also often analyzed by default. They are sometimes referred to as either minor or secondary constituents. Metallic elements that are usually found at concentrations less than 0.1 mg/L and less than 0.001 mg/L are sometimes called minor constituents and trace constituents, respectively. However, the significant concentrations and the relative importance of different groundwater constituents are site- and regulations-specific, and they do vary in different parts of the world, and in time.

A good example is arsenic, considered for a long time to be a "minor" groundwater constituent belonging to the concentration range between 0.0001 and 0.1 mg/L (e.g., Davis and DeWiest, 1991), or a "trace" constituent belonging to the concentration group of less than 0.1 mg/L (Domenico and Schwartz, 1990). However, as more and more analyses of groundwater used or considered for water supply are being made both in the United States and worldwide, arsenic emerges as the major groundwater constituent simply because of the new regulatory limit of 0.01 mg/L (10 ppb). In other words, many water supply utilities worldwide are discovering that arsenic is present at concentration levels greater than 10 ppb. About 35 or so important natural inorganic groundwater constituents, recognizing the relativity of word "important," can be divided into the following two practical analytical groups:

1. Primary constituents analyzed routinely
   Anions: Cl^-, SO_4^{2-}, HCO_3^-, CO_3^{2-}, NO_3^- (and other nitrogen forms)
   Cations: Ca^{2+}, Mg^{2+}, Na^+, K^+, Fe^{2+} (and other iron forms)
   Silica as SiO_2 (present mostly in uncharged form)
   pH

2. Secondary constituents analyzed as needed
   Elements/anions: boron (B), bromine/bromide (Br), fluorine/fluoride (F), iodine/iodide (I), phosphorus/phosphate (P)
   Metals, nonmetallic elements: aluminum (Al), antimony (Sb), arsenic (As), barium (Ba), beryllium (Be), cesium (Cs), chromium (Cr), copper (Cu), lead (Pb), lithium (Li), manganese (Mn), mercury (Hg), rubidium (Rb), selenium (Se), strontium (Sr), zinc (Zn)
   Radioactive elements: radium, uranium, alpha particles, beta particles
   Organic matter (total organic carbon [TOC])

Almost all primary and secondary inorganic constituents listed above are included in the list of primary or secondary drinking water standards by the U.S. Environmental Protection Agency (see Chapter 5.3.). Most of them, when in excess of a certain concentration, are considered contaminants and such groundwater is not suitable for human consumption. Groundwater contamination may be the result of both naturally occurring inorganic substances, and those introduced by human activities.

4.2.1 Graphical Presentation of Groundwater Analyses

Probably the clearest distinction between the primary and secondary constituents is that the first group always makes up more than 90% of total dissolved solids in a groundwater sample. Because of this, six ionic species in particular have been widely used to graphically present major chemical types of groundwater:

Cations: calcium (Ca^{2+}), magnesium (Mg^{2+}), and (expressed together) sodium (Na^+) + potassium (K^+)

Anions: chloride (Cl^-), sulfate (SO_4^{2-}), and (expressed together) hydrocarbonate (HCO_3^-) + carbonate (CO_3^{2-})

As explained in Section 4.1.1, in an analysis expressed in milliequivalents per liter, unit concentrations of all ions are chemically equivalent. This means that if all ions have been correctly determined, the total milliequivalents per liter of the four major cations listed above, (Ca, Mg, and Na + K) is equal to the total milliequivalents per liter of the four major anions (Cl, SO_4, and HCO_3 + CO_3). When this is not the case, in addition to possible analytical and computational errors, it may be that some of the ionic species usually considered minor and therefore not analyzed, are present in higher concentrations. For example, nitrate anion (NO_3^-)

in agricultural and otherwise contaminated areas is present in considerable amount, often in excess of 10 mg/L. Figure 4.5 shows the two most common graphical presentations of the contents of major ions in groundwater samples. The graphs are plotted using percentages of milliequivalents per liter for each ion within its respective group. Results of the chemical analysis for sample #1 expressed in additional units are given in Table 4.2. Piper diagram (Piper, 1944) is also the most convenient for plotting the results of multiple analyses on the same graph, which may reveal grouping of certain samples and indicate different types (origin) of groundwater. For example, samples 1, 2, and 3 in Figure 4.6 clearly belong to a different chemical type, compared to all other samples. Table 4.5 lists, in milligrams per liter, contents of major ions for the samples plotted on the Piper diagram in Figure 4.6. Concentrations expressed in milligrams per liter can be converted into milliequvalents per liter using conversion factors listed in Table 4.2. Circle diagrams are convenient for presentation of individual analyses on hydrogeologic maps: different ions can be shown in different colors, while the size of the circle may be used to represent the total dissolved solids in the sample. There are quite a few other graphical methods for representing the results of groundwater chemical analyses (e.g., see Alekin, 1962 and Hem, 1989). One of the more often used is the Stiff diagram, which gives an irregular polygonal shape that can help in recognizing possible patterns in multiple analyses (Figure 4.7).

The reason why ionic species shown in Figure 4.5 are the most prevalent in natural groundwaters is that the most important soluble salts occurring in relatively large quantitites in rocks are calcium carbonate ($CaCO_3$), magnesium carbonate ($MgCO_3$), their combination ($CaCO_3 \times MgCO_3$), sodium chloride ($NaCl$), calcium sulfate ($CaSO_4$), and hydrous calcium sulfate ($CaSO_4 \times 2H_2O$). Although aluminum is the third most abundant element in the Earth's crust, because of its chemical properties it rarely occurs in solution in natural water in concentrations greater than a few tenths or hundredths of a milligram per liter (Hem, 1989). The exceptions are waters with very low pH such as acidic mine drainage. Iron, the second most abundant metal in the Earth's crust, is also found in relatively small concentrations in groundwater, rarely exceeding 1 mg/L. The chemical behavior of iron and its solubility in water depend strongly on the redox potential and pH (see Figure 4.4). The forms of iron present are also strongly affected by microbial activity. Various ferrous complexes are formed by many organic molecules, and some of the complexes may be significantly more resistant to oxidation than free ferrous ions and insoluble in groundwater. For all of these reasons, iron is considered one of the primary groundwater constituents, even though its dissolved ionic forms are in most groundwaters found in smaller concentrations compared to the major ions.

Contrary to the examples of aluminum and iron, element silicon, being second only to oxygen in the Earth's crust, is found in appreciable quantities in most groundwaters, usually between 1 and 30 mg/L when expressed as silica. Relative abundance of silica in natural water is due to its many different chemical forms found in minerals and rocks. This fact goes contrary to the common belief that silica is not soluble in water and is therefore not present in groundwater. Most abundant forms of silica dissolved in water are thought not to form ions, although the complicated groundwater chemistry of silica is still not well understood. Consequently, when it is important to present the silica content in a groundwater sample graphically, a bar graph based on milliequivalents of millimoles per liter is often used (Figure 4.8).

Graphical presentation of relationships between individual ions, or one of the ions to the total concentration, is often helpful in understanding similarities and differences between samples, and may provide a good tool for the classification of groundwater types. Various classifications based on ratios between different ions and groups of ions have been proposed in the groundwater literature and are beyond the scope of this book (e.g., see Alekin, 1962; Matthess, 1982; Hem, 1989). An example illustrating the use of ratios is given in Table 4.6 where three hypothetical chemical analyses are compared using four different ratios (Hem, 1989). All three analyses could be considered representative of sodium bicarbonate waters, and

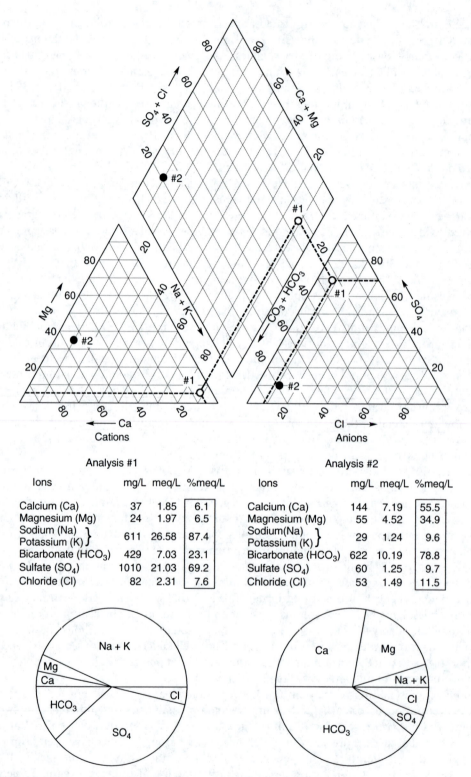

Analysis #1				Analysis #2			
Ions	mg/L	meq/L	%meq/L	Ions	mg/L	meq/L	%meq/L
Calcium (Ca)	37	1.85	6.1	Calcium (Ca)	144	7.19	55.5
Magnesium (Mg)	24	1.97	6.5	Magnesium (Mg)	55	4.52	34.9
Sodium (Na) } Potassium (K)	611	26.58	87.4	Sodium(Na) } Potassium (K)	29	1.24	9.6
Bicarbonate (HCO$_3$)	429	7.03	23.1	Bicarbonate (HCO$_3$)	622	10.19	78.8
Sulfate (SO$_4$)	1010	21.03	69.2	Sulfate (SO$_4$)	60	1.25	9.7
Chloride (Cl)	82	2.31	7.6	Chloride (Cl)	53	1.49	11.5

FIGURE 4.5 Piper and circle diagrams for two groundwater samples.

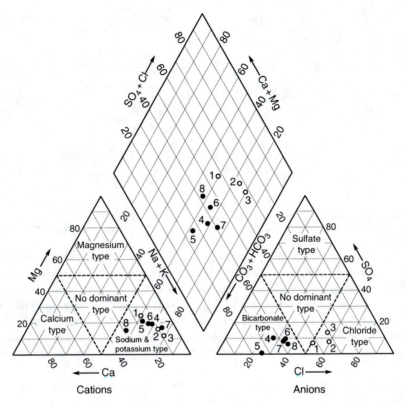

FIGURE 4.6 Piper diagram of multiple groundwater samples. (Modified from NAVFAC (Naval Facilities Engineering Command), Guidance for environmental background analysis; volume 3: groundwater. User's guide UG-2059-ENV, Engineering Service Center, Port Hueneme, CA, 177pp.)

have similar total concentration, specific conductance, and pH. The high proportion of silica in samples B and C, their similar Ca:Mg and Na:Cl ratios, and their similar proportions of SO₄ to total anions establish the close similarity of B and C, and the dissimilarity of both to A.

TABLE 4.5
Analytical Results (mg/L) for the Samples Plotted on the Piper Diagram in Figure 4.6

Sample	Ca^{2+}	Mg^{2+}	Na^+	K^+	Cl^-	HCO_3^-	SO_4^{2-}
MW-1	11.5	10.9	46.7	3.01	53	74	10
MW-2	11.6	11.2	92.4	3.45	100	91	18
MW-3	13.0	10.2	123	4.18	134	133	34
MW-4	26.7	25.4	162	9.86	94	375	47
MW-5	33.3	25.4	133	8.89	88	438	3
MW-6	30.9	25.8	159	7.26	150	468	58
MW-7	17.9	23.0	186	9.53	150	499	61
MW-8	113.0	20.0	184	9.65	200	500	56

Source: Modified from NAVFAC (Naval Facilities Engineering Command), Guidance for environmental background analysis; volume 3: groundwater. User's guide UG-2059-ENV, Engineering Service Center, Port Hueneme, CA, 2004, 177pp.

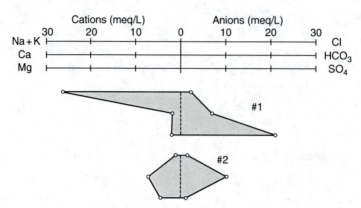

FIGURE 4.7 Stiff diagrams for water analyses shown in Figure 4.5.

Some useful ratios for establishing chemical types of groundwater are the ratio of calcium to magnesium for studying water from carbonate sediments (limestone and dolomite), and the ratio of silica to dissolved solids for identifying solutions of different silicate minerals in magmatic rock terrains. Other ratios may be useful in different terrains, as long as the mineral contents of the aquifer porous media is well understood. For example, a study of origin of groundwater at karst springs in an area composed of Triassic limestone, Jurassic Diabase-Chert Formation, and Jurassic ultramafic rocks (seprentinitie and peridotite) included interpretation of the Ca^{2+}/Mg^{2+} molar ratio and the silica content. The ophiolotic belt of southern Europe and Turkey is characterized by major karst aquifers, sometimes partially or completely covered by the younger Jurassic rocks. The main components of chemical composition of these rocks and the related groundwaters are shown in Figure 4.9. Groundwater in limestone has Ca^{2+}/Mg^{2+} molar ratio between 5 and 20, and the average silica content below 8 mg/L. Diabase rocks are composed of basic plagioclases, mainly anorthite, the major source of calcium in groundwater, and of monoclinic pyroxenes (augite), the source of calcium and magnesium. The origin of sodium in groundwater is from albite. Silica content

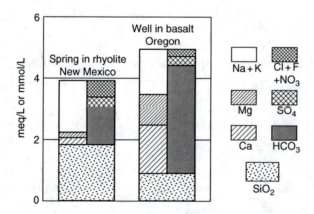

FIGURE 4.8 Graphs of major groundwater constituents including silica for two samples obtained from igneous rocks. (Modified from Hem, J.D., Study and interpretation of the chemical characteristics of natural water; 3rd edition. U.S. Geological Survey Water-Supply Paper 2254. Washington, D.C., 1989, 263pp.)

TABLE 4.6
Hypothetical Chemical Analyses Compared by Means of Ratios

Constituent	A		B		C	
	mg/L	meq/L	mg/L	meq/L	mg/L	meq/L
Silica (SiO$_2$)	12	—	33	—	30	—
Calcium (Ca)	26	1.30	12	0.60	11	0.55
Magnesium (Mg)	8.8	0.72	10	0.82	9.2	0.76
Sodium (Na)						
Potassium (K)	73	3.16	89	3.85	80	3.50
Bicarbonate (HCO$_3$)	156	2.56	275	4.51	250	4.10
Sulfate (SO$_4$)	92	1.92	16	0.33	15	0.31
Chloride (Cl)	24	0.68	12	0.34	12	0.34
Fluoride (F)	0.2	0.01	1.5	0.08	1.2	0.06
Nitrate (NO$_3$)	0.4	0.01	0.5	0.01	0.2	0.00
Calc. diss. solids	313	—	309	—	282	—
Hardness, CaCO$_3$	101	—	71	—	66	—
Spec. conductance						
(μmhos at 25°C)		475		468		427
pH		7.7		8.0		8.1

	$\dfrac{\text{SiO}_2\ (\text{mg/L})}{\text{Dissolved Solids (mg/L)}}$	$\dfrac{\text{Ca (mg/L)}}{\text{Mg (mg/L)}}$	$\dfrac{\text{Na (mg/L)}}{\text{Cl (mg/L)}}$	$\dfrac{\text{SO}_4\ (\text{mg/L})}{\text{Total Ions (mg/L)}}$
A	0.038	1.8	4.6	0.37
B	0.11	0.73	11.3	0.063
C	0.11	0.72	10.3	0.64

Source: Modified from Hem, J.D., Study and interpretation of the chemical characteristics of natural water; 3rd edition. U.S. Geological Survey Water-Supply Paper 2254. Washington, D.C., 1989, 263pp.

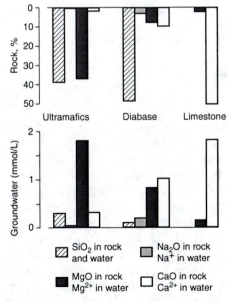

FIGURE 4.9 Main components of chemical composition of rocks and related groundwater in the ophiolitic belt of southern Europe. (From Kresic, N. and Papic, P., *Environ. Geol. Water Sci.*, 15(2), 131, 1990.)

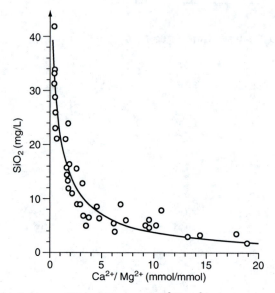

FIGURE 4.10 Relationship between silica content and Ca^{2+}/Mg^{2+} molar ratio at major karst springs in the ophiolitic belt of the Serbian Inner Dinarides. (From Kresic, N. and Papic, P., *Environ. Geol. Water Sci.*, 15(2), 131, 1990.)

in the rock, of about 48%, is dominant. The groundwater type (HCO₃-Ca-Mg) generally reflects the chemical composition of the rock, although the silica content is relatively small because of its lower solubility. Ca^{2+}/Mg^{2+} molar ratio is between 1 and 1.5. Major minerals in the ultramafics are olivines (forsterite and chrysolite) and rombic pyroxenes (enstatite–hypersthene series), i.e., magnesium silicates. The groundwater has distinct contents of Mg^{2+} and HCO_3^- ions, and pH of 8 or higher. The content of silica is much higher than in the diabase groundwater, a consequence of greater solubility of silica in the presence of magnesium. Ca^{2+}/Mg^{2+} molar ratio is about 0.15. Figure 4.10 shows that recharge of karst aquifers by water infiltrating through the overlying ultramafics, rich in both magnesium and silica, often plays a significant or even dominant role in the water balance of individual springs.

4.2.2 PRIMARY CONSTITUENTS

4.2.2.1 Calcium

Calcium is the most abundant of the alkaline-earth metals and is a major constituent of many common rock minerals and solutes in natural waters. Calcium has only one oxidation state, Ca^{2+}. It is an essential element for plant and animal life forms. Its presence in aqueous systems is due mainly to the more soluble solids containing calcium and is governed to a great extent by the equilibria between the solute and gaseous phases of the carbon dioxide species. Calcium also participates in cation-exchange equilibria at surfaces of aluminosilicates (clays) and other minerals (Hem, 1989). Although calcium is an essential constituent of many igneous rock minerals, especially of the chain silicates pyroxene and amphibole, and the feldspars, its concentration in the groundwater in magmatic terrains is generally low. This is mainly because of the slow rate of decomposition of most igneous-rock minerals. Calcium at low concentrations also occurs in groundwater that has been in contact with metamorphic rocks, which contain silicate minerals with calcium. Groundwater in silicate terrain often contains less than 100 mg Ca^{2+}/L, although this concentration may be elevated due to presence of soil–regolith sediments where the content of CO_2 in the ground air is consistently 10 to 100 times that of the atmosphere (Matthess, 1982). CO_2 forms carbonic acid in reaction with water and the acid accelerates dissolution of some calcium-bearing minerals.

Proportionally much higher concentrations of calcium are found in groundwaters in carbonate sedimentary rocks such as limestone, composed of calcite and aragonite (both have the formula of $CaCO_3$), dolomite, composed of calcite and mineral dolomite $CaMg(CO_3)_2$, calcium sulfates anhydrite ($CaSO_4$), and gypsum ($CaSO_4 \cdot 2H_2O$), and, more rarely, sedimentary rock fluorite composed of calcium fluoride (CaF_2). Calcium is also a component of some types of zeolites and montmorillonite (Hem, 1989). In sandstone and other intergranular rocks, calcium carbonate is commonly present in the form of cement between rock grains and may be dissolved by the flowing groundwater. The following equations express the general reactions between calcium carbonate and the CO_2–water system (see also Equation 4.9 through Equation 4.11):

$$CO_2 + H_2O = H_2CO_3 = H^+ + HCO_3^- \tag{4.17}$$

$$HCO_3^- = H^+ + CO_3^{2-} \tag{4.18}$$

$$CaCO_3 + H^+ = Ca^{2+} + HCO_3^- \tag{4.19}$$

The predominant influence of equilibria between the solute and gaseous phases of CO_2 on the calcite dissolution and precipitation processes greatly depends on pH and water temperature. Calculations determining if a particular solution is in thermodynamic equilibirium with calcite must therefore include pH (ionic activity of H^+), water temperature, concentration of free dissolved CO_2, and ionic activities of Ca^{2+} and HCO_3^- in the solution. Hem (1989) and Matthess (1982) provide a detailed discussion on calcite solubility including use of various reaction constants, nomograms, and ionic strength graphs. When groundwater undersaturated with respect to calcite flows through a porous medium rich in soluble calcite, the dissolution process commonly creates secondary porosity (voids) in the rock. Karst terrains are a typical example for this reaction. In time, the groundwater may become supersaturated with respect to calcite but may still keep it in solution (dissolved) due to various geochemical interactions and conditions that influence the carbonate equilibria. When the subsurface conditions abruptly change, calcium carbonate may precipitate at this geochemical barrier; typical examples are speleothems in caves, travertine (calc-sinter) deposits at springs, cementation of faults, or incrustation of well screens. The presence of sodium and potassium salts generally increases the solubility of calcium carbonate.

Concentrations of calcium in normal potable groundwater generally range between 10 and 100 mg/L, with some limestone and gypsum groundwaters exceeding this amount. The most commonly noticed effect of calcium in groundwater is hardness, or its reaction with soap demonstrated by the inability of soap to either cleanse or lather. Ions of magnesium, iron, manganese copper, barium, and zinc also cause a similar difficulty. Hardness is defined as the soap neutralizing power of these ions (Davis and DeWiest, 1991). Since all the ions except magnesium and calcium usually occur in low concentrations, hardness is attributed to the sum of effects of only calcium and magnesium dissolved in groundwater (the effect of alkaline-earth metals). It is calculated on the basis of calcium alone, as the concentration of equivalent calcium carbonate expressed in milligrams per liter. If the hardness exceeds the alkalinity (bicarbonate + carbonate in milligrams per liter of $CaCO_3$ or other equivalent units), the excess is termed "noncarbonated hardness" and may be reported as such. In some European countries, hardness values are reported in degrees of hardness, all in terms of calcium carbonate. One French degree is equivalent to 10 mg/L, one German degree to

17.8 mg/L, and one English or Clark degree to 14.3 mg/L (Hem, 1989). In the United States, it is now generally accepted that the "ideal" drinking water should not contain more than 80 mg/L of hardness.

4.2.2.2 Magnesium

Magnesium is a common alkaline-earth element essential in plant and animal nutrition, and has only one oxidation state of significance in water chemistry, Mg^{2+}. Although magnesium and calcium behave similarly in water solutions to some extent (such as when creating water hardness), the geochemical characteristics of magnesium are quite different because its ions are smaller than calcium ions (Hem, 1989). In magmatic rocks magnesium is typically a major constituent of the dark-colored ferromagnesian minerals such as olivine, pyroxenes, amphiboles, and dark-colored micas. In metamorphic rocks, magnesium is a common constituent of minerals chlorite and seprenitine.

Sedimentary rocks containing magnesium carbonate (magnesite), $MgCO_3$, and the double calcium–magnesium carbonate dolomite ($CaCO_3 \times MgCO_3$) are, similarly to calcite, soluble under the influence of CO_2 gas dissolved in groundwater. Although the solubility of $MgCO_3$ in pure water is greater than that of calcium carbonate, its actual solubility in the field conditions is much harder to determine because of the presence of various other hydrated magnesium carbonate forms such as nesquehonite, landsfordite, and basic hydromagnesite. Dolomite and magnesium hydroxide (brucite) in sediments are poorly soluble, while the solubility of magnesium chloride and sulfate is distinctly higher (Matthess, 1982). The magnesium ion, Mg^{2+}, will normally be the predominant form of magnesium in solution, while the complex $MgOH^+$ will not be significant below about pH 10 (Hem, 1989). Magnesium occurs in significant amounts in most limestones and especially in dolomites. The dissolution of these rocks brings magnesium into groundwater but the chemical reaction is not readily reversible. Magnesium concentration tends to increase along the flowpath as the precipitate formed may be nearly pure calcite. Mature groundwaters in carbonate sedimentary terrains therefore often have a high Mg:Ca ratio since the conditions for direct precipitation of dolomite from solution are not commonly found in normal groundwater (Hem, 1989).

In spite of the higher solubility of most of its compounds, the magnesium content in potable (low-mineralized) groundwater is generally below that of calcium, most probably because of the lower general abundance of magnesium. Occasional exceptions occur in magnesium-rich aquifers, such as olivine-basalts, serpentines, and dolomite rocks (Matthess, 1982).

4.2.2.3 Sodium and Potassium

Potassium is slightly less common than sodium in igneous rocks but more abundant in all the sedimentary rocks. However, in the ocean water, although notable, the concentration of potassium is far less than that of sodium. In most fresh water aquifers, if the sodium concentration substantially exceeds 10 mg/L, the potassium concentration is commonly half or a tenth that of sodium (Hem, 1989). In some dilute natural waters in which the sum of sodium and potassium is less than 10 mg/L, the potassium concentration may be equal or even exceeding that of sodium. These facts point out the very different behavior of the two alkali metals in natural systems. Sodium tends to remain in solution rather persistently once liberated from silicate-mineral structures. There are no important precipitation reactions that can maintain low sodium concentrations in water. Sodium may be retained by adsorption on mineral surfaces, especially by clays, which have high cation-exchange capacity. However, the interactions between surface sites and sodium (monovalent ion) are much weaker than the interactions with divalent ions such as calcium. Cation exchange processes therefore tend to extract divalent ions from the solution and to replace them with monovalent ions (such as sodium). Potassium feldspars orthoclase and microcline ($KAlSi_3O_8$) are less soluble than feldspars with sodium, and are less soluble than the sodium plagioclase albite ($NaAlSi_3O_8$). The soluble salts of sodium, such as halite, are readily dissolved and removed from sediments after environmental changes, such as sediment leaching by freshwater after sea level regression. Although potassium salts are also highly soluble and potassium is generally as abundant as sodium in rocks, it seldom occurs in concentrations equal to sodium in natural waters. However, reactions involving water chemistry of potassium are still not well understood so that quantifications of the different behavior of sodium and potassium are difficult. In addition, potassium has remained one of the more difficult ions to analyze accurately, which is the main reason why sodium and potassium are usually lumped together within the cation group of major groundwater constituents when reporting analytical results.

Although there are indications that biological factors may play an important role in controlling the availability of potassium for solution in groundwater, the following two generalizations are still used to explain the difference between the concentrations of sodium and potassium in most natural groundwaters: (1) the potassium concentrations are low because of the high degree of stability of potassium-bearing alumino-silicate minerals and (2) potassium from solution is incorporated strongly into some clay–mineral structures, such as in spaces between crystal layers of ilite and, unlike sodium, cannot be removed by further ion-exchange reactions (Hem, 1989).

Excessive sodium content in groundwater used for irrigation may damage soil structure as sodium replaces calcium and magnesium adsorbed on the soil clays and colloids. Two principal effects of this replacement are a reduction in soil permeability and a hardening of the soil. The U.S. Salinity Laboratory Staff (1954) defined the sodium-adsorption ratio (SAR) of a water as

$$SAR = (Na^+)/[0.5 \times (Ca^{2+}) \times (Mg^{2+})]^{-2} \qquad (4.20)$$

where ion concentrations (in parentheses) are expressed in milliequivalents per liter. The SAR predicts reasonably well the degree to which irrigation water tends to enter into cation-exchange reactions in soil (Hem, 1989). Values of SAR are therefore included in chemical analyses of irrigation water and water that might be considered for that use. The value is empirical and of otherwise limited geochemical significance. Figure 4.11 shows a diagram commonly used to assess suitability of water for irrigation. The graph relates SAR to the specific conductance (salinity) of water and divides water into various hazard classes. A detailed discussion on these classes, as they relate to crop tolerance for salinity and soil types, is given by the U.S. Salinity Laboratory Staff (1954).

4.2.2.4 Iron

Iron is one of the most abundant elements in magmatic rocks, particularly as part of dark-colored minerals such as pyroxenes, amphiboles, biotite, magnetite, and, especially, the olivine, which is a solid solution whose end members are forsterite (Mg_2SiO_4) and fayalite (Fe_2SiO_4). For the most part, iron in these minerals is in the ferrous (Fe^{2+}) form. Ferric iron (Fe^{3+}) is present in magnetite as Fe_3O_4 (Hem, 1989). When these minerals are exposed to water, the iron that is dissolved generally precipitates in the vicinity as sedimentary species. In reducing conditions, when sulfur is available, the precipitate is in form of ferrous polysulfides, or siderite ($FeCO_3$) when sulfur is less abundant. In oxidizing conditions, the precipitate (sediment) will contain ferric oxides or oxyhydroxides such as hematite (Fe_2O_3) or goethite ($FeOOH$). Ferric hydroxide, $Fe(OH)_3$, is freshly precipitated material with poorly developed

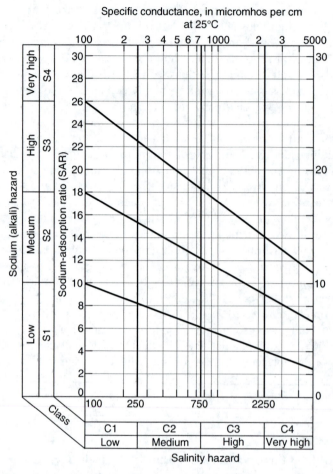

FIGURE 4.11 Diagram for use in interpreting the analysis of irrigation water. (After U.S. Salinity Laboratory Staff, Diagnosis and improvement of saline and alkali soils. *U.S. Department of Agriculture Handbook*, 60, 1954, 160 pp; From Hem, J.D., Study and interpretation of the chemical characteristics of natural water; 3rd edition. U.S. Geological Survey Water-Supply Paper 2254. Washington, D.C., 1989, 263pp.)

crystal structure, commonly refered to as gel. The availability of iron for aqueous solution is strongly affected by pH and redox conditions, where catalytic action of microorganisms plays the key role in oxidation to ferric (Fe^{3+}) iron under aerobic conditions, or the reduction to divalent, ferrous iron (Fe^{2+}) under anaerobic conditions. Figure 4.4 shows how small changes in pH and E_h affect the stability of various iron compounds. The most common form of iron in solution is the ferrous ion Fe^{2+}. The ferrous iron concentrations in reducing conditions are commonly between 1 and 10 mg/L. Groundwater with pH between 6 and 8 can be sufficiently reducing to retain as much as 50 mg/L of ferrous iron at equilibrium, when bicarbonate activity does not exceed 61 mg/L (Hem, 1989). This type of water is clear when first extracted from a well, but it may soon become cloudy and then brown from precipitating ferric hydroxide. Because of the unstable nature of ferrous iron in a sample, it should be preserved to prevent precipitation of original dissolved iron due to oxidation. An increase in dissolved ferrous iron may indicate the presence of organic substances, including pollution by human activities. Ferric iron can occur in acid solutions as Fe^{3+}, $FeOH^{2+}$, and $Fe(OH)_2^+$, and in polymeric hydroxide forms. Ferric iron can form various inorganic solution complexes with many anions beside OH^-, such as with chloride, fluoride, sulfate, and phosphate (Hem, 1989).

4.2.2.5 Bicarbonate and Carbonate

Bicarbonate (HCO_3^-) and carbonate (CO_3^{2+}) ions in groundwater are mostly derived from the atmospheric and soil CO_2, and dissolution of carbonate rocks such as calcium carbonate, as discussed earlier in the sections on solubility of gasses (Section 4.1.3), pH (Section 4.1.5), and alkalinity (Section 4.1.6). The term alkalinity is often used as a synonym for the measure of bicarbonate and carbonate content in groundwater. Note that this usage of the word alkalinity is contrary to the common chemical usage in which only water with a pH of more than 7.0 is considered to be alkaline (Davis and DeWiest, 1991). Groundwater generally contains more than 10 mg/L but less than 800 mg/L bicarbonate, usually between 50 and 400 mg/L (Matthess, 1982; Davis and DeWiest, 1991). Concentrations above 1000 mg/L can occur in water of low alkaline earth content, especially at high CO_2 concentrations caused by endogenetic or diagenetic processes such as in fault zones and regional metamorphic zones. When there is oversaturation with respect to partial pressure of atmospheric CO_2, precipitation of calcium carbonate (travertine, calc-sinter) will occur at the contact with atmosphere. The usual concentration of free dissolved (aggressive) CO_2 in groundwater, i.e., CO_2 not already bound to bicarbonate–carbonate, is 10–20 mg/L (Matthess, 1982). Under equilibrium conditions, the pH value of water indicates the fractions of different carbonate species (Figure 4.12). Bicarbonate ions are predominant between a pH of 6 and 8.5. Bicarbonate dissociation into carbonate starts at the pH of about 8.35, and carbonate dominates in highly alkaline waters. If the pH is less than 5, the solution contains only free CO_2.

4.2.2.6 Sulfate

Sulfate (SO_4^{2-}) ion is an oxidized form of sulfur, produced primarily when sulfide minerals undergo weathering in contact with aerated water. Hydrogen ions are also produced in this oxidation process in considerable quantity (Hem, 1989). Sulfur is widely distributed in reduced form in both magmatic and sedimentary rocks as metallic sulfides. Pyrite, in particular, constitutes a major source of both sulfate and ferrous iron in groundwater. Its

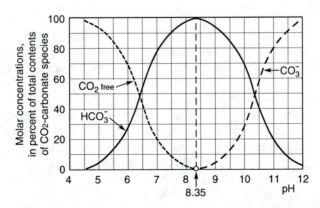

FIGURE 4.12 Relationship between free CO_2, bicarbonate, and carbonate species for various values of pH. (After Ovchinikov, M.A., *Obschaja gidrogeologia* (*General Hydrogeology*, in Russian). Nedra, Moskva, 1955; From Milojevic, N., *Hidrogeologija* (*Hydrogeology*, in Serbian). Univerzitet u Beogradu, Zavod za izdavanje udzbenika Socijalisticke republike Srbije, Beograd, 1967, 379pp.)

oxidation is also promoted by humans through combustion of fossil fuels and smelting of ores containing pyrite. Sulfate occurs in some magmatic-rock minerals of the feldspathoid group, but the most extensive and important occurrences are in evaporate sediments, most of which are soluble in water (Hem, 1989). Soluble calcium sulfate as gypsum, or as anhydrite, makes up a considerable part of many evaporate-rock sequences. Strontium sulfate is less soluble than calcium sulfate, and barium sulfate is almost insoluble, but these two salts are relatively rare. Soluble sodium sulfate is formed in some closed-basin lakes (Hem, 1989). It is believed that most sulfates are presently recycled from the atmosphere and from the solution of sulfate minerals in the sedimentary rocks, although in the history of the hydrosphere most sulfates probably originated from the oxidation of sulfides from magmatic rocks and gasses (Davis and DeWiest, 1991). All atmospheric precipitation contains sulfate, which is one of the major dissolved constituents of rain and snow. This sulfate is derived from the oxidation of sulfur dioxide and hydrogen sulfide gasses, and from dust particles containing sulfate minerals. Anthropogenic sulfur emissions, such as burning of fossil fuels and ore smelting, are a major factor in producing rain of low pH (acid rain) that has had many undesirable ecologic effects in northern Europe and parts of North America (Hem, 1989). In some areas, however, nitrogen oxides are equal or more important contributors to acid rain. One of the most prolific natural sources of hydrogen sulfide is from the bacterial reduction of organic material in tidal mud flats. Sulfate-reducing bacteria (sulfur bacteria) is active in groundwater where it derives energy from the oxidation of organic compounds and in the process obtains oxygen from the sulfate anions (Matthess, 1982). The resulting reduction of sulfate ions produces hydrogen sulfide as a by-product, most of which will remain in the subsurface as gas. If iron is present in the water under moderately reducing conditions, iron sulfide may be precipitated, thus removing both iron and sulfide from the water (Matthess, 1982). The rotten-egg odor of hydrogen sulfide can be detected by most people in waters that have only a few tenths of a milligram per liter H_2S. Groundwater associated with oil deposits, in areas of magmatic activity, and in deep reducing aquifers may attain high concentrations of hydrogen sulfide in excess of 100 mg/L. The sulfate content of normal groundwater in magmatic rocks and intergranular sedimentary rocks is usually less than 30 mg/L. In gypsum and anhydrite deposits SO_4^{2-} concentration may reach the saturation point of gypsum, corresponding to about 1360 mg/L. Very low or even zero sulfate concentrations are typical for groundwaters in which bacterial reduction has been taking place (Matthess, 1982).

4.2.2.7   Chloride

Chlorine is the most abundant element in the group of halogens, which includes fluorine, bromine, and iodine. More than three-fourths of the total amount of chlorine present in the Earth's outer crust, atmosphere, and hydrosphere is in the form of the ion chloride, Cl^-, dissolved in the ocean water. Average concentration of chloride in the ocean water is about 19,000 mg/L, by far the highest of all other constituents (the next highest is sodium, 10,500 mg/L, followed by sulfate, 2700 mg/L, magnesium, 1350 mg/L, and calcium, 410 mg/L; Hem, 1989). Based on the average contents of chloride in rocks, only a small part of its content in the ocean water is due to rock weathering. The bulk of it originates from degassing of the Earth's crust by volcanic emanations, which early in Earth's history gave rise to the chlorine in the primeval atmosphere and oceans (Matthess, 1982). Chlorine is volatile, dissolves readily in water, and is not stable in the atmosphere. Because of its strong and rapid oxidizing effect, chlorine has long been used as a disinfectant in purification of water supplies.

Chloride is present in the various rock types in concentrations lower than any of the other primary constituents of groundwater. Sedimentary rocks, particularly the evaporates, are considerably more important sources of chloride than magmatic rocks. In general, chloride in groundwater comes from four different sources: (1) chloride from ancient seawater entrapped in sediments, (2) solution of halite and related minerals in evaporates, (3) concentration by evaporation of chloride contributed by rain or snow, and (4) dry fallout from the atmosphere, particularly in arid regions (Davis and DeWiest, 1991). Chloride also enters the hydrologic cycle from liquid and solid waste materials, fertilizers, and highway salt. Values of Cl^- in normal potable groundwater are less than 30 mg/L, with higher values commonly indicating the admixture of mineralized waters or anthropogenic pollution (Matthess, 1982). Once dissolved in groundwater, chloride remains in solution and its behavior is often referred to as "conservative": chloride ions do not significantly enter into oxidation or reduction reactions, form no important solute complexes with other ions unless the chloride concentrations are extremely high, do not form salts of low solubility, are not significantly adsorbed on mineral surfaces, and play few vital biochemical roles (Hem, 1989). For all these reasons, chloride has been often used as a conservative tracer. In fact, Kauffman and Orlob (1956, from Hem, 1989) concluded that chloride ions move with the water through most soils tested with less retardation or loss than any of the other tracers tested, including tritium.

4.2.2.8 Nitrate (Nitrogen Forms)

Nitrogen occurs in groundwater as uncharged gas ammonia (NH_3), which is the most reduced inorganic form, nitrite and nitrate anions (NO_2^- and NO_3^- respectively), in cationic form as ammonium (NH_4^+), and at intermediate oxidation states as a part of organic solutes. Some other forms such as cyanide (CN^-) may occur in groundwater affected by waste disposal (Hem, 1989; Rees et al., 1995). Three gaseous forms of nitrogen may exist in groundwater: elemental nitrogen (oxidation state of zero), nitrous oxide (N_2O; slightly oxidized, +1), and nitric oxide (NO; +2). All three, when dissolved in groundwater, remain uncharged gasses (Rees et al., 1995). Ammonium cations are strongly adsorbed on mineral surfaces. Nitrate is readily transported by groundwater and stable over a considerable range of conditions. The nitrite and organic species are unstable in aerated water and easily oxidized. They are generally considered indicators of pollution by sewage or organic waste. The presence of nitrate or ammonium might be indicative of such pollution as well, but generally the pollution would have occurred at a site or time substantially removed from the sampling point. Ammonium and cyanide ions form soluble complexes with some metal ions, and certain types of industrial waste effluents may contain such species (Hem, 1989). Extensive application of nitrogen fertilizers has caused an increase in nitrate concentrations over wide agricultural areas in many countries. Nitrogen oxides, present in the atmosphere to a considerable extent due to combustion of fossil fuels, undergo various chemical alterations that produce H^+ and finally leave the nitrogen as nitrate. These processes can lower the pH of rain in the same way sulfur oxides do. Because of the many ways human activities influence various forms of nitrogen in the environment, and the public health concerns associated with the elevated concentrations of nitrite and nitrate in potable groundwater, a very large amount of scientific investigations of the sources of nitrogen, nitrogen cycle, and related groundwater impacts has been done (e.g., Feth, 1966; National Research Council, 1978; Zwirnmann, 1982; Keeney, 1990; Spalding and Exner, 1993; Puckett, 1994; Rees et al., 1995; Mueller et al., 1995).

Sources of various forms of nitrogen in groundwater are primarily derived from the biological activity of plants and microorganisms in the environment, from animal waste, and from many anthropogenic activities including agriculture, sewage disposal, and utilization of fossil fuels. Nonindustrially impacted rain may have a total nitrogen concentration of about 6 mg/L, and rainfall of 10 in./year would yield a nitrogen load to the soil column of about 13 pounds per acre per year in such a case. Significant evaporation of such rainwater could result in high concentrations of nitrogen in the infiltration water (Heaton, 1984; Rees et al., 1995). Geologic sources of nitrogen are much less significant. Lithologic units that typically have high nitrate concentrations include shales and Pleistocene-age loess that were deposited during periods favoring plant growth (Rees et al., 1995).

Nitrogen can undergo numerous reactions that can lead to storage in the subsurface, or conversion to gaseous forms that can remain in the soil for periods of minutes to many years. The main reactions include: (1) immobilization and mineralization, (2) nitrification, (3) denitrification, and (4) plant uptake and recycling (Keeney, 1990). Immobilization is the biological assimilation of inorganic forms of nitrogen by plants and microorganisms to form organic compounds such as amino acids, sugars, proteins, and nucleic acids. Mineralization is the inverse of immobilization. It is the formation of ammonia and ammonium ions during microbial digestion of organic nitrogen. Nitrification is the microbial oxidation of ammonia and ammonium ion first to nitrite, then ultimately to nitrate. Nitrification is a key reaction leading to the movement of nitrogen from the land surface to the water table because it converts the relatively immobile ammonium form (reduced nitrogen) and organic nitrogen forms to a much more mobile nitrate form. Chemosynthetic autotrophic soil bacteria of the family *Nitrobacteriaceae* is believed to be principally responsible for the nitrification process. Ammonium oxidizers, including the genera *Nitrosomonas*, *Nitrosospira*, *Nitrosolobus*, and *Nitrosvibrio*, oxidize ammonium to nitrite. The nitrite-oxidizing bacteria, which oxidize nitrite to nitrate, include the genus *Nitrobacter*. Nitrification can also be carried out by hetero-trophic bacteria and fungi (Rees et al., 1995). The nitrogen used by plants is largely in the oxidized form. Denitrification is the biological process that utilizes nitrate to oxidize (respire) organic matter into energy usable by microorganisms. This process converts the nitrate to more reduced forms, ultimately yielding nitrogen gas that can diffuse into the atmosphere. Uptake of nitrogen by plants also removes nitrogen from the soil column and converts it to chemicals needed to sustain the plants. Because the plants eventually die, the nitrogen incorporated into the plant tissues ultimately is released back to the environment, thus completing the cycle (Rees et al., 1995).

The concentrations of various nitrogen species in groundwater are determined and reported in different ways in published analyses. Most studies and laboratories engaged in analyses of groundwater contamination report ammonia, amino and organic nitrogen, and nitrite either separately or as a combined figure in terms of equivalent concentration of elemental nitrogen. Studies interested in general (inorganic) groundwater chemistry usually require reporting of nitrate ion (NO_3^-) only, since nitrite (NO_2^-) is rarely present in concentrations large enough to influence ionic balance to a noticeable degree (Hem, 1989). Nitrate concentrations in most unpolluted natural groundwaters in the United States are usually less than 2 mg/L (Mueller et al., 1995). Examples of much higher values include groundwaters often associated with arid climates and leaching of nitrate-bearing deposits in caves, caliche, and playas (Matthess, 1982). Shallow aquifers underlying areas impacted by anthropogenic activities, especially the application of fertilizers, may have nitrate concentrations of several hundred milligrams per liter or more. Nitrite concentrations in natural unpolluted groundwaters are usually in trace amounts, i.e., much less than 1 mg/L. Higher concentrations often indicate contamination by anthropogenic sources or animal waste.

4.2.2.9 Silica

The term "silica," meaning the oxide of silicon (SiO_2), is widely used in referring to silicon in natural water, but it should be understood that the actual form is an uncharged hydrated monomolecular silicic acid, H_4SiO_4 or $Si(OH)_4$, which is for the most part a true solution (Matthess, 1982; Hem, 1989). The bulk of silica occurring in groundwater comes from the weathering of silicate minerals. Amorphous silica also contributes to total dissolved silica, while crystalline silica, particularly quartz, is almost insoluble in water. In natural nonthermal groundwaters, silica is usually found in concentrations between 1 and 30 mg/L, and such waters are mostly undersaturated with respect to amorphous silica (Matthess, 1982). The solubility of silica is not affected by pH in the 0–9 range. However, high temperatures or pH > 9 may result in very high concentrations and supersaturation of groundwater with respect to silica. For example, Feth et al. (1961) report a pH of 11.6 and more than 3400 mg/L of dissolved SiO_2 in water of Aqua de Ney, a cold spring near the town of Mount Shasta in California (Hem, 1989).

4.2.3 SECONDARY CONSTITUENTS

Most secondary groundwater constituents occur in potable, nonpolluted natural groundwaters at very low concentrations (in trace amounts), and are not routinely analyzed as part of general hydrogeologic studies. However, since their natural concentrations have wide ranges depending on the local geochemical conditions in the aquifer, most of them are analyzed as part of water supply projects. This is because the aquifer-specific concentration of a naturally occurring constituent may be above a certain threshold considered safe for human consumption. This threshold is established by regulatory agencies based on extensive studies of the constituent's effect on human health, and is called maximum contaminant level (MCL) in the United States (see Chapter 5.3). A description of dozens of secondary groundwater constituents is beyond the scope of this book and the reader should consult hydrogeochemistry texts for more detail (e.g., Matthess, 1982; Hem, 1989). However, some constituents that are relatively widely distributed in both natural groundwaters and those associated with anthropogenic contamination, have in recent years gained unfavorable reputation due to their high toxicity or detrimental impact on the environment. This is the primary reason for including, in this chapter, more detail on phosphorus, fluoride, arsenic, chromium, and organic matter in general. Radioactive elements (radionuclides) are covered separately in the following section.

4.2.3.1 Phosphate (Phosphorus)

Although phosphorus is a common element in igneous and sedimentary rocks, its concentrations in natural potable groundwaters are usually less than a few tenths or hundredths of a milligram per liter, as a result of low solubility of most inorganic phosphorous compounds (Hem, 1989). Phosphorus is used by biota as an essential nutrient, which is another reason for its low content and mobility in natural groundwaters, and, at the same time, explains its extensive use as a fertilizer. However, unlike nitrogen in the form of nitrates, dissolved phosphorous compounds are easily precipitated or sorbed by clay minerals and metal oxides, so that phosphorous added to the subsurface through disposal of waste or leaching of fertilized fields may not remain available for longer periods. Reduced forms of phosphorous, present in some synthetic organic chemicals such as those used in insecticides, may persist in reducing environments long enough to be of some significance (Hem, 1989). Phosphate, or the fully oxidized form of phosphorous, is the only one of importance in natural waters. Dissociation of phosphoric acid, H_3PO_4, creates four possible solute species, the proportion of which depends on pH: $H_3PO_4(aq)$, $H_2PO_4^-$, HPO_4^{2-}, and PO_4^{3-}. The trivalent and neutral ions occur only outside the common pH range of natural fresh water (Hem, 1989).

The various forms of phosphate fertilizers, detergents, liquid and solid waste, and soil erosion from agricultural lands provide a major source of phosphate in surface waters, causing negative effects such as eutrophication. Phosphate mining, concentrating, and processing are sources of phosphate in river water in some areas, notably in Florida and Idaho.

4.2.3.2   Fluoride (Fluorine)

The element fluorine is used by higher life forms in the structure of bones and teeth. The importance of fluoride, its anion, in forming human teeth and the role of fluoride intake from drinking water in controlling the characteristics of tooth structure was recognized during the 1930s (Hem, 1989). Since that time the fluoride content of natural water has been studied extensively. Although intake of fluoride is necessary for promoting strong healthy teeth, at high concentrations it may cause bone disease and mottled teeth in children (MCL for fluoride in the United States is 4 mg/L). Although fluoride concentrations in most natural waters are small, less than 1 mg/L, groundwater exceeding this value has been found in many places in the United States in a wide variety of geologic terrains (Hem, 1989). Fluorite and apatite are common fluoride minerals in magmatic and sedimentary rocks, and amphiboles and micas may contain fluoride which has replaced part of the hydroxide. Rocks rich in alkali metals have higher fluoride content than most other magmatic rocks. Fresh volcanic ash may be rather rich in fluoride, and ash that is interbedded with other sediments could contribute significantly to fluoride concentrations. Fluoride is commonly associated with volcanic or fumarolic gases, and in some areas these may be important sources of fluoride in groundwater (Hem, 1989). Fluorine is the most electronegative of all the elements and its F^- ion forms strong solute complexes with many cations, particularly with aluminum, beryllium, and ferric iron. Anthropogenic sources of fluoride include fertilizers and discharge from ore processing and smelting operations, such as aluminum works.

4.2.3.3   Arsenic

Elemental arsenic is a steel gray metal-like substance rarely found naturally. As a compound with other elements such as oxygen, chlorine, and sulfur, arsenic is widely distributed throughout the Earth's crust, especially in minerals and ores that contain copper or lead. Natural arsenic in groundwater is largely the result of dissolved minerals from weathered rocks and soils. Principal ores of arsenic are sulfides (As_2S_3, As_4S_4, and $FeAsS$), which are almost invariably found with other metal sulfides. The hydrogen form of arsenic is arsine, a poisonous gas. Arsenic also forms oxide compounds. Arsenic trioxide (As_2O_3) is a transparent crystal or white powder that is slightly soluble in water and has a specific gravity of 3.74. Arsenic pentoxide (As_2O_5) is a white amorphous solid that is very soluble in water, forming arsenic acid. It has a specific gravity of 4.32 (USEPA, 2005a).

Dissolved arsenic in groundwater exists primarily as oxy-anions with formal oxidation states of 3+ and 5+. Either arsenate [As(V)] or arsenite [As(III)] can be the dominant inorganic form in groundwater. Arsenate ($H_nAs O_4^{n-3}$) generally is the dominant form in oxic (aerobic, oxygenated) waters with dissolved oxygen > 1 mg/L. Arsenite ($H_nAs O_3^{n-3}$) dominates in reducing conditions, such as sulfidic (dissolved oxygen < 1 mg/L with sulfide present) and methanic (methane present) waters. Aqueous and solid-water reactions, some of which are bacterially mediated, can oxidize or reduce aqueous arsenic. Both anions are capable of adsorbing to various subsurface materials, such as ferric oxides and clay particles. Ferric oxides are particularly important to the fate of arsenate and transport as ferric oxides are abundant in the subsurface and arsenate strongly adsorbs to these surfaces in acidic to neutral waters. An increase in the pH to an alkaline condition may cause both arsenite and arsenate to desorb, and they are expected to be mobile in an alkaline environment (Dowdle et al., 1996; Harrington et al., 1998, Welch et al., 2000; USEPA, 2005a).

All arsenic compounds consumed in the United States are imported. The arsenic has been used primarily for the production of pesticides, insecticides, and chromated copper arsenate (CCA), a preservative that renders wood resistant to rotting and decay. Increased environmental regulation, along with the decision of the wood-treating industry to eliminate arsenical wood preservatives from residential application by the end of 2003, caused arsenic consumption in the United States to decline drastically in 2004. Other industrial products containing arsenic include lead-acid batteries, light-emitting diodes, paints, dyes, metals, pharmaceuticals, pesticides, herbicides, soaps, and semiconductors. Anthropogenic sources of arsenic in the environment include mining and smelting operations, agricultural applications, and disposal of wastes that contain arsenic (USEPA, 2005a). Arsenic is a contaminant of concern at many remediation sites. Because arsenic readily changes valence states and reacts to form species with varying toxicities and mobilities, effective treatment of arsenic can be challenging. In addition, in January 2001, EPA published a revised MCL for arsenic in drinking water. This revised MCL for arsenic in drinking water has resulted in more stringent treatment goals for many public water supply systems throughout the United States, significantly affecting technology selection, and design and operation of treatment facilities.

Recent study of arsenic concentrations in the major U.S. aquifers by the USGS (accessible at http://water.usgs.gov/nawqa/trace/pubs/) shows wide regional variations of naturally occurring arsenic due to a combination of climate and geology. Although slightly less than half of 30,000 arsenic analyses of groundwater in the United States were equal or less than 1 µg/L, about 10% exceeded 10 µg/L. At a broad regional scale, arsenic concentrations exceeding 10 µg/L appear to be more frequently observed in the western United States than in the eastern half (USGS, 2004). Interestingly, more detailed recent investigations of groundwater in New England, Michigan, Minnesota, South Dakota, Oklahoma, and Wisconsin suggest that arsenic concentrations exceeding 10 µg/L are more widespread and common than previously recognized. Arsenic release from iron oxide appears to be the most common cause of widespread arsenic concentrations exceeding 10 µg/L in groundwater. This can occur in response to different geochemical conditions, including release of arsenic to groundwater through reaction of iron oxide with either natural or anthropogenic (i.e., petroleum products) organic carbon. Iron oxide also can release arsenic to alkaline groundwater, such as that found in some felsic volcanic rocks and alkaline aquifers of the western United States. Sulfide minerals in rocks may act both as a source and sink for arsenic, depending on local geochemistry. In oxic (aerobic, oxygenated) water, dissolution of sulfide minerals, most notably pyrite and arsenopyrite, contributes arsenic to groundwater and surface water in many parts of the United States. Other common sulfide minerals, such as galena, sphalerite, marcasite, and chalcopyrite, can contain 1% or more arsenic as an impurity.

As already mentioned, in addition to the natural geologic sources, there are many anthropogenic sources of arsenic. The most important are derived from agricultural practices, such as application of pesticides and herbicides. Inorganic arsenic was widely applied before it was banned for pesticide use in the 1980s and 1990s. Lead arsenate ($PbHAsO_4$) was the primary insecticide used in fruit orchards prior to the introduction of DDT in 1947. Inorganic arsenicals also have been applied to citrus, grapes, cotton, tobacco, and potato fields. For example, historic annual arsenic loading rates up to approximately 490 kg/ha (approximately 440 lb./acre) on apple orchards in eastern Washington led to arsenic concentrations in soil in excess of 100 mg/kg (Benson, 1976; Davenport and Peryea, 1991; from Welch et al., 2000). Agricultural soils in other parts of the United States also have high arsenic concentrations exceeding 100 mg/kg due to long-term application (20 to 40 years or more) of calcium and lead arsenate (Woolsen et al., 1971, 1973; Wauchope, 1983). Early studies suggested that arsenic in eastern Washington orchards was largely confined to the topsoil, although evidence for movement into the subsoil also has been cited (Peryea, 1991). This apparent movement of arsenic suggests a potential for contamination of shallow groundwater. The application of phosphate fertilizers creates the potential for releasing arsenic into groundwater. Laboratory studies suggest that phosphate applied to soils contaminated with lead arsenate can release arsenic to soil water. Increased use of phosphate at relatively high application rates has been adopted to decrease the toxicity of arsenic to trees in replanted orchards. Laboratory results suggest that this practice may increase arsenic concentrations in subsoil and shallow groundwater. The application of phosphate onto uncontaminated soil also may increase arsenic concentrations in groundwater by releasing adsorbed natural arsenic (Woolson et al., 1973; Davenport and Peryea, 1991; Peryea and Kammereck, 1997; Welch et al., 2000).

The adsorption (or coprecipitation) of arsenic on iron oxides has been cited as a concentration-limiting process in groundwater, and the related reactions have been investigated using a variety of approaches (Welch and Lico, 1998; Jain et al., 1999). The adsorption of arsenic onto $Fe(OH)_3$ is affected by a variety of factors, including pH, the chemistry and the amount of $Fe(OH)_3$ present, and concentrations of competing ions. A biologically mediated reaction that can release arsenic from iron oxide, commonly referred to as dissimilatory iron reduction, involves organic carbon and iron oxide (Lovley, 1991). The dissolution of iron oxide is the primary process responsible for high arsenic concentrations in some groundwater (Welch and Lico, 1998). Sources of organic carbon include sedimentary-organic matter and anthropogenic organic compounds. For example, the USGS studies on the geochemistry and mineralogy of the leachate-contaminated aquifer at the Saco Municipal Landfill in Saco, Maine have shown that the source of arsenic is not the landfill but the sediments the leachate is moving through (Stollenwerk and Colman, 2003). Much of the arsenic is present in hydrous ferric oxides that coat the aquifer sediments. The arsenic in these hydrous ferric oxides may have gradually accumulated over time because of adsorption of low concentrations of arsenic that occur naturally in the groundwater of this area. Results from laboratory experiments show that DOC in the leachate plume is promoting reductive dissolution of these hydrous ferric oxides, releasing arsenic to groundwater. Reductive dissolution occurs because the degradation of the DOC in the plume removes oxygen from the water and creates reducing conditions that favor the dissolution of hydrous ferric oxides and release of arsenic from the sediments (USGS, 2004).

Probably the most widely known groundwater contamination with arsenic is that of millions of shallow water supply wells in Bangladesh, installed by the government in an effort to prevent waterborne diseases caused by surface water supplies. Unfortunately, the arsenic concentrations in shallow water supply wells, sometimes measured in hundreds of micrograms per liter (ppb), have caused widespread arsenicosis, and are threatening to increase cancer rates. Shallow aquifer sediments typically contain only modest levels of arsenic, and the cause and timing of arsenic mobilization from these sediments is the subject of an ongoing debate. Based on an extensive multidisciplinary study, Harvey et al. (2002) concluded that the observed arsenic mobility is related to recent inflow of carbon through either organic carbon-driven reduction or displacement by carbonate. The distinctly older radiocarbon ages of DOC relative to dissolved inorganic carbon (DIC) and methane imply that mobilization is not driven by detrital organic carbon. Water budgets indicate that the advent of massive irrigation pumping has drawn relatively young water into the aquifer over the last several decades. This pumping may affect arsenic concentrations, but not by the oxidation of sulfides, or by slow reduction of iron oxyhydroxides or sorbed arsenate by detrital organic carbon, as has been previously proposed.

A recently completed study of 116 unlined construction and demolition debris (C&D) landfills in Florida demonstrates the complexity of chemical reactions in the subsurface, affecting arsenic mobility in groundwater. The occurrence of arsenic in Florida groundwater may result from both natural sources and use and disposal of arsenic-bearing materials such as arsenical pesticides in cattle dipping solutions, arsenical pesticides used for agricultural and golf course maintenance purposes, phosphate fertilizers, other arsenic-containing fertilizers, biosolids, and chicken manure. Irrespective of the possible sources, quantitative evaluation of the arsenic concentrations in groundwater from the Florida Department of Environmental Protection monitoring wells database for unlined C&D landfills resulted in the following findings: based on 4534 water quality samples collected in a 12-year period (February 1992 through July 2004), the average arsenic concentration from the background wells upgradient of the landfills exceeds the average concentration from both the detection and the compliance monitoring wells downgradient of the landfills. Background wells are located hydraulically upgradient from the waste disposal area, and are used to evaluate the quality of groundwater before it passes beneath the landfill; detection wells are located hydraulically downgradient from the disposal area and are used to evaluate the quality of groundwater after it passes beneath the landfill; compliance wells are located farther downgradient than the detection wells, generally at the property boundary, to determine whether any constituents that may have been introduced to, or elevated in the groundwater after passing under the disposal area may be migrating offsite. The average groundwater background concentration is 9.53 µg/L, compared to 5.82 and 7.31 µg/L for the detection and compliance monitoring wells, respectively (Kavanaugh et al., 2005). For comparison, MCL for total arsenic dissolved in groundwater is 10 µg/L or ppb (See Chapter 5.3). These results indicate that the C&D landfills themselves are not an overall additional source of dissolved arsenic mobile in groundwater. The results also indicate that the leachate from the unlined C&D landfills reduces the background concentration of arsenic and acts as a general geochemical barrier for the migration of arsenic dissolved in groundwater. Leachate from a C&D landfill is generally expected to have lower dissolved organic content than a municipal landfill leachate, and may contain a number of constituents, including iron, calcium, and sulfide–sulfate complexes that can precipitate arsenic. For example, data from monitoring wells at numerous unlined C&D landfills in Maine show nonarsenic groundwater impacts from disposal of concrete, gypsum wallboard, wood, and ash from brush burn piles (MDEP, 2005).

4.2.3.4 Chromium

Elemental chromium is a transition group metal found naturally in rocks, soil, and living organisms. It occurs in combination with other elements such as chromium salts, some of which are soluble in water. The pure metallic form does not occur naturally. Chromium can exist in several chemical forms with oxidation numbers ranging from −2 to +6. However, in the environment, it commonly exists in only two stable oxidation states, Cr(VI) and Cr(III), which have greatly contrasting toxicity and transport characteristics. Chromium speciation in the environment, particularly in groundwater, is affected primarily by E_h (oxidizing or reducing conditions) and pH (acidic or alkaline conditions). In general, hexavalent chromium, Cr(VI), predominates under oxidizing conditions, and trivalent chromium, Cr(III), predominates under more reducing conditions. It should be noted that the term hexavalent chromium is somewhat of a misnomer. This is because Cr(VI) is not present in the environment as a free cation (whereas several species of Cr(III) exist in the environment as cations). In fact, as all Cr(VI) species are oxides, they act like a divalent anion rather than a hexavalent cation (Kimbrough et al., 1999, from Stanin, 2005).

Chromium contamination of soil and groundwater is a significant problem worldwide. The extent of this problem is due primarily to its use in numerous industrial processes (i.e., metal plating and alloying, leather tanning, wood treatment, chemical manufacturing), but also its natural presence in rocks enriched in chromium. Chromite ore ($FeCr_2O_4$) is the most important commercial ore associated with ultramafic and serpentine rocks. Chromium is also associated with other ore bodies (e.g., uranium and phosphorites) and may be found in tailings and other beneficiation wastes from these mining operations where acid mine drainage can make the chromium available to the environment (USEPA, 2005b). Compared to the results of contamination of soil and groundwater by industrial and mining practices, the naturally occurring concentrations of chromium in soil and groundwater are low, commonly less than 10 µg/L (Hem, 1989). However, rather high concentrations of naturally occurring hexavalent chromium, in the range between 100 and 200 µg/L, have been observed in uncontaminated groundwater in Paradise Valley, north of Phoenix, Arizona (Robertson, 1975). Thus, both anthropogenic and natural sources of chromium can lead to locally elevated levels in soils and waters.

Chromium releases from anthropogenic sources are commonly from dumping of chromium-bearing liquid or solid wastes such as chromate byproducts (muds), ferrochromium slag, or chromium plating wastes. Such wastes can contain any combination of Cr(III) or Cr(VI) species with varying solubilities. The nature and behavior of different forms of chromium found in wastewaters can be quite variable. The presence, form, and concentration of chromium in discharged effluents depend mainly on the chromium compounds utilized in the industrial process, on the pH, and on the presence of other organic and inorganic processing wastes. In general, Cr(VI) dominates in wastewater from the metallurgical industry, metal finishing industry, refractory industry, and production or application of pigments (chromate color pigments and corrosion inhibition pigments). Cr(III) is found mainly in wastewaters of the tannery, textile (printing, dying), and decorative plating industries. However, there are exceptions to these generalities due to several factors. For example, in tannery wastewater where Cr(III) is the most expected form, the redox reactions occurring in sludge can increase the concentration of Cr(VI). Various chemical and biochemical transformations of chromium species are also common in the subsurface, including oxidation, reduction, sorption, precipitation, and dissolution.

The two different forms of chromium are quite different in their properties: charge, physiochemical characteristics, mobility in the environment, chemical and biochemical behavior, bioavailability, and toxicity. Most notably, Cr(III) is considered to be a trace element essential for the proper functioning of living organisms, whereas Cr(VI) exerts toxic effects on biological systems. In addition, the more toxic Cr(VI) compounds are generally more soluble, mobile, and bioavailable in the environment compared with Cr(III) compounds. Therefore, it is quite important to distinguish between the two forms of chromium rather than discussing this element as "total chromium."

Solubility can significantly limit the concentration of Cr(III) in groundwater at a pH above 4. The low solubility of the Cr(III) solid phases, Cr_2O_3 and $Cr(OH)_3$ (Hem, 1977), is likely the major reason why Cr(III) generally makes up a small percentage of the total chromium concentration in natural or contaminated groundwaters. Cr(III) tends to be essentially immobile in most groundwaters because of its low solubility (Calder, 1988). On the other hand, there are no significant solubility constraints on the concentrations of Cr(VI) in groundwater. The chromate (CrO_4^{2-}) and dichromate ions ($Cr_2O_7^{2-}$) are water-soluble at all pH. However, chromate can exist as a salt of a variety of divalent cations, such as Ba^{2+}, Sr^{2+}, Pb^{2+}, Zn^{2+}, and Cu^{2+}, and these salts have a wide range of solubilities. The rates of precipitation–dissolution reactions between chromate, dichromate anions, and these cations vary greatly and are pH-dependent. An understanding of the dissolution reactions is particularly important for assessing the environmental effects of chromium because Cr(VI) often enters the environment by dissolution of chromate salts (Rai et al., 1987, from Stanin, 2005).

Cr(VI) can be transported to considerable distances in groundwater due in part to its high solubility. For example, Hem (1989) cites an incident of shallow groundwater contamination by hexavalent chromium released from an industrial waste disposal pit in Long Island, New York, where Cr(VI) persisted in concentrations as high as 14 mg/L more than 3000 ft away from the original source, for more than 20 years after the release (Perlmutter et al., 1963). If, however, the transported Cr(VI) enters an area with relatively low E_h, it can be reduced readily to Cr(III) in the presence of organic matter, especially where pH is low. Cr(VI) can also be reduced by Fe(II) and dissolved sulfides (Stanin, 2005). Cr(III) generally is transported only short distances by groundwater because of its low solubility. If, however, the redox conditions along the transport pathway change from reducing to oxidizing, Cr(III) can be transformed to the more soluble Cr(VI). Under natural conditions, Cr(III) has been found to be oxidized to Cr(VI) by manganese (Bartlett and James, 1979).

4.2.3.5   Organic Matter

In addition to inorganic (mineral) substances, groundwater always contains natural organic substances, and almost always some living microorganisms (mainly bacteria), even at depths of up to 3.5 km in some locations (Krumholz, 2000). Organic matter in surface and groundwater is a diverse mixture of organic compounds ranging from macromolecules to low molecular weight compounds such as simple organic acids and short-chained hydrocarbons. In groundwater, there are three main natural sources of organic matter: organic matter deposits such as buried peat; kerogen and coal; soil and sediment organic matter; and organic matter present in waters infiltrating into the subsurface from rivers, lakes, and marine systems (Aiken, 2002). Various components of naturally occurring hydrocarbons (oil and gas) and their breakdown products formed by microbial activity are a significant part of groundwater chemical composition in many areas throughout the world. A very large number of artificial organic chemicals have become part of the groundwater reality in recent years because, if analyzed for using the latest available analytical methods, they are often detected. More on synthetic organic groundwater contaminants is given in Chapter 5.

Organic matter in groundwater plays important roles in controlling geochemical processes by acting as proton and electron donors and acceptors and pH buffers, by affecting the transport and degradation of pollutants, and by participating in mineral dissolution and precipitation reactions. Dissolved and particulate organic matter may also influence the availability of nutrients and serve as a carbon substrate for microbially mediated reactions. Numerous studies have recognized the importance of natural organic matter in the mobilization of hydrophobic organic species, metals (e.g., Pb, Cd, Cu, Zn, Hg, and Cr), and radionuclides (e.g., Pu, Am, U, and Co). Many contaminants that are commonly regarded as virtually immobile in aqueous systems can interact with DOC or colloidal organic matter, resulting in migration of hydrophobic chemicals far beyond distances predicted by the structure and activity relationships (Aiken, 2002).

Historically, organic matter in natural waters has been arbitrarily divided into DOC and particulate organic carbon (POC), based on filtration through a 0.45 m filter. No natural cutoff exists between these two fractions and the distinction is arbitrary, based on the filtration of the sample. The definition of terms, therefore, is operational. Overlapping the dissolved and particulate fractions is the colloidal fraction, which consists of suspended solids that are operationally considered solutes. Colloidal organic matter in natural waters is composed of living and senescent organisms, cellular exudates, and partially-to-extensively degraded detrital material, all of which may be associated with mineral phases. Generally, DOC is in greater abundance than POC, accounting for approximately 90% of the TOC of most waters (Aiken, 2002).

Microbial degradation of organic matter results in the formation of many of the compounds that comprise DOC, especially nonvolatile organic acids that dominate the DOC in most aquatic environments. Many of these organic acids are considered refractory because the rates of subsequent biodegradation are slower than for other fractions or classes of organic matter. Organic matter derived from different source materials has distinctive chemical characteristics associated with those source materials.

A number of significant, although poorly understood, mechanisms can be responsible for the transport or retention of organic molecules in the subsurface. Once in the system, organic compounds, whether of anthropogenic or natural origin, can be truly dissolved, associated with immobile particles or associated with mobile particles. Mobile particles include DOC, DOC–iron complexes, and colloids. For an organic compound, each state is related to the other states through equilibrium partitioning and air–water exchange. The magnitude of the partitioning coefficients and the abundance of sorbents determine the mechanisms and enhancement of transport for a particular organic compound. Regardless of environment, chemical reactivity and speciation will be controlled by thermodynamics and reaction kinetics (Aiken, 2002).

Positively charged organic solutes are readily removed from the dissolved phase by cation exchange, which can be a significant sorption mechanism. Organic solutes that may exist as cations in natural waters include amino acids and polypeptides. Hydrophilic neutral (e.g carbohydrates, alcohols) and low molecular weight anionic organic compounds (e.g., organic acids) are retained the least by aquifer solids. Hydrophobic organic compounds interact strongly with the organic matter associated with the solid phase. These interactions are controlled, in part, by the nature of the organic coatings on solid particles, especially with respect to its polarity and aromatic carbon content. Interactions of hydrophobic organic compounds with stationary particles can result in strong binding and slow release rates of these compounds (Aiken, 2002).

The common way to analyze the presence of organic substances in groundwater is to determine TOC. This is done by converting all carbon species present to carbon dioxide and correcting for subsequent determination of CO_2 for any dissolved CO_2 species that were initially present. Organic carbon concentrations in groundwater normally are smaller than in surface water, but information on actual ranges in various natural groundwaters is scarce (Hem, 1989). TOC is typically analyzed in groundwater affected by pollution. Chemical oxygen demand (COD), a criterion more often used in surface water and wastewater analyses, can also be used to quantify the amount of organic matter present in groundwater. COD is measurement of the oxygen necessary to remove different groups of oxidizable constituents, and does not necessarily account for all dissolved organic matter. One common indicator of possible presence of high TOC in a groundwater sample is a notable color. This physical property has no direct chemical significance since many different substances, including inorganic solutes, at widely varying concentrations, may cause groundwater coloration.

Recently, as pointed out by Krumholz (2000), there has been an increased interest in the microbiology and microbial ecology of subsurface environments. One of the reasons for this interest in subsurface life lies in the fact that there is also an increased interest in possible life on Mars, backed up by a very significant funding from the U.S. Government for several NASA missions to Mars and the related scientific research. Life probably does not now exist on the surface of Mars, due to the cold temperatures and high levels of solar radiation. However, one could imagine the possibility of life below the surface, where water may be present, temperatures may be more favorable, and solar radiation nonexistent. Krumholz (2000) provides an overview of microbial communities in the Earth's deep subsurface, including examples of their existence in extreme conditions which, up until recently, were thought it to be unsupportive of any life forms.

4.2.4 RADIONUCLIDES

The arrangements of energetic particles within nuclei of certain elements are not stable and spontaneously break down to form more stable energy and particle configurations. Energy released during this process is called radioactive energy and such elements are called radioactive elements or radionuclides. The most unstable configurations disintegrate very rapidly and some of them do not exist in the Earth's crust anymore (e.g., chemical elements 85 and 87, astatine and francium). Other radioactive elements, such as rubidium-87, have a slow rate of decay and are still present in significant quantity (Hem, 1989). The decay of a radionuclide is a first-order kinetic process, usually expressed in terms of a rate constant (λ) given as:

$$\lambda = \frac{\ln 2}{t_{1/2}}$$ (4.21)

where $t_{1/2}$ is half-life of the element, i.e., the length of time required for half the quantity present at time 0 to disintegrate.

Radioactive energy is released in various ways, with the following three types being of interest in water chemistry: (1) alpha radiation, consisting of positively charged helium nuclei, (2) beta radiation, consisting of electrons or positrons, and (3) gamma radiation, consisting of electromagnetic wave-type energy similar to x-rays (Hem, 1989). Potential effects from radionuclides depend on the number of radioactive particles or rays emitted (alpha, beta, or gamma) and not the mass of the radionuclides (USEPA, 1981). As such, it is essential to have a unit that describes the number of radioactive emissions per time period. The activity unit is used to describe the nuclear transformations or disintegrations of a radioactive substance, which occur over a specific time interval. The activity is related to the half-life; longer half-lives mean lower activity. The radioactivity of water is expressed in one of the following two units: Curie per liter and Becquerel per liter. One Curie (Ci) is defined as 3.7×10^{10} atomic disintegrations per second, which is the approximate specific activity of 1 g of radium in equilibrium with its disintegration products. By comparison, 1 g of uranium-238 has an activity of 0.36 millionth of a Curie. The unit is named after Pierre Curie and Marie Curie, the discoverers of radium. Curie is a large unit for expressing natural radioactivity levels and picoCurie (pCi) is usually used instead (1 pCi = 2.7×10^{-12} Ci). Becquerel (Bq) is the unit for radioactivity in the International System (SI) of units, defined as the radiation caused by one disintegration per second; this is equivalent to approximately 27.0270 pCis. The unit is named after a French physicist, Antoine-Henri Becquerel, the discoverer of radioactivity. Where possible, radioactivity is reported in terms of concentration of specific nuclides, as is commonly the case with uranium, which is conveniently analyzed by chemical means. For some elements, radiochemical analytical techniques permit the detection of concentrations much lower than what can be analyzed by any current chemical method. This fact is of special significance when performing tracing with radioactive isotopes, which can be introduced into the groundwater in very small quantities.

The effect of radioactivity depends not only on the activity (disintegrations/time) but also on the type of radiation (alpha, beta, or gamma) and its energy. These two properties, activity and type of radiation, collectively determine the absorbed dose to the tissue when decay occurs internally and the internal organs are the target (USEPA, 1991). A dose unit, absorbed dose, reflects the amount of radiation, or how much energy, was imparted to the tissue. The total amount of energy imparted is related to the number of particles emitted by the radioisotope per second, and the respective energies of the particles (USEPA, 1991). A common unit used to measure absorbed dose is a rad. One rad is equivalent to the amount of ionizing radiation that deposits 100 ergs (metric unit of energy) in 1 g of matter or tissue. For a perspective on the size of an erg, 10 million erg/s is equivalent to 1 W. In general, rad units are quite large (USEPA, 1991). Because of the particle mass and charge, 1 rad deposited in tissue by alpha particles creates more concentrated biological damage than 1 rad of gamma rays. To compensate for the difference in damage between different types of radiation particles and their subsequent effect, a new unit was created—the rem. Rem is the unit of measurement for the dose equivalent from ionizing radiation to the total body or any internal organ or organ system. It is equal to the absorbed dose in rads multiplied by a quality factor (to account for different radiation types). This relationship is written as follows:

$$\text{Quantity in rems} = [\text{Quantity in rads}] \times [Q] \tag{4.22}$$

$Q = 1$, for beta particles and all electromagnetic radiation (gamma rays and x-rays);
$Q = 10$, for neutrons from spontaneous fission and protons; and
$Q = 20$, for alpha particles and fission fragments.

An example of dose equivalent: an alpha particle is a heavy form of radiation and, in a relatively short distance, imparts a great amount of energy to the human body. Alpha exposure produces approximately 20 times the effect of a beta particle, which is a smaller form of radiation and travels faster, consequently imparting its energy over a longer path. The difference in energy is accounted for by a quality factor (Q) and the result is expressed in rems or dose units. While dose equivalent is sometimes not an exact measurement, it nevertheless can be a useful administrative unit (USEPA, 1981). To reiterate, the absorbed dose is measured in rads and the dose equivalent is measured in rems (USEPA, 1991).

There are approximately 2000 known radionuclides, which are all isotopes of elements that break down by emitting radiation. Radionuclides can be categorized as follows:

- As naturally occurring or anthropogenic (artificial)
- By the type of radioactive decay (alpha, beta, or gamma emission)
- By radioactive decay series

The natural radionuclides include the primordial elements that were incorporated into the Earth's crust during its formation, the radioactive decay products of these primordial elements, and radionuclides that are formed in the atmosphere by cosmic ray interactions. Anthropogenic radionuclides are produced through the use of nuclear fuels, radio pharmaceuticals, and other nuclear industry activities. Anthropogenic radionuclides have also been released into the atmosphere as the result of atmospheric testing of nuclear weapons and in rare cases, accidents at nuclear fuel stations, and discharge of radio-pharmaceuticals. The two types of radioactive decays that carry the most health risks due to ingestion of water are alpha emitters and beta or photon emitters. Many radionuclides are mixed emitters, with each radionuclide having a primary mode of disintegration. The naturally occurring radionuclides are largely alpha emitters, though many of the short-lived daughter products emit beta particles. Anthropogenic radionuclides are predominantly beta or photon emitters and include those that are released to the environment as the result of activities of the nuclear industry, but also include releases of alpha-emitting plutonium from nuclear weapon and nuclear reactor facilities (USEPA, 2000). The natural radionuclides involve three decay series, which start with uranium-238, thorium-232, and uranium-235, and are known collectively as the uranium, thorium, and actinium series. Each series decays through stages of various nuclides, which emit either an alpha or a beta particle as they decay, and terminates with a stable isotope of lead. Some of the radionuclides also emit gamma radiation, which accompany the alpha or beta decay. The uranium series contains uranium-238 and -234, radium-226, lead-210, and polonium-210. The thorium series contains radium-228 and radium-224. The actinium series contains uranium-235 (USEPA, 2000).

Partly in response to rising public concerns, including several lawsuits, the U.S. Environmental Protection Agency has revisited radionuclides regulation, which has been in effect since 1977, by requiring new monitoring provisions and by promulgating the following current standards: combined radium 226/228 of 5 pCi/L; a gross alpha standard for all alphas of 15 pCi/L, not including radon and uranium; a combined standard of 4 mrem/year for beta emitters. The new MCL for uranium is 30 μg/L (http://www.epa.gov/safewater/standard/pp/radnucpp.html, accessed on January 19, 2006). As part of the new standard promulgation, the USEPA, in cooperation with the USGS, issued a technical document (USEPA, 2000), which includes sections on the fundamentals of radioactivity in drinking water, overview of natural occurrence of major radionucludes in groundwater, and the results of a nationwide survey of selected wells in all hydrostratigraphic provinces in the U.S. performed by the USGS (Focazio, 2001). Excerpts from these two documents, together with references to various other publications on radionuclides published by the USGS, are presented below.

4.2.4.1 Radium

Radium isotopes occur as decay products in each of the three decay series previously mentioned. The two isotopes with an MCL include radium-226, a decay product in the uranium series, and radium-228, a decay product in the thorium series. Uranium and thorium are ubiquitous components of rocks and soils; therefore, radium radionuclides are also ubiquitous trace elements in rocks and soils. The occurrence of radionuclides in groundwater depends first on the presence and solubility of the parent products. Each radioactive decay product has its own unique chemical characteristics that differ from the radionuclide parent. Consequently, the occurrence and distribution of a parent radionuclide in solution does not necessarily indicate the presence of a daughter radionuclide in solution. For example, uranium (parent of radium-226) tends to be the least mobile in oxygen-poor groundwater, and tends to be strongly adsorbed onto humic substances. Conversely, radium tends to be most mobile in reducing groundwater that is chloride-rich with high concentrations of total dissolved solids (Kramer and Reid, 1984; Zapecza and Szabo, 1986). Radium behaves similarly to other divalent cations such as calcium, strontium, and barium. Therefore, in aquifers with limited sorption sites, radium solubility can be enhanced by the common ion effect in which competing cations are present in abundance and occupy sorption sites, keeping the radium in solution. Recently, high concentrations of radium were found to be associated with groundwater that was geochemically affected by agricultural practices in the recharge areas by strongly enriching the water with competing ions such as hydrogen, calcium, and magnesium (Szabo and dePaul, 1998). Radium-228 was detected in about equivalent concentrations as radium-226 in the aquifer study in New Jersey (Szabo and dePaul, 1998). The process of "alpha recoil" of alpha-emitting radionuclides also enhances their solubility. When an alpha particle is ejected from the nucleus of a radionuclide during decay, the newly created progeny radionuclide recoils in the opposite direction. The energy associated with this recoil is 104 to 106 times larger than typical chemical bond energies (Cothern and Rebers, 1990) and can cause atoms on the surface of a grain to be recoiled directly into the water in pore spaces. To date, measured radium in surface waters has been traced back to discharge from groundwater (Elsinger and Moore, 1983).

4.2.4.1.1 Radium-226

Radium-226 is the fifth member of the uranium-238 series, has a half-life of approximately 1622 years, and decays by alpha-particle emission. Uranium forms soluble complexes under oxygen-rich conditions, particularly with carbonates. Uranium precipitates from groundwater under oxygen-poor conditions and can be concentrated in secondary deposits (Cothern and Rebers, 1990). Consequently, uranium can be expected to range widely with different geologic environments. Because the chemical behavior of uranium and radium are vastly different, the degree of mobilization of the parent and product are different in most chemical environments. Szabo and Zapecza (1991) detail the differences in the occurrence of uranium and radium-226 in oxygen-rich and oxygen-poor areas of aquifers. The most significant radium-226 occurrence in groundwater in the nation is concentrated in the north-central states, including southern Minnesota, Wisconsin, northern Illinois, Iowa, and Missouri (Gilkeson et al., 1983; Zapecza and Szabo, 1987; Kay, 1999). In these states, the drinking water wells tap deep aquifers of Cambrian and Ordovician sandstones and dolomites and Cretaceous sandstones. These aquifers tend to have limited sorption sites, and radium solubility is enhanced by the common-ion effect wherever the concentration of total dissolved solids are high (Gilkeson et al., 1983). In some areas, such as northern Illinois, reduction of sulfate decreases coprecipitation of Ra-226 with barium sulfate, another mechanism which, if present, tends to limit dissolved Ra-226 (Gilkeson et al., 1983).

Ra-226 is also found in high concentrations in water derived from aquifers that straddle the Fall Line of the southeastern States from Georgia to New Jersey (King et al., 1982; Zapecza and Szabo, 1987; Szabo and dePaul, 1998). These aquifers are composed of unconsolidated sands that contain fragments of uranium-bearing minerals derived from the crystalline rocks of the Blue Ridge and Piedmont provinces. Sands of the Coastal Plain where uranium and radium concentrations tend to be the highest (Zapecza and Szabo, 1986) were directly derived from these crystalline rocks as fluvial deposits. Furthermore, in areas with saltwater intrusion or brackish water, the common-ion effect again is likely the cause for high concentrations of dissolved Ra-226 (Kraemer and Reid, 1984; Miller and Sutcliffe, 1985).

4.2.4.1.2   Radium-228

Radium-228 is the second member of the thorium-232 series, has a half-life of about 5.7 years, and decays by beta-particle emission. Thorium is extremely insoluble (Cothern and Rebers, 1990) and thus is not subject to mobilization in most groundwater environments. The relatively short half-life of this isotope limits the potential for transport of unsupported radium-228 relative to that of the longer-lived radium-226 isotope. Consequently, although radium-228 is chemically similar to radium-226, its occurrence distribution can be different. Michel and Cothern (1986) developed a national model to depict the occurrence of radium-228. There was little available occurrence data on radium-228; the model and associated data showed, however, that radium-228 activities tend to be the highest in arkosic sand and sandstone aquifers. Generally, the areas associated with the highest potential for radium-228 include the Coastal Plain aquifers that straddle the Fall Line from Georgia to Pennsylvania. Large areas of northern Illinois, Iowa, Minnesota, and Wisconsin were also ranked as having a high potential for radium-228 occurrence due to the sandstone aquifers and presence of high total dissolved solids. In contrast, aquifers that were mostly alluvial or glacial sand and gravel aquifers have a low potential for radium-228 in those same states. Other parts of the country such as areas in Colorado, Montana, and California are underlain by granitic rock that have the prerequisite geochemical characteristics for radium-228 occurrence but the aquifers are not used extensively for water supplies. Other areas such as the High Plains (Ogallala) aquifer and other locations where alluvial valley and sandstone aquifer material was derived from feldspathic minerals are considered to have medium potential for radium-228 occurrence.

4.2.4.2   Uranium

Uranium is the heaviest element in nature. Natural uranium consists of three isotopes: uranium-238, uranium-235, and uranium-234. The predominant isotope is uranium-238, which has a long half-life (4.5 billion years) and is relatively abundant in rocks and soils.

Uranium is predominately found in groundwater in the Colorado Plateau, the Western Central Platform, the Rocky Mountain System, Basin and Range, and the Pacific Mountain System (Zapecza and Szabo, 1987). Concentrations of uranium in groundwater in the eastern United States is typically low. Uranium is found in concentrated amounts in granite, metamorphic rocks, lignites, monazite sand, and phosphate deposits, as well as in the uranium-rich minerals of uraninite, carnotite, and pitchblende. Uranium must be oxidized before it is transported into groundwater but once in solution, it can travel great distances. The three natural uranium isotopes have the following proportions: uranium-234 (0.006% by weight), uranium-235 (0.72%), and uranium-238 (99.27%). The activity-to-mass ratio of the sum of the three radioisotopes in rock is 0.68 pCi/µg. The crustal abundances of uranium are not duplicated in groundwater. Uranium-234 is enriched in water relative to rock when standardized to uranium-238 in the water. Uranium-234 activity to uranium mass ratio in water varies from that in rock. In order to convert µg/l to pCi/L a ratio of U-234/U-238 of 0.9 to 1.3 is typically observed to account for the excess alpha-particle activity from uranium-234 in the water as opposed to the expected activity if crustal abundance were preserved. This uranium isotopic ratio may vary regionally, as well as seasonally. Uranium concentrations in groundwater are reported in micrograms per liter; 1 µg/L of uranium is equal to an activity of approximately 1.5 pCi/L when the isotopes of uranium-238 and uranium-234 are present in a 1:1 ratio.

4.2.4.3   Radon

Radon is a colorless, odorless radioactive gas that is soluble in water. The most abundant isotope of radon is radon-222. It is produced through the radioactive decay of uranium-238 and radium-226, which are naturally present in rock and soil. Radon-222 has a half-life of 3.8 days. Groundwater radon concentrations are highly variable, even within individual geologic units. The observed wide range of radon concentrations probably reflects the variable distribution of uranium or radium in the aquifer and the variable distribution of aquifer properties. Radon concentrations can range up to three orders of magnitude in water from a single geologic unit and can differ significantly from well to well locally (Sloto, 1994; Senior and Vogel, 1995; Senior, 1998).

The Surgeon General of the United States has recognized exposure to radon gas as being second only to cigarette smoking as a cause of lung cancer (USEPA, 1992). Radon gas can cause lung cancer if inhaled because its decay products can accumulate in the lungs and damage tissue. Radon moves from its source in rocks and soils through voids and fractures. It can enter buildings as a gas through foundation cracks, or it can dissolve in groundwater and be carried to buildings through the use of water supply wells. Most of the radon escapes from the water at the faucet or other point of use, leaving little in the water itself. The radon that escapes from the water adds to the radon that enters the home through the basement, and, in some cases, the water may contribute a large part of the radon that is present in a home. A study by Mose et al. (1990) found that cancer occurrences increase as the amount of radon in household water increases.

USEPA has not yet established an MCL for radon in drinking water. Waterborne radon commonly is a concern only for those who use well water for individual water supply. Because of its short half-life, radon in some public water supplies utilizing groundwater may decay to low concentrations before the water is delivered to users, especially if the water has been treated (Zapecza and Szabo, 1988). As a recognized public health hazard, radon is receiving a continuing attention from the government agencies. Various information regarding radon, including mitigation of its negative impacts, can be found on the following dedicated web pages maintained by the USEPA (http://www.epa.gov/iaq/radon), and the USGS (http://sedwww.cr.usgs.gov:8080/radon) and in USEPA (1992), USEPA and CDC (1992), and USEPA (1993).

4.2.5 Environmental Isotopes and Groundwater Age Dating

Environmental tracers commonly used to estimate age of groundwater formed less than 50–70 years ago are the chlorofluorocarbons (CFCs) and tritium–helium 3, i.e., $^3H/^3He$ ratio. Because of various uncertainties and assumptions that are associated with sampling, analysis, and interpretation of the environmental tracer data, groundwater ages estimated by use of the CFC and ^3H–^3He methods are regarded as apparent ages and must be carefully reviewed to ensure that they are geochemically consistent and hydrologically realistic (Rowe et al., 1999). Isotopes typically used for the determination of groundwater that was formed long ago are carbon-14, chlorine-36, oxygen-18, and deuterium, while many other isotopes are increasingly studied for their applicability (e.g., see Geyh, 2000).

4.2.5.1 Chlorofluorocarbons

Chlorofluorocarbons (CFCs) are manufactured VOCs that have been used extensively as refrigerants, aerosol propellants, cleaning solvents, and blowing agents in a variety of industries. Since their first introduction in the 1930s, the concentrations of CFCs have been steadily increasing. However, because CFCs are also believed to catalyze the destruction of atmospheric ozone, worldwide controls in early 1990s have been instituted in an effort to reduce global atmospheric CFC concentrations. Data reported through the mid-1990s indicate that this effort resulted in leveling-off or declining of CFC-11 and CFC-113 concentrations in air. Therefore, CFC dating of modern groundwater will become less precise in the future as concentrations of CFC compounds used for dating decline further: the use of CFCs as a dating tool is based on the more or less steady increase in northern troposphere CFC concentrations that has accompanied the large-scale historic use of CFC compounds in various industries (Rowe et al., 1999).

CFCs released into the atmosphere are partitioned into rainwater. The equilibrium solubility of individual CFC compounds in water (C_{CFC}) is governed by gas–liquid exchange equilibria, which are expressed in terms of Henry's law:

$$C_{CFC} = K_{CFC(T,S)} \times P_{CFC} \tag{4.23}$$

where K is Henry's law constant for the individual CFC compound at a known temperature (T) and salinity (S), and P is the atmospheric partial pressure of the CFC compound. Henry's law constants for CFC-11, CFC-12, and CFC-113 as a function of temperature and salinity have been compiled (Warner and Weiss, 1985; Bu and Warner, 1995; from Rowe et al., 1999). Concentrations of CFC-11, CFC-12, and CFC-113 in continental U.S. air from 1940 to the present have been reconstructed from CFC production records (Chemical Manufacturers Association, 1992) and atmospheric measurements that began in the mid-1970s (Busenberg et al., 1993). An example of concentrations of CFC-11, CFC-12, and CFC-113 in air and water in equilibrium at 10°C, for a study area near Dayton, southwestern Ohio is shown in Figure 4.13.

To calculate the age of a groundwater sample (or recharge year, if age is subtracted from the date of sampling), concentrations of the individual CFC compounds in groundwater are divided by the appropriate Henry's law constant to give the partial pressure of the CFC compound in air at the time the sample was isolated from the atmosphere. Because Henry's law constants are temperature dependent, an estimate of the recharge temperature is needed.

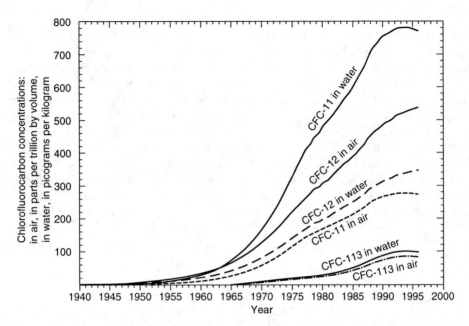

FIGURE 4.13 Concentrations of chlorofluorocarbon compounds CFC-11, CFC-12, and CFC-113 in air and water at equilibrium at 10°C and 760 feet above sea level; study area near Dayton, southwestern Ohio, the U.S. (From Rowe, G. L., et al., Ground-water age and water-quality trends in a Buried-Valley aquifer, Dayton area, Southwestern Ohio. U.S. Geological Survey Water Resources Investigations Report 99-4113. Columbus, OH, 1999, 81pp; Data from Busenberg, E., U.S. Geological Survey, written communication, 1996).

The recharge temperature is the temperature at the base of the unsaturated zone and is usually close to the mean annual air temperature. Recharge temperatures can be estimated independently by a variety of techniques, such as by use of oxygen and deuterium isotope data, dissolved-gas (N_2/Ar) ratios, and long-term air temperature records for the study area (Rowe et al., 1999). The calculated partial pressure is then compared with the atmospheric concentration curves (Figure 4.13) to derive the age of the sample. It is assumed that CFC concentrations in recharge waters are in equilibrium with the soil atmosphere and, more importantly, that CFC concentrations in the soil gas are equal to those found in air (Busenberg et al., 1993).

Groundwater ages derived by the CFC technique are considered to be minimum ages because trace-level contamination by small amounts of CFC compounds introduced during sampling or mixing of groundwater of different ages can never be completely excluded. Trace-level contamination affects the reliability of older recharge ages to a greater degree because of lower CFC concentrations in older waters (Figure 4.13). Other processes that can affect the reliability of CFC ages include presence of other chlorinated VOCs, cross contamination during drilling or sampling, sorption, microbial degradation, hydrodynamic dispersion, and diffusion in the unsaturated zone (Busenberg and Plummer, 1992; Busenberg et al., 1993; Plummer et al., 1993; Reilly et al., 1994; and Katz et al., 1995). Uncertainty in the recharge temperature affects the recharge age to a varying extent. Uncertainty of several degrees celsius causes an uncertainty of less than a year in waters recharged prior to 1975. For waters recharged in the 1980s, an uncertainty of 2°C results in an uncertainty of 2 to 3 years. For post-1989 waters, a 2°C uncertainty in the recharge temperature can cause errors of several years or more (Busenberg and Plummer, 1992; Plummer et al., 1993; from Rowe et al., 1999).

In their study of environmental tracers to determine the fate and transport of recycled water in Los Angeles County, California, Anders and Schroeder (2003) note that the deviation from a 1:1 ratio of CFC-11 and CFC-12-determined recharge ages indicates that there are processes altering the CFC-11 and CFC-12 concentrations to different extents. These processes may include a more rapid anaerobic microbial degradation of CFC-11 than that of CFC-12 (Lovley and Woodward, 1992) and a stronger sorption of CFC-11 by soils than that for CFC-12 (Russell and Thompson, 1983; from Anders and Schroeder, 2003). Since modeled CFC-11 recharge ages are greater than modeled CFC-12 recharge ages for post-1970 water, sorption and degradation likely occur during the initial stage of recharge. The greater modeled CFC-12 recharge ages for pre-1970 water suggests greater amounts of CFC-11 were reaching the groundwater in the past than in the present. Furthermore, all three CFCs are known to degrade in environments where methane is present (Dunkle et al., 1993), a condition possibly present in the Central Basin prior to the use of recycled water. Anders and Schroeder (2003) therefore conclude that CFCs do not provide reliable groundwater ages in most water samples collected from the multiple-well monitoring sites along the flow path of the recycled water.

4.2.5.2   Tritium–Helium-3 (^{3}H–^{3}He)

Tritium (^{3}H), the radioactive isotope of hydrogen with the half-life of 12.43 years, has been used extensively as a hydrologic tracer and dating tool. It is produced naturally in the upper atmosphere by bombardment of nitrogen with cosmic radiation and, although few measurements are available, it is estimated that the natural concentration of tritium in precipitation is between 5 and 20 tritium units (TU) (Kauffman and Libby, 1954). The use of tritium as a hydrologic tracer is related to the release of large quantities of tritium into the atmosphere during atomic-weapons testing in the 1950s and early 1960s. Peak tritium activities in various parts of the world can be estimated from data published by the International Atomic Energy Agency (IAEA) in Vienna (ftp.iaea.org) or by agencies of individual countries. For example, Figure 4.14 shows a graph for southwestern Ohio rainwater in which the peak tritium activities are believed to have approached 2000 TU (1 TU equals 3.24 pCi/L, or 1.185 Bq/L, or 1 tritium atom per 10^{18} hydrogen atoms). The atmospheric-testing peak therefore provides an absolute time marker from which to estimate groundwater age. However, because radioactive decay and hydrodynamic dispersion have greatly reduced maximum tritium concentrations in groundwater, identification of the 1960s atmospheric testing peak has become increasingly difficult. The interpretation of ages from tritium data alone is further complicated by the fact that monitoring and extraction wells are commonly screened over intervals that represent a wide range of groundwater ages. Questions relating to whether the groundwater at depth contains tritium solely from the falling limb (postbomb peak, or after about 1965) of the tritium curve can also complicate the interpretation. Therefore, tritium by itself is used only as a qualitative indicator of groundwater age (Rowe et al., 1999; Anders and Schroeder, 2003).

The most accurate use of the tritium data is to indicate pre- or post-1952 groundwater recharge. Assuming that piston-flow conditions (no dispersion or mixing) are applicable, Clark and Fritz (1997, from Kay et al., 2002) provide the following guidelines for using the tritium data: (1) groundwater that contains less than 0.8 TU and is underlying regions with continental climates has recharged the water table prior to 1952; (2) water with 0.8 to 4 TU may represent a mixture of water that contains components of recharge from before and after 1952; (3) tritium concentrations from about 5 to 15 TU may indicate recharge after about 1987; (4) tritium concentrations between about 16 and 30 TU are indicative of recharge since 1953 but cannot be used to provide a more specific time of recharge; (5) water with more

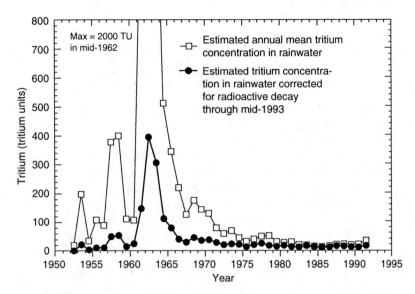

FIGURE 4.14 Estimated annual mean tritium concentration in rainwater, 1952–1991, southwestern Ohio. (From Rowe, G. L., et al., Ground-water age and water-quality trends in a Buried-Valley aquifer, Dayton area, Southwestern Ohio. U.S. Geological Survey Water-Resources Investigations Report 99-4113. Columbus, OH, 1999, 81pp; Michel, R.L., Tritium deposition in the continental United States, 1953–1983. U.S. Geological Survey Water-Resources Investigations Report 89-4072, 1989, 46pp; Michel, R.L., U.S. Geological Survey, written communication, 1991.)

than 30 TU probably is from recharge in the 1960s or 1970s; and (6) water with more than 50 TU predominately is from recharge in the 1960s. The continuing depletion of the artificial tritium in the environment will likely reduce its future usefulness in the groundwater studies.

The tritium/helium-3 (^{3}H/^{3}He) method is used to exclude artifacts of mixing of water with various ages in a well. This dating method is based on the decay of tritium to its daughter product, the noble gas helium-3 (^{3}He$_{trit}$, or tritiogenic helium). Because these substances virtually are inert in groundwater, unaffected by groundwater chemistry, and not derived from contamination from most of the typical anthropogenic sources, ^{3}H/^{3}He dating can be applied to a wide range of hydrologic investigations since the input function of ^{3}H does not have to be known (Geyh, 2000; Kay et al., 2002). By measuring both the mother and daughter activities, actual water ages can be calculated, provided the samples were unmixed and collected from an aquifer with a piston-flow groundwater movement. Incorporating the effects of dispersion and diffusion, which is a much more realistic scenario, requires application of modeling techniques.

The activity of ^{3}H in the sample (^{3}H$_{spl}$) is given as (Geyh, 2000):

$$^3H_{spl} = {^3H_{init}}e^{-\lambda t} \tag{4.24}$$

where ^{3}H$_{init}$ is the activity of initial tritium, λ is the radioactive decay constant, and t is the time since decaying started (absolute age). The growth of ^{3}He in a sample is given by

$$^3He_{spl} = {^3H_{init}}(1 - e^{-\lambda t}) \tag{4.25}$$

Combining Equation 4.24 and Equation 4.25, the unknown and variable initial ^{3}H activity (^{3}H$_{init}$) is eliminated and the age of water (t) is obtained from

$$^3He_{spl} = {}^3H_{spl}(e^{-\lambda t} - 1) \tag{4.26}$$

$$t = -\frac{\ln\left(1 + \dfrac{{}^3He_{spl}}{{}^3H_{spl}}\right)}{\lambda} \tag{4.27}$$

The ^3He concentration in the sample has to be corrected for admixed ^3He from the Earth's crust and from the atmosphere.

The sum of tritium and its daughter product ($^3H + {}^3He_{trit}$) represents a conservative quantity equivalent to the amount of tritium in rainwater at the time of recharge, assuming that helium produced by tritium decay is not lost by upward diffusion to the unsaturated zone. This assumption is considered valid for aquifers where vertical flow velocities exceed approximately 1.5 ft/yr (Poreda et al., 1988; Schlosser et al., 1988, 1989). Procedures used to evaluate and correct the helium data for nontritiogenic sources of ^3He are described by Schlosser et al. (1988, 1989) and Shapiro et al. (1998). However, in most shallow groundwater systems, subsurface sources of ^3He are minor and the following equation can be used to calculate $^3He_{trit}$ (Jenkins, 1987, from Anders and Schroeder, 2003):

$$^3He_{trit} = 4.021 \times 10^{14} \times \left[{}^4He_{tot}(R_{tot} - R_{atm}) + {}^4He_{eq}R_{atm}(1 - \beta)\right] \tag{4.28}$$

where $^4He_{tot}$ is measured ^4He concentration of the sample [cm^3 STP/g H_2O], R_{tot} is measured $^3He/{}^4He$ ratio of the water sample [dimensionless], R_{atm} is $^3He/{}^4He$ ratio of atmospheric helium [1.384×10^{-6}; dimensionless], and $^4He_{eq}$ is ^4He concentration in air-equilibrated water [cm^3 STP/g H_2O].

The factor 4.021×10^{14} is used to convert cm^3 STP/g H_2O into TU and β is the effect of the difference in solubility of the two isotopes of helium (0.983). Furthermore, the use of this age-dating technique assumes that the system is closed (does not allow ^3He to escape) and is characterized by piston flow (no hydrodynamic dispersion) (Scanlon et al., 2002).

Once the amount of tritiogenic ^3He is known, the ^3H–^3He age is then calculated from the daughter/parent ratio ($^3He_{trit}/{}^3H$) by the use of standard decay equation (Rowe et al., 1999):

$$\tau = \frac{T_{1/2}}{\ln 2} \ln\left(1 + \frac{{}^3He_{trit}}{{}^3H}\right) \tag{4.29}$$

where τ is the ^3H–^3He age, in years; $^3He_{trit}$ is the amount of ^3He derived from tritium decay, in TU; ^3H is the measured tritium concentration, in TU; and $T_{1/2}$ is the half-life of tritium, i.e., 12.43 years (Schlosser et al., 1988, 1989; Solomon et al., 1993). In simple terms, the concentration of tritiogenic ^3He will increase as tritium decays; thus, older waters will have higher $^3He_{trit}/{}^3H$ ratios. The age derived by use of this technique represents the time since the water became isolated from the atmosphere and traveled to the sampling point, and is independent of the tritium source function; it is further assumed that diffusion, dispersion, contamination, or mixing at the well screen have not affected the concentrations of ^3H or $^3He_{trit}$ in the sample (Rowe et al., 1999). Apart from radioactive decay of tritium, ^3H and ^3He are both chemically inert and therefore are unaffected by microbial degradation or sorption, processes known to affect the reliability of the CFC dating method. Figure 4.15 shows an example of groundwater age determination using tritium/helium method at various depths below aquifer recharge basins in the Central Basin, Los Angeles County, California.

4.2.5.3 Oxygen-18 and Deuterium

The most abundant isotopes of oxygen, ^{16}O (99.7%) and ^{18}O (0.2%), and those of hydrogen, ^1H and ^2H (deuterium, about 0.016% of hydrogen atoms), combine to produce water

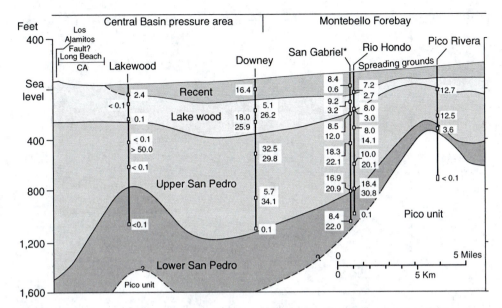

FIGURE 4.15 Tritium concentration in five multiple-well monitoring sites and tritium/helium-3 ages in 16 monitoring wells along groundwater flow path, Los Angeles County, California. Numbers in single boxes and upper numbers represent tritium content in tritium units (TU); lower numbers represent tritium/helium-3 age determination, in years, for the sampling year 1998. (Modified from Anders, R. and Schroeder, R.A., Use of water-quality indicators and environmental tracers to determine the fate and transport of recycled water in Los Angeles County, California. U.S. Geological Survey Water-Resources Investigations Report 03-4279, Sacramento, CA, 2003, 104pp.)

molecules of differing molecular mass between 18 and 22, of which the most abundant are $^1H_2^{16}O$, $^1H^2H^{16}O$, and $^1H_2^{18}O$ (Geyh, 2000). These isotopes of oxygen and hydrogen are stable and do not disintegrate by radioactive decay. As constituents of the water molecules they can therefore act as conservative tracers in groundwater. The natural atomic ratios are $^2H/^1H = {}^2R = 1.5 \times 10^{-4}$ and $^{18}O/^{16}O = {}^{18}R = 2 \times 10^{-3}$. These ratios are expressed in delta units (δ) relative to a reference standard (Kay et al., 2002):

$$\delta x = [(R_x/R_{STD}) - 1] \times 1000 \qquad (4.30)$$

where R_x and R_{STD} are the $^2H/^1H$ and $^{18}O/^{16}O$ ratios of the sample and reference standard, respectively. The delta units are given in parts per thousand (per mil, written as ‰). Ocean water has $\delta^{18}O$ and δ^2H values of $\pm 0‰$, and has been chosen as the Vienna standard mean ocean water (V-SMOW) standard. Most freshwaters have negative delta values (Geyh, 2000). For example, an oxygen sample with an $\delta^{18}O$ value of $-50‰$ is depleted in ^{18}O by 5% or 50‰ relative to the standard. The difference in the mass of oxygen and hydrogen isotopes in water results in distinct partitioning of the isotopes (fractionation) caused by evaporation, condensation, freezing, melting, or chemical and biological reactions. For example, δ^2H and $\delta^{18}O$ values in precipitation are isotopically lighter in areas with lower mean annual temperature. Strong seasonal variations are expected at any given location (Dansgaard, 1964). Average annual values of δ^2H and $\delta^{18}O$ in precipitation, however, show little variation at any one location. The IAEA provides δ^2H and $\delta^{18}O$ precipitation data measured at various locations throughout the world (accessible at: ftp://ftp.iaea.org). For example, values of δ^2H with $\delta^{18}O$ in precipitation at Midway Airport (Figure 4.16) show a significant linear correlation

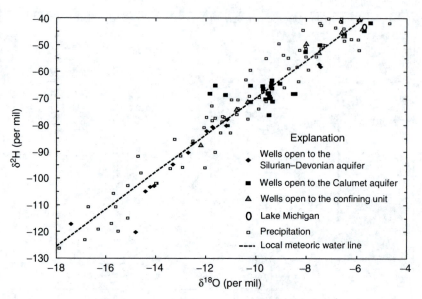

FIGURE 4.16 Isotopic composition of groundwater, precipitation, and lake water, Calumet region of northwestern Indiana and northeastern Illinois, July 1997–February 2001. (From Kay, R.T., et al., Use of isotopes to identify sources of ground water, estimate ground-water-flow rates, and assess aquifer vulnerability in the Calumet Region of Northwestern Indiana and Northeastern Illinois. U.S. Geological Survey Water-Resources Investigation Report 02-4213, Indianapolis, IN, 2002, 60pp.)

between the hydrogen and oxygen isotopic composition of precipitation in the study area, described by the equation (Kay et al., 2002):

$$\delta^2H = 6.9 \times {}^{18}O + 0.08 \text{ [given in ‰]} \tag{4.31}$$

This equation describes the local meteoric water line (LMWL) and provides a reference with which the isotopic composition of the groundwater in the region of interest can be evaluated. The strong relationship between the $\delta^{18}O$ and δ^2H values of precipitation is also reflected in the global meteoric water line (MWL). The slope is 8 and the so-called deuterium excess is +10‰. The deuterium excess (d) is defined as

$$d_{excess} = \delta^2H - 8\delta^{18}O \tag{4.32}$$

The deuterium excess near the coast is smaller than +10‰ and approximately 0‰ only in Antarctica. In areas where, or during periods in which, the relative humidity immediately above the ocean is or was below the present mean value, d is greater than +10‰. An example is the deuterium excess of +22‰ in the eastern Mediterranean. The value of d is primarily a function of the mean relative humidity of the atmosphere above the ocean water. The coefficient d can therefore be regarded as a palaeoclimatic indicator (Gat and Carmi, 1970; Merlivat and Jouzel, 1979; Geyh, 2000).

In groundwater systems with temperatures less than 50°C, the isotopic compositions of δ^2H and $\delta^{18}O$ in water are usually not significantly affected by water–rock interactions (Perry et al., 1982). Differences in the isotopic composition of groundwater, therefore, can be used to detect differences in the source water, which may include recent precipitation. Geyh (2000) discusses various processes and factors that affect local isotopic compositions of precipitation and groundwater, as well as their deviations from the global and LMWLs (e.g., see Figure 4.17). This includes groundwater mixing, reactions, evaporation, temperature, altitude, and continental effects.

4.2.5.4 Carbon-14

Radiocarbon (carbon-14 or ^{14}C) is the radioactive isotope of carbon with a half-life of 5730 years. It occurs in atmospheric CO_2, living biosphere, and the hydrosphere after its production by cosmic radiation. Underground production is negligible. The ^{14}C activity is usually given as an activity ratio relative to a standard activity, about equal to the activity of recent or modern carbon. Therefore, the ^{14}C content of carbon-containing materials is given in percent modern carbon (pMC): 100 pMC (or 100% modern carbon) corresponds by definition to the ^{14}C activity of carbon originating from (grown in) 1950 AD (Geyh, 2000). In addition to the radioactive isotope ^{14}C, two other stable carbon isotopes, ^{13}C and ^{12}C, are important for understanding the origin of CO_2 involved in the dissolved carbonate-CO_2 system in groundwater, and for correcting the age-dating results obtained from the ^{14}C isotope.

The time since water from a sampling well has been isolated from the atmosphere (i.e., the apparent age of groundwater) can be estimated using its ^{14}C composition. In general, water is isolated from the atmosphere when it moves from the unsaturated zone to the groundwater flow system. The ^{14}C composition of groundwater is the result of two main factors: (1) chemical reactions that affect the inorganic carbon concentrations and ^{14}C composition in the water as the water infiltrates through the unsaturated zone or moves through the aquifer and (2) radioactive decay of ^{14}C in the water (Anderholm and Heywood, 2002). The isotopic composition of carbon in the dissolved inorganic carbon (DIC) constituents of groundwater is very variable. The sources of carbon dissolved in groundwater are soil CO_2, CO_2 of geogenic origin or from magmatic CO_2 (from deep crustal or mantle sources or in fluid inclusions), living and dead organic matter in soils and rocks, methane, and carbonate

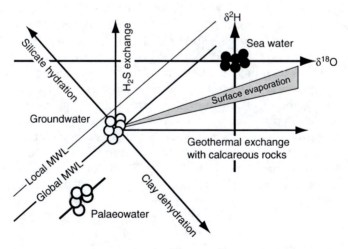

FIGURE 4.17 Various processes which shift the $\delta^{18}O$ and δ^2H values from the MWL (meteoric water line): evaporation shifts both $\delta^{18}O$ and δ^2H values; the former are displaced as a result of isotopic exchange with volcanic CO_2 and limestone, the latter due to exchange with H_2S and silicate hydration. Local MWL corresponds to the Mediterranean precipitation; the MWL of Pleistocene palaeowater may be apart from the MWL. (Modified from Geyh, M., In: Mook, W.G. (ed.), *Environmental Isotopes in the Hydrological Cycle; Principles and Applications.* IHP-V, Technical Documents in Hydrology, No. 39, Vol. IV, UNESCO, Paris, 2000, 196pp.)

minerals. Each of these sources has a different carbon isotopic composition and contributes to total dissolved carbon in various proportions. Therefore, the isotopic composition of DIC compounds in groundwater has a wide range of $\delta^{13}C$ values. Soil carbon dioxide usually has a value of about $-22‰$, and in tropical soils it may be more positive to about $-11‰$. Carbon dioxide of an endogenous or magmatic origin has $\delta^{13}C$ values of about $-6‰$, and metamorphic carbon from sedimentary rocks is usually close to zero if it is derived from marine carbonates. The organic carbon of terrestrial plants has $\delta^{13}C$ values between $-30‰$ and $-20‰$. The heaviest carbon isotopic composition is found in evaporate carbonates with $+10‰$. Such carbonates occur in sedimentary basins where the $\delta^{13}C$ values of DIC of fresh groundwater might have elevated $\delta^{13}C$ values (Geyh, 2000).

Two common chemical reactions affecting the inorganic carbon concentrations and ^{13}C and ^{14}C compositions in groundwater include dissolution of carbon dioxide and dissolution of carbonate minerals. These reactions could occur in the unsaturated zone during infiltration of water or as groundwater moves through the aquifer. The dissolution of carbon dioxide gas, most of which generally occurs in the unsaturated zone by infiltrating water, is an important reaction that affects the ^{14}C composition of groundwater. Recently infiltrating water and dissolved carbon dioxide gas in the unsaturated zone would have a ^{14}C composition of about 100% modern because carbon dioxide gas diffuses from the atmosphere and because plants respire carbon dioxide gas to the soil zone or unsaturated zone that is 100% modern. Water that has dissolved carbon dioxide gas and is infiltrating through the unsaturated zone or moving through the aquifer can also dissolve carbonate minerals, which increases DIC concentrations and can significantly reduce the ^{14}C composition of the water (Anderholm and Heywood, 2002). Groundwater with a long resident time in the aquifer was subject to similar processes at the time of its origin, and has undergone various transformations through reactions with the aquifer porous media since that time. Because of all these factors, ^{14}C is not a conservative tracer and its use for groundwater dating studies is not straightforward.

Groundwater dating using ^{14}C is generally considered applicable for water up to 30,000 years old, although the original dating technique proposed for organic carbon samples is applicable to the 45,000–50,000-year range (Libby, 1946). Before determining how long groundwater has been isolated from the atmosphere or from the modern ^{14}C reservoir, the effect of chemical reactions on the ^{14}C composition of groundwater needs to be determined. Various models have been used to adjust or estimate the ^{14}C composition of water, resulting from processes in the unsaturated zone and aquifer (Mook, 1980, from Anderholm and Heywood, 2002, and Geyh, 2000). These models range from simple ones that require little data, to complex models that require much information about the carbon isotopic composition (ratio of stable carbon isotopes $^{13}C/^{12}C$ and ^{14}C) of gas in the unsaturated zone, carbon isotopic composition of carbonate minerals in the unsaturated zone and aquifer, and reactions that occur as water moves through the aquifer. The following equation can be used to estimate the apparent age of groundwater (Anderholm and Heywood, 2002):

$$t = \frac{5730}{\ln 2} \cdot \ln\left(\frac{A_0}{A_S}\right) \tag{4.33}$$

where t is apparent age, in years, A_0 is the ^{14}C composition of water before radioactive decay and after chemical reactions, in percent modern, and A_S is the ^{14}C composition measured in the sample, in percent modern.

Experience with various correction models shows that, using the same hydrochemical and isotope information, different models may produce corrections varying by many thousands of years (Geyh, 2000).

4.2.5.5 Chlorine-36

Chlorine-36 (^{36}Cl) is the radioactive isotope of chlorine and has a half-life of 300,000 years. It is produced cosmogenically, by natural underground production and by nuclear weapon tests. The subsurface production by radionuclide-derived neutron fluxes on ^{38}Cl depend heavily of the variable geological settings. As radioactive decay and underground production on the one hand, and dilution by dissolution of salt or salt enrichment by evaporation on the other interfere, the interpretation of ^{36}Cl data in terms of groundwater ages remains ambiguous. If uranium-rich or chlorine-bearing minerals are present, apparently too low ^{36}Cl groundwater ages may be obtained. The relative ^{36}Cl abundance in groundwater is not changed by evaporation of the water, mineral interactions, or secondary mineral formation, but it is influenced by chloride dissolution. Therefore, the total chloride concentration has to be always determined. The plot of $^{36}Cl/Cl$ ratio versus the chloride concentration yields information on groundwater mixing, evaporation, remobilization of chloride, and radioactive decay and subsurface production of ^{36}Cl. Dating techniques using ^{36}Cl are applied to groundwater with residence times between 40,000 and 3,000,000 years. As ^{36}Cl was also produced during testing of nuclear bombs, it has been used for dating young groundwater in the unsaturated zone and in unconfined aquifers recharged since the 1950s and 1960s (Bentley et al., 1986; Florkowski et al., 1988; Andrews and Fontes, 1992; Mazor, 1992; Geyh, 2000).

The best source for numerous studies on the application of the above discussed and various other environmental isotopes, as well as artificially introduced radioactive isotopes, are the proceedings of the international conferences organized by the IAEA, and related monographs published by IAEA (www.iaea.org).

REFERENCES

Aiken, G.R, 2002. Organic matter in ground water. In: Aiken, G.R. and E.L. Kuniansky (eds.), U.S. Geological Survey Open-File Report 02-89, pp. 21–23.

Alekin, O.A., 1953. *Osnovi gidrohemii* (*Principles of Hydrochemistry*, in Russian). Gidrometeoizdat, Leningrad.

Alekin, O.A., 1962. *Grundlages der wasserchemie. Eine einführung in die chemie natürlicher wasser.* VEB Deutsch. Verl., Leipzig, 260 pp. (originally published in Russian in 1953).

Anderholm, S.K. and Heywood, C.E., 2003. Chemistry and age of ground water in the southeastern Hueco Bolson, New Mexico and Texas. U.S. Geological Survey Water-Resources Investigations Report 02-4237, Albuquerque, New Mexico, p. 16.

Anders, R. and Schroeder, R.A., 2003. Use of water-quality indicators and environmental tracers to determine the fate and transport of recycled water in Los Angeles County, California. U.S. Geological Survey Water-Resources Investigations Report 03-4279, Sacramento, CA, 104pp.

Andrews, J.N. and Fontes, J.C., 1992. Importance of the *in situ* production of ^{36}Cl, ^{36}Ar and ^{14}C in hydrology and hydrogeochemistry. In: *Isotope Techniques in Water Resources Development*, IAEA, Vienna, pp. 245–269.

Banerjee, S., 1984. Solubility of organic mixtures in water. *Environmental Science and Technology*, 18(8), 587–591.

Bartlett, R.J. and James, B., 1979. Behavior of chromium in soils: III. Oxidation, *Journal of Environmental Quality*, 8, 31–35.

Benson, N.R., 1976. Retardation of apple tree growth by soil arsenic residues from old insecticideal treatments. *Journal American Society of Horticultural Science*, 101(3), 251–253.

Bentley, H.W., Phillips, F.M., and Davis, S.N., 1986. Chlorine-36 in the terrestrial environment. In: Fritz, P. and Fontes, J.C. (eds.), *Handbook of Environmental Isotope Geochemistry*, Vol. IIB, The Terrestrial Environment, Elsevier, Amsterdam, pp. 427–480.

Brownlow, A.H., 1996. *Geochemistry*. Prentice-Hall, Upper Saddle River, New Jersey, 580pp.

Bu, X. and Warner, M.J., 1995. Solubility of chlorofluorocarbon 113 in water and seawater. *Deep Sea Research*, 42(7), 1151–1161.

Busenberg, E. and Plummer, L.N., 1992. Use of chlorofluorocarbons (CCl_3F and CCl_2F_2) as hydrologic tracers and age-dating tools—the alluvium and terrace system of central Oklahoma. *Water Resources Research*, 28(9), 2257–2283.

Busenberg, E., et al., 1993. Age dating ground water by use of chlorofluorocarbons in the unsaturated zone, Snake River Plain aquifer, Idaho National Engineering Laboratory, Idaho. U.S. Geological Survey Water-Resources Investigations Report 93-4054, 47pp.

Calder, L.M., 1988. Chromium contamination of groundwater. In: Nriagu, J.O. and Nieboer, E. (eds.), *Chromium in the Natural and Human Environments*, John Wiley & Sons, New York, Chapter 8, pp. 215–231.

Chemical Manufacturers Association, 1992. Alternative fluorocarbons environmental acceptability study—production and atmospheric release data for CFC-11 and CFC-12 (through 1991), Chemical Manufacturers Association, Washington D.C., 34pp.

Clark, I.D., and Fritz, P., 1997. *Environmental Isotopes in Hydrogeology*. Lewis Publishers, New York, 311pp.

Cohen, R.M., and Mercer, J.W., 1993. *DNAPL Site Evaluation*. C.K. Smoley—CRC Press, Boca Raton, FL.

Cothern, R.C., and Rebers, P.A., 1990. *Radon, Radium and Uranium in Drinking Water*. Lewis Publishers,. Chelsea, MI, 286pp.

Dansgaard, W., 1964. Stable isotopes in precipitation. *Tellus*, 16(4), 437–468.

Davenport, J.R. and Peryea, F.J., 1991. Phosphate fertilizers influence leaching of lead and arsenic in a soil contaminated with lead and arsenic in a soil contaminated with lead arsenate. *Water, Air and Soil Pollution*, 57–58, 101–110.

Davis, S.N. and DeWiest, J.M., 1991. *Hydrogeology*. Krieger Publishing Company, Malabar, FL, 463pp.

Dimitrijevic, N.S., 1988. *Hidrohemija* (*Hydrochemistry*, in Serbian). Univerzitet u Beogradu, Beograd, 313pp.

Domenico, P.A. and Schwartz, F.W., 1990. *Physical and Chemical Hydrogeology*. John Wiley & Sons, New York, 824pp.

Dowdle, P.R., Laverman, A.M., and Oremland, R.S., 1996. Bacterial dissimilatory reduction of arsenic (V) to arsenic (III) in anoxic sediments. *Applied Environmental Microbiology*, 62(5), 1664–1669.

Dunkle, S.A., et al., 1993. Chlorofluorocarbons (CCl3F and CCl2F2) as dating tools and hydrologic tracers in shallow ground water of the Delmarva Peninsula, Atlantic coastal plain, United States. *Water Resources Research*, 29, 3837–3860.

Elsinger, R.J., and Moore, W.S., 1983. Radium-224, Ra-228, and Ra-226 in Winyah Bay and Delaware Bay. *Earth and Planetary Science Letters*, 64, 430–436.

Feth, J.H., 1966. Nitrogen compounds in natural water—a review. *Water Resources Research*, 2, 41–58.

Feth, J.H., Rogers, S.M., and Roberson, C.E., 1961. Aqua de Ney, California, a spring of unique chemical character. *Geochimica et Cosmochimica Acta*, 22, 75–86.

Florkowski, T., Morawska, L., and Rozanski, K., 1988. Natural production of radionuclides in geological formations. *Nuclear Geophysics*, 2, 1–14.

Focazio, M.J., 2001. Occurrence of selected radionuclides in ground water used for drinking water in the United States: a targeted reconnaissance survey, 1998. U.S. Water-Resources Investigations Report 00-4273, Reston, VI, 40pp.

Fountain, J.C., 1998. Technologies for dense nonaqueous phase liquid source zone remediation. Technology Evaluation Report TE-98-02, Ground-Water Remediation Technologies Analysis Center (GWRTAC), Pittsburgh, Pennsylvania, 62 pp.

Freeze, R.A., and Cherry, J.A., 1979. *Groundwater*. Prentice-Hall, Englewood Cliffs, NJ, 604pp.

Gat, J.R. and Carmi, I., 1970. Evolution of the isotopic composition of atmospheric waters in the Mediterranean Sea area. *Journal of Geophysical Research*, 75, 3039–3048.

Geyh, M., 2000. Groundwater, saturated and unsaturated zone. In: Mook, W.G. (ed.), *Environmental Isotopes in the Hydrological Cycle; Principles and Applications*. IHP-V, Technical Documents in Hydrology, No. 39, Vol. IV, UNESCO, Paris, 196pp.

Harrington, J.M., Fendorf, S.B., and Rosenzweig, R.F., 1998. Biotic generation of arsenic(III) in metal(loid)-contaminated sediments. *Environmental Science and Technology*, 32(16), 2425–2430.

Harvey, C.F., et al., 2002. Arsenic mobility and groundwater extraction in Bangladesh. *Science*, 298, 1602–1606.

Hem, J.D., 1989. Study and interpretation of the chemical characteristics of natural water; 3rd edition. U.S. Geological Survey Water-Supply Paper 2254. Washington, D.C., 263pp.

Jain, A., Raven, K.P., and Loeppert, R.H., 1999. Arsenite and arsenate adsorption on ferrihydrite: Surface charge reduction and net OH-release stoichiometry. *Environmental Science and Technology*, 33, 1179–1184.

Jenkins, W.J., 1987. ^{3}H and ^{3}He in the Beta Triangle: observations of gyre ventilation and oxygen utilization rates. *Journal of Physical Oceanography*, 17, 763–783.

Johnson, R.L., Palmer, C.D., and Fish, W., 1989. Subsurface chemical processes. In: USEPA, Transport and fate of contaminants in the subsurface, seminar publication. EPA/625/4-89/019, pp. 41–56.

Katz, B.G., et al., 1995. Chemical evolution of groundwater near a sinkhole lake, northern Floridan—1. Flow patterns, age of groundwater, and influence of lake water leakage. *Water Resources Research*, 31(6), 1549–1564.

Kauffman, S. and Libby, W.S., 1954. The natural distribution of tritium. *Physical Review*, 93(6), 1337–1344.

Kauffman, W.J. and Orlob, G.T., 1956. Measuring ground water movement with radioactive and chemical tracers. *American Water Works Association Journal*, 48, 559–572.

Kavanaugh, M.C., Kresic, N., Schwarz, S.C., and Wright, A.P., 2005. Analysis of data and research concerning CCA-treated wood and economic impact & practicability issues. RedOak Consulting, a Division of Malcolm Pirnie, 36 pp. (Accessed January 17, 2006 at http://www.woodpreservative-science.org/disposal.shtml).

Kay, R., 1999, Radium in ground water from public-water supplies in northern Illinois. U.S. Geological Survey Fact Sheet 137–99, 4 pp.

Kay, R.T., Bayless, E.R., and Solak, R.A., 2002. Use of isotopes to identify sources of ground water, estimate ground-water-flow rates, and assess aquifer vulnerability in the Calumet Region of Northwestern Indiana and Northeastern Illinois. U.S. Geological Survey Water-Resources Investigation Report 02-4213, Indianapolis, IN, 60pp.

Keeney, D., 1990. Sources of nitrate to ground water. *Critical Reviews in Environmental Control*, 16, 257–304.

Kimbrough, D.E., et al., 1999. A critical assessment of chromium in the environment. *Critical Reviews in Environmental Science and Technology*, 29, 1–46.

King, P.T., Michel, J., and Moore, W.S., 1982. Ground-water geochemistry of Ra-228, Ra-226, and Rn-222. *Geochimica Cosmochimica Acta*, 46, 1173–1182.

Kraemer, T.F., and Reid, D.F., 1984. The occurrence and behavior of radium in saline formation water of the U.S. Gulf Coast Region. *Isotope Geoscience*, 2, 153–174.

Kresic, N. and Papic, P., 1990. Specific chemical composition of karst groundwater in the ophiolite belt of the Yugoslav Inner Dinarides: a case for covered karst. *Environ. Geol. Water Sci.*, vol. 15, no. 2, pp. 131–135.

Krumholz, L.R., 2000. Microbial communities in the deep subsurface. *Hydrogeology Journal*, 8(1), 4–10.

Leinonen, P.J. and Mackay, D., 1973. The multicomponent solubility of hydrocarbons in groundwater. *Canadian Journal of Chemical Engineering*, 51, 230–233.

Libby, W.F., 1946. Atmospheric helium three and radiocarbon from cosmic radiation. *Physics Review*, 69, 671–672.

Lide, D.R., 2005. *CRC Handbook of Chemistry and Physics*, 86th ed. CRC Press, Boca Raton, FL, 2544pp.

Lovley, D.R., 1991. Dissimilatory Fe(III) and Mn(IV) reduction. *Microbiological Reviews*, 55(2), 259–287.

Lovley, D.R. and Woodward, J.C., 1992. Consumption of freons CFC-11 and CFC-12 by anaerobic sediments and soils. *Environmental Science and Technology*, 26, 925–929.

MDEP (Maine Department of Environmental Protection), 2005. Report to the Joint Standing Committee on Natural Resources concerning the safe management of arsenic-treated wood wastes. Bureau of Remediation and Waste Management, 10 p. (Accessed on January 17, 2006 at: http://www.rookscommunications.com/nelma/january/Maine%20Arsenic%20Report.pdf).

Matthess, G., 1982. *The Properties of Groundwater*. John Willey & Sons, New York, 406pp.

Mazor, E., 1992. Reinterpretation of Cl-36 data: physical processes, hydraulic interconnections, and age estimates in groundwater systems. *Applied Geochemistry*, 7, 351–360.

Merlivat, L. and Jouzel, J., 1979. Global climatic interpretation of the deuterium-oxygen 18 relationship for precipitation. *Journal of Geophysical Research*, 84, 5029–5033.

Michel, R.L., 1989. Tritium deposition in the continental United States, 1953–1983. U.S. Geological Survey Water-Resources Investigations Report 89-4072, 46pp.

Michel, J., and Cothern, C.R., 1986. Predicting the occurrence of Ra-228 in ground water. *Health Physics*, 51(6), 715–721.

Miller, R.L., and Sutcliffe, H., Jr., 1985. Occurrence of natural radium-226 radioactivity in ground water of Sarasota County, Florida. *U.S. Geological Survey Water-Resources Investigations Report*, 84-4237, 34pp.

Milojevic, N., 1967. *Hidrogeologija* (*Hydrogeology*, in Serbian). Univerzitet u Beogradu, Zavod za izdavanje udzbenika Socijalisticke republike Srbije, Beograd, 379pp.

Mook, W.G., 1980, Carbon-14 in hydrogeological studies. In: Fritz, P., and Fontes, J. Ch., (eds.), Handbook of environmental isotope geochemistry, Volume 1—the terrestrial environment, A, chap. 2. Elsevier Scientific Publishing Co., New York, pp. 49–74.

Mose, D.G., Mushrush, G.W., and Chrosinak, C., 1990. Radioactive hazard of potable water in Virginia and Maryland. *Bulletin of Environmental Contamination and Toxicology*, 44(4), 508–513.

Mueller, D.K., et al., 1995. Nutrients in ground water and surface water of the United States—an analysis of data through 1992. U.S. Geological Survey Water Resources Investigations Report 95-4031, 1995.

National Research Council, 1978. Nitrates: an environmental assessment. National Academy of Sciences, Washington, D.C., 723pp.

NAVFAC (Naval Facilities Engineering Command), Guidance for environmental background analysis; volume 3: groundwater. User's guide UG-2059-ENV, Engineering Service Center, Port Hueneme, CA, 177pp.

Ovchinikov, M.A., 1955. *Obschaja gidrogeologia* (*General Hydrogeology*, in Russian). Nedra, Moskva.

Ovchinikov, M.A., 1970. *Gidrogeohimia* (Hydrogeochemistry, in Russian). Nedra, Moskva.

Palmer, C.D., and Puls, R.W., 1994. Natural attenuation of hexavalent chromium in groundwater and soils. EPA/540/5-94/505. U.S. EPA, Office of Solid Waste and Emergency Response and Office of Research and Development.

Pauling, L., and Pauling, P., 1975. *Chemistry*. W.H. Freeman, San Francisco, 767pp.

Parkhurst, D.L. and Appelo, C.A.J., 1999. User's guide to PHREEQC (Version 2)—a computer program for speciation, batch-reaction, one-dimensional transport, and inverse geochemical calculations. U.S. Geological Survey Water-Resources Investigations Report 99-4259, 310pp.

Perlmutter, N.M., et al., 1963. Movement of waterborne cadmium and hexavalent chromium wastes in South Farmingdale, Nassau County, Long Island, New York. In: Short papers in geology and hydrology, U.S. Geological Survey Professional Paper 475-C, pp. C179–C184.

Peryea, F.J., 1991. Phosphate-induced release of arsenic from soils contaminated with lead arsenate. *Soil Science Society America Journal*, 55, 1301–1306.

Peryea, F.J. and Kammereck, R., 1997. Phosphate-enhanced movement of arsenic out of lead arsenate-contaminated topsoil and through uncontaminated subsoil. *Water, Air, and Soil Pollution*, 93(1–4), 243–254.

Perry, E.C., Grundl, T., and Gilkeson, R.H., 1982. H, O, and S isotopic study of the ground water in the Cambrian-Ordovician aquifer system of northern Illinois, In: Isotope studies of hydrologic processes, Northern Illinois University Press, DeKalb, Illinois, pp. 35–45.

Piper, A.M., 1944. A graphic procedure in the geochemical interpretation of water analyses. *American Geophysical Union Transactions*, 25, 914–923.

Plummer, L.N., et al., 1993. Environmental tracers for age dating young ground water. In: Alley, W.M. (ed.), *Regional Ground-Water Quality*. Van Nostrand Reinhold, New York, pp. 255–294.

Poreda, R.J., Cerling, T.E., and Solomon, D.K., 1988. Tritium and helium isotopes as hydrologic tracers in shallow aquifers. *Journal of Hydrology*, 103, 1–9.

Puckett, L.J., 1994. Nonpoint and point sources of nitrogen in major watersheds of the United States. U.S. Geological Survey Water-Resources Investigations Report 94-4001, 9pp.

Rai, D., Sass, B.M., and Moore, D.A., 1987. Chromium (III) hydrolysis constants and solubility of chromium (III) hydroxide. *Inorganic Chemistry*, 26, 345–349.

Reilly, T., et al., 1994. The use of simulation and multiple environmental tracers to quantify ground-water flow in a shallow aquifer. *Water Resources Research*, 30(2), 421–433.

Rees, T.F., et al., 1995. Geohydrology, water quality, and nitrogen geochemistry in the saturated and unsaturated zones beneath various land uses, Riverside and San Bernardino Counties, California, 1991–1993. U.S. Geological Survey Water-Resources Investigations Report 94-4127, Sacramento, CA, 267pp.

Robertson, F.N., 1975. Hexavalent chromium in the groundwater in the Paradise Valley, Arizona, *Ground Water,* 13, 516–527.

Rowe, G.L. Jr., Shapiro, S.D., and Schlosser, P., 1999. Ground-water age and water-quality trends in a Buried-Valley aquifer, Dayton area, Southwestern Ohio. U.S. Geological Survey Water-Resources Investigations Report 99-4113. Columbus, OH, 81pp.

Rowland, E.S., 1991. Stratospheric ozone in the 21st century—the chlorofluorocarbon problem. *Environmental Science & Technology*, 25, 622–628.

Russell, A.D. and Thompson, G.M., 1983. Mechanisms leading to enrichment of the atmospheric fluorocarbons CCl3F and CCL2F2 in groundwater. *Water Resources Research*, 19, 57–60.

Sander, R, 1999. Compilation of Henry's law constants for inorganic and organic species of potential importance in environmental chemistry (Version 3). http://www.mpch-mainz.mpg.de/~sander/res/henry.html; accessed on January 6, 2006.

Scanlon, B.R., Healy, R.W., and Cook, P.G., 2002. Choosing appropriate techniques for quantifying groundwater recharge. *Hydrogeology Journal*, 10, 18–39.

Schlosser, P., et al., 1988. Tritium/^{3}He dating of shallow ground water. *Earth and Planetary Science Letters*, 89, 353–362.

Schlosser, P., et al., 1989. Tritiogenic ^{3}He in shallow ground water. *Earth and Planetary Science Letters*, 94, 245–256.

Senior, L.A., 1998. Radon-222 in the ground water of Chester County, Pennsylvania. U.S. Geological Survey Water-Resources Investigations Report 98-4169, 79pp.

Senior, L.A., and Vogel, K.L., 1995. Radium and radon in ground water in the Chickies quartzite, southeastern Pennsylvania. U.S. Geological Survey Water-Resources Investigations Report 92-4088, 145pp.

Sloto, R.A., 1994. Geology, hydrology, and ground-water quality of Chester County, Pennsylvania. *Chester County Water Resources Authority Water-Resource Report*, 2, 127 pp.

Shapiro, S.D., et al., 1998. Utilization of the ^{3}H–^{3}He dating technique under complex conditions to evaluate hydraulically stressed areas of a buried-valley aquifer. *Water Resources Research*, 34(5), 1165–1180.

Solomon, D.K., et al., 1993. A validation of the ^{3}H/^{3}He method for determining ground-water recharge. *Water Resources Research*, 29(9), 2951–2962.

Spalding, R.F., and Exner, M.E., 1993. Occurrence of nitrate in groundwater—a review. *Journal of Environmental Quality*, 22, 392–402.

Stanin, F.T., 2005. The transport and fate of chromium (VI) in the environment. In: Guertin, J., Jacobs, J.A., and Avakian, C.P. (eds.), *Chromium(VI) Handbook*, CRC Press, Boca Raton, FL, pp. 165–214.

Stollenwerk, K.G., and Colman, J.A., 2003. Natural remediation potential of arsenic-contaminated ground water. In: Welch, A.H., and Stollenwerk, K.G. (eds.), *Arsenic in Ground Water—Geochemistry and Occurrence*. Kluwer, Boston, MA, pp. 351–379.

Szabo, Z., and Zapecza, O.S., 1991. Geologic and geochemical factors controlling uranium, radium-226, and radon-222 in ground water, Newark Basin, New Jersey. In: Gundersen, L.C.S. and Wanty, R.B. (eds.), Field studies of radon in rocks, soils, and water, U.S. Geological Survey Bulletin 1971, pp. 243–266.

Szabo, Z., and dePaul, V.T., 1998. Radium-228 and radium-228 in shallow ground water, southern New Jersey. U.S. Geological Survey Fact Sheet FS-062-98, 6pp.

USGS, 2002. PHREEQCI—a graphical user interface to the geochemical model PHREEQC. U.S. Geological Survey Fact Sheet FS-031-02, 2pp.

USGS, 2004. Natural remediation of arsenic contaminated ground water associated with landfill leachate. U.S. Geological Survey Fact Sheet 2004-3057, 4pp.

USEPA, 1981. Radioactivity in drinking water. Glossary. EPA 570/9-81-002, U.S. Environmental Protection Agency, Health Effects Branch, Criteria and Standards Division, Office of Drinking Water.

USEPA, 1985. Nationwide occurrence of radon and other natural radioactivity in public water supplies. National Air and Environmental Radiation Laboratory, EPA 550-5-85-008. 230pp.

USEPA, 1991. National Primary Drinking Water Regulations; Radionuclides; Proposed Rule. Appendix A—Fundamentals of Radioactivity in Drinking Water. Federal Register. Vol. 56, No. 138, p. 33050.

USEPA and CDC (Centers for Disease Control), 1992. A citizen's guide to radon—the guide to protecting yourself and your family from radon (2nd ed.): EPA 402-K92-001, 15pp.

USEPA, 1992. Consumers guide to radon reduction—how to reduce radon levels in your home. EPA 402-K92-003, 17pp.

USEPA, 1993. Home buyer's and seller's guide to radon: EPA 402-R-93-003, 32pp.

USEPA, 2000. Radionuclides notice on data availability: technical support document. Targeting and Analysis Branch, Standards and Risk Management Division, Office of Ground Water and Drinking Water (www.epa.gov/safewater/rads/tsd.pdf).

USEPA, 2005a. http://www.clu-in.org/contaminantfocus/default.focus/sec/arsenic/.Accessed on November 21, 2005.

USEPA, 2005b. http://www.clu-in.org/contaminantfocus/default.focus/sec/chromium/. Accessed on November 21, 2005.

U.S. Salinity Laboratory Staff, 1954. Diagnosis and improvement of saline and alkali soils. *U.S. Department of Agriculture Handbook*, 60, 160 pp.

Warner, M.J., and Weiss, R.E., 1985. Solubilities of chlorofluorocarbons 11 and 12 in water and seawater. *Deep Sea Research*, 32, 1485–1497.

Wauchope, R.D. 1983. Uptake, translocation and phytotoxicity of arsenic in plants. In: Lederer, W.H. and Fensterheim, R.J. (eds.), *Arsenic: Industrial, Biomedical, Environmental Perspectives*, Van Nostrand Reinhold, New York, pp. 348–375.

Welch, A.H., and Lico, M.S., 1998. Factors controlling As and U in shallow ground water, southern Carson Desert, Nevada. *Applied Geochemistry*, 13(4), 521–539.

Welch, A.H., et al., 2000. Arsenic in ground water of the United States—occurrence and geochemistry. *Ground Water*, 38(4), 589–604.

Woolsen, E.A., Axley, J.H., and Kearney, P.C., 1971. The chemistry and phytotoxicity of arsenic in soils: I. Contaminated field soils. *Soil Science Society America Proceedings*, 35(6), 938–943.

Woolsen, E.A., Axley, J.H., and Kearney, P.C., 1973. The chemistry and phytotoxicity of arsenic in soils: II. Effects of time and phosphorous. *Soil Science Society America Proceedings*, 37(2), 254–259.

Zapecza, O.S. and Szabo, Z., 1986. Source of natural radioactivity in ground water in the Kirkwood-Cohansey aquifer system, southwestern Coastal Plain, New Jersey. In: Geological Society America, Abstracts with Programs, 21(2), 78 (abstract).

Zapecza, O. S., and Szabo, Z., 1987. Natural radioactivity in ground water—a review. U.S. Geological Survey National Water Summary 1986, Ground-Water Quality: Hydrologic Conditions and Events. U.S. Geological Survey Water Supply Paper 2325, pp. 50–57.

Zwirnmann, K.H., 1982. Nonpoint nitrate pollution of municipal water supply sources: issues of analysis and control. IIASA Collaborative Proceedings Series CP-82-S4, International Institute for Applied Systems Analysis, Laxenburg, Austria, 303pp.

5 Groundwater Contamination

5.1 REGULATORY OVERVIEW

In one of its useful fact sheets, the United States Environmental Protection Agency (USEPA) explains the history of drinking water treatment to an average citizen with the following narrative:

Ancient civilizations have always established themselves around water sources. While the importance of ample water quantity for drinking and other purposes was apparent to our ancestors, an understanding of drinking water quality was not well known or documented. Although historical records have long mentioned aesthetic problems (an unpleasant appearance, taste or smell) with regard to drinking water, it took thousands of years for people to recognize that their senses alone were not accurate judges of water quality.

Water treatment originally focused on improving the aesthetic qualities of drinking water. Methods to improve the taste and odor of drinking water were recorded as early as 4000 B.C. Ancient Sanskrit and Greek writings recommended water treatment methods such as filtering through charcoal, exposing to sunlight, boiling, and straining. Visible cloudiness (later termed turbidity) was the driving force behind the earliest water treatments, as many source waters contained particles that had an objectionable taste and appearance. To clarify water, the Egyptians reportedly used the chemical alum as early as 1500 B.C. to cause suspended particles to settle out of water. During the 1700s, filtration was established as an effective means of removing particles from water, although the degree of clarity achieved was not measurable at that time. By the early 1800s, slow sand filtration was beginning to be used regularly in Europe.

During the mid to late 1800s, scientists gained a greater understanding of the sources and effects of drinking water contaminants, especially those that were not visible to the naked eye. In 1855, epidemiologist Dr. John Snow proved that cholera was a waterborne disease by linking an outbreak of illness in London to a public well that was contaminated by sewage ('Broad Street well'). In the late 1880s, Louis Pasteur demonstrated the 'germ theory' of disease, which explained how microscopic organisms (microbes) could transmit disease through media like water.

During the late nineteenth and early twentieth centuries, concerns regarding drinking water quality continued to focus mostly on disease-causing microbes (pathogens) in public water supplies. Scientists discovered that turbidity was not only an aesthetic problem; particles in source water, such as fecal matter, could harbor pathogens. As a result, the design of most drinking water treatment systems built in the U.S. during the early 1900s was driven by the need to reduce turbidity, thereby removing microbial contaminants that were causing typhoid, dysentery, and cholera epidemics. To reduce turbidity, some water systems in U.S. cities (such as Philadelphia) began to use slow sand filtration. While filtration was a fairly effective treatment method for reducing turbidity, it was disinfectants like chlorine that played the largest role in reducing the number of waterborne disease outbreaks in the early 1900s. In 1908, chlorine was used for the first time as a primary disinfectant of drinking water in Jersey City, New Jersey. The use of other disinfectants such as ozone also began in Europe around this time, but were not employed in the U.S. until several decades later. (USEPA, 2000)

In the early 1900s, reacting to the large number of typhoid and other disease outbreaks, the United States and local governments began establishing public health programs to protect water supplies. The first were water pollution control programs, which focused on keeping surface water supplies safe by identifying and limiting sources of contamination. Early water pollution control programs concentrated on keeping raw sewage out of surface waters used for drinking water. Efforts were also made to site intakes used to collect drinking water upstream from sewage discharges (USEPA, 2003a).

At the same time, public agencies such as U.S. Geological Survey (USGS) were educating citizens, rural homeowners, and farmers on groundwater contamination issues as illustrated with Figure 5.1 through Figure 5.3, and the following excerpts from Fuller (1910):

Farms, which are generally remote from towns, cities, or other areas of congested population, seem to be almost ideally situated for obtaining pure and wholesome water. In reality, however, polluted water is exceedingly common on them (Figure 5.1 in this book) and typhoid-fever rates are usually greater in country districts than in cities. Typhoid fever is now almost universally believed to be transmitted solely through drink or food taken into the stomach, and is especially liable to be communicated by polluted waters obtained from shallow wells near spots where the discharges of typhoid patients have been thrown upon the ground and subsequently carried down through the soil and into the wells, and it is doubtless principally this fact that makes the disease so common in farming regions.

Protection of sink holes. It has already been pointed out that much of the water in limestones, the springs of which are frequently used for drinking and domestic purposes, enters the rock through open sink holes, into which in some places manure and other refuse have been dumped or sewage drained. Plate VI, A, (Figure 5.2 in this book) shows a small but continuous stream of sewage from a large college building discharging into a sink from which it finds its way to the underground water channels. Such practices are very dangerous. Cases of typhoid fever have resulted from drinking water from springs or wells, which have become polluted by such matter entering the sinks; and even where specific pollution is absent, undesirable slimes and rubbish often render the water highly objectionable. Instead of discharging refuse or sewage into sinks, every care should be taken to protect them against their access.

An example of danger from refuse of a more disgusting type is shown in Plate X, A. (Figure 5.3 in this book). Located in the middle of a well-traveled street, only a few inches above a gutter filled with paper and refuse, a part of which is sure to enter whenever a heavy rain occurs; open to the rain which washes into it from the steps leading down to it such dirt from the street as is brought in by the feet of the users; subject to the dipping of all sorts of more or less dirty buckets and utensils; receiving the underground drainage and presumably more or less sewage from the buildings on the slopes above; and containing in its bottom several inches of decaying paper and other refuse, this spring is on the whole one of the worst and most dangerously located sources of drinking water in the United States.

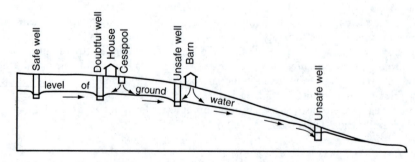

FIGURE 5.1 Diagram showing ordinary location of farm wells. (From Fuller, M.L., 1910. Underground waters for farm use. US Geological Survey Water-Supply Paper 255, Washington, DC.)

FIGURE 5.2 Pollution of ground water: sewage discharging into sinkhole. (From Fuller, M.L., 1910. Underground waters for farm use. US Geological Survey Water-Supply Paper 255, Washington, DC.)

Typhoid deaths dropped rapidly with the development of water quality and drinking water programs at the state and local levels in the early 1900s. In particular, chlorination and slow and rapid sand filtration had a significant impact. For example, in Albany, New York, prior to filtration of the public water supply in 1899, the typhoid death rate was 110 per 100,000. From 1900 to 1910, filtration was used and the typhoid death rate dropped to 20 per 100,000. In 1910, chlorination was introduced and the typhoid death rate for 1924 to 1929 dropped to zero (USEPA, 2003a).

In 1970, USEPA was established as an independent agency. A major factor in its establishment was an implicit understanding of the need for a federal enforcement authority. The drinking water, air pollution control, and solid waste programs were moved from the public health service to USEPA. Water pollution control moved from the Department of Interior to USEPA. A 1972 study detected 36 chemicals in treated water taken from treatment plants that drew water from the Mississippi River in Louisiana. In New Orleans, cancer was found to be present at higher rates in the population using the public water supply than in the population using private wells. These surveys raised concerns and prompted USEPA to

FIGURE 5.3 Spring in center of city street. (From Fuller, M.L., 1910. Underground waters for farm use. US Geological Survey Water-Supply Paper 255, Washington, DC.)

conduct a national survey to detail the quality of drinking water. The survey showed that drinking water was widely contaminated on a national scale, particularly with synthetic organic chemicals. Contamination was especially alarming in large cities. This survey raised concerns about drinking water in the public health community and in the general public. Increased concern and awareness of contamination of drinking water supplies prompted Congress to enact the Safe Drinking Water Act (SDWA) in 1974. The purpose of SDWA was to establish national enforceable standards for drinking water quality and to guarantee that water suppliers monitor water to ensure that it meets the national standards (USEPA, 2003a).

National Interim Drinking Water Regulations (renamed to National Primary Drinking Water Regulations in the 1986 SDWA amendments) established either the maximum concentration of pollutants allowed in drinking water or the minimum treatment required for water that is delivered to customers. Maximum contaminant level goal (MCLG) is the maximum level of a contaminant in drinking water at which no known or anticipated adverse health effects would occur. A maximum contaminant level (MCL) is an enforceable requirement. It is the maximum permissible level of a contaminant in water that can be delivered to any user of a public water system. An MCL is set as close to an MCLG as possible, taking into account the costs and benefits and feasible technologies. For some contaminants, there is not a reliable method that is economically and technologically feasible to measure the contaminant, particularly at low concentrations. In these cases, USEPA establishes a treatment technique (TT). The TT is an enforceable procedure or level of technological performance that public water systems must follow to ensure control of a contaminant. The hazardous waste (RCRA) and Superfund programs also use MCLs to define acceptable cleanup levels for contaminated water (USEPA, 2003a,b).

The public water system supervision (PWSS) program implements the National Primary Drinking Water Regulations. The PWSS program also implements programs to enhance water system operation. The underground injection control (UIC) program regulates discharges of fluids into underground sources of drinking water. The act provides USEPA with the authority to limit the concentrations of contaminants discharged by wells or to close wells that endanger drinking water sources. From 1974 until 1986, the UIC program was USEPA's major tool for protecting groundwater resources. Today, the injection into the subsurface is one of the primary means of disposing of liquid wastes. Nationwide, over 814,000 wells are used for disposal of hazardous and nonhazardous wastes. The sole source aquifer program provides special status to aquifers that represent the primary source of drinking water in a particular area. Such designation gives USEPA the ability to review and comment on federally funded projects, which results in project design and practices that focus greater attention on ground water protection.

The 1986 SDWA amendments were prescriptive and required USEPA to issue drinking water regulations for 83 specified contaminants by 1989. Further, USEPA was required to regulate an additional 25 contaminants (to be specified by USEPA) every 3 years and to designate a best available treatment technology for each contaminant regulated. As of 2006, USEPA has the total of 87 contaminants included in the list of National Primary Drinking Water Standards (see Chapter 5.3).

The 1986 amendments also initiated the groundwater protection program, including the wellhead protection program. The law specified that certain program activities, such as delineation, contaminant source inventory, and source management, be incorporated into state wellhead protection programs, which are approved by USEPA prior to implementation. In addition, the sole source aquifer demonstration program was added to the existing sole source aquifer provision. This program provides funding to identify and provide the special protections needed for sole source aquifers (USEPA, 2003a).

Increased monitoring requirements and monitoring for organic chemicals at a greater number of water systems led to increased detection of chemicals and the identification of potential problems from the widespread presence of organic chemicals. In addition, increased monitoring detected previously unidentified microbial problems. The increased detection of previously unknown water system contaminant problems created a need for water system operators and states to develop risk communication skills to inform the public of impacts of contaminants on their health. Increased knowledge of *Giardia* improved methods for detecting the pathogen, and continuing outbreaks of the disease prompted tightened requirements for surface water treatment. This included lowered turbidity standards, disinfectant contact time (CT) calculations, and strict criteria to avoid filtration.

Along with increased treatment requirements for surface water systems, some groundwater supplies were recognized as providing water of essentially surface water quality. These sources are recharged by surface water to the extent that pathogens, such as *Giardia* cysts, can contaminate the water. These sources are known as groundwater under the direct influence (of surface water) or GWUDI. Identification of GWUDI sources and regulation as surface water systems was required. Public notification requirements increased the communication between water systems and consumers, further increasing awareness of contamination of drinking water. Public notification requirements were strictly prescribed and included broadcast and printed notices depending on the severity of the contamination problem. More stringent *coliform monitoring* requirements in the 1986 amendments increased the frequency of coliform detection. Increased requirements for follow-up monitoring after initial detection revealed even more problems. This led to greater awareness of the inadequacy of some sources of water, even after treatment (USEPA, 2003a,b).

Among the key provisions, the 1996 SDWA amendments authorized a drinking water state revolving loan fund (DWSRF) program to help public water systems finance projects needed to comply with SDWA rules. The amendments also established a process for selecting contaminants for regulation based on health risk and occurrence, gave USEPA some added flexibility to consider costs and benefits in setting most new standards, and established schedules for regulating certain contaminants (such as *Cryptosporidium*, arsenic, and radon). The law added several provisions aimed at building the capacity of water systems (especially small systems) to comply with SDWA regulations, and it imposed many new requirements on the states including programs for source water assessment, operator certification and training, and compliance capacity development. The amendments also required that community water suppliers provide customers with annual consumer confidence reports that provide information on contaminants found in the local drinking water (Tiemann, 2005).

Two programs established by USEPA in the 1980s have had the largest impact on the rapid development of hydrogeology and many scientific and engineering disciplines related to groundwater remediation. Comprehensive Environmental Response, Compensation, and Liability Act (CERCLA or Superfund), provides for the cleanup of inactive and abandoned waste sites. Resource Conservation and Recovery Act (RCRA) regulates hazardous waste generation, storage, transportation, treatment, and disposal, including remediation of contaminated groundwater at active industrial and other facilities (USEPA, 1986, 1994).

5.1.1 Health Effects

Various substances in drinking water can adversely affect or cause disease in humans, animals, and plants. These effects are known as toxic effects. Below are the general categories of toxicity, based on the organs or systems in the body affected (USEPA, 2003a):

- Gastrointestinal: affecting the stomach and intestines
- Hepatic: affecting the liver
- Renal: affecting the kidneys
- Cardiovascular or hematological: affecting the heart, circulatory system, or blood
- Neurological: affecting the brain, spinal cord, and nervous system (in nonhuman animals, behavior changes can result in lower reproductive success and increased susceptibility to predation)
- Respiratory: affecting the nose, trachea, and lungs or the breathing apparatus of aquatic organisms
- Dermatological: affecting the skin and eyes
- Reproductive or developmental: affecting the ovaries or testes, or causing lower fertility, birth defects, or miscarriages. This includes contaminants with genotoxic effects, i.e., capable of altering deoxyribonucleic acid (DNA). This can have mutagenic effects (changes in the genetic materials causing cells to misfunction), which can cause cancer or birth defects (teratogens)

Substances that cause cancer are known as carcinogens and are classified as such based on evidence gathered in studies. USEPA has a three-category approach to classifying compounds as carcinogenic, based on evidence of carcinogenicity, pharmacokinetics (the absorption, distribution, metabolism, and excretion of substances from the body), potency, and exposure:

1. Category I compounds are carcinogens.
2. Category II compounds exhibit limited evidence of carcinogenic endpoints and also exhibit noncarcinogenic endpoints.
3. Category III compounds are noncarcinogenic.

The effects a contaminant has on various life forms depend not only on its potency and the exposure pathway, but also on the temporal pattern of exposure. Short-term exposure (minutes to hours) is referred to as acute. Longer term exposure (days, weeks, months, years) is referred to as chronic. An example of an acute exposure with widespread health effects and a very high associated cost of mitigation is ingestion of water from Milwaukee's river contaminated by *Cryptosporidium*, which sickened hundreds of people in 1993 and required the city to upgrade its water system. The cost of the system improvements, along with costs to the water utility, city, and health department associated with the disease outbreak were $89 million (USEPA, 2003a).

The constancy of exposure is also a factor in how the exposure affects an organism. For instance, the effects of 7 days of exposure may differ depending on whether the exposure was on seven consecutive days or seven days spread over a month, a year, or several years. In addition, some organisms may be more susceptible to the effects of contaminants. If evidence shows that a specific subpopulation is more sensitive to a contaminant than the population at large, then safe exposure levels are based on that population. If no such scientific evidence exists, pollution standards are based on the group with the highest exposure level. Some commonly identified sensitive subpopulations include infants and children, the elderly, pregnant and lactating women, and immunocompromised individuals (USEPA, 2003a).

5.2 SOURCES OF CONTAMINATION AND GROUNDWATER CONTAMINANTS

In the broadest sense, all sources of groundwater contamination and contaminants themselves can be grouped into two major categories: naturally occurring and artificial (human-made). Although some natural contaminants, such as arsenic and radionuclides, may have significant local or regional impacts on groundwater supplies depending on geology, numerous human-made sources and contaminants have disproportionately greater negative effects on quality of groundwater resources. Arguably, almost every human activity has a potential to impact groundwater to some extent. An exponential advancement of analytical laboratory techniques in the last decade or so has demonstrated that many synthetic organic chemicals are widely distributed in the environment, including in groundwater (Hamilton et al., 2004), and that a considerable number of them can now be found in human tissue and organs of people living across the world. At the same time, a similar advancement in water treatment technologies, understanding of contaminant fate and transport, and groundwater remediation technologies is (arguably) making this fact somewhat less alarming. Strongly related to the ever-increasing public awareness of the environmental pollution is a very rapid growth in consumption of bottled drinking water, also across the world. Many consumers are ready to pay premium for brands marketed as "pure spring water" or "water coming from deep pristine aquifers," so that major multinational corporations are frantically looking for groundwater resources that can be marketed as such. In general, there is still a lot of truth in the following statement, very much appreciated by many hydrogeologists: groundwater in general is much less vulnerable to contamination than surface water, it is of better quality and thus requires much less investment in water supply development. It is, however, also true that, in general, it takes more time and it is more difficult to "cleanup" groundwater than surface water once it becomes contaminated.

For the most part, groundwater contamination results from the following activities, with the note that the list is far from all-inclusive:

- Misuse and improper disposal of liquid and solid wastes and chemicals at commercial, industrial, agricultural, and governmental facilities, and in households
- Illegal dumping or abandonment of household, commercial, or industrial chemicals

- Accidental spilling of chemicals from trucks, railways, aircrafts, handling facilities, and storage tanks
- Use of road salt in winter
- Land application (disposal of wastewater treatment effluent and sludge through infiltration basins, and on land and farmland)
- Urban runoff from parking lots, streets, and construction sites
- Improper location, design, construction, operation, or maintenance of agricultural, residential, municipal, commercial, and industrial drinking-water wells and liquid and solid waste disposal facilities
- Application of fertilizers, pesticides, and insecticides in agriculture, on household lawns and gardens, and on golf courses
- Livestock feeding operations
- Atmospheric pollutants, such as airborne sulfur and nitrogen compounds, which are created by smoke, flue dust, aerosols, and automobile emissions, fall as acid rain, and percolate through the soil (modified from: www.wrds.uwyo.edu/wrds/deq/whp/; accessed November 12, 2005)

Contaminants can reach groundwater from activities occurring on the land surface, such as industrial waste storage, from sources below the land surface but above the water table, such as septic systems, from structures beneath the water table, such as wells, and from contaminated artificial recharge water.

5.2.1 Point and Nonpoint Sources

Probably the most frequently used definition of a point source of groundwater contamination is that it occupies a small (limited) area at the land surface, or in the shallow subsurface, as in the case of a leaky underground storage tank (UST). Such source, by the same definition, creates contaminant plume of a limited extent. However, some people may have a very different understanding of the word "limited" in this context, which may cause confusion. For example, a rather large complex of closely spaced, unlined municipal landfill cells can hardly be equated in size to a gasoline station with a leaky UST. However, either may create contaminant plumes which, given enough monitoring wells, should be easily definable in terms of their three-dimensional spatial extent (horizontal and vertical extent of the plume). At the same time, an agricultural parcel of a size similar to that of a large landfill, with a continuing application of fertilizers and pesticides, may also create contaminant plume of similar spatial characteristics to the one originating from the landfill. However, according to most definitions, the landfill plume, which in reality may consist of multiple small plumes of different contaminants, emanating from multiple small sources such as buried drums of "nasty stuff," would be collectively described as the point-source plume, while the agricultural parcel would be called a nonpoint source of groundwater contamination. The preceding discussion illustrates the importance of both the scale of observation and the nature of contaminant introduction into the subsurface, when defining point and nonpoint sources. In general, nonpoint sources refer to a widespread introduction of potential contaminants into the subsurface such as due to application of fertilizers over large agricultural areas. Point sources result from both unintentional and intentional hazardous waste disposal practices, spills, leaks, or otherwise limited in extent introduction of contaminants into the subsurface.

Many RCRA and Superfund sites are examples of multiple point sources of groundwater contamination. These sources may form individual plumes of individual contaminants, individual plumes of mixed contaminants from identifiable sources, or, in the most complicated cases, commingled (merged) plumes of various contaminants from multiple sources, some of which are not easily, or not at all identifiable. Figure 5.4 shows just a few possible cases from the plume world. Sites on military installations, large industrial complexes, and multiple chemical manufacturing plants, are likely to have groundwater contaminated by multiple constituents, which may be distributed at various depths in the underlying aquifers and form plumes with complicated shapes. Complex sites of groundwater contamination are often a nightmare for groundwater professionals trying to characterize possible contaminant sources and "attach" to them their own plumes. This, however, is the favorite topic of attorneys working for various potentially responsible parties (PRPs). Heavy involvement of attorneys in groundwater contamination and remediation issues is understandable as costs associated with groundwater remediation may be astronomical, and the question of who is responsible for the plume becomes of utmost importance. In addition, remediation of certain contaminants in certain hydrogeologic settings may not be practically or technically feasible, which is often very hard to convey to various stakeholders. Karst and highly fractured aquifers are examples of such settings. Figure 5.5 illustrates this fact by showing the unique nature of heterogeneous karst porosity. In this particular case, the waste visible in the photograph was eventually removed by dedicated spelunkers; in the vast majority of potentially similar cases, however, even the most dedicated and skilled spelunkers would not be able to reach all possible cavities, conduits, and open fractures present in a karst aquifer.

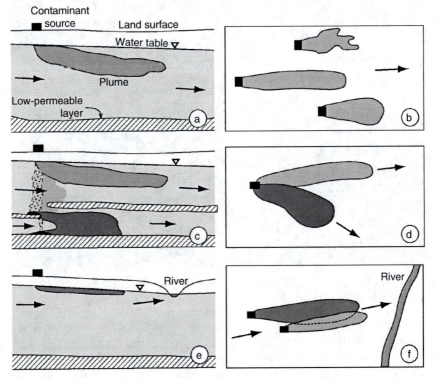

FIGURE 5.4 Some possible cases of dissolved contaminant plumes originating from point sources; cross section views on the left and map views on the right. a: Plume diving due to significant recharge from land surface; b: Various shapes of plumes caused by aquifer heterogeneities; c: Plumes at different depths originating from primary and secondary sources (deeper plume in this case is caused by DNAPL penetrating deep into the saturated zone; d: Plumes at different depths may flow in different directions when separated by aquitards; e: Shallow, nondiving plume in the absence of significant recharge, flowing toward a discharge zone; f: Plumes from multiple sources may commingle (mix, overlap) as they flow in the same general directions toward a common discharge zone.

FIGURE 5.5 Spelunker examines trash around a flowstone waterfall at the bottom of the entrance to Midnight Cave near Austin, Texas, on November 20, 1993. The trash includes household garbage, used oil filters, corroded 55-gallon drums, glass pesticide bottles, partially filled turpentine cans, and automobile parts. Note the trash on the higher ledges of the cave. (Photo: Nico M. Hauwert; from Hauwert, N.M. and Vickers, S. 1994. Barton Springs/Edwards Aquifer: Hydrogeology and groundwater quality. Barton Springs/Edwards Aquifer Conservation District, Austin, Texas, 91 p. + Appendices.)

Table 5.1 and Table 5.2 show the results of an analysis of contaminant sources and wastes most commonly found at Superfund sites (Reisch and Bearden, 1997). Soil contamination occurs at 80% and groundwater contamination occurs at nearly 79% of Superfund sites with records of decisions (RODs) yet to be implemented. Manufacturing operations contribute the largest share of the waste, while mining activities contribute the smallest portion. Liquid waste is present at 92.4% of all Superfund sites, solid waste at 58.3%, and sludge at 49.2%.

TABLE 5.1
Common Sources of Waste at Superfund Sites

Source of Waste	Share of Waste (%)
Manufacturing operations	38.90 %
Municipal landfills	16.50 %
Recyclers	8.50 %
Industrial landfills	6.50 %
Department of Energy and Department of Defense	5.00 %
Mining	2.00 %
Other sources	22.50 %

Source: Reisch, M. and D.M. Bearden, 1997. Superfund fact book. National Council for Science and the Environment, Congressional Research Service Reports, 97-312 ENR, 33 p. http://ncseonline.org/NLE/CRSreports/Waste/ (accessed on January 24, 2006).

Potential impacts on large public water-supply wells from various contaminant sources are obviously of most concern due to often extensive capture zones of high-yielding wells. The screened or open intervals of such wells are commonly from tens to hundreds of feet in length; therefore, water from these wells is generally a mixture of waters of different ages that enter the well at different depths and are associated with different potential sources of contamination, both point and nonpoint (Eberts et al., 2005). For example, Figure 5.6 and Figure 5.7 illustrate a case where water entering the well may be coming from two distinct areas: (1) water from the urban area may contain contaminants from point sources, such as chlorinated solvents from dry-cleaners and machine shops, and will enter the top portion of the well screens; (2) water that has traveled from the more distant agricultural area, where recharge water may contain contaminants such as agricultural pesticides and fertilizers, will enter the bottom portion of the well screens.

TABLE 5.2
Types of Contaminants Commonly Found at Superfund Sites

Contaminant	Frequency of Occurrence (%)
Organic chemicals	71.40 %
Metals	64.30 %
Oily wastes	35.10 %
Inorganic chemicals	30.90 %
Municipal waste	27.30 %
Acids or bases	24.50 %
PCBs (Polychlorinated biphenyls)	20.30 %
Pesticides or herbicides	18.40 %
Paints or pigments	17.70 %
Solvents	6.30 %

Source: Reisch, M. and D.M. Bearden, 1997. Superfund fact book. National Council for Science and the Environment, Congressional Research Service Reports, 97-312 ENR, 33 p. http://ncseonline.org/NLE/CRSreports/Waste/ (accessed on January 24, 2006).

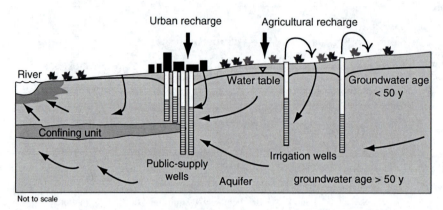

FIGURE 5.6 An aquifer system and public-water system in an urban setting. The water entering the well screens of the public-supply wells is of different ages and from different areas because of their long screened intervals, which commonly make public-supply wells vulnerable to contamination from multiple sources. In this example, sources of contaminants may include those associated with urban and agricultural land-use activities. Aquifer materials may also serve as sources of natural contaminants such as arsenic. (From Eberts, S.M., et al., 2005. Assessing the vulnerability of public-supply wells to contamination from urban, agricultural, and natural sources. US Geological Survey Fact Sheet 2005–3022, 4 p.)

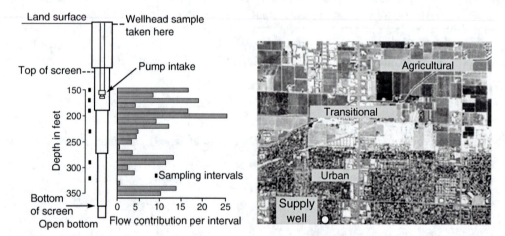

FIGURE 5.7 Inflow at different depths within a public-supply well. The areal photo shows an approximately 63-square-mile area near the well. Water entering the well screen is associated with different potential sources of contaminants because of the different land-use activities in the areas contributing recharge to various intervals along the well screen, as well as the different aquifer materials through which water flows between the recharge areas and the well. The amount of contamination that might be contributed by any given interval is related to the volume of water that flows into the well along the interval and the concentration of any associated contaminants. Depth-dependent samples are a composite of all intervals beneath the sampling point; these samples are being analyzed for chemical quality and groundwater age and then compared to samples collected from the wellhead. (From Eberts, S.M., et al., 2005. Assessing the vulnerability of public-supply wells to contamination from urban, agricultural, and natural sources. US Geological Survey Fact Sheet 2005–3022, 4 p.)

In some cases, the same contaminant may originate from very different sources, both point and nonpoint, impact the same aquifer, and cause real headache even to seasoned environmental attorneys used to various improbable cases. Nitrogen in different forms is the best example of such a contaminant as discussed by Seiler of the USGS (1996): potential sources of the nitrogen in groundwater flow systems in Nevada include domestic sewage, animal wastes, fertilizers, natural sources, and munitions constituents. Partially based on a detail study of sources of nitrogen in groundwater underlying valleys in Washoe County, Nevada, Seiler provides a very informative general overview of nitrogen presence in groundwater, excerpts of which are included throughout the following discussion.

5.2.1.1 Domestic Sewage and Animal Wastes

Domestic sewage in sparsely populated areas of the United States is disposed of primarily in on-site septic systems. In 1980, 20.9 million residences (about 24% of the total in the United States) disposed of about 4 million acre-feet of domestic sewage in on-site septic systems (Reneau et al., 1989). In rural areas of Nevada, central waste-treatment plants are not used because of their cost. Rather, domestic sewage is treated and disposed of in on-site septic systems. Inherent in this method is the discharge of effluent to the local groundwater. To avoid contamination problems in an area, treated sewage effluent can be removed from a basin and discharged elsewhere if wastewater treatment is centralized. Unfortunately, removal is not possible with on-site septic systems, and even properly designed and constructed on-site septic systems frequently cause nitrate concentrations to exceed the MCL in the underlying groundwater (Wilhelm et al., 1994).

Because of the significant amounts of nitrogen excreted by large animals, confined animal operations have been identified as sources of nitrogen contamination in many areas of the United States. Grazing ruminants excrete 75%–90% of the nitrogen they ingest, mostly as urea in urine (Ball and Ryden, 1984). Van Vuren (1949) states that the average cow, horse, and pig excretes 156, 128, and 150 pounds of nitrogen per year, respectively (Seiler, 1996). Compared to domestic sewage, wastes produced by one horse contain twice as much nitrogen as that produced by a family of four during one year (Hantzsche and Finnemore, 1992). However, because the ammonia in animal wastes is exposed to atmospheric effects such as wind and drying, the loss of ammonia through volatilization is undoubtedly much greater from animal wastes than from human wastes, which is disposed of in underground septic systems and is thus isolated from atmospheric effects. Therefore, even though domestic animals produce more nitrogen than humans do, probably more animal nitrogen than human nitrogen is lost to the air.

Total-nitrogen concentrations in septic-tank effluent range from 25 mg/L to as much as 100 mg/L, and the average is in the range of 35–45 mg/L (USEPA, 1980), of which about 75% is ammonium and 25% is organic. Wilhelm et al. (1994) report that nitrate concentrations in the effluent below a septic field can be two to seven times the MCL, and distinct plumes of nitrate-contaminated groundwater may extend from the septic system. On the basis of detailed analytical data available for several studies in Wisconsin, Seiler (1996) estimates that septic systems contribution of nitrogen to groundwater in the East Lemmon Subarea of Washoe County, Nevada, is between 16,500 to 42,000 kg (18 to 46 tons) of nitrogen annually.

In cattle feeding lots, animal wastes may lose much of the nitrogen by ammonia volatilization, particularly in corrals that are not subject to water application; water can transport the nitrogen to the subsurface before substantial volatilization has occurred. The amount of nitrate from animal wastes that percolates to the groundwater depends on the amount of nitrate formed from the wastes, the infiltration rate, the frequency of manure removal, the animal density, the soil texture, and the ambient temperature (National Academy of Sciences, 1978).

5.2.1.2 Fertilizers and Natural Sources

Nitrogen fertilizers are obvious potential sources of groundwater contamination with nitrogen. The amount of nitrogen contributed depends on several factors, including the type and amount of fertilizer applied, the acidity (pH) of the soil, the air temperature at the time of application, and the amount of water applied after fertilization. Hantzsche and Finnemore (1992) estimate that the nitrogen load to groundwater from fertilization of lawns in residential subdivisions and rural communities is in the range of 0.5–1.5 kg (1.1 to 3.3 pounds) per developed acre per year.

The decay of natural organic material in the ground can contribute substantial amounts of nitrogen to groundwater. For example, in the late 1960s in West Central Texas several cattle died from drinking groundwater containing high concentrations of nitrate; the source of the nitrate was determined to be naturally occurring organic material in the soil (Kreitler and Jones, 1975, from Seiler, 1996). The average nitrate concentration (as NO_3) for 230 wells was 250 mg/L, and the highest concentration exceeded 3000 mg/L. Native vegetation, which included a nitrogen-fixing plant, was destroyed by plowing of the soil for dryland farming. This increased oxygen delivery to the soil, and the nitrate causing the contamination was formed by oxidation of the naturally occurring organic material in the soil.

5.2.1.3 Munitions

Dynamite and other explosives contain nitrogen, and breakdown of the explosives can contribute nitrogen to the groundwater. Nitrogen contamination of the groundwater has been found at facilities in Nevada where munitions have been processed (Van Denburgh et al., 1993).

5.2.1.4 Identification of Nitrogen Sources

A variety of chemical constituents and groundwater characteristics may be indicative of potential sources of nitrogen, sometimes definitively as described by case studies in Seiler (1996). Stable isotopes of nitrogen and oxygen have been used for this purpose as well, and are particularly useful when combined with other hydrochemical methods. Probably the simplest and analytically the least expensive chemical method is the use of major cations and anions to differentiate the contaminant plumes from native groundwater. Robertson et al. (1991) successfully characterized the chemical composition of two contaminant plumes from on-site septic systems near Cambridge, Ontario, Canada, by using concentrations of sodium, calcium, potassium, and chloride, which were all elevated in the plumes compared with background levels. Of all the constituents they measured, sodium was the best indicator of contamination from the septic system because its concentrations in the plume were 10–20 times greater than the background level. Total concentration of dissolved solids (TDS) and its surrogate, specific conductance, are also simple indicators of contamination. Domestic water use increases the TDS concentration through the addition of salts, detergents, and other household byproducts. Aerobic oxidation of domestic-waste effluent causes acidity, which can dissolve minerals in the subsurface and hence can increase the TDS concentration of the water. Over a 20-year period, this process could dissolve 225–900 kg (about 500 to 2000 pounds) of $CaCO_3$ below a typical septic system (Wilhelm et al., 1994). Examples of more exotic indicators that have been, or have potential to be used for plume fingerprinting, are given below.

Methylene blue active substances (MBAS) are anionic surfactants that commonly are added to fertilizers, to pesticides, and to some household consumer products (such as detergents, shampoos, and toothpastes). One major type of MBAS used in detergents is branched-chain alkylaryl sulfonates (ABS). Because ABS are resistant to biodegradation, their use in detergents was discontinued in the 1960s. Another type of MBAS, linear alkylaryl sulfonates (LAS), are biodegraded more easily but still persist in groundwater. On Cape Cod, Thurman et al. (1984) could identify the location of the sewage-contaminant plume by the presence of MBAS and could distinguish contaminants discharged before the mid-1960s by the presence of ABS. Because MBAS are used in fertilizers also, they may not be suitable to differentiate between fertilizer and sewage sources of nitrogen.

It may be possible to identify in wastewater some chemicals derived from products that are consumed by humans but not by animals. Beverages containing caffeine (such as coffee, tea, and soft drinks) can be disposed of directly into septic systems, and some caffeine may pass unmetabolized through the digestive system. Caffeine was found to be a significant component of wastewater effluent that has undergone secondary treatment (Sievers et al., 1977). Other consumer chemicals, such as aspirin, ibuprofen, or nicotine (from tobacco products), also may be identifiable in domestic sewage.

One group of chemicals that might be used as identifiers of animal wastes are veterinary pharmaceuticals used for control of internal parasites such as worms and flies. For these chemicals to be useful as identifiers, they must migrate with the animal wastes and not be bound to soil particles. Although the potential exists for veterinary pharmaceuticals to be useful identifiers of animal wastes, the method has not been reported in the literature by 1996 (Seiler, 1996).

Because of its widespread use, boron may generally be a good tracer of sewage in ground water (Thurman et al., 1984). Boron has numerous industrial and household uses—as a cleaning agent; as a water softener in washing powders; and in insecticidal, antibacterial, and antifungal agents. Boron concentrations in effluent at the Cape Cod site were about 500 mg/L and were less than 30 mg/L in the native groundwater.

5.2.1.5   Use of Stable Isotopes

As explained in Section 4.2.5, stable isotopes are nonradioactive forms of an element; isotopes of a given element have the same number of protons but a different number of neutrons. The nitrogen isotopes ^{15}N and ^{14}N constitute an isotope pair where the lighter isotope ^{14}N is more abundant; there is one atom of ^{15}N per 273 atoms of ^{14}N in the atmosphere (Drever, 1988; from Seiler, 1996). Measurements of the ratio of the abundance of the heavier isotope to that of the lighter isotope in a substance can provide useful information because the slight differences in the mass of the isotopes cause slight differences in their behavior. Isotope values for nitrogen and their elements are presented in the delta notation (Seiler, 1996):

$$\delta^{15}N = \left[\frac{(^{15}N/^{14}N)_{sample}}{(^{15}N/^{14}N)_{air}} - 1\right] \times 1000 \qquad (5.1)$$

Several steps in the nitrogen cycle result in fractionation. In other words, changes in the stable-isotope composition of a nitrogen-containing chemical occur as a result of chemical reactions. Isotopic effects, caused by slight differences in the mass of two isotopes, tend to cause the heavier isotope to remain in the starting material of a chemical reaction. Denitrification, for example, causes the nitrate of the starting material to become isotopically heavier. Volatilization of ammonia results in the lighter isotope preferentially being lost to the atmosphere, and the ammonia that remains behind becomes isotopically heavier. These isotopic effects mean that, depending on its origin, the same compound may have different isotopic compositions. For stable isotopes to provide a useful tool in identifying sources of nitrogen contamination, the isotopic composition of the potential source materials must be distinguishable. The major potential sources of nitrogen contamination in the hydrosphere commonly have characteristic $^{15}N/^{14}N$ ratios. Typical $\delta^{15}N$ values for important sources of nitrogen contamination, given in permil (parts per million) are as follows (Heaton, 1986, from Seiler, 1996): (1) for commercial fertilizer between –4 and +4; (2) for animal wastes >10; (3) for precipitation −3; and (4) for organic nitrogen in soil between +4 and +9.

Even when the stable-isotope composition of the source material is known, what reactions occur after its deposition and how they affect its isotopic composition also must be known if the source of nitrate in groundwater is to be identified. Because fractionation after deposition blurs the isotopic signatures of the source materials, the use of ^{15}N data alone may not be sufficient to differentiate among sources. Predicting isotopic changes in nature is difficult; Heaton (1986) concluded that the unpredictable magnitude of fractionation in natural settings restricts the use of nitrogen isotope data to semiquantitative interpretations.

5.2.2 Microbiological Contaminants

Microorganisms that can cause disease in humans, animals, and plants are called pathogens. They can cause an adverse effect after an acute (short-term) exposure such as ingestion of just one glass of water. Pathogens include bacteria, viruses, or parasites and are found in sewage, in runoff from animal farms or rural areas populated with domestic and wild animals, and in water used for drinking and swimming. Fish and shellfish contaminated by pathogens, or the contaminated water itself, can cause serious illnesses (USEPA, 2003a). A virus is the smallest form of microorganism capable of causing disease. A virus of fecal origin that is infectious to humans by waterborne transmission is of special concern for drinking water regulators. Many different waterborne viruses can cause gastroenteritis, including Norwalk virus, and a group of Norwalk-like viruses (Kaplan et al., 1982). Bacteria are microscopic living organisms usually consisting of a single cell. Waterborne disease-causing bacteria include *E. coli* and *Shigella*. *Protozoa* or parasites are also single cell organisms. Examples include *Giardia lamblia* and *Cryptosporidium*.

Giardia (see photograph in Figure 5.8) was only recognized as a human pathogen capable of causing waterborne disease outbreaks in the late 1970s. Its occurrence in relatively pristine

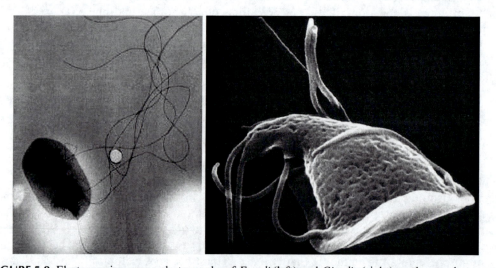

FIGURE 5.8 Electron microscope photographs of *E. coli* (left) and *Giardia* (right), pathogens known to cause waterborne disease. (Photo of *E. coli* 0157:H7: Center for Disease Control; photo of *Giardia* trophozoite: S. Erlandsen; USEPA, 2003a. Overview of the Clean Water Act and the Safe Drinking Water Act. http://www.epa.gov/OGWDW/dwa/electronic/ematerials/ [accessed September 12, 2005]; USEPA, 2003c. Regulating microbial contamination; unique challenge, unique approach. http://www.epa.gov/OGWDW/dwa/electronic/ematerials/ [accessed September 12, 2005].)

water as well as wastewater treatment plant effluent called into question water-system definitions of "pristine" water sources. This parasite, now recognized as one of the most common causes of waterborne disease in humans in the United States, is found in every region of the United States and throughout the world. In 1995, outbreaks in Alaska and New York were caused by *Giardia*. The outbreak of giardiasis in Alaska affected ten people, and was associated with untreated surface water. The outbreak in New York affected an estimated 1449 people, and was associated with surface water that was both chlorinated and filtered. The symptoms of giardiasis include diarrhea, bloating, excessive gas, and malaise.

Cryptosporidium (often called "crypto"), which cannot be seen without a very powerful microscope, is so small that over 10,000 of them would fit on the period at the end of this sentence (Tiemann, 1996). The infectious dose for crypto is less than 10 organisms and, presumably, one organism can initiate an infection. As late as 1976, it was not known to cause disease in humans. In 1993, 403,000 people in Milwaukee, Wisconsin, became ill with diarrhea after drinking water contaminated with the parasite, resulting in the largest water-borne disease outbreak ever documented in the United States (Tiemann, 1996). For the 2 y period 1993–1994, the Centers for Disease Control reported that 17 states identified 30 disease outbreaks associated with drinking water. Since then attention was focused on determining and reducing the risk of cryptosporidiosis from public water supplies. Crypto is commonly found in lakes and rivers and is highly resistant to disinfection. Groundwater under influence of surface water, and groundwater in highly transmissive karst and gravel aquifers, may also be susceptible to contamination with parasites such as *Giardia* and *Cryptosporidium*. People with severely weakened immune systems are likely to have more severe and more persistent symptoms than healthy individuals.

As already mentioned, viruses are among the smallest of the disease-causing microorganisms found in the aquatic environment. More than 120 different types of potentially harmful enteric viruses are excreted in human feces, and are widely distributed in type and number in domestic sewage, agricultural wastes, and septic drainage systems (Gerba, 1988; from Banks and Batiggelli, 2002). Many of these viruses are stable in natural waters and have long survival times with half-lives ranging from weeks to months (Gerba, 1999). Because they may cause disease even when just a few virus particles are ingested, low levels of environmental contamination may affect water consumers. From 1971 to 1979, approximately 57,974 people in the United States were affected by outbreaks of waterborne pathogens (Craun, 1986; from Banks and Batiggelli, 2002). Outbreaks of waterborne disease attributed to enteric viruses are poorly documented, even though viruses are commonplace in natural waters contaminated with human feces. Illnesses in humans caused by waterborne viruses range from severe infections such as myocarditis, hepatitis, diabetes, and paralysis to relatively mild conditions such as self-limiting gastroenteritis. It has not been possible to identify the etiologic agent or agents responsible for community illness in approximately half of the reported water-borne outbreaks because the isolation and identification of the causative agent was either unsuccessful or not attempted (Craun, 1986; Sobsey, 1989; from Banks and Batiggelli, 2002). Additional analyses indicate that caliciviruses such as the Norwalk virus and other enteric viruses may be responsible for as much as 60% of the reported waterborne outbreaks of gastroenteritis as the clinical features of the cases in many of these epidemics are consistent with viral infections, and bacterial pathogens were ruled out as disease agents (Keswick and Gerba, 1980; Kaplan et al., 1982; from Banks and Batiggelli, 2002).

There are no viruses currently included in the national primary drinking water standards issued by USEPA, and studies of possible groundwater contamination with viruses are very rare. In 1999, the USGS, in cooperation with the Maryland Department of the Environment and the Wisconsin State Laboratory of Hygiene, began to assess the occurrence and distribution of viral contamination in small (less than 10,000 gallons per day) public water-supply wells in Baltimore and Harford Counties, Maryland. Ninety sites were selected based on a method that used an inclusion probability proportional to an arbitrary weight for each county. An additional site, selected randomly in Baltimore County, also was sampled. Forty-six sampling sites were in Baltimore County, and 45 sampling sites were in Harford County. None of the 91 environmental samples contained culturable viruses; however, viral ribonucleic acid for rotavirus was detected at one site in Harford County. These data indicate that viruses are not frequently found in small public water-supply wells in the Piedmont Physiographic Province of Baltimore and Harford counties, Maryland (Banks and Batiggelli, 2002). The potential for viral contamination of groundwater is significantly elevated in highly permeable and transmissive aquifers, where resident time of newly infiltrated, contaminated water, before it discharges from the aquifer (either via water wells or springs) is short. Unconfined karst and gravel aquifers are particularly vulnerable to viral (bacterial) contamination.

5.2.3 INORGANIC CONTAMINANTS

Inorganic substances, which are naturally occurring in all groundwaters to some degree, sometimes can have elevated concentrations harmful to human health. Such concentrations may be the result of the underlying geology (naturally occurring contaminants) or may be caused by industry, mining, waste disposal, or agricultural activities. Inorganic contaminants include radionuclides, metals and metallic elements such as lead, chromium, and arsenic, and inorganic nutrients such as nitrogen and phosphorus in various forms. At elevated concentrations inorganic substances can cause a variety of damaging effects to the liver, kidney, nervous system, circulatory system, gastrointestinal system, bones, and skin, depending upon the element and level of exposure (USEPA, 2003a). Pregnant women and infants are especially susceptible to harm from inorganic chemicals.

An exposure to radionuclides results in an increased risk of cancer. Certain elements accumulate in specific organs. For example, radium accumulates in the bones and iodine accumulates in the thyroid. For uranium, there is also the potential for kidney damage. Many water sources have very low levels of naturally occurring radioactivity usually low enough not to be considered a public health concern. In some parts of the United States, however, the underlying geology causes elevated concentrations of some radionuclides in aquifers used for water supply (see Chapter 4.2.4). Contamination of water from human-made radioactive materials occurs primarily as the result of improper waste storage, leaks, or transportation accidents. These radioactive materials are used in various ways in the production of commercial products (such as television and smoke detectors), electricity, nuclear weapons, and in nuclear medicine in therapy and diagnosis.

Nitrogen and phosphorus are widely used in fertilizers. Nitrogen is used to promote green, leafy, vegetative growth in plants. Phosphorus promotes root growth, root branching, stem growth, flowering, fruiting, seed formation, and maturation. Excessive phosphorus from runoff and erosion can fertilize surface waters. In this process, called eutrophication, algae multiply rapidly when fertilized by phosphorus. These algae cloud the water making it difficult for larger submerged aquatic vegetation (SAV) to obtain enough light. The SAV may die, reducing available habitat of aquatic animals. When the algae themselves eventually die, they decompose. During decomposition, dissolved oxygen is removed from the water. Lowered oxygen levels make it difficult for other aquatic organisms to survive. As discussed in Chapter 4.2.3.1, phosphorus does not appear to be a widely present groundwater contaminant despite its extensive use in agriculture. This is primarily due to its low mobility in the subsurface. Different forms of nitrogen, on the other hand, collectively constitute the most widely spread groundwater contaminant coming from many different sources and present in all environmental settings, rural, urban, and industrial. More information on nitrogen is given in Section 4.2.2.8 and Section 5.2.1.

5.2.4 ORGANIC CONTAMINANTS

Concern about synthetic organic chemicals in the drinking water supplies of some cities was a significant force in the passage of the 1974 SDWA, even though data were scarce. In 1981, USEPA conducted the Groundwater Supply Survey to determine the occurrence of volatile organic chemicals (VOCs) in public drinking water supplies using groundwater. The survey showed detectable levels of these chemicals in 28.7% of public water systems serving more than 10,000 people and in 16.5% of smaller systems. Other USEPA and state surveys also revealed VOCs in public water supplies (USEPA, 1999). The Agency has used these surveys to support regulation of numerous organic chemicals, many of which are carcinogenic.

Synthetic organic chemicals (SOCs) are human-made compounds that are used for a variety of industrial and agricultural purposes. Adverse health effects from exposure to synthetic organic chemicals include damage to the nervous system and kidneys, and cancer risks. SOCs can be divided into two groups: nonvolatile (semivolatile) and volatile organic chemicals:

- Nonvolatile and semivolatile organic chemicals include pesticides, insecticides, and herbicides, such as atrazine, aroclor, and DDT. Atrazine has the potential to cause weight loss, cardiovascular damage, retinal and some muscular degeneration; and cancer. Aroclor can cause eye, liver, kidney, or spleen problems, anemia, and an increased risk of cancer. Herbicides can harm aquatic plants.
- VOCs are synthetic compounds used for a variety of industrial and manufacturing purposes. Among the most common VOCs are degreasers and solvents such as benzene, toluene, and trichloroethylene (TCE), insulators and conductors such as polychlorinated biphenyls (PCBs), and dry-cleaning agents such as tetrachloroethylene (PCE). VOCs have the potential to cause chromosome aberrations, cancer, nervous system disorders, and liver and kidney damage (USEPA, 2003a). There are 53 organic chemicals included on the USEPA primary drinking water standards list (see Section 5.3).

Widespread detections of agricultural chemicals (generally a subset of organic chemicals) in groundwater and surface water have prompted considerable public and governmental concern in recent years. Although concentrations of most of the detected pesticides have been very low, health officials note that little is known about the long-term health effects of low-level exposures to pesticides. In areas of heavy agricultural–chemical use, pesticides have been detected more frequently and at higher levels. In 1992, EPA issued the pesticides in groundwater database (1971–1991), which showed that nearly 10,000 of 68,824 tested wells contained pesticides at levels that exceeded drinking water standards or health advisory levels. Almost all the data were from drinking water wells. EPA has placed restrictions on 54 pesticides found in groundwater, 28 of which are no longer registered for use in the United States but may still be present in soils and groundwater due to the widespread historic use (Tiemann, 1996).

Disinfection of drinking water is one of the major public health advances of the twentieth century. Disinfection is a major factor in reducing the typhoid and cholera epidemics that were common in the United States and European cities in the nineteenth century and the beginning of the twentieth century. While disinfectants are effective in controlling many microorganisms, certain disinfectants (notably chlorine) react with natural organic and inorganic matter in source water and distribution systems to form disinfection byproducts (DBPs). A large portion of the U.S. population is potentially exposed to DBPs through its drinking water. More than 240 million people in the United States are served by public water systems that apply a disinfectant to water to protect against microbial contaminants. Results from toxicology studies have shown several DBPs (e.g., bromodichloromethane, bromoform, chloroform, dichloroacetic acid, and bromate) to be carcinogenic in laboratory animals. Other DBPs (e.g., chlorite, bromodichloromethane, and certain haloacetic acids) have also been shown to cause adverse reproductive or developmental effects in laboratory animals. Epidemiological and toxicological studies involving DBPs have provided indications that these substances may have a variety of adverse effects across the spectrum of reproductive and developmental toxicity: early-term miscarriage; stillbirth; low birth weight; premature babies; and congenital birth defects (USEPA, 2003a). DBPs are of special concern when studying potential for artificial aquifer recharge using treated wastewater. Three disinfectants and four DBPs are currently on the USEPA primary drinking water standards list (see Section 5.3).

5.2.4.1 Light Nonaqueous Phase Liquids

Nonaqueous phase liquids (NAPLs) are hydrocarbons that exist as a separate, immiscible phase when in contact with water or air. Differences in the physical and chemical properties of water and NAPL result in the formation of a physical interface between the liquids that prevents the two fluids from mixing. NAPLs are typically classified as either light nonaqueous phase liquids (LNAPLs), which have densities less than that of water, or dense nonaqueous phase liquids (DNAPLs), which have densities greater than that of water. LNAPLs affect groundwater quality at a variety of sites. The most common LNAPL related groundwater contamination problems result from the release of petroleum products. Leaking of underground storage tanks (USTs) at gas stations and other facilities is arguably the most widely spread point-source contamination of groundwater in developed countries. LNAPL products are typically multicomponent organic mixtures composed of chemicals with varying degrees of water solubility. Some gasoline additives (e.g., methyl tertiary-butyl ether (MTBE), and alcohols such as ethanol) are highly soluble. Other components (e.g., benzene, toluene, ethylbenzene, and xylenes, collectively known as BTEX) are slightly soluble. Many components (e.g., *n*-dodecane and *n*-heptane) have relatively low water solubility under ideal conditions (Newell et al., 1995).

When LNAPLs (e.g., fuel) are released to the subsurface, the components of the fuel may remain in the original NAPL, dissolve into and migrate with any water present in the vadose zone, adsorb onto the solid material in the soil, or volatilize into soil gas. This partitioning between NAPL, water, solids, and gas can be largely described by four chemical properties: solubility in water, solids–water partition coefficient, vapor pressure, and Henry's law constant. The movement of fuel components in the saturated zone is a function of the ground-water flow velocity and the retardation factor, and their persistence in the environment as affected by the biodegradation rate (Nichols et al., 2000; see Chapter 6 for more details on the contaminant fate and transport processes).

As a spilled LNAPL enters the unsaturated zone, it flows through the central portion of the unsaturated pores because LNAPL is a nonwetting fluid relative to water and air. Water, as the wetting phase relative to both air and LNAPL, tends to line the edges of the pores and cover the soil grains. If the amount of product released is small, the product flows until residual saturation is reached (Figure 5.9a). Therefore, a three-phase system consisting of water, product, and air is created within the vadose zone. Infiltrating water dissolves the components within the LNAPL (e.g., benzene, toluene, xylene and others) and carries them to the water table. These dissolved constituents then form a contaminant plume emanating from the area of the residual product, where LNAPL phase is immobile (trapped by the porous media). Many of the components commonly found in LNAPLs are volatile and can divide

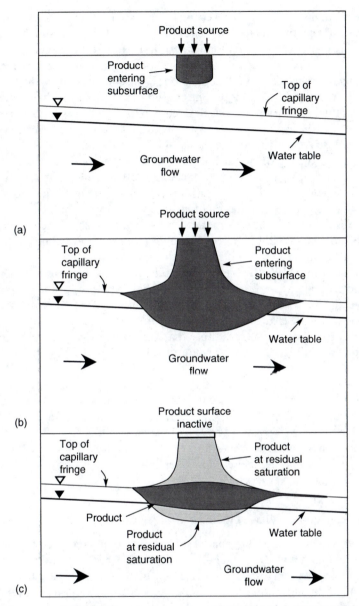

FIGURE 5.9 Movement of LNAPLs into the subsurface: a: distribution of LNAPL after small volume has been spilled; b: depression of the capillary fringe and water table; c: rebounding of the water table as LNAPL drains from overlying pore space. (From Palmer, C.D. and Johnson, R.L., 1989. Physical processes controlling the transport of nonaquesous phase liquids in the subsurface, in: *USEPA, Seminar Publication: Transport and Fate of Contaminant in the Subsurface*, EPA/625/4–89/019, US Environmental Protection Agency, p. 23–27.)

into the soil air and be transported by molecular diffusion in all directions within the vadose zone and away from the area of the residual mass. These vapors may partition back into the water phase and spread contamination over a wider area. They can also diffuse across the land surface boundary into the atmosphere (Palmer and Johnson, 1989).

If large volumes of product are spilled (Figure 5.9b), the product flows through the pore space to the top of the capillary fringe. The dissolved components of the infiltrating product precede the product and may change the wetting properties of the water, causing a reduction in the residual water content and collapse of the capillary fringe (Johnson and Palmer, 1989). The LNAPL product is lighter (less dense) than water and tends to float on top of the capillary fringe. As the head created by the infiltrating product increases, the water table is depressed and the product begins to accumulate in the depression. If the source of the spilled product is then turned off, the LNAPL within the vadose zone continues to flow under the influence of gravity until it reaches residual saturation. As this drained product continues to recharge the product pool accumulated on the capillary fringe, it spreads laterally. The draining of the upper portions of the vadose zone also reduces the total head at the interface between the product and the groundwater, causing the water table to rebound slightly. The rebounding water can only displace a portion of the product because the latter remains at residual saturation. Groundwater passing through this area of residual saturation dissolves the components within the residual product, creating a contaminant plume. Water infiltrating from the surface also can dissolve the residual product and vapors within the vadose zone, thereby contributing to the overall dissolved contaminant load to the aquifer (Johnson and Palmer, 1989).

Fluctuations of the water table, combined with the downgradient migration of the three-contaminant phases (product pool, phase dissolved in groundwater, vapor phase), may create quite complex horizontal and vertical distributions (redistribution) of the contaminant in the subsurface, especially if the porous media is heterogeneous (presence of clay lenses and layers).

Accumulations of LNAPL at or near the water table are susceptible to "smearing" from changes in water-table elevation such as those that occur due to seasonal changes in recharge and discharge, or tidal influence in coastal environments. Mobile LNAPL floating above the water-saturated zone will move vertically as the groundwater elevation fluctuates. As the water table rises or falls, LNAPL will be retained in the soil pores, leaving behind a residual LNAPL "smear zone." If smearing occurs during a decline in groundwater elevations, residual LNAPL may be trapped below the water table when groundwater elevations rise. A similar situation may develop during product recovery efforts. LNAPL will flow toward a recovery well or trench in response to the gradient induced by water-table depression. LNAPL residual will be retained below the water table as the water-table elevation returns to prepumping conditions (Newell et al., 1995).

5.2.4.2 Dense Nonaqueous Phase Liquids

The major types of DNAPLs are halogenated solvents, coal tar, creosote-based wood-treating oils, polychlorinated biphenyls (PCB), and pesticides. As a result of widespread production, transportation, utilization, and disposal practices, particularly since 1940s, there are numerous DNAPL contamination sites in North America and Europe. The potential for serious long-term contamination of groundwater by some DNAPL chemicals at many sites is high due to their toxicity, limited solubility (but much higher than drinking water limits), and significant migration potential in soil gas, groundwater, and as a separate phase. DNAPL chemicals, especially chlorinated solvents, are among the most prevalent groundwater contaminants identified in groundwater supplies and at waste disposal sites (Huling and Weaver, 1991; Cohen and Mercer, 1993). Some of the more common DNAPLs found in groundwater at various industrial and governmental sites are included in Table 5.3. As shown in Figure 5.10, pure chlorinated solvents are generally much more mobile than creosote and coal tar and PCB oil mixtures due to their relatively high density and viscosity ratios (Cohen and Mercer, 1993).

5.2.4.2.1 Halogenated Solvents

Halogenated solvents, particularly chlorinated hydrocarbons, and brominated and fluorinated hydrocarbons to a much lesser extent, are DNAPL chemicals encountered at numerous

TABLE 5.3
Some of the More Common DNAPLs Found in Groundwater at Various Industrial and Governmental Sites

Halogenated Volatiles	Type	Nonhalogenated Semivolatiles	Type
Chlorobenzene	Liquid solvent	Phenol	Nonchlorinated phenol
1,2-Dichloropropane	Liquid solvent	2,4-Dinitrophenol	Nonchlorinated phenol
1,1-Dichloroethane (DCA)	Liquid solvent	o-Cresol	Nonchlorinated phenol
1,2-Dichloroethane	Liquid solvent	p-Cresol	Nonchlorinated phenol
1,1-Dichloroethylene (DCE)	Liquid solvent	m-Cresol	Nonchlorinated phenol
trans-1,2 Dichloroethylene	Liquid solvent	Anthracene	PAH
cis-1,2 Dichloroethylene	Liquid solvent	Benzo(a)anthracene	PAH
1,1,1-Trichloroethane (TCA)	Liquid solvent	Benzo(a)pyrene	PAH
1,1,2-Trichloroethane	Liquid solvent	Chrysene	PAH
Trichloroethylene (TCE)	Liquid solvent	Fluorene	PAH
Tetrachloroethylene (PCE)	Liquid solvent	2-Methyl napthalene	PAH
1,1,2,2,-Tetrachloroethane	Liquid solvent	Naphthalene	PAH
Methylene chloride	Liquid solvent	Phenathrene	PAH
Chloroform	Liquid solvent	Pyrene	PAH
Carbon tetrachloride	Liquid solvent		
Ethylene dibromide	Liquid solvent		

Halogenated Semivolatiles	Type	Miscellaneous	Type
1,2-Dichlorobenzene	Chlorinated benzene	Coal tar	Mixture
1,4-Dichlorobenzene	Chlorinated benzene	Cresosote	Mixture
Aroclor 1242, 1254, 1260	PCB		
Chlordane	Pesticide	Chemicals associated with DNAPLs:	
DDD	Pesticide		
DDE	Pesticide	Chloroethane	Gas
DDT	Pesticide	Vinyl chloride	Gas
Dieldrin	Pesticide		
2,3,4,6-Tetrachlorophenol	Chlorinated phenol		
Pentachlorophenol	Chlorinated phenol		

Note: Many of these chemicals are found mixed with other chemicals or carrier oils.

contamination sites. These halocarbons are produced by replacing one or more hydrogen atoms with chlorine (or another halogen) in petrochemical precursors such as methane, ethane, ethene, propane, and benzene. Many bromocarbons and fluorocarbons are manufactured by reacting chlorinate hydrocarbon intermediates (such as chloroform or carbon tetrachloride) with bromine and fluorine compounds, respectively. DNAPL halocarbons at ambient environmental conditions include: chlorination products of methane (methylene chloride, chloroform, carbon tetrachloride), ethane (1,1-dichloroethane, 1,2-dichloroethane, 1,1,1-trichloroethane, 1,1,2,2-tetrachloroethane), ethene (1,1-dichloroethene, 1,2-dichloroethene isomers, trichloroethene, tetrachloroethene), propane (1,2-dichloropropane, 1,3-dichloro-propene isomers), and benzene (chlorobenzene, 1,2-dichlorobenzene, 1,4-dichlorobenzene); fluorination products of methane and ethane such as 1,1,2-trichlorofluormethane (Freon-11) and 1,1,2-trichlorotrifluorethane (Freon-113); and, bromination products of methane

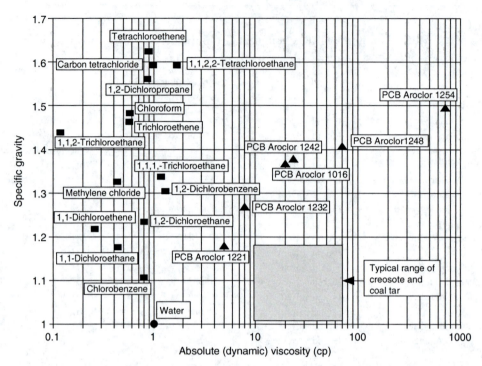

FIGURE 5.10 Specific gravity versus absolute viscosity for some DNAPLs. DNAPL mobility increases with increasing density and viscosity ratios. (From Cohen, R.M., and Mercer, J.W., in *DNAPL Site Evaluation*, C.K. Smoley, Boca Raton, Florida, 1993.)

(bromochloromethane, dibromochloromethane, dibromodifluoromethane, bromoform), ethane (bromoethane, 1,1,2,2-tetrabromoethane), ethene (ethylene dibromide), and propane (l,2-dibromo-3-chloropropane).

Although most chlorinated solvents were first synthesized during the 1800s, large-scale production generally began around the mid-1900s primarily for use as solvents and chemical intermediates. They account for 10 of the 20 organic contaminants detected most frequently at contamination sites (Cohen and Mercer, 1993). Trichloroethylene (also known as trichloroethene or TCE), methylene chloride (DCM), and tetrachloroethylene (tetrachloroethene or PCE) are ranked number 2, 3, and 5, respectively. Toluene and benzene, which are LNAPLs, are ranked number 1 and 4, respectively. Subsurface contamination derived from halogenated solvents are associated with industries that produce or use these DNAPLs, and with waste disposal sites. Their subsurface presence is caused by leakage from tanks, pipelines, drums, and other containers; spillage during filling operations; and intentional discharge to landfills, pits, ponds, and sewers. Halogenated solvents are frequently present at mixed DNAPL sites (Cohen and Mercer, 1993).

5.2.4.2.2   Coal Tar and Creosote

Coal tar and creosote are complex chemical mixture DNAPLs derived from the destructive distillation of coal in coke ovens and retorts. Historically, coal tar has been produced by coal tar distillation plants and as a byproduct of manufactured gas plant and steel industry coking operations. Creosote consists of various coal tar distillates (primarily the 200°C to 400°C fraction) that are blended to meet American Woodpreservers' Association (AWPA) product standards. These creosote blends are then used to treat wood alone or diluted with coal tar, petroleum, or, to a very limited extent, pentachlorophenol. In addition to wood preservation, coal tar is used for road, roofing, and waterproofing solutions. Considerable use of coal tar is also made for fuels (Cohen and Mercer, 1993).

Creosote and coal tar contamination of the subsurface is associated with wood-treating plants, former manufactured gas plants, coal tar distillation plants, and steel industry coking plants. Prior to the 1970s, liquid wastes (including creosote and coal tar) from wood-treating plants were typically discharged to ponds, sumps, or streams. Many plants had small (1–4 acres) unlined ponds to trap the DNAPL wastes as effluent discharged to streams or public water treatment facilities (McGinnis, 1989). As a result, many wood-treating sites have large volumes of DNAPL-contaminated soils in the vicinity of former discharge ponds. Soil contamination at wood-treating plants is also prevalent in the wood treating, track, and storage areas due to preservative drippage from wood as it is moved and stored, and around preservative tanks and pipelines due to spillage and leaks.

Creosote and coal tar are complex mixtures containing more than 250 individual compounds. Creosote is estimated to contain 85% polycyclic aromatic hydrocarbons (PAHs), 10% phenolic compounds, and 5% N-, S-, and O-heterocyclic compounds. The composition of creosote and coal tar are quite similar, although coal tar generally includes a light oil component (<5% of the total) consisting of monocyclic aromatic compounds such as benzene, toluene, ethylbenzene, and xylene (BTEX). Consistent with the composition of creosote and coal tar, PAHs (in addition to BTEX compounds) are common contaminants detected in groundwater at wood-treating sites (Rosenfeld and Plumb, 1991; Cohen and Mercer, 1993).

5.2.4.2.3 Polychlorinated Biphenyls

PCBs are extremely stable, nonflammable, dense, and viscous liquids that are formed by substituting chlorine atoms for hydrogen atoms on a biphenyl (double benzene ring) molecule. In the United States, the only large producer of PCBs was Monsanto Chemical Co., which sold them between 1929 and 1977 under the Aroclor trademark for use primarily as dielectric fluids in electrical transformers and capacitors. PCBs were also sold for use in oil-filled switches, electromagnets, voltage regulators, heat transfer media, fire retardants, hydraulic fluids, lubricants, plasticizers, carbonless copy paper, dedusting agents, and other products. Prior to use PCBs were frequently mixed with carrier fluids (Cohen and Mercer, 1993).

Due to environmental concerns, Monsanto ceased production of Aroclor 1260 in 1971; restricted the sale of other PCBs to totally enclosed applications (transformers, capacitors, and electromagnets) in 1972; and ceased all production and sale of PCBs in 1977. In 1979, USEPA issued final rules under the 1976 Toxic Substances Control Act restricting the manufacture, processing, use, and distribution of PCBs to specifically exempted and authorized activities. Due to their widespread use and persistence, PCBs are often detected in the environment at very low concentrations. The potential for DNAPL migration is greatest at sites where PCBs were produced, utilized in manufacturing processes, stored, reprocessed, or disposed in quantity (Cohen and Mercer, 1993).

5.2.4.2.4 Miscellaneous and Mixed DNAPL Sites

Miscellaneous DNAPLs refer to dense, immiscible fluids that are not categorized as halogenated solvents, coal tar, creosote, or PCBs. These include various exotic compounds and some insecticides, herbicides, and pesticides such as chlordane, dieldrin, and disulfoton, and phthalate plasticizers. Mixed DNAPL sites refer to landfills, lagoons, chemical waste handling or reprocessing sites, and other facilities where various organic chemicals were released to the environment and DNAPL mixtures are present. Many mixed DNAPL sites derive from the disposal of off-specification products and process residues in landfills and lagoons by chemical manufacturers. Typically, these mixed DNAPL sites include a significant component of chlorinated solvents.

As already mentioned, DNAPLs can have great mobility in the subsurface because of their relatively low solubility, high density, and low viscosity. Hydrophobic DNAPLs do not readily mix with water (they are immiscible) and tend to remain as separate phases (i.e., nonaqueous). The relatively high density of these liquids provides a driving force that can carry product deep into aquifers. DNAPL infiltrating from the land surface because of a spill or leak, may encounter two general conditions: (1) in the presence of moisture (water) within the vadose zone, DNAPL exhibits viscous fingering during infiltration (when a high-density, low-viscosity fluid, i.e., DNAPL, displaces a lower density, higher viscosity fluid, i.e., water, the flow is unstable resulting in the occurrence of viscous fingering), and (2) If the vadose zone is dry, viscous fingering is generally not observed (Palmer and Johnson, 1989). When the spill of DNAPL is small, it will flow through the vadose zone until it reaches the residual saturation, i.e., until all of the mobile DNAPL is trapped in the porous media. This residual DNAPL may still cause the formation of a dissolved plume in the underlying saturated zone (aquifer) through one of the following two mechanisms:

1. The DNAPL can partition into the vapor phase and these dense vapors may sink to the capillary fringe where they are eventually dissolved in water and transported downgradient.
2. Infiltrating water can dissolve the residual DNAPL and transport it down to the water table.

When a greater amount of DNAPL is spilled and reaches the water table, it starts to penetrate into the saturated zone. However, in order to do this, the DNAPL must displace the water by overcoming the capillary forces between the water and the porous medium. The critical height of DNAPL (z_c) required to overcome these capillary forces can be calculated from (Villaume et al., 1983, from Palmer and Johnson, 1989)

$$z_c = 2\gamma \cos(\theta) \times \frac{(1/r_t - 1/r_p)}{\Delta\rho g} \tag{5.2}$$

where γ is the interfacial tension between the water and the DNAPL, θ is the contact angle between the fluid boundary and the solid surface, r_t is the radius of the pore throat, r_p is the radius of the pore, $\Delta\rho$ is the difference in the density between the water and the DNAPL, and g is the acceleration due to gravity.

For example, the calculated critical heights required for TCE to penetrate saturated porous media of different grain size range from a few centimeters for coarse grain sediments (sand and gravel) to tens of meters for clays. Thus, unfractured, saturated clays and silts can be effective barriers to the migration of DNAPLs, provided the critical heights are not exceeded (Palmer and Johnson, 1989).

Once DNAPL enters the saturated zone, its further migration will depend on the amount (mass) of product and the aquifer heterogeneity, such as presence of low-permeable lenses and layers of clay. Figure 5.11 shows two examples of free-phase (mobile) DNAPL accumulation over impermeable layers. This free-phase (pooled) DNAPL serves as a continuing source of dissolved-phase contamination, which is carried downgradient by the flow of groundwater. As it migrates, the DNAPL may leave residual phase in the porous media of the saturated (aquifer) zone along its path. This residual phase also serves as a source of dissolved-phase contamination. Being denser than water, free-phase DNAPL moves because of gravity, and not because of the hydraulic gradients normally present in natural aquifers. As a consequence, DNAPL encountering a low-permeable layer may flow along its slope, in a different (including opposite) direction from the dissolved plume, as shown in Figure 5.11. This

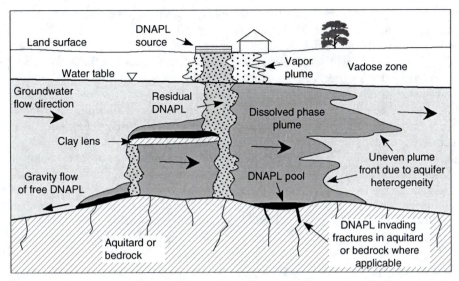

FIGURE 5.11 Schematic of possible migration pathways of free-phase DNAPL and the derived dissolved contaminant plumes in the subsurface.

free-phase DNAPL would also create its own dissolved-phase plume. As can be seen, migration of DNAPL in the unsaturated and saturated zones may create a rather complex pattern of multiple secondary sources of free-phase and residual-phase DNAPL, and multiple dissolved plumes at various depths within the aquifer. This pattern may not be definable based on the hydraulics of groundwater flow alone and, in any case, presents a great challenge when attempting to restore the aquifer to its beneficial use: "Once in the subsurface, it is difficult or impossible to recover all of the trapped residual" (USEPA, 1992).

The complexity of DNAPL groundwater contamination is especially pronounced in fractured rock and karst aquifers. DNAPL introduced into a fractured rock or fractured clay system follows a complex pathway based on the distribution of fractures in the original matrix. The number, density, size, and direction of the fractures usually cannot be determined due to the extreme heterogeneity of a fractured system and the lack of economical aquifer characterization technologies (USEPA, 1993). Relatively small volumes of DNAPL can penetrate deeply into fractured systems due to the low retention capacity of the fractures and the ability of some DNAPLs to migrate through very small ($< 20 \mu$m) fractures. The following assessment by USEPA, made more than a decade ago, still holds true for the majority of DNAPL-contaminated sites in fractured rock and karst aquifers: "Although many DNAPL removal technologies are currently being tested, to date there have been no field demonstrations where sufficient DNAPL has been successfully recovered from the subsurface to return the aquifer to drinking water quality. The DNAPL that remains trapped in the soil or aquifer matrix acts as a continuing source of dissolved contaminants to ground water, preventing the restoration of DNAPL-affected aquifers for many years" (USEPA, 1993).

Recognizing the many difficulties associated with the characterization and remediation of groundwater contaminated by DNAPLs, especially in fractured rock and karst aquifers, in 1993 USEPA issued a directive entitled "Guidance for Evaluating the Technical Impracticability of Ground-Water Restoration." Following are the key excerpts from the directive:

- "The long-term remediation objectives for a DNAPL zone should be to remove the free-phase, residual and vapor phase DNAPL to the extent practicable and contain DNAPL sources that cannot be removed. EPA recognizes that it may be difficult to locate and remove all of the subsurface DNAPL within a DNAPL zone. Removal of DNAPL mass should be pursued wherever practicable and, in general, where significant reduction of current or future risk will result. Where it is technically impracticable to remove subsurface DNAPLs, EPA expects to contain the DNAPL zone to minimize further release of contaminants to the surrounding ground water, wherever practicable (footnote: As DNAPLs may be remobilized during drilling or ground-water pumping, caution should be exercised where such activities are proposed for DNAPL zone characterization, remediation, or containment)."

- "Where it is technically practicable to contain the long-term sources of contamination, such as the DNAPL zone, USEPA expects to restore the aqueous contaminant plume outside the DNAPL zone to required cleanup levels. Effective containment of the DNAPL zone generally will be required to achieve this long-term objective because ground-water extraction remedies (e.g., pump-and-treat) or *in situ* treatment technologies are effective for plume restoration only where source areas have been contained or removed."

From these two key points by USEPA it is obvious that, in same cases, it may not be possible to contain the DNAPL zone or remove free product from it because it may be impossible to define where such zone (product) is. It is also obvious that, in some cases, it may not be possible to restore the "aqueous contaminant plume" to drinking water standards because it may not be technically practicable to contain the flux of dissolved contaminants flowing through a highly heterogeneous, dual-porosity aquifer, and coming off unknown DNAPL zones. Figure 5.12 and Figure 5.13 (see also Figure 1.17) show some of the features in the unsaturated and saturated zones of karst aquifers, which illustrate an unpredictable aquifer heterogeneity and types of spaces where DNAPL may be accumulating and flowing. These and many other karst-specific types of porosity often make DNAPL contamination impossible to adequately characterize and remediate within reasonable limits of cost and time (Kresic et al., 2005).

DNAPL Site Evaluation by Cohen and Mercer (1993) and *Dense Chlorinated Solvents and Other DNAPLs in Groundwater* by Pankow and Cherry (1996) are two invaluable technical resources for the general study and characterization of DNAPLs in the subsurface.

5.3 DRINKING WATER STANDARDS

There are two categories of drinking water regulations in the United States:

- National Primary Drinking Water Regulations (NPDWRs or Primary Standards) are legally enforceable standards that apply to public water systems. Primary standards protect drinking water quality by limiting the levels of specific contaminants that can adversely affect public health and are known or anticipated to occur in water. They take the form of MCLs or TTs. These standards must be met at the discharge point from the distribution system or, in some cases, at various points throughout the distribution system (USEPA, 2003a).
- National Secondary Drinking Water Regulations (NSDWRs or Secondary Standards) are nonenforceable guidelines regarding contaminants that may cause cosmetic effects (such as skin or tooth discoloration) or have esthetic effects (such as affecting the taste, odor, or color of drinking water). USEPA recommends secondary standards to water systems but does not require systems to comply. However, states may choose to adopt them as enforceable standards. NSDWRs are intended to protect "public welfare" (USEPA, 2003a).

FIGURE 5.12 Caver rafting in the Martin's Creek section of Sloans Valley Cave. Pulaski County, Kentucky. (Photo: K.L. Day; from Dougherty, P.H. (ed.), in *Caves and Karst of Kentucky*, Special Publication 12, Series XI, Kentucky Geological Survey, Lexington, Kentucky, 1985, 196 p.)

Once USEPA has selected a contaminant for regulation, it examines the contaminant's health effects and sets a MCLG. This is the maximum level of a contaminant in drinking water at which no known or anticipated adverse health effects would occur, and which allows an adequate margin of safety. MCLGs do not take cost and technologies into consideration. MCLGs are nonenforceable public health goals. In setting the MCLG, USEPA examines the size and nature of the population exposed to the contaminant, and the length of time and concentration of the exposure. Because MCLGs consider only public health and not the limits of detection and treatment technology, they are sometimes set at a level that water systems cannot meet. For most carcinogens (contaminants that cause cancer) and microbiological contaminants, MCLGs are set at zero because a safe level often cannot be determined (USEPA, 2003a).

MCLs, which are enforceable limits that finished drinking water must meet, are set as close to the MCLG as feasible. SDWA defines "feasible" as the level that may be achieved with the

FIGURE 5.13 Turner Avenue cave passage, in the Flint Ridge section of the Mammoth Cave system. (Photo: Arthur N. Palmer; from Palmer, A.N., The Mammoth Cave region and Pennyroyal Plateau, in Dougherty, P.H. (ed.), *Caves and Karst of Kentucky*. Special Publication 12, Series XI, Kentucky Geological Survey, Lexington, Kentucky, 1985, p. 97–118.)

use of the best available technology (BAT), TT, or other means specified by USEPA, after examination for efficacy under field conditions (that is, not solely under laboratory conditions) and taking cost into consideration (USEPA, 2003a).

For some contaminants, especially microbiological contaminants, there is no reliable method that is economically and technically feasible to measure a contaminant at particularly low concentrations. In these cases, USEPA establishes TTs. A TT is an enforceable procedure or level of technological performance that public water systems must follow to ensure control of a contaminant. Examples of rules with TTs are the surface water treatment rule and the lead and copper rule.

To support a drinking water contaminant rulemaking, USEPA

- evaluates occurrence of the contaminant (number of systems affected by a specific contaminant and concentrations of the contaminant);
- evaluates the number of people exposed and the ingested dose; and
- characterizes choices for water systems to meet regulatory standards (treatment technologies).

In developing an MCL or TT, USEPA assesses multiple possible MCL or TT alternatives in terms of costs (for example, the cost of installing new treatment equipment). USEPA also assesses benefits resulting from the various regulatory alternatives. Some of the benefits can be quantified (for example, cost of illness avoided), but some are unquantifiable (for example, cost savings associated with the removal of other contaminants, gaining economies of scale by merging with other water systems).

When thinking about how MCLs and TTs are established, it is important to understand how acute and chronic health effects are addressed under SDWA. USEPA determines whether adverse health effects from a given contaminant are generally from acute or chronic exposures based on information about the occurrence of the contaminant in drinking water. MCLs are then set to protect humans from those exposures. There must be evidence of the presence of a contaminant at acute levels to set a standard at such a level. In addition, in meeting the levels for chronic concerns, drinking water would automatically be treated to a level that would prevent acute effects. It is also important to note that a one-time exceedance of an MCL is not necessarily considered a violation of the MCL. Some MCLs distinguish between an exceedance and a violation and others provide a method for calculating, over a period of time, whether a water system is in violation or not (USEPA, 2003a).

As of January 2006, USEPA has set MCLs or TTs for 87 contaminants included in the national primary drinking water standards list (Table 5.4). There are 15 contaminants included in the national secondary drinking water standards list (Table 5.5).

5.4 EMERGING CONTAMINANTS

The 1986 SDWA Amendments required USEPA to select contaminants for regulation from an existing list of contaminants with known health effects. However, this approach did not take into account how often a contaminant occurred in drinking water, and it did not provide a means to prioritize contaminants for regulation. The approach outlined in the 1996 amendments for developing new standards requires broad public and scientific input to ensure that contaminants posing the greatest risk to public health will be selected for future regulation. A contaminant's presence in drinking water and public health risks associated with a contaminant must be considered to determine whether a public health risk is evident. In addition, the new contaminant selection approach explicitly takes into account the needs of sensitive populations such as children and pregnant women. Under the 1996 amendments, the contaminant candidate list (CCL) guides scientific evaluation of new contaminants. Contaminants on the CCL are prioritized for regulatory development, drinking water research (including studies of health effects, treatment effects, and analytical methods), and occurrence monitoring. The unregulated contaminant monitoring rule (UCMR) guides collection of data on contaminants not included in the national primary drinking water standard. The data are used to evaluate and prioritize contaminants that USEPA is considering for possible new drinking water standards. Figure 5.14 illustrates the process of regulating drinking water contaminants by USEPA.

The CCL must be updated every 5 years, providing a continuing process to identify contaminants for future regulations or standards, and for prevention activities. To prioritize contaminants for regulation, USEPA considers peer-reviewed science and data to support an "intensive technological evaluation," which includes many factors: occurrence in the environment; human exposure and risks of adverse health effects in the general population and sensitive subpopulations; analytical methods of detection; technical feasibility; and impacts of regulation on water systems, the economy and public health (USEPA, 2003a).

USEPA published first CCL of 60 contaminants in March 1998 and the second CCL in February 2005 after deciding to continue research on the list of contaminants on the first CCL. The second CCL carries forward 51 (of the original 60) unregulated contaminants from the first CCL, including 9 microbiological contaminants and 42 chemical contaminants or contaminant groups (see Table 5.6). In July 2003, EPA announced its final determination for a subset of nine contaminants from the first CCL, which concluded that sufficient data and information was available to make the determination not to regulate Acanthamoeba, aldrin, dieldrin, hexachlorobutadiene, manganese, metribuzin, naphthalene, sodium, and sulfate. These nine contaminants were not carried forward to the 2005 CCL2 (USEPA, 2005b).

It is important to note that USEPA has not limited itself to making regulatory determinations for only those contaminants on the CCL. The agency can also decide to regulate other unregulated contaminants if information becomes available showing that a specific contaminant presents a public health risk. Some of these "other" contaminants have already been

TABLE 5.4
National Primary Drinking Water Standards

	Contaminant	MCL or TT[a] (mg/L)[b]	Potential Health Effects from Exposure above the MCL	Common Sources of Contaminant in Drinking Water	Public Health Goal
OC	Acrylamide	TT[c]	Nervous system or blood problems;	Added to water during sewage or wastewater increased risk of cancer treatment	0
OC	Alachlor	0.002	Eye, liver, kidney, or spleen problems; anemia; increased risk of cancer	Runoff from herbicide used on row crops	0
R	Alpha particles	15 picocuries per Liter (pCi/L)	Increased risk of cancer	Erosion of natural deposits of certain minerals that are radioactive and may emit a form of radiation known as alpha radiation	0
IOC	Antimony	0.006	Increase in blood cholesterol; decrease in blood sugar	Discharge from petroleum refineries; fire retardants; ceramics; electronics; solder	0.006
IOC	Arsenic	0.01	Skin damage or problems with circulatory systems, and may have increased risk of getting cancer	Erosion of natural deposits; runoff from orchards, runoff from glass and electronics production wastes	0
IOC	Asbestos (fibers >10 mm)	7 million fibers per Liter (MFL)	Increased risk of developing benign intestinal polyps	Decay of asbestos cement in water mains; erosion of natural deposits	7 MFL
OC	Atrazine	0.003	Cardiovascular system or reproductive problems	Runoff from herbicide used on row crops	0.003
IOC	Barium	2	Increase in blood pressure	Discharge of drilling wastes; discharge from metal refineries; erosion of natural deposits	2
OC	Benzene	0.005	Anemia; decrease in blood platelets; increased risk of cancer	Discharge from factories; leaching from gas storage tanks and landfills	0
OC	Benzo(a)pyrene (PAHs)	0.0002	Reproductive difficulties; increased risk of cancer	Leaching from linings of water	0

IOC	Beryllium	0.004	Intestinal lesions	Discharge from metal refineries and coal-burning factories; discharge from electrical, aerospace, and defense industries	0.004
R	Beta particles and photon emitters	4 millirems per year	Increased risk of cancer	Decay of natural and man-made deposits of certain minerals that are radioactive and may emit forms of radiation known as photons and beta radiation	0
DBP	Bromate	0.010	Increased risk of cancer	Byproduct of drinking water disinfection	0
IOC	Cadmium	0.005	Kidney damage	Corrosion of galvanized pipes; erosion of natural deposits; discharge from metal refineries; runoff from waste batteries and paints	0.005
OC	Carbofuran	0.04	Problems with blood, nervous system, or reproductive system	Leaching of soil fumigant used on rice and alfalfa	0.04
OC	Carbon tetrachloride	0.005	Liver problems; increased risk of cancer	Discharge from chemical plants and other industrial activities	0
D	Chloramines (as Cl_2)	MRDL = 4^a	Eye or nose irritation; stomach discomfort, anemia	Water additive used to control microbes	MRDLG = 4^a
OC	Chlordane	0.002	Liver or nervous system problems; increased risk of cancer	Residue of banned termiticide	0
D	Chlorine (as Cl_2)	MRDL = 4^a	Eye or nose irritation; stomach discomfort	Water additive used to control microbes	MRDLG = 4^a
D	Chlorine dioxide (as ClO_2)	MRDL = 0.8^a	Anemia; infants and young children: nervous system effects	Water additive used to control microbes	MRDLG = 0.8^a
DBP	Chlorite	1.0	Anemia; infants and young children: nervous system effects	Byproduct of drinking water disinfection	0.8
OC	Chlorobenzene	0.1	Liver or kidney problems	Discharge from chemical and agricultural chemical factories	0.1
IOC	Chromium (total)	0.1	Allergic dermatitis	Discharge from steel and pulp mills; erosion of natural deposits	0.1

continued

TABLE 5.4 (continued)
National Primary Drinking Water Standards

	Contaminant	MCL or TT[a] (mg/L)[b]	Potential Health Effects from Exposure above the MCL	Common Sources of Contaminant in Drinking Water	Public Health Goal
IOC	Chromium (total)	0.1	allergic cermatitis	Discharge from steel and pulp mills; erosion of natural deposits	0.1
IOC	Copper	TT[d]; Action Level = 1.3	Short-term exposure: Gastrointestinal distress. Long-term exposure: liver or kidney damage. People with Wilson's disease should consult their personal doctor if the amount of copper in their water exceeds the action level	Corrosion of household plumbing systems; erosion of natural deposits	1.3
M	*Cryptosporidium*	TT[e]	Gastrointestinal illness (e.g., diarrhea, vomiting, and cramps)	Human and animal fecal waste	0
IOC	Cyanide (as free cyanide)	0.2	Nerve damage or thyroid problems	Discharge from steel and metal factories; discharge from plastic and fertilizer factories	0.2
OC	2,4-D	0.07	Kidney, liver, or adrenal gland problems	Runoff from herbicide used on row crops	0.07
OC	Dalapon	0.2	Minor kidney changes	Runoff from herbicide used on rights of way	0.2
OC	1,2-Dibromo-3-chloropropane (DBCP)	0.0002	Reproductive difficulties; increased risk of cancer	Runoff or leaching from soil fumigant used on soybeans, cotton, pineapples, and orchards	0
OC	*o*-Dichlorobenzene	0.6	Liver, kidney, or circulatory system problems	Discharge from industrial chemical factories	0.6
OC	*p*-Dichlorobenzene	0.075	Anemia; liver, kidney or spleen damage; changes in blood	Discharge from industrial chemical factories	0.075
OC	1,2-Dichloroethane	0.005	Increased risk of cancer	Discharge from industrial chemical factories	0
OC	1,1-Dichloroethylene	0.007	Liver problems	Discharge from industrial chemical factories	0.007

	Contaminant		Health Effects	Sources	
OC	*cis*-1,2-Dichloroethylene	0.07	Liver problems	Discharge from industrial chemical factories	0.07
OC	*trans*-1,2-Dichloroethylene	0.1	Liver problems	Discharge from industrial chemical factories	0.1
OC	Dichloromethane	0.005	Liver problems; increased risk of cancer	Discharge from drug and chemical factories	0
OC	1,2-Dichloropropane	0.005	Increased risk of cancer	Discharge from industrial chemical factories	0
OC	Di(2-ethylhexyl)adipate	0.4	Weight loss, liver problems, or possible reproductive difficulties	Discharge from chemical factories	0.4
OC	Di(2-ethylhexyl)phthalate	0.006	Reproductive difficulties; liver problems; increased risk of cancer	Discharge from rubber and chemical factories	0
OC	Dinoseb	0.007	Reproductive difficulties	Runoff from herbicide used on soybeans and vegetables	0.007
OC	Dioxin (2,3,7,8-TCDD)	0.00000003	Reproductive difficulties; increased risk of cancer	Emissions from waste incineration and other combustion; discharge from chemical factories	0
OC	Diquat	0.02	Cataracts	Runoff from herbicide use	0.02
OC	Endothall	0.1	Stomach and intestinal problems	Runoff from herbicide use	0.1
OC	Endrin	0.002	Liver problems	Residue of banned insecticide	0.002
OC	Epichlorohydrin	TT[c]	Increased cancer risk, and over a long period of time, stomach problems	Discharge from industrial chemical factories; an impurity of some water treatment chemicals	0
OC	Ethylbenzene	0.7	Liver or kidneys problems	Discharge from petroleum refineries	0.7
OC	Ethylene dibromide	0.00005	Problems with liver, stomach, reproductive system, or kidneys; increased risk of cancer	Discharge from petroleum refineries	0
IOC	Fluoride	4.0	Bone disease (pain and tenderness of the bones); children may get mottled teeth	Water additive which promotes strong teeth; erosion of natural deposits; discharge from fertilizer and aluminum factories	4.0
M	*Giardia lamblia*	TT[c]	Gastrointestinal illness (e.g., diarrhea, vomiting, and cramps)	Human and animal fecal waste	0
OC	Glyphosate	0.7	Kidney problems; reproductive difficulties	Runoff from herbicide use	0.7

continued

TABLE 5.4 (continued)
National Primary Drinking Water Standards

	Contaminant	MCL or TT^a (mg/L)^b	Potential Health Effects from Exposure above the MCL	Common Sources of Contaminant in Drinking Water	Public Health Goal
DBP	Haloacetic acids (HAA5)	0.060	Increased risk of cancer	Byproduct of drinking water disinfection	n/a^f
OC	Heptachlor	0.0004	Liver damage; increased risk of cancer	Residue of banned termiticide	0
OC	Heptachlor epoxide	0.0002	Liver damage; increased risk of cancer	Breakdown of heptachlor	0
M	Heterotrophic plate count (HPC)	TT^e	HPC has no health effects; it is an analytic method used to measure the variety of bacteria that are common in water. The lower the concentration of bacteria in drinking water, the better maintained the water system is.	HPC measures a range of bacteria that are naturally present in the environment	n/a
OC	Hexachloro-cyclopentadiene	0.05	Kidney or stomach problems	Discharge from chemical factories	0.05
IOC	Lead	TT^d, Action Level = 0.015	Infants and children: delays in physical or mental development; children could show slight deficits in attention span and learning abilities; adults: kidney problems; high blood pressure	Corrosion of household plumbing systems; erosion of natural deposits	0
M	*Legionella*	TT^e	Legionnaire's disease, a type of pneumonia	Found naturally in water; multiplies in heating systems	0
OC	Lindane	0.0002	Liver or kidney problems	Runoff or leaching from insecticide used on cattle, lumber, gardens	0.0002
IOC	Mercury (inorganic)	0.002	Kidney damage	Erosion of natural deposits; discharge from refineries and factories; runoff from landfills and croplands	0.002
OC	Methoxychlor	0.04	Reproductive difficulties	Runoff or leaching from insecticide used on fruits, vegetables, alfalfa, livestock	0.04

Category	Contaminant	MCL	Potential health effects from exposure	Sources of contaminant in drinking water	MCLG
IOC	Nitrate (measured as Nitrogen)	10	Infants below the age of six months who drink water containing nitrate in excess of the MCL could become seriously ill and, if untreated, may die. Symptoms include shortness of breath and blue-baby syndrome	Runoff from fertilizer use; leaching from septic tanks, sewage; erosion of natural deposits	10
IOC	Nitrite (measured as Nitrogen)	1	Infants below the age of six months who drink water containing nitrite in excess of the MCL could become seriously ill and, if untreated, may die. Symptoms include shortness of breath and blue-baby syndrome	Runoff from fertilizer use; leaching from septic tanks, sewage; erosion of natural deposits	1
OC	Oxamyl (Vydate)	0.2	Slight nervous system effects	Runoff or leaching from insecticide used on apples, potatoes, and tomatoes	0.2
OC	Pentachlorophenol	0.001	Liver or kidney problems; increased cancer risk	Discharge from wood preserving factories	0
OC	Picloram	0.5	Liver problems	Herbicide runoff	0.5
OC	Polychlorinated biphenyls (PCBs)	0.0005	Skin changes; thymus gland problems; immune deficiencies; reproductive or nervous system difficulties; increased risk of cancer	Runoff from landfills; discharge of waste chemicals	0
R	Radium 226 and Radium 228 (combined)	5 pCi/L	Increased risk of cancer	Erosion of natural deposits	0
IOC	Selenium	0.05	Hair or fingernail loss; numbness in fingers or toes; circulatory problems	Discharge from petroleum refineries; erosion of natural deposits; discharge from mines	0.05
OC	Simazine	0.004	Problems with blood	Herbicide runoff	0.004
OC	Styrene	0.1	Liver, kidney, or circulatory system problems	Discharge from rubber and plastic factories; leaching from landfills	0.1
OC	Tetrachloroethylene (PCE)	0.005	Liver problems; increased risk of cancer	Discharge from factories and dry-cleaners	0

continued

TABLE 5.4 (continued)
National Primary Drinking Water Standards

	Contaminant	MCL or TT[a] (mg/L)[b]	Potential Health Effects from Exposure above the MCL	Common Sources of Contaminant in Drinking Water	Public Health Goal
IOC	Thallium	0.002	Hair loss; changes in blood; kidney, intestine, or liver problems	Leaching from ore-processing sites; discharge from electronics, glass, and drug factories	0.0005
OC	Toluene	1	Nervous system, kidney, or liver problems	Discharge from petroleum factories	1
M	Total Coliforms (including fecal coliform and *E. coli*)	5.0%[g]	Not a health threat in itself; it is used to indicate whether other potentially harmful bacteria may be present[h]	Coliforms are naturally present in the environment as well as feces; fecal coliforms and *E. coli* only come from human and animal fecal waste	0
DBP	Total Trihalomethanes (TTHMs)	0.080	Liver, kidney, or central nervous system problems; increased risk of cancer	Byproduct of drinking water disinfection	n/a[f]
OC	Toxaphene	0.003	Kidney, liver, or thyroid problems; increased risk of cancer	Runoff or leaching from insecticide used on cotton and cattle	0
OC	2,4,5-TP (Silvex)	0.05	Liver problems	Residue of banned herbicide	0.05
OC	1,2,4-Trichlorobenzene	0.07	Changes in adrenal glands	Discharge from textile finishing factories	0.07
OC	1,1,1-Trichloroethane	0.2	Liver, nervous system, or circulatory problems	Discharge from metal degreasing sites and other factories	0.20
OC	1,1,2-Trichloroethane	0.005	Liver, kidney, or immune system problems	Discharge from industrial chemical factories	0.003
OC	Trichloroethylene (TCE)	0.005	Liver problems; increased risk of cancer	Discharge from metal degreasing sites and other factories	0
M	Turbidity	TT[e]	Turbidity is a measure of the cloudiness of water. It is used to indicate water quality and filtration effectiveness (e.g., whether disease-causing organisms are present) Higher turbidity levels are often associated with higher levels of disease-causing microorganisms such as viruses, parasites, and some bacteria	Soil runoff	n/a

	Contaminant	MCL	Potential health effects	Sources of contaminant	MCLG
R	Uranium	30 ug/L	Increased risk of cancer, kidney toxicity	Erosion of natural deposits	0
OC	Vinyl chloride	0.002	Increased risk of cancer	Leaching from PVC pipes; discharge from plastic factories	0
M	Viruses (enteric)	TT^e	These organisms can cause symptoms such as nausea, cramps, diarrhea, and associated headaches; Gastrointestinal illness (e.g. diarrhea, vomiting, and cramps)	Human and animal fecal waste	0
OC	Xylenes (total)	10	Nervous system damage	Discharge from petroleum factories; discharge from chemical factories	10

Note: OC: Organic Chemical; IOC: Inorganic Chemical; D: Disinfectant; DBP: Disinfection Byproduct; M: Microorganism; R: Radionuclides; n/a: Not available.

[a]Definitions:

Maximum Contaminant Level Goal (MCLG)

Maximum Contaminant Level (MCL)

Maximum Residual Disinfectant Level Goal (MRDLG)

Maximum Residual Disinfectant Level (MRDL)

Treatment Technique (TT)

[b]Units are in milligrams per liter (mg/L) unless otherwise noted. Milligrams per liter are equivalent to parts per million (ppm).

[c]Each water system must certify, in writing, to the state (using third-party or manufacturers certification) that when it uses acrylamide or epichlorohydrin to treat water, the combination (or product) of dose and monomer level does not exceed the levels specified, as follows:

Acrylamide = 0.05% dosed at 1 mg/L (or equivalent); Epichlorohydrin = 0.01% dosed at 20 mg/L (or equivalent).

[d]Lead and copper are regulated by a TT that requires systems to control the corrosiveness of their water. For copper, the action level is 1.3 mg/L, and for lead is 0.015 mg/L. To exceed the action level, water systems must take additional steps.

[e]EPA's surface water treatment rules require systems using surface water or groundwater under the direct influence of surface water to:

(1) disinfect their water, and (2) filter their water or meet criteria for avoiding filtration so that the following contaminants are controlled at these levels:

Cryptosporidium 99% removal; *Giardia lamblia:* 99.9% removal or inactivation; Viruses: 99.99% removal or inactivation

Turbidity: may never exceed 1 NTU, and must not exceed 0.3 NTU in 95% of daily samples in any month.

HPC: No more than 500 bacterial colonies per milliliter

[f]Although there is no collective MCL for this contaminant group, there are individual MCLGs for some of the individual contaminants:

– Haloacetic acids: dichloroacetic acid (zero): trichloroacetic acid (0.3 mg/L)

– Trihalomethanes: bromodichloromethane (zero); bromoform (zero); dibromochloromethane (0.06 mg/L). If more than 10% of tap water samples

[g]No more than 5.0% samples total coliform-positive in a month. (For water systems that collect fewer than 40 routine samples per month, no more than one sample can be total coliform-positive per month.) Every sample that has total coliform must be analyzed for either fecal coliforms or *E. coli* if two consecutive TC-positive samples, and one is also positive for *E. coli* fecal coliforms, system has an acute MCL violation.

[h]Fecal coliform and *E. coli* are bacteria whose presence indicates that the water may be contaminated with human or animal wastes.

Source: USEPA, 2005a. National Primary and Secondary Drinking Water Standards. http://www.epa.gov/safewater/mcl.html#mcls (accessed on December 18, 2005).

TABLE 5.5
National Secondary Drinking Water Standards

Contaminant	Secondary Standard
Aluminum	0.05 to 0.2 mg/L
Chloride	250 mg/L
Color	15 (color units)
Copper	1.0 mg/L
Corrosivity	noncorrosive
Fluoride	2.0 mg/L
Foaming agents	0.5 mg/L
Iron	0.3 mg/L
Manganese	0.05 mg/L
Odor	3 threshold odor number
pH	6.5–8.5
Silver	0.10 mg/L
Sulfate	250 mg/L
Total dissolved solids	500 mg/L
Zinc	5 mg/L

Source: USEPA, 2005a. National Primary and Secondary Drinking Water Standards. http://www.epa.gov/safewater/mcl.html#mcls (accessed on December 18, 2005).

regulated by the various states, which often react faster to widely expressed public concerns than the federal government. Examples include MTBE (an infamous gasoline additive), which was regulated by quite a few states before it finally made it on the CCL2, and 1,4-dioxane (solvent stabilizer), which is increasingly detected in association with 1,1,1-TCA plumes, is regulated by some states, but it is not on the CCL2. Some examples of true emerging contaminants (those that neither are on the CCL2 list nor regulated by individual states), include pharmaceuticals and endocrine disrupting compounds (EDCs) such as birth control pills, estrogen replacement products, or steroids (Masters et al., 2004). Some pharmaceuticals

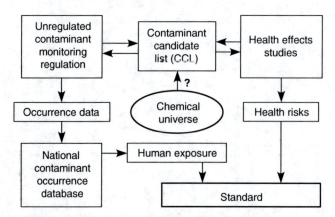

FIGURE 5.14 Schematic presentation of the USEPA process for developing drinking water standards for emerging contaminants.

TABLE 5.6
Drinking Water Contaminant Candidate List 2—CCL2

Microbial Contaminant Candidates		Chemical Contaminant Candidates	CASRN
Adenoviruses		Aluminum	7429-90-5
Aeromonas hydrophila		Boron	7440-42-8
Caliciviruses		Bromobenzene	108-86-1
Coxsackieviruses		DCPA mono-acid degradate	887-54-7
Cyanobacteria (blue-green algae), other freshwater algae, and their toxins		DCPA di-acid degradate	2136-79-0
Echoviruses		DDE	72-55-9
Helicobacter pylori		Diazinon	333-41-5
Microsporidia (Enterocytozoon and Septata)		Disulfoton	298-04-4
Mycobacterium avium intracellulare (MAC)		Diuron	330-54-1
Chemical Contaminant Candidates	**CASRN**	EPTC (s-ethyl-dipropylthiocarbamate)	759-94-4
1,1,2,2-Tetrachloroethane	79-34-5	Fonofos	944-22-9
1,2,4-Trimethylbenzene	95-63-6	p-Isopropyltoluene (p-cymene)	99-87-6
1,1-Dichloroethane	75-34-3	Linuron	330-55-2
1,1-Dichloropropene	563-58-6	Methyl bromide	74-83-9
1,2-Diphenylhydrazine	122-66-7	Methyl-t-butyl ether (MTBE)	1634-04-4
1,3-Dichloropropane	142-28-9	Metolachlor	51218-45-2
1,3-Dichloropropene	542-75-6	Molinate	2212-67-1
2,4,6-Trichlorophenol	88-06-2	Nitrobenzene	98-95-3
2,2-Dichloropropane	594-20-7	Organotins	n/a
2,4-Dichlorophenol	120-83-2	Perchlorate	14797-73-0
2,4-Dinitrophenol	51-28-5	Prometon	1610-18-0
2,4-Dinitrotoluene	121-14-2	RDX	121-82-4
2,6-Dinitrotoluene	606-20-2	Terbacil	5902-51-2
2-Methyl-Phenol (o-cresol)	95-48-7	Terbufos	13071-79-9
Acetochlor	34256-82-1	Vanadium	7440-62-2
Alachlor ESA and other acetanilide pesticide degradation products	n/a	Triazines and degradation products of triazines including, but not limited to Cyanazine	21725-46-2 6190-65-4

Note: n/a: not avaialble.

Source: USEPA, 2005b. Fact sheet: the drinking water contaminant candidate list—the source of priority contaminants for the Drinking Water Program. Office of Water, 6 p. (Available at http://www.epa.gov/safewater/ccl/ccl2_list.html.)

are not completely metabolized after consumption by humans or animals and are excreted in their original form, while others are transformed into different compounds (conjugates). Thanks to continuously advancing analytical methods, dozens of these compounds, mostly at very low levels, have been detected in recent years in municipal wastewater and drinking water systems. Releases of pharmaceuticals and EDCs to the environment are likely to continue as the human population increases and ages; the pharmaceutical industry formulates new prescription and nonprescription drugs, and promotes their use; and more wastewater is generated that enters the hydrologic cycle and may impact groundwater resources (Masters et al., 2004).

REFERENCES

Ball, E.R. and J.C. Ryden, 1984. Nitrogen relationships in intensively managed temperate grassland. *Plant Soil*, 76: 23–33.

Banks, W.S.L. and D.A. Battigelli, 2002. Occurrence and distribution of microbiological contamination and enteric viruses in shallow groundwater in Baltimore and Harford counties, Maryland. US Geological Survey Water-Resources Investigations Report 01–4216, Baltimore, Maryland, 39 p.

Cohen, R.M. and J.W. Mercer, 1993. *DNAPL Site Evaluation*. C.K. Smoley, Boca Raton, Florida.

Craun, G.F., 1986. *Waterborne Diseases in the United States*. CRC Press, Boca Raton, Florida, 192 p.

Dougherty, P.H. (ed.), 1985. *Caves and Karst of Kentucky*. Special Publication 12, Series XI, Kentucky Geological Survey, Lexington, Kentucky, 196 p.

Drever, J.I., 1988. *The Geochemistry of Natural Waters*. Prentice-Hall, Englewood Cliffs, N.J., 437 p.

Eberts, S.M., Erwin, M.L., and P.A. Hamilton, 2005. Assessing the vulnerability of public-supply wells to contamination from urban, agricultural, and natural sources. US Geological Survey Fact Sheet 2005–3022, 4 p.

Focazio, M.J., et al., 2002. Assessing ground-water vulnerability to contamination—Providing scientifically defensible information for decision makers. US Geological Survey Circular 1224, 33 p.

Fuller, M.L., 1910. Underground waters for farm use. US Geological Survey Water-Supply Paper 255, Washington, DC.

Gerba, C.P., 1988. Methods for virus sampling and analysis of ground water. In: Collins, A.G., and Johnson, A.I. (eds.), *Groundwater Contamination Field Methods*. American Society for Testing and Materials, ASTM STP 963, p. 343–348.

Gerba, C.P., 1999. Virus survival and transport in groundwater. *J. Ind. Microbiol. Biotechnol.*, 22(4): 247–251.

Hamilton, P.A., Miller, T.L., and D.N. Myers, 2004. Water quality in the nation's streams and aquifers—Overview of selected findings, 1991–2001. US Geological Survey Circular 1265, 20 p.

Hantzsche, N.N. and E.J. Finnemore, 1992. Predicting groundwater nitrate–nitrogen impacts. *Ground Water*, 30(4): 490–499.

Hauwert, N.M. and S. Vickers, 1994. Barton Springs/Edwards Aquifer: Hydrogeology and groundwater quality. Barton Springs/Edwards Aquifer Conservation District, Austin, Texas, 91 p. + Appendices.

Heaton, T.H.E., 1986. Isotopic studies of nitrogen pollution in the hydrosphere and atmosphere—A review. *Chem. Geol.*, 59: 87–102.

Huling, S.G. and J.W.Weaver, 1991. Dense nonaqueous phase liquids. Ground Water Issue, EPA/540/4-91-002, Robert S. Kerr Environmental Research Laboratory, Ada, Oklahoma, 28 p.

Izbicki, J.A., 2004. A small-diameter sample pump for collection of depth-dependent samples from production wells under pumping conditions. US Geological Survey Fact Sheet 2004–3096, 2 p.

Kaplan J.E., et al., 1982. Epidemiology of Norwalk gastroenteritis and the role of Norwalk virus in outbreaks of acute nonbacterial gastroenteritis. *Ann Intern. Med.*, 96: 756–761.

Keswick, B.H. and C.P. Gerba, 1980. Viruses in groundwater. *Environ. Sci. Technol.*, 14: 1, 290–291, 297.

Kreitler, C.W. and D.C. Jones, 1975. Natural soil nitrate—The cause of nitrate contamination of ground water in Runnels County, Texas. *Ground Water*, 13(1): 53–61.

Kresic, N., et al., 2005. Technical impracticability (TI) of DNAPL remediation in karst. In: Stevanovic, Z., and Milanovic P. (eds.), *Water Resources and Environmental Problems in Karst–Cvijić 2005*, Proceedings of the International Conference, University of Belgrade, Institute of Hydrogeology, Belgrade, p. 63–65.

Masters, R.W., Verstraeten, I.M., and T. Heberer, 2004. Fate and transport of pharmaceuticals and endocrine disrupting compounds during ground water recharge. *Ground Water Monit. Rem.*, 54–57. [Special Issue: Fate and transport of pharmaceuticals and endocrine disrupting compounds during ground water recharge.]

McGinnis, G.D., 1989. Overview of the wood-preserving industry, *Proceedings Technical Assistance to USEPA Region IX: Forum on Remediation of Wood Preserving Sites,* San Francisco, California.

National Academy of Sciences, 1978. *Nitrates—An Environmental Assessment*. National Academy of Sciences, Washington, DC, 723 p.

Newell, C.J., et al., 1995. Light nonaqueous phase liquids. Ground Water Issue, EPA/540/S-95/500, Robert S. Kerr Environmental Research Laboratory, Ada, Oklahoma, 28 p.

Nichols, E.M., Beadle, S.C., and M.D. Einarson, 2000. Strategies for characterizing subsurface releases of gasoline containing MTBE. American Petroleum Institute (API), Regulatory and Scientific Affairs Publication Number 4699, Washington, DC.

Palmer, A.N., 1985. The Mammoth Cave region and Pennyroyal Plateau. In: Dougherty, P.H. (ed.), *Caves and Karst of Kentucky*. Special Publication 12, Series XI, Kentucky Geological Survey, Lexington, Kentucky, p. 97–118.

Palmer, C.D. and R.L. Johnson, 1989. Physical processes controlling the transport of nonaquesous phase liquids in the subsurface. In: *USEPA, Seminar Publication: Transport and Fate of Contaminant in the Subsurface*, EPA/625/4–89/019, US Environmental Protection Agency, p. 23–27.

Pankow, J.F. and J.A. Cherry, 1996. *Dense Chlorinated Solvents and Other DNAPLs in Groundwater*. Waterloo Press, Guelph, Ontario, Canada, 522 p.

Reisch, M. and D.M. Bearden, 1997. Superfund fact book. National Council for Science and the Environment, Congressional Research Service Reports, 97-312 ENR, 33 p. http://ncseonline. org/NLE/CRSreports/Waste/ (accessed on January 24, 2006).

Reneau, R.B., Jr., et al., 1989. Fate and transport of biological and inorganic contaminants from on-site disposal of domestic wastewaters. *J. Environ. Qual.*, 18: 135–144.

Robertson, W.D., Cherry, J.A., and E.A. Sudicky, 1991. Groundwater contamination from two small septic systems on sand aquifers. *Ground Water*, 29(1): 82–92.

Rosenfeld, J.K. and R.H. Plumb, Jr., 1991. Groundwater contamination at wood treatment facilities. *Ground Water Monit. Rev.*, 11(1): 133–140.

Seiler, R.L., 1996. Methods for identifying sources of nitrogen contamination of groundwater in valleys in Washoe County, Nevada. US Geological Survey Open-File Report 96–461, Carson City, Nevada, 20 p.

Sievers, R.E., et al., 1977. Environmental trace analysis of organics in wastewater by glass capillary column chromatography and ancillary techniques. *J. Chromatogr.*, 142: 745–754.

Sobsey, M.D., 1989. Inactivation of health related microorganisms in water by disinfecting processes. *Water Sci. Technol.*, 21(3): 179–195.

Thurman, E.M., et al., 1984. Sewage contaminants in ground water. In: LeBlanc, D.R. (ed.), *Movement and Fate of Solutes in a Plume of Sewage-Contaminated Ground water*, Cape Cod, MA. US Geological Survey Open-File Report 84-475, p. 47–64.

Tiemann, M., 1996. *91041*: Safe Drinking Water Act: Implementation and reauthorization. National Council for Science and the Environment, Congressional Research Service Reports. http://ncseonline.org/nle/crsreports (accessed January 21, 2006).

Tiemann, M., 2005. Safe Drinking Water Act: Implementation and issues. CRS Issue Brief for Congress, received through the CRS Web. Congressional Research Service, The Library of Congress, 16 p. http://ncseonline.org/nle/crsreports (accessed January 21, 2006).

USEPA, 1980. Design manual—On-site wastewater treatment and disposal systems. US Environmental Protection Agency Report EPA-6625/1-77-008, 480 p.

USEPA, 1986. Solving the hazardous waste problem: EPA's RCRA Program. EPA/530-SW-86-037, Office of Solid Waste, Washington, D.C., 34 p.

USEPA, 1992. Estimating potential for occurrence of DNAPL at Superfund sites. Publication 9355.4-D7FS, R.S. Kerr Environmental Research Laboratory, Office of Solid Waste and Emergency Response, US Environmental Protection Agency, 9 p.

USEPA, 1993. Guidance for evaluating the technical impracticability of ground-water restoration; Interim Final. Directive 9234.2-25, Office of Solid Waste and Emergency Response, US Environmental Protection Agency, Washington, DC, 26 p.

USEPA, 1994. This is Superfund. A citizen's guide to EPA's Superfund Program. EPA 540-K-93-008, Office of Emergency and Remedial Response, Washington, DC, 15 p.

USEPA, 1999. A review of contaminant occurrence in public water systems. EPA 816-R-99-006, US Environmental Protection Agency, Office of Water, 78 p.

USEPA, 2000. The history of drinking water treatment. EPA-816-F-00-006, Office of Water, 4 p.

USEPA, 2003a. Overview of the Clean Water Act and the Safe Drinking Water Act. http://www.epa.gov/OGWDW/dwa/electronic/ematerials/ (accessed September 12, 2005).

USEPA, 2003b. An overview of the Safe Water Drinking Act. http://www.epa.gov/OGWDW/dwa/electronic/ematerials/ (accessed September 12, 2005).

USEPA, 2003c. Regulating microbial contamination; unique challenge, unique approach. http://www.epa.gov/OGWDW/dwa/electronic/ematerials/ (accessed September 12, 2005).

USEPA, 2005. http://www.epa.gov/OGWDW/dwh/t-soc/pcbs.html (accessed October 25, 2005).

USEPA, 2005a. National Primary and Secondary Drinking Water Standards. http://www.epa.gov/safewater/mcl.html#mcls (accessed on December 18, 2005).

USEPA, 2005b. Fact sheet: the drinking water contaminant candidate list—the source of priority contaminants for the Drinking Water Program. Office of Water, 6 p. (Available at http://www.epa.gov/safewater/ccl/ccl2_list.html.)

Van Denburgh, A.S., Goerlitz, D.F., and E.M. Godsy, 1993. Depletion of nitrogen-bearing explosives wastes in a shallow ground-water plume near Hawthorne, Nevada. In: Morganwalp, D.W., and Aronson, D.A. (compilers), *US Geological Survey Toxic Substances Hydrology Program—Abstracts of the Technical Meeting*, Colorado Springs, CO, September 20–24, 1993. US Geological Survey Open-File Report 93–454, p. 172.

Van Vuren, J.P.J., 1949. *Soil Fertility and Sewage—An Account of Pioneer Work in South Africa in the Disposal of Town Wastes*. Dover, New York, 236 p.

Villaume, J.F., Lowe, P.C., and D.F. Unites, 1983. Recovery of coal gasification wastes: An innovative approach. *Proceedings of the Third National Symposium on Aquifer Restoration and Ground-Water Monitoring*. National Water Well Association, Worthington, OH, p. 434–445.

Wilhelm, S.R., Schiff, S.L., and J.A. Cherry, J.A., 1994. Biogeochemical evolution of domestic waste water in septic systems, Pt. 1, Conceptual model. *Ground Water*, 32(6): 905–916.

6 Fate and Transport of Contaminants

Potential groundwater contaminants can enter the subsurface soils in the source area in two basic states: (1) as already dissolved in the infiltrating water and (2) as liquid, either readily miscible with water, or hydrophobic (immiscible) such as various nonaqueous phase liquids (NAPLs; see Section 5.2.4). In any case, before reaching the water table (saturated zone, aquifer), the potential contaminant first has to move through the vadose, or unsaturated zone, where it is subject to various physical and bio-geochemical processes collectively called contaminant fate and transport (F&T) processes. Figure 6.1 and Figure 6.2 illustrate the general concept of contaminant movement from the land surface into the subsurface and further downward through the vadose zone. One or more F&T processes may result in the potential contaminant never reaching the water table, which would be an ideal outcome. If the contaminant does reach the saturated zone, it would continue its "journey" through the aquifer and it will be affected by most of the same F&T processes as in the vadose zone. Although distinction between the terms "fate" and "transport" is not always clear, it is generally understood that fate refers to various bio-geochemical processes acting upon the contaminant, whereas transport refers to physical movement of the contaminant. An example of a fate process would be complete mineralization of an organic contaminant, i.e., its conversion into inorganic substances such as carbon dioxide and water. An example of a pure transport process would be advection, or movement of the dissolved contaminant together with groundwater. On the other hand, some critical processes that affect contaminant movement, without changing its chemical nature, may be the result of various chemical interactions among the three media: contaminant, water, and aquifer solids. An example would be processes collectively described as sorption of the dissolved contaminant particles (molecules) onto solid surfaces of the porous media (grains). This general term is used to describe immobilization of the contaminant particles by the porous media, irrespective of the actual mechanism. It may be the result of various more specific processes caused by geochemical interactions (forces) between the solids and the dissolved contaminant. Cation exchange would be one example of sorption where the contaminant is immobilized by the mineral (usually clay) surfaces. This immobilization may not be permanent, and the contaminant may be released back into the water solution by the reverse process when geochemical conditions in the aquifer change (e.g., change of pH or inflow of another chemical species with the greater affinity for cation exchange with the mineral surfaces). Adsorption is the term often used to describe a process of contaminant particles or molecules "sticking" to aquifer materials simply because of the affinity for each other. For example, many hydrophobic organic contaminants are adsorbed onto particles of organic carbon present in the aquifer, and can be desorbed if conditions change. Adsorption is commonly used interchangeably with sorption, a more generic term, which sometimes may cause confusion. Absorption, a rather vague term, usually refers to contaminant incorporation "deep" into the solid particle

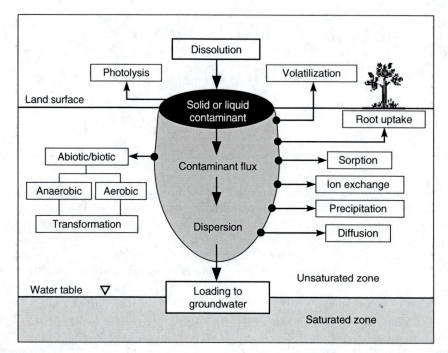

FIGURE 6.1 Physical, chemical, and biological processes affecting contaminant fate and transport in the unsaturated zone.

structure and it has chemical connotation. The term, however, is seldom used as its net effect would be equal to a complete destruction of the contaminant, i.e., its permanent removal from the flow system. Precipitation is another mechanism that completely removes the contaminant from the flow system. It is a well-understood chemical complexation reaction in which the complex formed by two or more aqueous species is a solid. When the precipitate

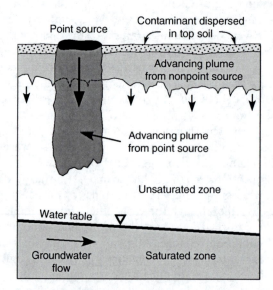

FIGURE 6.2 Migration of dissolved contaminants through vadose zone from point and nonpoint sources.

is insoluble, the contaminant is permanently (irreversibly) removed from the flow system. Precipitation is particularly important to the behavior of heavy metals in soil or groundwater systems, and it is heavily influenced by pH and redox potential. Dissolution is the reverse of precipitation, and it may reintroduce the same contaminant back into the flow system, or it may introduce a new chemical species into the solution based on the changing geochemical conditions.

It should be noted that most commonly applied analytical equations of contaminant F&T do not incorporate various actual chemical reactions between the contaminants, groundwater, and aquifer solids. When it is important to more accurately represent the contaminant fate in that regard, the flow equations have to be coupled with geochemical models, which substantially increases the complexity of calculations and data requirements. In many practical applications, however, various processes that immobilize or completely remove the contaminant from the flow system can be sufficiently accurately described by couple of generic parameters, as explained in more detail in the following sections of this chapter.

After it has been subject to a variety of complex and rapid interactions in the source zone, the dissolved, mobile contaminant phase flowing away from the source is usually characterized by much slower reactions and the flow system is often described as being in (quasi) equilibrium for practical purposes. Although this assumption helps to greatly simplify F&T calculations, it is not entirely correct because whenever the front of a moving contaminant plume encounters uncontaminated groundwater, the system enters into nonequilibrium conditions. The exception would be a stable, nonexpanding plume. Figure 6.3 illustrates main types of contaminant plumes with respect to various possible F&T processes influencing their development. Each plume type, expanding, stable or shrinking, is to a varying degree subject to all of the F&T processes discussed further in this chapter. The bullets in Figure 6.3 list only those that may have the greatest net effect on the particular plume type.

Three key questions when evaluating the F&T of a contaminant are:

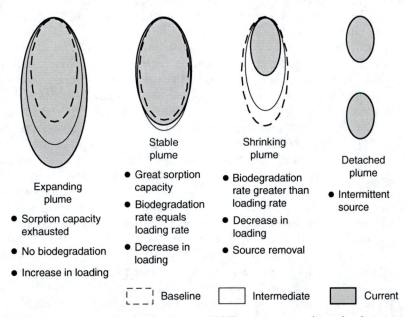

FIGURE 6.3 Influence of various fate and transport (F&T) processes on plume development; while most F&T processes may be present in any give case, the bullets list only those with the possibly greatest net effect. (Modified from USEPA, The report to Congress: Waste disposal practices and their effects on ground water–Report. EPA 570977001, 1977, pp. 531.)

- Will there be an impact on the human health and the environment (or, rephrased, is there a pathway between the contaminant source zone and the potential receptor)?
- How long would it take the contaminant to reach the potential receptor?
- What would the concentration of the contaminant be when it reaches the receptor?

An answer to the first question requires a thorough understanding of the underlying geology and hydrogeology, and development of a defensible conceptual site model (CSM). The remaining two questions, in addition to having a defensible CSM, may be answered based on the thorough understanding of the contaminant release mechanism (the timing of the release, and the contaminant mass introduced into the subsurface), and the contaminant F&T processes applicable to the CSM. The accuracy of the quantitative answers involving the time and the concentration will depend on the site-specific data available for estimating (determining) quantitative parameters that describe the F&T processes. Regardless of the data availability, the answers will always, by default, have a degree of uncertainty. This uncertainty should be discussed with various stakeholders whenever possible and addressed in any report that may be accompanying F&T calculations or modeling in general. The main reason for the quantitative uncertainty is the inevitable heterogeneity and anisotropy of any natural porous media, as discussed throughout Chapter 1 and Chapter 2.

6.1 FREE-PHASE AND RESIDUAL-PHASE CONTAMINATION

NAPLs may be present in the subsurface as continuous (contiguous) bodies of a relatively significant extent (volume), which occupy all the pore space in the aquifer material. In such a case, they are referred to as free-phase NAPLs. Detecting free-phase NAPL in the saturated zone is in most cases impracticable and when it happens, it is often a matter of either luck or a very significant lateral extent of the NAPL in the form of a pool. At residual saturation, NAPL occurs as disconnected singlet and multipore globules (ganglia) within the larger pore spaces that have been cutoff and disconnected from the continuous NAPL body by the invading water (Cohen and Mercer, 1993; Pankow and Cherry, 1996). The actual saturation of pore space with NAPL can vary between 0 and 1, with the individual saturations of all fluids present (NAPLs and water) always summed to 1 in the saturated zone. Determining the percentage of the NAPL saturation, even from the actual core samples, is difficult and requires application of complicated, laborious techniques, and care in obtaining, preserving, and transporting the samples. In any case, numeric saturation data from discrete sampling points would have to be extrapolated and interpolated (contoured) to estimate the volume of dense nonaqueous phase liquid (DNAPL) present in the corresponding volume of the aquifer. Cohen and Mercer (1993) present laboratory and field values of residual saturation for various NAPL fluids in intergranular porous media (clays, silts, sands, gravels, and their mixtures). They conclude that values of residual saturation generally range between 0.10 and 0.50 in saturated porous media, and tend to be higher in the preferential pathways of NAPL transport. Residual saturation also tends to increase with increasing pore aspect ratios and pore size heterogeneity, and with decreasing porosity, probably due to reduced pore connectivity and a decrease in mobile nonwetting fluid (NAPL) in smaller pore throats. Values of residual saturation in the vadose zone are generally smaller than in the saturated zone, and range between 0.10 and 0.20. Residual saturation and retention capacity in the vadose zone increase with decreasing intrinsic permeability, effective porosity, and moisture content (Cohen and Mercer, 1993).

Relatively small residual sources of DNAPL may contaminate groundwater for very long periods of time because of the low solubility and mobility of the product trapped in pore spaces by capillary forces. For this reason, the main goal during characterization of sites suspected to be contaminated by DNAPLs is to find their source zones, try to remove DNAPL, and contain the dissolved phase contaminant plume leaving the source area. U.S. Environmental Protection Agency (USEPA) defines the DNAPL zone as "that portion of the subsurface where immiscible liquids (free-phase or residual DNAPL) are present either above or below the water table" (USEPA, 1996). Unfortunately, however, the interactions among DNAPL, porous media, and other subsurface fluids are complex and affect the ability to detect DNAPLs and find their source zones. DNAPL migration tends to follow small preferential pathways that are very difficult to delineate. The more complex the geology, the more difficult it is to characterize a DNAPL release. A site with complicated geology will generally have a greater number of preferential pathways of varying size as well as more confining layers to trap large and small amounts of DNAPL. Characterizing a DNAPL release can also be difficult even in formations that appear to be homogeneous. Column tests by Schwille (1988) and field tests by Poulsen and Kueper (1992) demonstrated the control on DNAPL migration exerted by preferential flow pathways in what appeared to be homogeneous materials. This is exacerbated in heavily heterogeneous geologic environments, making it even more difficult to determine the location of residual and potentially mobile (free-phase) DNAPL (ITRC, 2003). Many field techniques for detecting the presence and estimating the volume of DNAPL in the subsurface have been proposed and tested, and their description is beyond the scope of this book. Interested readers should refer to Cohen and Mercer (1993), Pankow and Cherry (1996), and ITRC (2003). The same authors also discuss in detail various mechanisms that control DNAPL saturation and migration in both the vadose and saturated zones.

One potential indication of the presence of DNAPL in the saturated zone in a monitoring well is that the concentration of the dissolved contaminant is greater than 1% to 10% of the compound's effective solubility (Feenstra and Cherry, 1988; Pankow and Cherry, 1996). Figure 6.4 illustrates how concentrations in monitoring wells may be used to delineate aquifer zones with potential presence of DNAPL. One reasoning behind this widely

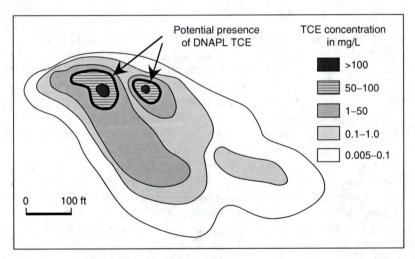

FIGURE 6.4 Delineation of potential aquifer zones with DNAPL based on the 1%–10% solubility rule of thumb. The contaminant of concern is trichloroethene (TCE), which has aqueous solubility approximately between 1100 and 1400 mg/L. The aquifer area that may contain residual DNAPL is assumed to be within the 50 mg/L concentration contour, or approximately 4% of the pure-phase solubility. TCE is not the only compound present in DNAPL. In case of a DNAPL mixture, the effective solubility of TCE would be less than the pure-phase solubility, and would have to be determined based on Raoult's law.

accepted "rule of thumb" is that if DNAPL is present, it will generally be present either as a small lens in a small preferential pathway, as residual-phase ganglia, or diffused from a preferential pathway into a fine-grained matrix. If a 10 ft well screen is close to or intersects one of these areas, the area where the DNAPL is present will likely be thin when compared to the full length of the well screen. This is mainly because groundwater flow is generally laminar and will not mix quickly with the larger interval of the formation over short distances. As a consequence, the aqueous phase contamination dissolving from the DNAPL into groundwater at a concentration close to its solubility limit will remain contained within a narrow (thin) interval some distance downgradient of the source zone. This contamination will be diluted in the monitoring well during sampling by the larger screened interval of the formation. Therefore, concentrations of a small percentage of solubility may indicate presence of DNAPL in the vicinity of the monitoring well. If well screens are short, there will be less dilution and the contaminant concentration will be a higher percentage of solubility before it indicates DNAPL. This technique is subjective and must be applied very carefully because if used alone it may grossly overestimate or underestimate presence and volumes of DNAPL in the aquifer. It should be considered only a part of the process used to determine if DNAPL is present, and not a method that by itself will indicate the presence or absence of DNAPL (ITRC, 2003). A very useful discussion on the behavior and dissolved concentrations expected to be found in DNAPL source zones is given by Anderson et al. (1987, 1992).

Another indirect method for detecting potential presence of residual DNAPL is to calculate the hypothetical pore-water concentration from the measured total soil concentration by assuming equilibrium chemical partitioning between the solid phase, the pore water, and the soil gas, and assuming that no DNAPL is present in the collected sample. This pore-water concentration (C_w, in mg/L or $\mu g/cm^3$) can be expressed in terms of the total soil concentration (C_t, in $\mu g/g$ dry weight) as (Pankow and Cherry, 1996)

$$C_w = \frac{C_t \rho_b}{K_d \rho_b + \theta_w + H_c \theta_a} \tag{6.1}$$

where ρ_b is the dry bulk density of the soil sample (g/cm^3), θ_w is water-filled porosity (volume fraction), θ_a is air-filled porosity (volume fraction), K_d is partition coefficient between pore water and solids for the compound of interest (cm^3/g) (see Section 6.2.3 for explanation of K_d), and H_c is dimensionless Henry's gas law constant for the compound of interest.

If no DNAPL is present, there is a maximum amount of chemical, which can be contained in the soil sample at equilibrium with the soil pore water and air. In other words, for true aqueous solute (dissolved phase) in equilibrium, the calculated pore-water concentration (C_w) has to be equal to the solubility concentration (S_w) of the chemical: $C_w = S_w$. If the calculated pore-water concentration is higher than the solubility concentration, some DNAPL phase of the chemical has to be present in the sample. Note that for a DNAPL mixture, effective solubilities of individual compounds will be lower than their pure-phase aqueous solubilities (see Section 4.1 and application of Raoult's Law). Pankow and Cherry (1996) provide the following example of the application of Equation 6.1:

- Measured TCE concentration (C_t) in the soil sample taken from the saturated zone (where θ_a equals zero) was 3100 mg/kg, or 3100 $\mu g/g$.
- The partition coefficient (K_d) for TCE was calculated from the fraction of organic carbon in the soil sample ($f_{oc} = 0.001$) and the so-called organic–carbon partition coefficient for TCE ($K_{oc} = 126$) as follows: $K_d = f_{oc} \times K_{oc} = 0.001 \times 126 = 0.126$.

- The bulk density (ρ_b) was estimated to be 1.86 g/cm^3.
- The total porosity, equal to water-filled porosity (θ_w), was estimated to be 0.3.

Inserting the above values into Equation 6.1 gives the calculated value for the pore-water concentration of 10,790 mg/L, which is much higher than the TCE pure aqueous solubility of approximately 1100 mg/L. The conclusion is that residual liquid TCE DNAPL is present in the sample.

6.1.1 DISSOLUTION RATES AND RATE-LIMITED DISSOLUTION

When a contaminant is highly soluble or completely miscible in water, such as many salts (e.g., sodium chloride, perchlorate) and a considerable number of organic compounds (e.g., ethanol, 1,4-dioxane), the rate of dissolution by groundwater flowing through the source zone is not limited. The time required to completely deplete the source zone is theoretically instantaneous: as soon as the contaminant comes in contact with water, it is dissolved and carried away by the groundwater flux (flow). The same is true when sources of miscible constituents in the vadose zone or at the land surface are exposed to the infiltrating water (precipitation). The flux of the contaminant entering the subsurface will depend only on the infiltration rate. This is illustrated schematically in Figure 6.5. If the contaminant is not retarded by any of the F&T processes in the unsaturated zone, and if the loading of the

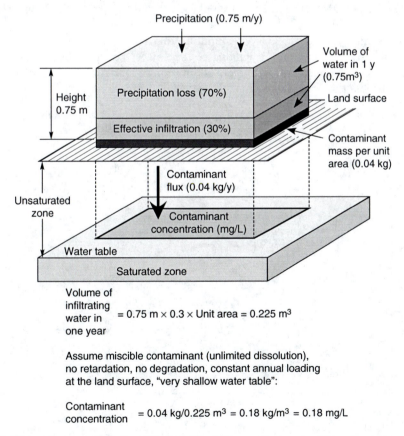

FIGURE 6.5 Determination of the flux of infinitely soluble contaminant entering the vadose zone, as it is being dissolved by the infiltrating water.

contaminant at the land surface is constant (continuous source of the same strength), the concentration arriving at the water table will eventually become constant as well. In such a case, the concentration can be calculated by dividing the mass of the contaminant at the surface with the volume of the infiltrating water.

In case of low-soluble, immiscible compounds such as NAPLs, the dissolution rate is limited and it depends on various factors including groundwater flow velocity, effective solubility, contact area between the NAPL body (e.g., pool, ganglia) and the flowing water, dispersivity, diffusion, effective porosity, and DNAPL density. Even when assuming that some of these factors can be estimated rather accurately at a certain site, it is almost impossible to reasonably accurately determine the distribution of free-phase and residual DNAPL, and the actual geometric shapes of various DNAPL bodies that are needed to determine their contact areas with water. When comparing two DNAPL bodies with the same volume of product, all other factors being equal, the time needed for the complete source depletion will be longer for the DNAPL body with the smaller contact area. This is schematically illustrated in Figure 6.6. DNAPL pool resting on an impermeable layer is exposed to the groundwater flux only at the top, while irregular, suspended, and disconnected DNAPL bodies have proportionally much greater contact areas and are more quickly dissolved (depleted).

One seemingly intuitive way of estimating the time needed for a DNAPL source zone depletion is to base the calculation on the pure aqueous phase solubility of the single DNAPL compound, the estimated mass of DNAPL in the subsurface, and the groundwater flux through the source zone. For example, it is assumed that there is mass (m) of 1000 kg (1 million mg) of TCE DNAPL, either as free-phase, or residual, or both, "somewhere" in a particular volume of the aquifer. Knowing that the pure-phase aqueous solubility (C_w) of TCE is approximately 1100 mg/L, and estimating that the effective groundwater flux (Q_{ef}) through the effective porosity in the volume of the aquifer occupied by DNAPL is 2 m^3/d (2000 L/d), the time needed for the complete dissolution of DNAPL TCE can be misleadingly calculated to be less than one day:

$$t = \frac{m}{Q_{ef} C_w} = \frac{1,000,000 \text{ mg}}{2000 \text{ L/d} \times 1100 \text{ mg/L}} = 0.45 \text{ d}$$

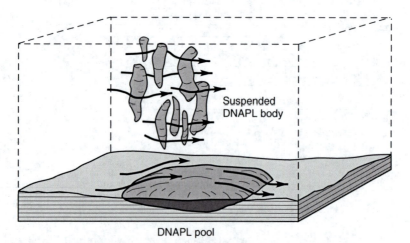

Suspended
DNAPL body

DNAPL pool

FIGURE 6.6 DNAPL pool resting on an impermeable layer is exposed to the groundwater flux only at the top, while irregular, suspended, and disconnected DNAPL bodies have proportionally greater contact areas and are dissolved (depleted) faster. Arrows show directions of groundwater flow.

Although the most simple and the fastest, this "flush volume method" is the least desirable and potentially the most erroneous, because it completely ignores various other factors that are collectively much more important for the source depletion than the pure-phase TCE solubility and the mass of TCE present "somewhere" in the aquifer. Figure 6.7 illustrates one of the reasons why residual DNAPL can persist in the aquifer for a longtime and is dissolved in the groundwater at a limited rate. DNAPL blobs in dead-end or otherwise restricted pores are surrounded by the stagnant water and cannot dissolve as quickly as the blobs that are being constantly flushed by the flowing groundwater (groundwater advective flux). This flowing water may be either "clean" or have a much lower concentration of the dissolved constituent than water around the DNAPL blob in the dead-end pore. As a result, the dissolution rate of the blobs being constantly flushed by the flowing groundwater will be much higher compared to the diffusion-driven dissolution of the dead-end blobs. One significant consequence of the residual DNAPL persistency on groundwater remediation of the source zones is that even after a significant mass of product has been removed from the aquifer, the impact of the remaining mass on downgradient dissolved concentrations may still be great. In other words, this mass removal may not result in a concentration decrease down to maximum contaminant level (MCL) for the contaminant of concern, which is typically a very low number (e.g., 5 ppb for TCE). It is therefore not surprising that there is an ongoing (and often heated) debate among various stakeholders as to the benefits of mass removal in cases where there is a high level of uncertainty as to the locations and mass of DNAPL present in all potential source zones. Some stakeholders argue that removing some, but not all DNAPL, sources that may be creating dissolved plumes would not necessarily reduce contaminant concentrations to the levels not posing risk to human health and the environment. This risk is always calculated based on the dissolved concentrations at the points of compliance and exposure.

The complicated dissolution process of NAPLs, as suggested by some researchers, is likely dependent on at least 10 different dimensionless variables, which in turn are functions of numerous flow and transport parameters (Miller et al., 1990; see also Clement et al., 2003, and

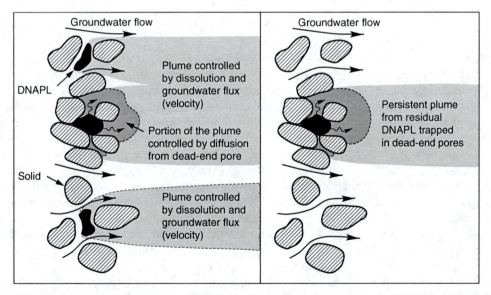

FIGURE 6.7 Schematic presentation of DNAPL dissolution at a pore scale. DNAPL blobs being constantly flushed by the flowing groundwater will dissolve faster than those trapped in dead-end pores with stagnant water where dissolution is driven by slow diffusion.

Pankow and Cherry, 1996 for discussion on various related equations). A commonly accepted simplified approach for quantifying the rate of mass transfer from the DNAPL phase to the dissolved constituent phase is to lump various processes into one parameter called mass transfer coefficient. In general terms, this parameter is expressed as (Pankow and Cherry, 1996)

$$
\begin{matrix}
\text{Rate of} & & \text{Mass transfer} & & \text{Concentration} & & \text{Contact} \\
\text{mass transfer} & & \text{coefficient} & & \text{difference} & & \text{area} \\
& = & & \times & & \times & \\
(M/T) & & (L/T) & & (M/L^3) & & (L^2)
\end{matrix}
\tag{6.2}
$$

where units for mass (M), length (L), and time (T) are in parentheses. The mass transfer coefficient is usually first estimated by some of the proposed equations, and then calibrated to match field data if the related analysis is performed with the aid of numeric models. In practice, the majority of projects that include a more thorough quantitative analysis of contaminant F&T associated with NAPL source zones use numeric modeling techniques to account for the rate-limited dissolution processes. For example, MT3DMS and RT3D, two widely used F&T models based on the Modflow groundwater flow solution (see Chapter 7) include options for modeling a constituent transfer from the immobile phase (e.g., DNAPL) to the mobile phase (dissolved phase) by using the mass transfer coefficient as one of the model parameters.

6.2 DISSOLVED PHASE CONTAMINATION

The characterization and quantification of the dissolved phase groundwater contamination can be performed with the following two approaches, which make the most sense when applied together: (1) measurement of the contaminant concentration and (2) measurement of the contaminant flux. Advective contaminant flux (Q_c) is simply the amount of contaminant, dissolved in groundwater and expressed with its concentration (C_c), that flows through a certain cross section of the aquifer (A) driven by the linear (effective) groundwater velocity (v_{ef}):

$$
Q_c = C_c \times v_{ef} \times A \text{ (kg/d)}
\tag{6.3}
$$

Whatever the investigative approach is, it is very important to collect three-dimensional, site-specific information on most (if not all) physical and chemical parameters needed to quantify the contaminant F&T. As emphasized throughout this book, groundwater flow takes place in a three-dimensional space, which is heterogeneous and anisotropic by default. F&T of dissolved contaminants takes place in the same three-dimensional heterogeneous space, and it is even more important not to represent it as a "sand box": a contaminant may move through a preferential, narrow, and convoluted flow zone at some high concentration and it may even remain undetected, causing various negative impacts at distances far from the source. For this and other reasons, costs and efforts associated with contaminant F&T characterization, and subsequent groundwater-remediation at an average "contaminated site," far outweigh most groundwater supply projects. Figure 6.8 illustrates one of the more complicated "sites" and difficulties in calculating (estimating) contaminant fluxes, and measuring representative dissolved concentrations. For example, relatively high concentrations measured in a highly transmissive flow zone, such as an open fracture or karst conduit, may indicate the existence of a DNAPL source zone in the bedrock, strong enough to create such a high contaminant flux. At the same time, comparable or higher concentrations in the residuum sediments may be created by similarly strong or stronger DNAPL zones, but the contaminant flux in the residuum would normally be much lower because of its low hydraulic conductivity compared to karst

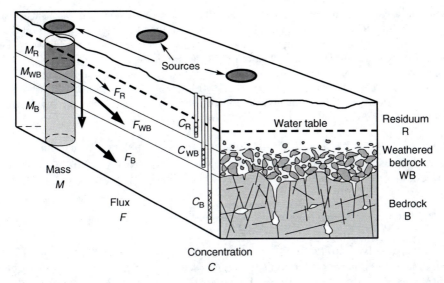

FIGURE 6.8 Schematic presentation of contaminant mass distribution in karst subsurface below DNAPL spill zones, including pathways of dissolved phase fate and transport. M_R, M_{WB}, and M_B—immobile contaminant mass in the residuum, weathered bedrock, and fractured (karst) bedrock, respectively; F_R, F_{WB}, and F_B—flux of dissolved contaminants through the residuum, weathered bedrock, and fractured (karst) bedrock, respectively; C_R, C_{WB}, and C_B—dissolved contaminant concentration measured at the point of compliance (e.g., monitoring well) in the residuum, weathered bedrock, and fractured (karst) bedrock, respectively. (From Kresic et al., Technical impractability (TI) of DNAPL remediation in karst. In: Stevanovic, Z. and P. Milanovic (eds.), *Water Resources and Environmental Problems in Karst-Cvijic 2005, Proceedings of International Conference, University of Belgrade*, Institute of Hydrogeology, Belgrade, pp. 63–65, 2005.)

features. To make things more complicated, the bulk of the bedrock volume may have low hydraulic conductivity as well, and thus relatively low overall advective contaminant flux; at the same time, DNAPL in the bedrock source zones may drive the contaminant into the rock matrix by diffusion, and this new "reservoir" of the contaminant may act as a secondary source for long periods of time, as it slowly diffuses back into the bedrock fractures and becomes available for the advective flow (flow with groundwater). As can be seen, substantial resources and a lot of luck would be required to monitor dissolved concentrations in various portions of the described "site" (aquifer system), to first understand, then estimate, and finally remediate (eliminate) all components of the contaminant flux that are causing elevated contaminant concentrations at some downgradient groundwater "receptors."

The following excerpt from Meinzer (1932) illustrates a visionary understanding of the importance of determining both the groundwater and the tracer (contaminant) flux by early researchers in the United States:

> In 1921 and subsequent years uranin dye was successfully used by Stiles and his assistants in an investigation to determine the extent to which bacteria are carried through formations of sand. This investigation, which was made at an experiment station near Fort Caswell, N.C., by the United States Public Health Service, with the cooperation of the United States Geological Survey, involved a minute three-dimensional survey of the direction and rate of movement of the groundwater and made a notable contribution in demonstrating the use of the dye method in fine-grained materials and in providing means for detailed study of movement of ground water.
>
> The use of dye probably affords the most accurate method of studying in detail the movements of groundwater. It may be rather easily applied in some creviced rocks that have relatively definite

underground streams, but in porous rocks with small interstices the dye may be very elusive, and the method may be found to be much more laborious and difficult than would appear on casual consideration. In the Fort Caswell experiment about 550 test wells were sunk in addition to some trenches, and the exact distribution of the uranin was determined.

To estimate the flow of ground water, whether by electrolytic, chemical, or dye methods, it is necessary to ascertain with some degree of accuracy the cross section through which the water flows and the velocity of flow through each unit of the cross section. This is accomplished by sinking test wells and to some extent by studying the records of existing wells.

Contaminant F&T processes, and their quantitative parameters, are explained in more detail in the following chapters. To set the stage, here is the general equation of contaminant F&T in one dimension (e.g., along horizontal X-axis), which brings them all together:

$$\frac{\partial C}{\partial t} = \frac{D_x}{R}\frac{\partial^2 C}{\partial x^2} - \frac{v_x}{R}\frac{\partial C}{\partial x} \pm Q_s \tag{6.4}$$

where C is the dissolved contaminant concentration (kg/m^3, or mg/L), t is time (d), D_x is hydrodynamic dispersion in x-direction (m^2/d), R is retardation coefficient (dimensionless), x is the distance from the source along X-axis (m), v_x is linear groundwater velocity in x-direction (m/d), and Q_s is the general term for source or sink of contaminant, such as due to biodegradation ($kg/m^3/d$). This term can also be expressed using the first rate degradation constant, λ (1/d) which gives

$$\frac{\partial C}{\partial t} = \frac{D_x}{R}\frac{\partial^2 C}{\partial x^2} - \frac{v_x}{R}\frac{\partial C}{\partial x} - \lambda C \tag{6.5}$$

The above equation does not have an explicit solution and various approximate solutions, based on simplifying assumptions, have been proposed by various authors. One of the more widely and simple analytical equations used for an instantaneous (such as a chemical spill) point source, which takes into account hydrodynamic dispersion in all three Cartesian directions, X, Y, and Z, and calculates the concentration at the x, y, z coordinate, is the Baestle equation, formulated by Domenico and Schwartz (1990) as

$$C(x, y, z, t) = \left[\frac{C_0 V_0}{8(\pi t)^{3/2}(D_x D_y D_z)^{1/2}}\right] \exp\left[-\frac{X - v_c t}{4 D_x t} - \frac{Y^2}{4 D_y t} - \frac{Z^2}{4 D_z t} - \lambda t\right] \tag{6.6}$$

where X, Y, Z are distances in the x, y, z directions from the center of the gravity of the contaminant mass, t is time, C_0 is the initial concentration of the contaminant at the source, V_0 is the initial volume of the released contaminant, such that $C_0 V_0$ equals the mass of contaminant which has entered the saturated zone, D_x, D_y, D_z are coefficients of hydrodynamic dispersion in all three main directions, x, y, and z, v_c is the velocity of contaminant in one dimension, equal or lower than that of water (it is lower when the contaminant is retarded due to sorption for example), and λ is the degradation constant.

The maximum contaminant concentration caused by a point source spill occurs at the center of the spreading plume, where $y = z = 0$, and $x = v_c t$:

$$C_{max} = \frac{C_0 V_0 e^{-\lambda t}}{8(\pi t)^{3/2}(D_x D_y D_z)^{1/2}} \tag{6.7}$$

Analytical solutions for other source types and geometries are given in Domenico and Schwartz (1990), and Chapter 18 and Chapter 19. Table 6.1 and Table 6.2 list various physical and chemical properties of common NAPLs, which are used in F&T calculations.

TABLE 6.1
Selected Physical and Chemical Properties of Common DNAPL Chemicals

Compound	Density (g/cm^3)	Dynamic Viscosity (Centipoise)	Water Solubility (mg/L)	Henry's Law Constant (atm-m^3/mol)	Vapor Pressure (mm Hg)	Log K_{ow}	Log K_{oc} (mL/g)
Halogenated semivolatiles							
1,2-Dichlorobenzene	1.306	1.302	1.0E+02	1.9E−03	9.6E−01	3.38	3.06
1,4-Dichlorobenzene	1.247	1.258	8.0E+01	1.6E−03	6.0E−01	3.39	3.07
Aroclor 1242	1.385	24	4.5E−01	3.4E−04	4.1E−04	5.58	5.00
Aroclor 1254	1.538	700	1.2E−02	2.8E−04	7.7E−05	6.03	5.61
Aroclor 1260	1.440		2.7E−03	3.4E−04	4.1E−05	7.15	6.42
Chlordane	1.600	1.104	5.6E−02	2.2E−04	1.0E−05	5.48	4.58
Dieldrin	1.750		1.9E−01	9.7E−06	1.8E−07	5.34	3.23
Pentachlorophenol	1.978		1.4E+01	2.8E−06	1.1E−04	5.12	4.80
Halogenated volatiles							
Chlorobenzene	1.106	0.756	4.9E+02	3.5E−03	8.8E+00	2.84	2.20
1,2-Dichloropropane	1.158	0.840	2.7E+03	3.6E−03	3.95E+01	2.02	1.71
1,1-Dichloroethane	1.175	0.377	5.5E+03	5.5E−04	1.82E+02	1.79	1.48
1,2-Dichloroethane	1.253	0.840	8.7E+03	1.1E−03	6.37E+01	1.48	1.15
1,1-Dichloroethylene	1.214	0.330	4.0E+02	1.5E−03	5.0E+02	2.13	1.81
trans-1,2-Dichloroethylene	1.257	0.404	6.3E+03	5.3E−03	2.65E+02	2.09	1.77
cis-1,2-Dichloroethylene	1.248	0.467	3.5E+03	7.5E−03	2.0E+02	1.86	1.50
1,1,1-Trichloroethane	1.325	0.858	9.5E+02	4.1E−03	1.0E+02	2.49	2.18
1,1,2-Trichloroethane	1.444	0.119	4.5E+03	1.2E−03	1.88E+01	2.17	1.75
Trichloroethylene (TCE)	1.462	0.570	1.0E+03	8.9E−03	5.87E+01	2.42	2.10
Tetrachloroethylene (PCE)	1.625	0.890	1.5E+02	2.3E−02	1.4E+01	3.14	2.82
Carbon tetrachloride	1.595	0.965	8.0E+02	2.0E−02	9.13E+01	2.83	2.64
Chloroform	1.485	0.563	8.2E+03	3.8E−03	1.6E+02	1.97	1.64
Methylene chloride	1.325	0.430	1.3E+04	2.6E−03	3.5E+02	1.25	0.94
1,1,2,2-Tetrachloroethane	1.600	1.770	2.9E+03	5.0E−04	4.9E+00	2.39	2.34
Ethylene dibromide	2.172	1.676	3.4E+03	3.2E−04	1.1E+01	1.76	1.45
Vinyl chloride (gas)	0.912		1.1E+03	1.2E−00	2.6E+03	0.60	0.91
Chloroethane (gas)	0.941		5.7E+03	4.5E−01	7.7E+02	1.43	1.17

continued

TABLE 6.1 (continued)
Selected Physical and Chemical Properties of Common DNAPL Chemicals

Compound	Density (g/cm^3)	Dynamic Viscosity (Centipoise)	Water Solubility (mg/L)	Henry's Law Constant (atm-m^3/mol)	Vapor Pressure (mm Hg)	Log K_{ow}	Log K_{oc} (mL/g)
Nonhalogenated semivolatiles							
2-Methyl napthalene	1.006		2.5E+01	5.1E−02	6.80E−02	3.86	3.93
o-Cresol	1.027		3.1E+04	4.7E−05	2.45E−01	1.95	1.23
p-Cresol	1.035		2.4E+04	3.5E−04	1.08E−01	1.94	1.28
m-Cresol	1.038	21. 0	2.4E+04	3.8E−05	1.53E−01	1.96	1.43
Phenol	1.058		8.4E+04	7.8E−07	5.29E−01	1.46	1.15
Naphthalene	1.162		3.1E+01	1.3E−03	2.34E−01	3.30	3.11
Benzo(a)anthracene	1.174		1.4E−02	4.5E−06	1.16E−09	5.61	6.14
Flourene	1.203		1.9E+00	7.6E−05	6.67E−04	4.18	3.90
Acenaphthene	1.225		3.9E+00	1.2E−03	2.31E−02	3.92	3.70
Anthracene	1.250		7.5E−02	3.4E−05	1.08E−05	4.45	4.10
Dibenz(a,h)anthracene	1.252		2.5E−03	7.3E−08	1.0E−10	6.80	6.52
Pyrene	1.271		1.5E−01	1.2E−05	6.67E−06	4.88	4.58
Chrysene	1.274		6.0E−03	1.1E−06	6.3E−09	5.61	5.30
2,4-Dinitrophenol	1.680		6.0E+03	6.5E−10	1.49E−05	1.54	1.22

Sources: From Huling, S.G. and J.W. Weaver, 1991. Dense nonaqueous phase liquids. *Ground Water Issue*, EPA/540/4-91-002; Robert S. Kerr Environmental Research Laboratory, Ada, Oklahoma, 28pp.; Knox, R.C., Sabatini, D.A., and L.W. Canter, *Subsurface Transport and Fate Processes*, Lewis Publishers, Boca Raton, FL, 1993, 430pp.; Cohen, R.M. and J.W. Mercer, *DNAPL Site Evaluation*, C.K. Smoley, Boca Raton, FL, 1993.

TABLE 6.2
Selected Properties of Some Common LNPAL Chemicals and Gasoline Additives

Chemical	Solubility* (mg/L)	log K_{oc} (L/kg)	Vapor Pressure (mm Hg)	Henry's Constant (Dimensionless)
Benzene	1780	1.5–2.2	76–95.2	0.22
Toluene	535	1.6–2.3	28.4	0.24
Ethylbenzene	161	2.0–3.0	9.5	0.35
m-Xylene	146	2.0–3.2	8.3	0.31
Ethanol	Miscible	0.20–1.21	49–56.5	0.00021–0.00026
Methanol	Miscible	0.44–0.92	121.6	0.00011
TBA	Miscible	1.57	40–42	0.00048–0.00059
MTBE	43,000–54,300	1.0–1.1	245–256	0.023–0.12
ETBE	26,000	1.0–2.2	152	0.11
TAME	20,000	1.3–2.2	68.3	0.052
DIPE	2039–9000	1.46–1.82	149–151	0.195–0.41

*Pure-phase solubility; solubility in mixture is lower (Raoult's Law).

Source: Nichols et al., 2000. Strategies for characterizing subsurface releases of gasoline containing MTBE. American Petroleum Institute (API), Regulatory and Scientific Affairs Publication Number 4699, Washington, D.C. Copyright American Petroleum Institute, all rights reserved.

6.2.1 ADVECTION

Advection is the movement of the dissolved contaminant with (in) groundwater, and it refers to the average linear flow velocity of the "bulk" of contaminant. Jumping a bit ahead (see next section), it should be immediately noted that some dissolved contaminant particles, just like some water particles, will also jump ahead of the rest (the "bulk") of the dissolved contaminant mass. This spreading, or contaminant dispersion, happens because some particles will move faster than the others in the actual pore space, but the bulk of the dissolved contaminant will move with the average linear groundwater velocity (v_L), which is given as

$$v_L = \frac{K \times i}{n_{ef}} \tag{6.8}$$

where K is the hydraulic conductivity, which has units of length per time (m/d; ft/d), i is the hydraulic gradient, which is dimensionless, and n_{ef} is effective porosity of the porous aquifer material, also dimensionless.

Although the definition of the above three main parameters of groundwater flow in the saturated zone is part of any hydrogeology textbook (including this one—see Section 1.3), it is crucial to clearly understand various implications of using their interrelationships, as quantitative parameters, when assessing the flow of dissolved contaminants. In other words, all applicable groundwater flow, and F&T models, from the simplest ones such as a single analytical equation, to the most complex three-dimensional models, include calculation of groundwater velocity by default. The linear groundwater velocity value may sometimes be used alone to answer the following common question:

- How long would it take for a groundwater (contaminant) particle to reach the potential receptor?

This question obviously ignores all other F&T processes, and assumes that there is a direct groundwater pathway between the point of release and the potential receptors. Figure 6.9

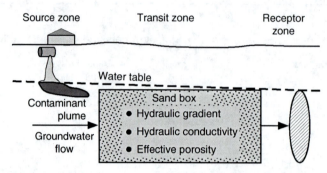

FIGURE 6.9 Three conceptual zones of groundwater flow between the source and the receptor, all represented as one "sand box."

illustrates a commonly applied approach for quickly obtaining a preliminary answer to the above question. This approach (screening) is potentially also the most misleading because it assumes the following simple conditions:

- The hydraulic gradient between the two points of interest (e.g., the point of release and the receptor) is uniform, i.e., constant.
- The hydraulic conductivity of the porous medium is constant (it does not vary spatially), which then also implies only one sediment (rock) type through which the flow takes place.
- Effective porosity is constant (same implications as with the hydraulic conductivity).
- The contaminant is not retarded (slowed down) or degraded in any way.

However, in most real field situations, especially over distances of hundreds or thousands of feet, the above assumptions do not hold. Figure 6.10 shows a schematic representation of far more realistic subsurface conditions present at most sites with unconsolidated sediments. The question then becomes: given the expected heterogeneity of the porous media within the aquifer, what values of the effective porosity and the hydraulic conductivity (assuming the hydraulic gradient is known from the monitoring well measurements) should one use to quickly perform the screening in the absence of more detailed field (site-specific) data? Figure 1.21 and Figure 1.35 (Chapter 1) illustrate that there may be more than one acceptable answer if the values are selected "from literature." Consider the following scenario: point of release

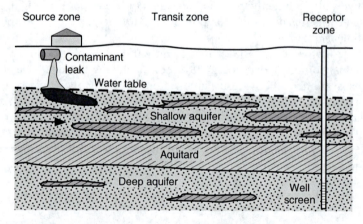

FIGURE 6.10 Most sites in geologic setting with unconsolidated sediments have various proportions of gravel, sand, silt, and clay in layers and lenses, including their various spatial configurations.

and a potential receptor are 2500 ft apart; regional hydraulic gradient in the shallow aquifer (which consists "predominantly" of fine sands), is estimated from the available monitoring well data to be 0.002. How long would it take a water (contaminant) particle to travel between the two points, assuming there is a direct flow path between them, within the same aquifer?

As shown in Figure 1.31, fine sand (our sand box) can have hydraulic conductivity anywhere between about 0.5 and 50 ft/d. Effective porosity (specific yield) can vary between 10% and 30% (Figure 1.21). Assuming the lowest values from the two ranges, the linear velocity of a groundwater particle, using Equation 6.8, is

$$v_L = \frac{K \times i}{n_{ef}} = \frac{0.5\text{ft/d} \times 0.002}{0.1} = 0.01 \text{ ft/d}$$

On the basis of this velocity, the time of travel between the two points of interest would be 250,000 days or about 685 years (2500 ft distance is divided by the velocity of 0.01 ft/d). Assuming the highest values from the two ranges (50 ft/d and 30%) the time of travel would be about 20 y, which is a very significant difference, to say the least. This simple quantitative example shows inherent uncertainties in quantifying linear groundwater flow velocities, even when assuming that the sand box approach is appropriate and the groundwater flow is strictly horizontal.

The time of travel is particularly important when considering F&T of contaminants dissolved in groundwater if they undergo biodegradation. It is obvious that longer groundwater resident times would result in a more significant decrease in the contaminant concentration as more time will be available for its degradation before it reaches the receptor.

All other things in a porous medium being equal (hydraulic gradient, hydraulic conductivity), theoretically speaking, a change in effective porosity will change the groundwater velocity through that medium (e.g., higher effective porosity will decrease the groundwater velocity). However, in reality, if effective porosity changes the porous material also changes, which then means that the hydraulic conductivity changes as well. Therefore, it is very important to understand that one cannot arbitrarily assume a change in effective porosity without considering the related change of the hydraulic conductivity. Because the hydraulic conductivity can change couple of orders of magnitude for the seemingly same material, it is a much more "sensitive" parameter (i.e., it has greater impact on the calculated velocity) than the effective porosity, which can only change within a limited range. However, changing the effective porosity by a factor of two will simply double or cut in half the time of travel. Finally, using the same effective porosity of, say 25% for "all" porous media, silt or gravel, would be completely erroneous, regardless of the intended level of effort (e.g., it is just for the screening purposes).

6.2.2 Dispersion and Diffusion

Very broadly stated, dispersion in the aquifer is a process of mixing between the advancing fluid, such as a contaminant dissolved in groundwater and being carried by the flow of groundwater, and the fluid being displaced (e.g., clean groundwater). Dispersion always takes place and the main result of this mixing is a decrease in concentration of the dissolved contaminant. Dispersion is porous medium-specific and considered independent of the flowing fluid. In a more narrow sense, scientists and engineers define dispersion as the sum of two processes: mechanical dispersion and diffusion. Defined in this way, dispersion is often called hydrodynamic dispersion (D) and, for the main direction of groundwater flow (in x-direction), it is given as

$$D_x = \alpha_x v_x + D_e \tag{6.9}$$

where a_x is the longitudinal dispersivity, in dimension of length (m or ft), v_x is linear groundwater velocity (m/d), and D_e is the coefficient of effective diffusion (cm^2/s), described further in this chapter.

Mixing due to mechanical dispersion occurs along the main direction of groundwater flow (this is called longitudinal dispersion), as well as perpendicular to the main flow direction (transverse dispersion). There is also a third main direction of mixing called vertical dispersion. Dispersion causes some solute particles to advance faster than the bulk of contamination, thus creating a halo (cloud) of low concentrations around the main portion of the plume. For the purposes of quantifying the effects of three-dimensional dispersion along the main (horizontal) direction of groundwater flow, the following equations relating linear groundwater velocity (v) and parameter called dispersivity (α) are used in practice:

$$D_L = \alpha_L \times v_x; \quad D_T = \alpha_T \times v_x; \quad D_V = \alpha_V \times v_x \qquad (6.10)$$

where D_L, D_T, and D_V are longitudinal, transverse, and vertical (mechanical) dispersions, respectively, and α_L, α_T, and α_V are longitudinal, transverse, and vertical dispersivities, respectively. Dispersivity has unit of length. There is a general practice to use transverse dispersivity 10 times smaller than the longitudinal dispersivity, and the vertical dispersivity one to two orders of magnitude smaller than the longitudinal dispersivity.

Most researchers and practitioners, as well as government agencies such as USEPA (e.g., see Wiedemeier et al., 1998; Aziz et al., 2000), have concluded that the dispersion is dependent on the length of solute flow being observed: the greater the plume length, the larger the value of longitudinal dispersivity. There are still discussions regarding the nature of this relationship (e.g., is it linear or nonlinear, is there a practical limit after which the dispersion stabilizes and does not increase with the increasing plume length) but there is a general agreement that dispersion depends on both the scale and the time of flow. Various graphs based on various experiments at different scales conducted in the field conditions, and various model calibrations, show that values of longitudinal dispersivity could range from 0.01 to 5500 m or more. However, as seen in Figure 6.11, which is one example of experimental dispersivity graphs, the reliability of the available data varies greatly, especially for large plume lengths. It is also important to understand that the dispersivities determined in the field are larger than dispersivities determined in the laboratory experiments using small column samples. The rule of thumb, suggested by USEPA, is that the longitudinal dispersivity in most cases could be initially estimated from the plume length as being 10 times smaller (Wiedemeier, 1998; Aziz et al., 2000). This means that, for example, if the plume length is 300 ft, the initial estimate of the longitudinal dispersivity is about 30 ft. There are also other suggested empirical relationships relating the plume length and the longitudinal dispersivity (Wiedemeier, 1998; Aziz et al., 2000). Recognizing the limitations of the available and reliable field-scale data on dispersivity, the agency also suggests that the final values of dispersivities used in F&T calculations should be based on calibration to the site-specific (field) concentration data. The main reason why very few (if any) projects for practical groundwater remediation purposes consider field determinations of dispersivity, is that it would require a large number of monitoring wells and application of large-scale tracer tests. Such studies are expensive by default and usually not feasible due to generally slow movement of tracers in intergranular porous media over long distances.

The graph in Figure 6.11 shows how problematic it is to simply (and blindly) use "literature" values, without considering site-specific conditions. Unfortunately, this is sometimes done because of the desired outcome of calculations by some stakeholders, rather than being based on sound science (or logic for that matter). Consider this situation from an actual lawsuit: an expert for the plaintiff has produced the final, definitive, and calibrated F&T

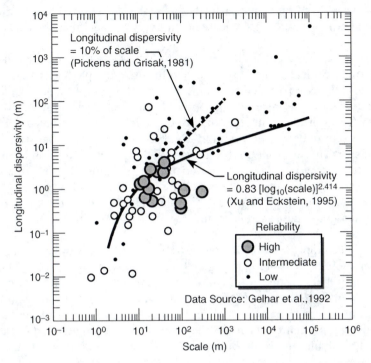

FIGURE 6.11 Longitudinal dispersivity vs. scale data reported by Gelhar et al. (1992). Data includes Gelhar's reanalysis of several dispersivity studies. Size of circle represents general reliability of dispersivity estimates. Location of 10% of the scale linear relation plotted as dashed line (Pickens and Grisak 10% rule of thumb). Xu and Eckstein's regression shown as solid line. (From Aziz et al., BIOCHLOR: Natural Attenuation Decision Support System: User's manual, version 1.0. EPA/600/R-00/008, U.S. Environmental Protection Agency, Office of Research and Development, Washington, D.C, 46 p., 2000.)

model, which predicted certain future impact on a water-supply well from a chemical added to gasoline, which was spilled at a gas station. The distance between the gas station and the well is thousands of feet, and the extent of the contaminant plume is limited to the immediate area of the gas station, a couple of hundred feet or so. With his model, the expert predicted that the contaminant would start arriving at the well, with the concentration of 0.2 parts per billion, 17 years from the present date (date of depositions). At two depositions, the expert stated that the accuracy of his model is very high, and that he has a very high level of confidence that the concentration at the supply well will be detected in 17 years. In addition, because the detection limit of the chemical at the time of the depositions was exactly 0.2 ppb, the expert was very confident in his model. And, because of all this, the plaintiff was asking $9 million for future damages to the public water-supply well. However, because of some legal and other issues, the judge in the case ruled in the meantime (not knowing what any expert in the lawsuit would say in his courtroom in front of the jury) that the future damages can be considered only if they were certain to occur by a certain date, which was set by the judge to be several months into the ongoing lawsuit. Having learned that, the plaintiff's expert changed the single most important parameter in his already calibrated, definitive, and very accurate model. He increased the longitudinal dispersivity by 1100%, which then enabled the contaminant to reach the public water-supply well, at very low but still detectable concentrations (0.2 ppb), in just about the right time to satisfy the new requirement for future damages of $9 million. When he was asked, at his last-minute deposition one day before the continuation of the trial and his appearance in front of the jury, why did he do that, he referred to the graph in Figure 6.11 and said that he made that decision based on his very considerable practical experience of more than 25 years.

There are at least several points to be made from this case, but the author will discuss only those related to the technical "rules of thumb" and the use of graphs such as the one in Figure 6.11. Most, if not all, F&T parameters commonly used in hydrogeologic calculations act to decrease a contaminant concentration as it travels dissolved in groundwater. Dispersion is especially important if the contaminant does not degrade or sorb to aquifer solids. Scientists and agencies that set standards and rules of thumb almost always do that with caveats, as is the case with dispersivity. Explanations that come with these rules of thumb should be read carefully by practicing hydrogeologists. In our dispersivity case, the 10% rule refers to the existing plumes and their lengths, not some future plumes that are yet to be developed. Consider this scenario, which defies the logic, but may be handy for winning $9 million when referring to a graph published in quite a few books (including this one):

1. There is a public supply well 40,000 m away (that is right, 40 km away) from this particular gas station. Note that the graph in Figure 6.11 has horizontal scale of 1000 km.

2. There was a spill of gasoline at this station that happened yesterday (but no one will do anything about it, so the free NAPL product will somehow find its way to the water table, which happens to be at a considerable depth of 25 ft below ground surface).

3. From the graph in Figure 6.11 it may be concluded that longitudinal dispersivity for a plume that is 40 km long could be as much as 1 km (there are two data points on the graph suggesting that, and it does not matter that they are not that reliable; I will not share my thoughts with anyone in the scientific and regulatory community and will not suggest that no scientific or regulatory graph should include data that has "low reliability").

4. Although there is currently no 40 km long plume, I am expecting it to develop because my receptor (water-supply well) is that far.

5. I will create a F&T numeric model for the saturated zone (because I am not that familiar with the unsaturated zone processes and models, I will assume that all of the spilled gasoline will reach the water table and be dissolved at very, very high concentrations).

6. I will run the model and show that there will be a measurable impact 40 km away in, say, $4\frac{1}{2}$ years. I will not show the associated (predicted by the model) plume map; rather, I will show a graph of concentration vs. time (*C* vs. *T*) at the 40 km distant well because that is my focus in this legal case.

7. If someone (say, expert for the other side) tells his attorney to ask me if the model was calibrated for some data in between the gasoline station and the 40-km distant water-supply well, I will answer that the model accurately represents F&T of the contaminant as expected, based on my experience.

8. I will keep for myself the somewhat unnerving fact that the model-generated plume (which I am not going to share with anyone) shows contamination in some water-supply and other wells, which are about couple of thousand feet away from the gas station, in just several months after the spill (after all, the dispersivity in my model is 1 km so I am not that surprised). Incidentally, the plume map also shows that the contaminant traveled almost 900 ft upgradient from the gas station (not surprised). If the attorney for the other side starts asking me all these questions, I may say something about numeric dispersivity, salmon effect, or something similarly incomprehensible. I will, however, stick to the graph shown in Figure 6.11 and keep mentioning my experience. (Author's note to the reader: salmon is a fish that can jump over some very high waterfalls; in other words, it can jump upgradient the river, i.e., "uphill"; the phrase is sometimes used to describe incorrect numeric modeling results caused by incorrect dispersion).

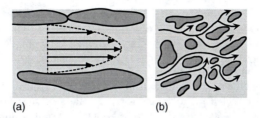

FIGURE 6.12 (a) Approximate fluid velocity distribution in a single pore, and (b) tortuous paths of fluid movement in an unconsolidated porous medium. (From Franke, O.L., et al., Study guide for a beginning course in ground-water hydrology; Part 1, Course participants. *U.S. Geological Survey Open-File Report* 90–183, Reston, Virginia, 184pp., 1990.)

Following is one of the most illustrative discussions on dispersivity and its use in hydro-geologic calculations, by Franke et al. (1990) of the USGS; it is accompanied by Figure 6.12 and Figure 6.13:

At the microscopic (pore) scale, velocity varies from a maximum along the centerline of each pore to zero along the pore walls, as shown in Figure 6.12a; both the centerline velocity and the velocity distribution differ in pores of different size. In addition, flow direction changes as the fluid moves through the tortuous paths of the interconnecting pore structure, as shown in Figure 6.12b. At a larger (macroscopic) scale, local heterogeneity in the aquifer causes both the magnitude and direction of velocity to vary as the flow concentrates along zones of greater permeability or diverges around pockets of lesser permeability. In this discussion, the term

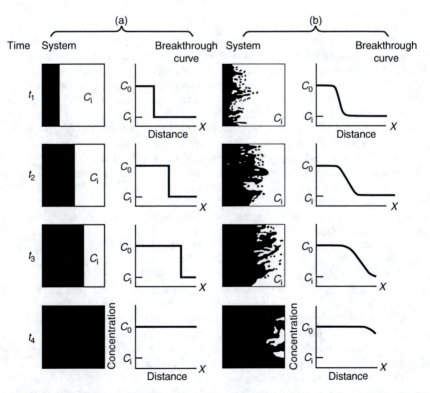

FIGURE 6.13 Advance of a tracer for (a) a sharp front and (b) an irregular front (From Franke, O.L., et al., Study guide for a beginning course in ground-water hydrology; Part 1, Course participants. *U.S. Geological Survey Open-File Report* 90–183, Reston, Virginia, 184pp., 1990.)

'macroscopic heterogeneity' is used to suggest variations in features large enough to be readily discernible in surface exposures or test wells, but too small to map (or to represent in a mathematical model) at the scale at which we are working. For example, in a typical problem involving transport away from a landfill or waste lagoon, macroscopic heterogeneities might range from the size of a baseball to the size of a building.

Using the velocity from the model, a tracer front introduced at the left side of region (shown in Figure 6.13) would be predicted to traverse this region (R) as a sharp front moving with the average linear velocity of the water. In reality, however, a tracer front becomes progressively more irregular and diffuse as it moves through a porous medium. If we consider a vertical plane through the aquifer at the left edge of region R, the actual velocity varies in both magnitude and direction from one point to another; the same is true in the flow direction. Thus each tracer particle enters R at a velocity that generally is different from that of its neighbors, and each particle experiences a different sequence of velocities as it crosses R from left to right. Instead of a sharp front of advancing tracer as shown in Figure 6.13a, we see an irregular advance as in this figure, with the forward part of the tracer distribution becoming broader and more diffuse with time. The pore-scale or microscopic velocity variations contribute only slightly to this overall dispersion; macroscopic variations contribute more significantly, whereas 'mappable' variations generally have the largest effect.

If it were possible to generate a model or a computation that could account for all of the variations in velocity in natural aquifers, dispersive transport would not have to be considered (except for molecular diffusion); sufficiently detailed calculations of advective transport theoretically could duplicate the irregular tracer advance observed in the field. In practice, however, such calculations are impossible. Field data at the macroscopic scale never are available in sufficient detail, information at the 'mappable' scale rarely is complete, and descriptions of microscopic scale variations are impossible except in a statistical sense. Even if complete data were available, however, an unreasonable computational effort would be required to define completely the natural velocity variations in an aquifer.

The more closely we represent the actual permeability distribution of an aquifer, the more closely our calculations of advective transport match reality; the finer the scale of simulation, the greater is the opportunity to match natural permeability variations. In most situations, however, when both data collection and computational capacity have been extended to their practical limits, calculations of advective transport fail to match field observation; therefore, we must find a tractable method of adjusting or correcting such calculations. Historically, the effort to develop such a method of correction followed the diffusion model. Diffusion had been analyzed successfully as a process of random particle movement which, in the presence of concentration change, results in a net transport proportional to the concentration gradient in the direction of decreasing concentration. In the case of a moving fluid, the random movement ascribed to diffusion was viewed as superimposed on the motion caused by the fluid velocity. Thus, the net movement of any solute particle could be regarded as the vector sum of an advective component and a random diffusive component.

By analogy, it was assumed that solute transport through porous media could be viewed in the same way—as the sum of an advective component in which solutes move with the average linear velocity of the fluid, and a random 'dispersive' component superimposed on the advective motion (Saffman, 1959). In effect, dispersion was seen as the net transport with respect to a point moving with the average linear velocity of the fluid. Because the dispersive motion of solute particles was assumed to be random, the flux was taken to be proportional to the concentration gradient.

While many difficulties have been perceived with the concentration-gradient approach, no satisfactory alternative has yet been found. Currently, we know that some method is required to adjust and correct the results of advective-transport calculations. The method commonly employed is to postulate an additional transport that is proportional to the concentration gradient in the direction of decreasing concentration; however, the coefficient of proportionality is treated as a function of the average flow velocity.

This approach can be derived or justified mathematically if assumptions similar to those used in the analysis of molecular diffusion in moving liquids are made—that is, if the actual velocity of particles through the system can be described as the sum of two components: (1) the average velocity used in advective calculation, and (2) a random deviation from the average velocity. To the extent that scale variations in velocity represent random deviation from the velocity used in advective transport calculation, and to the extent that these variations occur on a scale which is significantly smaller than the size of the region used for advective calculation (for example, region R in Figure 6.13), dispersion theory may describe adequately the differences between advective calculation and field observation. However, if the velocity variations are not random, or if they are large relative to the region used for advective calculation, the suitability of the dispersion approach is questionable. Moreover, even when this approach appears to be justified, the determination of the necessary coefficients usually must be approached empirically (for example, through model calibration). The range of validity of the quantities determined in this manner is uncertain.

Variations in velocity most often are caused by variations in the permeability and effective porosity of the porous medium on all three of the relevant scales. In theory, therefore, it should be possible to describe the dispersive-transport process through statistical analysis of variations in aquifer permeability. Gelhar and Axness (1983) have attempted to do this by using a stochastic analysis of permeability variation at the macroscopic scale to generate dispersivity values. The utility of this approach currently is limited by the difficulty in obtaining the necessary data on the statistics of permeability variation. However, Gelhar has demonstrated that in the limit, as distances of transport become large, a concentration-gradient approach is justified on theoretical grounds.

Because dispersive transport actually represents an aggregate of the deviations of actual particle velocities from the velocity used in advective-transport calculation, coefficients of dispersion must vary as the overall velocity of flow varies to create agreement between computed and observed results. As overall flow velocities in the system increase, the magnitude of velocity deviations from the average velocity used in advective-transport calculation must increase as well; therefore, dispersive transport is dependent on average flow velocity.

The description of dispersion in terms of velocity variation implies that problem scale must be a factor in any calculation of dispersive effects. As the size of the region used in advective-transport calculation (for example, region R in Figure 6.13) increases, more heterogeneities are included in that region. If a small region of calculation is chosen (for example, corresponding to the size of a laboratory column), the dominant heterogeneities within it are those at the pore scale; dispersive effects and dispersion coefficients are correspondingly small. As the region R becomes larger, macroscopic and ultimately 'mappable' heterogeneities dominate. Thus, as larger regions of calculation are taken, the dispersive effects tend to increase in magnitude, the determination of the coefficients required for their description becomes more difficult, and the applicability of the conventional concentration-gradient approach becomes questionable. In general, the scale at which advective-transport calculations are made (for example, the scale of discretization in a model analysis) ideally reflects the existing level of knowledge of heterogeneities in the system. The scale is chosen to be fine enough so that the effects of all recognized heterogeneities can be accounted for by advective transport, yet coarse enough so that individual regions of advective-transport calculation are large with respect to their unknown internal heterogeneities, which must be described by dispersive terms. Thus, in any calculation of the physical mechanisms of solute transport, advection and dispersion are interrelated, and the appropriate values of dispersion depend on the scale at which the advective field is quantified.

One hundred years ago, in 1905, Charles Slichter of the USGS published *Field Measurements of the Rate of Movements of Underground Waters*. This exceptional work includes results of several very detail laboratory experiments addressing, among other things, the issue of dispersion of solutes in a simulated aquifer (Figure 6.14). Here are several excerpts from Slichter's work pertaining to dispersion:

The results of the experiments are best shown by the series of diagrams figs. 8 to 20, in which the strength of the electrolyte found at each test well is shown by a circle of appropriate size. Among the various electrolytes tested were ammonium chloride (sal ammoniac), sodium chloride

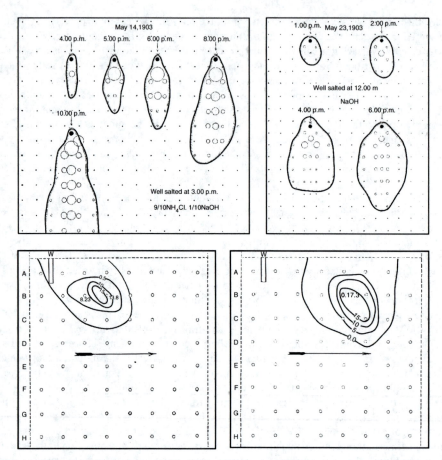

FIGURE 6.14 Examples of numerous graphs showing the results of Slichter's experiments with solute flow in a simulated aquifer. Top: map views of the solute "plumes" as they travel through saturated gravel in a horizontal tank, showing the effect of dispersion. The size of the circles indicates magnitude of the measured concentrations at well ports (each dot represents one measuring port, which are spaced 3 in. apart). The solute (tracer) is injected at port or well "W." Bottom: flow of solutes in cross-sectional view, analyzed in a vertical tank. (From Slichter, C.S., Field measurements of the rate of movement of underground waters. *U.S. Geological Survey water-supply and Irrigation Paper* 140, Washington, D.C., 122pp., 1905.)

(common salt), concentrated ammonia water, and mixtures of ammonium chloride and caustic soda, or lye. One of the most remarkable conclusions from the experiment, was that diffusion plays but a very small part in the spread of the electrolyte through the ground water. In none of the experiments was it found that the electrolyte extended more than about 3 in. upstream from the large well W. This fact can be seen by consulting the series of diagrams illustrating the distribution of electrolyte. In general, it can be seen that the electrolyte moves downstream in a pear-shaped mass, the width of the stream varying somewhat with the nature of the electrolyte used. The high velocities always gave a stream of electrolyte which was quite narrow and the low velocities gave broader streams.

It seems to be conclusively shown by these experiments, as has been already stated (pp. 22–23), that the diffusion of the dissolved salt plays a very small part in the way in which the electrolyte is distributed in the moving current of ground water, but, as already stated, that the central thread of water in each capillary pore of the soil moves faster than the water in contact with the walls of the capillary pore. Likewise the spread of the electrolyte, as shown by these experiments, is not to be explained by the diffusion of the salt, but must be explained by the continued branching and subdivision of the capillary pores around the individual grains of the sand. The stream of electrolyte issuing from the salt well W will gradually broaden as it passes downstream, because each thread of it must divide and divide again and again as it meets with each succeeding grain of soil. If diffusion had much to do with its rate of spread, it would also make itself apparent by causing an upstream motion to the electrolyte against the current of ground water. As before stated; in no case did the electrolyte succeed in moving upstream a distance as great as 3 inches.

The excerpts explaining the results of experiments in the vertical tank (Figure 6.14, bottom):

As will be observed by consulting the diagrams, the dissolved salt entered the ground water and passed to the right with the moving stream, at the same time moving slightly downward, as shown by the contour curves. The velocity of water through the gravel during this experiment was about 17 feet for twenty-four hours. The elliptical outline of the contour curves is due to the two components of motion, one component being the velocity of ground water to the right, and the other being the downward motion, due to the high density of the solution of sal ammoniac. It will be noticed that the elliptical contour lines have their longest dimension sloping downward to the right, as they should if they represent the resultant of these two motions. It should also be noted (consult the diagrams) that after an interval of an hour nearly all of the electrolyte had left well W, and the water in the well had become fresh again.

One of the most interesting experiments with the vertical tank was made for the purpose of determining the amount of diffusion of the electrolyte. For this purpose the electrolyte was introduced into the well W and the ground water was permitted to remain stationary, no water being run into or out of the tank during the eight hours covered by the experiment . . . well W was placed directly over column 2. A charge of salt was introduced into the well W at 9 a.m., and samples were taken at the end of one-half hour and at the end of each hour thereafter until 5 p.m. The salt was found to drop vertically with a rapidity equal to the vertical component of motion noted in the experiments in which flow took place. In the eight hours of the test no portion of the charge could be detected in the test wells of columns 1 or 3. This experiment showed that the electrolyte had not diffused sufficiently to reach the wells of columns 1 and 3, while dropping a vertical distance of about 3 feet.

6.2.2.1   Diffusion

As demonstrated by Slichter's experiments, and many other similar research efforts that followed, diffusion does not play a significant role in the advective–dispersive transport of contaminants dissolved in groundwater, as they move freely through the effective porosity of the aquifer. It is when the groundwater velocity becomes very low, due to small pore sizes and very convoluted pore-scale pathways (e.g., see Figure 1.13 and Figure 1.14), that diffusion may become an important F&T process. Porosity that does not readily allow advective groundwater flow (flow under the influence of gravity), but it does allow movement of the contaminant due to diffusion, is sometimes called diffusive porosity. Dual-porosity media has one type of porosity which allows preferable advective transport through it; it also has another type of porosity that does not allow free gravity flow, or flow in it is significantly smaller than the flow taking place through the higher effective (advective) porosity. Examples of dual-porosity media include fractured rock, where advective flow takes place, preferably, through fractures, while the advective flow rate through the rest of the rock mass, or rock matrix, is comparably lower, much lower, or does not exist for all practical purposes. This gradation depends on the nature of matrix porosity; in some rocks such as sandstones and young limestones, matrix porosity may be fairly high and it may allow a very significant rate of advective flow, often as high as or higher than through the fractures. In most consolidated, "hard" rocks, matrix porosity is usually low, less than 5%–10%, and it does not provide for significant advective flow. Other examples of dual-porosity media, of particular interest for the migration of DNAPLs, include fractured clay and residuum sediments. In some cases, various discontinuities and fractures in such media may serve as pathways for some advective contaminant transport, while the bulk of the sediments may have a very high matrix porosity, which does not allow advective transport. Flow of solutes with high concentration through the fractures may result in the solute diffusion into the surrounding matrix (Figure 6.15).

The preceding discussion explains one of the two factors that have to be present for any significant diffusion to take place—diffusive, or matrix porosity where the contaminant can move. The other factor is existence of a high-enough concentration gradient, for enough time, that would cause the contaminant to start diffusing into the rock matrix (diffusive porosity). A good example is a persistent body of NAPL, resting on, or suspended in, a low-permeable clay sediment and creating high-concentration gradients, for long periods of time. Another example would be a DNAPL body "sitting" in fractures in limestone for a long time, thus driving high contaminant concentrations into the rock matrix. If, for any reason, the concentration gradient reverses itself, such as due to the final dissolution of the free-phase (residual) DNAPL in fractures and flushing of the fractures with the incoming uncontaminated groundwater, the contaminant that was diffused into the matrix will start diffusing back into the fractures. In some cases, this back-diffusion may act as a significant secondary source of groundwater contamination.

To summarize, diffusion is a micro-scale process (it happens at the molecular level), which causes movement of a solute in water from the area of its higher concentration to the area of its lower concentration (including, of course, where the solute is absent, i.e., where its concentration in groundwater is zero). This difference in concentration is called concentration gradient. Regardless of the rate of advective transport taking place, the solute in groundwater will diffuse into the diffusive porosity as long as there is a concentration gradient.

The rate of diffusion for different chemicals (solutes) in water depends on the concentration gradient and the coefficient of diffusion, which is solute-specific (different solutes have different coefficients of diffusion). The diffusion coefficients for electrolytes, such as major ions in groundwater (Na^+, K^+, Mg^{2+}, Ca^{2+}, Cl^-, HCO_3^-, SO_4^{2-}) range between 1×10^{-9}

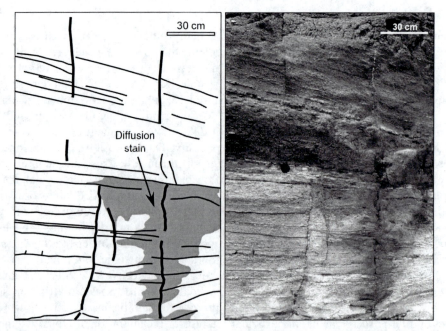

FIGURE 6.15 Diffusion stain around a vertical fracture in saprolite, Piedmont, North Carolina (detail from Figure 1.14).

TABLE 6.3

Aqueous Diffusion Coefficients (D_0) for Some Common Organic Chemicals

Chemical	Water Diffusion Coefficient (cm^2/s)	Chemical	Water Diffusion Coefficient (cm^2/s)
Benzene	1.0E−5	1,1-Dichloroethene	9.5E−6
Ethylbenzene	8.3E−6	trans-1,1-Dichloroethene	9.5E−6
Toluene	9.0E−6	Dichloromethane	9.1E−6
Xylenes	8.4E−6	Methylene chloride	1.1E−6
Carbon tetrachloride	7.1E−6	Nitrobenzene	7.6E−6
Carbon disulfide	1.1E−5	Trichloroethene (TCE)	1.0E−5
Chlorobenzene	7.9E−6	Tetrachloroethylene (PCE)	1.0E−6
Chloroform	9.1E−6	1,1,1-Trichloroethane	9.4E−6

Source: From Cohen, R.M. and Mercer, J.W., *DNAPL Site Evaluation*. C.K. Smoley, Boca Raton, FL., 1993; Pankow, J.F. and Cherry, J.A., *Dense Chlorinated Solvents and Other DNAPLs in Ground water*. Waterloo Press, Guelph, Ontario, Canada, 522pp., 1996.

and 2×10^{-9} m^2/s at 25°C (Robinson and Stokes, 1965; from Freeze and Cherry, 1979). The coefficient of diffusion is temperature-dependent and decreases with the decreasing temperature (e.g., at 5°C these coefficients are about 50% smaller than at 25°C). Table 6.3 lists aqueous diffusion coefficients for some common organic contaminants.

Flux (F) of a contaminant moving due to diffusion in the porous medium, is described by Fick's first law:

$$F = -D_e \frac{\partial C}{\partial x} \tag{6.11}$$

The Fick's second law describes the change in concentration of a nonsorbing contaminant due to diffusion:

$$\frac{\partial C}{\partial t} = D_e \frac{\partial^2 C}{dx^2} \tag{6.12}$$

and if the contaminant is also subject to sorption as it moves through the porous media:

$$\frac{\partial C}{\partial t} = \frac{D_e}{R} \frac{\partial^2 C}{\partial x^2} \tag{6.13}$$

where D_e is the effective coefficient of diffusion in the porous medium, R is the coefficient of retardation (see next section on sorption and retardation), and C is the contaminant concentration in groundwater. In porous aquifer materials, the apparent diffusion coefficients are smaller than in free water. This is because solute molecules (or ions) have to travel longer paths because of solid particles in the soil, which act as obstacles. These convoluted paths of travel are described with the term tortuosity. The apparent effective diffusion coefficient is therefore used to calculate rate of solute diffusion in the subsurface, rather than the aqueous phase diffusion coefficient. The effective diffusion coefficient (D_e) can be determined by using the known (experimentally determined) tortuosity of the porous media (rock) in question, or by multiplying the aqueous diffusion coefficient (D_0) with an empirical coefficient, called apparent tortuosity factor (τ), which can range between 0 and 1. This empirical coefficient is related to the aqueous (D_0) and effective (D_e) diffusions, and the rock matrix porosity (θ_m) through the following expression (Parker et al., from Pankow and Cherry, 1996):

$$\frac{D_e}{D_0} = \tau \cong \theta_m^p \tag{6.13}$$

where the exponent p varies between 1.3 and 5.4 depending on the type of porous geologic medium. Low porosity values result in small τ values and low D_e values. Laboratory studies of nonadsorbing solutes show that apparent tortuosity usually has values between 0.5 and 0.01. For example, for generic clay τ is estimated at 0.33, for shale and sandstone it is 0.10, and for granite it is quite small: 0.06 (Parker et al., from Pankow and Cherry, 1996).

The movement of the vapors emanating from an NAPL residual in the vadose zone is typically controlled by molecular diffusion and described by Fick's second law (Equation 6.12), where the tortuosity factor needed for determination of the free-air diffusion coefficient in partially saturated (moist) media is given as (Millington, 1959)

$$\tau = \frac{\theta_a^{2.33}}{\theta_t^2} \tag{6.14}$$

where θ_a is the air-filled porosity of the medium and θ_t is the total porosity. Thus, diffusion through the soil air is significantly reduced in high-water content (low θ_a) soils (Johnson et al., 1989).

The concentration profile of a nonsorbing solute in subsurface, moving only due to diffusion and in one direction (x) from the high-concentration layer into the zero-concentration layer, can be analytically calculated for various times (t) based on Fick's second law using Crank's equation (Freeze and Cherry, 1979):

$$C_i(x, t) = C_0 \operatorname{erfc}\left(\frac{x}{2} \times \sqrt{D_e t}\right) \tag{6.14}$$

where C_0 is the initial concentration on the high-concentration side of the contact between two layers, and erfc is complimentary error function.

6.2.3 Sorption and Retardation

Sorption results in distribution of a solute between the solution (groundwater where it is dissolved) and the solid phase (where it is held by the solids of the aquifer). This distribution is called partitioning and it is quantitatively described with the term distribution coefficient (or adsorption coefficient, or partition coefficient), and denoted with K_d. Because of sorption, the contaminant movement in groundwater is slowed down relative to the average groundwater velocity. This effect of sorption is called retardation, and the affinity of different solutes (chemicals dissolved in groundwater) to be retarded is quantified with a parameter called retardation factor, denoted with R. The overall effect of sorption is decrease in dissolved contaminant concentration.

When a contaminant is associated with the solid phase, it is not known if it was adsorbed onto the surface of a solid, absorbed into the structure of a solid, precipitated as a three-dimensional molecular structure on the surface of the solid, or partitioned into the organic matter (Sposito, 1989). Dissolution or precipitation is more likely to be the key process where chemical nonequilibrium exists, such as at a point source, an area where high contaminant concentrations exist, or where steep pH or redox gradients exist. Adsorption or desorption will likely be the key process controlling contaminant migration in areas where chemical equilibrium exists, such as in areas far from the point source. K_d is a generic term devoid of any particular mechanism and used to describe the general partitioning of aqueous phase constituents to a solid phase due to sorption.

The K_d parameter is very important in estimating the potential for the sorption of dissolved contaminants in contact with soil. As typically used in F&T transport calculations, the K_d is defined as the ratio of the contaminant concentration associated with the solid (C_s) to the contaminant concentration in the surrounding aqueous solution (C_w) when the system is at equilibrium:

$$K_d = \frac{C_s}{C_w} \qquad (6.15)$$

More generally, the distribution coefficient, K_d, is defined as the ratio of the quantity of the adsorbate adsorbed per mass of solid to the amount of the adsorbate remaining in solution at equilibrium. For the reaction

$$A + C_i = A_i \qquad (6.16)$$

the mass action expression for K_d is

$$K_d = \frac{\text{Mass of adsorbate sorbed}}{\text{Mass of adsorbate in solution}} = \frac{A_i}{C_i} \qquad (6.17)$$

where A is free or unoccupied surface adsorption sites, C_i is total dissolved adsorbate remaining in solution at equilibrium, and A_i is amount of adsorbate on the solid at equilibrium (USEPA, 1999b).

The K_d is typically given in units of mL/g. Describing the K_d in terms of this simple reaction assumes that A is in great excess with respect to C_i and that the activity of A_i is equal to 1.

Retardation due to sorption, R, is defined as

$$R = \frac{v_w}{v_c} \qquad (6.18)$$

where v_w is linear velocity of groundwater through a control volume; and v_c is velocity of contaminant through a control volume.

The sorption (chemical) retardation term does not equal unity when the solute interacts with the soil; almost always the retardation term > 1 due to solute sorption to soils. In rare cases, the retardation factor is actually < 1, and such circumstances are thought to be caused by anion exclusion (USEPA, 1999a,b).

Knowledge of the K_d and of media bulk density and porosity for porous flow, or of media fracture surface area, fracture opening width, and matrix diffusion attributes for fracture flow, allows calculation of the retardation factor. For porous flow with saturated moisture conditions, the R is defined as (USEPA, 1999b)

$$R = 1 + \frac{\rho_b K_d}{n} \qquad (6.19)$$

where ρ_b is bulk density of aquifer porous media (mass/length^3), K_d is distribution coefficient (length^3/mass), and n is porosity of the media at saturation (length^3/length^3).

Soil and geochemists knowledgeable of sorption processes in natural environments have long known that generic or default K_d values can result in significant error when used to predict the absolute impacts of contaminant migration or site-remediation options. Therefore, for site-specific calculations, K_d values measured at site-specific conditions are absolutely essential (USEPA, 1999a). To address some of this concern when using generic or default K_d values for screening calculations, modelers often incorporate a degree of conservatism into their calculations by selecting limiting or bounding conservative K_d values. For example, the most conservative estimate from an off-site risk perspective of contaminant migration through the subsurface natural soil is to assume that the soil has little or no ability to slow (retard) contaminant movement (i.e., a minimum bounding K_d value). Consequently, the contaminant would migrate in the direction and, for a K_d value of 0, travel at the rate of water. Such an assumption may in fact be appropriate for certain contaminants such as tritium, but may be too conservative for other contaminants, such as thorium or plutonium, which react strongly with soils and may migrate 10^2 to 10^6 times more slowly than the water (USEPA, 1999a). On the other hand, to estimate the maximum risks (and costs) associated with on-site remediation options, the bounding K_d value for a contaminant will be a maximum value (i.e., maximize retardation). USEPA has issued two extensive volumes on understanding variation in partition coefficient values, including theoretical background on various partitioning processes (adsorption being only one of them), methods for determining K_d, and a number of other very useful data and information essential for a practicing hydrogeologist.

The K_d value is usually a measured parameter that is obtained from laboratory experiments. The general methods used to measure K_d values include the laboratory batch method, *in-situ* batch method, laboratory flow-through (or column) method, field modeling method, and K_{oc} method. The ancillary information needed regarding the adsorbent (soil), solution (contaminated groundwater or process wastewater), contaminant (concentration, valence state, speciation distribution), and laboratory details (spike addition methodology, phase separation techniques, contact times) are explained in this USEPA document. The advantages, disadvantages, and, perhaps more importantly, the underlying assumptions of each method are also summarized in this USEPA reference text, which also includes conceptual overview of geochemical modeling calculations and computer codes as they pertain to evaluating K_d values and modeling of adsorption processes. The use of geochemical codes in evaluating aqueous speciation, solubility, and adsorption processes associated with contaminant fate studies is reviewed. This approach is compared to the traditional calculations that rely on the constant K_d construct. The use of geochemical modeling to address quality assurance and technical defensibility issues concerning available K_d data and the measurement of K_d values is also discussed (USEPA, 1999a).

In many natural systems, the extent of sorption is controlled by the electrostatic surface charge of the mineral phase. Most soils have net negative charges. These surface charges originate from permanent and variable charges. The permanent charge results from the substitution of a lower valence cation for a higher valence cation in the mineral structure, whereas the variable charge results from the presence of surface functional groups. Permanent charge is the dominant charge of 2:1 clays, such as biotite and montmorillonite. Permanent charge constitutes a majority of the charge in unweathered soils, such as those existing in temperate zones in the United States, and it is not affected by solution pH. Permanent positive charge is essentially nonexistent in natural rock and soil systems. Variable charge is the dominant charge of aluminum, iron, and manganese oxide solids and organic matter. Soils dominated by variable charge surfaces are primarily located in semitropical regions, such as Florida, Georgia, and South Carolina, and tropical regions. The magnitude and polarity of the net surface charge changes with a number of factors, including pH. As the pH increases, the surface becomes increasingly negatively charged. The pH where the surface has a zero net charge is referred to as the pH of zero-point-of-charge, pH_{ZPC} (Table 6.4). At the pH of the majority of natural soils (pH 5.5–8.3), calcite, gibbsite, and goethite, if present, would be expected to have some, albeit little, positive charge and therefore some anion sorption capacity (USEPA, 1999a).

6.2.3.1 Adsorption

Adsorption, in general, can be considered as the net accumulation of matter at the interface between a solid phase and an aqueous-solution phase. It differs from precipitation because it

TABLE 6.4
pH of Zero-Point-of-Charge, pH$_{ZPC}$

Material	pH$_{ZPC}$
Gibbsite [Al(OH)$_3$]	5.0
Hematite (α-Fe$_2$O$_3$)	6.7
Goethite (α-FeOOH)	7.8
Silica (SiO$_2$)	2
Feldspars	2–2.4
Kaolinite [Al$_2$Si$_2$O$_5$(OH)$_4$]	4.6
COOH	1.7–2.6*
NH$_3$	9.0–10.4*

*These values represent range of pK_a values for amino acids.
Source: Data from Stumm, W. and Morgan, J.J., *Aquatic Chemistry: An Introduction Emphasizing Chemical Equilibria in Natural Waters*. John Wiley & Sons, New York, 1981; Lehninger, A.L., *Biochemistry*. Worth Publishers, Inc., New York, 1970.

does not include the development of a three-dimensional molecular structure. The matter that accumulates in two-dimensional molecular arrangements at the interface is the adsorbate. The solid surface on which it accumulates is the adsorbent. Adsorption on clay particle surfaces can take place via three mechanisms. In the first mechanism, an inner-sphere surface complex is in direct contact with the adsorbent surface and lies within the so-called Stern layer. As a rule, the relative affinity of a contaminant to sorb will increase with its tendency to form inner-sphere surface complexes. The tendency for a cation to form an inner-sphere complex in turn increases with increasing valence (i.e., more specifically, ionic potential) of a cation (Sposito, 1984). The second mechanism creates an outer-sphere surface complex that has at least one water molecule between the cation and the adsorbent surface. If a solvated ion (i.e., an ion with water molecules surrounding it) does not form a complex with a charged surface functional group but instead neutralizes surface charge only in a delocalized sense, the ion is said to be adsorbed in the diffuse-ion swarm, and these ions lie in a region called the diffuse sublayer. The diffuse-ion swarm and the outer-sphere surface complex mechanisms of adsorption involve exclusively ionic bonding, whereas inner-sphere complex mechanisms are likely to involve ionic as well as covalent bonding (USEPA, 1999a).

The mechanisms by which anions adsorb are inner-sphere surface complexation and diffuse-ion swarm association. Outer-sphere surface complexation of anions involves coordination to a protonated hydroxyl or amino group or to a surface metal cation (e.g., water-bridging mechanisms) (Gu and Schulz, 1991). Almost always, the mechanism of this coordination is hydroxyl-ligand exchange (Sposito, 1984). In general, ligand exchange is favored at pH levels less than the zero-point-of-charge (Table 6.4). The anions CrO$_4^{2-}$, Cl$^-$, and NO$_3^-$, and to lesser extent HS$^-$, SO$_4^{2-}$, and HCO$_3^-$, are considered to adsorb mainly as diffuse-ion and outer-sphere complex species (USEPA, 1999a).

Adsorption of dissolved contaminants is very dependent on pH. As noted previously in the discussion of pH$_{ZPC}$ (Table 6.4.), the magnitude and polarity of the net surface charge of a mineral changes with pH (Langmuir, 1997; Stumm and Morgan, 1981). At pH$_{ZPC}$, the net charge of a surface changes from positive to negative. Mineral surfaces become increasingly more negatively charged as pH increases. At pH < pH$_{ZPC}$, the surface becomes protonated, which results in a net positive charge and favors adsorption of contaminants present as dissolved anions. Because adsorption of anions is coupled with a release of OH^- ions, anion adsorption is greatest at low pH and decreases with increasing pH. At $pH > pH_{ZPC}$, acidic dissociation of surface hydroxyl groups results in a net negative charge, which favors adsorption of contaminants present as dissolved cations. Because adsorption of cations is coupled with a release of H^+ ions, cation adsorption is greatest at high pH and decreases with decreasing pH. It should be noted that some contaminants may be present as dissolved cations or anions depending on geochemical conditions (USEPA, 1999a).

6.2.3.2   Ion Exchange

One of the most common adsorption reactions in soils is ion exchange. In its most general meaning, an ion-exchange reaction involves the replacement of one ionic species on a solid phase by another ionic species taken from an aqueous solution in contact with the solid. As such, a previously sorbed ion of weaker affinity is exchanged by the soil for an ion in aqueous solution. Most metals in aqueous solution occur as charged ions and thus metal species adsorb primarily in response to electrostatic attraction. In the cation-exchange reaction

$$CaX(s) + Sr^{2+} = SrX(s) + Ca^{2+}$$ (6.20)

Sr^{2+} replaces Ca^{2+} from the exchange site, X. The equilibrium constant (K_{ex}) for this exchange reaction is defined by the equation

$$K_{ex} = \frac{\{SrX(s)\} \times \{Ca^{2+}\}}{\{CaX(s)\} \times \{Sr^{2+}\}}$$ (6.21)

There are numerous ion-exchange models that are described by Sposito (1984) and Stumm and Morgan (1981). The ranges of cation-exchange capacity (CEC, in milliequivalents/100 g) exhibited by several clay minerals and based on values tabulated in Grim (1968) are listed in Table 6.5.

TABLE 6.5
Cation-Exchange Capacities (CEC) for Several Clay Minerals

Mineral	CEC (Milliequivalents/100 g)
Chlorite	10–40
Halloysite · $2H_2O$	5–10
Halloysite · $4H_2O$	40–50
Illite	10–40
Kaolinite	3–15
Sepiolite-attapulgite-palygorskite	3–15
Smectite	80–150
Vermiculite	100–150

Source: From USEPA, Understanding variation in partition coefficient, K_d, values; Volume I: The K_d model, methods of measurements, and applications of chemical reaction codes, EPA 402-R-99-004A, U.S. Environmental Protection Agency, Office of Air and Radiation, Washington, D.C., 1999a; Data from Grim, R.E., *Clay Mineralogy*. McGraw-Hill Book Company, New York, 1968.

6.2.3.3 Sorption of Organic Solutes

Organic solutes tend to preferably adsorb onto organic carbon present in the aquifer porous media. This organic carbon in the soil is of various forms—as discrete solids, as films on individual soil grains, or as stringers of organic material in soil grains. Distribution coefficient (K_d) can be calculated for different organic solutes based on fraction soil organic carbon contents (f_{oc}) and partition coefficient with respect to the soil organic carbon (K_{oc}), using the following equation:

$$K_d = f_{oc} \times K_{oc} \tag{6.22}$$

Values of K_{oc} for common organic contaminants are given in Table 6.1 and Table 6.2.

The coefficient of retardation (R) for linear sorption is determined from the distribution coefficient using the relationship (same as Equation 6.19)

$$R = 1 + \frac{\rho_b K_d}{n} \tag{6.23}$$

where ρ_b is bulk density of aquifer porous media (mass/length^3), K_d is distribution coefficient (length^3/mass), and n is porosity of the media at saturation (length^3/length^3).

This equation can also be written as

$$R = 1 + \left(\frac{1-n}{n}\right)\rho_s \times K_d \tag{6.24}$$

where ρ_s is the particle density of porous media (mass/length^3), which is often assumed to be 2.65 g/cm^3 for most mineral soils (in the absence of actual site-specific information).

Table 6.6 illustrates the relative mobility of some common organic contaminants as calculated with Equation 6.22 and Equation 6.24 for several different contents of organic carbon in the aquifer porous media. Values of K_{oc} are from USEPA (1996c). Lower R values indicate higher mobility (less retardation due to sorption). Note that chlordane is practically

---

TABLE 6.6
Comparison of Retardation Coefficients Calculated Using Equation 6.22 and Equation 6.24 for Some Common Organic Contaminants and Several Values of Percent Organic Carbon in the Porous Medium

Chemical	K_{oc} (mL/g)	Fraction Organic Carbon (Dimensionless)					
		0.05		0.01		0.001	
		K_d (mL/g)	R	K_d (mL/g)	R	K_d (mL/g)	R
TCE	166	8.3	60.6	1.66	12.9	0.166	2.2
PCE	155	7.75	56.6	1.55	12.1	0.155	2.1
Chloroform	39.8	1.99	15.3	0.398	3.9	0.0398	1.3
Naphthalene	2000	100	719.0	20	144.6	2	15.4
Benzene	58.9	2.945	22.1	0.589	5.2	0.0589	1.4
Chlordane	120,000	6,000	43,081.0	1,200	8,617.0	120	862.6
Vinyl chloride	18.6	0.93	7.7	0.19	2.3	0.02	1.1

Note: Soil particle density is 2.65 g/cm^3, and porosity is 30%.
Values of K_{oc} are from USEPA Soil Screening Guidance (USEPA, 1996c)

TABLE 6.7
Comparison of Retardation Coefficients Calculated Using Equation 6.16 for Some Common Metal Contaminants and Three Values of pH

Metal	pH = 4.9		pH = 6.8		pH = 8.0	
	K_d (mL/g)	R	K_d (mL/g)	R	K_d (mL/g)	R
Arsenic (+3)	25	180.5	29	209.2	31	223.6
Barium	11	80.0	41	295.4	52	374.4
Cadmium	15	108.7	75	539.5	4300	30,875.0
Chromium (+3)	1,200	8,617.0	1,800,000	Infinite	4,300,000	Infinite
Chromium (+6)*	31	223.6	19	137.4	14	101.5
Mercury (+2)	400	2,873.0	52	374.4	200	1,437.0
Selenium	18	130.2	5	36.9	2	16.8
Zinc	16	115.9	62	446.16	530	3,806.4

Note: Soil bulk density is 2.65 g/cm^3 and porosity is 30% in all calculations.
Values of distribution coefficient, K_d, are from USEPA (1996c)
*See also Table 6.8.

immobile in groundwater due to its extremely high K_{oc}, while vinyl chloride (VC) is the most mobile. For comparison, Table 6.7 shows calculation of R for some common metal contaminants using Equation 6.24 and generic K_d values listed in the *USEPA Soil Guidance Manual* (USEPA, 1996c). As can be seen, the metals are disproportionally less mobile than the organic chemicals because they strongly sorb to soil particles. Note, however, that the mobility of most metals and metallic compounds, in addition to pH, is dependent on redox potential and presence of other compounds, and is therefore site-specific. Table 6.8 illustrates this point: distribution coefficient for chromium (VI) can range between 1770 mL/g (highly immobile) to zero (not sorbed). This again emphasizes the need for site-specific determination of K_d, as some literature values can be highly misleading.

In areas with high clay concentrations and low total organic carbon concentrations, the clay minerals become the dominant sorption sites. Under these conditions, the use of K_{oc} to compute K_d might result in underestimating the importance of sorption in retardation calculations, a source of error that will make retardation calculations more conservative based on the total organic carbon content of the aquifer matrix (Wiedemeier et al., 1998). However, there is a "critical level of organic matter" (f_{oc}^*) below which sorption onto mineral surfaces is the dominant sorption mechanism (McCarty et al., 1981):

$$f_{oc}^* = \frac{A_s}{200} \times \frac{1}{K_{ow}^{0.84}} \tag{6.24}$$

where f_{oc}^* is the critical level of organic matter (mass fraction), A_s is the surface area of mineralogical component of the aquifer matrix (m^2/g), and K_{ow} is the octanol–water partitioning coefficient.

From this relationship, it is apparent that the total organic carbon content of the aquifer matrix is less important for solutes with low octanol–water partitioning coefficients (K_{ow}). Also apparent is the fact that the critical level of organic matter increases as the surface area of the mineralogic fraction of the aquifer matrix increases. This surface area of the aquifer matrix is most strongly influenced by the amount of clay. For compounds with low K_{ow} values in materials with a high clay content, sorption to mineral surfaces could be an

TABLE 6.8
Estimated Range of K_d Values for Chromium (VI) as a Function of Soil pH, Extractable Iron Content, and Soluble Sulfate

Soluble Sulfate Conc (mg/L)	K_d (mL/g)	pH											
		4.1–5.0			5.1–6.0			6.1–7.0			≥7.1		
		DCB Extractable Fe (mmol/g)			DCB Extractable Fe (mmol/g)			DCB Extractable Fe (mmol/g)			DCB Extractable Fe (mmol/g)		
		≤0.25	0.26–0.29	≥0.30	≤0.25	0.26–0.29	≥0.30	≤0.25	0.26–0.29	≥0.30	≤0.25	0.26–0.29	≥0.3
0–1.9	Min	25	400	990	20	190	390	8	70	80	0	0	1
	Max	35	700	1770	34	380	920	22	180	350	7	30	60
2–18.9	Min	12	190	460	10	90	180	4	30	40	0	0	1
	Max	15	330	820	15	180	430	10	80	160	3	14	30
19–189	Min	5	90	210	4	40	80	2	15	20	0	0	0
	Max	8	150	380	7	80	200	5	40	75	2	7	13
≥190	Min	3	40	100	2	20	40	1	7	8	0	0	0
	Max	4	70	180	3	40	90	2	20	35	1	3	6

Source: From USEPA, 1999b. Understanding variation in partition coefficient, K_d, values; volume II: Review of geochemistry and available K_d values for cadmium, cesium, chromium, lead, plutorium, radon, strontium, thorium, tritium (^{3}H), and uranium. EPA 402-R-99-004B, U.S. Environmental Protection Agency, Office of Air and Radiation, Washington, D.C.

important factor causing retardation of the chemical (Wiedemeier et al., 1998). For example, specific surface area in clay mineral montmorillonite ranges between 50 and 120 m^2/g, for ilite between 65 and 100 m^2/g, and for kaolinite between 10 and 20 m^2/g (Knox et al., 1993).

6.2.3.4 Sorption Isotherms

The relationship between the concentration of chemical sorbed onto solid surfaces (C_s) and the concentration remaining in aqueous solution (C_w) at equilibrium is referred to as the sorption isotherm because laboratory experiments for determining distribution coefficient values are performed at constant temperature. Sorption isotherms generally exhibit one of three characteristic shapes depending on the sorption mechanism. These isotherms are referred to as the Langmuir isotherm, the Freundlich isotherm, and the linear isotherm, which is a special case of the Freundlich isotherm.

The Langmuir isotherm model describes sorption in solute transport systems in which the sorbed concentration increases linearly with increasing solute concentration at low concentrations and approaches a constant value at high concentrations (experimental line flattens out). The sorbed concentration approaches a constant value because there are a limited number of sites available on the aquifer matrix for contaminant sorption. The Langmuir equation is described mathematically as (Devinny et al., 1990, from Wiedemeier et al., 1998)

$$C_s = \frac{KC_w b}{1 + KC_w} \tag{6.25}$$

where C_s is the sorbed contaminant concentration (mass contaminant/mass soil), K is the equilibrium constant for the sorption reaction (μg/g), C_w is the dissolved contaminant concentration (μg/mL), and b is the maximum sorptive capacity of the solid surface.

The Langmuir isotherm model is appropriate for highly specific sorption mechanisms where there are a limited number of sorption sites. This model predicts a rapid increase in the amount of sorbed contaminant as contaminant concentrations increase in a previously pristine area. As sorption sites become filled, the amount of sorbed contaminant reaches a maximum level equal to the number of sorption sites, b.

The Freundlich isotherm is a modification of the Langmuir isotherm model for the case when the number of sorption sites is large (assumed infinite) relative to the number of contaminant molecules. This is generally a valid assumption for dilute solutions (e.g., downgradient from a petroleum hydrocarbon spill in the dissolved BTEX plume) where the number of unoccupied sorption sites is large relative to contaminant concentrations. The Freundlich isotherm is expressed mathematically as (Devinny et al., 1990, modified from Wiedemeier et al., 1998)

$$C_s = K_d C_w^n \tag{6.26}$$

where K_d is the distribution coefficient, C_s is the sorbed contaminant concentration (mass contaminant/mass soil, mg/g), C_w is the dissolved concentration (mass contaminant/volume of solution, mg/mL), and n is the chemical-specific coefficient.

The value of n in this equation is a chemical-specific quantity that is determined experimentally. Values of n typically range from 0.7 to 1.1, but may be as low as 0.3 and as high as 1.7 (Lyman et al., 1992, from Wiedemeier et al., 1998).

The simplest expression of equilibrium sorption is the linear sorption isotherm, a special form of the Freundlich isotherm that occurs when the value of n is 1. The linear isotherm is valid for a dissolved species that is present at a concentration less than one half of its solubility (Lyman et al., 1992). This is a valid assumption for BTEX compounds partitioning from fuel mixtures into groundwater. Dissolved BTEX concentrations resulting from this type of partitioning are significantly less than the pure compound's solubility in pure water. The linear sorption isotherm is expressed as (Jury et al., 1991; from Wiedemeier et al., 1998)

$$C_s = K_d C_w \qquad (6.27)$$

where the notification is the same as in Equation 6.26. Distribution coefficient, K_d, is the slope of the linear isotherm plotted using experimental laboratory data.

6.2.4 VOLATILIZATION

Volatilization refers to mass transfer from liquid and soil to the gaseous phase. Chemicals in the vadose zone gas may be derived from either the presence of NAPL dissolved chemicals, or adsorbed chemicals. Chemical properties affecting volatilization include vapor pressure and aqueous solubility (see Table 6.1 and Table 6.2). Other factors influencing volatilization rate are concentration of contaminant in soil, soil moisture content, soil air movement, sorptive and diffusive characteristics of the soil, soil temperature, and bulk properties of the soil such as organic carbon content, porosity, density, and clay content (Lyman et al., 1982, from Cohen and Mercer, 1993). Volatile organic compounds (VOCs) in soil gas can (1) migrate and ultimately condense, (2) sorb onto soil particles, (3) dissolve in groundwater, (4) degrade, or (5) escape to the atmosphere. Volatilization of flammable organic chemicals in soil can create a fire or explosion hazard if vapors accumulate in combustible concentrations in the presence of an ignition source (Fussell et al., 1981).

Volatilization losses from subsurface NAPL are expected where NAPL is close to the ground surface, or in dry pervious sandy soils, or where NAPL has a very high vapor pressure (Feenstra and Cherry, 1988). Estimating volatilization from soil involves (1) estimating the organic partitioning between water and air, and NAPL and air; and (2) estimating the vapor transport from the soil. Henry's law and Raoult's law are used to determine the partitioning between water and air, and between NAPL and air, respectively (see Section 4.1). Vapor transport in the soil is usually described by the diffusion equation and several models have been developed where the main transport mechanism is macroscopic diffusion (e.g., Lyman et al., 1982). More complex models are also available (Falta et al., 1989; Sleep and Sykes, 1989; Brusseau, 1991).

Volatilization represents a source to subsurface vapor transport. Density-driven gas flow can be an important transport mechanism in the vadose zone that may result in contamination of the underlying groundwater and significant depletion of residual NAPL. Density-driven gas flow is a function of the gas-phase permeability, the gas-phase retardation coefficient, and the total gas density, which depends on the NAPL molecular weight and saturated vapor pressure (Falta et al., 1989). Density-driven gas flow will likely be significant where the total gas density exceeds the ambient gas density by more than 10% and the gas-phase permeability exceeds 1×10^{-11} m^2 in homogeneous media such as coarse sands and gravel (Falta et al., 1989; Mendoza and Frind, 1990a,b). Dense gas emanating from NAPL in the vadose zone will typically sink to the water table where it and the gas that has volatilized from the saturated zone will spread outward, with the pattern of soil gas migration strongly influenced by subsurface heterogeneities (Cohen and Mercer, 1993).

6.2.5 BIOTIC AND ABIOTIC TRANSFORMATIONS

As defined in the USEPA directive, on applying monitored natural attenuation (MNA) at contaminated sites (1999c), natural attenuation processes in general include

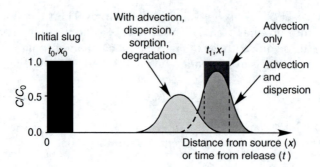

FIGURE 6.16 Changes in concentration of a contaminant slug from an instantaneous release, as it travels away from the source; the graph is applicable to both the time and the distance scales. C_0 is initial concentration; C/C_0 is relative concentration. (Modified from Wiedemeier, T.H., et al., Technical protocol for evaluating natural attenuation of chlorinated solvents in groundwater. EPA/600/R-98/128, U.S. Environmental Protection Agency, Office of Research and Development, Washington, D.C., 1998).

a variety of physical, chemical, or biological processes that, under favorable conditions, act without human intervention to reduce the mass, toxicity, mobility, volume, or concentration of contaminants in soil or groundwater. These *in-situ* processes include biodegradation; dispersion; dilution; sorption; volatilization; radioactive decay; and chemical or biological stabilization, transformation, or destruction of contaminants. When relying on natural attenuation processes for site remediation, EPA prefers those processes that degrade or destroy contaminants.... Natural attenuation processes are typically occurring at all sites, but to varying degrees of effectiveness depending on the types and concentrations of contaminants present and the physical, chemical, and biological characteristics of the soil and groundwater.... Where conditions are favorable, natural attenuation processes may reduce contaminant mass or concentration at sufficiently rapid rates to be integrated into a site's soil or groundwater remedy. Following source control measures, natural attenuation may be sufficiently effective to achieve remediation objectives at some sites without the aid of other (active) remedial measures.

Figure 6.16 and Figure 6.17 illustrate the impact of various processes discussed in the preceding chapters, as well as of biodegradation, on contaminant concentrations as it travels downgradient from the source zone. In the most general sense, processes that do not involve microorganisms are called abiotic. These include sorption, dispersion, diffusion, dilution, and volatilization, for example. There is, however, some confusion as to the "correct" meaning of the term, as many chemical reactions in the subsurface are facilitated by microorganisms,

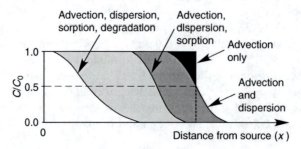

FIGURE 6.17 Graph of concentration vs. distance from the source, in direction of groundwater flow, in case of a continuous source of constant strength; C_0 is initial concentration; C/C_0 is relative concentration. (Modified from Wiedemeier, T.H., et al., Technical protocol for evaluating natural attenuation of chlorinated solvents in groundwater. EPA/600/R-98/128, U.S. Environmental Protection Agency, Office of Research and Development, Washington, D.C., 1998.)

including very important inorganic redox reactions. Some redox-controlled reactions chemically transform contaminants, both organic and inorganic, and may completely remove them from the flow system. An example is precipitation of an insoluble mineral compound, which most people would not define as "degradation or destruction." However, the net F&T effect of this biochemical transformation of the contaminant may be the same as complete destruction. The following discussion by USEPA (1999c) probably best describes the Agency's concerns as to the F&T of inorganic contaminants, which are generally thought not to be biodegradable:

> MNA (monitored natural attenuation) may, under certain conditions (e.g., through sorption or oxidation–reduction reactions), effectively reduce the dissolved concentrations and/or toxic forms of inorganic contaminants in groundwater and soil. Both metals and non-metals (including radionuclides) may be attenuated by sorption reactions such as precipitation, adsorption on the surfaces of soil minerals, absorption into the matrix of soil minerals, or partitioning into organic matter. Oxidation–reduction (redox) reactions can transform the valence states of some inorganic contaminants to less soluble and thus less mobile forms (e.g., hexavalent uranium to tetravalent uranium) and/or to less toxic forms (e.g., hexavalent chromium to trivalent chromium). Sorption and redox reactions are the dominant mechanisms responsible for the reduction of mobility, toxicity, or bioavailability of inorganic contaminants. It is necessary to know what specific mechanism (type of sorption or redox reaction) is responsible for the attenuation of inorganics so that the stability of the mechanism can be evaluated. For example, precipitation reactions and absorption into a soil's solid structure (e.g., cesium into specific clay minerals) are generally stable, whereas surface adsorption (e.g., uranium on iron-oxide minerals) and organic partitioning (complexation reactions) are more reversible. Complexation of metals or radionuclides with carrier (chelating) agents (e.g., trivalent chromium with EDTA) may increase their concentrations in water and thus enhance their mobility. Changes in a contaminant's concentration, pH, redox potential, and chemical speciation may reduce a contaminant's stability at a site and release it into the environment. Determining the existence, and demonstrating the irreversibility, of these mechanisms is important to show that a MNA remedy is sufficiently protective.
>
> Inorganic contaminants persist in the subsurface because, except for radioactive decay, they are not degraded by the other natural attenuation processes. Often, however, they may exist in forms that have low mobility, toxicity, or bioavailability such that they pose a relatively low level of risk. Therefore, natural attenuation of inorganic contaminants is most applicable to sites where immobilization or radioactive decay is demonstrated to be in effect and the process/mechanism is irreversible.

In groundwater contamination and remediation studies, the term biotic is now almost exclusively applied to biochemical reactions, which "degrade" (transform) or completely "destroy" organic chemical compounds due to microbial activity and involvement of enzymes. Hydrolysis is one example of a direct chemical reaction between water and some organic compounds that frequently results in the formation of alcohols and alkenes (Johnson et al., 1989). Although hydrolysis may sometimes be indirectly facilitated by microorganisms, it is generally considered an abiotic reaction because it does not involve enzymes, which can be produced only by microorganisms. For practical (simplification) purposes, the term biotic is therefore used as a synonym to biodegradation.

6.2.5.1 Biodegradation

For almost a decade now, MNA has been the favorite first option for consideration of at most (if not all) groundwater contamination sites. Although at a relatively few sites, percentage wise, it has been approved as the sole remedy, it is always very attractive as a supplemental remedy for three main reasons: it is noninvasive, it does not involve operations and maintenance (O&M) costs, which are often very substantial for various engineered groundwater remediation systems, and it usually costs less to implement, particularly because it does not require energy source or elaborate equipment. However, installation of monitoring wells, which is a necessary part of any MNA remedy, may involve a significant initial cost. One of the potentially most attractive aspects of MNA to general public is that it is "noninvasive": unlike many elaborate engineered site cleanup facilities, it is "quietly" working below ground so that the land surface above ground may continue to be used. In its effort to educate the public on the benefits of natural attenuation and bioremediation in general, and alleviate concerns that MNA is not a "do-nothing" groundwater remedial alternative, USEPA has published various pamphlets (e.g., USEPA, 1996a,b), which include general explanations such as the following:

> Bioremediation, is a process in which naturally occurring microorganisms (yeast, fungi, or bacteria) break down, or *degrade*, hazardous substances into less toxic or nontoxic substances. Microorganisms, like humans, eat and digest organic substances for nutrition and energy. (In chemical terms, 'organic' compounds are those that contain carbon and hydrogen atoms.) Certain microorganisms can digest organic substances such as fuels or solvents that are hazardous to humans. Biodegradation can occur in the presence of oxygen (aerobic conditions) or without oxygen (anaerobic conditions). In most subsurface environments, both aerobic and anaerobic biodegradation of contaminants occur. The microorganisms break down the organic contaminants into harmless products—mainly carbon dioxide and water in the case of aerobic biodegradation. Once the contaminants are degraded, the microorganism populations decline because they have used their food sources. Dead microorganisms or small populations in the absence of food pose no contamination risk.
>
> Many organic contaminants, like petroleum, can be biodegraded by microorganisms in the underground environment. For example, biodegradation processes can effectively cleanse soil and groundwater of hydrocarbon fuels such as gasoline and the BTEX compounds—benzene, toluene, ethylbenzene, and xylenes. Biodegradation also can break down chlorinated solvents, like trichloroethylene (TCE), in groundwater but the processes involved are harder to predict and are effective at a smaller percentage of sites compared to petroleum-contaminated sites. Chlorinated solvents, widely used for degreasing aircraft engines, automobile parts, and electronic components, are among the most often-found organic ground-water contaminants. When chlorinated compounds are biodegraded, it is important that the degradation be complete, because some products of the breakdown process can be more toxic than the original compounds.

The microorganisms most responsible for bioremediation are bacteria. Bacteria are found in all environmental media, including air, water, and soil. When indigenous (native to the site) bacteria degrade compounds under existing subsurface conditions, the degradation is called passive bioremediation or natural attenuation. Enhanced bioremediation involves stimulating indigenous bacteria by adding electron donors (substrates) or nutrients to the subsurface to increase bacterial growth yielding faster degradation rates. The compounds injected are determined by the type of bacteria that are stimulated to degrade the specific contaminants. The type of bacteria that dominate the subsurface are heterotrophic forms that require organic substrates (carbon based compounds) to serve as a source of energy. Bacteria also are categorized according to the use of oxygen as aerobes, anaerobes, and facultative anaerobes. Aerobes require oxygen, anaerobes require an environment devoid of oxygen, and facultative anaerobes can survive in both aerobic and anaerobic environments. Enhanced bioremediation also refers to the addition of exogenous microorganisms to a contaminated site (bioaugmentation) specifically for the purpose of bioremediating a polluted area (ITRC, 1998, 1999).

Detailed explanation of various biodegradation processes and associated bioremediation technologies is beyond the scope of this book. Because of the ever-increasing interest in bioremediation of contaminated groundwater, there are many excellent publications and resources available in the public domain and accessible for free download at various web sites maintained by the U.S. Government agencies. A good starting point are works by Wiedemeier et al. (1998, 1999), Azdapor-Keeley et al. (1999), and ASTM (1998). To set the stage for solving a quantitative problem in Section 19.2 involving sequential degradation, a short description of one of the biodegradation processes that may be successful in reducing or completely eliminating contamination of groundwater by chlorinated solvents is included, accompanied by Figure 6.18 and Figure 6.19.

Reductive anaerobic reduction of chlorinated solvents, one of the better-understood biodegradation processes utilized in their bioremediation, involves the substitution of H^+ for Cl^- in the chlorinated solvent structure (R–Cl in the equation below):

$$R - Cl + H^+ + 2e \rightarrow R - H + Cl^- \tag{6.28}$$

Chlorinated solvents undergo a series of reductions through dechlorination reactions. For example, perchloroethylene (PCE) degrades to trichloroethylene (TCE), which degrades primarily to *cis*-1,2-dichloroethylene (cDCE), which in turn degrades to VC, which is

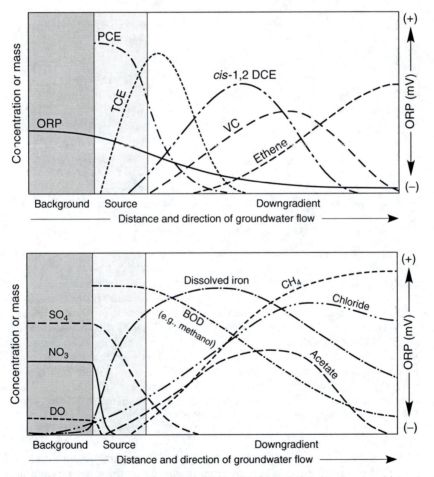

FIGURE 6.18 Common patterns of chlorinated solvent biodegradation in an anaerobic system. Interpretation: BOD (e.g., methanol) is supporting the growth of anaerobic bacteria as shown by the production of methane and acetate, and depletion of sulfate. Sulfate-reducing and possibly iron-reducing bacteria appear responsible for the initial dechlorination of PCE to DCE. As the sulfate concentration decreases, the activity of methanogenic bacteria increases. Under methanogenic or acetogenic conditions 1,2-DCE and VC are dechlorinated to ethane. (ITRC [The Interstate Technology & Regulatory Cooperation Work Group], 1999. Natural attenuation of chlorinated solvente in groundwater: Principles and practices. Technical/Regulatory Guidelines; graphs by S. Jamal, Beak International, Inc., 1997.)

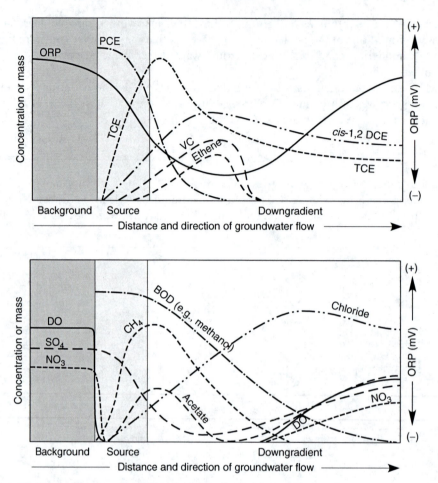

FIGURE 6.19 Common patterns of chlorinated solvent biodegradation in a sequential aerobic or anaerobic system. Interpretation: An anaerobic zone has developed in an aerobic groundwater system due to metabolism of the BOD (e.g., methanol) in the source area. In the anaerobic zone PCE is dechlorinated to RCE, DCE, VC, and finally ethane. Methanogenic, sulfate-reducing, iron-reducing, and acetogenic bacteria are active, and their interactions are responsible for the dechlorination. However, the dechlorination rate is insufficient to cause all of the TCE and DCE to be dechlorinated in the anaerobic zones. These chemicals along with methane, ethane, and vinyl chloride migrate into the transition and aerobic zones. In the transition zone the TCE and DCE are partially cometabolized by methanotrophs growing on methane. Ethene and VC are mineralized to CO_2 by aerobic bacteria in the aerobic zone. (ITRC [The Interstate Technology & Regulatory Cooperation Work Group], 1999. Natural attenuation of chlorinated solvents in groundwater: Principles and practices. Technical/Regulatory Guidelines; graphs by S. Jamal, Beak International, Inc., 1997.)

dechlorinated to ethene. This series of reductions is called sequential degradation. The higher compound is called parent and the lower daughter (e.g., TCE is the parent compound for *cis*-1,2-DCE). Each step requires a lower redox potential than the previous one. PCE degradation occurs in a wide range of-reducing conditions, whereas VC is reduced to ethene only under sulfate-reducing and methanogenic conditions. During each of these transformations, the parent compound (R–Cl) releases one chloride ion and gains one hydrogen. Two electrons are transferred during the process, which may provide a source of energy for the microorganism. The ultimate source for the hydrogen and electrons in this reaction is some sort of organic substrate. Hydrogen (H_2) released during fermentation of the substrate acts as the actual electron donor for respiration (e.g., Distefano et al., 1991). Complete reductive dechlorination was first documented in the laboratory by Freedman and Gossett (1991) and in the field by Major and Cox (1992) (ITRC, 1998, 1999). Some chlorinated solvent compounds and degradation products can be biodegraded by other mechanisms as well, including aerobic degradation (e.g., DCE and VC) and cometabolism. (Cometabolism is a reaction in which microbes transform a contaminant even though the contaminant cannot serve as an energy source for the organisms. To degrade the contaminant, the microbes require the presence of other compounds [primary substrates] that can support their growth.)

One quantitative parameter needed to quantify and model sequential degradation or a complete destruction of an organic contaminant is the stoichiometric conversion ratio. Governed by the chemical structure of contaminants and their molecular weights, this ratio determines the relative masses of daughter products produced in the transformation processes. A compound degradation may be documented to occur at a particular site, by directly measuring the presence of its degradation products, or it may be assumed indirectly by measuring changes in concentrations of various groundwater constituents that indicate such degradation is likely occurring (see Figure 6.18 and Figure 6.19). In either case, and regardless of the actual mechanism and reactions involved, the mass of the daughter product produced has to follow the exact chemical balance between the parent and the daughter products. Yield of the specific transformation, i.e., parameter required by models simulating sequential degradation (decay), is the ratio between molecular weights of the daughter product (e.g., 4A-DNT) and its parent chemical (TNT) multiplied by the mass conversion factor (MCF) for the specific transformation:

$$\left[\frac{4A-DNT}{TNT}\right] \times MCF = \left[\frac{197}{227}\right] \times 0.67 = 0.868 \times 0.67 = 0.582 \qquad (6.29)$$

Table 6.9 shows an example of yield calculations for degradation of TNT and RDX, two common munition constituents. The MCF is 1 (or 100%) when the entire mass of the parent

TABLE 6.9
Molecular Weights of TNT, RDX and Their Transformation Products, and Stoichiometric Conversion Ratios for Yield of 100%

Component	Molecular Formula	Molecular Weight (g)	Transformation	Yield of 100%
TNT	$C_7H_5N_3O_6$	227	TNT > 2A-DNT	0.868
2A-DNT	$C_7H_7N_3O_4$	197	2A-DNT > 2,6-DANT	0.736
4A-DNT	$C_7H_7N_3O_4$	197	2A-DNT > 2,4-DANT	0.736
2,4-DANT	$C_7H_9N_3O_2$	167	2,6-DANT > TAT	0.604
2,6-DANT	$C_7H_9N_3O_2$	167	2,4-DANT > TAT	0.604
TAT	$C_7H_{11}N_3$	137	TNT > 4A-DNT	0.868
TNB	$C_6H_3N_3O_9$	213	4A-DNT > 2,4-DANT	0.736
DNB	$C_6H_4N_2O_6$	168	2,4-DANT > TAT	0.604
NB	$C_6H_5NO_3$	123	TNT > TNB	0.938
RDX	$C_3H_6N_6O_6$	222	TNB > DNB	0.740
MNX	$C_3H_6N_6O_5$	206	DNB > NB	0.542
DNX	$C_3H_6N_6O_4$	190	RDX > MNX	0.928
TNX	$C_3H_6N_6O_3$	174	MNX > DNX	0.856
			DNX > TNX	0.784

Source: Courtesy of Malcolm Pirnie, Inc.

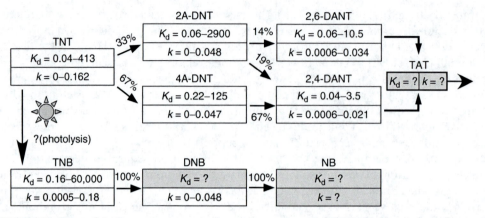

FIGURE 6.20 Some of the more common transformation pathways of TNT with mass balance in percent, based on the current state of knowledge. K_d—distribution coefficient in L/kg; k—transformation rate constant in h^{-1}. (Courtesy of Malcolm Pirnie, Inc.)

chemical transforms into only one daughter product. However, under site-specific conditions, the parent chemical may transform into two or more daughter products. For example, TNT may transform into both 4A-DNT and 2A-DNT with the MCFs of 0.67 and 0.33, respectively, as shown in Figure 6.20.

6.2.5.1.1 Degradation Rate
Many chemicals will undergo zero- or first-order degradation in the dissolved, solid or gaseous phase. A first-order degradation is described by the following equation:

$$C = C_0 e^{-kt} \qquad (6.30)$$

where C is concentration at time t, C_0 is initial concentration at time $t = 0$, k is first-order rate constant (units $=$ time^{-1}), and t is time.

First-order degradation can also be stated in terms of a chemical's half-life. The half-life is the time it takes for half the contaminant to degrade:

$$\frac{C}{C_0} = 0.5 = e^{-kt} \qquad (6.31)$$

$$t_{1/2} = \frac{\ln 2}{k} = \frac{0.693}{k} \qquad (6.32)$$

As can be seen, the contaminant half-life and the first-order degradation constant can both be used in quantitative analyses, but consistently, so that there is no confusion. For example, a 2-year half-life is equivalent to a first-order rate constant of 0.35/y. Most practitioners, and most analytical equations, consistently use the first-order degradation constant in the same units of time as all other time-dependent parameters.

Some chemicals undergo zero-order degradation, which is described by the following equation:

$$C = C_0 - k_0 t \qquad (6.33)$$

where k is the zero-order rate constant (units $=$ mass/volume-time^{-1}). Zero-order kinetics does not typically occur for most common organic compounds found in groundwater.

Obviously, the key parameter in all calculations involving contaminant degradation and its "half-life," is the degradation constant k. Unfortunately, it is also one of the most difficult parameters that can be accurately determined in the field, and it may change in time as geochemical conditions at the site change. It is also the least likely parameter to be accepted by regulatory agencies in case all other abiotic parameters, such as advection, recharge (very important when considering effects of potential dilution), sorption, and dispersion, are not reasonably accurately established for the particular site. In case of sequential decay reactions, such as those involving chlorinated solvents and munition constituents, degradation constants are certain to vary for different daughter products, and may not even be applicable for some of them. In other words, some daughter products may be less biodegradable or not biodegradable at all in certain conditions.

USEPA published a very informative work on different methods commonly used to determine the degradation rate constant k, including discussions on associated uncertainties and applicability of individual methods (Newell et al., 2002). The key point is that constant k may mean different things to different people and in different contexts. It is therefore very important to make a clear distinction between the general attenuation constant and the biodegradation constant. Although either constant can be used to quantitatively describe a general decrease in contaminant concentration in time (a first-order decay process), the degradation constant should be used only if the associated biological process is confirmed, and its rate quantified. The more general attenuation constant includes both abiotic and biotic processes, without making distinction between them. It is relatively determined in an easy manner from the measured contaminant concentrations at monitoring wells, at multiple times, by establishing a quantitative relationship between the concentration decrease and the time. This is illustrated in Figure 6.21. However, it would be completely erroneous to include, in a mathematical equation of contaminant F&T, this generic constant and any (or all) of the processes that may have caused the concentration decrease. By doing this, one would account for the same process twice (e.g., sorption included as a separate parameter, and also through the apparent attenuation parameter), therefore overestimating the actual natural attenuation.

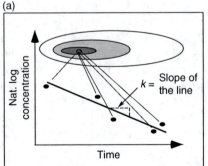
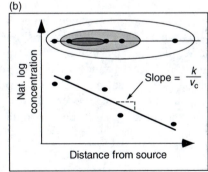

FIGURE 6.21 Determining rate constant from the monitoring well concentration data. (a) Concentration vs. time rate constant for individual wells. (b) Concentration vs. distance for a line of wells along groundwater flow direction; v_c is contaminant velocity, equal to linear groundwater velocity divided by the retardation factor. (Modified from Newell, C.J., et al., 2002. Calculation and use of first-order rate constants for monitored natural attenuation studies. *Ground Water Issue.* EPA/540/S-02/500, U.S. Environmental Protection Agency, National Risk Management Research Laboratory, Cincinnati, Ohio, 27pp.)

6.2.5.1.2   Concentration vs. Time Rate Constant (Figure 6.21a)

Rate constant derived from a C vs. T plot at a single monitoring location provides information regarding the potential plume lifetime at that location, but cannot be used to evaluate the distribution of contaminant mass within the groundwater system. The C vs. T rate constant at a location within the source zone represents the persistence in source strength over time and could be used to estimate the time required to reach a remediation goal at that particular location. To adequately assess an entire plume, monitoring wells that adequately delineate the entire plume must be available, and an adequate time record of monitoring data must be available to calculate a C vs. T plot for each well. At most sites, the rate of attenuation in the source area (due to weathering of residual source materials such as NAPLs) is slower than the rate of attenuation of materials in groundwater, and concentration profiles in plumes tend to retreat back toward the source over time. In this circumstance, the lifecycle of the plume is controlled by the rate of attenuation of the source, and can be predicted by the C vs. T plots in the most contaminated wells. At some sites, the rate of attenuation of the source is rapid compared to the rate of attenuation in groundwater. This pattern is most common when contaminants are readily soluble in groundwater and are not biodegraded. In this case, the rate of attenuation of the source as predicted by a C vs. T plot will underestimate the lifetime of the plume (Newell et al., 2002).

6.2.5.1.3   Concentration vs. Distance Rate Constant (Figure 6.21b)

Attenuation rate constants derived from concentration vs. distance (C vs. D) plots serve to characterize the distribution of contaminant mass within space at a given point in time. A single C vs. D plot provides no information with regard to the variation of dissolved contaminant mass over time and, therefore, cannot be used to estimate the time required for the dissolved plume concentrations to be reduced to a specified remediation goal. This rate constant incorporates all attenuation parameters (sorption, dispersion, biodegradation) for dissolved constituents after they leave the source. Use of the rate constant derived from a C vs. D plot (i.e., characterization of contaminant mass over space) for this purpose (i.e., to characterize contaminant mass over time) will provide erroneous results. The C vs. D based rate constant indicates how quickly dissolved contaminants are attenuated once they leave the source but provides no information on how quickly a residual source zone is being attenuated. Note that most sites with organic contamination will have some type of continuing residual source zone, even after active remediation (Wiedemeier et al., 1999), making the C vs. D rate constant inappropriate for estimating plume lifetimes for most sites (Newell et al., 2002).

The attenuation coefficient for the case shown in Figure 6.21b is the slope, k, of the straight line on the semilog graphs, multiplied by the contaminant velocity, v_c. The contaminant velocity equals the linear groundwater velocity divided by the retardation factor; see Equation 6.18, Equation 6.19, Equation 6.23, and Equation 6.24. The slope (k) is found from the regression equation describing the semilog straight line; note that the vertical scale is a natural logarithm of concentration, and the horizontal scale (x) is distance:

$$y = C_0 e^{-kx}$$

where C_0 is the intercept of the straight line with the y-axis.

6.2.5.1.4   Biodegradation Rate Constant

The biodegradation rate constant applies to both space and time, but only to one degradation mechanism. Quantification of this parameter is arguably the most critical part of contaminant F&T studies and remediation projects that consider biodegradation, in any form, as a potentially viable alternative. It can be performed in the laboratory, using controlled, time-consuming microcosm studies with soil and groundwater samples from the site, or in the field using extensive (and expensive) tracer studies. It can also be estimated during model calibration of all other abiotic F&T parameters. Whatever the case may be, selecting a literature value should be the last option, and only for the "screening" purposes. Because every site will have a very specific degradation rate for any particular contaminant, and this rate may change in time, the author deliberately did not include any table of "literature values" of degradation rate constants. This is, again, simply because every organic chemical may or may not degrade at any particular site. Every effort should be made to avoid using literature values because it may result in very misleading results, whatever any particular stakeholders were hoping for.

REFERENCES

Anderson, M.R., Johnson, R.L., and J.F. Pankow, 1987. The dissolution of residual dense nonaqueous phase liquid (DNAPL) from a saturated porous medium. In: *Proceedings of the NWWA/API Conference on Petroleum Hydrocarbons and Organic Chemicals in Ground Water: Prevention, Detection, and Restoration, National Water Well Association-American Petroleum Institute*, Houston, Texas, pp. 409–428.

Anderson, M.R., Johnson, R.L., and J.F. Pankow, 1992. Dissolution of dense chlorinated solvents into groundwater. 3. Modeling of contaminant plumes from fingers and pools of solvent. *Environ. Sci. Technol.*, 26: 901–908.

ASTM (American Society for Testing and Materials), 1998. Standard guide for remediation of groundwater by natural attenuation at petroleum release sites. E 1943–98, West Conshohocken, PA. www.astm.org

Azdapor-Keeley, A., Russell, H.H., and G.W. Sewell, 1999. Microbial processes affecting monitored natural attenuation of contaminants in the subsurface. *Ground Water Issue*, U.S. Environmental Protection Agency, Office of Research and Development, EPA/540/S-99/001, 18pp.

Aziz, C.E., et al., 2000. BIOCHLOR: Natural Attenuation Decision Support System; User's manual, version 1.0. EPA/600/R-00/008, U.S. Environmental Protection Agency, Office of Research and Development, Washington, D.C., 46pp.

Brusseau, M., 1991. Transport of organic chemicals by gas advection in structured or heterogeneous porous media: Development of a model and application of column experiments. *Water Resour. Res.*, 27(12): 3189–3199.

Cohen, R.M. and J.W. Mercer, 1993. *DNAPL Site Evaluation*. C.K. Smoley, Boca Raton, FL.

Devinny, J.S., et al., 1990. *Subsurface Migration of Hazardous Wastes*. Van Nostrand Reinhold, New York, 387pp.

Distefano, T.D., Gossett, J.M., and S.H. Zinder, 1991. Reductive dechlorination of high concentrations of tetrachloroethene to ethene by an anaerobic enrichment culture in the absence of methanogenesis. *Appl. Environ. Microbiol.*, 57: 2287–2292.

Falta, R.W., et al., 1989. Density-drive flow of gas in the unsaturated zone due to evaporation of volatile organic chemicals, *Water Resour. Res.*, 25(10): 2159–2169.

Feenstra, S. and J.A. Cherry, 1988. Subsurface contamination by dense nonaqueous phase liquid (DNAPL) chemicals. In: *Proceedings of International Groundwater Symposium, International Association of Hydrogeologists*, May 1–4, 1988, Halifax, Nova Scotia, pp. 62–69.

Franke, O.L., et al., 1990. Study guide for a beginning course in ground-water hydrology; Part 1, Course participants. *U.S. Geological Survey Open-File Report* 90–183, Reston, Virginia, 184pp.

Freedman, D.L. and J.M. Gossett, 1991. Biodegradation of dichloromethane in a fixed-film reactor under methanogenic conditions. In: R. Hinchee and R. Olfenbuttel (eds.), *In Situ and On Site Bioreclamation*, Buttersworth-Heineman, Stoneham, MA.

Fussell, D.R., et al., 1981. *Revised Inland Oil Spill Clean-Up Manual*. CONCAWE Report No. 7/81, Management of manufactured gas plant sites, GRI-87/0260, Gas Research Institute, Den Haag, 150pp.

Gelhar, L.W. and C.L. Axness, 1983. Three-dimensional stochastic analysis of macrodispersion in aquifers. *Water Resour. Res.*, 19(1): 161–180.

Gelhar, L.W., Welty, C., and K.R. Rehfeldt, 1992. A critical review of data on field-scale dispersion in aquifers. *Water Resour. Res.*, 28(7): 1955–1974.

Grim, R.E., 1968. *Clay Mineralogy*. McGraw-Hill Book Company, New York.

Gu, B. and R.K. Schulz, 1991. *Anion Retention in Soil: Possible Application to Reduce Migration of Buried Technetium and Iodine*. NUREG/CR-5464, U.S. Nuclear Regulatory Commission, Washington, D.C.

Huling, S.G. and J.W. Weaver, 1991. Dense nonaqueous phase liquids. *Ground Water Issue*, EPA/540/4-91-002, Robert S. Kerr Environmental Research Laboratory, Ada, Oklahoma, 28pp.

ITRC (The Interstate Technology & Regulatory Cooperation Work Group), 1998. Technical and regulatory requirements for enhanced *in situ* bioremediation of chlorinated solvents in groundwater—final. *In Situ* Bioremediation Subgroup.

ITRC (The Interstate Technology & Regulatory Cooperation Work Group), 1999. Natural attenuation of chlorinated solvents in groundwater: Principles and practices. Technical/Regulatory Guidelines.

ITRC (The Interstate Technology & Regulatory Council), 2003. An introduction to characterizing sites contaminated with DNAPLs. Technology Overview, Dense Nonaqueous Phase Liquids Team, Washington, D.C.

Johnson, R.L., Palmer, C.D., and W. Fish, 1989. Subsurface chemical processes. In: *USEPA, Seminar Publication; Transport and Fate of Contaminant in the Subsurface*, EPA/625/4-89/019, U.S. Environmental Protection Agency, pp. 41–56.

Jury, W.A., Gardner, W.R., and W.H. Gardner, 1991. *Soil Physics*. John Wiley & Sons, New York, 328pp.

Knox, R.C., Sabatini, D.A., and L.W. Canter, 1993. *Subsurface Transport and Fate Processes*. Lewis Publishers, Boca Raton, FL, 430pp.

Kresic, N., et al., 2005. Technical impracticability (TI) of DNAPL remediation in karst. In: Stevanovic, Z. and P. Milanovic (eds.), *Water Resources and Environmental Problems in Karst–Cvijić 2005, Proceedings of International Conference*, University of Belgrade, Institute of Hydrogeology, Belgrade, pp. 63–65.

Langmuir, D., 1997. *Aqueous Environmental Geochemistry*. Prentice Hall, Upper Saddle River, NJ.

Lehninger, A.L., 1970. *Biochemistry*. Worth Publishers, Inc., New York.

Lyman, W.J., Reehl, W.F., and D.H. Rosenblatt, 1982. *Handbook of Chemical Property Estimation Methods: Environmental Behavior of Organic Compounds*. McGraw-Hill Book Co., New York.

Lyman, W.J., Reidy, P.J., and B. Levy, 1992. *Mobility and Degradation of Organic Contaminants in Subsurface Environments*. C.K. Smoley, Chelsea, MI, 395pp.

Major, D.W. and E.E. Cox, 1992. Field and laboratory evidence of *in situ* biotransformation of chlorinated ethenes at two district sites: Implications for bioremediation. In: S. Lesage (ed.), *In Situ Bioremediation Symposium '92*. Environment Canada, Niagara-on-the-Lake, Canada.

McCarty, P.L., Reinhard, M., and B.E. Rittmann, 1981. Trace organics in groundwater. *Environ. Sci. Technol.*, 15(1): 40–51.

Mendoza, C.A. and E.O. Frind, 1990a. Advective–dispersive transport of dense organic vapors in the unsaturated zone, 1. Model development. *Water Resour. Res.*, 26(3): 379–387.

Mendoza, C.A. and E.O. Frind, 1990b. Advective–dispersive transport of dense organic vapours in the unsaturated zone, 2. Sensitivity analysis. *Water Resour. Res.*, 26(3): 388–398.

Millington, R.J., 1959. Gas diffusion in porous media. *Science*, 130: 100–102.

Newell, C.J., et al., 1995. Light nonaqueous phase liquids. *Ground Water Issue*, EPA/540/S-95/500, Robert S. Kerr Environmental Research Laboratory, Ada, Oklahoma, 28pp.

Newell, C.J., et al., 2002. Calculation and use of first-order rate constants for monitored natural attenuation studies. *Ground Water Issue*, EPA/540/S-02/500, U.S. Environmental Protection Agency, National Risk Management Research Laboratory, Cincinnati, Ohio, 27pp.

Nichols, E.M., Beadle, S.C., and M.D. Einarson, 2000. Strategies for characterizing subsurface releases of gasoline containing MTBE. American Petroleum Institute (API), Regulatory and Scientific Affairs Publication Number 4699, Washington, D.C.

Pankow, J.F. and J.A. Cherry, 1996. *Dense Chlorinated Solvents and Other DNAPLs in Groundwater.* Waterloo Press, Guelph, Ontario, Canada, 522pp.

Poulsen, M. and B.H. Kueper, 1992. A field experiment to study the behavior of tetrachloroethylene in unsaturated porous media. *Environ. Sci. Technol.*, 26: 889–895.

Robinson, R.A. and R.H. Stokes, 1965. *Electrolyte Solutions*, 2nd ed. Butterworth, London.

Saffman, P.G., 1959. A theory of dispersion in a porous medium. *J. Fluid Mech.*, 6: 321–349.

Schwille, F., 1988. *Dense Chlorinated Solvents in Porous and Fractured Media—Model Experiments* (translated by J.F. Pankow). Lewis Publishers, Chelsea, MI.

Sleep, B.E. and J.F. Sykes, 1989. Modeling the transport of volatile organics in variably saturated media. *Water Resour. Res.*, 25(1): 81–92.

Slichter, C.S., 1905. Field measurements of the rate of movement of underground waters. *U.S. Geological Survey Water-Supply and Irrigation Paper* 140, Washington, D.C., 122pp.

Sposito, G., 1984. *The Surface Chemistry of Soils.* Oxford University Press, New York.

Stumm, W. and J.J. Morgan, 1981. *Aquatic Chemistry: An Introduction Emphasizing Chemical Equilibria in Natural Waters.* John Wiley & Sons, New York.

USEPA, 1996. Presumptive Response Strategy and *Ex-Situ* Treatment Technologies for Contaminated Groundwater at CERCLA Sites. OSWER Directive 9283. pp. 1–12.

USEPA, 1996a. A citizen's guide to bioremediation, EPA 542-F-96-007, Solid Waste and Emergency Response, 4pp.

USEPA, 1996b. A citizen's guide to natural attenuation, EPA 542-F-96-015, Solid Waste and Emergency Response, 4pp.

USEPA, 1996c. *Soil Screening Guidance: User's Guide*, 2nd ed. Publication 9355.4–23, Office of Emergency and Remedial Response, Washington, D.C.

USEPA, 1999a. Understanding variation in partition coefficient, K_d, values; Volume I: The K_d model, methods of measurements, and application of chemical reaction codes, EPA 402-R-99-004A, U.S. Environmental Protection Agency, Office of Air and Radiation, Washington, D.C.

USEPA, 1999b. Understanding variation in partition coefficient, K_d, values; Volume II: Review of geochemistry and available K_d values for cadmium, cesium, chromium, lead, plutonium, radon, strontium, thorium, tritium (^3H), and uranium, EPA 402-R-99-004B, U.S. Environmental Protection Agency, Office of Air and Radiation, Washington, D.C.

USEPA, 1999c. Use of monitored natural attenuation at Superfund, RCRA corrective action, and underground storage tank sites. OSWER Directive 9200.4-17P, April 21, 1999, 32 p. http://www.epa.gov:80/ordntrnt/ORD/WebPubs/biorem/D9200417.pdf

Wiedemeier, T.H., et al., 1998. Technical protocol for evaluating natural attenuation of chlorinated solvents in groundwater, EPA/600/R-98/128, U.S. Environmental Protection Agency, Office of Research and Development, Washington, D.C.

Wiedemeier, T.H., et al., 1999. Technical protocol for implementing intrinsic remediation with long-term monitoring for natural attenuation of fuel contamination dissolved in groundwater; Volume I (Revision 0), Air Force Center for Environmental Excellence (AFCEE), Technology Transfer Division, Brooks Air Force Base, San Antonio, Texas.

7 Groundwater Modeling

Groundwater modeling in some form is now a major part of most projects dealing with groundwater development, protection, and remediation. As computer hardware and software continue to be improved and become more affordable, the role of models in highly quantitative earth sciences such as hydrogeology will continue to increase accordingly. It is essential, however, that for any groundwater model to be interpreted and used properly, its limitations should be clearly understood. In addition to strictly "technical" limitations, such as accuracy of computations (hardware and software), the following is true for any model:

- It is based on various assumptions regarding the real natural system being modeled.
- Hydrogeologic and hydrologic parameters used by the model are always just an approximation of their actual field distribution, which can never be determined with 100% accuracy.
- Theoretical differential equations describing groundwater flow are replaced with systems of algebraic equations that are more or less accurate.

It is therefore obvious that a model will have a varying degree of reliability, and that it could not be "misused" as long as all the limitations involved are clearly stated, the modeling process follows industry-established procedures and standards, and the modeling documentation and any generated reports are transparent, and also follow the industry standards.

7.1   TYPES OF GROUNDWATER MODELS

In general, a model simulates the areal and temporal properties of a system, or one of its parts, in either a physical (real) or mathematical (abstract) way. An example of a physical model in hydrogeology would be a tank filled with sand and saturated with water—the so-called "sandbox," an equivalent to a miniature aquifer of limited extent. This aquifer can be subject to miniature stresses such as pumping from a perforated tube placed into the sand thus representing a water well. An obvious question when considering similar models is how feasible it is to build a multilayer "aquifer" exposed to various stresses such as precipitation, surface streamflow, leakage from deep underlying strata, and then change some of its geometric and hydrogeologic properties as needed. Consequently, the application of real physical models has been limited to educational and demonstration purposes. A groundwater system can also be simulated using the analogy between groundwater flow and some other similar physical process such as the flow of electrical currents through conductors. Such models are called analog and were often used in hydrogeologic practice before the rapid development of numeric computer modeling (see Rushton and Redshaw, 1979). Models that use mathematical equations to describe elements of groundwater flow are called mathematical. Depending upon the nature of equations involved, these models can be:

- Empirical (experimental)
- Probabilistic
- Deterministic

Empirical models are derived from experimental data that are fitted to some mathematical function. A good example is Darcy's law. (Note that Darcy's law was later found to be theoretically grounded and actually became a physical or deterministic law.) Although empirical models are limited in scope and are usually site or problem-specific, they can be an important part of a more complex numeric modeling effort. For example, the behavior of a certain pollutant in porous media can be studied in the laboratory or in controlled field experiments, and the derived experimental parameters can then be used for developing numeric models of groundwater transport.

Probabilistic models are based on laws of probability and statistics. They can have various forms and complexity starting with a simple probability distribution of a hydrogeological property of interest, and ending with complicated stochastic, time-dependent models. The main limitations for a wider use of probabilistic (stochastic) models in hydrogeology are that: (1) they require large data sets needed for parameter identification and (2) they cannot be used to answer (predict) many of the most common questions from hydrogeologic practice such as effects of a future pumping, for example.

Deterministic models assume that the stage or future reactions of the system (aquifer) studied are predetermined by physical laws governing groundwater flow. An example is the flow of groundwater toward a fully penetrating well in a confined aquifer as described with the Theis equation. Most problems in traditional hydrogeology are solved using deterministic models, which can be as simple as the Theis equation or as complicated as a multiphase flow through a multilayered, heterogeneous, anisotropic aquifer system. There are two large groups of deterministic models depending upon the type of mathematical equations involved:

- Analytical
- Numeric

Simply stated, analytical models solve one equation of groundwater flow at a time and the result can be applied to one point or "line of points" in the analyzed flow field (aquifer). For example, if we want to find (i.e., to model) what the drawdown at 50 m from the pumping well would be after 24 h of pumping, we would apply one of the equations describing flow toward a well depending upon the aquifer and well characteristics (confined, unconfined, leaky aquifer; fully or partially penetrating well). To find the drawdown at 1000 m from the well, we would have to solve the same equation (say, the Theis equation) for this new distance. If the aquifer is not homogeneous, these solutions would be applicable just for a limited radial distance of 50 or 1000 m within the same distribution of aquifer transmissivity. Obviously, if our aquifer is quite heterogeneous, and we want to know drawdown at "many" points, we might spend a rather long period of time solving the same equation (with slightly changed variables) again and again. If the situation gets really complicated, such as when there are several boundaries, more pumping wells, and several hydraulically connected aquifers, the feasible application of analytical models terminates.

Numeric models describe the entire flow field of interest at the same time, providing solutions for as many data points as specified by the user. The area of interest is subdivided into many small areas (referred to as cells or elements) and a basic groundwater flow equation is solved for each cell usually considering its water balance (water inputs and outputs). The solution of a numeric model is the distribution of hydraulic heads at points representing individual cells. These points can be placed at the center of the cell, at intersections between adjacent cells, or elsewhere. The basic differential flow equation for each cell is replaced (approximated) by an algebraic equation so that the entire flow field is represented by x equations with x unknowns, where x is the number of cells. This system of algebraic equations is solved numerically, through an iterative process, thus the name numeric models. Based on various methods of approximating differential flow equations, and methods used for numerically solving the resulting system of algebraic equations, numeric models are divided into several groups. The two most widely applied groups are

- finite differences (numeric models) and
- finite elements (numeric models).

Both types of models have their advantages and disadvantages and for certain problems one may be more appropriate than the other. However, because they are easier to design and understand, and require less mathematical involvement, finite-difference models have prevailed in hydrogeologic practice. In addition, several excellent finite-difference modeling programs have been developed by the Unites States Geological Survey (USGS) and are in public domain, which ensures their widest possible use. One of these is Modflow, probably the most widely used, tested, and verified modeling program today and it has become the industry standard thanks to its versatility and open structure: independent subroutines called "modules" are grouped into "packages," which simulate specific hydrologic features. New modules and packages can be easily added to the program without modifying the existing packages or the main code. For these reasons, Modflow has been chosen as the standard for explaining modeling principles and for solving numeric modeling problems in this book. USGS has recently made public a significantly upgraded version of the finite-element model SUTRA, now capable of simulating three-dimensional flow (Voss and Provost, 2002). This computer program can simulate both unsaturated and saturated flow, heat and contaminant transport, as well as variable density flow, which makes it a powerful tool for modeling just about any imaginable condition. Unfortunately, unlike Modflow, SUTRA3D is not yet part of any of the most widely used user-friendly commercial programs for processing model input and output data, which severely limits its greater application. For a thorough explanation of finite-element and finite-difference models, and their various applications, the reader should consult the excellent work by Anderson and Woessner (1992).

Groundwater models can be used for three general purposes:

- To predict or forecast expected artificial or natural changes in the system (aquifer) studied. The term predict is more appropriately applied to deterministic (numeric) models as it carries a higher degree of certainty, while forecasting is the term used with probabilistic (stochastic) models. Predictive models are by far the largest group of models built in hydrogeologic practice.
- To describe the system in order to analyze various assumptions about its nature and dynamics. Descriptive models help to better understand the system and plan future investigations. Although not originally planned as a predictive tool, they often grow to be full predictive models.
- To generate a hypothetical system that will be used to study principles of groundwater flow associated with various general or more specific problems. Generic models are used for training and are often created as part of a new computer code development.

Predictive numeric models are divided into two main groups: (1) models of groundwater flow and (2) models of contaminant fate and transport. The latter models cannot be developed without first solving the groundwater flow field of the system studied, i.e., they use the solution of the groundwater flow model as the base for fate and transport calculations. Some of the more common questions that fully developed and calibrated groundwater flow, and fate and transport models may help answer, are:

- What is the safe (sustainable) yield of the aquifer portion targeted for groundwater development?
- At what locations and how many wells are needed to provide a desired flow rate?
- What is the impact of current or planned groundwater extraction on the environment (e.g., on surface streamflows, wetlands)?
- Is there a potential for saltwater intrusion from an increased groundwater pumpage?
- Where is the contaminant flowing to, and where is it coming from?
- How long would it take the contaminant to reach potential receptors?
- What would the contaminant concentration be once it reaches a receptor?

Once these questions are addressed by the models, many new ones may pop-up, which is exactly what the purpose of a well-documented and calibrated groundwater model should be: to answer "all kinds" of possible questions related to groundwater flow, and fate and transport of contaminants. Here are just two of the common "big" questions, often with a multimillion dollar price tag and possibility of a protracted lawsuit: Who is responsible for the groundwater contamination? What is the most feasible groundwater remediation option?

7.2 NUMERIC MODEL SETUP

Numeric groundwater model setup consists of the following main stages:

- Development of the conceptual site model (CSM), which is the most important part of modeling and the basis for all further related activities
- Selection of a computer code that can most effectively simulate the concept and meet the purpose of modeling
- Definition of the model geometry: lateral and vertical extent of the area to be modeled defined by model boundaries, grid layout, and position and number of layers
- Input of hydrogeologic parameters, and fate and transport parameters when required, for each model cell such as horizontal and vertical hydraulic conductivities, including possible anisotropy, storage properties, effective porosity, dispersivity, distribution coefficient, and others
- Definition of model boundary conditions that influence the flux of water, or are directly causing it; this flux of water (and contaminant when required) both enters and leaves the model domain and has to be provided for in the model design (e.g., boundaries with known hydraulic head, known flux, or head-dependent flux)
- Definition of initial conditions, such as estimated distribution of the hydraulic head in the model domain, and distribution of contaminant concentration when required
- Definition of external and internal hydraulic stresses acting upon the system, in addition to those assigned along model boundaries, such as areal recharge, evapotranspiration (ET), well pumpage, outflow through springs, drains, inflow of water from other sources (recharge wells, adjacent aquifers)

After the model has been set up, it is run and then adjusted (calibrated) to match the hydraulic (and chemical or contaminant where required) information collected in the field, which served as the "water" basis for the CSM:

- The hydraulic head measured in monitoring wells
- The flux along model boundaries measured or calculated externally to the model; all such fluxes comprise the water budget of the model
- The contaminant concentration measured at monitoring wells and at model boundaries, in case of fate and transport models

Model calibration and other major aspects of model development and utilization, such as sensitivity analysis, model validation, prediction, and documentation, are described in the respective subchapters.

7.2.1 CONCEPTUAL MODEL

Developing a modeling concept is the initial and the most important part of every modeling effort. It requires a thorough understanding of hydrogeology, hydrology, and dynamics of groundwater flow in and around the area of interest. The result is a computerized database, and simplified electronic maps and cross sections that will be used in model design and could effortlessly be incorporated into the actual electronic model. The most efficient way to organize all the information required for model development is to utilize a geographic information system (GIS) environment. This enables all interested stakeholders, including nonmodelers, to provide invaluable input as to the validity of certain hydrogeologic assumptions, spatial information related to contaminant fate and transport, and other aspects of the model (Kresic and Rumbaugh, 2000). Most commercial GIS programs offer free software for viewing and sharing electronic files and maps, which is arguably the most efficient way (other than meeting face-to-face) to quickly exchange visual information. Modern development of groundwater models is a highly visual process, greatly enhanced by various graphical programs, which facilitate quick and accurate input of data into models and visualization of model results (such programs are called graphic user interface, or GUI programs).

Probably the single most important initial assurance that the model will be developed in a technically sound manner and efficiently is the involvement of the "computer modeler" from the very beginning of the concept development. Ideally, the leading modeler and the leading hydrogeologist on the project should be the same person since there is no valid excuse why any practicing hydrogeologist would not be intimately knowledgeable in groundwater modeling. Unfortunately, in many cases nonhydrogeologists (or even worse—nongeologists) may end up developing groundwater models and calibrating them, without realizing that some (or many) parts of such models simply do not make hydrogeologic sense. It cannot be emphasized enough that every numeric groundwater model is a nonunique solution of the underlying flow field. In other words, various combinations of various model parameters may produce very similar or identical results. The opposite is also true; what may seem a "slight parameter change" to some may result in a dramatically different model output. It is therefore not enough to simply match a few hydraulic heads measured in the field. It is infinitely more important that the model makes hydrogeologic sense, and that all its uncertainties and (inevitable) errors be fully documented. Then and only then, the model will be hard to misuse and it may be useful to most (if not all) stakeholders that have to make some decisions based on the modeling results.

Figure 7.1 is a hydrogeologic map of a rather simple study area that will be used throughout this chapter to illustrate various aspects of model development. It shows water table

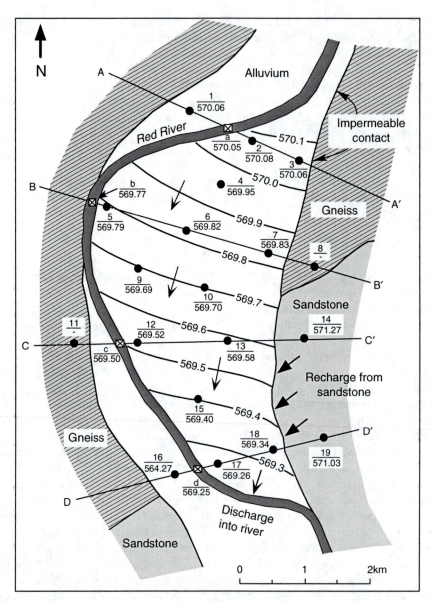

FIGURE 7.1 Map showing water table contour lines in meters above mean sea level (amsl) and remarks on the boundaries of the flow field in an alluvial aquifer. Arrows indicate general direction of groundwater flow.

contours for an alluvial aquifer and general directions of groundwater flow. The Red River is in most part a gaining stream except in the far north where water enters the aquifer from the river. The alluvium is in contact with practically impermeable gneiss in the northeast and does not receive any significant lateral recharge. Gneiss is the aquifer base for the most part except in the south where sandstone is also present, as illustrated with cross sections in Figure 7.2.

Where there is contact with the sandstone, the aquifer receives lateral recharge, which increases southward as the area of contact increases (see cross-sections CC′ and DD′ in Figure 7.2). The hydraulic gradient of the water table is uniform indicating homogeneous transmissivity of the alluvial deposits. The borehole data show that the aquifer has similar

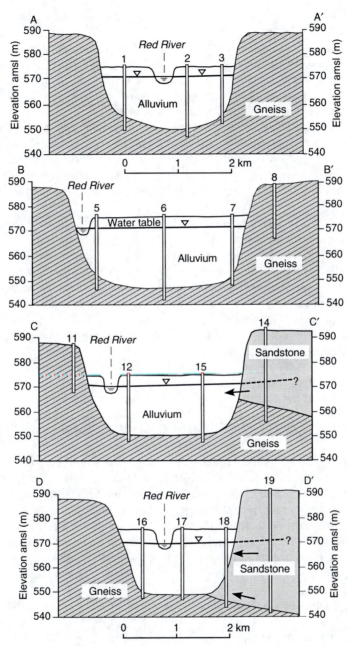

FIGURE 7.2 Hydrogeologic cross-sections of the alluvial aquifer shown in Figure 7.1 (the vertical scale is exaggerated).

thickness throughout its central part, which means that the hydraulic conductivity is fairly uniform. The average hydraulic conductivity from slug tests performed at several monitoring wells is 2.4×10^{-3} m/s. The porous media is mainly fine sand, and the assumed effective porosity is 25%. The water table elevation is recorded in two wells completed in sandstone aquifer (wells 14 and 19). Based on these two data and the projected water table in the alluvial aquifer, the groundwater flow in sandstone is oriented toward the alluvium aquifer.

The hydraulic gradient in the sandstone aquifer is steeper than in the alluvial aquifer indicating a lower hydraulic conductivity.

According to the borehole logs, there is no significant silty or clayey cover over the alluvial sands and the aquifer is directly exposed to recharge from precipitation. The exact amount of precipitation in the area is not known. Data for a 23-miles distant gauging station in slightly different orographic conditions show that the average annual precipitation is 32 inches. The site is located in a relatively humid moderate climate in the northern hemisphere. The water table and the river stage measurements were taken only once during the month of May. Although there is no information about the river flow and stage before the measurements, it could be concluded, based on the climate characteristics, that the collected data are representative for the season of a relatively high water table and major aquifer recharge. A relatively simple initial modeling concept can be developed based on the above facts:

- The groundwater flow system to be modeled consists of a single unconfined aquifer.
- There are five well-defined boundaries of the flow field: (1) equipotential boundary along the contact between the river and the alluvial aquifer, (2) impermeable lateral boundary between the alluvial aquifer and gneiss, (3) impermeable base boundary, (4) water table boundary directly exposed to the surface, and (5) lateral boundary between the alluvial and the sandstone aquifers.
- The alluvial aquifer is of relatively uniform thickness, transmissivity (hydraulic conductivity), and effective porosity and can be modeled with one layer where all model cells will have the same values of these parameters.
- The alluvial aquifer receives recharge from precipitation and from the adjacent sandstone aquifer. In both cases, the recharge will have to be estimated and calibrated. We will initially assume that the areal recharge is 10% of the total annual precipitation.
- Since there is only one set of water table data, the model will have to be calibrated for that set assuming steady-state conditions and it cannot be verified (validated).

It is obvious that any model built on the above assumptions would be, in reality, susceptible to criticism since we have spent a considerable amount of time discussing, in all preceding chapters, the inherent heterogeneity of aquifer porous media. In addition, the fact that there is only one set of water table and river stage measurements available, severely limits the utility of this model for long-term predictions of any kind. However, we will proceed with building this model since it will not be subject to any serious criticism, being only an academic exercise. In reality, if we were to consider this assignment more seriously and had appropriate funds to do so, we would, ideally

- acquire more rounds of seasonal synoptic water level measurements in the two aquifers and the river;
- conduct aquifer pumping test;
- carefully examine the boring logs to facilitate design of a multilayer model;
- analyze land cover and land use in the aquifer area to better estimate potential aquifer recharge; and
- continuously and simultaneously measure precipitation and water table fluctuations in several "strategically" located monitoring wells (adjacent to the river, in the central aquifer portion, and adjacent to the low-permeable aquifer boundary in the east).

As we implement this data collection program, we may conclude that some additional information would also be useful, and may try to obtain additional funds to collect the new data (we will argue that it was not our oversight not including this new data requirement in the initial program; rather, we will try to convince the client that "all aquifers" are unpredictable to a certain extent, and that we are trying to best serve his, her, or their interest).

In conclusion, the CSM should have the following straightforward questions covered:

- Where is the water in the aquifer coming from, approximately how much of it, and how does this change in time?
- Where (in which directions) is the water in the aquifer flowing, at approximately what rate, and how does this change in time?
- Where is the water leaving the aquifer, approximately how much of it, and how does this change in time?

In case we are interested in developing a fate and transport model for certain contaminants, the word *water* in the above questions would be replaced with word *contaminants* in the new set of questions.

7.2.2 GOVERNING EQUATIONS

A computer program for groundwater modeling numerically solves a system (matrix) of algebraic equations. This matrix represents an approximation of the mathematical model formulated by partial differential equations of groundwater flow. Finite-difference and finite-element methods are two common ways in which this approximation is made. (Note that there are some other methods yet to be widely accepted such as integrated finite differences, the boundary integral equation method and analytic elements; see Ligget and Liu, 1983; Ligget, 1987; Strack, 1987, 1988). Which method will be used depends largely on the type of problem and the knowledge of the modeler. Finite elements more easily describe irregular model boundaries and internal boundaries such as faults. They are also more appropriate in handling point sources and sinks and large variations in the water table. However, finite-difference models are easier to program, require less data, and are friendlier for data input. Modflow (A modular three-dimensional finite-difference groundwater flow model) developed at the USGS by McDonald and Harbaugh (1988), Harbaugh and McDonald (1996), and Harbaugh et al. (2000) is considered by many to be the most reliable, verified, and utilized groundwater flow computer program available. There are several integrated, user-friendly pre- and postprocessing graphical software packages (GUIs) for Modflow that provide easy data input and visualization of modeling results. Author's favorites are Processing Modflow Pro (Chiang and Kinzelbach, 2001; Chiang, 2005, WebTech360) and Groundwater Vistas Advanced Version (Rumbaugh and Rumbaugh, 2004; Environmental Simulations, Inc.). Other widely used GUIs are Groundwater Modeling System (GMS) initially developed for the U.S. Department of Defense, and Visual Modflow developed by Waterloo Hydrogeologic.

Modflow utilizes a numeric solution for the equation governing groundwater flow through porous media:

$$\frac{\partial}{\partial x}\left(K_{xx}\frac{\partial h}{\partial x}\right) + \frac{\partial}{\partial y}\left(K_{yy}\frac{\partial h}{\partial y}\right) + \frac{\partial}{\partial z}\left(K_{zz}\frac{\partial h}{\partial z}\right) - W = S_s\frac{\partial h}{\partial t} \tag{7.1}$$

where K_{xx}, K_{yy}, and K_{zz} are values of hydraulic conductivity along the x, y, and z coordinate axes, which are assumed to be parallel to the major axes of hydraulic conductivity (LT^{-1}); h is the hydraulic head (L), W is the volumetric flux per unit volume and represents sources or sinks of water (T^{-1}), S_s is the specific storage of the porous material (L^{-1}), and t is time.

For derivation of Equation 7.1 see for example Rushton and Redshaw (1979). In general, S_s, K_{xx}, K_{xy}, and K_{zz} may be functions of space ($S_s = S_s(x, y, z)$, $K_{xx} = K_{xx}(x, y, z)$, etc.) and

W may be a function of space and time $W = (W(x, y, z, t))$. Equation 7.1 describes groundwater flow under nonequilibrium conditions in a heterogeneous and anisotropic medium, provided the principal axes of hydraulic conductivity are aligned with the coordinates directions (McDonald and Harbaugh, 1988). The following discussion on the numeric solution implemented in Modflow for solving Equation 7.1 is from the original publication by McDonald and Harbaugh (1988).

Figure 7.3 shows a spatial discretization of an aquifer system with a mesh of blocks called cells, the locations of which are described in terms of rows, columns, and layers. An i, j, k indexing system is used. For a system consisting of "nrow" rows and "ncol" columns, and "nlay" layers, i is the row index, $i = 1, 2, \ldots,$ nrow; j is the column index, $j = 1, 2, \ldots,$ ncol; and k is the layer index, $k = 1, 2, \ldots,$ nlay. For example, Figure 7.3 shows a system with nrow = 5, ncol = 9, and nlay = 5. In formulating the equations of the model, an assumption was made that layers would generally correspond to horizontal geohydrologic units or intervals. Thus in terms of Cartesian coordinates, the k index denotes changes along the vertical, z; because the convention followed in this model is to number layers from the top-down, an increment in the k index corresponds to a decrease in elevation. Similarly, rows would be considered parallel to the x-axis, so that increments in the row index, i, would correspond to decreases in y; and columns would be considered parallel to the y-axis, so that increments in the column index, j, would correspond to increases in x. These conventions were followed in constructing Figure 7.3; however, application of the model requires only that rows and columns fall along consistent orthogonal directions within the layers, and does not require the designation of x, y, or z coordinate axes.

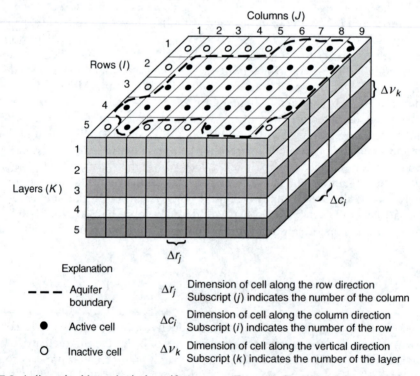

FIGURE 7.3 A discretized hypothetical aquifer system. (From McDonald, M.G. and Harbaugh, A.W., 1988. A modular three-dimensional finite-difference ground-water flow model. *U.S. Geological Survey Techniques of Water-Resources Investigations*, Book 6, Chap. A1, 586pp.)

Following the conventions used in Figure 7.3, the width of cells in the row direction, at a given column, j, is designated Δr_j; the width of cells in the column direction at a given row, i, is designated Δc_i; and the thickness of cells in a given layer, k, is designated Δv_k. Thus a cell with coordinates $(i, j, k) = (4, 8, 3)$ has a volume of $\Delta r_8 \Delta c_4 \Delta v_3$. Within each cell there is a point called a "node" at which head is to be calculated. Figure 7.4 illustrates, in two dimensions, two conventions for defining the configuration of cells with respect to the location of the nodes—the block-centered formulation and the point-centered formulation. Both systems start by dividing the aquifer with two sets of parallel lines that are orthogonal. In the block-centered formulation, which is adopted in Modflow, the blocks formed by the sets of parallel lines are the cells; the nodes are at the center of the cells. In the point-centered formulation, the nodes are at the intersection points of the sets of parallel lines, and cells are drawn around the nodes with faces halfway between nodes. In either case, spacing of nodes should be chosen so that the hydraulic properties of the system are, in fact, generally uniform over the extent of a cell. In Equation 7.1, the head, h, is a function of time as well as space so that, in the finite-difference formulation, discretization of the continuous time domain is also required.

Development of the groundwater flow equation in finite-difference form follows from the application of the continuity equation: the sum of all flows into and out of the cell must be equal to the rate of change in storage within the cell. Under the assumption that density

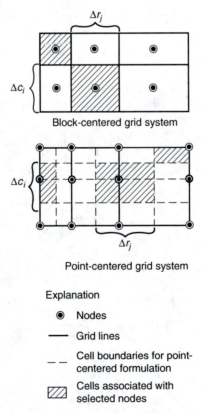

FIGURE 7.4 Grids showing the difference between block-centered and point-centered grids. (From McDonald, M.G. and Harbaugh, A.W., 1988. A modular three-dimensional finite-difference ground-water flow model. *U.S. Geological Survey Techniques of Water-Resources Investigations*, Book 6, Chap. A1, 586pp.)

of groundwater is constant, the continuity equation expressing the balance of flow for a cell is

$$\sum Q_i SS^{\Delta h} \Delta V \tag{7.2}$$

where Q_i is a flow rate into the cell (L^3/t); SS has been introduced as the notation for specific storage in the finite-difference formulation; its definition is equivalent to that of S_s in Equation 7.1, i.e., it is the volume of water that can be injected per unit volume of aquifer material per unit change in head (L^{-1}); ΔV is the volume of the cell (L^3); and Δh is the change in head over a time interval of length Δt.

The term on the right-hand side is equivalent to the volume of water taken into storage over a time interval Δt given a change in head of Δh. Equation 7.2 is stated in terms of inflow and storage gain. Outflow and loss are represented by defining outflow as negative inflow and loss as negative gain.

Figure 7.5 depicts a cell i, j, k and six adjacent aquifer cells $i-1, j, k$; $i+l, j, k$; $i, j-1, k$; $i, j+1, k$; $i, j, k-1$; and $i, j, k+1$. To simplify the following development, flows are considered positive if they are entering the cell i, j, k; and the negative sign usually incorporated in Darcy's law has been dropped from all terms. Following these conventions, flow into cell i, j, k, in the row direction from cell $i, j-1, k$ (Figure 7.6) is given by Darcy's law as

$$q_{i,j-1/2,k} = KR_{i,j-1/2,k} \Delta c_i \Delta v_k \frac{(h_{i,j-1,k} - h_{i,j,k})}{\Delta r_{j-1/2}} \tag{7.3}$$

where $h_{i,j,k}$ is the head at node i, j, k and $h_{i,j-1,k}$ that at node $i, j-1, k$; $q_{i,j-1/2,k}$ is the volumetric fluid discharge through the face between cells i, j, k and $i, j-1, k$ (L^3/t^{-1}); $KR_{i,j-1/2,k}$ is the hydraulic conductivity along the row between nodes i, j, k and $i, j-1, k$ (L/t^{-1}); $\Delta c_i \Delta v_k$ is the area of the cell faces normal to the row direction; and $\Delta r_{j-1/2}$ is the distance between nodes i, j, k and $i, j-1, k$ (L).

Similar expressions can be written approximating the flow into the cell through the remaining five faces. Although the discussion is phrased in terms of flow into the central cell, it can be misleading to associate the subscript $j-1/2$ of Equation 7.3 with a specific point

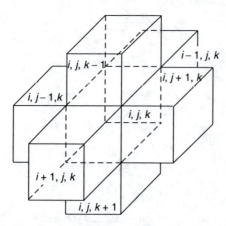

FIGURE 7.5 Cell i, j, k and indices for the six adjacent cells. (From McDonald, M.G. and Harbaugh, A.W., 1988. A modular three-dimensional finite-difference ground-water flow model. *U.S. Geological Survey Techniques of Water-Resources Investigations*, Book 6, Chap. A1, 586pp.)

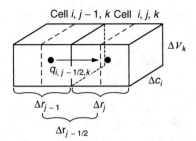

FIGURE 7.6 Flow into cell i, j, k from cell $i, j-1, k$. (From McDonald, M.G., and Harbaugh, A.W. A modular three-dimensional finite-difference ground-water flow model. U.S. Geological Survey Techniques of Water-Resources Investigations, Book 6, Chap. A1, 586 pp., 1988.)

between the nodes. Rather, the term $KR_{i,j-1/2,k}$ of Equation 7.3 is the effective hydraulic conductivity for the entire region between the nodes, normally calculated as a harmonic mean. If this is done, Equation 7.3 gives the exact flow, for a one-dimensional steady-state case, through a block of aquifer extending from node $i, j-1, k$ to node i, j, k and having a cross-sectional area $\Delta c_j \Delta v_k$.

7.2.2.1 Conductance

Each of the six equations expressing the inflow through the six faces of cell i, j, k in terms of heads, grid dimensions, and hydraulic conductivity can be simplified by combining grid dimensions and hydraulic conductivity into a single constant, the hydraulic conductance or, more simply, the conductance. For example,

$$CR_{i,j-1/2,k} = KR_{i,j-1/2,k} \Delta c_i \Delta v_k / \Delta r_{j-1/2} \tag{7.4}$$

where $CR_{i,j-1/2,k}$ is the conductance in row i and layer k between nodes $i, j-1, k$ and i, j, k (L^2/t^{-1}).

Conductance is thus the product of hydraulic conductivity and cross-sectional area of flow divided by the length of the flow path (in this case, the distance between the nodes). Substituting conductance from Equation 7.4 into Equation 7.3 yields

$$q_{i,j-1/2,k} = CR_{i,j-1/2,k}(h_{i,j-1,k} - h_{i,j,k}) \tag{7.5}$$

Equations for the remaining five cell faces are similar and together with Equation 7.5 describe internal inflow and outflow into cell i, j, k from the six adjacent cells.

An example of the final finite-difference equation for each cell, which receive both internal and external fluxes (e.g., recharge from river, well pumpage and or injection) is

$$CR_{i,j-1/2,k}(h^m_{i,j-1,k} - h^m_{i,j,k}) + CR_{i,j+1/2,k}(h^m_{i,j+1,k} - h^m_{i,j,k}) + CC_{i-1/2,j,k}(h^m_{i-1,j,k} - h^m_{i,j,k})$$
$$+ CC_{i+1/2,j,k}(h^m_{i+1,j,k} - h^m_{i,j,k}) + CV_{i,j,k-1/2}(h^m_{i,j,k-1} - h^m_{i,j,k}) + CV_{i,j,k+1/2}(h^m_{i,j,k+1} - h^m_{i,j,k})$$
$$+ P_{i,j,k}h^m_{i,j,k} + Q_{i,j,k} = SS_{i,j,k}\frac{(\Delta r_j \Delta c_i \Delta v_k)(h^m_{i,j,k} - h^{m-1}_{i,j,k})}{t_m - t_{m-1}} \tag{7.6}$$

where $P_{i,j,k} \, h_{i,j,k} + Q_{i,j,k}$ is the general known external flow term for cell i, j, k; SS is the specific storage; $\Delta r_j \Delta c_i \, \Delta v_k$ is the volume of the cell i, j, k; t_m is the time at which the flow terms are evaluated; t_{m-1} is the time which precedes t_m; h^m is the hydraulic head at each node, denoted with subscripts, and has to be calculated for time t_m; and h^{m-1} is the initial hydraulic head at the beginning of calculation for the current time step, calculated for the preceding time step t_{m-1}.

Equation 7.6 is a backward-difference equation, which can be used as the basis for the simulation of partial differential equation of groundwater flow, Equation 7.1. Like the term $Q_{i,j,k}$, the coefficients of various head terms in Equation 7.6 are all known, as is the head at the beginning of the time step. The seven heads at time t_m, the end of time step, are unknown; that is, they are part of the head distribution to be predicted. Thus Equation 7.6 cannot be solved independently, since it represents a single equation in seven unknowns. However, an equation of this type can be written for each active cell in the mesh; and, since there is only one unknown head for each cell, we are left within a system of n equations in n unknowns. Such a system can be solved simultaneously (McDonald and Harbaugh, 1988).

Numeric models of fate and transport start where the groundwater flow models have stopped—they use the flow and the groundwater velocity fields computed by the flow model to create their own solution. Two of the most widely used fate and transport models based on Modflow are MT3DMS (modular three-dimensional transport model of multi-species) by Zheng and Wang (1999), and RT3D (modular computer code for simulating reactive multispecies transport in three-dimensions) by Clement (1997). RT3D is based on solutions of both Modflow and MT3DMS, so that these three models constitute the so-called Modflow suite of models, which is a very powerful tool for modeling various fate and transport processes in the saturated zone. MT3DMS and RT3D can accommodate multiple sorbed and aqueous phase species with any reaction framework that the user wishes to define. With the flexibility to insert user-specific kinetics, these two reactive transport models can simulate a multitude of scenarios. For example, the models can simulate and evaluate natural attenuation processes or an active remediation. Simulations could be applied to scenarios involving contaminants such as heavy metals, explosives, petroleum hydrocarbons, and chlorinated solvents. Both models also have dual-porosity option, which is useful for simulating dissolution of nonaqueous phase liquids (NAPLs), i.e., transfer of mass from immobile (residual NAPL) to mobile (dissolved constituent) phases.

General equation of fate and transport in three dimensions solved by MT3DMS is (Zheng and Wang, 1999)

$$\frac{\partial(\theta C^k)}{\partial t} = \frac{\partial}{\partial x_i}\left(\theta D_{ij}\frac{\partial C^k}{\partial x_j}\right) - \frac{\partial}{\partial x_{i,j}}(\theta v_i C^k) + q_s C_s^k + \sum R_n \qquad (7.7)$$

where C^k is the dissolved concentration of the k species (ML^{-3}), θ is the porosity (dimensionless), t is time (T), $x_{i,j}$ is the distance along the respective Cartesian coordinate axis (L), D_{ij} is the hydrodynamic dispersion tensor, v_i is the seepage or linear pore water velocity, given as $v_i = q_i/\theta$, q_i is the volumetric flow rate per unit volume of aquifer representing fluid sources (positive) and sinks (negative) (T^{-1}), C_s^k is the concentration of the source or sink flux for species k (ML^{-3}); ΣR_n is the chemical reaction term ($ML^{-3}\,T^{-1}$).

The left-hand side of Equation 7.7 can be expanded into two terms:

$$\frac{\partial(\theta C^k)}{\partial t} = \theta\frac{\partial C^k}{\partial t} + C^k\frac{\partial \theta}{\partial t} = \theta\frac{\partial C^k}{\partial t} + q_s' C^k \qquad (7.8)$$

where $q_s' = \partial\theta/\partial t$ is the rate of change in transient groundwater storage (unit, T^{-1}).

The chemical reaction term in Equation 7.7 can be used to include the effect of general biochemical and geochemical reactions on contaminant fate and transport. Considering only two basic types of chemical reactions, i.e., aqueous-solid surface reaction (sorption) and first-order rate reaction, the chemical reaction term can be expressed as follows:

$$\sum R_n = -\rho_b \frac{\partial C_A^k}{\partial t} - \lambda_1 \theta C^k - \lambda_2 \rho_b C_A^k \tag{7.9}$$

where ρ_b is the bulk density of the subsurface medium (ML^{-1}); C_A^k is the concentration of species k (ad)sorbed on the subsurface solids (MM^{-1}); λ_1 is the first-order reaction rate for the dissolved phase (T^{-1}); and λ_2 is the first-order reaction rate for the sorbed (solid) phase (T^{-1}).

Substituting Equation 7.8 and Equation 7.9 into Equation 7.7 and dropping the species index for simplicity of presentation, Equation 7.7 can be rearranged and rewritten as

$$\theta \frac{\partial C}{\partial t} + \rho_b \frac{\partial C_A}{\partial t} = \frac{\partial}{\partial x_i} \left(\theta D_{ij} \frac{\partial C}{\partial x_j} \right) - \frac{\partial}{\partial x_i} (\theta v_i C) + q_s C_s - q_s' C_s - \lambda_1 \theta C - \lambda_2 \rho_b C_A \tag{7.10}$$

Equation 7.10 is essentially a mass balance statement, i.e., the change in the mass storage (both dissolved and sorbed phases) at any given time is equal to the difference in the mass inflow and outflow due to dispersion, advection, sink or source, and chemical reactions.

Local equilibrium is often assumed for the various sorption processes (i.e., sorption is sufficiently fast compared to the transport timescale). When the local equilibrium assumption (LEA) is invoked, it is customary to express Equation 7.10 in the following form:

$$R\theta \frac{\partial C}{\partial t} = \frac{\partial}{\partial x_i} \left(\theta D_{ij} \frac{\partial C}{\partial x_j} \right) - \frac{\partial}{\partial x_i} (\theta v_i C) + q_s C_s - q_s' C_s - \lambda_1 \theta C - \lambda_2 \rho_b C_A \tag{7.11}$$

where R is referred to as the retardation factor, which is a dimensionless factor defined as

$$R = 1 + \frac{\rho_b}{\theta} \frac{\partial C_A}{\partial C} \tag{7.12}$$

When the LEA is not appropriate, the sorption processes are typically represented through a first-order kinetic mass transfer equation between the immobile (sorbed) and mobile (dissolved) phases. Not that this mass transfer coefficient can be used to represent the effects of rate-limited dissolution of NAPLs as discussed in Section 6.1.1.

7.2.3　Model Domain and Model Geometry

7.2.3.1　External Boundaries

There are two types of external model boundaries: physical (real) and hydraulic (artificial). Physical boundaries are well-defined geologic and hydrologic features that permanently influence the pattern of groundwater flow. Examples are impermeable contact between two geologic units, a fault (either as a preferential flow path or a low-permeable barrier), contact between the porous medium and a large body of surface water, or contact between the porous medium and a human-made structure such as a slurry wall. It is preferable to have real physical boundaries as external model boundaries. If that is not possible because of the model scale limitations (i.e., the real boundaries are too far and it is not feasible to include them), the hydraulic boundaries need to be defined. Hydraulic boundaries are derived from the groundwater flow net and are therefore "artificial" boundaries set by the model designer. They can be no-flow boundaries represented by chosen streamlines (remember that there is no flow across the streamline), or boundaries with known hydraulic head represented by equipotential lines (see Figure 7.7). It is less desirable to model with hydraulic boundaries since they are not permanent features and can change in time. In addition, their location, although guided by general characteristics of flow nets, is subjective. The main requirement (and difficulty) in setting hydraulic boundaries is that they have to be far enough from the area of interest within the model in order not to influence the future flow pattern created by projected activity (e.g., pumping, injection). The major consequence of this requirement is that the hydraulic boundaries very often have to be placed arbitrarily because of the lack of data beyond the area of an immediate interest. For example, many contaminated sites are relatively small in terms of regional groundwater flow pattern and the related hydrogeologic investigations, for various reasons, focus almost exclusively on the site itself.

As seen from Figure 7.1 and Figure 7.2, most boundaries for the groundwater flow field in the case of our little alluvial aquifer are real physical boundaries: the large perennial stream (Red River), and the impermeable contact between alluvial deposits and gneiss. The contact between the alluvial aquifer and the sandstone aquifer is well defined and represents the boundary between the two distinct flow fields that are connected. This is also a logical boundary of the model since the main interest is in the alluvial aquifer. The influence of the sandstone aquifer and the role of the boundary between the two aquifers can be modeled with an appropriate boundary condition. Only one hydraulic boundary (equipotential line) must be inferred in the far north where the river runs parallel to the contact between the aquifer and gneiss, which leaves the flow field open.

Surface streams in general should be carefully analyzed before deciding what their role will be in the model design. Fully penetrating perennial streams offer a straightforward answer; if the flow in the river is large enough to sustain an extensive potential groundwater withdrawal in the adjacent aquifer, they are ideal equipotential boundaries. If the river is partially penetrating, and the amount of groundwater withdrawal is such that the flow in it might be

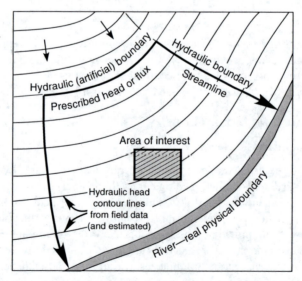

FIGURE 7.7 In the absence of real (physical) boundaries, the artificial (hydraulic) boundaries should be placed far enough from the area of interest in order not to influence the future flow pattern created by the anticipated activity (e.g., well pumpage). The artificial no-flow boundary corresponds to the selected flow line, and the artificial known-head boundary corresponds to the selected equipotential line.

affected to the point of drying, the river cannot be set as an external boundary. It will still, however, be an important internal boundary and must be described by an appropriate boundary condition.

7.2.3.2 Model Grid

Laying out the grid is the starting point in the actual computer model design. In Modflow, which is a finite-difference model, the grid is formed by two sets of parallel lines that are orthogonal. The blocks formed by these lines are called cells. In the center of each cell is the node—the point at which the model calculates hydraulic head. It is assumed that hydraulic and hydrogeologic properties are uniform over the extent of a cell so that the cell is represented by its node. This type of grid utilized in Modflow is called block-centered grid (Figure 7.3). Although application of Modflow does not require designation of x, y, or z model axes in terms of Cartesian coordinates (the only requirement is that rows and columns fall along consistent orthogonal coordinates within the layers), the following association is commonly assumed:

- Columns (J) correspond to x-coordinate
- Rows (I) correspond to y-coordinate
- Layers (K) correspond to z-coordinate

There are two important differences between the Modflow and the Cartesian coordinate systems: (1) in Modflow the origin is in the upper left corner of the grid and (2) the k index corresponds to a decrease in elevation, i.e., the layers are numbered from the top-down.

The grid mesh can be uniform, when all cells have the same dimensions, and custom, when cell size varies. Although the uniform grid is preferred from a mathematical standpoint, it will often be necessary to design a custom grid. A uniform grid is a better choice when

- available data on aquifer (system) characteristics are evenly distributed over the model area,
- there is not more interest in some parts of the model area than others, i.e., the entire flow field is equally important, and
- number of cells (size of the model) is not an issue either from computer memory or the modeling time standpoints.

The custom grid is more appropriate when there is less or no data available for certain parts of the model area and there is specific interest in one or more smaller areas within the flow field. For example, a model for design of a groundwater extraction system for containment and cleanup of a plume is a "perfect candidate" for a custom grid because the following is usually true:

- Extent of the plume is much smaller than the required model area—remember that if there are hydraulic (artificial) model boundaries, they must be far enough away to not influence future pumping within the plume
- Data on aquifer (system) characteristics beyond the extent of the plume are limited or not available
- A small cell size (fine resolution) within the plume is needed in order to optimize the multiple-well pump-and-treat system, i.e., analyze well interaction and interference

In general, a custom grid can greatly reduce computational time by decreasing the model size (number of cells). The areas of interest are discretized into smaller cells, which give them more weight and modeling accuracy, while distant portions of the flow field are described with fewer and larger cells thus keeping the model size reasonable. Custom grids can also more accurately describe model boundaries. This is very important in the case of real physical boundaries (e.g., surface streams, impermeable contacts), especially when they are close to areas of interest.

The grid is laid out keeping in mind the modeling concept and particularly the position and shape of the areas of interest and model boundaries. Figure 7.8 shows a custom grid laid over the alluvial aquifer in our case. The smallest cell size is in the west-central part of the aquifer closest to the river. This is based on the assumption that the highest pumping rates can be achieved by placing wells closer to the river and away from impermeable boundaries. The cell size increases somewhat outside this area of interest but not significantly since all boundaries are relatively close. Cell size in the row direction (y-axis direction if viewed in Cartesian coordinate system) could have been larger in the far south if there were no data on aquifer characteristics. However, the grid remains relatively fine since the water table measurements are available in wells 15 through 19 and will be useful during model calibration. In addition,

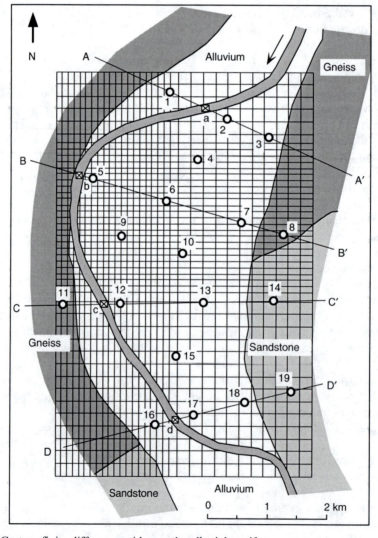

FIGURE 7.8 Custom finite-difference grid over the alluvial aquifer.

the model has only one layer (unconfined aquifer) and its overall size of 31 columns × 45 rows does not pose any problem for the major versions of Modflow and common PC platforms. (Note that most current versions on the market are limited only by available computer memory, which means that a standard desktop PC with 1 GB of RAM can handle tens of millions of cells.)

A rule of thumb when designing a custom grid is that the size of a cell, in all three directions (row, column, layer), cannot be more than 1.5 times larger (or smaller) than the size of the adjacent cells (see Figure 7.9). This is necessary in order to preserve the mathematical stability of the numerical solution. In case of fate and transport models, two additional requirements are very important. The Peclet number constraint requires that the grid cell size in either of the two horizontal directions (denoted as Δx) is less than two times the longitudinal dispersivity (α_L):

$$\Delta x \leq 2\frac{D_L}{v_L} \leq 2\alpha_L \tag{7.13}$$

where D_L is the longitudinal dispersion (see also Equation 6.10, Chapter 6) and v_L is the linear groundwater velocity.

This rule of thumb is very important because it greatly reduces possibility of numeric model dispersion, which is the "process" by which contaminant particles are moving between the cells faster than the actual physical mixing process (i.e., dispersion). If the distance between grid nodes is too large, particles can numerically "jump" between the adjacent grid cells in successive time steps and this numerical dispersion may become the dominant mixing process. One indicator of the inadequate grid size is contaminant transport upgradient from the source or calculated concentrations elsewhere in the aquifer exceeding that at the source.

The Courant number constraint helps with selecting an appropriate time step for advective transport so that the contaminant particle does not travel more than one cell length in one time step (Δt):

$$\frac{\Delta x}{v_L} \geq \Delta t \tag{7.14}$$

Comparing Equation 7.13 and Equation 7.14 it becomes apparent that sometimes it may be quite difficult to satisfy both criteria at once since a finer grid designed to reduce numeric dispersion results in a smaller time step required to reduce numeric oscillations. It should be noted that these problems are not trivial because fate and transport calculations involving large models, multiple layers, and several parameters may last "many" hours for one model run, even when using most powerful desktop computers currently available.

It is a good practice to dimension the overall grid size larger than the present estimated model area since first model runs may show that, for example, hydraulic boundaries are

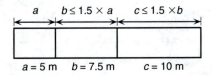

FIGURE 7.9 Size of adjacent cells in a custom grid cannot increase or decrease by more than 1.5 times in any of the three directions (i.e., along column, row, and layer directions).

placed too close and adjacent aquifers should be included in the model. It is important to remember that, once the grid size (i.e., the total lengths in row and column directions) is set, it cannot be changed later without having to reinput all other parameters.

7.2.3.2.1 Grid Orientation

When the aquifer is isotropic, i.e., when its hydraulic conductivity is the same in all directions, orientation of the grid is not critical. In such cases it is still recommended to orient the grid so that the number of inactive cells is minimized. Figure 7.10a shows an isotropic system of irregular shape without any dominant direction. Here a common south–north orientation seems the most "natural": columns and rows are aligned with x- and y-axes, respectively. Figure 7.10b shows an isotropic system that is elongated in one direction. In this case it is recommended that one of the model axes (either row or column) be aligned with the predominant direction. This reduces the number of inactive cells and the size of the model. Note that inactive cells take space in the computer memory even though the model does not calculate hydraulic heads for them. When the aquifer is anisotropic, the model coordinate axes must be aligned with the main axes of the hydraulic conductivity tensor. These axes, i.e., directions of maximum and minimum hydraulic conductivities, are always perpendicular to each other. Finding the main axes of anisotropy and assigning the corresponding values of hydraulic conductivity is explained further in this chapter.

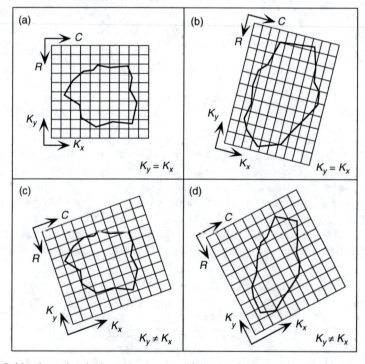

FIGURE 7.10 Grid orientation is dependent on aquifer's shape and degree of anisotropy. (a) Isotropic aquifer with irregular shape. (b) Isotropic aquifer elongated in one direction. (c) Anisotropic aquifer with irregular shape. (d) Anisotropic aquifer elongated in direction, which is different from two main directions of hydraulic conductivity anisotropy.

7.2.3.2.2   Layer Type

There are four basic types of layers in Modflow:

- Type 0. This layer type is used to simulate confined conditions (layers and aquifers) in which transmissivity of each cell remains constant for the entire simulation time. For transient simulations, this layer type requires the confined storage coefficient (specific storage times layer thickness), which is used to calculate the rate of change in storage.
- Type 1. This layer type is used strictly for unconfined conditions and is valid for the first (uppermost) layer only. It requires specific yield for transient conditions. Transmissivity of the layer varies as saturated thickness of the aquifer changes during simulation. The alluvial aquifer in our case will be modeled as one layer of type 1.
- Type 2. This layer type is used when the aquifer alternates between confined and unconfined. However, it is assumed that the saturated thickness remains everywhere a high fraction of the layer thickness so that recalculation of transmissivity is not necessary (it is constant throughout the period of simulation). In transient simulations, it is needed to specify both the storage coefficient for fully saturated confined conditions, and the specific yield for unconfined flow. If the layer completely desaturates, vertical leakage from above ceases.
- Type 3. A layer of this type is also used for confined–unconfined transitions. It includes varying transmissivity, which is recalculated at each iteration using hydraulic conductivity and new saturated thickness. Confined storage coefficient and specific yield are both needed for transient simulations. Vertical leakage from above terminates when the layer is completely dry.

Note that a semiconfining unit does not have to be simulated as active in Modflow. Its influence can be described by a term named vertical leakance between two active model layers. Models, which do not directly include semiconfining layers are called quasi-three-dimensional models. In the original 1988 Modflow and some commercially available processors, the leakance must be calculated directly by the user. Most widely used GUIs can calculate the leakance term from the layer thickness and the vertical hydraulic conductivity, or it can be left to the user if desired.

Although a quasi-three-dimensional model may sometimes seem easier to design and more appropriate (particularly when there is no data on the semiconfining layer), it is better to use the full potential of Modflow and design a real three-dimensional model. For example, the model may be used later to analyze flow paths between aquifers and the exact role of the separating layer; or it may become part of a groundwater transport model. In both cases, it is necessary to fully include the semiconfining layer and assign realistic values of effective porosity and other transport parameters. It should be noted that, if necessary, one hydrostratigraphic unit (aquifer) can be modeled with more layers of the same hydraulic conductivity. This is typically done when steep hydraulic gradients are present, such as in aquitards, or expected to occur due to an imposed stress such as well pumpage or drain installation. More layers are also recommended when a well is partially penetrating thus allowing for the accurate placement of the well screen.

7.2.3.2.3   Boundary Array and Cell Type

The next step after laying out the grid is to describe model boundaries, i.e., to determine the type of each cell in the model. There are two types of cells in Modflow used to describe boundaries: inactive (no-flow) cells and constant-head cells. Inactive cells (designated with number 0 in Modflow's boundary array—IBOUND) are those for which no flow into or out of the cell occurs during the entire time of simulation. The model does not calculate hydraulic head for these cells and they are used to deactivate model domain beyond the boundary of the flow field. Figure 7.11a shows the principle of delineating an impermeable boundary with inactive cells. If more than one-half of a cell is within the impermeable unit, it is designated as an inactive cell. This estimate is quickly done by focusing on the cell side closest to the boundary. As seen in Figure 7.11a, all other cells beyond the boundary cells are also designated as inactive (no-flow). The remaining cells are called active, or variable-head cells. They are designated with number 1 in Modflow's boundary array (IBOUND). The hydraulic heads for active cells are calculated by the model and are free to vary in time. Constant-head cells are used to describe model boundaries with known heads such as aquifer contacts with major surface water features. The hydraulic head at these cells is specified in advance and does not change during one simulation time period. Constant-head cells are designated with number −1 in the IBOUND array. Figure 7.11b shows the principle of delineating the contact between an aquifer and a surface stream, which is at the same time the external model boundary (as in our case). This is carried out simply by noting cell nodes that are closest to the contact. Only the line of the contact is important when designating constant-head cells. In other words, the width of the river is not important and all cells beyond the riverbank of interest, including cells in the river, are designated as inactive. When a river is an internal model boundary, the described principle is applied to both riverbanks and all cells in between (cells within the river) are designated as constant-head cells. Note that it is not always recommended to simulate the contact between the aquifer and the surface water body with constant-head cells since they are an inexhaustible source of water. Such boundaries can be simulated with several Modflow packages depending upon actual conditions as explained further in this chapter. Having three cell types allows for accurate description of irregularly shaped flow fields (aquifers). This is important since the model array is always rectangular in outline, whereas the active flow domain is often of irregular shape, which may vary from layer to layer as well.

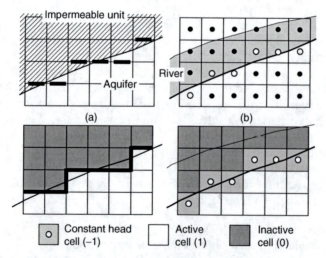

FIGURE 7.11 (a) Delineation of an impermeable boundary with inactive (no-flow) cells. If more than half of the cell is in the impermeable unit, the cell is designated as inactive. (b) Delineation of a surface stream boundary with constant-head cells. The cell whose center is the closest to the contact between the stream and the aquifer is designated as a constant-head cell.

7.2.4 Input Parameters

Model parameters are divided into four groups: (1) time, (2) space, as defined with layer top and bottom, (3) hydrogeologic characteristics such as hydraulic conductivity, transmissivity, storage parameters, and effective porosity, and (4) fate and transport parameters such as dispersivity, sorption isotherms, retardation coefficient, diffusion, chemical reactions, and degradation rate constant.

7.2.4.1 Time

Time parameters are specified when modeling transient (time-dependent) conditions. They include time unit, the length and number of time periods, and the number of time steps within each time period. During one time (stress) period all model parameters associated with boundary conditions and various stresses remain constant. Having more time periods allows these parameters to change in time. For example, a pumping well can change its pumping rate in different time periods, and the river stage or recharge from precipitation can vary according to seasons. A time period is further divided into time steps, which are useful for analyzing changes in hydraulic head and drawdown. Time steps do not have to be of same length. This option, controlled by the time step multiplier, is particularly useful when simulating a well pumping test: each time step is longer than the preceding one and the results can be plotted over several logarithmic cycles. Having more time steps increases the accuracy of iterative computations. The length of stress periods is not relevant for steady-state simulations.

7.2.4.2 Layer Top and Bottom

The elevation of the layer top and bottom is required if the user wants the program to calculate aquifer (layer) transmissivity, vertical leakance, or confined storage coefficient, which is recommended. Original Modflow reads the top elevation only for layers of type 2 and 3, and the bottom elevation for layers of type 1 and 3. Hydraulic conductivity is needed only when modeling unconfined aquifers, i.e., layers of type 1 and 3, in which case Modflow calculates transmissivity by multiplying K with the saturated thickness. In all other conditions, the transmissivity and storage coefficient (for transient modeling) are specified by the user. This is convenient when the actual layer thickness is unknown or the layer shape is too complicated. Newer versions of Modflow (Modflow-96 and Modflow-2000) have the option for assigning actual elevation of layer top and bottom by default. The model uses the actual layer thickness and vertical conductivity to calculate leakance between the layers. Whenever possible, it is best to input the top and bottom for all layers since it is important to relate hydraulic heads in different layers with their respective geometry (see Figure 7.12).

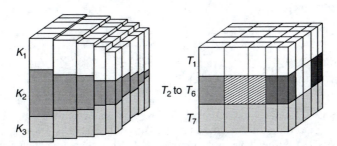

FIGURE 7.12 Modeling layers of varying thickness. *Left*: With hydraulic conductivities, and actual top and bottom. *Right*: With varying transmissivity, and horizontal top and bottom.

This is particularly useful when analyzing possible hydraulic connection and migration of contaminants between layers (aquifers).

When setting up the geometry of the layers it is important to remember that the layer top for unconfined aquifers is the water table and not the land (topographic) surface. Also (very) important is to make sure that the bottom of one layer is not lower than the top of the layer below, which can sometimes happen when importing surfaces interpolated by an external program. If this is the case, the model, i.e., some GUIs, will report the error. In any case, the model should not start running by default.

One common problem when defining layer top and bottom is the question of interpolation. Numerical groundwater models require that these and other areally distributed model parameters be assigned to each cell. Since available data are usually limited to irregularly scattered locations in the model domain, it will be necessary to estimate (interpolate) the parameter values for each cell. A nightmare for every modeler using original Modflow was the manual input of model parameters for large multilayered models of highly heterogeneous aquifers. Today, most major Modflow GUIs, such as Processing Modflow Pro and *Groundwater Vistas*, offer several convenient options for data input including zones and various interpolators that generate data for each cell. Parameter arrays can also be defined using some external interpolator such as SURFER and then imported in the model. This greatly reduces modeling time and increases accuracy.

One common situation from the "real world" is that the actual geologic layers within the model domain are not continuous. A discontinuous layer in the area where it is not present can be modeled by adjusting its thickness and hydraulic conductivity as shown in Figure 7.13. A realistic hydraulic connection between the two aquifers in the "window" area is simulated by assigning the hydrogeologic parameters (horizontal and vertical hydraulic conductivities, storage coefficient, effective porosity) of the layer above or below the aquitard. Alternatively, this portion of layer 2 (aquitard) can be given the average values of the parameters in layers 1 and 3.

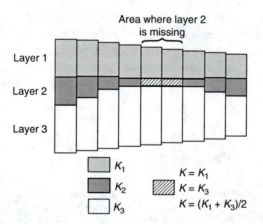

FIGURE 7.13 Principle of modeling discontinuous layers. The area where the actual layer is missing is still modeled as having a certain thickness (usually similar to adjacent cells where it is present) but the hydraulic conductivity (K) is the same as in layer 1 or 2. Note that, following the rule of thumb, the thickness of successive cells should not increase by more than 1.5 times in order to avoid possible model instability.

7.2.4.3 Horizontal Hydraulic Conductivity and Transmissivity

The hydraulic conductivity (K) is in many cases the most critical and sensitive modeling parameter. Every attempt should be made to design a model with realistic values of K obtained in the field, preferably by pumping tests. By varying the spatial distribution of hydraulic conductivity arbitrarily, an experienced modeler can almost always produce a modeling result that closely resembles the field data set used for calibration. Another question is if that distribution is justified. Only after all other calibration avenues such as adjusting boundaries, boundary conditions, and stresses acting upon the system are exhausted, should the user engage in changing K distribution. There are already enough models designed with oddly shaped "K rectangles" scattered throughout the flow domain without any hydrogeologic support or explanation. Although some of them may look interesting, especially when K shapes are colored, and may resemble a nice southwestern rug created by some imaginative and skillful craftsmen, it would still not qualify them as being anything logical.

In Modflow, the horizontal hydraulic conductivity is the conductivity along grid rows (K_x). If the aquifer (layer) is isotropic, K along model columns (K_y) will be the same. Otherwise, the horizontal anisotropy is specified with the anisotropy factor, which multiplies K_x to obtain K_y. The original Modflow and most GUIs require input of the anisotropy factor (1 for isotropic conditions) while some programs allow for direct input of K_y. During field work, it is not uncommon to discover two dominant directions of the hydraulic conductivity anisotropy that are at a certain angle to each other. This information can be used to define the two axes of K that are perpendicular to each other—important when orienting the model grid (see Figure 7.14). The following equation relates the principal conductivity components K_x and K_y to the resultant K_r in any angular direction θ (adapted from Freeze and Cherry, 1979, pp. 35–36):

$$K_y = \frac{K_r K_x \sin^2 \theta}{K_x - K_r \cos^2 \theta} \tag{7.15}$$

The hydraulic conductivity is required for layers of type 1 and 3 (unconfined conditions), and the user can specify transmissivity for layers of type 0 and 2 directly (confined conditions). Most Modflow processors can calculate transmissivity for a layer of any type by multiplying the hydraulic conductivity with the layer thickness derived from the layer top and bottom elevations, which is recommended.

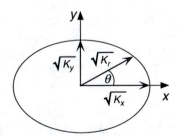

FIGURE 7.14 Hydraulic conductivity ellipse for determining the second major axis of hydraulic conductivity anisotropy from two distinct directions found in the field that are not perpendicular to each other. Note that the directions of maximum and minimum hydraulic conductivities are always perpendicular to each other in an anisotropic flow field by modeling convention, which is not necessarily true in field conditions.

7.2.4.4   Vertical Hydraulic Conductivity and Leakance

For quasi-three-dimensional models with more than one layer and for full three-dimensional models, Modflow requires the input of the vertical leakance between two adjacent layers. As already mentioned, most GUIs can calculate the vertical leakance for each layer from the information specified by the user: layer top and bottom elevations and the vertical hydraulic conductivity. Unless accurately determined from pumping tests, the vertical hydraulic conductivity is usually assumed and calibrated if, for some reason, the modeler does not want the program to perform this task. For example, it is common to assume that the vertical hydraulic conductivity is one magnitude lower than the horizontal conductivity for most stratified sedimentary rocks. If the vertical grid axis is not aligned with the principal axis of vertical hydraulic conductivity (such as in dipping strata), a correction should be made as it was in the case of horizontal anisotropy (see Figure 7.14 and Equation 7.15).

7.2.4.5   Storage Terms

Storage terms, i.e., storage coefficient for confined layers (layer type 0, 2, and 3) and specific yield for unconfined layers (layer type 1, 2, and 3) are required only for transient simulations. The storage coefficient (S) is the product of layer thickness (b) and specific storage (S_s):

$$S = bS_s \quad \text{(dimensionless)} \tag{7.16}$$

Note that S is the storage term determined by pumping tests in confined aquifers as explained in Chapter 2. If S is assigned directly to the layer, the layer thickness should be accounted for. Some GUIs allow selection between the two parameters and will calculate S based on the layer thickness and the assigned S_s. Pumping tests in unconfined aquifers give specific yield (also dimensionless), which is several orders of magnitude larger than the storage coefficient. Storativity and specific yield concepts are explained in detail in Chapter 1 and Chapter 2.

7.2.4.6   Effective Porosity

Effective porosity is required only if the model will be used in conjunction with other programs that use results generated by Modflow to calculate the average linear velocity of groundwater flow. This velocity is needed to track water particles as they move through the porous medium, and to calculate contaminant concentrations.

7.2.5   MODEL BOUNDARY CONDITIONS

After assigning time parameters, and space and hydrogeologic parameters for each model cell, the next step is to accurately describe boundary conditions using the most appropriate Modflow packages. Since the choice is not straightforward, the package that will be used for a particular boundary will often be decided during model calibration.

7.2.5.1   Boundary with Known Head

A boundary with known head, such as an equipotential line or a contact between the aquifer and a large surface water body, can be simulated by constant-head cells as explained earlier. However, such simulation should be done with great caution since constant-head cells represent an inexhaustible source of water, when higher than the aquifer head, regardless of changes at the boundary and in the aquifer. The head at constant-head cells remains the same throughout the time period of simulation. For example, a small stream may dry out because of the nearby well pumpage but the constant-head cells simulating it will continue to supply water to the aquifer.

An option developed by Leake and Prudic (1991; Appendix C), called the Time-Variant Specified-Head Package (CHD1) and used in transient simulations, allows constant-head cells to take on different head values for each time step during a simulation time period. The user specifies the value of the head in the cell at the start of the stress period (h_s), and at the last time step in the stress period (h_e). The package then linearly interpolates boundary heads (h) for each cell using the following equation:

$$h = h_s + (h_e - h_s)\frac{\text{PERTIM}}{\text{PERLEN}} \tag{7.17}$$

where PERTIM is the starting time of a time step in a stress period and PERLEN is the length of the whole stress period.

The Time-Variant Specified-Head Package (CHD1) is useful when, for example, data on weekly, monthly, or seasonal variations in river stages are available.

Leake and Lilly (1997) introduced another package for simulating transient specified-head boundaries, FHB1, also capable of simulating transient specified-flow boundaries.

Figure 7.15 shows boundaries and boundary conditions for our example. The Red River is simulated by constant-head cells since there is no data on time variations in the river stage, river flows, or river bed characteristics. In addition, it is certain that the river is perennial with a substantial flow during the summer-fall period. If some or all of the above-mentioned data were available, it would be recommended to apply the Modflow packages describing head-dependent flux boundaries.

7.2.5.2 Boundary with Head-Dependent Flux

The boundary, for which the flow rate is calculated, based on the head difference between the boundary cells and the adjacent aquifer cells, is called the head-dependent flux boundary. Three Modflow packages can simulate this condition: River Package, Drain Package, and General-Head Boundary (GHB) Package. Common to all three packages is that the user must specify a conductance term, which can often be completely defined only during the calibration process. Despite this "inconvenience," all three packages offer versatility in describing various external and internal model boundaries.

7.2.5.2.1 River Package

The River Package simulates the flow between an aquifer and a surface water body such as a river or a lake. The rate of flow is calculated from the difference in hydraulic heads at the boundary cell and the adjacent aquifer cell using the following equation (McDonald and Harbaugh, 1988; see Figure 7.16 and Figure 7.17):

$$\text{QRIV} = \frac{KLW}{M}(\text{HRIV} - h_{i,j,k}) \tag{7.18}$$

$$\text{QRIV} = \text{CRIV}(\text{HRIV} - h_{i,j,k}) \tag{7.19}$$

where QRIV is the flow between the stream and the aquifer, taken as positive if it is directed into the aquifer, HRIV is the head in the stream, h is the head at the node in the cell underlying the stream reach, and CRIV is the hydraulic conductance of the stream–aquifer interconnection:

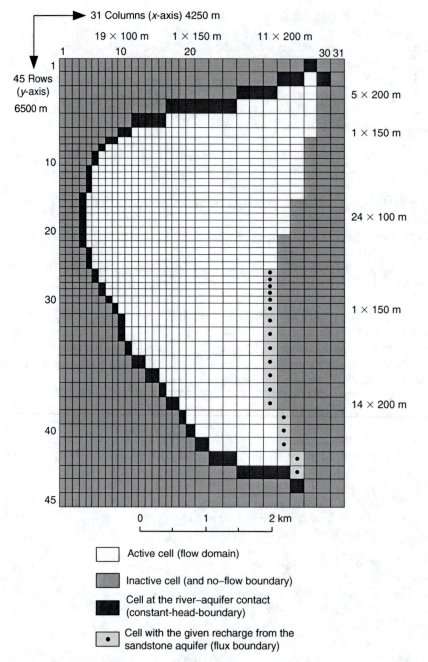

FIGURE 7.15 Boundaries and boundary conditions for the flow model in our case (alluvial aquifer shown in Figure 7.1 and Figure 7.8).

$$\text{CRIV} = \frac{KLW}{M} \tag{7.20}$$

where K is the hydraulic conductivity of the streambed material, L is the length of the river channel in the cell, W is the width of the river channel in the cell, and M is the thickness of the riverbed material.

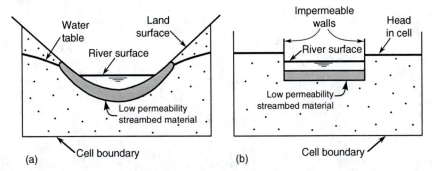

FIGURE 7.16 (a) Cross section of an aquifer containing a stream and (b) conceptual representation of stream–aquifer interconnection in simulation. (From McDonald, M.G. and Harbaugh, A.W., 1988. A modular three-dimensional finite-difference ground-water flow model. *U.S. Geological Survey Techniques of Water-Resources Investigations*, Book 6, Chap. A1, 586pp.)

When the hydraulic head in the aquifer falls below the river channel bottom, the flow remains constant and is given by

$$QRIV = CRIV(HRIV - RBOT) \qquad (7.21)$$

where RBOT is the elevation of the river channel bottom. As can be seen from Equation 7.21, the River Package should be used with care since there is no adjustment for river flow and stage, so that the supply of water to the aquifer from a losing stream can be more than the flow in the stream.

7.2.5.2.2 Drain Package
The Drain Package simulates both closed (buried) and open drains. It is different from the River Package in that the flow is directed only from the aquifer toward the drain and it stops when the head in the aquifer drops below the elevation of the drain. This makes the Drain Package a useful tool for modeling springs. The rate of flow entering the drain (QD) is calculated from the following equations (modified from McDonald and Harbaugh, 1988):

$$QD = CD(h - d) \quad \text{for } h > d \qquad (7.22)$$

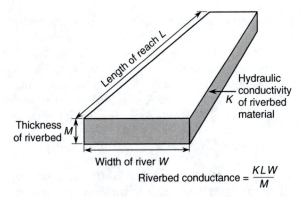

FIGURE 7.17 Idealization of streambed conductance in an individual cell. (From McDonald, M.G. and Harbaugh, A.W., 1988. A modular three-dimensional finite-difference ground-water flow model. *U.S. Geological Survey Techniques of Water-Resources Investigations*, Book 6, Chap. A1, 586pp.)

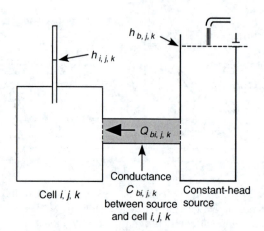

FIGURE 7.18 Schematic diagram illustrating principle of General-Head Boundary package. (From McDonald, M.G. and Harbaugh, A.W., 1988. A modular three-dimensional finite-difference groundwater flow model. *U.S. Geological Survey Techniques of Water-Resources Investigations*, Book 6, Chap. A1, 586pp.)

$$QD = 0 \quad \text{for } h \leq d \tag{7.23}$$

where CD is the drain hydraulic conductance, h is the hydraulic head in the aquifer cell, and d is the elevation of the drain in the drain cell.

CD depends on the aquifer hydraulic conductivity near the drain, distribution and characteristics of the fill material, number and size of the drainpipe openings, and characteristics of the clogging materials. In practice, many of these data are not available and the drain hydraulic conductivity has to be estimated and adjusted during model calibration. For open drains, CD can be calculated using Equation 7.18 for the conductance of riverbed material.

7.2.5.2.3 General-Head Boundary Package

The head-dependent flow (Q_b) through a boundary simulated by the GHB package is calculated from (modified from McDonald and Harbaugh; see Figure 7.18):

$$Q_b = C_b(h_b - h) \tag{7.24}$$

where C_b is the conductance of the boundary analogous to the streambed conductance, h_b is the hydraulic head at the boundary cell, and h is the hydraulic head at the aquifer cell adjacent to the boundary.

The GHB package assumes unlimited linear flow between the boundary and the aquifer and should be used with care. It is useful for transient simulations of known head boundaries since it allows changes in C_b and h_b for different time periods. When a large value of conductance is used, the cell simulated by this package is equivalent to a constant-head cell.

7.2.5.3 Boundary with Known Flux

Typical boundaries with known flux are water tables in unconfined aquifers, inflow and outflow through lateral contacts between different aquifers, seepage to or from adjacent layers, spring flow, and recharge to or from surface streams. Water table flux and a lateral inflow are boundary conditions present in our case as shown in Figure 7.15. Whenever possible, the model should have boundary conditions described with known (prescribed) head or head-dependent flux rather than with flux only. This is important for the following reasons—the model result is a distribution of hydraulic heads and the iteration process is more accurate if it starts with known heads, and it is easier to measure hydraulic head in the field.

In general, known flux conditions can be simulated by one of the following packages.

7.2.5.3.1 Recharge Package

Recharge to a model cell is specified as recharge intensity and must have dimensions of velocity expressed in same length-time units as other parameters. The model calculates the recharge flux by multiplying the intensity with the cell area. The cell within each vertical column to which the recharge is applied is specified through the recharge option menu. The options include (Figure 7.19):

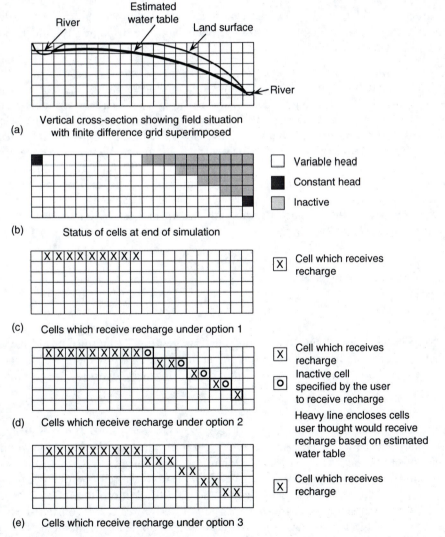

FIGURE 7.19 Hypothetical problem showing which cells receive recharge under the three options available in the Recharge Package. (From McDonald, M.G. and Harbaugh, A.W., 1988. A modular three-dimensional finite-difference ground-water flow model. *U.S. Geological Survey Techniques of Water-Resources Investigations*, Book 6, Chap. A1, 586pp.)

- Application of the recharge to the first model layer. If the cell is inactive, it will not receive recharge.
- Application of the recharge to any cell in the vertical column as specified by layer numbers. If the cell is inactive, it will not receive recharge.
- Application of the recharge to the uppermost active cell in the vertical column, provided there is no constant cell above it in the column. If there is a constant cell in a vertical column of cells and there is no active cell above, then no recharge is applied to this column because it is assumed that any recharge would be intercepted by the constant-head source.

In our problem, the recharge is estimated as 10% of the average annual precipitation of 630 mm. This number is based on the fact that the climate in the area is semiarid with high temperatures causing a high rate of ET most of the year. In general, the exact recharge rate is hard to define accurately in the absence of more site-specific information; recharge depends on various factors such as climate, land cover and land use, topography, and underlying geology (see Section 2.6).

7.2.5.3.2 Evapotranspiration Package

The ET Package simulates the effects of plant transpiration and direct evaporation from a shallow water table. The maximum ET occurs at the user-specified water table elevation, usually the elevation of the land surface. ET ceases when the water table drops below the user-specified "extinction depth." Between these two extreme elevations the ET rate varies linearly. ET is drawn from only one cell in the vertical column, usually from the uppermost cell (layer 1). However, there is an option to specify the cell within the vertical column from which ET will withdraw water. In both cases, ET is set automatically to zero by the model if the designated cell is either a no-flow (inactive) cell or a constant-head cell.

The model calculates the volume of water extracted from a cell by ET using the following information provided by the user:

- Maximum ET rate (dimensions of velocity)
- Elevation of the ET surface
- ET extinction depth
- Layer indicator

Similar to recharge, it is difficult to accurately determine the ET rate since it is influenced by many factors. For more information about the nature of ET and methods for its determination, the reader should consult some of the textbooks on general hydrology (e.g., Maidment, 1993).

7.2.5.3.3 Well Package

The Well Package simulates pumping or injection wells using rates (Q) specified by the user for each time period of the simulation. Negative values of Q indicate a withdrawal well, while positive values of Q indicate a recharging well. Since the rate of withdrawal and injection is independent of both the cell area and the head in the cell, the well package is the ideal tool for simulating boundaries with known flux. In addition, the inflow rate can be assigned to any cell in any layer (with the exception of no-flow cells and constant-head cells). In our example, the influx from the sandstone aquifer is simulated by recharge wells placed in the cells along the boundary between the alluvium and the sandstone. Figure 7.20 illustrates the principle of assigning injection rate to a well that simulates the boundary with known flux. This rate can vary for each stress period thus simulating seasonal changes in the lateral recharge.

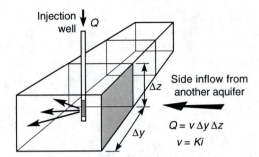

FIGURE 7.20 Determination of injection (recharge) rate for a well that simulates boundary with known flux. The rate must be calculated externally using (or assuming) the hydraulic conductivity of the adjacent aquifer (K), the hydraulic gradient (i; usually determined from the well data), and the area of the cell side.

When simulating several real wells that fall within a one-cell area, the user must sum the individual pumping rates. If the well penetrates more layers, it should be represented by a group of single-layer wells, each open to one of the layers tapped by the multilayer well, and each having an individual Q term specified for each time period of pumping. In such cases, the discharge rate for individual-layer wells must be determined externally to the program. Some GUIs, such as Groundwater Vistas, can calculate pumping rates of multilayer wells automatically, by assigning pumping rates for the individual (one-layer) portions of the well proportionally to the transmissivity of individual layers in which the well is screened.

7.2.5.3.4 Stream Package

The Stream Package (or the Streamflow-Routing Package; Prudic, 1989) is a combination of a known flux and a head-dependent flux boundary. It is similar to the River Package since it allows flow into and from the stream. However, it is more sophisticated than the River Package because it considers the flow rate in the stream and limits the leakage between the aquifer and the stream accordingly. The package increases streamflow in areas of gaining stream segments (reaches) and reduces the flow by taking water out through riverbed seepage in losing reaches. The package can also calculate the stream stage needed for other calculations by using a set of required data instead of a preassigned stage value. Because of its versatility, the Stream Package requires intensive data preparation and more information than any other Modflow package.

7.2.5.4 No-Flow Boundaries

Boundaries with the flux set to zero are called no-flow boundaries. Figure 7.11A shows the principle of delineation of an impermeable boundary with inactive (no-flow) cells. Modflow automatically assumes a no-flow boundary around the perimeter of the grid. If the user's intention is to align one or more model edges with some other boundary type, this must be done by assigning appropriate boundary conditions to the last cells along the grid edge.

7.2.6 STEADY-STATE AND TRANSIENT MODELS

Modflow is designed to simulate steady-state or transient conditions. For steady state, the storage term in the groundwater flow equation is set to zero. This is the only part of the flow equation that depends on length of time, so the stress-period length does not affect the calculated heads in a steady-state simulation. However, it is required in Modflow-2000, as in earlier versions of Modflow, that the length of a steady-state stress period be specified, partly so that the same input mechanism for all stress periods can be used. A single time step is all that is required for steady-state stress periods. Unlike earlier versions of Modflow, the stress period length can be zero, but care is needed because a nonzero length may be important for other processes such as transport. The biggest differences in the way stress periods are implemented compared to previous versions of Modflow is that Modflow-2000 allows individual stress periods in a single simulation to be either transient or steady state instead of requiring the entire simulation to be either steady-state or transient. Steady-state and transient stress periods can occur in any order. Commonly the first stress period is steady state and produces a solution that is used as the initial condition for subsequent transient stress periods (Harbaugh et al., 2000).

7.2.6.1 Initial Conditions

Initial conditions are values of the hydraulic head for each active and constant-head cell in the model. A very common error associated with setting the top and bottom elevation of model cells is when the initial head (assigned to every model cell by the user) is set to be below the bottom of a cell (or more cells) in the layer populated by the data. Although quite a few modelers, when working on a steady-state model, do not worry about this problem (i.e., the model should theoretically run and create the hydraulic head solution regardless of the initially assigned head, except in the case of the constant-head cells) there is potentially a big problem with possible model instability and such an approach. In the case of transient models, the initial head has to be higher than the cell bottom by default and the model would not run if that were not the case. In any event, it is advisable to set the initial heads in each model layer, for both steady-state and transient models, close to the heads estimated from the field data. This will decrease the time for model run and should result in quicker convergence of the solution, provided other model parameters are reasonable and close to the field-observed data.

Once the model is calibrated (in either steady-state or transient conditions), the resulting distribution of hydraulic heads will become the new set of initial heads for the prediction phase. In other words, these heads will be imported into the initial heads array over the old estimated heads. This can also be done several times during the calibration phase to reduce the calculation time.

7.2.7 WATER AND CONTAMINANT FLUXES

In addition to boundary conditions, all Modflow packages described earlier are used to simulate various stresses (water fluxes) acting upon the aquifer. An aquifer stress is defined as any supply to or withdrawal of water from the system. This broad statement also shows that various packages can be used for various stresses regardless of their name as seen fit by the user. For example, the Drain Package can be used to simulate a river or a spring; the River Package can simulate a leaky confined aquifer, and so on. Following is a brief summary of the most widely used Modflow packages that can be selected to model various water fluxes:

River Package. Simulates the flow between an aquifer and a surface water feature in both directions. Includes riverbed conductance for simulating fine sediment clogging of a river channel. Should not be used for intermittent streams since there is no adjustment for the streamflow and stage once it drops below the channel bottom. Very versatile because of two adjustable parameters (conductance and head) and the elevation of the channel bottom which also plays an important role. Often used to simulate wetlands.

Recharge Package. Typically used to simulate a real infiltration from precipitation or irrigation. Can be used to simulate local recharge from ponds. Flexible in assigning vertical flux to different layers along the vertical column.

Well Package. Simulates both extraction and recharge wells. Assumes full penetration of the layer. Partial penetration should be modeled with more layers within the same hydrostratigraphic unit. Good tool for modeling lateral inflow or outflow flux in any layer.

Drain Package. Simulates both closed and open drains. Best tool to model springs because the inflow into the drain ceases when the head in the aquifer drops below drain elevation. Requires intensive calibration.

Evapotranspiration Package. Typically used in irrigation for simulating the effects of plant water intake. Varies linearly with the water table elevation accounting for the user-specified maximum and minimum (zero) depth of ET.

General-Head Boundary Package. Used for transient simulations of constant-head boundaries as it allows the user to change the head from one time period to another. It can be used to simulate permanent surface water features but does not allow for any limit to discharge depending upon actual flow and stage in surface streams. Conductance term can be used to fine-tune the flow of water into the domain. As the name implies, it can be used for a variety of conditions.

Stream Package (Prudic, 1989). Simulates interaction between an aquifer and a surface stream accounting for the flow rate in the stream. Flow into or out of the stream stops when the stream dries out. Requires intensive data preparation and more data input than any other package.

Horizontal-flow Barrier Package (Hsieh and Freckleton, 1993). Simulates thin low-permeable features such as vertical slurry walls and faults.

BCF2 Package (McDonald et al., 1992). Allows for simulations in which a water table rises into unsaturated model layers. A typical application is the simulation of the recovery of overstressed aquifers, such as after heavy pumpage, either through artificial recharge or through the reduction of stress.

Transient Leakage Package (Leake et al., 1994). Simulates transient leakage from confining units.

RES1 Package (Fenske et al., 1996). Simulates leakage from reservoirs.

Contaminant fluxes in the model may come from two general types of contaminant sources: (1) point and (2) nonpoint. For every contaminant source, the user has to assign the initial contaminant concentration dissolved in groundwater, including zero concentrations (note that the starting values for each model cell are set to zero by default). There are various options for assigning the initial concentration. It is not uncommon to select an "unusual" mechanism for this, similarly to the water fluxes where various packages can simulate conditions not implied by their name (e.g., river package simulating wetlands). Imaginary injection well can be used to deliver the dissolved contaminant to the saturated zone, acting as a continuous point source. The user has to externally calculate what the concentration assigned to the injected water should be, so that a realistic net effect is accurately simulated. In other words, the transient (including constant) contaminant flux estimated to be reaching the saturated zone should be equal to the flux injected by the imaginary well. One potential problem with this approach is that the imaginary well has to have a very low negative pumping rate not to impact the natural groundwater flow pattern. This also means that the contaminant concentrations would often have to be unrealistically (or impossibly) high to match the estimated contaminant flux. However, as soon as this flux is injected into the cell, it will be immediately diluted by the volume of water already present in the source cell. In the absence of any monitoring wells and the actual measured concentration in the source zone over some period of time, that could be used to develop reasonable, external-to-the-model, flux estimates, this approach may not be defensible.

Continuous or decaying sources can also be simulated by assigning constant contaminant concentration in any model cell. This concentration may or may not vary in different time periods. Constant concentration boundaries, similar to the constant-head boundaries, can be used to fix the contaminant input into the model domain. Water entering the model domain through recharge package, can also be used as a contaminant carrier by assigning the concentration to this areally distributed flux. Finally, the mass transfer coefficient can be used as a generic term, for simulating various reactions that transfer contaminant mass, through modeled concentrations, between the immobile and mobile flow domains. As already mentioned, this approach is often used in modeling rate-limited dissolution of NAPLs, as well as contaminant back-diffusion phenomena.

7.3   MODEL CALIBRATION AND SENSITIVITY ANALYSIS

Before starting model calibration, which is arguably the most lengthy and the most important part of the modeling process, and assuming that all the above-explained model parameters and boundary conditions are correctly assigned to each cell, and before actually running the model, the user must choose one of the solving packages (Solvers). There are several solvers in Modflow to select from and each one requires specification of certain calculation parameters. Common for most solvers are the head change criterion for convergence and the maximum allowed number of iterations. When the maximum absolute value of hydraulic head change in any model cell is less than or equal to the head change criterion, the iteration process stops. Three of the most widely used Modflow solvers are the Strongly Implicit Procedure (SIP), the Slice-Successive Overrelaxation method (SSOR), and the Preconditioned Conjugate-Gradient 2 method (PCG2). It is a good practice to try all three solvers and see which one gives better results for the particular problem. For a more detailed explanation of these three solvers and their applicability, see McDonald and Harbaugh (1988) and Hill (1990).

The first model run is the fear (or joy) of every model designer. When dealing with a "real-life" model, it is almost certain that the first result will not be a satisfactory match between the calculated and the measured hydraulic heads. The number of model runs during calibration will depend on the quantity and quality of available data, desirable accuracy of the model results, and often the patience of the user.

Although often explained separately in modeling reports, calibration and sensitivity analysis are inseparable and are part of the same process. While performing calibration, which is comprised of numerous single and multiple changes of model parameters, every user determines quickly which parameters are more sensitive to changes with regard to the final model result. By carefully recording all of the changes made during calibration and commenting on their results, the model designer is already engaged in the sensitivity analysis and can effortlessly finalize this part of the modeling effort later. Calibration is the process of finding a set of boundary conditions, stresses, and hydrogeologic parameters, which produces the result that most closely matches field measurements of hydraulic heads and flows. Calibration of every model should have the target of an acceptable error set beforehand. Its range will depend mainly on the model purpose. For example, a groundwater flow model for evaluation of a regional aquifer system can sometimes "tolerate" a difference between calculated and measured heads of up to several feet. This, however, would be an unacceptable error in the case of a model for design of containment and cleanup of a contaminant plume spread over, say, 50 acres.

In many instances, the quality of calibration will depend on the amount and reliability of available field data. It is therefore crucial to assess these field data, or calibration data set, for their consistency, homogeneity, and measurement error. Such assessment is the basis for setting the calibration target.

Model calibration can be performed for steady-state conditions, transient conditions, or both. Although steady-state calibration has prevailed in modeling practice, every attempt should be made to have a transient calibration as well for the following reasons:

- Groundwater flow is transient by its nature, and is often subject to artificial (man-made) changes.
- The usual purpose of the model is prediction, which is by definition time-related.
- Steady-state calibration does not involve aquifer storage properties, which are critical for a viable (transient) prediction.

A limited field data set does not leave choice and predetermines the steady-state calibration. In such a case, an appropriate approach would be to define boundary conditions and stresses that are representative for the period in which the field data are collected.

When a transient field data set of considerable length is available, some meaningful average measure should be derived from it for a steady-state calibration. For example, this can be the mean annual water table elevation or the mean water table for the dry season, the average annual groundwater withdrawal, the mean annual precipitation (recharge), the average baseflow in a surface stream, and so on.

Transient calibration typically involves water levels recorded in wells during pumping tests or long-term aquifer exploitation. An ideal set that incorporates all common relevant boundary conditions and stresses would be

- Monthly water table (hydraulic head) elevations
- Monthly precipitation (recharge)
- Average monthly river stage
- Average monthly groundwater withdrawal

Transient calibration based on monthly values is preferred over daily or weekly data since groundwater systems usually react with a certain delay to surface stresses. In addition, monthly data enable accurate analysis of seasonal influences, which is very important for long-term predictions.

There are two methods of calibration: (1) trial-and-error (manual) and (2) automated calibration. Trial-and-error calibration was the first technique applied in groundwater modeling and is still preferred by most users. Although it is heavily influenced by the user's experience, it is always recommended to perform this type of calibration, at least in part. By changing parameter values and analyzing the corresponding effects, the modeler develops a better feeling for the model and the assumptions on which its design is based. During manual calibration boundary conditions, parameter values and stresses are adjusted for each consecutive model run until calculated heads match the preset calibration targets.

It will sometimes be necessary to change input values and run the model tens of times before reaching the target. The worst-case scenario involves a complete redesign of the model with new geometry, boundaries, and boundary conditions.

The first phase of calibration typically ends when there is a good visual match between calculated and measured hydraulic heads as seen on a contour map (Figure 7.21). The next step involves quantification of the model error with various statistical parameters such as standard deviation and distribution of model residuals, i.e., differences between calculated and measured values (see Section 7.4). Once this error is minimized (through a lengthy process of calibration), and satisfies a preset criterion, the model is ready for predictive use.

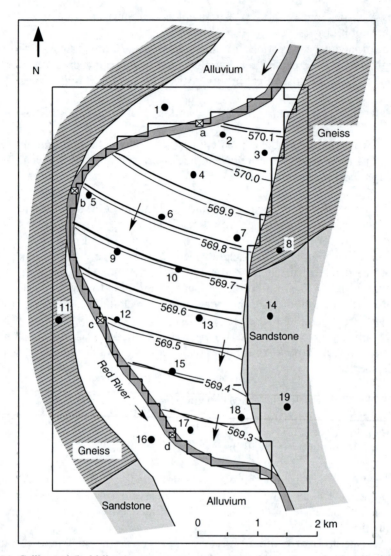

FIGURE 7.21 Calibrated (bold lines) versus measured water table contours in the alluvial aquifer. The calibrated hydraulic conductivity is 3.5×10^{-3} m/s, compared to the one initially estimated from the slug tests, 2.4×10^{-3} m/s. The largest discrepancy between calculated and estimate contours is along the contact with sandstone since it was assumed that the inflow from the sandstone aquifer would influence the water table contours in the alluvial aquifer. However, even unrealistically high inflow rates do not significantly change the shape of the calculated contours.

During calibration, the user should focus on parameters that are determined with less accuracy or assumed, and change only slightly those parameters that are more certain. For example, hydraulic conductivity determined by several pumping tests should be the last parameter to change "freely" because it is usually the most sensitive. Figure 7.22 shows how changing of K by one order of magnitude can significantly alter distribution of the calculated hydraulic head. Most other parameters are less sensitive and can be changed only within a certain realistic range: it is obviously not possible to increase precipitation infiltration rate ten times from 10% to 100%. In general, hydraulic conductivity and recharge are two parameters with equivalent quality: an increase in hydraulic conductivity creates the same

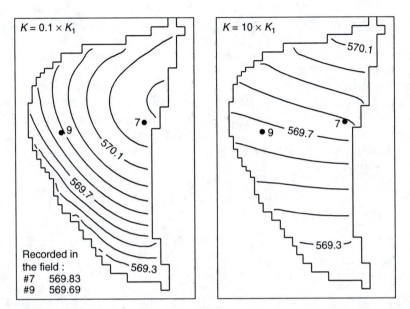

FIGURE 7.22 Effects of varying uniformly distributed hydraulic conductivity on the calculated water table in the aquifer. Monitoring wells 7 and 9 are shown for numeric comparison of individual data points; data from the same wells are used for creating the calibration graph in Figure 7.23.

effect as a decrease in recharge. Since different combinations of parameters can yield similar, or even the same results, trial-and-error calibration is not unique. During calibration, it is recommended to plot calculated versus recorded heads at several control points (i.e., nodes with monitoring wells) as shown in Figure 7.23. This enables more accurate determination of the parameter value that produces the best fit.

A very effective measure of calibration is the analysis of water budget calculated by Modflow. This model provides flows across boundaries, flows to and from all sources and

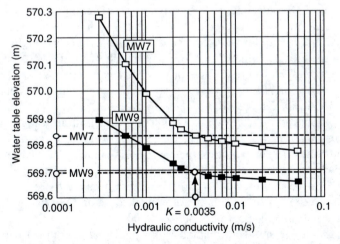

FIGURE 7.23 Water table elevation at wells 7 and 9 versus hydraulic conductivity distribution tested during calibration. The calibrated (chosen) hydraulic conductivity is found as an intersect between the calculated curve and the water table elevation recorded in the field (569.83 m for 7, and 569.69 m for 9).

sinks, and flows generated by the storage. These calculated values should be compared with measurements and estimates made during concept development. Unrealistic components of the water balance should be analyzed in order to calibrate the parameters and conditions that are causing them.

Arguably, it is much more difficult to calibrate a contaminant fate and transport model because aquifer heterogeneities and biochemical reactions usually have a much greater effect on contaminant flow pathways and concentrations, compared to "bulk" flow rate of ground-water. However, the process of calibration of F&T models is exactly the same as for groundwater flow models: various F&T parameters are being changed, within reasonable bounds, until a satisfactory match between the field-measured and model-predicted contaminant concentrations is achieved. The most critical parameter that should be "freely" adjusted the last is the rate of biodegradation. Following is the citation from a USEPA publication that speaks to that fact (Azdapor-Keeley et al., 1999):

> Many times during calibration, if a model does not fit observed concentrations, it is assumed that the biodegradation coefficient is the proper parameter to be adjusted. Using biodegradation to adjust a model without supporting field data should not be done until all abiotic mechanisms for reduction are explored. When using a model which incorporates a biodegradation term, care should be taken to verify that assumptions made about degradation rates and the amount and activity of biomass are valid for the site in question. Degradation rates are sensitive to a wide array of field conditions which have been discussed previously. Extrapolation of laboratory derived rates to a site can also lead to significant errors. Likewise, using models to derive degradation rates from limited field data where abiotic variables are not well defined can be misleading. . . . Kinetic constants derived from laboratory microcosms or other sites are generally not useful on a wide scale to predict overall removal rates. Site specific degradation rates should be developed and incorporated into a model.

Automated calibration is gaining in popularity since several powerful computer programs became widely available and are now incorporated in most GUIs by default. It is a technique developed in order to minimize uncertainties associated with the user's subjectivity. As with any relatively new approach, it has been criticized, particularly because of its nonuniqueness (see Anderson and Woessner, 1992). However, this is the case with any calibration, including manual, and it is up to the modeler to use it wisely, as an aid, rather than some final solution that has to be accepted because of its "objectivity." Most computer codes for automated calibration search an optimal parameter set for which the sum of squared deviations between calculated and measured values is reduced to a minimum. Two of the highly regarded codes for parameter estimation developed for Modflow are PEST by Doherty et al. (1994) and UCODE by Poeter and Hill (1998). Version 4 of Groundwater Vistas (Rumbaugh and Rumbaugh, 2004) now includes an easy-to-use and streamlined parameter optimization code developed specifically for this GUI.

For many data input quantities, Modflow-2000 allows definition using parameter values, each of which can be applied to data input for many grid cells. In combination with new multiplication and zone array capabilities, the parameters make it much easier to modify data input values for large parts of a model. Defined parameters also can have associated sensitivities calculated and be modified to attain the closest possible fit to measured hydraulic heads, flows, and advective travel. This is accomplished using the Observation, Sensitivity, and Parameter-Estimation Processes of Modflow-2000.

In conclusion, the efficiency of automated calibration codes, coupled with the trial-and-error input from the user, is arguably the most appropriate calibration method available. Readers interested in learning more about automated calibration should consult Hill (1998; "Methods and guidelines for effective model calibration") and Hill et al. (2000; "User guide to the observation, sensitivity, and parameter-estimation processes and three post-processing programs").

7.4 MODEL ERROR

Quantitative techniques for comparing model results (simulations) to site-specific information include calculating residuals, assessing correlation among the residuals, and plotting residual on maps and graphs (ASTM, 1999). Individual residuals are calculated by subtracting the model-calculated values from the targets (values recorded in the field, *not* extrapolated or otherwise assumed). They are calculated in the same way for hydraulic heads, drawdowns, concentrations, or flows; for example, the *hydraulic head residuals* are differences between the computed heads and the heads actually measured in the field:

$$r_i = h_i - H_i \tag{7.25}$$

where r_i is the residual, H_i is the measured hydraulic head at point i, and h_i is the computed hydraulic head at the approximate location where H_i was measured.

If the residual is positive, the computed value was too high; if negative, the computed value was too low (ASTM, 1999).

Residual mean is the arithmetic mean of the residuals computed from a given simulation:

$$R = \frac{\sum_{i=1}^{n} r_i}{n} \tag{7.26}$$

where R is the residual mean and n is the number of residuals.

Of two simulations, the one with the residual mean closest to zero has a better degree of correspondence, with regard to this criterion, and assuming there is no correlation among residuals (ASTM, 1999). It is possible that large positive and negative residuals could cancel each other, resulting in a small residual mean. For this reason, the residual mean should never be considered alone, but rather always in conjunction with the other quantitative and qualitative comparisons (ASTM, 1999).

The weighted residual mean can be used to account for differing degrees of confidence in the measured heads:

$$R = \frac{\sum_{i=1}^{n} r_i}{n \sum_{i=1}^{n} w_i} \tag{7.27}$$

where w_i is the weighting factor for the residual at point i. The weighting factors can be based on the modeler's judgment or statistical measures of the variability in the water level measurements. A higher weighting factor should be used for a measurement with a high degree of confidence than for one with a low degree of confidence.

Second-order statistics give measures of the amount of spread of the residuals about the residual mean. The most common second-order statistics is the standard deviation of residuals:

$$s = \left\{ \frac{\sum_{i=1}^{n} (r_i - R)^2}{n - 1} \right\}^{1/2} \tag{7.28}$$

where s is the standard deviation of residuals and R is given with Equation 7.26. Smaller values of the standard deviation indicate better degrees of correspondence than larger values (ASTM, 1999).

Correlation among Residuals. Spatial or temporal correlation among residuals can indicate systematic trends or bias in the model. Correlation among residuals can be identified through listings, scattergrams, and spatial and temporal plots. Of two simulations, the one with less correlation among residuals has a better degree of correspondence, with regard to this criterion (ASTM, 1999). Spatial correlation is evaluated by plotting residuals, with their sign (negative or positive) on a site map or cross sections. If applicable, the residuals can also be contoured. Apparent trends or spatial correlations in the residuals may indicate a need to refine aquifer parameters or boundary conditions, or even to reevaluate the CSM. For example, if all of the residuals near a no-flow boundary are positive, then the recharge may need to be reduced or the hydraulic conductivity increased (ASTM, 1999). For transient simulations, plot of residuals at a single point versus time may identify temporal trends. Temporal correlations in residuals can indicate the need to refine input aquifer storage properties or initial conditions (ASTM, 1999).

Figure 7.24 shows a mandatory graph of calculated versus actually measured heads at monitoring wells. Only field-measured hydraulic heads, not those estimated (interpolated) by the user for the purposes of creating the initial CSM hydraulic head, can be used to plot such a graph. If there were no calculation error for any of the control monitoring wells, all data would fall on the straight 1:1 ratio line, which can never happen (even if one were to engage in creating a nice-looking southwestern rug of hydraulic conductivity). Deviations of points or clusters of points, such as several monitoring wells in the same portion of the aquifer, from this line can reveal certain patterns and point toward the need for additional calibration or adjustment of the CSM.

7.5 MODEL VERIFICATION

A model is verified if its accuracy and predictive capability have been proven to lie within acceptable limits of error by tests independent of the calibration data (Konikow, 1978, from Anderson and Woessner, 1992). A typical verification is performed for an additional field data set in either steady-state or transient simulation. If the calibrated model parameters such as boundary conditions, stresses and distribution of hydrogeologic characteristics (e.g., conductivity, transmissivity, effective porosity) are correct, this additional independent field

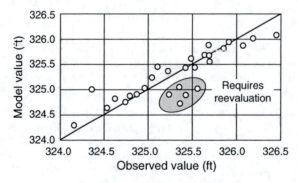

FIGURE 7.24 Calculated versus observed hydraulic heads for hypothetical modeling results, showing a cluster of errors that points toward the need for additional calibration. Such graphs, together with maps of posted residuals, can help in reevaluating the CSM assumptions.

data set should be closely matched by the model for the new boundary conditions and stresses. For example, the calibration data set may be representative of a wet period with high water table, high stage in the nearby river, and high lateral influx from the adjacent aquifer. If the verification data set is collected in the field during a recession period, the corresponding boundary conditions (river stage, lateral influx) and stresses (recharge) must be changed accordingly. If the model run with these changed conditions produces a head distribution that closely matches the new monitoring well data set, the model is verified. Often, however, it will be necessary to change some of the hydrogeologic parameters to obtain an acceptable match. In such cases, the verification actually becomes another calibration and a new independent data set is needed for model verification.

Since models are often calibrated in steady state, it is necessary to estimate storage parameters during the transient verification which somewhat decreases its accuracy. Despite this limitation, transient verification is preferred as it offers a better insight into the aquifer responses to time-dependent stresses. Examples of data sets typically used for transient verification are pumping tests, long-term pumpage, and seasonal changes in water table. Unfortunately, in modeling practice often only one data set is available in which case the model cannot be verified.

7.6 MODEL RESULTS AND PREDICTION

Once the model is calibrated and verified, preferably for both steady-state and transient conditions, it can be used for prediction, which is the purpose of most modeling projects. Prediction is the reward for the effort invested into all the preceding modeling phases and it is the stage, which model designers, more or less patiently, wait for. It is recommended that the user carry out a "mock prediction" during very early stages of model development. This means that, for example, anticipated wells are placed in the area of interest and pumped heavily in order to test placement of hydraulic (artificial) boundaries. If this pumping is influenced by the boundaries, the model area should be expanded. Doing this at the beginning of the modeling effort, rather than after lengthy calibration and sensitivity analyses, may save a lot of time, nerves, and spare a colleague or two of some inappropriate words.

Unless the prediction is made for some very general purposes, in which case the steady-state approach may be justified, it is very important to anticipate changes in all boundary conditions and stresses during the prediction period. For example, seasonal variations of recharge, fluctuations in river stage and other equipotential boundaries, or seasonal groundwater withdrawal, should be simulated with an appropriate number of stress periods. Transient prediction is very important for successful water resources management since it allows for more accurate design of well fields and pumping schedules.

Figure 7.25 and Figure 7.26 show traditional ways of presenting model predictions with maps and graphs. Thanks to the continuous interest of various software developers in the groundwater modeling market, there are now quite a few computer programs that can create attractive visualizations of modeling results, including in three dimensions and with animations. Some examples are shown in Figure 7.27 through Figure 7.30.

7.7 MODEL DOCUMENTATION

This is the final phase of the modeling effort and arguably the most important from the client's standpoint. A poorly documented and confusing report can ruin days of work and an otherwise excellent model. Every effort should be made to produce an attractive and user-friendly document that will convey clearly all previous phases of the model design. Special attention should be paid to clearly state the model's limitations and uncertainties

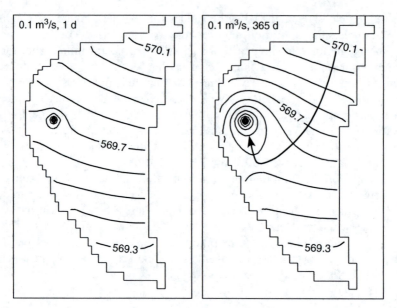

FIGURE 7.25 Transient effects of pumping 0.1 m³/s with one well placed close to the river. The capture zone after 1 and 365 d of pumping is shown with the boundary streamline. Water table contours are in meters amsl, and the contour interval is 0.1 m.

associated with calibrated parameters. Electronic modeling files (model input and output) and GIS files, if applicable, will have to be made available to the client in most cases. All modeling documentation should strictly follow widely accepted industry practices, guidelines and standards for groundwater modeling as detailed in the following section. And, needless to say, if the modeling process followed the recommendations in this book, section-by-section, it would be very easy (in all likelihood) to prepare a defensible modeling report.

7.8   STANDARDS FOR GROUNDWATER MODELING

The following industry standards, created by the leading industry experts for the groundwater modeling community under the auspices of ASTM, cover all major aspects of groundwater

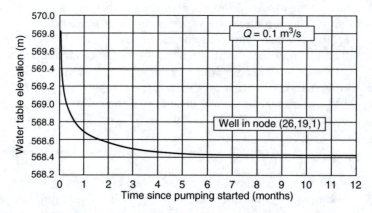

FIGURE 7.26 Hydraulic head versus time for a well pumping near the eastern impermeable boundary of the alluvial aquifer shown in Figure 7.20. A near steady-state condition is established after approximately 6 months of pumping.

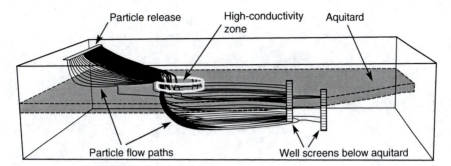

FIGURE 7.27 Results of a three-dimensional particle-tracking model showing the effects of a high-conductivity zone in an aquitard on particle flow paths. (Modified from Chiang, W.H., Chen, J., and Lin, J., 2002. *3D Master—A Computer Program for 3D Visualization and Real-time Animation of Environmental Data*, Excel Info Tech, Inc., 146pp.)

modeling and should be followed when attempting to create a defensible groundwater model that can be used for predictive purposes:

- Guide for application of groundwater flow model to a site-specific problem (D 5447–93)
- Guide for comparing groundwater flow model simulations to site-specific information (D 5490–93)
- Guide for defining boundary conditions in groundwater flow modeling (D 5609–94)
- Guide for defining initial conditions in groundwater flow modeling (D 5610 94)
- Guide for conducting a sensitivity analysis for a groundwater flow model application (D 5611–94)
- Guide for documenting a groundwater flow model application (D 5718–95)
- Guide for subsurface flow and transport modeling (D 5880–95)
- Guide for calibrating a groundwater flow model application (D 5981–96)

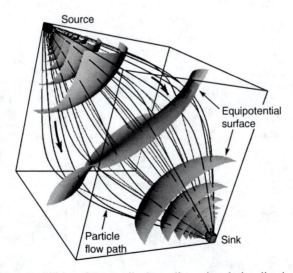

FIGURE 7.28 One of the capabilities of a versatile three-dimensional-visualization program, 3D Master, for displaying groundwater modeling results. (From Chiang, W.H., Chen, J., and Lin, J., 2002. *3D Master—A Computer Program for 3D Visualization and Real-time Animation of Environmental Data*, Excel Info Tech, Inc., 146pp.)

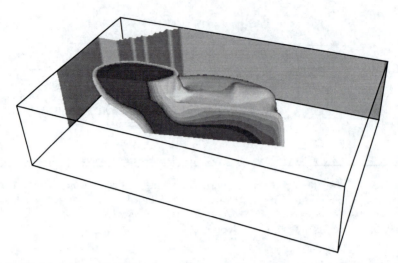

FIGURE 7.29 Modeled three-dimensional contaminant plume with a cut-through showing different concentrations inside the plume. (From Chiang, W.H., Chen, J., and Lin, J., 2002. *3D Master—A Computer Program for 3D Visualization and Real-time Animation of Environmental Data*, Excel Info Tech, Inc., 146pp.)

- Practice for evaluating mathematical models for the environmental fate of chemicals (E 978–92)
- Guide for developing conceptual site models for contaminated sites (E 1689–95)

The following language accompanies the USEPA OSWER Directive #9029.00 (USEPA, 1994):

> The purpose of this guidance is to promote the appropriate use of ground-water models in EPA's waste management programs. More specifically, the objectives of the "Assessment framework for ground-water model applications" are to

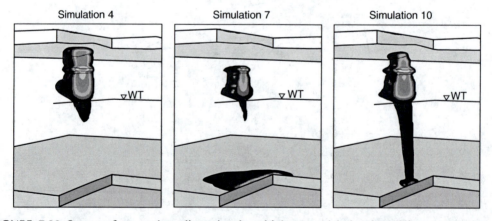

FIGURE 7.30 Outputs from a three-dimensional multiphase model showing different scenarios of DNAPL migration through unsaturated and saturated zones. (From Oostrom, M. et al., 2004. Three-dimensional modeling of DNAPL in the subsurface of the 216-Z-9 trench at the Hanford site. Prepared for the U.S. Department of Energy under Contract DE-AC05–76RL01830, Pacific Northwest National Laboratory, Richland, Washington).

- support the use of ground-water models as tools for aiding decision-making under conditions of uncertainty;
- guide current or future modeling;
- assess modeling activities and thought processes; and
- identify model application documentation needs.

Following is the introduction to "Guidelines for evaluating groundwater flow models" published by the U.S. Geological Survey (Reilly and Harbaugh, 2004):

Ground-water flow modeling is an important tool frequently used in studies of ground-water systems. Reviewers and users of these studies have a need to evaluate the accuracy or reasonableness of the ground-water flow model. This report provides some guidelines and discussion on how to evaluate complex ground-water flow models used in the investigation of ground-water systems. A consistent thread throughout these guidelines is that the objectives of the study must be specified to allow the adequacy of the model to be evaluated.

REFERENCES

Anderson, M.P. and Woessner, W.W., 1992. *Applied Ground Water Modeling: Simulation of Flow and Advective Transport*, Academic Press, San Diego, 381pp.

ASTM (American Society for Testing and Materials), 1999. *ASTM Standards on Determining Subsurface Hydraulic Properties and Ground Water Modeling*, 2nd ed., 320pp.

Azdapor-Keeley, A., Russell, H.H., and Sewell, G.W., 1999. Microbial processes affecting monitored natural attenuation of contaminants in the subsurface. *Ground Water Issue*, U.S. Environmental Protection Agency, Office of Research and Development, EPA/540/S-99/001, 18pp.

Chiang, W.H., 2005. Processing Modflow Pro^{TM}, WebTech360, 409pp.

Chiang, W.H. and Kinzelbach, W., 2001. *3D-Groundwater Modeling with PMWIN*, Springer, New York, 346pp.

Chiang, W.H., Chen, J., and Lin, J., 2002. *3D Master—A Computer Program for 3D Visualization and Real-time Animation of Environmental Data*, Excel Info Tech, Inc., 146pp.

Clement, T.P., 1997. *RT3D—A Modular Computer Code for Simulating Reactive Multi-species Transport in 3-Dimensional Groundwater Aquifers*, Pacific Northwest National Laboratory, Richland, WA, PNNL-11720.

Doherty, J., Brebber, L., and Whyte, P., 1994. *PEST—Model-independent Parameter Estimation. User's Manual*, Watermark Computing. Australia.

Fenske, J.P., Leake, S.A., and Prudic, D.E., 1996. Documentation of a computer program (RES1) to simulate leakage from reservoirs using the modular finite-difference ground-water flow model (Modflow), U.S. Geological Survey Open-File Report 96–364, 51pp.

Freeze, R.A. and Cherry, J.A., 1979. *Groundwater*, Prentice-Hall, Englewood Chiffs, NJ, 604pp.

Harbaugh, A.W. and McDonald, M.G., 1996. User's documentation for Modflow-96, an update to the U.S. Geological Survey modular finite-difference ground-water flow model. U.S. Geological Survey Open-File Report 96–485, Reston, VA, 56pp.

Harbaugh, A.W. et al., 2000. Modflow-2000, the U.S. Geological Survey modular ground-water model—user guide to modularization concepts and the ground-water flow process. The U.S. Geological Survey Open-File Report 00–92, Reston, VA, 121pp.

Hill, M.C., 1990. Preconditioned conjugate-gradient 2 (PCG2), a computer program for solving ground-water flow equations. U.S. Geological Survey Water-Resources Investigations Report 90–4048, 43pp.

Hill, M.C. 1998. Methods and guidelines for effective model calibration. U.S. Geological Survey Water-Resources Investigations Report 98–4005.

Hill, M.C. et al., 2000. Modflow-2000, the U.S. Geological Survey modular ground-water model—user guide to the observation, sensitivity, and parameter-estimation processes and three post-processing programs. U.S. Geological Survey Open-File Report 00–184, 210pp.

Hsieh, P.A. and Freckleton, J.R., 1993. Documentation of a computer program to simulate horizontal-flow barriers using the U.S. Geological Survey's modular three-dimensional finite-difference ground-water flow model. U.S. Geological Survey Open-File Report 92–477, 32pp.

Konikow, L.F., 1978. Calibration of ground-water models. In: *Verification of Mathematical and Physical Models in Hydraulic Engineering*, American Society of Civil Engineers, NY, pp. 87–93.

Kresic, N. and Rumbaugh, J. 2000. GIS and Data Management for Ground Water Modeling, National Ground Water Education Foundation Course, June 2001, San Francisco, National Ground Water Association, Westerville, OH.

Leake, S.A. and Lilly, M.R., 1997. Documentation of a computer program (FHB1) for assignment of transient specified-flow and specified-head boundaries in applications of the modular finite-difference ground-water flow model (Modflow). U.S. Geological Survey Open-File Report 97–571, 50pp.

Leake, S.A. and Prudic, D.E. 1991. Documentation of a computer program to simulate aquifer-system compaction using the modular finite-difference ground-water flow model. *U.S. Geological Survey Techniques of Water-Resources Investigations*, Book 6, Chap. A2, 68pp.

Leake, S.A., Leahy, P.P., and Navoy, A.S. 1994. Documentation of a computer program to simulate transient leakage from confining units using the modular finite-difference ground-water flow model. U.S. Geological Survey Open-File Report 94–59, 70pp.

Ligget, J.A., 1987. Advances in the boundary integral equation method in subsurface flow. *Water Resources Bulletin*, 23(4), 637–651.

Ligget, J.A. and Liu, P.L.F. 1983. *The Boundary Integral Equation Method for Porous Media Flow*, Allen & Unwin, London, 255pp.

McDonald, M.G. and Harbaugh, A.W., 1988. A modular three-dimensional finite-difference ground-water flow model. *U.S. Geological Survey Techniques of Water-Resources Investigations*, Book 6, Chap. A1, 586pp.

McDonald, M.G. et al., 1992. A method of converting no-flow cells to variable-head cells for the U.S. Geological Survey Modular Finite-Difference Ground-Water Flow Model. U.S. Geological Survey Open-File Report 91–536, Reston, VA, 99pp.

Maidment, D.R. (Ed.), 1993. *Handbook of Hydrology*, McGraw Hill, New York 1.1 to I.48.

Oostrom, M. et al., 2004. Three-dimensional modeling of DNAPL in the subsurface of the 216-Z-9 trench at the Hanford site. Prepared for the U.S. Department of Energy under Contract DE-AC05–76RL01830, Pacific Northwest National Laboratory, Richland, Washington.

Poeter, E.P. and Hill, M.C. 1998. Documentation of UCODE, a computer code for universal inverse modeling. U.S. Geological Survey Water-Resources Investigations Report 98–4080.

Prudic, D.E., 1989. Documentation of a computer program to simulate stream–aquifer relations using a modular, finite difference, ground-water flow model. U.S. Geological Survey Open-File Report 88–729, 113pp.

Reilly, T.E. and Harbaugh, A.W., 2004. Guidelines for evaluating ground-water flow models. U.S. Geological Survey Scientific Investigations Report 2004–5038, Version 1.01, Reston, VA, 30pp.

Rumbaugh, J.O. and Rumbaugh, D.B. 2004. Guide to using Groundwater Vistas, Version 4. Environmental Simulations, Inc., Reinholds, PA, 366pp (http://www.groundwatermodels.com).

Rushton, K.R. and Redshaw, S.C. 1979. *Seepage and Groundwater Flow; Numerical Analysis by Analog and Digital Methods*, John Wiley & Sons, New York.

Strack, O.D.L., 1987. The analytic element method for regional groundwater modeling. In: *Solving Ground Water Problems with Models*, National Water Well Association, Columbus, OH, pp. 929–941.

Strack, O.D.L., 1988. *Groundwater Mechanics*, Prentice-Hall, Englewood Cliffs, NJ, 732pp.

USEPA, 1994. Assessment framework for ground-water model applications, OSWER Directive 9029.00. U.S. Environmental Protection Agency, Office of Solid Waste and Emergency Response, Washington, D.C.

Voss, C.I. and Provost, A.M. 2002. SUTRA: a model for saturated-unsaturated, variable-density ground-water flow with solute or energy transport. U.S. Geological Survey Water-Resources Investigations Report 02–4231, Reston, VA, 250pp.

Zheng, C. and Wang P.P., 1999. MT3DMS: a modular three-dimensional multispecies model for simulation of advection, dispersion and chemical reactions of contaminants in groundwater systems; Documentation and Users Guide, Contract Report SERDP-99–1, U.S. Army Engineer Research and Development Center, Vicksburg, MS.

Part II

Solved Problems in
Hydrogeology and
Groundwater Modeling

8 Porosity and Related Parameters

8.1 POROSITY, EFFECTIVE POROSITY, AND SATURATION

Formulation of problem: The volume of a moist sand specimen is 72.5 cm^3 and its weight is 152 g. The volume is determined by submerging the sample in a container with mercury and measuring the weight of the spilled mercury. After drying in the oven for 24 h at 105°C, the weight of the specimen is 145 g and its volume is 71.2 cm^3 (a 1.8% decrease). Determine the following parameters for the sand sample: porosity, specific weights and densities, void ratio, moisture content, degree of saturation, and effective porosity.

Note: Confusion is often present when the mass and the weight of a sample are expressed with the same unit, such as gram (g) or kilogram (kg; 1 kg = 1000 g). The unit of mass in the SI system is the kilogram, while the unit of weight is the Newton (N), which is also the unit of force (weight is a force). However, in laboratory and field engineering practices the weight is measured by scales that are usually calibrated in kilograms (grams). The obtained value for weight is then equaled to the value of mass, which means, for example, that a sample that has a "weight" of 152 g has a mass of 152 g in the gravitational field of the earth (weight = mass × acceleration of gravity). This means that the mass of a body (sample) is a constant quantity, while its weight will change if the acceleration of gravity changes. For practical purposes the weight will here be expressed in grams.

The porosity is given as the ratio of volume of voids to the total initial volume of the specimen before drying, which includes both voids and solids (as can be seen, a sample can change the volume upon drying in the oven):

$$n = \frac{V_v}{V}(\times 100\%)$$

(8.1)

The volume of voids can be determined using the three-phase diagram approach, or a water displacement test, which is used to find the effective porosity.

8.1.1 THREE-PHASE DIAGRAM

Weight–volume relationships of a soil sample, i.e., its specific weight, density, porosity, void ratio, and saturation, can be determined using three-phase diagrams as shown in Figure 8.1. If one of the volumes or weights is unknown, others should be indicated on the diagram in terms of that unknown. The weight (moist and dry) of the sample is easily measured in the laboratory. The total volume is determined by measuring the weight of the spilled mercury of known specific weight after the sample is submerged in a container with mercury. Specific weight and gravity of the sample, sample solids and water are calculated and are found in tables. When the specific weight of water is assumed to be unity (1 g/cm^3), then knowing the weight (g) of water lost during oven-drying, which is the difference between the sample weights before and after drying, gives the same number for the volume of that water in cm^3 (see Figure 8.2).

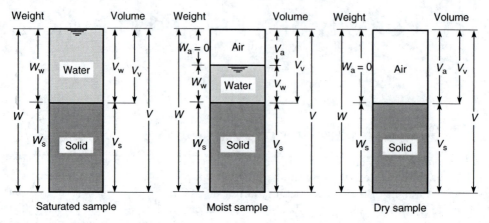

FIGURE 8.1 Three-phase diagrams for the samples with different degrees of saturation.

The specific weight of the sample (sand specimen) is

$$\gamma = \frac{W}{V} = \frac{152 \text{ g}}{72.5 \text{ cm}^3} = 2.09 \text{ g/cm}^3 \tag{8.2}$$

where W is the weight of the sample, i.e., the weight of the specimen before oven-drying and V is the volume of the sample. The specific weight is numerically equal to the sample density (ρ) if both weight and mass are expressed in the same units, for example, grams. Density is defined as the sample mass divided by the sample volume and its (physically correct) unit in the SI system is g/cm^3.

The dry specific weight is the weight of the sample after oven-drying divided by the total initial volume of the sample before oven-drying (a sample can change its volume during drying in the oven, as is the case with our sand specimen; these changes are usually small for

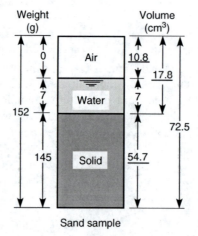

FIGURE 8.2 Weights and volumes of the three phases of the sand sample.

sand and gravel, but may be significant for clayey sediments and clays). The weight of the sample after drying equals the weight of the solids or sample mineral matter (W_s).

$$\gamma_d = \frac{W_s}{V}(g/cm^3) \tag{8.3}$$

The dry specific weight numerically equals the bulk density (ρ_b), which is defined as the mass of the oven-dried sample divided by the sample volume. The bulk density is always smaller than the overall density of a moist sample since the mass of water present in the voids is excluded from the calculation. The bulk density depends on the total pore space and varies accordingly—the fine texture materials, such as silts and clays, have more pore space and lower bulk densities than sand and gravel.

The specific weight of the solids (mineral particles), which numerically equals the particle density (ρ_s), is the ratio of the weight of solids to the volume of the sample solids excluding pore space:

$$\gamma_s = \frac{W_s}{V_s}(g/cm^3) \tag{8.4}$$

It is very difficult, if practically impossible, to measure directly the volume of solids in a soil or aquifer material sample. Rather, it is determined indirectly by assuming the particle density and using Equation 8.4. Particle density of mineral soils and unconsolidated aquifer materials vary a great deal. They are not dependent on moisture content, particle size or void space, and are a material constant property. Sample particle density is an average of the densities of all present minerals. Most soils and unconsolidated aquifer materials reflect densities of the most abundant minerals—quartz (2.65 g/cm^3) and feldspars (2.56–2.76 g/cm^3). Therefore, the particle density of 2.65 g/cm^3 is usually assumed unless very precise data is required. The specific weight of the solids (particle density) can be expressed also as

$$\gamma_s = G\gamma_w \tag{8.5}$$

where G is the specific gravity of the sample solids (dimensionless) and γ_w is the specific weight of water. If the specific weight of water is considered as unity (the average γ_w is 1 g/cm^3), the specific weight of solids (given in g/cm^3) equals the specific gravity the value of which can be found in tables. The terms specific weight of solids, specific gravity, and particle density are used as synonyms throughout literature and are numerically equal. For this problem, it is assumed that the sand specimen has the specific gravity of 2.65, which gives the following value of the volume of solids using Equation 8.4:

$$V_s = \frac{W_s}{\gamma_s} = \frac{145\ g}{2.65\ g/cm^3} = 54.7\ cm^3 \tag{8.6}$$

The volume of voids is then

$$V_v = V - V_s = 72.5\ cm^3 - 54.7\ cm^3 = 17.8\ cm^3 \tag{8.7}$$

and the volume of voids filled with air is

$$V_a = V_v - V_w = 17.8\ cm^3 - 7.0\ cm^3 = 10.8\ cm^3 \tag{8.8}$$

Again, note that the volume of water numerically equals the weight of water if they are expressed in cubic centimeters (cm^3) and grams (g), respectively, and if we assume that the specific weight of water is unity, i.e., 1 g/cm^3.

The phase diagram in Figure 8.2 shows all the measured and calculated (bold underlined) values of the weights (masses) and volumes for this problem.

8.1.2 Porosity

The porosity of the sample is calculated using Equation 8.1:

$$n = \frac{V_v}{V} = \frac{17.8 \text{ cm}^3}{72.5 \text{ cm}^3} = 0.245 \text{ or } 24.5\%$$

8.1.3 Void Ratio

The void ratio is given as

$$e = \frac{V_v}{V_s} = \frac{17.8 \text{ cm}^3}{54.7 \text{ cm}^3} = 0.325 \tag{8.9}$$

or

$$e = \frac{n}{1 - n} = \frac{0.245}{1 - 0.245} = 0.324 \tag{8.10}$$

where V_v is the volume of voids in the sample, V_s is the volume of solids in the sample, and n is the porosity of the sample. The void ratio is inversely related to the particle density (specific gravity): the greater the particle density, the smaller the void ratio, and vice versa.

Note: Unlike porosity, the void ratio can have values greater than 1, in which case it means that the percentage of voids in the unit volume of the sample is greater than the percentage of solids (mineral particles).

8.1.4 Moisture Content

The moisture content (w) is the ratio of the weight of water (W_w) in the sample to the weight of solids (W_s):

$$w = \frac{W_w}{W_s} = \frac{W - W_s}{W_s} = \frac{7 \text{ g}}{145 \text{ g}} = 0.0483 \text{ or } 4.8\% \tag{8.11}$$

8.1.5 Degree of Saturation (Saturation Index)

The degree of saturation is usually expressed with the saturation index, which is the percentage of sample voids filled with water:

$$S_I = \frac{V_w}{V_v} = \frac{7 \text{ cm}^3}{17.8 \text{ cm}^3} = 0.3933 \text{ or } 39.3\% \tag{8.12}$$

8.1.6 EFFECTIVE POROSITY

A water displacement test is used to determine the effective porosity (n_{ef}) of the sample. Effective porosity is the volume of interconnected voids that allow free water (fluid) flow divided by the total sample volume. The oven-dried sample of the known volume ($V = 72.5\,\text{cm}^3$) is immersed in the chamber containing $500\,\text{cm}^3$ of water, and left until it becomes saturated (the chamber is sealed to prevent water evaporation loss). It is assumed that during the oven-drying all the free water contained in the voids is expelled. The volume of water used to completely fill the voids, i.e., saturate the sample, is therefore equal to the volume of voids. However, this volume will yield the value of effective porosity, which is smaller than the overall porosity since it excludes micropores, which do not allow the free flow of water, and the pores that are not interconnected. The volume of the interconnected voids (V_{vi}) is determined as the difference between the initial water volume in the chamber ($500\,\text{cm}^3$) and the water volume after the saturated sample is removed ($483.5\,\text{cm}^3$):

$$V_{vi} = V_{w1} - V_{w2} = 500\ \text{cm}^3 - 483.5\ \text{cm}^3 = 16.5\ \text{cm}^3 \tag{8.13}$$

The effective porosity is

$$n_{ef} = \frac{V_{vi}}{V} = \frac{16.5\ \text{cm}^3}{72.5\ \text{cm}^3} = 0.2276\ \text{or}\ 22.8\% \tag{8.14}$$

As mentioned earlier, the effective porosity is smaller than the overall porosity. In the case of homogeneous medium-to-coarse sand and gravel, this difference is only a few percent. It dramatically increases, however, with an increase in the percentage of silt and clay particles in the rock. Some clays that have a total porosity of more than 60% may have an effective porosity of less than 1–3%. Values of the porosity of various rocks are given in Appendix C.

9 Laboratory Methods for Determining Hydraulic Conductivity

9.1 CONSTANT HEAD PERMEAMETER

Formulation of problem: Determine the hydraulic conductivity and permeability of a coarse sand sample that was tested in the laboratory with a constant head permeameter. The results of the test are given in Table 9.1. The fluid used for testing was deaired water with a temperature of 23°C. The apparatus used for testing was Darcy's constant head permeameter, explained in Section 1.4.2 (Figure 1.33). The inside diameter of the sample tubing was 1 7/8 in. (4.8 cm) and the length of the sand sample was 6 in. (15.2 cm). The experiment consisted of 10 individual tests each with a different position of the constant head (the head difference between the constant head and the head of the permeameter outlet is given in Table 9.1). The flow rate for each test was determined by measuring the volume of water collected at the outlet and the corresponding time. After determining the hydraulic conductivity for water as the test fluid, calculate what the hydraulic conductivity of the sample would be if, instead of water, the test fluid was gasoline at 18°C (kinematic viscosity of gasoline at 18°C is 6.5×10^{-7} m^2/s).

9.1.1 HYDRAULIC CONDUCTIVITY

The permeameter (shown in Figure 1.33) is based on Darcy's apparatus named after Henry Darcy, a French civil engineer who formulated the equation of flow of water through intergranular porous media. The rate of a fluid flow (Q) through a sand sample is directly proportional to the cross-sectional area of the flow (A) and the loss of the hydraulic head between two points of measurements (Δh), and it is inversely proportional to the length of the sample (l):

$$Q = KA\frac{\Delta h}{l} \quad (\text{m}^3/\text{s}) \tag{9.1}$$

where K is the proportionality constant of the law called hydraulic conductivity and has units of velocity. The change in hydraulic head between two measuring ports, separated by distance l, is hydraulic gradient, i. The form of Darcy's law that we will use, expressed in terms of the hydraulic conductivity, is

$$K = \frac{v}{i} \tag{9.2}$$

557

TABLE 9.1
Results of the Constant Head Permeameter Test of the Sand Sample

Test number	1	2	3	4	5	6	7	8	9	10
Head difference (cm)	5	6	7	8	9	10	12	14	16	18
Volume (cm^3)	4	3.5	5	5.5	3	6	3.5	3	6	4.5
Time (s)	125	92	106	112	46	92	48	34	61	45

where v is the flow velocity determined from measuring the volumetric flow rate through the cross-sectional area of sample and using the flow continuity equation:

$$v = \frac{Q}{A}$$ (9.3)

The sand sample inserted into the permeameter is a cylindrical specimen of the cross-sectional area A and length l (Figure 1.33). It is maintained in place by two porous stones. A steady-state flow of water through the sample is established by maintaining the upgradient and downgradient hydraulic heads. The upgradient hydraulic head is adjustable (it can be changed) while the downgradient hydraulic head is fixed by the position of the outlet. For each position of the upgradient hydraulic head, the flow rate (Q) is determined by measuring the volume of water (V) collected at the outlet through time t ($Q = V/t$). Measurements for at least 10 head differences should be performed in order to plot a graph velocity versus hydraulic gradient. The graph in Figure 9.1 is used to check the validity of the experiment and to calculate the representative value of the hydraulic conductivity. Measurements and calculations needed for plotting the graph are shown in Table 9.2.

The cross-sectional area of the sample, i.e., the fluid flow, is determined using the following equation:

$$A = r^2 \pi = \left(\frac{d}{2}\right)^2 \pi$$ (9.4)

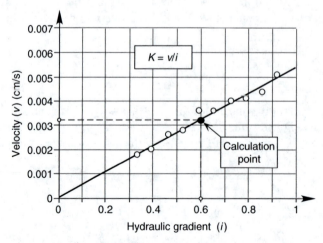

FIGURE 9.1 Graph velocity (v) versus hydraulic gradient (i) for the constant head permeameter test, which is used to determine the hydraulic conductivity of the sand.

TABLE 9.2
Elements for Plotting the Graph Velocity versus Hydraulic Gradient

Test #	1	2	3	4	5	6	7	8	9	10
V (cm^3)	4	3.5	5	5.5	3	6	3.5	5.5	7	5
t (s)	125	92	106	112	46	92	48	74	87	54
$Q = V/t$	0.032	0.038	0.047	0.049	0.065	0.065	0.072	0.074	0.08	0.092
$v = Q/A$	0.0018	0.0021	0.0026	0.0027	0.0036	0.0036	0.0040	0.0041	0.0044	0.0051
Δh (cm)	5	6	7	8	9	10	11	12	13	14
$i = \Delta h/l$	0.329	0.395	0.460	0.526	0.592	0.658	0.724	0.789	0.855	0.921

where r is the radius of the sample and d is the diameter of the sample.

$$A = (2.4 \text{ cm})^2 \times 3.14 = 18.1 \text{ cm}$$

If the experiment is correctly set, the individual tests for various head differences should group along a straight line on the graph velocity versus hydraulic gradient as shown in Figure 9.1. In addition, it should be possible to start the best-fit straight line drawn through the experimental data at the graph's origin. This would indicate that the assumptions of Darcy's law, i.e., the flow is laminar (linear) in steady state, are satisfied and that Equation 9.2 can be applied. The representative hydraulic conductivity of the sample is determined by choosing a point anywhere on the straight line and by substituting its coordinates in Equation 9.2:

$$K = \frac{0.0032 \text{ cm/s}}{0.6} = 5.3 \times 10^{-3} \text{ cm/s}$$

9.1.2 PERMEABILITY

The permeability of a porous medium is the ease with which a fluid can flow through that medium. In other words, permeability characterizes the ability of a porous medium to transmit a fluid. In hydrogeology, the permeability is also referred to as the intrinsic permeability. It is dependent only on the physical properties of the porous medium: grain size, grain shape and arrangement, pore interconnections, etc. On the other hand, hydraulic conductivity is dependent on properties of both the porous medium and the fluid. The relationship between the permeability (K_i) and the hydraulic conductivity (K) is expressed through the following formula:

$$K_i = K \frac{\mu}{\rho g} \quad (\text{m}^2) \tag{9.5}$$

where μ is the absolute viscosity of the fluid (also called dynamic viscosity or simply viscosity), ρ is the density of the fluid, and g is the acceleration of gravity.

The viscosity and the density of the fluid are related through the property called kinematic viscosity (v):

$$v = \frac{\mu}{\rho} \quad (\text{m}^2/\text{s}) \tag{9.6}$$

Inserting the kinematic viscosity into Equation 9.6 somewhat simplifies the calculation of the permeability since only one value (that of v) has to be obtained from tables or graphs (for most practical purposes, the value of the acceleration of gravity (g) is 9.81 m/s^2):

$$K_i = K \frac{v}{g} \tag{9.7}$$

The kinematic viscosity of water used in the constant head permeameter test is shown in the graph in Figure 1.34; the water temperature is 23°C and the kinematic viscosity is 9.4×10^{-7} m^2/s. Substituting these values and the value of the hydraulic conductivity into Equation 9.7 gives

$$K_i = 5.3 \times 10^{-5} \text{ m/s} \times \frac{9.4 \times 10^{-7} \text{ m}^2/\text{s}}{9.81 \text{ m/s}^2} = 5.08 \times 10^{-12} \text{ m}^2$$

It is much better to express the permeability in the units of area (m^2 or cm^2) for the reasons of consistency and easier use in other formulas. However, in practice, it is more commonly given in darcys (which is a "tribute" to the oil industry):

$$1 \text{ darcy (D)} = 9.87 \times 10^{-9} \text{ cm}^2 \quad \text{(or approximately } 1 \times 10^{-8} \text{ cm}^2\text{)}$$

Knowing that 1 m^2 = 10,000 cm^2, the permeability of our sand sample is

$$K_i = \frac{5.08 \times 10^{-12} \times 10,000 \text{ cm}^2}{9.87 \times 10^{-9} \text{ cm}^2} = \frac{5.08 \times 10^{-8} \text{ cm}^2}{9.87 \times 10^{-9} \text{ cm}^2}$$
$$K_i = 5.15 \text{ D}$$

9.1.3 HYDRAULIC CONDUCTIVITY WITH GASOLINE AS A TEST FLUID

As already mentioned, the permeability of a porous medium is independent of the characteristics of the fluid flowing through that medium. This means that, once determined, the permeability can be used to estimate the hydraulic conductivity of the medium for various fluids or temperatures using Equation 9.7 in the following form:

$$K = K_i \frac{g}{v} \tag{9.8}$$

Knowing that the kinematic viscosity of gasoline at 18°C is 6.5×10^{-7} m^2/s and that the permeability of our sand sample is 5.08×10^{-12} m^2, the hydraulic conductivity is easily calculated:

$$K = 5.08 \times 10^{-12} \text{ m}^2 \times \frac{9.81 \text{ m/s}^2}{6.5 \times 10^{-7} \text{ m}^2/\text{s}} = 7.67 \times 10^{-5} \text{ m/s}$$

The result shows that the hydraulic conductivity of the sand sample is higher for gasoline than for water: 7.7×10^{-5} m/s versus 5.3×10^{-5} m/s.

9.2 FALLING HEAD PERMEAMETER

Formulation of problem: Determine the hydraulic conductivity and permeability of a silty sand sample that was tested in the laboratory with a falling head permeameter. The results of the test are given in Table 9.3. The fluid used for testing was deaired water with a temperature of 26°C. Schematic drawing of the falling head laboratory permeameter, which is used to test low-permeable intergranular porous media, such as silts and clays, is shown in Figure 9.2. The inside diameter of the sample tubing is 1 7/8 in. (4.8 cm) and the length of the silty sand sample is 6 in. (15.2 cm). The experiment consisted of 10 individual tests each with a different time period and position of the initial head (h_0).

The sand sample inserted into the permeameter is a cylindrical specimen of the diameter $2r_s$ and length L (Figure 9.2). It is maintained in place by two porous stones. The flow of water through the sample is established by adding water to the tall, narrow burette. In time t, the head falls from the position h_0 to the position h_1. Measurements for at least 10 changes of head and corresponding times should be performed in order to plot a graph time versus ratio between the initial and final heads. The graph in Figure 9.3 is used to check the validity of the experiment and to calculate the representative value of the hydraulic conductivity. Measurements and calculations needed for plotting the graph are given in Table 9.3.

9.2.1 HYDRAULIC CONDUCTIVITY

The hydraulic conductivity of the sample is found to be started with the general hydraulic equation of the continuity of flow; fluid flow through the burette equals flow through the sample:

$$Q_{burette} = Q_{sample} \tag{9.9}$$

The flow rate is the product of the cross-sectional area of flow (A) and the velocity of flow (v):

$$Q_{burette} = A_{burette} \times v_{burette} \tag{9.10}$$

Knowing the radius of burette (r_b) and understanding that the flow velocity equals the change of head in time gives

$$Q_{burette} = r_b^2 \pi \left(-\frac{dh}{dt} \right) \tag{9.11}$$

The minus sign indicates that the hydraulic head, which causes the flow in the burette, changes from the higher one to the lower one.

TABLE 9.3
**Elements for Plotting the Graph Time versus Ratio between the Initial Head (h_0)
and the Final Head (h_1)**

H_0 (cm)	48	50.5	59	60	50	49	55	52	60	65
h_1 (cm)	42.5	44.0	50.5	47.0	35.0	32.0	37.0	45.0	49.0	47.0
h_0/h_1	1.13	1.16	1.17	1.28	1.42	1.52	1.48	1.15	1.22	1.38
Time (s)	422	461	505	781	1162	1423	1270	421	644	1000

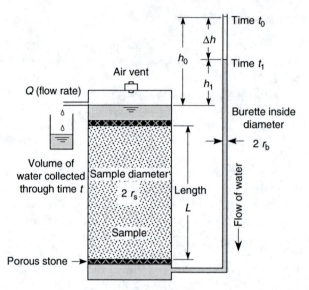

FIGURE 9.2 Schematic presentation of the falling head permeameter used for testing silty sand sample.

The fluid flow through the sand sample is

$$Q_{sample} = A_{sample} \times v_{sample} \tag{9.12}$$

or

$$Q_{sample} = r_s^2 \pi K \frac{h}{L} \tag{9.13}$$

where r_s is the sample radius, L is the sample length, h is the change in hydraulic head, and K is the hydraulic conductivity of the sample.

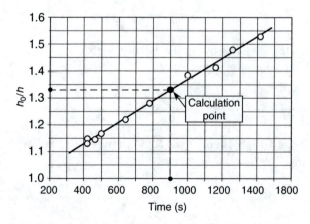

FIGURE 9.3 Graph time versus ratio of the initial and the final heads for calculating the hydraulic conductivity using falling head permeameter.

Equation 9.11 and Equation 9.13 give

$$r_b^2 \pi \left(-\frac{dh}{dt} \right) = r_s^2 \pi K \frac{h}{L} \tag{9.14}$$

This equation is then integrated with the following boundary conditions: at time t_0, the hydraulic head is h_0 and at time t_1, the hydraulic head is h_1 (see also Figure 9.2):

$$r_b^2 L \left(-\int_{h_0}^{h_1} \frac{dh}{h} \right) = r_s^2 K \int_{t_0}^{t_1} dt \tag{9.15}$$

Since both integrals are elementary, they are easily solved (see Appendix B):

$$r_b^2 L [-(\ln h_1 - \ln h_0)] = r_s^2 K (t_1 - t_0) \tag{9.16}$$

$$r_b^2 L \left(\ln \frac{h_0}{h_1} \right) = r_s^2 K t \tag{9.17}$$

Finally, the hydraulic conductivity is

$$K = \frac{r_b^2 L}{r_s^2 t} \ln \frac{h_0}{h_1} \tag{9.18}$$

The value of the hydraulic conductivity is calculated by inserting known values of the burette's radius (r_b), length (L), and radius of the sample (r_s), and the coordinates of a point on the best-fit line drawn through the experimental data as shown in Figure 9.3:

$$K = \frac{(1 \text{ cm})^2 \times 15.2 \text{ cm}}{(2.4 \text{ cm})^2 \times 900 \text{ s}} \times \ln (1.338)$$

$$K = 0.00085 \text{ cm/s} = 8.5 \times 10^{-4} \text{ cm/s} = 8.5 \times 10^{-6} \text{ m/s}$$

9.2.2 PERMEABILITY

As already mentioned in Section 9.1, the permeability of a porous medium is the ease with which a fluid can flow through that medium. Permeability characterizes the ability of a porous medium to transmit a fluid and its value is independent of the fluid characteristics. Permeability is dependent only on the physical properties of the porous medium such as grain size, grain shape and arrangement, pore interconnections, etc. On the other hand, hydraulic conductivity is dependent on properties of both the porous medium and the fluid flowing through it.

The relationship between the permeability (K_i) and the hydraulic conductivity (K) is expressed through the following formula:

$$K_i = K \frac{\mu}{\rho g} \tag{9.19}$$

where μ is the absolute viscosity of the fluid (also called dynamic viscosity or simply viscosity), ρ is the density of the fluid, and g is the acceleration of gravity.

Remember that the viscosity and the density of the fluid are related through the property called kinematic viscosity (ν):

$$\nu = \frac{\mu}{\rho} \tag{9.20}$$

Inserting the kinematic viscosity into Equation 9.19 simplifies the calculation of the permeability since only one value (that of ν) has to be obtained from tables or graphs (for most practical purposes, the value of the acceleration of gravity is 9.81 m/s^2):

$$K_i = K \frac{\nu}{g} \tag{9.21}$$

The dynamic viscosity of water used in the falling head permeameter test is shown in the graph (Figure 1.34, Section 1.4.2); the water temperature is 26°C and the dynamic viscosity is 8.8×10^{-4} Ns/m^2. The density of water at 26°C is found in the graph in Figure 1.34 and it is found to be 996.56 kg/m^3 (or 0.99656 g/cm^3). The values of dynamic viscosity and density can be inserted in Equation 9.20 to find kinematic viscosity (if a graph like the one shown in Figure 1.34 is not readily available). Together with the acceleration of gravity, they can also be inserted into Equation 9.19 to give the permeability:

$$K_i = 8.5 \times 10^{-6} \text{ m/s} \times \frac{8.8 \times 10^{-4} \text{ Ns/m}^2}{996.56 \text{ kg/m}^3 \times 9.806 \text{ m/s}^2}$$

$$K_i = 7.65 \times 10^{-13} \text{ m}^2$$

Note: Equations with physical properties that have complex derived units can often lead to dimensionally inconsistent (erroneous) results if conversions between various units are not properly applied. In our case the key for obtaining permeability in square meters is

$$1 \text{ Newton (N)} = 1 \text{ (kg m)/s}^2$$

It is much better to express the permeability in the units of area (m^2 or cm^2) for reasons of consistency and easier use in other formulas. However, it is more commonly given in darcys:

$$1 \text{ darcy (D)} = 9.87 \times 10^{-9} \text{ cm}^2 \text{ (or approximately } 1 \times 10^{-8} \text{ cm}^2)$$

Knowing that 1 m^2 = 10,000 cm^2, the permeability of our sand sample is

$$K_i = \frac{7.65 \times 10^{-13} \times 10,000 \text{ cm}^2}{9.87 \times 10^{-9} \text{ cm}^2} = \frac{7.65 \times 10^{-9} \text{ cm}^2}{9.87 \times 10^{-9} \text{ cm}^2}$$

$$K_i = 0.775 \text{ D}$$

9.3 GRAIN SIZE ANALYSIS

Formulation of problem: Three sets of standard American Society for Testing Materials (ASTM) sieves were used to determine grain size distribution of three aquifer samples. Using the test results from Table 9.4, the following has to be done for each sample:

TABLE 9.4
Results of Laboratory Sieve Tests for Three Porous Media Samples

Sample 1 (500 g)

Retained at sieve #	6	10	12	16	20	30	40	60	100	Bottom pan
Diameter of sieve (mm)	3.350	2.000	1.700	1.180	0.850	0.600	0.425	0.250	0.150	—
Weight of fraction (mg)	45.2	95.1	204.0	97.8	39.5	13.6	2.2	1.9	1.0	0.1
Content of fraction (%)	9.04	19.02	40.80	19.56	7.90	2.72	0.44	0.38	0.19	0.28
% Coarser than sieve diameter	9.04	28.06	68.86	88.42	96.32	99.04	99.48	99.86	100.05	100.33

Sample 2 (500 g)

Retained at sieve #	6	10	16	20	30	40	60	100	140	Bottom pan
Diameter of sieve (mm)	3.350	2.000	1.180	0.850	0.600	0.425	0.250	0.150	0.106	—
Weight of fraction (mg)	25.1	52.4	58.3	72.9	124.7	89.3	52.2	18.8	4.6	2.8
Content of fraction (%)	5.02	10.48	11.66	14.58	24.94	17.86	10.44	3.76	0.92	0.56
% Coarser than sieve diameter	5.02	15.50	27.16	41.74	66.68	84.54	94.98	98.74	99.66	100.22

Sample 3 (500 g)

Retained at sieve #	10	16	20	30	40	60	100	140	200	Bottom pan
Diameter of sieve (mm)	2.000	1.180	0.850	0.600	0.425	0.250	0.150	0.106	0.075	—
Weight of fraction (mg)	18.7	69.3	47.9	45.4	34.2	72.5	39.9	52.6	26.2	91.7
Content of fraction (%)	3.74	13.86	9.58	9.08	6.84	14.50	7.98	10.52	5.24	18.34
% Coarser than sieve diameter	3.74	17.6	27.18	36.26	43.1	57.6	65.58	76.1	81.34	99.68

1. Draw the grain size distribution curve.
2. Determine the uniformity coefficient and the effective grain size.
3. Calculate hydraulic conductivity of the aquifer material using appropriate empirical formulas.

9.3.1 Grain Size Distribution Curve

Porosity, hydraulic conductivity, and permeability are hydrogeological parameters that all greatly depend on the size of sediment grains and the percentage of various sediment fractions. The most widely used method to determine grain size distribution is by laboratory sieve analysis. As shown in Figure 9.4, a sediment sample of known weight is placed in the top sieve of a set of sieves with decreasing mesh openings and diameters. The set is then placed in a mechanical shaker for up to 10 min to separate the sample into fractions, which are retained at individual sieves. The weight of each fraction is measured and expressed as percent of the initial (total) weight of the sample. The grain size distribution curve is plotted on a semilogarithmic paper in which the cumulative percent coarser by weight (as shown in the last column in Figure 9.4) is plotted on the vertical arithmetic scale, while sieve openings diameter is plotted on the horizontal logarithmic scale (Figure 9.5). Alternatively, the cumulative percent finer by weight can be used to plot the curve as well. As shown in Figure 9.5, it is a common practice to have two vertical arithmetic scales: one (usually on the right) with the cumulative percent coarser than the sieve diameter and the other (usually on the left) with the cumulative percent finer than the sieve diameter. Logarithmic scale of the grain size decreases from left to right, but the opposite direction is not uncommon, especially in foreign literature.

Sieve #	Sieve Diameter (mm)		Weight of the Fraction (mm)	Content of the Fraction (%)	Percent Coarser than the Sieve Diameter
6	3.35		45.2	9.04	9.04
10	2.00		95.1	19.02	28.06
12	1.70		204.0	40.80	68.86
16	1.18		97.8	19.56	88.42
20	0.850		39.5	7.90	96.32
30	0.600		13.6	2.72	99.04
40	0.425		2.2	0.44	99.48
60	0.250		1.9	0.38	99.86
100	0.150		1.0	0.19	100.05
Bottom pan	—		0.1	0.38	—

FIGURE 9.4 Schematic presentation of the calculations for plotting the grain size distribution curve for sample 1. Due to weighing imprecision, the final cumulative percent may be different from 100.

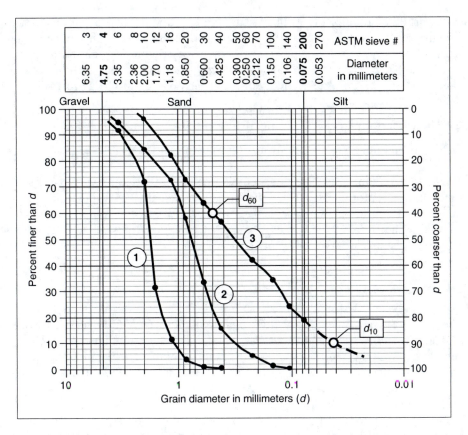

Curve #.	Coefficient of uniformity $U = d_{60}/d_{10}$	Effective grain size $d_e = d_{10}$ (mm)
1	1.67	1.12
2	2.78	0.32
3	11.63	0.043

FIGURE 9.5 Grain size distribution curves for three samples.

The lower limit of the applicability of sieve analysis is for grains retained at ASTM sieve #200 with an opening size (diameter) of 0.075 mm. This is also the boundary between sand and "fines" (silt and clay) according to engineering classification of sediment grain sizes adopted by ASTM. Grains smaller than 0.075 mm are separated and analyzed by several methods and the most common one is the hydrometer method. After removal of the coarser fraction by sieving, the remaining silt and clay particles are suspended in a vertical cylinder of water and left to settle. A hydrometer is submerged in the cylinder at various time intervals to measure the density of the suspension at given elevations. The diameter of the grains is determined from the suspension density, which progressively decreases over time at a given elevation. The relationship is explained by Stoke's law, which, simply put, states that grains of different sizes have different fall velocities in a suspension, which cause them to separate (for a detailed explanation of the hydrometer method, see Lambe, 1951).

If the percentage of the fines in the sample is smaller than 10%, the hydrometer method is usually not performed in common hydrogeological studies. When the fines constitute a significant portion of a sample, as is the case in sample 3, and the hydrometer analysis is not readily available, the grain size curve may be extrapolated following the shape established by the sieve method. For more precise studies (especially for groundwater protection and remediation) and when the hydraulic conductivity of low-permeable porous media is important to determine, both sieve and hydrometer methods should be performed.

9.3.1.1   The Uniformity Coefficient, Effective Grain Size

Grain size distribution curves serve four main purposes in hydrogeological practice.

1. First is to determine the range of grain size present in the sample, i.e., if the sample is more or less uniform. This could be done just visually, or using some numerical parameters from the curve, which also allows classification and comparison of different materials. The slope of the curve, or range of the grains present, is most commonly defined by the uniformity coefficient (U):

$$U = \frac{d_{60}}{d_{10}} \tag{9.22}$$

   where d_{60} is the sieve opening size (diameter) that allows 60% of the sample by weight to pass and therefore retains 40% of the sample; as shown in Figure 9.5, d_{60} is easily determined from the grain size curves (1.87 mm for sample 1, 0.89 mm for sample 2, and 0.5 mm for sample 3), d_{10} is the sieve diameter that allows 10% of the sample to pass, thus retaining 90%. Values of d_{10} for samples 1, 2, and 3 are 1.12, 0.32, and 0.043 mm, respectively (Figure 9.5).

   The lower the uniformity coefficient, the more uniform (well-graded, well-sorted) the material. Large values of U represent a wide range of grain sizes and generally poorly graded porous material. A somewhat arbitrary set criterion for well-sorted material is $U < 5$. In our example, curves 1 and 2 indicate well-sorted (uniform) samples while curve 3 represents a poorly graded material since the coefficient of uniformity is more than 5.

2. Second is to determine the effective grain size, or the grain size that is most important for groundwater flow in a sampled porous medium. On the basis of numerous experimental data, it is now common practice to use d_{10} as the effective grain size ($d_{ef} = d_{10}$), although some other values can still be found in the literature (e.g., d_{17} and d_{20}). The effective grain size expresses the equivalent permeability of a porous material; the smallest 10% of grains, which also fill the pore spaces between larger grains, actually determines the material's permeability. In our example, values of $d_{ef} = d_{10}$ as determined from the grain size distribution curves, are given in Figure 9.5. The value for sample 3 is taken from the extrapolated part of the curve since the hydrometer method was not available for determining the grain size of the fraction passing through #200 sieve (grains smaller than 0.075 mm). Such extrapolation, however, is not recommended since fines (silt and clay) constitute almost 20% of the sample and the slope of the curve may vary considerably.
3. Third is to design a gravel pack around well screens and drains. A detailed explanation of gravel pack calculations is given in Chapter 15.
4. Fourth is to determine the hydraulic conductivity and permeability of the sampled porous medium. The calculation of hydraulic conductivity using grain size curve is explained in the next section. Once the hydraulic conductivity is determined, it is easy to find permeability following the example in Section 9.1 and Section 9.2 (and *vice versa*).

9.3.2 Hydraulic Conductivity

Quite a few empirical formulas for calculating hydraulic conductivity using grain size analysis can be found in today's literature. The vast majority, however, are not dimensionally homogeneous and represent various modifications of the few initial works such as those of Hazen or Kozeny. It should be clearly understood that all these empirical formulas have various limits of application and give just approximate values of hydraulic conductivity for small point samples. Since they are derived for different experimental materials and conditions, it is very common that several formulas applied to the same sample will yield several different values of hydraulic conductivity.

The following dimensionally correct general formula of hydraulic conductivity can be used for expressing most empirical equations currently in use (Vukovic and Soro, 1992):

$$K = \frac{g}{\nu} Cf(n)d_e^2 \tag{9.23}$$

where g is the gravitational acceleration, ν is the kinematic viscosity of the fluid, C is a dimensionless coefficient, which depends on various parameters of the porous medium such as grain shape, structure, and heterogeneity, $f(n)$ is a function of porosity n, and d_e is the effective grain size (usually d_{10}).

Three empirical formulas, all having specific application limits, are expressed in the form of Equation 9.23 and used to calculate the hydraulic conductivity of the three samples (adapted from Vukovic and Soro, 1992).

The Hazen equation is

$$K = \frac{g}{\nu} C_h f(n)d_{10}^2 \tag{9.24}$$

where

$$C_h = 6 \times 10^{-4} \tag{9.24a}$$

$$f(n) = [1 + 10(n - 0.26)] \tag{9.24b}$$

The Hazen equation is applicable for sediments with the coefficient of uniformity less than 5 ($U < 5$) and the effective grain size between 0.1 and 3 mm (0.1 mm $< d_{10} <$ 3 mm). Equation 9.24 is a more complex form of the initial Hazen formulas. The simplified form is

$$K = Cd_{10}^2 \tag{9.24c}$$

where the values of C are arbitrarily taken from the range of 1200 for clean uniform sand to 400 for silty and clayey heterogeneous sand. This formula was later modified by Lange (1958) to account for the sample porosity (Equation 9.24).

The Kozeny equation is

$$K = \frac{g}{\nu} C_k f(n)d_{10}^2 \tag{9.25}$$

where

$$C_k = 8.3 \times 10^{-3} \tag{9.25a}$$

TABLE 9.5
Hydraulic Conductivity (in m/s) of the Three Samples
Calculated Using Three Empirical Formulas

Formula	Sample 1	Sample 2	Sample 3
Hazen	1.23×10^{-2}	5.81×10^{-4}	4.77×10^{-6}
Kozeny	9.09×10^{-3}	2.70×10^{-4}	1.96×10^{-6}
Breyer	1.60×10^{-3}	1.19×10^{-3}	1.56×10^{-5}

$$f(n) = \frac{n^3}{(1-n)^2} \tag{9.25b}$$

The formula is applicable for coarse sand.

The Breyer equation is

$$K = \frac{g}{\nu} C_b d_e^2 \tag{9.26}$$

where

$$C_b = 6 \times 10^{-4} \log \frac{500}{U} \tag{9.26a}$$

The formula does not express the hydraulic conductivity as a function of porosity. It is applicable for $1 < U < 20$, and $0.06 \text{ mm} < d_{10} < 0.6 \text{ mm}$, which makes it useful for analyzing heterogeneous porous media with poorly sorted grains, such as sample 3 in our example.

For comparison, the hydraulic conductivities of all three samples are calculated using all three equations and the results are shown in Table 9.5. Gravity acceleration (g) is assumed to be 9.81 m/s^2 and the kinematic viscosity (ν) is $1.14 \times 10^{-6} \text{ m}^2/\text{s}$ for the groundwater having a temperature of $15°C$. Porosities of samples 1, 2, and 3, as determined in the laboratory, are 0.35, 0.27, and 0.21, respectively.

Table 9.5 shows that the difference between calculated hydraulic conductivities increases with the increasing uniformity coefficient. When choosing hydraulic conductivity that will be used for further calculations, one should always have in mind specific limitations of these formulas. Having in mind the applicability of each equation, the following values should be chosen:

Sample 1 (coarse sand): 9.09×10^{-3} m/s (Kozeny formula)
Sample 2 (medium sand): 5.81×10^{-4} m/s (Hazen formula)
Sample 3 (silty sand): 1.56×10^{-5} m/s (Breyer formula)

REFERENCES

Lambe, T.W., 1951. *Soil Testing for Engineers*, John Wiley & Sons, New York, 165 pp.

Lange, O.K., 1958. *Basic Hydrogeology* (in Russian: *Osnovi gidrogeologii*), Moskovskii Gosudarstvenii Univerzitet, Moscow.

Vukovic, M. and Soro, A., 1992. *Determination of Hydraulic Conductivity of Porous Media from Grain-Size Composition*, Water Resources Publications, Littleton, 86 pp.

10 One-Dimensional Steady-State Flow

10.1 CONFINED HOMOGENEOUS AQUIFER

Calculate groundwater flow rate per unit width between the river and the drainage channel as shown in Figure 10.1. The average elevation of the water surface in the river is 144.60 m, and in the channel 142.20 m. The hydraulic conductivity of the confined intergranular aquifer developed in medium alluvial sands is 3.5×10^{-4} m/s. The hydrogeologic cross section shows the relationship between the aquifer, the overlying silty clay (aquitard), and the underlying dense (impermeable) clay. Also calculate the position of the potentiometric surface at a midpoint between the river and the channel.

10.1.1 GROUNDWATER FLOW PER UNIT WIDTH

On the basis of the hydrogeologic cross-section A-A' in Figure 10.1 and the available information, a scheme for the required calculation is developed as shown in Figure 10.2. The equation of the continuity of groundwater flow is

$$Q = Av \tag{10.1}$$

where Q is the flow rate in cubic meters per second, A is the cross-sectional area of flow in square meters, and v is Darcy's velocity of groundwater in meters per second. The cross-sectional area of flow is

$$A = ab \tag{10.2}$$

where a is the width of the flow in meters and b is the aquifer thickness in meters.

Derivation of the following basic equation of groundwater flow rate per unit width of a confined homogeneous aquifer is given in 1.4.3.1. of Chapter 1 as

$$q = T\frac{h_1 - h_2}{L} \quad (\text{m}^2/\text{s}) \tag{10.3}$$

where q is the flow per unit width, h_1 and h_2 are the hydraulic heads measured in two monitoring wells separated by the distance L, and T is aquifer transmissivity, given as the product of the hydraulic conductivity (K) and the aquifer thickness (b).

From Figure 10.1 it can be seen that the aquifer thickness remains fairly constant and that an average value for b of 3.5 m can be adopted. If, however, aquifer thickness and slope of the impermeable base vary significantly, they should be integrated (see Section 10.4). Substituting the values from the calculation scheme (Figure 10.2) into equation gives

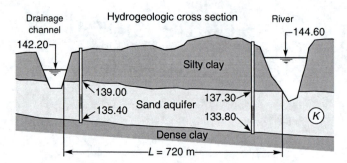

FIGURE 10.1 Hydrogeologic cross section between the river and the drainage channel as determined by field investigations.

$$q = 0.00122 \text{ m}^2/\text{s} \frac{144.60 \text{ m} - 142.20 \text{ m}}{720 \text{ m}} = 0.000004067 \text{ m}^2/\text{s}$$

$$q = 4.07 \times 10^{-6} \text{ m}^2/\text{s}$$

i.e., groundwater flow rate is $4.07 \times 10^{-6} \text{ m}^3/\text{s}$ per meter of aquifer width.

10.1.2 POSITION OF THE HYDRAULIC HEAD

The position of the hydraulic head surface at distance x from the higher hydraulic head (river) using Equation 10.3 is

$$h_x = h_1 - \frac{q}{T}x \qquad\qquad (10.4)$$

Substituting the values from the calculation scheme (Figure 10.2) into Equation 10.4 gives the final solution:

$$h_x = 144.60 \text{ m} - \frac{4.07 \times 10^{-6} \text{ m}^2/\text{s}}{1.22 \times 10^{-3} \text{ m}^2/\text{s}} \times 360 \text{ m} = 143.40 \text{ m}$$

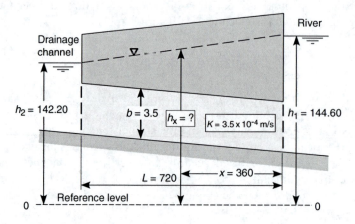

FIGURE 10.2 Calculation scheme for the problem.

It can be seen that the drop of the hydraulic head surface at midpoint is exactly half the difference between the water surface elevations in the river and the channel. Decrease of the hydraulic gradient in a confined, homogeneous, isotropic aquifer is linear, i.e., the potentiometric surface shown in the cross-section is a straight line.

10.2 CONFINED HETEROGENEOUS AQUIFER

Calculate the groundwater flow per unit width of a confined aquifer in the section between two parallel rivers shown in Figure 10.3. The aquifer can be divided into three homogeneous areas with different hydraulic conductivities: $K_1 = 6.2 \times 10^{-5}\,\text{m/s}$, $K_2 = 7.9 \times 10^{-4}\,\text{m/s}$, $K_3 = 4.1 \times 10^{-4}$ m/s. Draw a calculation scheme showing all parameters necessary for the solution. Determine the position of the hydraulic head (potentiometric) surface on the cross section. Determine the "equivalent hydraulic conductivity" and compare it with the average hydraulic conductivity. Also compare the results of groundwater flow calculation using these two hydraulic conductivities with the result obtained using the original calculation scheme.

10.2.1 Unit Groundwater Flow

Groundwater flow rate per unit width of a confined, homogeneous, and isotropic aquifer is given as:

$$q = bK\frac{h_1 - h_2}{L} \tag{10.5a}$$

$$q = T\frac{h_1 - h_2}{L} \tag{10.5b}$$

The total loss of groundwater potential along a flow path is

$$h_1 - h_2 = \Delta h = \frac{qL}{T} \tag{10.6}$$

The loss of hydraulic head in each section is

$$\Delta h_1 = \frac{q_1 L_1}{T_1} \tag{10.7a}$$

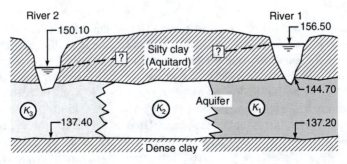

FIGURE 10.3 Hydrogeologic cross section of a confined aquifer with three homogeneous areas of different hydraulic conductivity.

$$\Delta h_2 = \frac{q_2 L_2}{T_2} \tag{10.7b}$$

$$\Delta h_3 = \frac{q_3 L_3}{T_3} \tag{10.7c}$$

The total loss is the sum of the three (all) partial losses:

$$\Delta h = \Delta h_1 + \Delta h_2 + \Delta h_3 \tag{10.8}$$

$$\Delta h = \sum_{i=1}^{i=3} \Delta h_i \tag{10.9}$$

The flow rate through each section is the same because the cross-sectional area of the flow is the same: $q_1 = q_2 = q_3$, which gives the following form for the Equation 10.9:

$$\Delta h = q \sum_{i=1}^{i=3} \frac{L_i}{T_i} \tag{10.10}$$

The unit flow rate (flow rate per unit width of the aquifer) is then

$$q = \frac{\Delta h}{\sum_{i=1}^{i=3} \frac{L_i}{T_i}} \tag{10.11}$$

$$q = \frac{b \Delta h}{\sum_{i=1}^{i=3} \frac{L_i}{K_i}} \tag{10.12}$$

where b is the thickness of the aquifer assumed to be constant as the elevations of the aquifer top and bottom do not change significantly, i.e., they remain roughly parallel to each other and horizontal. If this was not the case, the aquifer thickness would have to be integrated. Substituting information from the calculation scheme (Figure 10.4) into Equation 10.12 gives the final solution:

$$q = \frac{7.50 \text{ m} \times 6.40 \text{ m}}{\frac{290 \text{ m}}{6.2 \times 10^{-5} \text{ m/s}} + \frac{320 \text{ m}}{7.9 \times 10^{-4} \text{ m/s}} + \frac{210 \text{ m}}{4.1 \times 10^{-4} \text{ m/s}}}$$

$$q = 8.58 \times 10^{-6} \text{ m}^2/\text{s}$$

Note: In Equation 10.5 and Equation 10.6 one can use either actual values for the hydraulic heads in the rivers ($h_1 = 56.5$ m; $h_2 = 50.1$ m) or some "new" values if a changing of the reference level is preferred. For example, the reference level can be placed at the base of the aquifer below the River 1 because it is approximately horizontal. This would give the following values for the hydraulic heads: $h_1 = 9.3$ m; $h_2 = 12.9$ m. In both cases, the head difference is the same: $\Delta h = 6.4$ m. In the case of unconfined aquifers, one should be very careful in choosing a reference level because the related equation for calculating flow rate is not linear (see Section 10.5). In other words, the hydraulic head used for the calculation should correspond to the actual saturated thickness of the aquifer; using water table elevations (heights above the sea level) for the hydraulic heads will yield incorrect results.

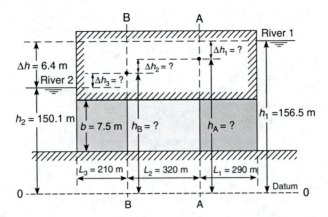

FIGURE 10.4 Calculation scheme with information necessary for the solution.

10.2.2 Position of the Hydraulic Head

The position (elevation) of the hydraulic head surface at verticals A and B is calculated using Equation 10.7a–c:

$$\Delta h_1 = \frac{q_1 L_1}{T_1} = 5.35 \text{ m}$$

$$\Delta h_2 - \frac{q_2 L_2}{T_2} - 0.46 \text{ m}$$

$$\Delta h_3 = \frac{q_3 L_3}{T_3} = 0.59 \text{ m}$$

$$h_A = h_1 - \Delta h_1 = 151.15 \text{ m}$$

$$h_B = h_A - \Delta h_2 = 150.69 \text{ m}$$

Figure 10.5 is a schematic presentation of the solution. The largest hydraulic loss of groundwater flow is in the least permeable section of the aquifer (the slope of the hydraulic head is the steepest in the section with the hydraulic conductivity K_1). Accordingly, this loss is the smallest in the central, most permeable section (K_2).

Note that, in general, the position (slope) of the hydraulic head surface is dependent on the aquifer transmissivity. It means that, in addition to hydraulic conductivity, the thickness of the aquifer is also important. For the same aquifer material, the slope will be steeper where the aquifer is thin and less steep where the aquifer is thicker. This is true for both confined and unconfined aquifers.

10.2.3 Equivalent and Average Hydraulic Conductivities

The equivalent hydraulic conductivity represents the aquifer as homogeneous throughout the calculation domain. It gives the same value for the rate of groundwater flow as the varying conductivities. However, the distribution (slope) of the hydraulic head in such a case is uniform and does not reflect the reality. Introducing the equivalent hydraulic conductivity (K_{eqv}) into the equation for the unit discharge rate (Equation 10.5a) gives

$$q = bK_{eqv}\frac{h_1 - h_2}{L} \tag{10.13}$$

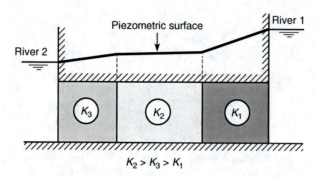

$$K_2 > K_3 > K_1$$

FIGURE 10.5 Schematic presentation of the position of the hydraulic head in the aquifer. Note that the linear loss of the hydraulic head is the largest in the least permeable section of the aquifer (K_1).

The unit discharge rate (discharge rate per unit width) is also given by Equation 10.12 as

$$q = \frac{b\Delta h}{\sum\limits_{i=1}^{i=3} \dfrac{L_i}{K_i}} \tag{10.14}$$

Relating Equation 10.13 and Equation 10.14 gives the final expression for the equivalent hydraulic conductivity:

$$K_{eqv} = \frac{L}{\sum\limits_{i=1}^{i=3} \dfrac{L_i}{K_i}} \tag{10.15}$$

Substitution of the values from the calculation scheme into Equation 10.15 leads to the final solution for the equivalent hydraulic conductivity:

$$K_{eqv} = 1.47 \times 10^{-4} \text{ m/s}$$

The average (mean) hydraulic conductivity is given as

$$K^* = \frac{\sum\limits_{i=1}^{i=3} K_i}{n} \tag{10.16}$$

or, by using actual values,

$$K^* = \frac{6.2 \times 10^{-5} \text{ m/s} + 7.9 \times 10^{-4} \text{ m/s} + 4.1 \times 10^{-4} \text{ m/s}}{3}$$

$$K^* = 4.2 \times 10^{-4} \text{ m/s}$$

The unit groundwater flow rate, if calculated with the mean hydraulic conductivity, is

$$q^* = bK^* \frac{\Delta h}{L} \tag{10.17}$$

or, after substituting numeric values,

$$q^* = 2.46 \times 10^{-5} \text{ m}^2/\text{s}$$

This result is considerably higher than the real value of the groundwater flow rate per unit width that is calculated using all three hydraulic conductivities (Equation 10.12).

10.3 CONFINED ONE-DIMENSIONAL STEADY-STATE FLOW, AQUIFER WITH FORKS

Calculate groundwater flow per unit width downgradient from the monitoring well 3 (MW3) that is shown on the hydrogeologic cross section in Figure 10.6. Hydraulic conductivity of the confined sand aquifer is 3.7×10^{-5} m/s and it may be considered homogeneous (uniform) throughout the study area. Water levels registered in the monitoring wells MW1, MW2, and MW3 are 144.6, 145.0, and 142.4 m above sea level, respectively. The general slope of the "upper fork" of the aquifer is 20°, and the "lower fork" slopes at approximately 15°. Monitoring wells MW1 and MW2 are clustered, i.e., very close to each other (about 1 m apart). The distance between the MW3 and the MW1 and MW2 cluster is 170 m. Draw a calculation scheme for solving the problem.

10.3.1 UNIT GROUNDWATER FLOW

Groundwater flow per unit width in a confined aquifer of uniform thickness is a product of the aquifer transmissivity and the hydraulic gradient:

$$q = T \frac{h_1 - h_2}{L} \tag{10.18}$$

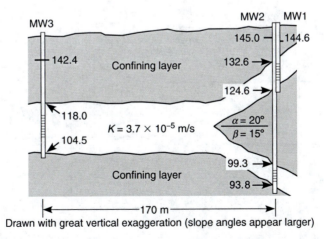

Drawn with great vertical exaggeration (slope angles appear larger)

FIGURE 10.6 Simplified hydrogeologic cross section of the confined aquifer with forks. Note that, because of the vertical exaggeration, sloping angles α and β appear bigger than their true values of 20° and 15°, respectively.

where h_1 is a higher hydraulic head and h_2 is a lower hydraulic head measured in a homogeneous part of the aquifer along the same flow line, and L is the distance between these two heads.

The aquifer shown in Figure 10.6 has three sections: (1) the horizontal portion, (2) the "upper fork," and (3) the "lower fork," all with different uniform thickness, and with only one known hydraulic head per section. Calculation scheme in Figure 10.7 shows all the elements needed to determine the flow downgradient from MW3. According to the general hydraulic equation of the continuity of flow, the unit groundwater flow through the horizontal section of the confined aquifer remains constant becasue groundwater neither enters nor leaves the aquifer. The same is true for the flows in two forks which combine to generate the flow in the horizontal section:

$$q_1 + q_2 = q_3 \tag{10.19}$$

To calculate all three unit flows, the following elements should be determined for each section:

- One additional hydraulic head
- Distance between this new calculated head and the known head
- Thickness of the aquifer

As shown in Figure 10.7, the hydraulic head h_x is common for all three sections of the aquifer: it is a lower (downgradient) head for both forks and, at the same time, it is a higher (upgradient) head for the horizontal section. Its position is determined by the distance x, i.e., the distance between the MW1–MW2 cluster and the beginning of the horizontal section where the two forks meet. The unit groundwater flows for the three sections of the aquifer are given with the following equations:

Upper fork:

$$q_1 = T_1 \frac{h_1 - h_x}{x} \tag{10.20}$$

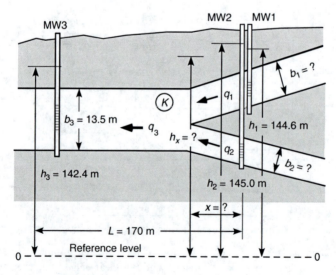

FIGURE 10.7 Calculation scheme for the problem.

Lower fork:

$$q_2 = T_2 \frac{h_2 - h_x}{x} \tag{10.21}$$

Horizontal section:

$$q_3 = T_3 \frac{h_x - h_3}{L - x} \tag{10.22}$$

where T_1, T_2, and T_3 are the transmissivities determined from the aquifer hydraulic conductivity and the aquifer thickness ($T_1 = Kb_1$; $T_2 = Kb_2$; $T_3 = Kb_3$). The thicknesses of the three sections, and the position of the unknown hydraulic head h_x (distance x), can be easily found by performing simple trigonometric calculations. Once these values are found, Equation 10.19 through Equation 10.22 can be solved as a system of four equations with four unknowns (q_1, q_2, q_3, and h_x).

10.3.2 AQUIFER THICKNESS

In calculations of groundwater flow rate and aquifer transmissivity it is very important to use true thickness of the aquifer and not its apparent thickness (which is often the case in practice). Aquifer true thickness is the perpendicular distance between its top and bottom, whereas aquifer apparent thickness is the vertical distance between its top and bottom as determined by exploratory drilling. If the aquifer sloping angle is equal to or less than $10°$, it is a common practice to use the apparent thickness as the error introduced is only 1.5% or less. However, for steeper slopes the errors can be significant. For example, the angles of $20°$, $30°$, and $40°$ degrees would yield errors of 6.0%, 13.4%, and 23.4%, respectively, if the apparent thickness is used instead of the aquifer true thickness. Figure 10.8 shows the trigonometric relationship between the angle of slope (α), the aquifer true thickness (b), and the aquifer apparent thickness (Δh).

True thickness of the "upper fork" is

$$\begin{aligned} b_1 &= \Delta h_1 \cos \alpha \\ b_1 &= (132.6 \text{ m} - 124.6 \text{ m}) \times \cos 20° = 7.5 \text{ m} \end{aligned} \tag{10.23}$$

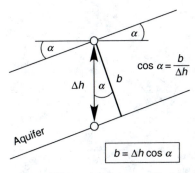

FIGURE 10.8 Determination of the aquifer true thickness (b) from the angle of slope (α) and the apparent thickness (Δh).

True thickness of the "lower fork" is

$$b_2 = \Delta h_2 \cos \beta$$
$$b_1 = (99.3 \text{ m} - 93.8 \text{ m}) \times \cos 15° = 5.3 \text{ m} \qquad (10.24)$$

The thickness of the horizontal section, as determined from the MW3 log, is

$$118.0 \text{ m} - 104.5 \text{ m} = 13.5 \text{ m}$$

10.3.3 Position of the Hydraulic Head (h_x)

Figure 10.9 shows the elements needed for the trigonometric determination of the distance x. Combining basic equations for tangents of angles α and β leads to

$$x \tan \alpha + x \tan \beta = y + z \qquad (10.25)$$

and

$$x = \frac{y + z}{\tan \alpha + \tan \beta} \qquad (10.26)$$

Substitution of the known values into Equation 10.26 gives the distance x:

$$x = \frac{25.3 \text{ m}}{\tan 20° + \tan 15°} = \frac{25.3 \text{ m}}{0.27 + 0.36} = 40 \text{ m}$$

10.3.4 Hydraulic Head (h_x)

After finding the thickness of each of the three sections, and the position of the hydraulic head h_x, its value is easily determined by combining Equation 10.19 through Equation 10.22:

$$Kb_1 \frac{h_1 - h_x}{x} + Kb_2 \frac{h_2 - h_x}{x} = Kb_3 \frac{h_x - h_3}{L} \qquad (10.27)$$

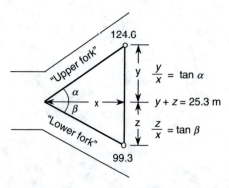

FIGURE 10.9 Elements for trigonometric determination of the distance x (position of the hydraulic head h_x).

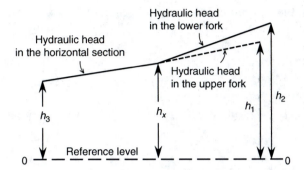

FIGURE 10.10 Schematic presentation of the position of the hydraulic head in all three sections of the aquifer.

Taking out the hydraulic conductivity (K) and solving for h_x as the only unknown gives:

$$h_x = \frac{b_1 L h_1 + b_2 L h_2 + b_3 x h_3}{b_1 L + b_2 L + b_3 x}$$ (10.28)

$$h_x = 144.3 \text{ m}$$

Figure 10.10 is a schematic presentation of the position of the hydraulic head in all three sections of the aquifer (note that all lines are straight indicating the linear decrease of the hydraulic head characteristic for confined aquifers).

10.3.5 UNIT GROUNDWATER FLOWS

Unit groundwater flow downgradient from MW3, or flow in the aquifer's horizontal section (which is the same), is determined by Equation 10.22:

$$q_3 = T_3 \frac{h_x - h_3}{L} = K b_3 \frac{h_x - h_3}{L}$$

$$q_3 = 3.7 \times 10^{-5} \text{ m/s} \cdot 13.5 \text{ m} \cdot \frac{144.3 \text{ m} - 142.4 \text{ m}}{170 \text{ m}}$$

$$q_3 = 5.58 \times 10^{-6} \text{ m}^2/\text{s}$$

Similarly, unit groundwater flows (i.e., groundwater flow rates per 1 m of aquifer width) in the upper and lower forks are calculated using Equation 10.20 and Equation 10.21, respectively:

$$q_1 = 3.7 \times 10^{-5} \text{ m/s} \cdot 7.5 \text{ m} \cdot \frac{144.6 \text{ m} - 144.3 \text{ m}}{40 \text{ m}}$$

$$q_1 = 2.08 \times 10^{-6} \text{ m}^2/\text{s}$$

and

$$q_2 = 3.7 \times 10^{-5} \text{ m/s} \cdot 5.3 \text{ m} \cdot \frac{145.0 \text{ m} - 144.3 \text{ m}}{40 \text{ m}}$$

$$q_2 = 3.43 \times 10^{-6} \text{ m}^2/\text{s}$$

As mentioned earlier, the unit groundwater flow downgradient from MW3 is the sum of unit groundwater flows in two forks:

$$q_3 = q_1 + q_2$$
$$q_3 = 2.08 \times 10^{-6} \ \text{m}^2/\text{s} + 3.43 \times 10^{-6} \ \text{m}^2/\text{s} = 5.51 \times 10^{-6} \ \text{m}^2/\text{s}$$

This result is very close to the result obtained using Equation 10.22 and the hydraulic heads h_x and h_3.

10.4 CONFINED ONE-DIMENSIONAL FLOW, AQUIFER WITH VARYING THICKNESS

Calculate unit groundwater flow in a confined aquifer that is shown in Figure 10.11. Also find the hydraulic head surface in the monitoring well MW2. The hydraulic conductivity of the aquifer, which can be considered as homogeneous and isotropic, is 5.6×10^{-5} m/s.

As can be seen in Figure 10.11, the aquifer has a varying thickness and therefore a varying transmissivity. This means that a simple equation describing one-dimensional groundwater flow in a confined aquifer (e.g., Equation 10.3) cannot be applied directly as the aquifer thickness has to be integrated. In many cases, the varying thickness can be described with a simple linear function or a combination of several linear functions with different slopes. For the situation shown in Figure 10.11, the linear function is easily derived by setting the coordinate system $X-Y$ so that it corresponds to the length-aquifer thickness system in the field $(L-b)$ as shown in Figure 10.12.

The equation of the straight line in the $X-Y$ coordinate system is

$$y = ax + b \tag{10.29}$$

where a is the slope of the line $(a = \Delta y/\Delta x)$ and b is the Y-axis intercept. The corresponding equation of the top of the confined aquifer is

$$b = \frac{b_3 - b_1}{L}x + b_1 \tag{10.30}$$

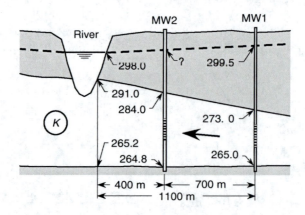

FIGURE 10.11 Schematic hydrogeologic cross section of a confined aquifer with varying thickness.

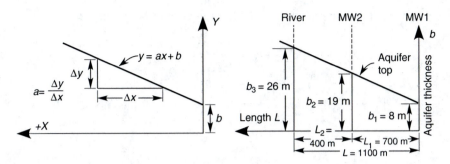

FIGURE 10.12 Comparison of the X–Y coordinate system (left) and the Length–Thickness coordinate system (right) used to derive equation of the varying thickness of the confined aquifer shown in Figure 10.11.

where $(b_3 - b_1)/L$ is the slope and b_1 is the intercept. The bottom of the aquifer can be considered horizontal and the reference level is placed at the elevation 265.0 m (which is the elevation of the contact between the sand aquifer and dense clay at monitoring well MW1).

The general equation of the unit groundwater flow (groundwater flow per unit width of the aquifer) is given as the product of the aquifer thickness (b) and the flow velocity (v):

$$q = b \cdot v \tag{10.31}$$

Groundwater velocity (Darcy's velocity) is

$$v = -Ki \tag{10.32}$$

where K is the hydraulic conductivity of the aquifer material in meters per second, and i is the hydraulic gradient. The minus sign denotes that Darcy's velocity is oriented in the direction of the hydraulic head decrease, i.e., downgradient, and it should be used only when the equation is expressed in a differential form. In other words, the hydraulic gradient is an infinitesimally small change of hydraulic head (dh) along an infinitesimally small length of the flow path (dx):

$$i = \frac{dh}{dx} \tag{10.32a}$$

In the case of one-dimensional groundwater flow in an isotropic, homogeneous, confined aquifer, the flow rate per unit width is then

$$q = -bK\frac{dh}{dx} \tag{10.33}$$

The unit flow in the aquifer with the changing thickness, as shown in Figure 10.11 and Figure 10.12, is expressed by combining Equation 10.30 and Equation 10.33:

$$q = -\left(\frac{b_3 - b_1}{L}x + b_1\right) \cdot K\frac{dh}{dx} \tag{10.34}$$

To solve the differential Equation 10.34, the variables x and h first have to be separated:

$$-dh = \frac{q}{K} \cdot \frac{dx}{\frac{b_3 - b_1}{L} x + b_1} \tag{10.35}$$

Equation 10.35 is then solved for the known boundary conditions: for $x = 0$ the hydraulic head is h_1, and for $x = L$ the hydraulic head is h_2:

$$-\int_{h_1}^{h_3} dh = \frac{q}{K} \cdot \int_{0}^{L} \frac{dx}{\frac{b_3 - b_1}{L} x + b_1} \tag{10.36}$$

The integral on the left is an elementary integral, and the integral on the right is solved knowing that, in general notification (see also Appendix B):

$$\int \frac{dx}{ax + b} = \frac{1}{a} \ln (ax + b) + C$$

The solution for Equation 10.36 is

$$h_1 - h_3 = \frac{q}{K} \cdot \frac{1}{\frac{b_3 - b_1}{L}} \cdot \left[\ln \left(\frac{b_3 - b_1}{L} L + b_1 \right) - \ln \left(\frac{b_3 - b_1}{0} 0 + b_1 \right) \right] \tag{10.37}$$

$$h_1 - h_3 = \frac{q}{K} \cdot \frac{L}{b_3 - b_1} \cdot \ln \frac{b_3}{b_1} \tag{10.38}$$

and the unit flow rate is then

$$q = K \frac{h_1 - h_3}{L} \cdot \frac{b_3 - b_1}{\ln \frac{b_3}{b_1}} \tag{10.39}$$

The numeric value of the unit flow is found by inserting numbers from the calculation scheme shown in Figure 10.12 into Equation 10.39:

$$q = 5.6 \times 10^{-5} \text{ m/s} \cdot \frac{299.5 \text{ m} - 298.0 \text{ m}}{1100 \text{ m}} \cdot \frac{26 \text{ m} - 8 \text{ m}}{\ln \frac{26 \text{ m}}{8 \text{ m}}}$$

$$q = 1.16 \times 10^{-6} \text{ m}^2/\text{s}$$

Note that the volumetric flow rate is obtained when the unit flow is multiplied by aquifer width of interest. For example (see Figure 1.44 in Chapter 1), the total flow rate through an 800 m wide portion of the aquifer would be

$$Q = 1.16 \times 10^{-6} \text{ m}^2/\text{s} \cdot 800 \text{ m} = 9.28 \times 10^{-4} \text{ m}^3/\text{s}$$

10.4.1 POSITION OF THE HYDRAULIC HEAD

Hydraulic head (potentiometric) surface at MW2 is found by solving Equation 10.35 with the following boundary conditions (see Figure 10.12): for $x = 0$ the hydraulic head is h_1, and for $x = L_1$ the hydraulic head is h_2:

TABLE 10.1
Hydraulic Heads between MW1 and the River Calculated Using Equation 10.42

Distance from MW1 (m)	0	200	400	600	700	800	1000	1100
Aquifer thickness (m)	8	11.3	14.5	17.9	19	21	24.3	26
Hydraulic head (m asl)	299.5	299.1	298.7	298.5	298.4	298.3	298.1	298

$$-\int_{h_1}^{h_2} dh = \frac{q}{K} \cdot \int_0^{L_1} \frac{dx}{\frac{b_2 - b_1}{L_1} x + b_1} \qquad (10.40)$$

$$h_2 = h_1 - \frac{q}{K} \cdot \frac{L_1}{b_2 - b_1} \cdot \ln \frac{b_2}{b_1} \qquad (10.41)$$

Inserting values from the calculation scheme shown in Figure 10.12 gives

$$h_2 = 299.5 \text{ m} - \frac{1.16 \times 10^{-6} \text{ m}^2/\text{s}}{5.6 \times 10^{-5} \text{ m/s}} \cdot \frac{700 \text{ m}}{19 \text{ m} - 8 \text{ m}} \cdot \ln \frac{19 \text{ m}}{8 \text{ m}}$$

$$h_2 = 298.36 \text{ m}$$

The position of the hydraulic head at various distances from MW1 (h_x), and for known aquifer thickness at those distances (b_x), can be calculated using the following general equation based on Equation 10.41:

$$h_x = h_1 - \frac{q}{K} \cdot \frac{L_1}{b_x - b_1} \cdot \ln \frac{b_x}{b_1} \qquad (10.42)$$

Table 10.1 shows the hydraulic heads calculated for several verticals between MW1 and the river and used to draw the hydraulic head in Figure 10.13.

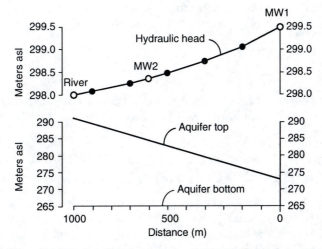

FIGURE 10.13 Position of the hydraulic head between MW1 and the river. Note a decrease in the hydraulic head gradient with the increasing thickness of the aquifer.

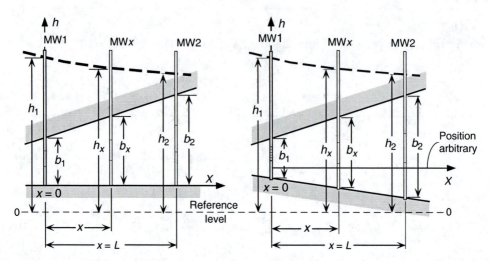

FIGURE 10.14 Schematic presentation of two confined aquifers with linearly varying thickness. In either case, the hydraulic head h_x can be calculated using Equation 10.43.

If two hydraulic heads (h_1 and h_2) in a confined aquifer that has a linearly varying thickness are known, the position of the hydraulic head at any point between these two heads can be found using the following equation:

$$h_x = h_1 - (h_1 - h_2) \cdot \frac{L_x}{L} \cdot \frac{b_2 - b_1}{b_x - b_1} \cdot \frac{\ln \dfrac{b_x}{b_1}}{\ln \dfrac{b_2}{b_1}} \tag{10.43}$$

Equation 10.43 is obtained by combining Equation 10.14 with a general form of Equation 10.39 in which both heads are known. It is useful and simple to use because one need not know the hydraulic conductivity and the unit flow in the aquifer. As shown in Figure 10.14, Equation 10.43 is valid for confined aquifers where the top changes linearly and the bottom is horizontal, as well as for aquifers where both the top and the bottom can be described with linear functions. This also means that the unit flow in both cases can be calculated using Equation 10.39 as the aquifer thickness function used in the integral in Equation 10.36 is the same:

$$b = \frac{b_1 - b_2}{L} x + b_1 \tag{10.44}$$

Note that X-axis can be placed anywhere along the vertical, but its origin should be at the first known hydraulic head (e.g., monitoring well MW1 as shown in Figure 10.14).

10.5 UNCONFINED HOMOGENEOUS AQUIFER WITH AND WITHOUT INFILTRATION, DRAINAGE GALLERY

Construction of a drainage gallery and an infiltration channel for aquifer flushing is proposed for the cleanup of the contaminant plume approaching the river as shown in Figure 10.15. The operating (pumping) water level in the gallery is designed at 275.3 m above mean sea level (amsl). Perform a preliminary calculation of the unit groundwater inflow (inflow per unit width) into the drainage gallery during the month of May, if the effective aquifer recharge in

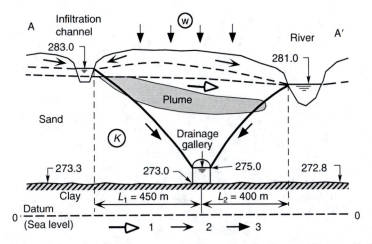

FIGURE 10.15 Cross-sectional view of the proposed alternative for the aquifer remediation. (1) Direction of groundwater flow in natural conditions. (2) Possible direction of groundwater flow between the infiltration channel and the river when the drainage gallery is not operating. (3) Direction of groundwater flow between the infiltration channel and the river when the drainage gallery is operating.

May is estimated to be 2.5 in. (i.e., 35% of the total May precipitation). The average elevation of the water surface in the river in May is 281.7 m amsl, and the water level in the infiltration channel is projected at 283.0 m amsl. The hydraulic conductivity of the aquifer material is 2.1×10^{-4} m/s. Other elements necessary for the calculation, as determined by field investigations, are shown in the hydrogeological cross section in Figure 10.15.

10.5.1 UNIT GROUNDWATER FLOW UNDER INFLUENCE OF INFILTRATION

Groundwater flow per unit width in an unconfined aquifer resting on a horizontal impermeable base and under the influence of infiltration, can be determined by analyzing an elementary aquifer volume (prism). Figure 1.45 and equations in Section 1.3.3.2 illustrate derivation of the following basic equation of unconfined steady-state flow under the influence of infiltration (Equation 1.44 in Section 1.4.3.2 in Chapter 1):

$$h^2 = h_1^2 - (h_1^2 - h_2^2)\frac{x}{L} + \frac{w}{K}(L - x)x \tag{10.45}$$

Equation 10.45 enables the calculation of the hydraulic head (water table elevation) for any given position between two points with known saturated thickness. Water table elevations at several points between the infiltration channel and the river, when the proposed drainage gallery is not operating, are calculated using Equation 10.45 and the results are shown in Table 10.2.

TABLE 10.2
Water Table Elevations between the Infiltration Channel and the River When the Aquifer Is Under the Influence of Infiltration

Distance from channel (x)	100	200	300	450	600	700	800
Saturated thickness (h)	10.21	10.31	10.29	9.95	9.57	9.07	8.40
Water table elevation (m)	283.21	283.31	283.29	282.95	282.57	282.07	281.40

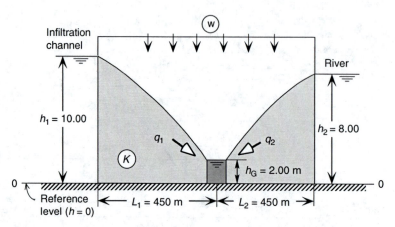

FIGURE 10.16 Calculation scheme for the problem.

The infiltration rate, which has dimension of velocity, was calculated from the effective precipitation data for May:

$$w = \frac{P_e}{t} = \frac{2.5 \text{ in.}}{31 \text{ d}} = \frac{0.0635 \text{ m}}{2.6784 \times 10^6 \text{ s}} = 2.371 \times 10^{-8} \text{ m/s}$$

The saturated thickness (h) was calculated from the base of the aquifer which is assumed to be horizontal and chosen for the new reference level (see Figure 10.15). The water table elevation or elevation above sea level (asl) was obtained by adding calculated saturated thickness to the elevation of the new reference level (273.0 m asl) placed at the bottom of the drainage gallery as shown in Figure 10.16.

If there is no gain of water due to infiltration of precipitation, or loss of water due to evapotranspiration, Equation 10.45 reduces to

$$h^2 = h_1^2 - (h_1^2 - h_2^2)\frac{x}{L} \tag{10.46}$$

Table 10.3 shows water table elevations at the same points as in Table 10.2 when there is no infiltration of precipitation, and the proposed drainage gallery is not operating; Equation 10.46 is used in this case.

Groundwater flow per unit width at any distance x from the coordinate beginning (i.e., the origin, where $x = 0$) is given as

$$q = -Kh\frac{dh}{dx} \tag{10.47}$$

TABLE 10.3
Water Table Elevations between the Infiltration Channel and the River When the Aquifer Is Not Under the Influence of Infiltration

Distance from channel (x)	100	200	300	450	600	700	800
Saturated thickness (h)	9.78	9.57	9.34	9.00	8.64	8.39	8.13
Water table elevation (m)	282.78	282.57	282.34	282	281.64	281.39	281.13

Differentiation of Equation 10.47 leads to

$$h\frac{dh}{dx} = -\frac{h_1^2 - h_2^2}{2L} - \frac{w}{K}\left(x - \frac{L}{2}\right) \tag{10.48}$$

The equation for the unit groundwater flow in an unconfined aquifer under the influence of infiltration is obtained by combining Equation 10.47 and Equation 10.48:

$$q = K\frac{h_1^2 - h_2^2}{2L} + w\left(x - \frac{L}{2}\right) \tag{10.49}$$

The unit groundwater flow at the point of contact between the infiltration channel and the aquifer is calculated for the vertical where $x = 0$ (the X-axis starts at this contact and has positive direction toward the river):

$$q_0 = K\frac{h_1^2 - h_2^2}{2L} - w\frac{L}{2} \tag{10.50}$$

$$q_0 = 2.1 \times 10^{-4} \text{ m/s}\frac{10^2 \text{ m} - 8^2 \text{ m}}{2(450 \text{ m} + 400 \text{ m})} - 2.371 \times 10^{-8} \text{ m/s}\frac{850 \text{ m}}{2}$$

$$q_0 = -5.64 \times 10^{-6} \text{ m}^2/\text{s}$$

The negative sign indicates that groundwater flow is in the negative direction of the X-axis, i.e., toward the infiltration channel. Unit groundwater flow entering the river is calculated for the vertical where $x = L$ (or 850 m):

$$q_L = K\frac{h_1^2 - h_2^2}{2L} + w\frac{L}{2}$$
$$q_L = 1.453 \times 10^{-5} \text{ m}^2/\text{s} \tag{10.51}$$

From the above analysis it can be seen that both the infiltration channel and the river are receiving groundwater flow generated by the infiltration of precipitation due to formation of a local groundwater divide (mound). The total groundwater flow per unit width in the section between the channel and the river is therefore the sum of the absolute values of two flows, and equals the rate of infiltration multiplied by the distance between the channel and the river:

$$q_{\text{total}} = q_0 + q_L = |-5.64 \times 10^{-6} \text{ m}^2/\text{s}| + |1.453 \times 10^{-5} \text{ m}^2/\text{s}| \tag{10.52}$$

$$q_{\text{total}} = 2.017 \times 10^{-5} \text{ m}^2/\text{s}$$

$$q_{\text{total}} = wL = 2.371 \times 10^{-8} \text{ m/s} \times 850 \text{ m} = 2.015 \times 10^{-5} \text{ m}^2/\text{s} \tag{10.53}$$

10.5.2 GROUNDWATER FLOW TOWARD DRAINAGE GALLERY (DRAIN) RESTING ON HORIZONTAL BASE

The unit groundwater inflow into the drainage gallery, as shown in Figure 10.16, is the sum of flows from two directions. From the infiltration channel direction, the flow is generated by the infiltration of precipitation and by the induced recharge from the channel due to the reversed hydraulic gradient:

$$q_1 = K \frac{h_1^2 - h_G^2}{2L_1} + w \frac{L_1}{2}$$

$$q_1 = 2.773 \times 10^{-5} \text{ m}^2/\text{s}$$

(10.54)

Similarly, the unit inflow from the river direction is generated by the infiltration of precipitation and by the induced recharge from the river:

$$q_2 = K \frac{h_2^2 - h_G^2}{2L_2} + w \frac{L_2}{2}$$

$$q_2 = 2.049 \times 10^{-5} \text{ m}^2/\text{s}$$

(10.55)

The total unit inflow into the drainage gallery, from both directions, is

$$q_{\text{total}} = q_1 + q_2 = 4.822 \times 10^{-5} \text{ m}^2/\text{s}$$

(10.56)

The calculated groundwater inflow per unit width of the drainage gallery can be used to estimate total inflow for a desired length. For example, for a 200 m long gallery, the total inflow would be

$$Q = q_{\text{total}} \times 200 \text{ m} = 9.644 \times 10^{-3} \text{ m}^3/\text{s}$$

(10.57)

11 One-Dimensional Transient Flow

11.1 SUDDEN CHANGE AT THE BOUNDARY, AQUIFER PARAMETERS

Find the hydrogeologic parameters of the unconfined aquifer shown in Figure 11.1. The aquifer is in a direct hydraulic connection with a large perennial river. The fluctuations of the water table are recorded in a monitoring well (MW) 30 m from the river and are shown in Figure 11.2. The hydrograph shows a sudden rise in the river stage and a consequent rise in the aquifer water table recorded at MW1. From the available data, find the hydraulic conductivity of the aquifer assuming it is homogeneous and isotropic. Also find what would be the rise in the water table 8 days into the flood at 100 m from the river.

The derivation of the general equation of transient (time-dependent) groundwater flow in two dimensions is given in Section 1.4.3.3, including its solution for the boundary and initial conditions applicable to this problem. The solution, in case of one-dimensional (cross-sectional) flow, is "borrowed" from the theory of heat transfer and has the following form:

$$\Delta H(x,t) = \Delta H_0 \, \text{erfc}(\lambda) \tag{11.1}$$

where $\Delta H(x,t)$ is a function of space and time, ΔH_0 is the change in hydraulic head at the boundary (river), and $\text{erfc}(\lambda)$ is the complementary error function given as $\text{erfc} = 1 - \text{erf}(\lambda)$, Error function (erf) is an integral of probability and is readily found in tables (see Appendix D). For our solution of transient one-dimensional flow with a sudden change at a boundary, it has the following form:

$$\text{erf}(\lambda) = \frac{2}{\sqrt{\pi}} \int_0^\lambda e^{-\lambda^2} d\lambda \tag{11.2}$$

where λ is a parameter given as

$$\lambda = \frac{x}{2\sqrt{at}} \tag{11.3}$$

where x is the distance from the boundary (river) at which the change in groundwater table (ΔH) is recorded, a is the hydraulic diffusivity, and t is the time of change. The hydraulic diffusivity is given as

$$a = \frac{K h_{\text{av}}}{S} = \frac{T}{S} \tag{11.4}$$

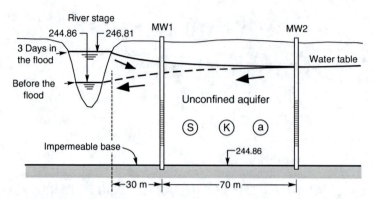

FIGURE 11.1 Hydrogeologic section of an unconfined aquifer in direct hydraulic connection with the river showing the water table position before the flood and 3 d into the flood.

The average aquifer thickness needed to find the hydraulic diffusivity is calculated as follows (see Figure 11.3):

$$h_{av} = \frac{(H + \Delta H_0) + (H_x + \Delta H)}{2} \tag{11.5}$$

Equation 11.1 can be used to predict a rise in water table at some distance x from the boundary after a sudden change of river stage such as during floods, or it can be used to estimate aquifer parameters if water levels in the river and the aquifer are recorded simultaneously. If two monitoring wells are placed along a stream line, the one closer to the boundary and having a larger change can act as a boundary (ΔH_0) thus avoiding measurements of the river stage.

The complementary error function for our case is found from Equation 11.1 and from a knowledge of the values of ΔH and ΔH_0 shown in Figure 11.3:

$$\text{erfc}(\lambda) = \frac{\Delta H(\text{at 30 m, after 3 d})}{\Delta H_0}$$

$$\text{erfc}(\lambda) = \frac{0.94 \text{ m}}{1.95 \text{ m}} = 0.48 \tag{11.6}$$

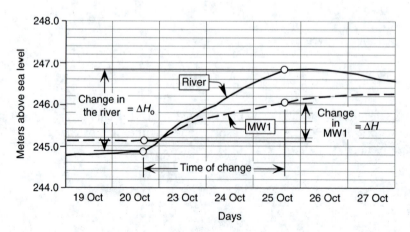

FIGURE 11.2 Hydrographs of the river stage and the water table recorded in the monitoring well MW1.

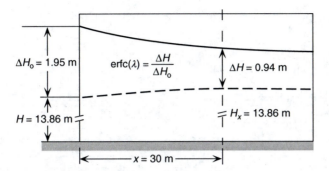

FIGURE 11.3 Calculation scheme for solving the problem.

The parameter λ is found from Table D.1 for the known value of erfc(λ): $\lambda = 0.50$. The hydraulic diffusivity is then calculated from Equation 11.3:

$$a = \frac{x^2}{4\lambda^2 t} \tag{11.7}$$

$$a = \frac{(30 \text{ m})^2}{4 \cdot (0 \cdot 50)^2 \cdot 3 \cdot 86400 \text{ s}} = \frac{900 \text{ m}^2}{259200 \text{ s}} = 3.47 \times 10^{-3} \text{ m}^2/\text{s}$$

If the specific yield of the aquifer is known (e.g., based on laboratory analyses of the effective porosity of aquifer samples), the aquifer transmissivity can be found using Equation 11.4 and vice versa. If both parameters are unknown, it is better to assume a value for the specific yield and find the transmissivity since the former has a smaller range and is easier to estimate. For example, if our aquifer consists of uniform medium-size sand, it is reasonable to assume specific yield of 0.25. The transmissivity is calculated from Equation 11.4:

$$T = aS = 3.47 \times 10^{-3} \text{ m}^2/\text{s} \cdot 0.25 = 8.675 \times 10^{-4} \text{ m}^2/\text{s}$$

The hydraulic conductivity then becomes

$$K = \frac{T}{h_{av}} = \frac{8 \cdot 675 \times 10^{-4} \text{ m}^2/\text{s}}{15.46 \text{ m}} = 5.61 \times 10^{-5} \text{ m/s}$$

where the aquifer average thickness is estimated using Equation 11.5 (see also Figure 11.3):

$$h_{av} = \frac{(H + \Delta H_0) + (H_x + \Delta H)}{2} = \frac{(13.86 \text{ m} + 1.95 \text{ m}) + (14.18 \text{ m} + 0.94 \text{ m})}{2} = 15.46 \text{ m}$$

11.1.1 CHANGE IN WATER TABLE

Once the hydraulic diffusivity is found, the change in water table at any distance from the boundary and at any time can be predicted using Equation 11.1. The rise in water table at 100 m from the river 8 d after the flood is found as follows:

$$\Delta H(x,t) = \Delta H_0 \text{ erfc}(\lambda)$$
$$\Delta H(x,t) = 1.95 \text{ m erfc}(\lambda)$$

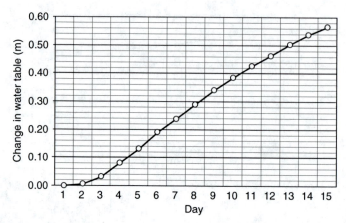

FIGURE 11.4 Change in water table at 100 m from the river calculated using Equation 11.1, Equation 11.3, calculated parameter a, and complementary error function, Table D.1.

$$\lambda = \frac{x}{2\sqrt{at}} = \frac{100 \text{ m}}{2\sqrt{3.47 \times 10^{-3} \text{ m}^2/\text{s} \times 8 \times 86400 \text{ s}}} = 1.02$$

For $\lambda = 1.02$, the complementary error function is approximately 0.148 (see Table D.1) and the change in water table is then

$$\Delta H(100 \text{ m}, 8 \text{ d}) = 1.95 \text{ m} \times 0.148 = 0.29 \text{ m}$$

Figure 11.4 shows the change in the water table at 100 m from the river for 15 d after the sudden change at the boundary (see also Table 11.1). The values are obtained using Equation 11.1 and Equation 11.3, the calculated hydraulic diffusivity (parameter a), and the complementary error function.

Note: it is somewhat arbitrary to choose between "a sudden" and "a gradual" change at the boundary. Theoretically, when solving the applicable general flow equation (Equation 1.86), a sudden change is considered to be instantaneous. In real field conditions, this should correspond to not more than 2–3 d of significant change in the water level at the boundary. A longer lasting change is better simulated with a gradual change at the boundary, especially if there is a significant precipitation in the area.

TABLE 11.1
Calculation of the Change in Water Table at 100 m from the River

Day	1	2	3	4	5	6	7	8	9	10	11	12	13	14	15
λ	2.888	2.042	1.667	1.444	1.291	1.179	1.091	1.021	0.963	0.913	0.871	0.834	0.801	0.772	0.746
erfc(λ)	0.000	0.004	0.017	0.042	0.067	0.098	0.122	0.148	0.174	0.197	0.219	0.238	0.258	0.275	0.291
ΔH (m)	0.000	0.007	0.033	0.082	0.131	0.191	0.238	0.289	0.339	0.384	0.427	0.464	0.503	0.536	0.567

Note: Calculated using Equations 11.1 and 11.3, calculated parameter a, and complementary error function, values of which are given in Table D.1 in Appendix D.

11.2   GRADUAL LINEAR CHANGE AT THE BOUNDARY, INFILTRATION RATE

Figure 11.5 shows a hydrogeologic cross section of an unconfined aquifer under direct influence of a large perennial river. Water table fluctuations are monitored by two wells placed along an average streamline of groundwater flow (which is roughly perpendicular to the river). The hydrographs in Figure 11.6, and Figure 11.7 show two episodes of a gradual increase in the water table that are consequences of the following:

- Change in the river stage during a period without precipitation in the study area (first episode)
- Change in the river stage accompanied with infiltration from precipitation during a 2-week period with frequent rains (second episode)

From the available data, find the hydrogeologic parameters of the sand aquifer whose effective porosity, determined from laboratory tests of core samples, is 0.20. Also, find the precipitation infiltration rate during the second episode.

Boundary and initial conditions of the flow field shown in Figure 11.5, which can be used to solve the general differential equation of unconfined transient flow in one dimension (Equation 1.106), are:

1. The change in the water table (ΔH) at the contact with the river, where $x = 0$, is given by the following expression:

$$\Delta H(0,t) = bt \qquad (11.8)$$

where b is the velocity of change and t is time.

2. This change in the water table equals zero far enough from the boundary:

$$\frac{\partial[\Delta H(x,t)]}{\partial x} = 0 \quad \text{for } x \to \infty \qquad (11.9)$$

i.e., the flow field is not disturbed when x approaches infinity.

3. There is a net infiltration across the water table: $w \neq 0$ (note that, in general, this net infiltration is the difference between the effective infiltration and the evaporation from

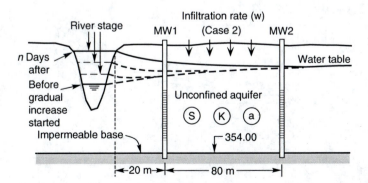

FIGURE 11.5 Hydrogeologic section of an unconfined aquifer in direct hydraulic connection with the river showing a general position of the water table before and at the end of a gradual increase (see Figure 11.6 and Figure 11.7 for the exact elevations of water table for the two cases discussed).

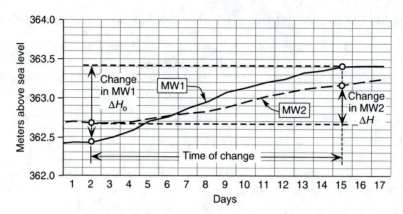

FIGURE 11.6 Hydrograph of the water table elevation recorded in two monitoring wells during an episode of a gradual linear change in river stage, and without significant precipitation in the study area.

the water table). The aquifer recharge by infiltration starts at time $t = 0$ and has constant intensity b during time t.

4. There is no leakage into the aquifer from below: $\varepsilon = 0$.

The solution of the basic differential equation for these initial and boundary conditions is

$$\Delta H(x,t) = \Delta H_0 \cdot R(\lambda) + \frac{wt}{S}[1 - R(\lambda)] \tag{11.10}$$

where $\Delta H(x,t)$ is the change in the water table at distance x from the boundary, at time t after the increase in the river stage began; ΔH_0 is the change in the river stage (change at the boundary); $R(\lambda)$ is a special function, values of which are found in tables (see Table D.3 in Appendix D); w is the intensity of infiltration; t is the time; S is the aquifer storativity (specific yield); and λ is a parameter given as

$$\lambda = \frac{x}{2\sqrt{at}} \tag{11.11}$$

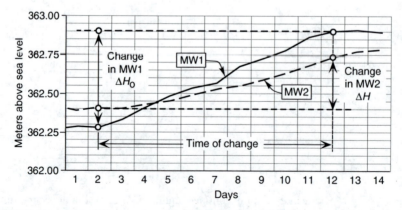

FIGURE 11.7 Hydrograph of the water table elevation recorded in two monitoring wells during an episode of gradual linear change in the river stage with significant precipitation in the study area.

where a is the hydraulic diffusivity, generally given as the ratio of transmissivity and storativity (specific yield): $a = T/S$. Since the saturated thickness changes in our case, the transmissivity is approximated with the average saturated thickness and the hydraulic conductivity: $T = h_{av}K$.

In the absence of infiltration and evapotranspiration, Equation 11.10 reduces to

$$\Delta H(x,t) = \Delta H_0 \cdot R(\lambda) \qquad (11.12)$$

Figure 11.6 shows hydrographs of the water table recorded in two monitoring wells (MW1 and MW2) during such an episode. The value of water-table change in both wells is easily determined from the hydrographs as the difference between the highest level recorded at the end of a noticeable increase, and the level recorded just before the change started. The time period between these two recordings is the time base for the calculations. For MW1 the change is

$$\Delta H_0 = 363.40 \text{ m} - 362.44 \text{ m} = 0.96 \text{ m}$$

and for MW2 the change is

$$\Delta H = 363.18 \text{ m} - 362.68 \text{ m} = 0.50 \text{ m}$$

The water table in MW2 lags behind MW1 by about 2 d. The time base for the calculation is 13 d. Once the change in water table is determined for both monitoring wells, the function $R(\lambda)$ is found using Equation 11.12:

$$R(\lambda) = \frac{\Delta H}{\Delta H_0} = \frac{0.50 \text{ m}}{0.96 \text{ m}} = 0.52$$

The parameter λ is then found from tables (such as Table D.3) from the known value of $R(\lambda)$: $\lambda = 0.27$. The relationship between hydraulic diffusivity (a) and parameter λ is given in Equation 11.11 and can be rewritten as

$$a = \frac{x^2}{4\lambda^2 t} \qquad (11.13)$$

where x is the distance between MW1 and MW2 (or, in general, the distance between the boundary and any point of water table measurement along the same streamline), and t is the time of change. Substituting known values for λ, x, and t into Equation 11.13 determines the hydraulic diffusivity:

$$a = \frac{(80 \text{ m})^2}{4 \cdot 0.52^2 \cdot 86400 \text{ s} \cdot 13} = 5.27 \times 10^{-3} \text{ m}^2/\text{s}$$

Once the hydraulic diffusivity of the aquifer is determined, it is possible to calculate the infiltration rates for the periods during which the rise in the water table can be partially attributed to aquifer recharge from precipitation.

The hydraulic diffusivity relates to the ability of the aquifer to transmit change imposed at the boundary to some distant location within the aquifer (or to transmit change imposed by a certain stress such as a pumping well). The time of response to this change is inversely proportional to the hydraulic diffusivity.

11.2.1 INFILTRATION RATE

Figure 11.7 shows hydrographs of the water table elevations recorded in two monitoring wells during an episode of aquifer recharge due to infiltration of precipitation, as well as the change in the river stage. The infiltration intensity (w) can be calculated using Equation 11.10 and the changes in the water table recorded in MW1 and MW2, the time of change, the aquifer hydraulic diffusivity, and the specific yield. The hydraulic diffusivity can be determined by analyzing water-table measurements in monitoring wells during a gradual change at the boundary. This procedure does not require a knowledge of aquifer hydraulic conductivity, transmissivity, or specific yield. Note, however, that if the value of hydraulic diffusivity determined in this way is to be used for calculations during some other boundary conditions, an assumption must be made that the average saturated thickness does not change significantly. In other words, the hydraulic diffusivity is assumed to be a constant property of the aquifer. If the hydraulic conductivity, transmissivity, and specific yield are determined previously by a pumping test or in some other way, the hydraulic diffusivity can then be calculated as

$$a = \frac{T}{S} = \frac{Kh_{av}}{S} \tag{11.14}$$

where

$$h_{av} = \frac{(H + \Delta H_0) + (H_x + \Delta H)}{2} \tag{11.15}$$

As determined earlier, the hydraulic diffusivity of the aquifer is

$$a = 5.27 \times 10^{-3} \text{ m}^2/\text{s}$$

and the other data necessary to apply Equation 11.10 are as follows:

Change at the boundary (MW1): $\Delta H_0 = 0.63$ m
Change in MW2: $\Delta H = 0.27$ m
Time of change: $t = 10$ d $= 86{,}400$ s $\times 10 = 864{,}000$ s
Specific yield: $S = 0.20$ (estimated from laboratory tests)
Distance between MW1 and MW2: $x = 80$ m

The parameter λ, which is also needed, is found using Equation 11.11:

$$\lambda = \frac{x}{2\sqrt{at}} = \frac{80 \text{ m}}{2\sqrt{5.27 \times 10^{-3} \text{ m}^2/\text{s} \cdot 864{,}000 \text{s}}} = 0.59$$

and the function $R(\lambda)$ is then found from Table D.3:

$$\text{for } \lambda = 0.59 \quad R(\lambda) = 0.22$$

Finally, the infiltration rate is calculated using Equation 11.10 and solving it for w:

$$w = \frac{S}{t} \cdot \frac{[\Delta H - \Delta H_0 \cdot R(\lambda)]}{[1 - R(\lambda)]}$$

$$w = \frac{0.20}{864{,}000 \text{ s}} \cdot \frac{[0.27 \text{ m} - 0.63 \text{ m} \cdot 0.22]}{[1 - 0.22]} = 3.89 \times 10^{-8} \text{ m/s} \tag{11.16}$$

It is assumed that the infiltration rate is constant during the 10 d period of gradual change in the water table. This assumption gives an equivalent height of effective precipitation (P_{ef}) over the study area in the 10 d period:

$$P_{ef} = w \times 10 \text{ d} = 0.034 \text{ m} = 34 \text{ mm}$$

If data on total or gross precipitation in the area are available, it is possible to determine the percentage of net infiltration using the following formula:

$$I = \frac{P_{ef}}{P_{total}} \times 100\% \tag{11.17}$$

where I is the infiltration in percent, P_{ef} is the effective precipitation recharging the aquifer, and P_{total} is the total, or gross precipitation recorded over the study area. In our case, the gross precipitation recorded during the 10 d period is 121.4 mm which gives the following value of infiltration:

$$I = \frac{34 \text{ mm}}{121.4 \text{ mm}} \times 100\% = 28\%$$

12 Two-Dimensional Steady-State Flow (Flow Nets)

12.1 CONSTRUCTION AND CHARACTERISTICS OF FLOW NETS AND GROUNDWATER FLOW RATE

The groundwater flow rate between the two lakes shown in Figure 12.1 is calculated using a flow net. The hydraulic conductivity of the confined aquifer is 7.3×10^{-6} m/s. The aquifer thickness at borehole B1 near the Upper Lake is 14.6 m.

As explained in detail in Section 1.5, a flow net is a set of streamlines and equipotential lines, which are perpendicular to each other (see Figure 1.64 and Figure 1.65). The flow line (or streamline) is an imaginary line representing the path of a groundwater particle as it flows (or would flow) through an aquifer. Two flow lines bound a flow segment of the flow field and never intersect, i.e., they are roughly parallel when observed in a relatively small portion of the aquifer. Similarly, two adjacent equipotential lines never intersect and can also be considered parallel within a small aquifer portion. These characteristics are the main reasons why a flow net in a homogeneous, isotropic aquifer is sometimes called the net of small (curvilinear) squares.

The groundwater flow field in the confined aquifer shown in Figure 12.1 has two types of hydrodynamic boundaries whose recognition is necessary for the flow net construction. These are:

1. Equipotential boundaries at the contact between the Upper and Lower Lakes and the aquifer
2. Impermeable or no-flow boundaries along the contact between the bedrock shales and the aquifer

It is best to start the flow net by drawing several flow lines perpendicular to one of the equipotential boundaries. The lines should be roughly equally spaced and parallel to each other (Figure 12.2). The next step is to draw an equipotential line parallel to the lake–aquifer contact (which is an equipotential boundary), and perpendicular to all the flow lines and the impermeable boundaries (remember that an impermeable boundary is also a flow line). The equipotential line and the flow lines should form "curvilinear squares," preferably of a similar size for this first attempt. After completing the first set of quadrangles, all flow lines are extended a bit further (keeping the outer ones parallel to the impermeable boundaries) and another equipotential line is drawn perpendicular to them. The process is repeated until the last set of squares is drawn at the lower equipotential boundary (Lower Lake). Alternatively,

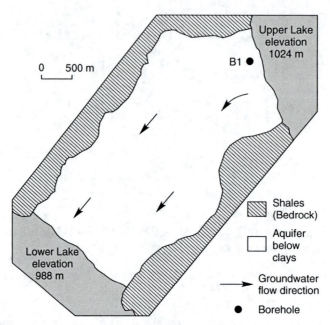

FIGURE 12.1 Simplified geologic map of the area between the two lakes.

a trial set of flow lines can be drawn through the entire flow field followed by the equipotential lines which should be spaced to form curvilinear squares. In any case, it will often be necessary to adjust the initial shape of both flow lines and equipotential lines, and change their number. The flow net is correct if the following tests are satisfied (see Figure 12.3):

- A circle touching all four sides can be drawn into each curvilinear square.
- Diagonals of the squares can be drawn to form continuous lines, i.e., another set of curvilinear squares.

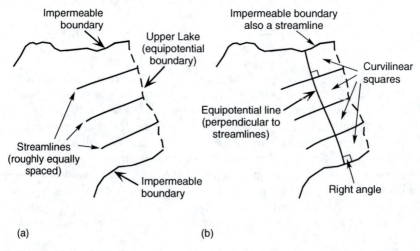

FIGURE 12.2 Beginning of the flow net construction by (a) drawing a trial set of streamlines and (b) continuing with the first equipotential line.

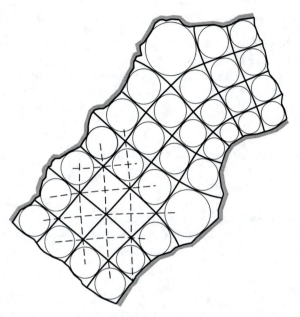

FIGURE 12.3 Tests of the flow net validity: circles touching all four sides can be drawn into each square; diagonals of the squares can be drawn to form continuous lines and another set of squares.

Once the flow net is finished, it can be used for the calculations of groundwater flow rates in the aquifer and the analysis of transmissivity distribution. The flow net can be made denser if a more precise analysis is required for certain portions of the aquifer (Figure 12.4).

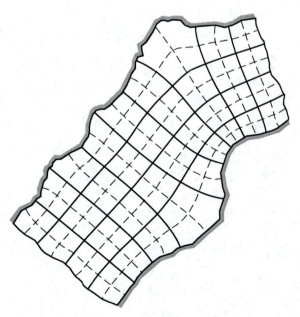

FIGURE 12.4 Once finished and tested, the flow net can be made denser for more precise calculations in the aquifer portions of interest.

12.1.1 GROUNDWATER FLOW RATE

Elements needed for the calculation of the total groundwater flow rate through the confined aquifer, from the Upper Lake to the Lower Lake, are shown in Figure 12.5. As usual, the flow rate per unit thickness of the aquifer (q) is first given as the product of the flow velocity (v) and the cross-sectional area of flow (A):

$$q = v \times A \tag{12.1}$$

The unit cross-sectional area of flow for the entire aquifer is

$$A = 1 \times \Delta W \times N \tag{12.2}$$

where 1 is the unit aquifer thickness, ΔW is the distance between two flow lines forming a flow segment (stream tube), and N is the number of segments. The average volumetric (Darcy's) groundwater velocity (v) through the entire aquifer is

$$v = K \times \frac{\Delta H}{M \times \Delta L} \tag{12.3}$$

where K is the hydraulic conductivity, ΔH is the head difference between the Upper and Lower Lakes, ΔL is the distance between two adjacent equipotential lines, and M is the number of squares along the flow path, i.e., between two lakes. Substituting Equation 12.2 and Equation 12.3 into Equation 12.1 gives

$$q = K\Delta H \times \frac{N}{M} \times \frac{\Delta W}{\Delta L} \tag{12.4}$$

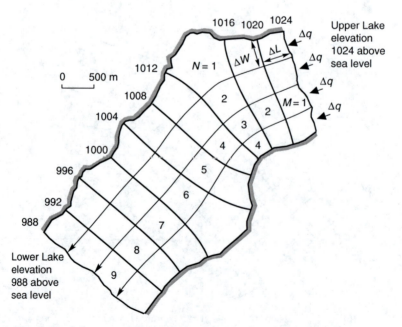

FIGURE 12.5 The finished flow net and the elements for the flow rate calculation.

Since the flow net in our case is a net of curvilinear squares (tests for the net validity are satisfied—see Figure 12.3), the distance between two adjacent equipotential lines can be considered equal to the distance between two flow lines: $\Delta L = \Delta W$, which reduces Equation 12.4 to

$$q = K\Delta H \times \frac{N}{M} \qquad (12.5)$$

As can be seen, the flow rate per unit thickness of the aquifer depends on the geometric characteristics of the flow net (i.e., number of squares along the flow direction and perpendicular to it), hydraulic conductivity, and the total head loss between the two equipotential boundaries. Inserting numeric values into Equation 12.5 gives the flow rate:

$$q = 7.3 \times 10^{-6} \text{ m/s } (1024 \text{ m} - 988 \text{ m})\frac{4}{9} = 1.168 \times 10^{-4} \text{ m}^2/\text{s}$$

To obtain the total flow rate through the aquifer, the flow rate per unit thickness has to be multiplied by the overall aquifer thickness. From the flow net in Figure 12.5, it can be seen that the squares first receiving the flow, i.e., those adjacent to the upper equipotential boundary (Upper Lake), are approximately of equal size. If the aquifer is homogeneous, i.e., if the hydraulic conductivity of the porous medium remains constant throughout (which is our assumption), then the aquifer thickness is roughly the same in the first four squares. The data from borehole B1 (see Figure 12.1) indicate an aquifer thickness of 14.6 m, which gives the total flow of

$$Q = q \times b = 1.168 \times 10^{-4} \text{ m}^2/\text{s} \times 14.6 \text{ m} = 1.705 \times 10^{-3} \text{ m}^3/\text{s}$$

12.1.2 TRANSMISSIVITY ANALYSIS

Groundwater contour maps and flow nets are useful tools for analyzing distribution of aquifer transmissivity and planning further investigations (e.g., location of monitoring and test wells). A portion of an aquifer with a wider spacing of equipotential lines (such as the $N = 1$ area in Figure 12.5) indicates a higher transmissivity, while closely spaced contour lines mean a lower transmissivity ($N = 4$ in Figure 12.5). From the definition of transmissivity

$$T = Kb$$

It follows that the higher transmissivity can be a consequence of higher hydraulic conductivity, greater aquifer thickness, or both. If, based on geologic and hydrogeologic studies, it is concluded that the aquifer hydraulic conductivity is uniform (as in our case), then the variation in the spacing of equipotential (contour) lines means that the aquifer thickness varies as well. Generally speaking, aquifer portions with widely spaced groundwater contours are better candidates for exploitation wells or other means of groundwater withdrawal. Note, however, that other hydrodynamic criteria must be considered as well. For example, even though area $N = 1$ in our case (see Figure 12.5) is the thickest portion of the aquifer, placing a water-supply well there would not be appropriate because of the closeness of the impermeable boundary. A well located near the aquifer center and equipotential boundaries would yield more.

12.2 GROUNDWATER CONTOUR MAPS AND INFLUENCE OF SURFACE STREAMS

The position of the water table in the unconfined alluvial aquifer of the Blue River is regularly recorded in 14 monitoring wells and 9 domestic wells as shown in Figure 12.6 (the river flood plain and the aquifer extend beyond the map). Two sets of measurements representing the summer recession period of 1994 and the wet spring period of 1995 are given in Table 12.1. Water table contour maps for both sets of data were constructed and analyzed. A short 8-h pumping test was conducted in the domestic well DW7 and the hydraulic conductivity of the aquifer was estimated to be 3.47×10^{-5} m/s. The aquifer impermeable base has an average elevation of 605 m above sea level in the area shown in Figure 12.6 (the base can be considered horizontal). The potential for a possible spread of groundwater contamination in the vicinity of DW7 is analyzed and presented graphically on a water table contour map.

The easiest way to produce a preliminary (draft) contour map is to let a computer program do it, understanding that the resulting first map will always need certain adjustments. However, in most cases, a computer program used would have great difficulties

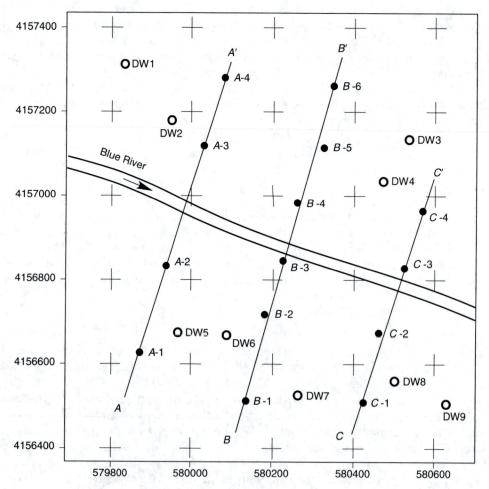

FIGURE 12.6 Monitoring wells (filled circles) and domestic wells (empty circles) used for measuring water table elevations in the Blue River alluvial aquifer.

TABLE 12.1
Alluvium Aquifer Water Table Elevations, in Feet above Sea Level, Measured in Monitoring and Domestic Wells on July 15, 1994 and May 24, 1995

Well	X-Coordinate	Y-Coordinate	July 15, 1994	May 24, 1995
A1	579,872.34	4,156,625.87	673.90	675.60
A2	579,938.25	4,156,833.42	672.60	675.55
A3	580,029.12	4,157,117.56	672.75	675.60
A4	580,080.49	4,157,282.19	674.65	675.65
B1	580,134.86	4,156,512.96	673.40	674.70
B2	580,180.35	4,156,717.51	672.70	674.40
B3	580,223.30	4,156,846.25	671.60	673.60
B4	580,259.27	4,156,984.09	672.35	674.15
B5	580,323.65	4,157,113.12	673.50	674.10
B6	580,350.45	4,157,268.09	674.70	675.00
C1	580,422.10	4,156,675.15	672.85	673.65
C2	580,463.54	4,156,675.23	672.10	673.05
C3	580,524.05	4,156,827.63	671.50	672.55
C4	580,570.72	4,156,966.07	672.70	672.70
DW1	579,832.14	4,257,313.89	674.75	676.45
DW2	579,949.16	4,157,180.20	673.45	675.90
DW3	580,533.25	4,157,134.51	673.75	673.80
DW4	580,470.81	4,157,036.78	672.90	673.35
DW5	579,968.06	4,156,674.83	673.45	675.30
DW6	580,087.19	4,156,669.12	673.20	674.85
DW7	580,270.34	4,156,527.45	673.10	674.20
DW8	580,500.18	4,156,558.78	672.65	673.10
DW9	580,629.02	4,156,506.19	672.80	672.85

UTM coordinates are in meters.

in appropriately interpreting the contours near the river and the hydraulic role of the river. One of the most important aspects of constructing contour maps in alluvial aquifers (as in our case) is to determine the hydraulic relationship between groundwater and the surface water stream. Figure 2.157 shows basic cases of this relationship, which can coexist to form more complicated situations (see next Section). In hydraulic terms, the contact between the aquifer and the surface stream is an equipotential boundary. This does not mean that everywhere along the contact the hydraulic head is the same. Since both the river water and the groundwater are flowing, there is a hydraulic gradient along their contact. If enough measurements of a river stage are available, it is relatively easy to draw the water table contours near the river and finish them along the river–aquifer contact at the same elevation. However, often few or no data are available and the river elevation, at the expense of precision, has to be estimated from a topographic map or inferred from the hydraulic gradient based on monitoring well data. River elevations estimated for the three cross sections (A, B, and C) from the aquifer hydraulic gradients are shown in Table 12.2. In addition, five more points were estimated based on presumed linear change in the hydraulic gradient of the river channel.

As explained in Section 1.5.1, contouring programs require that the groundwater level data be organized as XYZ files where X and Y are plane coordinates of the measuring points (monitoring wells) and Z is the elevation of the water table or hydraulic head surface above a

TABLE 12.2
**Estimated Elevation of the Water Surface in the Blue River
Channel, in Feet above Mean Sea Level, July 15, 1994**

Point	X-Coordinate	Y-Coordinate	Elevation
A	579,980.45	4,156,975.30	671.70
B	580,235.80	4,156,878.25	671.25
C	580,510.65	4,156,784.75	670.85
1	579,711.25	4,157,075.45	672.35
2	579,836.40	4,157,040.00	672.00
3	580,075.23	4,156,938.75	671.50
4	580,374.86	4,156,831.08	671.10
5	580,700.00	4,156,725.00	670.50

UTM coordinates of the estimation points are in meters.

chosen reference level, usually above mean sea level (amsl). The most widely used plane coordinate system in geosciences and engineering is The Universal Transverse Mercator system (UTM), which is based on the transverse Mercator projection. In our example, the lower left corner of the map shown in Figure 12.6 has the following UTM coordinates: X: 579,700 m east (the letters are omitted), Y: 4,156,370 m north (the letters are omitted). This means that the corner of the map is 79,700 m (79.7 km) east of the origin of the fifth UTM zone (number 5, the first number in the X-coordinate, designates the UTM zone to which the map area belongs), and that it is 4,156,370 m (4156.4 km) north of the equator. UTM coordinates are shown on standard USGS topographic maps usually with blue numbers and tick marks along the map boundaries. In case our monitoring well network were surveyed locally, and were not tied to any official coordinate system at the time (UTM or state), the lower left corner of the map would get an XY coordinate of (0,0) with distances along the X (horizontal) and Y (vertical) axes measured from the origin (Figure 12.7). Once the UTM or state coordinates of the origin (lower left corner) were determined, the local coordinates could be easily converted. Ultimately, all units for X, Y, and Z coordinates should be the same (meters or feet).

12.2.1 TRIANGULATION WITH LINEAR INTERPOLATION

This method is based on exact linear interpolation between three neighboring points as shown in Figure 1.67. Because the original data points are used to define the triangles, they are preserved (honored) very closely on the generated map. In general, when there are few data points (e.g., less than 20), or data are not evenly distributed, triangulation is not effective since it creates triangular facets and "holes" in the map. In our case, the method closely represents the expected water table in most of the map where the data are distributed evenly (Figure 12.8). The trend indicated by the map allows easy manual filling of the holes and completion of the contours. Triangulation is fast and accurate with 200 to 1000 data points evenly distributed over the map area. Note, however, that these numbers of original data are rarely available in groundwater investigations and one would have to estimate additional "auxiliary" data to satisfy the requirement. If enough data are available, the advantage of triangulation is that it preserves breaks in contours such as faults, geologic boundaries, and streams indicated in a data file.

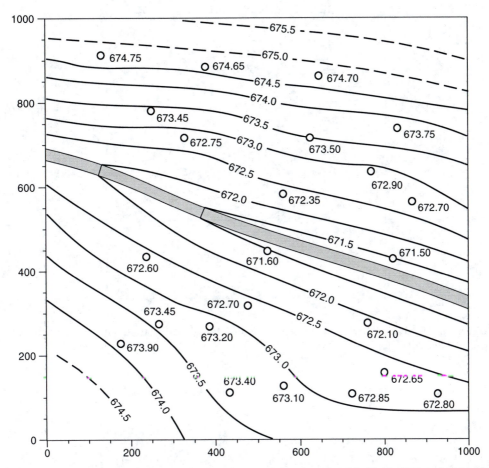

FIGURE 12.7 Manually drawn water table contour map for data listed in Table 12.1 and Table 12.2. The numbers next to well data points are recorded water table elevations in feet above sea level. Contour interval is 0.5 ft and the coordinate system is local, with the arbitrary origin placed at the lower left corner, with coordinates (0,0).

12.2.2 Inverse Distance to a Power

This interpolation method is usually present in various contouring packages because it is very fast and can handle large data sets. However, it does not produce good results when used to contour groundwater levels often represented with small data sets. Distance to a power method weights data points during interpolation so that the influence of one data point relative to another data point decreases with the power of distance. The result is the generation of "bull's eyes" pattern around the positions of data points as shown in Figure 12.9. Software smoothing of contours can only slightly reduce this effect but does not eliminate unnatural depressions and mounds in the water table.

12.2.3 Minimum Curvature

Contouring with minimum curvature is widely used in the earth sciences since this method generates the smoothest possible surface while still closely honoring the original data points. However, minimum curvature is not always accurate in preserving breaks in contours, which

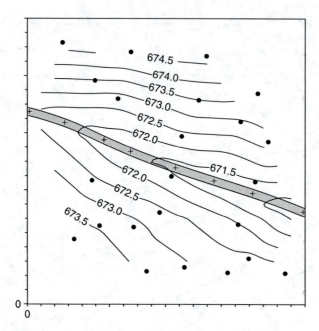

FIGURE 12.8 July 15, 1994 water table map generated using triangulation with linear interpolation method within SURFER.

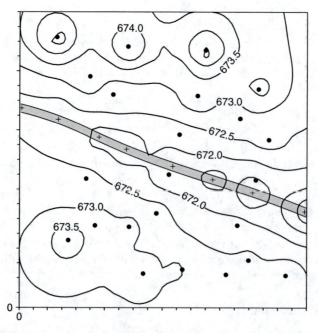

FIGURE 12.9 July 15, 1994 water table map generated using inverse distance to power method within SURFER. The power is two and the smoothing factor is one. Note a distinct "bull's eyes" pattern around the data points.

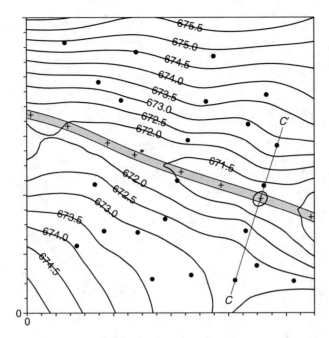

FIGURE 12.10 July 15, 1994 water table map generated using the minimum curvature method within SURFER.

is often most important when contouring piezometric surface near streams, faults, and geologic boundaries. Figure 12.10 shows some unnatural solutions by the method in our case (note a depression in the water table in the river at intersection C–C').

12.2.4 Radial Basis Functions (Splines)

A diverse group of radial basis functions is analogous to variograms in kriging and produces similar results (see Figure 12.11). They are all exact interpolators and closely preserve original data. Splines are frequently used radial basis functions in geosciences. The natural cubic spline function is (Golden Software, 2002):

$$B(h) = (h^2 + R^2)^{3/2} \qquad (12.6)$$

and the thin plate spline function is

$$B(h) = (h^2 + R^2) \times \log(h^2 + R^2) \qquad (12.7)$$

where h is the anisotropically rescaled, relative distance from the data point to the grid node and R^2 is the smoothing parameter specified by the user.

More detail on contouring with these and other radial basis functions can be found in the work of Carlson and Foley (1991).

Comparing all the maps shown in Figure 12.7 through Figure 12.11, it seems that the thin plate spline function most closely resembles the manually drawn map shown in Figure 12.7 and the expected natural look of the water table during the July 15, 1994 recession period. Since the computer-generated map (Figure 12.11) still shows some "irregularities" near the central-eastern edge (the contours bend away from the river) and in the southwestern section,

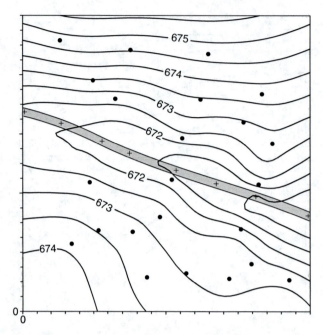

FIGURE 12.11 July 15, 1994 water table contour map of the Blue River alluvial aquifer generated using thin plate spline function within the radial basis functions contouring method available in SURFER.

further adjustments are needed to create the final *XYZ* file. This is achieved by introducing new auxiliary data points throughout the map, and particularly in the parts with few data points such as along the map edges. Table 12.3 shows coordinates and estimated elevations at 25 new locations. The elevations of the water table are estimated from the manually drawn map shown in Figure 12.7. The resulting final map (Figure 12.12) very closely resembles the expected groundwater flow conditions in the alluvial aquifer.

12.2.5 HYDRAULIC CONDUCTIVITY ANALYSIS

A contour map of an aquifer, accompanied by data on aquifer thickness, is an excellent tool for the analysis of transmissivity and hydraulic conductivity distributions. Wider spacing of water table contours indicates higher transmissivity of porous media, while closely spaced contours show parts of the aquifer with lower transmissivity. Since the transmissivity is the product of the hydraulic conductivity and the aquifer thickness ($T = Kb$), the following is true:

- In homogeneous aquifers with uniform hydraulic conductivity, widely spaced contours indicate larger aquifer thickness, while steeper hydraulic gradients (e.g., closely spaced contours) indicate smaller aquifer thickness.
- When an aquifer, or its portion of interest, has a uniform thickness, steeper hydraulic gradients indicate lower hydraulic conductivity of the porous media, while more widely spaced contours mean higher conductivity.

Based on the borehole data, it is concluded that the base of the Blue River alluvial aquifer is (sub)horizontal within the map area, with an average elevation of about 605 ft above sea level. The water table elevation varies between 675.0 and 670.6 ft, which is an amplitude of 4.4 ft.

TABLE 12.3
Estimated Elevation of the Water Table at Auxiliary Points, in Feet above Sea Level, July 15, 1994

ID	X-Coordinate	Y-Coordinate	Elevation
N1	579,750	4,156,700	674.00
N2	579,750	4,156,800	673.50
N3	579,700	4,157,000	672.50
N4	579,700	4,157,125	672.50
N5	579,750	4,157,150	673.00
N6	579,750	4,157,250	674.00
N7	579,850	4,157,100	672.50
N8	579,850	4,156,900	672.50
N9	579,800	4,156,850	673.00
N10	580,000	4,156,550	673.70
N11	580,000	4,156,450	674.00
N12	580,100	4,157,000	672.05
N13	580,300	4,156,750	672.00
N14	580,600	4,156,550	672.50
N15	580,600	4,156,850	672.00
N16	580,600	4,157,200	674.50
N17	580,600	4,157,250	675.00
N18	580,700	4,156,600	672.05
N19	580,700	4,156,775	671.50
N20	580,700	4,156,950	673.00
N21	580,700	4,157,050	673.50
N22	580,700	4,156,475	673.00
N23	580,000	4,157,400	675.50
N24	580,250	4,156,400	673.50
N25	580,700	4,156,825	672.00

UTM coordinates of the estimation points are in meters.

This amplitude is only about 6.5% of the average aquifer thickness (i.e., thickness of the saturated zone), which means that the variable spacing of the water table contours in Figure 12.12 is the result of the varying hydraulic conductivity rather than the (slight) change in thickness. The portion south of the Blue River, excluding a strip along the river, is generally more permeable than the portion north of the river where the hydraulic gradient is steeper. The highest hydraulic conductivity is expected around domestic well DW7, which is also the only well with the pumping test data available. However, keeping in mind the above-mentioned relationship between contour spacing and aquifer transmissivity, and knowing the numeric value of the hydraulic conductivity at one location, allows one to estimate the hydraulic conductivity for other portions of the aquifer. Figure 12.13 illustrates this procedure which consists of the following steps:

- A streamline through the aquifer portion of interest (in our case, the streamline is drawn through DW7) is drawn.
- Sections along the streamline with similar hydraulic gradients, i.e., with similar contour spacing (in our case, there are two such sections: A–B and B–C) is identified.

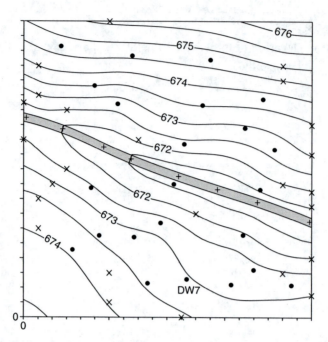

FIGURE 12.12 The final map of the water table on July 15, 1994 generated using the thin plate spline contouring method within SURFER. New auxiliary data points with estimated water table elevations are shown with × marks, while the original recorded data points (monitoring and domestic wells) are shown with circles.

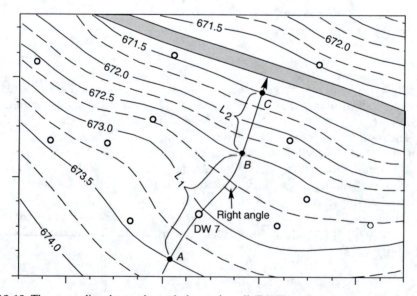

FIGURE 12.13 The streamline drawn through domestic well (DW7), a suspected source of groundwater contamination. The streamline is perpendicular to the water table contours. Note the auxiliary contours (dashed lines), which make drawing the streamline easier.

- Groundwater flow rate per unit aquifer width along the streamline in the section with the known hydraulic conductivity (A–B in our case) is calculated and then this calculated value is used to find the hydraulic conductivity in the other section (B–C).

Groundwater flow rate per unit width in an unconfined aquifer without infiltration and with the horizontal base is described with the following equation:

$$q = K \frac{h_1^2 - h_2^2}{2L} \tag{12.8}$$

where q is the unit flow, K is the hydraulic conductivity, h_1 is the higher hydraulic head, h_2 is the lower hydraulic head, and L is the distance between the two heads measured along the streamline.

The unit flow in section A–B is

$$q_1 = K_1 \frac{h_A^2 - h_B^2}{2L_1} \tag{12.9}$$

The hydraulic head h_A is the height of the water table above the reference level, which is set at the horizontal base:

$$h_A = 673.5 \text{ ft} - 605 \text{ ft} = 68.5 \text{ ft} = 20.88 \text{ m}$$

The hydraulic head h_B is

$$h_B = 672.5 \text{ ft} - 605 \text{ ft} = 67.5 \text{ ft} = 20.57 \text{ m}$$

Inserting all numeric values into Equation 12.9 gives the unit flow along the streamline in section A–B:

$$q_1 = 3.47 \times 10^{-5} \text{ m/s} \times \frac{(20.88 \text{ m})^2 - (20.57 \text{ m})^2}{2 \cdot 281 \text{ m}} = 7.93 \times 10^{-7} \text{ m}^2/\text{s}$$

From the principle of the continuity of flow along the streamline, it follows that the unit flow in section A–B equals the unit flow in section B–C:

$$q_1 = q_2 = K_2 \frac{h_B^2 - h_C^2}{2L_2} \tag{12.10}$$

which gives the following expression for the hydraulic conductivity in section B–C:

$$K_2 = \frac{q_1 2L_2}{h_B^2 - h_C^2} \tag{12.11}$$

$$K_2 = \frac{7.93 \times 10^{-7} \text{ m}^2/\text{s} \times 2 \times 130 \text{ m}}{(20.57 \text{ m})^2 - (20.27 \text{ m})^2} = 1.68 \times 10^{-5} \text{ m/s}$$

The hydraulic conductivity in the section B–C can also be found by inserting Equation 12.9 into Equation 12.11, which eliminates the unit flow q:

$$K_2 = K_1 \times \frac{L_2}{L_1} \times \frac{(h_A^2 - h_B^2)}{(h_B^2 - h_C^2)} \tag{12.12}$$

12.2.6 Influence of Surface Streams

The importance of understanding possible influences of a surface stream on a water table is illustrated with the situation recorded in the Blue River alluvial aquifer on May 24, 1995 following a 5-week period of intensive precipitation. Figure 12.14 shows a computer-generated contour map of the water table that looks quite different from the map shown in Figure 12.12, which is representative of the recession period recorded on July 1994. Although the May map needs further adjustments (it is based on the recorded data only), it is clear that the changes in river stage cause significant changes in the water table far into the aquifer. This is explained by a general example shown in Figure 12.15.

Cross section 1 is characteristic for a gaining stream and groundwater flow directed toward the stream in the whole aquifer. In alluvial aquifers of large perennial streams, this situation is present most of the time. Case 2 shows a reverse direction of groundwater flow close to the stream following a rapid rise in the river stage during and immediately following major wet periods. Groundwater flow in most of the aquifer is still oriented toward the stream. After the river stage returns to normal, a rather complex groundwater pattern may develop as case 3 shows. The aquifer area affected by these changes depends on their duration and the hydrogeologic characteristics of the porous media. The most pronounced influence is illustrated with the map view for case 3. Water table contours show two "ridges" parallel to the river that cause groundwater to flow in two opposite directions (small arrows). There are also two major (new) directions of groundwater flow parallel to the stream as indicated with large arrows. This picture may become even more complex if there are several close periods of significant changes in river stage. If the hydrometeorologic and hydrologic conditions in the area for the preceding few months are not known, it may, at times, be quite difficult to correctly interpret the water table recordings, especially if there are few data.

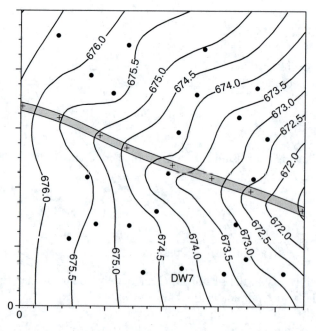

FIGURE 12.14 Computer-generated contour map of the water table in the Blue River alluvial aquifer for May 24, 1995. Contouring software is SURFER and the gridding method is thin plate spline.

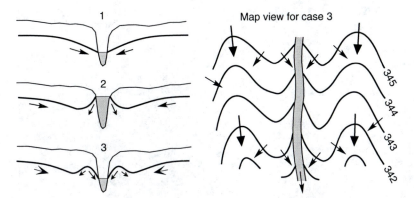

FIGURE 12.15 *Left*: Cross-sectional views of some common cases of the influence of surface streams on the water table in adjacent alluvial aquifers. (1) Basic case of a gaining (effluent) stream; (2) position of water table after a significant rise in river stage; and (3) water table soon after the river stage returns to normal. *Right*: Map view for case 3; small arrows indicate local directions of groundwater flow while large arrows show two major directions parallel to the river.

The computer-generated map shown in Figure 12.14 is adjusted manually so that the contours along the northern and southern edges band more toward the river in Figure 12.16. Although there are no supporting data, this interpretation is based on the contour map developed for the recession period of July 1994.

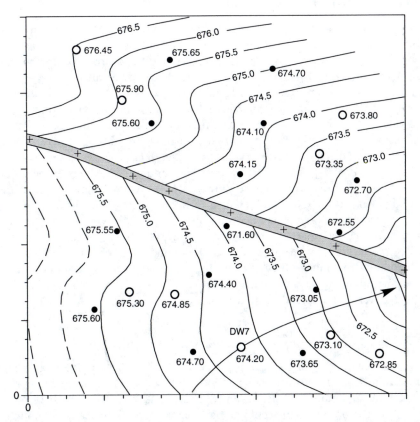

FIGURE 12.16 Manually adjusted water table contour map for May 24, 1995 with the flow line through domestic well DW7. Numbers next to the wells are water table elevations in feet above sea level.

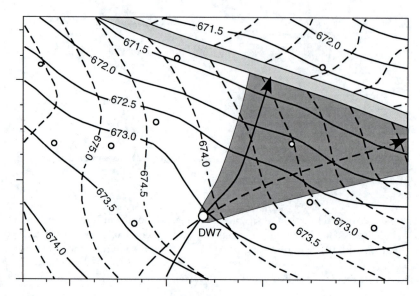

FIGURE 12.17 Estimated spread of a possible groundwater contamination (hatched area) around domestic well DW7 if no prevention or remediation were conducted. Solid contours are for July 15, 1994 and dashed contours are for May 24, 1995.

12.2.7 SPREAD OF GROUNDWATER CONTAMINATION

In order to correctly assess the spread of possible groundwater contamination, it is best to analyze water table contour maps for all distinct hydrologic seasons in the area of interest. Why one, or even several maps for similar hydrologic conditions, are not enough is illustrated with the situation shown in Figure 12.17. The streamline drawn through DW7 (the area of suspected contamination) is perpendicular to the river during the recession period of July 15, 1994. If these were the only data set available, and an assumption was made that, more or less, the groundwater flow direction is the same throughout the year, an area affected by the contamination would be a relatively narrow strip around the flow line (assuming that the contaminant is transported primarily by groundwater, i.e., due to advection). A similar mistake of underestimating the spread of possible contamination would be made if the situation recorded on May 24, 1995 (Figure 12.16) were considered representative for the aquifer. Knowing that both data sets are recorded during two distinct and relatively extreme hydrologic periods (summer recession and spring high water stage), the estimated spread of possible contamination shown in Figure 12.17 seems reasonable. It includes anticipated common lateral dispersion so that the plume limits are somewhat further than the two controlling streamlines. Of course, depending on how wet was the wet period of May 24, 1995, the corresponding streamline may approach the river at a sharper angle (after higher floods), causing even wider spread of pollution.

12.3 KRIGING

Kriging is one of the most robust and widely used methods for interpolation and contouring in different fields. It is a geostatistical method that takes into consideration the spatial variance, location, and sample distribution in data. It is useful in heterogeneous porous media when used to contour hydrogeologic parameters such as hydraulic conductivity and storage. There are various interpolation (gridding) models within the kriging method and the most appropriate one for the existing data set can be chosen after determining the semivariogram model, which best describes the underlying experimental (sampled) data. Although all major contouring programs on the market include kriging, few allow for user-friendly, visual generation of a semivariogram by interactively adjusting all of its key components. Therefore, in order to select an appropriate kriging method, it may be necessary to generate the semivariogram using some external program. SURFER by Golden Software includes very powerful and simple-to-use options for generating experimental and theoretical semivariograms, and several kriging contouring methods.

12.3.1 SEMIVARIOGRAM OR VARIOGRAPHY

Variography is the term often used to describe the process of determining if there is spatial correlation among data measured in the field. If such correlation is evident, it includes determining which geostatistical model is the most appropriate to describe it quantitatively. The selected model is then used to predict (interpolate) the variable in question at the unsampled locations, which is part of another process, called kriging. Variography includes the following steps:

- Testing spatial correlation between the field measurements of the same variable (spatial autocorrelation).
- Computing spatial covariance (autocovariance) between the data, including calculating semivariance for each pair of observations.
- Presenting the calculated semivariances graphically as a variogram cloud or (semi)variogram. Note that terms variogram and semivariogram are interchangeably used through out the geostatistical literature and the only real difference is the factor of two: semivariogram is one half of the sample variogram, just like semivariance is one half of the sample variance.

Covariance (S_{XY}) between two variables X and Y that are not "regionalized variables" (i.e., are not spatially correlated) is

$$S_{XY} = \frac{1}{n-2} \sum_{i=1}^{n} (x_i - x_{\text{av}})(y_i - y_{\text{av}}) \qquad (12.13)$$

where n is the number of paired data, x_i and y_i are individual values of each variable, and x_{av} and y_{av} are average values of each variable.

Semivariance (γ_i) for each pair of observation points of the spatially correlated (regionalized) variable is (see Figure 12.18)

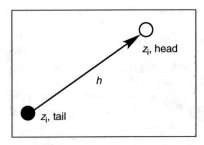

FIGURE 12.18 Semivariance is calculated for each pair of observations, separated by distance h.

$$\gamma_i = \frac{1}{2}(z_{i,\text{head}} - z_{i,\text{tail}})^2 \tag{12.14}$$

and it is calculated for all possible data pairs, which is often a fairly large number:

$$\text{Total number of pairs} \ = \frac{n(n-1)}{2} \tag{12.15}$$

where n is the number of measured data points. This information is then presented in two ways:

1. As a variogram cloud where all semivariances for all pairs are plotted against their respective individual separation distances. This graph contains a large number of data and does not reveal much in terms of spatial correlation among the data.
2. As experimental (semi)variogram where all semivariances within one separation distance or lag (also called "bin") are averaged and plotted against the average separation distance between the data pairs within that lag.

Figure 12.19 illustrates the process of calculating semivariances within one lag and then moving to the next lag. The average variance for one lag is

$$2\gamma^*(h) = \frac{1}{n}\sum_{i=1}^{i=n}[g(x)_i - g(x+h)_i]^2 \tag{12.16}$$

where $\gamma^*(h)$ is the experimental semivariance for given distance (lag) h, h is the separation distance between the two data points, g is the value of the sample (data point), x is the

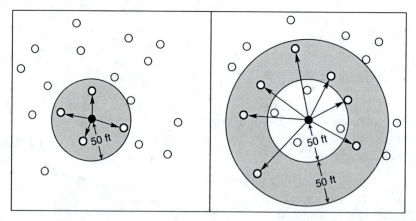

FIGURE 12.19 Calculation pattern for omnidirectional (isotropic) variogram, looking at one data point at the time. *Left*: Separation distance between the black data and the other data inside the shaded circle is between 0 and 50 ft, i.e., the lag width is 50 ft. This smallest separation distance (chosen by the user arbitrarily) is called lag 1 (Bin 1). As the calculation window moves throughout the sampled area from one point to another, each new step produces a number of pairs, which are all added to lag 1. *Right*: Separation distance between the black data and the data inside the shaded area is between 50 and 100 ft, i.e., the lag width is also 50 ft. This separation distance is called lag 2 (Bin 2). As the calculation window moves throughout the sampled area from one point to another, each new step produces a number of pairs, which are all added to lag 2.

position of one sample in the pair, $x + h$ is the position of the other sample in the pair, and n is the number of pairs in calculation.

The calculated value (divided by two) is the measure of the difference between the two data points, which are h distance apart. The semivariogram measure plotted on the vertical graph axis is in squared units of data (in case of the hydraulic head it is squared distance, or ft^2). The lag is usually given a tolerance (say, 40 ± 10 ft) so that more spatial information is included; note that most groundwater information in the field is collected from irregularly spaced sampling points so the exact lag distance (say, 40 ft) may lead to fewer calculated values for plotting the experimental semivariogram graph. This tolerance should not be greater than one half of the basic distance. For example, if the basic distance h is 40 ft, the tolerance should be maximum ± 20 ft. This means that all the pairs that fall between 20 and 60 ft will be included in the calculation of the semivariogram. The process in then repeated for as many new lags as possible. Each new distance (lag) is increased by the basic interval (say, $h = 40 + 40 = 80$ ft, $h = 120$ ft, $h = 160$ ft, etc.) and the results are plotted on the same graph. The maximum separation distance should not be larger than one-half distance between two points in the field data set that are farthest apart. Figure 12.20 shows one experimental semivariogram with the number of data pairs per lag (also called "bin"), and the variance of all data in the sample. The sample variance (s^2) is given as

$$s^2 = \frac{1}{n-1} \sum_{i=1}^{i=n} (g_i^2 - g_{av}^2) \tag{12.17}$$

where g_i are the values of individual data and g_{av} is the sample mean given as

$$g_{av} = \frac{1}{n} \sum_{i=1}^{i=n} g_i \tag{12.18}$$

Note that the X-coordinates of the plotted points are not regularly spaced because each coordinate is the average separation distance for all pairs included in the individual bin; these average numbers vary from bin to bin by default since the data are irregularly spaced.

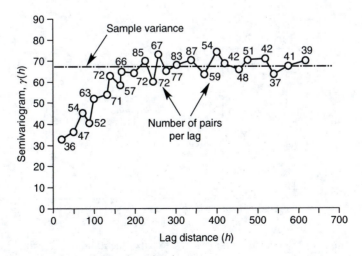

FIGURE 12.20 Example of experimental semivariogram.

Semivariograms can be plotted for various directions (so-called directional semivario-grams), which is recommended, especially if a certain degree of anisotropy in data is expected. Directions that produce noticeably different (while still meaningful) semivario-grams may indicate actual anisotropy in the data. Note, however, that there is a substantial degree of subjectivity in interpreting semivariograms. For a statistically valid semivario-gram, there should be at least 30 pairs of data, which is often not the case in groundwater studies. It is up to the users to know their data well and make an evaluation as to the reasonable lower limit of data per bin. When the data set is small, the directional calcula-tion (e.g., all pairs falling within a 30° window $\pm$ x degrees of tolerance) may not produce a meaningful semivariogram even if there is an underlying anisotropy. In such cases, the calculation is performed by default for all possible directions and for all sample pairs that fall within the calculation interval. This will produce a global, omnidirectional experimental semivariogram.

Once the experimental semivariogram shows a certain recognizable structure, the next step is to fit a theoretical curve that best describes the experimental data. The shape of the experimental variogram should indicate which theoretical model is appropriate for the given data, and what are the values of the three important parameters of the semivariogram (see Figure 12.21):

1. a, range of influence
2. C, sill
3. C_0, nugget effect

The range of influence (a) is the distance at which data become independent of one another. After this point, the graph is horizontal and the corresponding value of γ is called the sill of the semivariogram (C). Arguably, the scale, range, and nugget (where applicable) should be selected such that the sill of the theoretical function is as close to the sample variance as possible. Figure 12.21 shows a semivariogram with the nugget effect—the graph's intercept at the vertical axis. Its presence may indicate a potential error in data collection, or any other unexplainable (random) component in the data at the scale smaller than the separation

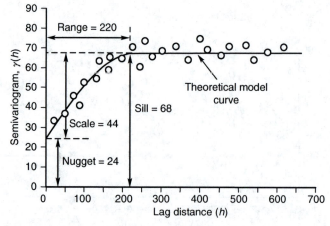

FIGURE 12.21 Example of theoretical function fitted to experimental data, showing three main elements: range, sill, and nugget.

distances between the field data. The nugget effect is a constant that raises a theoretical semivariogram C_0 units along the vertical:

$$\gamma(h) = C_0 + \gamma'(h) \tag{12.19}$$

where $\gamma'(h)$ is one of the common theoretical semivariograms shown in Figure 12.22.

Specifying a nugget effect causes kriging to become more of a smoothing interpolator, implying less confidence in individual data points versus the overall trend of the data. The higher the nugget effect, the smoother is the resulting grid (Golden Software, 2002).

Figure 12.23 shows three most common theoretical semivariogram models fitted to the May 1995 water table data for the Blue River aquifer: spherical, exponential, and linear. The so-called "ideal shape" for the semivariogram, which is derived theoretically and known as the spherical or Matheron model, is given as

$$\gamma(h) = C\left(1.5\frac{h}{a} - 0.5\frac{h^3}{a^3}\right) \tag{12.20}$$

where the sill value is equal to the sample variance. The exponential model is

$$\gamma(h) = C\left[1 - e^{-\frac{h}{a}}\right] \tag{12.21}$$

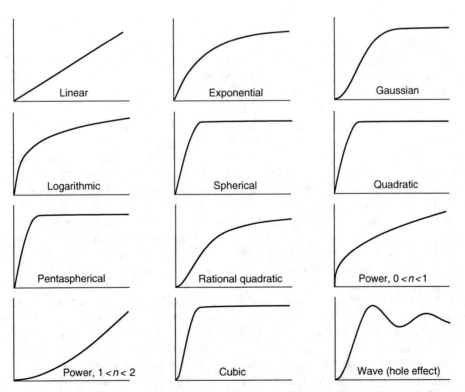

FIGURE 12.22 Some of the most common theoretical semivariograms. (From Golden Software, Inc., *SURFER: User's Guide. Contouring and 3D Surface Mapping for Scientists and Engineers*, Golden Software, Inc., Golden, Colorado, 2002, 640 p.)

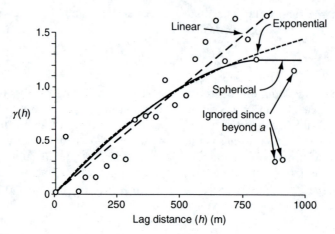

FIGURE 12.23 Fitted standardized theoretical semivariogram models for the July 15, 1994 water table recorded in the Blue River alluvial aquifer.

and the linear model is

$$\gamma(h) = bh \tag{12.22}$$

where b is the slope of the straight line.

Once the "right" semivariogram model and its parameters are selected, the kriging contouring method will use them to interpolate a map. Again, the selection of a model is rather subjective and requires experience, including the possibility of combining several models. For more information on semivariograms and kriging, see for example, Clark (1979), Isaaks and Srivastava (1989), Pannatier (1996), Kitanidis (1997), Webster and Oliver (2001), and Cressie (1991).

Figure 12.24 shows a map of the water table in the Blue River alluvial aquifer generated with the linear kriging model within SURFER contouring software (published by Golden Software, Golden, Colorado). The map does not accurately represent the role of the Blue River as a gaining stream because there is not enough data on the stream elevations. In addition, contours along the edges of the map, where there are less data (especially in northeast and southwest), tend to bend away from the river, which does not seem "natural" for the base flow period. (Note that during base flow conditions, i.e., during extended periods without significant precipitation, the river stage is at its lowest and the supply of water comes from the aquifer only. This means that groundwater is finding the shortest way to reach the river, i.e., the streamlines are more perpendicular to the river, and the contour lines are then more parallel to the river.)

Figure 12.25 shows a much more acceptable map after five more data points are estimated along the river (see Table 12.2). However, the contours remain bent and questionable in the corners of the map indicating that more adjustments have to be done to generate the final *XYZ* computer file. Even though kriging is widely considered as a reliable contouring method, it is recommended to try other methods available within the chosen software package since some may work better with a particular set of data. Interpolation is not an exact process and can yield often very different results. It is iterative, with the final accepted result assessed qualitatively.

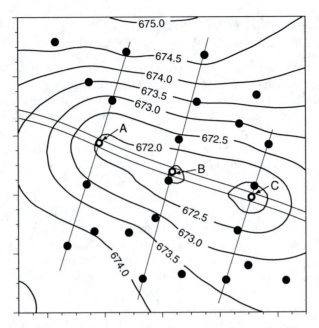

FIGURE 12.24 July 15, 1994 water table contour map of the Blue River alluvial aquifer generated using the linear kriging model; three river elevations at intercepts with three cross sections are estimated.

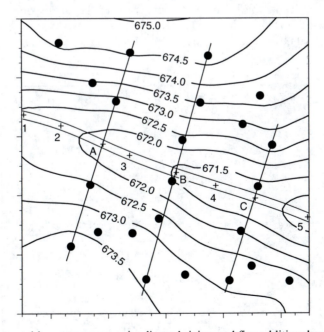

FIGURE 12.25 Water table contour map using linear kriging and five additional estimated data points along the river.

12.4 FLOW NETS IN HETEROGENEOUS AQUIFERS, REFRACTION OF FLOW LINES, AND GROUNDWATER VELOCITY

Estimate the path of groundwater contamination that would develop from the spill site shown in Figure 12.26 if no remedial action were taken. Three monitoring wells, MW1, MW2, and MW3, were drilled around the spill to determine the immediate direction of groundwater flow. Water table elevations recorded in the unconfined silty sand aquifer were 467.98, 467.82, and 467.53 m above sea level, respectively. The hydraulic conductivity of the porous medium was estimated to be 4.3×10^{-6} m/s based on laboratory analyses of core samples. Two short pumping tests were performed in the nearby domestic wells DW1 and DW2 and yielded hydraulic conductivities of 1.45×10^{-5} and 6.11×10^{-6} m/s, respectively. The water table elevations recorded in DW1 and DW2 before the test were 467.32 and 467.23 m. The hydraulic conductivity of the main sandy gravel aquifer in the area, which has been alternately and discontinuously pumped by two municipal wells, is 2.81×10^{-5} m/s. The water table recorded in W1 during a long pumping break was 466.91 m. If it was shown that the main aquifer could be threatened by the spill, the time needed for the contamination to reach the aquifer is estimated assuming that there is no significant retardation of the contaminant.

12.4.1 GROUNDWATER FLOW DIRECTION

The first step in dealing with an accidental spill of hazardous materials that have the potential for groundwater contamination is to determine the direction of groundwater flow in the immediate vicinity of the spill. This is done by recording the water table elevations in the nearby existing water wells or, if they are not available, in newly drilled monitoring wells as in our case. A minimum of three wells placed around the spill is needed to determine the direction of groundwater flow as shown in Figure 12.27. The water table contour lines are drawn using exact linear interpolation among the three points. The flow line (i.e., the direction of groundwater flow) through the center of the spill is drawn perpendicular to the contours and is directed downgradient, toward north–northwest. This first analysis shows that, in general, the flow is indeed oriented toward the municipal wells.

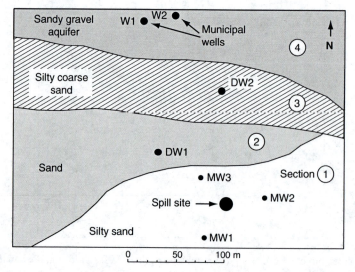

FIGURE 12.26 Geologic map of the spill area with the location of new monitoring wells (MW), domestic wells (DW), and municipal wells pumping the main unconfined aquifer (W).

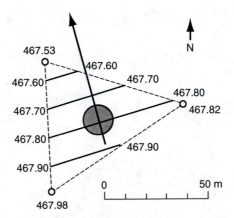

FIGURE 12.27 Water table contours and direction of groundwater flow determined by the triangular linear interpolation of data recorded in monitoring wells MW1, MW2, and MW3.

The flow direction determined in this way is valid only for the silty sand hydrogeologic unit. It would change at the boundary between two different porous media having different hydraulic conductivities—silty sand and sand. In other words, the flow line representing the flow in the spill area will refract at the boundary and a new direction of flow will have to be determined for the sand unit. Similarly, the flow will change direction at each subsequent boundary between two porous media with different hydraulic conductivities.

12.4.2 Estimated Path of Contamination

Flow lines representing the direction of groundwater flow refract at the boundary between porous media with different hydraulic conductivities. The relationship between the angle of incidence (α_1) and the angle of refraction (α_2), and the two related hydraulic conductivities (K_1 and K_2) is (see also Equation 1.126):

$$\frac{K_1}{K_2} = \frac{\tan \alpha_1}{\tan \alpha_2} \tag{12.23}$$

Since the spill is localized in our case (less than 20 m in diameter), it is enough to analyze just one flow line drawn through the spill center (see Figure 12.28). The actual spread of contamination will depend largely on the amount of transverse dispersion and will have more or less diverging outer limits. The flow line in the first hydrogeologic unit (silty sand) in the spill area is determined as shown in Figure 12.27. After reaching the first boundary, the flow line refracts with the angle of refraction calculated using Equation 12.23:

$$\tan \alpha_2 = \frac{K_2 \times \tan \alpha_1}{K_1} \tag{12.24}$$

$$\alpha_2 = \arctan \left(\frac{K_2 \times \tan \alpha_1}{K_1} \right) \tag{12.25}$$

where α_1 is the angle of incidence measured on the map and K_1 and K_2 are the hydraulic conductivities of the two porous media.

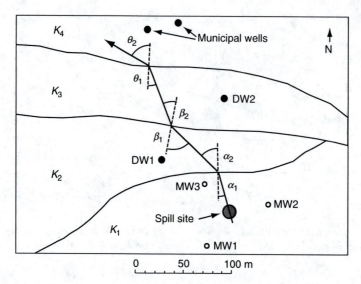

FIGURE 12.28 Flow line representing the path of possible contamination from the spill site. The angles of incidence at the boundaries (α_1, β_1, and θ_1) are measured on the map, and the angles of refraction (α_2, β_2, and θ_2) are calculated using general Equation 12.23 for each boundary (where β_2 and θ_2 replace α_2 in the notification). Note that an assumption was made that the groundwater flow is not disturbed by pumping wells or other sinks or sources.

The angle of refraction is found after substituting known values into Equation 12.25:

$$\alpha_2 = \arctan\left(\frac{1.45 \times 10^{-5}\ \text{m/s} \times \tan 16^\circ}{4.30 \times 10^{-6}\ \text{m/s}}\right)$$

$$\alpha_2 = \arctan 0.96569 = 44^\circ$$

The flow line is extended at the 44° angle until the next boundary. Next, the refraction angle β_2 is found by applying Equation 12.23 again (note that the angle of incidence, β_1, is measured on the map):

$$\beta_2 = \arctan\left(\frac{K_3 \times \tan\ \beta_1}{K_2}\right)$$

$$\beta_2 = \arctan\left(\frac{6.11 \times 10^{-5}\ \text{m/s} \times \tan\ 56^\circ}{1.45 \times 10^{-5}\ \text{m/s}}\right)$$

$$\beta_2 = 32^\circ \tag{12.26}$$

Finally, after further extending the streamline, the angle of refraction at the last boundary is found as (θ_1 is measured on the map)

$$\theta_2 = \arctan\left(\frac{K_4 \times \tan\ \theta_1}{K_3}\right)$$

$$\theta_2 = \arctan\left(\frac{2.81 \times 10^{-5}\ \text{m/s} \times \tan\ 25^\circ}{6.11 \times 10^{-6}\ \text{m/s}}\right)$$

$$\theta_2 = 65^\circ \tag{12.27}$$

The flow line shown in Figure 12.28 is drawn assuming no disturbance of groundwater flow by pumping wells or other sinks or sources. If there was a significant groundwater withdrawal in the area such as a continuous pumping in two municipal wells, this streamline and the corresponding equipotential lines would look different.

12.4.3 FLOW VELOCITY

Linear groundwater velocity (v_L) is found as

$$v_L = \frac{K \times i}{n_{eff}}$$

(12.28)

where K is the hydraulic conductivity, i is the hydraulic gradient, and n_{eff} is the effective porosity. Note that the linear groundwater velocity is accurate when used to estimate the average travel time of groundwater, and Darcy's velocity is accurate for calculating flow rates. Neither, however, is the real groundwater velocity, which is the time of travel of a water particle along its actual path through the voids. It is obvious that, for practical purposes, the real velocity cannot be measured or calculated.

As seen from Equation 12.28, there are three elements needed to calculate the linear groundwater velocity in each of the four hydrogeologic sections shown in Figure 12.26: the hydraulic gradient (i), the hydraulic conductivity (K), and the effective porosity (n_{eff}). In our case, the hydraulic conductivities are known, the hydraulic gradients can be determined from the drawn streamline and the water table contours, while the effective porosities have to be estimated. The hydraulic gradient in Section 1 on the map in Figure 12.29 is:

$$i_1 = \frac{h_A - h_B}{L_{AB}}$$

$$i_1 = \frac{467.78 \text{ m} - 467.47 \text{ m}}{40 \text{ m}} = 0.00775$$

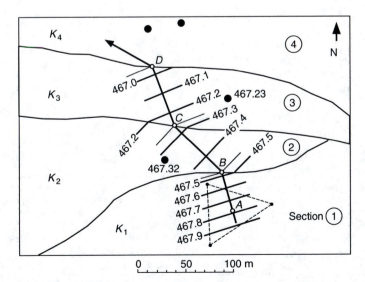

FIGURE 12.29 The streamline representing the path of potential contamination and the water table contour lines drawn perpendicular to it. The contours are constructed by the exact linear interpolation using field data and following the refraction rules. Contour interval is 0.1 m.

where h_A and h_B are the hydraulic heads at points A and B, respectively, whose values are determined from the water table contour map and L_{AB} is the distance between points A and B measured on the map.

As discussed in Chapter 1.3, the effective porosity is always smaller than the overall porosity of a porous medium. This difference is up to several percent for clean, uniform sediments that are medium-to-coarse grained. When a sedimentary deposit is a mixture of various grain sizes this difference becomes larger, especially if there is a significant amount of fine sediments such as silt and clay. Some uniform clays have the overall porosity above 80% while their effective porosity is only 1 to 2%. The overall porosity of various rocks is given in Appendix C and similar tables can be found throughout literature. Because of the wide possible ranges and much less experimental data available, tables of the effective porosity are not common and are less accurate. Effective porosity can be estimated from the specific yield information (see Figure 1.21 and Figure 1.22), or simple rules can be followed when estimating it from the tables of overall porosity (see Figure 1.19):

- For clean uniform sands and gravels, the difference between the two porosities is up to 5%, usually less than that.
- For nonuniform sands and gravels, and their mixtures, the effective porosity is up to 10% smaller than the overall porosity.
- Presence of fines, i.e., silts and clays, rapidly reduces the effective porosity: a 50:50 mixture of uniform sand and clay can have the overall porosity of 50%, while the effective porosity can be less than 5%.
- The effective porosity of clean clays is always smaller than 5%, often only 1% or less.

Using these rules and information from Figure 1.19, Figure 1.21, and Figure 1.22, the following values of the effective porosity are estimated for the hydrogeologic units in our case:

- Silty sand: 20%
- Sand: 35%
- Silty coarse sand: 15%
- Sandy gravel: 25%

Finally, the linear groundwater velocity in Section 1 is determined by inserting all known, calculated, and estimated values into Equation 12.28:

$$v_{L1} = \frac{K_1 \times i_1}{n_{eff1}}$$

$$v_{L1} = \frac{4.30 \times 10^{-6} \text{ m/s} \times 0.00775}{0.20} = 1.67 \times 10^{-7} \text{ m/s}$$

$$v_{L1} = 0.0144 \text{ m/d}$$

The linear velocity in Section 2 is

$$v_{L2} = \frac{K_2 \times \dfrac{h_B - h_C}{L_{BC}}}{n_{eff2}}$$

$$v_{L2} = \frac{1.45 \times 10^{-5} \text{ m/s} \times \dfrac{467.47 \text{ m} - 467.27 \text{ m}}{67 \text{ m}}}{0.35} = 1.24 \times 10^{-7} \text{ m/s}$$

$$v_{L2} = 0.0107 \text{ m/d}$$

The linear velocity in Section 3 is

$$v_{L3} = \frac{K_3 \times \dfrac{h_C - h_D}{L_{CD}}}{n_{eff3}}$$

$$v_{L3} = \frac{6.11 \times 10^{-6} \text{ m/s} \times \dfrac{467.27 \text{ m} - 466.96 \text{ m}}{63 \text{ m}}}{0.15} = 2.00 \times 10^{-7} \text{ m/s}$$

$$v_{L3} = 0.0173 \text{ m/d}$$

12.4.4 TRAVEL TIME

The time needed for the potential groundwater contamination caused by the spill to reach the sandy gravel aquifer (Section 4) is determined using the basic velocity equation:

$$v_L = \frac{L}{t} \tag{12.29}$$

where v_L is the linear velocity, L is the distance traveled, and t is the travel time. The total time is the sum of travel times in each of the three sections:

$$t = t_1 + t_2 + t_3 \tag{12.30}$$

$$t = \frac{L_{AB}}{v_{L1}} + \frac{L_{BC}}{v_{L2}} + \frac{L_{CD}}{v_{L3}}$$

$$t = \frac{40 \text{ m}}{0.0144 \text{ m/d}} + \frac{67 \text{ m}}{0.0107 \text{ m/d}} + \frac{63 \text{ m}}{0.0173 \text{ m/d}} \tag{12.31}$$

$$t = 7.6 \text{ y} + 17.2 \text{ y} + 10.0 \text{ y} = 34.8 \text{ y}$$

The time of approximately 35 y is calculated on the assumption that the flow pattern shown in Figure 12.29 is not influenced by groundwater withdrawal at the two municipal wells. A significant pumping in the wells would alter the shape of the guiding flow line and the contour lines. It would also likely induce steeper hydraulic gradients and reduce the travel time.

REFERENCES

Carlson, R.E. and Foley, T.A., 1991. *Radial Basis Interpolation on Track Data*, Lawrence Livermore National Laboratory, UCRL-JC-1074238.

Clark, I., 1979. *Practical Geostatistics*. Applied Science Publishers, London, 129 p.

Cressie, N., 1991. *Statistics for Spatial Data*. John Wiley & Sons, New York, 900 p.

Golden Software, Inc., 2002. *SURFER: User's Guide. Contouring and 3D Surface Mapping for Scientists and Engineers*. Golden Software, Inc., Golden, Colorado, 640 p.

Isaaks, E. and Srivastava, M., 1989. *An Introduction to Applied Geostatistics*. Oxford University Press, New York, 562 p.

Kitanidis, P.K., 1997. *Introduction to Geostatistics, Applications in Hydrogeology*. Cambridge University Press, Cambridge, 249 p.

Pannatier, Y., 1996. *VARIOWIN—Software for Spatial Data Analysis in 2D*. Springer-Verlag, New York, 91 p.

Webster, R. and Oliver, M.A., 2001. *Geostatistics for Environmental Scientists*. John Wiley & Sons, Chichester, England, 271 p.

13 Steady-State Flow to Water Wells

13.1 CONFINED AQUIFER

An extraction (pumping) well and three monitoring wells were used to conduct a pumping test in a confined aquifer. The location of the wells and a simplified hydrogeologic cross section are shown in Figure 13.1 and Figure 13.2, respectively. Monitoring wells MW1, MW2, and MW3 are 8, 35, and 120 m from the pumping well PW, respectively. The radius of the pumping well in its screen section is 0.3 m. The test conducted was a three-step pumping test with each step lasting 6 h. The pumping rates were 10, 15, and 20 L/s. Table 13.1 lists the drawdown data for all wells recorded at the end of each step. From the available data, assuming quasi–steady-state conditions, determine the transmissivity and the hydraulic conductivity of the aquifer.

Figure 13.3 is the plot of drawdown versus distance for all three steps for all the wells. The graph shows that the data for the monitoring wells MW1, MW2, and MW3 form straight lines for all three steps. Note that the drawdown recorded in the pumping well does not fall on the straight line in any of the steps; it is below the straight lines indicating that there is an additional drawdown in the well because of the well loss. This well loss increases for each step due to the increasing pumping rates. The well loss is explained in detail in Section 15.2. Theory behind the following equation describing drawdown at any radial distance from the pumping well is given in Section 2.5.1:

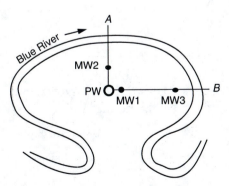

FIGURE 13.1 Map view of the locations of the pumping well PW and three monitoring wells used for aquifer testing. Hydrogeologic cross section in Figure 13.2 is along *AB*.

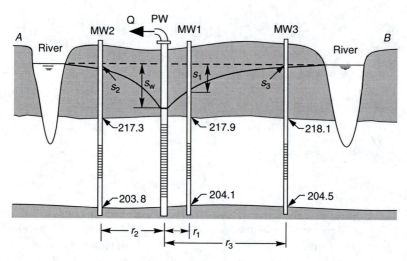

FIGURE 13.2 Simplified hydrogeologic cross section AB.

$$s_r = \frac{Q}{2\pi T} \ln \frac{R}{r} \tag{13.1}$$

Noting all constant terms in Equation 13.1, the drawdown can be expressed as the function of distance only:

$$s_r = \frac{Q}{2\pi T} \ln R - \frac{Q}{2\pi T} \ln r \tag{13.2}$$

or, when the constant terms are replaced with general numbers:

$$s_r = a - b \ln r \tag{13.3}$$

Equation 13.3 means that the recorded data of drawdown versus the radial distance from the well would form a straight line when plotted on a semilogarithmic graph paper.

TABLE 13.1
Pumping Rates and Recorded Drawdowns at the End of Each Step, for the Pumping Well (PW) and Three Monitoring Wells (MW1, MW2, MW3)

Well	Drawdown (m)		
	$Q = 10$ L/s	$Q = 15$ L/s	$Q = 20$ L/s
PW	1.10	1.84	2.72
MW1	0.54	0.80	1.08
MW2	0.32	0.48	0.65
MW3	0.17	0.24	0.33

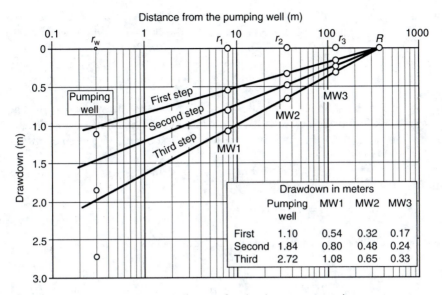

FIGURE 13.3 Graph of drawdown versus distance for the three-step pumping test.

The transmissivity of the aquifer can be determined from the drawdown versus distance graph for any of the three steps. This is done by relating the coordinates of any two points on the same straight line (line of the same step):

$$s_1 = \text{const} - \frac{Q}{2\pi T} \ln r_1 \tag{13.4}$$

$$s_2 = \text{const} - \frac{Q}{2\pi T} \ln r_2 \tag{13.5}$$

$$s_1 - s_2 = \Delta s = \frac{Q}{2\pi T} \ln \frac{r_2}{r_1} \tag{13.6}$$

The transmissivity is then found from this difference in drawdown for the chosen points:

$$T = \frac{Q}{2\pi \Delta s} \ln \frac{r_2}{r_1} \tag{13.7}$$

For practical purposes it is better to express transmissivity in terms of the logarithm to the base 10 (log) as the corresponding graph is easier to use. Knowing that

$$\log x = 0.4343 \ln x \tag{13.8}$$

and replacing π with its value 3.14, Equation 13.8 becomes

$$T = \frac{0.366Q}{\Delta s} \log \frac{r_2}{r_1} \tag{13.9}$$

If two points on the straight line are chosen to be one log cycle apart (i.e., the ratio of the distance coordinates r_2 and r_1 is 10), Equation 13.9 reduces to the following simple form:

$$T = \frac{0.366Q}{\Delta s} \tag{13.10}$$

First step: For the monitoring wells MW1 and MW2, registered drawdown at the end of the first step is 0.54 and 0.32 m, respectively. The transmissivity is calculated using Equation 13.9 and with the knowledge that the pumping rate for the first step is 10 L/s (0.01 m³/s):

$$T = \frac{0.366 \times 0.01 \text{ m}^3/\text{s}}{(0.54 \text{ m} - 0.32 \text{ m})} \log \frac{35 \text{ m}}{8 \text{ m}}$$
$$T = 0.0107 \text{ m}^2/\text{s} = 1.07 \times 10^{-2} \text{ m}^2/\text{s}$$

The hydraulic conductivity is calculated from the equation

$$K = \frac{T}{b} \tag{13.11}$$

and from the knowledge that the average thickness of the aquifer determined from the logs for MW1, MW2, and MW3 (see Figure 13.2) is 13.6 m:

$$K = \frac{1.07 \times 10^{-2} \text{ m}^2/\text{s}}{13.6 \text{ m}} = 7.87 \times 10^{-4} \text{ m/s}$$

Choosing any two points on the straight line, other than the actual measurements, and applying Equation 13.9 or Equation 13.10 would yield the same result for T in this case because the actual measurements fall exactly on the straight line. However, if the actual data do not all exactly fall along the straight line (which is usually the case, especially if there are more monitoring wells), choosing two points on the best-fit line is more appropriate.

Second step: For the monitoring wells MW2 and MW3, registered drawdown at the end of the second step is 0.48 and 0.24 m, respectively (Figure 13.3). The transmissivity is calculated using Equation 13.9 and from the knowledge that the pumping rate for the second step is 15 L/s (0.015 m³/s):

$$T = \frac{0.366 \cdot 0.015 \text{ m}^3/\text{s}}{(0.48 \text{ m} - 0.24 \text{ m})} \log \frac{120 \text{ m}}{35 \text{ m}}$$
$$T = 0.0122 \text{ m}^2/\text{s} = 1.22 \times 10^{-2} \text{ m}^2/\text{s}$$
$$K = 9.04 \times 10^{-4} \text{ m/s}$$

Third step: For the monitoring wells MW1 and MW3, registered drawdown at the end of the third step is 1.08 and 0.33 m, respectively (Figure 13.3). The transmissivity is calculated using Equation 13.9 and with the knowledge that the pumping rate for the third step is 20 L/s (0.02 m³/s):

$$T = \frac{0.366 \times 0.02 \text{ m}^3/\text{s}}{(1.08 \text{ m} - 0.33 \text{ m})} \log \frac{120 \text{ m}}{8 \text{ m}}$$
$$T = 0.0115 \text{ m}^2/\text{s} = 1.15 \times 10^{-2} \text{ m}^2/\text{s}$$
$$K = 8.46 \times 10^{-4} \text{ m/s}$$

13.1.1 ALL THREE STEPS TOGETHER (DRAWDOWN/PUMPING RATE METHOD)

Another graphical method for determining aquifer transmissivity and hydraulic conductivity combines all three steps or drawdown into a single straight line on a semilogarithmic paper. This method is preferred over individual steps as it averages the response of the aquifers to several different stresses. The equation that gives the drawdown at any radial distance from the pumping well can be divided by the pumping rate:

$$s_r = \frac{Q}{2\pi T} \ln \frac{R}{r} \tag{13.12}$$

$$\frac{s_r}{Q} = \frac{1}{2\pi T} \ln \frac{R}{r} \tag{13.13}$$

The above equation rewritten to include logarithm to the base 10 (note that $\ln x = 2.3 \log x$, and $\pi = 3.14$) is

$$\frac{s_r}{Q} = \frac{0.366}{T} \log \frac{R}{r} \tag{13.14}$$

The ratio between the drawdown and the pumping rate can be expressed as the function of distance only because R and T are constant terms (assuming steady-state conditions and homogeneity of the aquifer):

$$\frac{s_r}{Q} = \frac{0.366}{T} \log R - \frac{0.366}{T} \ln r \tag{13.15}$$

$$\frac{s_r}{Q} = a - b \log r \tag{13.16}$$

The graph that shows points with the coordinates r (distance from the pumping well) and s/Q (registered drawdown over pumping rate) for all the wells and all three drawdowns is shown in Figure 13.4. As can be seen, the points for the individual monitoring wells have practically the same s/Q ratio for all three steps. However, this is not the case with the pumping well because of the increasing well loss. Note that, because of the well loss, the actual points for the pumping well are below the straight line formed by the monitoring wells.

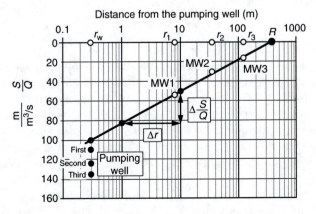

FIGURE 13.4 Graph of distance versus drawdown and pumping rate ratio for the three-step pumping test.

The transmissivity is obtained by relating coordinates of any two points on the straight line formed by the monitoring wells:

$$\left(\frac{s}{Q}\right)_1 = \frac{0.366}{T} \log \frac{R}{r_1} \tag{13.17}$$

$$\left(\frac{s}{Q}\right)_2 = \frac{0.366}{T} \log \frac{R}{r_2} \tag{13.18}$$

$$T = \frac{0.366}{(s/Q)_2 - (s/Q)_1} \log \frac{r_2}{r_1} \tag{13.19}$$

Again, if two points are chosen so that the ratio between their distance coordinates is 10 (note that log 10 equals 1), Equation 13.19 becomes

$$T = \frac{0.366}{\Delta(s/Q)} \tag{13.20}$$

In our example the two points chosen have distance coordinates 2 and 20 m, respectively (see Figure 13.4), which gives the ratio of 10. Reading from the graph that $(s/Q)_2 - (s/Q)_1 = 31.5$ s/m^2, the transmissivity is calculated using Equation 13.20:

$$T = \frac{0.366}{31.5 \text{ s/m}^2} = 1.16 \times 10^{-2} \text{ m}^2/\text{s}$$

The hydraulic conductivity is $K = T/b$, or $K = 8.53 \times 10^{-4}$ m/s.

13.1.2 RADIUS OF WELL INFLUENCE

The radius of well influence is determined from the graphs in Figure 13.3 or Figure 13.4. It is the intercept of the straight line formed by the monitoring wells and the zero drawdown. As seen from the graph in Figure 13.3, this value remains constant for all three steps indicating the position of the meandering river (strong equipotential boundary). Although the pumping well's location may not be in the center of the meander, and the meander itself may be of a more or less irregular shape, the radius of well influence determined this way corresponds to a circle hydraulically equivalent to the actual meander. The radius of well influence determined from both graphs is $R = 375$ m.

13.2 UNCONFINED AQUIFER

A fully penetrating well is pumping from an unconfined aquifer as shown in Figure 13.5. Drawdowns registered in the pumping well (PW) and two monitoring wells (MW1 and MW2) reflect steady-state conditions and are 7.73, 2.74, and 1.63 m, respectively. Find the hydraulic conductivity of the aquifer and the radius of well influence if the pumping rate is 17.5 L/s (0.0175 m^3/s). Diameter of the pumping well is 300 mm and the monitoring wells MW1 and MW2 are at 8 and 30 m from PW, respectively. Also find the approximate position of the water table in the vicinity of the well, as well as the distance from the well at which Dupuit's assumption is valid.

Section 2.5.2 provides detailed derivation of the equation describing flow toward a fully penetrating well in an unconfined aquifer. The hydraulic conductivity is found from the

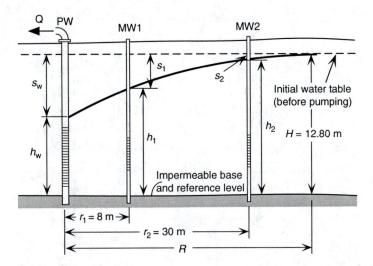

FIGURE 13.5 Hydrogeologic cross section of the unconfined aquifer with the pumping well PW and two monitoring wells MW1 and MW2.

following equation relating two hydraulic heads (h_1 and h_2) measured at two radial distances from the pumping well (r_1 and r_2):

$$K = \frac{Q \ln(r_2/r_1)}{\pi(h_2^2 - h_1^2)} \tag{13.21}$$

The hydraulic heads in MW1 and MW2 to be substituted into Equation 13.21 can be found from the data on recorded drawdown and the initial hydraulic head (see Figure 13.5):

$$h_2 = H - s_2 = 12.80 \text{ m} - 1.63 \text{ m} = 11.17 \text{ m}$$
$$h_1 = H - s_1 = 12.80 \text{ m} - 2.74 \text{ m} = 10.06 \text{ m}$$

and the hydraulic conductivity is then

$$K = \frac{0.0175 \text{ m}^3/\text{s} \cdot \ln\dfrac{30 \text{ m}}{8 \text{ m}}}{3.14 \cdot \left[(11.17 \text{ m})^2 - (10.06 \text{ m})^2\right]} = 1.34 \times 10^{-3} \text{ m/s}$$

If there are more than two monitoring wells, or there are several pumping tests, and it is required to find an average hydraulic conductivity representative of the tested aquifer, the graphoanalytical method presented below is recommended.

From the general equation for groundwater flow toward a fully penetrating well in an unconfined aquifer

$$H^2 - h^2 = \frac{Q}{\pi K} \ln \frac{R}{r} \tag{13.22}$$

it can be seen that the recorded cone of depression would appear as a straight line if plotted on a semilogarithmic graph $H^2 - h^2 = f(\log r)$ if all known terms are replaced with a constant:

$$H^2 - h^2 = \text{const} - \frac{0.733 \cdot Q}{K} \log r \tag{13.23}$$

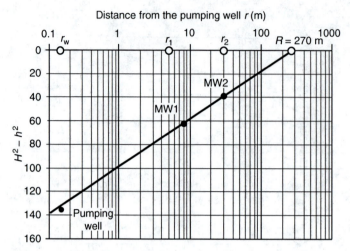

FIGURE 13.6 Semilogarithmic graph of $H^2 - h^2$ versus radial distance from a pumping well (r) for determining radius of well influence.

Figure 13.6 shows data for PW, MW1, and MW2 plotted on such a graph. Coordinates of the points are:

$$\text{PW:} \quad H^2 - h_w^2 = (12.80 \text{ m})^2 - (5.28 \text{ m})^2 = 135.96 \text{ m}^2; \quad r_w = 0.150 \text{ m}$$

$$\text{MW1:} \quad H^2 - h_w^2 = (12.80 \text{ m})^2 - (10.06 \text{ m})^2 = 62.6 \text{ m}^2; \quad r_1 = 8 \text{ m}$$

$$\text{MW2:} \quad H^2 - h_w^2 = (12.80 \text{ m})^2 - (11.17 \text{ m})^2 = 39.07 \text{ m}^2; \quad r_2 = 30 \text{ m}$$

The radius of well influence is determined by the interception of the straight line drawn through the monitoring-well points with the line of zero drawdown, i.e., for $H^2 - h^2 = 0$. In our case, the radius of well influence is $R = 270$ m. Note that the straight line should be drawn only through the monitoring-well points because the actual drawdown recorded in the pumping well includes an additional, undetermined component and the pumping-well point in most cases is below the theoretical straight line. This additional drawdown is a consequence of well losses, as well as the formation of a seepage face along the well perimeter (see Section 15.2).

Graphs, such as the one shown in Figure 13.6, can also be used to determine the aquifer's hydraulic conductivity. The coordinates of any two points on the straight line, chosen to be one log cycle apart (i.e., the ratio of the distance coordinates r_2 and r_1 is 10), can be used to calculate the hydraulic conductivity using Equation 13.23:

$$K = \frac{0.733 \cdot Q}{\Delta(H^2 - h^2)} \quad (13.24)$$

For the pumping test shown in Figure 13.6 and for the pumping rate of 17.5 L/s (0.0175 m³/s) the hydraulic conductivity is

$$K = \frac{0.733 \times 0.0175 \text{ m}^3/\text{s}}{40 \text{ m}^2} = 3.21 \times 10^{-4} \text{ m/s}$$

13.2.1 POSITION OF WATER TABLE NEAR THE WELL

An equation, based on Dupuit's assumption, that describes the position of the water table (h) at a distance r from the pumping well is

$$h = \sqrt{h_w^2 + \frac{Q}{\pi K} \ln \frac{r}{r_w}} \tag{13.25}$$

where h_w is the hydraulic head recorded in the well and r_w is the well radius. This equation is also known as Dupuit's parabola. The well pumping rate is given as

$$Q = \pi K \frac{(H^2 - h_w^2)}{\ln(R/r_w)} \tag{13.26}$$

By inserting Equation 13.26 into Equation 13.25, the hydraulic head is expressed only in terms of other known heads and distances at which they are recorded. This is a more convenient method of calculating Dupuit's parabola as one does not need to know the hydraulic conductivity and the well pumping rate. Therefore

$$h = \sqrt{h_w^2 + (H^2 - h_w^2) \frac{\ln(r/r_w)}{\ln(R/r_w)}} \tag{13.27}$$

where H is the initial hydraulic head (before pumping) and R is the radius of well influence. The position of the actual water table is above the Dupuit's parabola and several experimental (approximation) equations have been proposed for its calculation. One of the more common is the Babbitt–Caldwell equation (1948, adapted from Kashef, 1987), which estimates the actual hydraulic head at the well perimeter as

$$h_0 = H - \frac{0.6}{H} \times \frac{H^2 - h_w^2}{\ln(R/r_w)} \times \ln \frac{R}{0.1H} \tag{13.28}$$

In our case this gives the following value:

$$h_0 = 12.8 \text{ m} - \frac{0.6}{12.8 \text{ m}} \times \frac{(12.8 \text{ m})^2 - (5.07 \text{ m})^2}{\ln(270 \text{ m}/0.15 \text{ m})} \times \ln \frac{270 \text{ m}}{0.1 \times 12.8 \text{ m}} = 8.18 \text{ m}$$

which means that the seepage face at the well perimeter is

$$\Delta h = h_0 - h_w = 8.18 \text{ m} - 5.07 \text{ m} = 3.11 \text{ m}$$

Another equation, proposed by Schneebeli, is also used for determining the height of the seepage face (Δh) at the well perimeter (adapted from Vukovic and Soro, 1984):

$$\Delta h = \sqrt{h_w^2 + \frac{Q}{\pi K} \left[0.4343 \ln \frac{Q}{\pi K r_w^2} - 0.4 \right]} - h_w \tag{13.29}$$

TABLE 13.2
Calculation of Distance $r*$ by Trial, Using Equation 13.32

$r*$ (m)	30	20	10	5	4.5	4.9	4.8
Calculated C	333	215	101.5	47.52	42.31	46.47	45.43
Actual C	46.07	46.07	46.07	46.07	46.07	46.07	46.07

or

$$\Delta h = \sqrt{h_w^2 + \frac{H^2 - h_w^2}{\ln(R/r_w)}\left[0.4343\ln\frac{H^2 - h_w^2}{r_w^2 \cdot \ln(R/r_w)} - 0.4\right]} - h_w \qquad (13.30)$$

which in our case gives 3.41 m after substituting the known values for the hydraulic heads and the radial distances. The position of the actual water table (h_0) at the well perimeter is then

$$h_0 = h_w + \Delta h = 5.07 \text{ m} + 3.41 \text{ m} = 8.48 \text{ m} \qquad (13.31)$$

The position of the water table in the vicinity of the pumping well is estimated based on the calculated seepage face and a distance from the well at which Dupuit's assumption is valid, i.e., where Dupuit's parabola and the actual water table can be considered close enough for practical purposes. This distance ($r*$) is found for the condition that $dh/dr = 0.2$, which, for most practical purposes, provides that the actual water table and the Dupuit's parabola are close enough to each other (Vukovic and Soro, 1984):

$$\frac{2.5(H^2 - h_w^2)}{\ln(R/r_w)} = r* \times \sqrt{h_w^2 + (H^2 - h_w^2)\frac{\ln(r*/r_w)}{\ln(R/r_w)}} \qquad (13.32)$$

$$\text{Const} = r* \times \sqrt{h_w^2 + (H^2 - h_w^2)\frac{\ln(r*/r_w)}{\ln(R/r_w)}} \qquad (13.32a)$$

As can be seen from Equation 13.32, distance $r*$ is expressed implicitly and its value of 4.9 m (in our case) was found by trial (see Table 13.2). As practice shows, distance $r*$ is usually within a radius of 10 m from the pumping well. In cases of extremely high well losses, $r*$ can be placed at greater distances.

Table 13.3 shows calculations of the water table around pumping well PW using Equation 13.27 (Dupuit's parabola). The results are plotted in Figure 13.7 together with the estimated

TABLE 13.3
Values of the Calculated Water Table Using Equation 13.27 (Dupuit's Parabola) in the Vicinity of PW

Distance from PW (m)	0.15	1	2	3	4	5	6	7	8	9	10	15
Dupuit's parabola (m)	5.07	7.79	8.57	8.99	9.28	9.5	9.68	9.82	9.95	10.06	10.15	10.51
Measured in wells (m)	5.07										10.06	

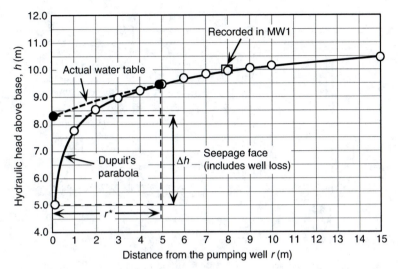

FIGURE 13.7 Calculated (Dupuit's parabola) and estimated real water table in the vicinity of pumping well PW.

position of the actual water table. The latter one is constructed by connecting the upper limit of the seepage face (found as the average value of Equation 13.28 and Equation 13.31: 8.33 m) with the point on the Dupuit's parabola at distance $r^* = 4.9$ m. This line should follow the trend of Dupuit's parabola beyond the radius of 4.9 m. Also plotted in Figure 13.7 is the recorded water table in MW1, which is 8 m from the pumping well. As can be seen, the actual water table is very close to the line calculated using Equation 13.27. Again, this confirms common experience that a significant difference between the actual water table and Dupuit's parabola occurs only in the immediate vicinity of a pumping well. Finding (calculating) this difference is important in cases of construction dewatering, drainage works, and other related well operations.

13.3   UNCONFINED AQUIFER WITH INFILTRATION

A water supply well at a paper factory is tapping an unconfined aquifer that is exposed to significant infiltration from precipitation. An incomplete record shows a fairly constant drawdown in the well and in the two nearby abandoned wells (see Figure 13.8) of 5.80, 1.65, and 0.95 m, respectively. At the time of well completion, the static water table was at 100.80 m above mean sea level. The diameter of the well is 400 mm and the hydraulic conductivity of the aquifer was determined to be 9.3×10^{-5} m/s. According to the production record, the installed water supply system has been consistently using 137,500 gal of water per day for the last 3 y. Although there is no exact data on the well pumping rate, it is suspected that there is a loss of water somewhere in the system. The average annual precipitation in the area was 29 in. for the last 3 y. From the available information find the average pumping rate of the well and the average annual infiltration rate.

13.3.1   Radius of Well Influence According to Dupuit

The general equation of groundwater flow toward a fully penetrating well in an unconfined aquifer is

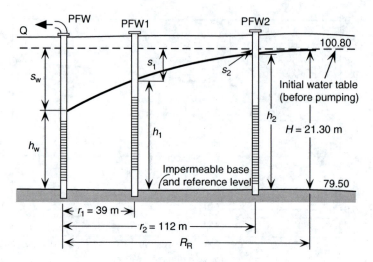

FIGURE 13.8 Hydrogeologic cross section of the unconfined aquifer with the pumping well PFW and two abandoned wells PFW1 and PFW2.

$$H^2 - h^2 = \frac{Q}{\pi K} \ln \frac{R_D}{r} \qquad (13.33)$$

where Q is the well pumping rate, h is the hydraulic head at distance r from the pumping well, H is the initial hydraulic head and R_D is the radius of well influence. In our case this formula cannot be applied directly as the pumping rate is not known. In addition, although the infiltration plays an important role in the system, Equation 13.33 does not include it. However, one can assume that the pumping rate is generated entirely by the infiltration rate as the system is in equilibrium, i.e., in steady state. This means that Equation 13.33 can be used to determine the radius of well influence as explained with Equation 13.34 through Equation 13.37, i.e., using a graph of $H^2 - h^2$ versus distance r (see Figure 13.9). Note, however, that this radius of well influence is not the real (actual) one. Rather, it is calculated, i.e., a graphically determined radius of influence according to Dupuit's assumptions. The real radius of influence is somewhat larger and it depends on the infiltration.

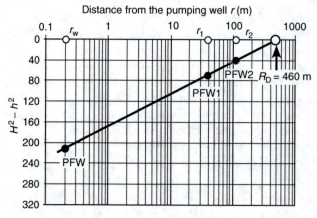

FIGURE 13.9 Semilogarithmic graph of $H^2 - h^2$ versus radial distance from a pumping well (r) for determining radius of well influence according to Dupuit.

From the basic rules of logarithmic calculus (see Appendix B), Equation 13.33 can be written as

$$H^2 - h^2 = \frac{Q}{\pi K} \ln R_D - \frac{Q}{\pi K} \ln r \qquad (13.34)$$

Since all the factors of the first term on the right-hand side are constant values (note that the radius of well influence does not change in time for steady-state conditions), Equation 13.34 can be expressed as

$$H^2 - h^2 = \text{const} - \frac{Q}{\pi K} \ln r \qquad (13.35)$$

or, when the natural logarithm is replaced with the logarithm to the base 10:

$$H^2 - h^2 = \text{const} - \frac{Q}{\pi K} \times 2.303 \log r \qquad (13.36)$$

$$H^2 - h^2 = \text{const} - \frac{0.733 Q}{K} \log r \qquad (13.37)$$

From Equation 13.37 it can be seen that the recorded cone of depression would appear as a straight line if plotted on a semilogarithmic graph, i.e., $H^2 - h^2 = f(\log r)$. Figure 13.9 shows data for PFW, PFW1, and PFW2 plotted on such a graph. Coordinates of the points are

PFW: $H^2 - h_w^2 = (21.30 \text{ m})^2 - (15.50 \text{ m})^2 = 213.4 \text{ m}^2$; $r_w = 0.200 \text{ m}$

PFW1: $H^2 - h_w^2 = (21.30 \text{ m})^2 - (19.65 \text{ m})^2 = 67.6 \text{ m}^2$; $r_1 = 39 \text{ m}$

PFW2: $H^2 - h_w^2 = (21.30 \text{ m})^2 - (20.35 \text{ m})^2 = 39.6 \text{ m}^2$; $r_2 = 112 \text{ m}$

The radius of well influence according to Dupuit (R_D) is determined by the interception of the straight line drawn through the data points with the line of zero drawdown, i.e., for $H^2 - h^2 = 0$. In our case, Dupuit's radius of well influence is $R_D = 460$ m. Note that the straight line should be drawn only through the monitoring-well points, as the actual drawdown recorded in the pumping well includes an additional undetermined component and the pumping-well point in most cases is below the theoretical straight line. This additional drawdown is a consequence of well losses, as well as the formation of a seepage face along the well perimeter. It appears, however, that in our case the pumping-well loss is not significant.

Once the radius of influence according to Dupuit is determined, it is possible to find the actual radius of influence and apply the differential equation of groundwater flow toward a fully penetrating well in an unconfined aquifer under the influence of infiltration.

Figure 13.10 shows elements of groundwater flow toward a fully penetrating well in an unconfined aquifer that is exposed to infiltration from the surface. The flow, or pumping rate at the well (Q_w), is the sum of two flows:

- the flow from the aquifer (Q_r) that enters the side of the cylinder with radius r, and
- the flow generated by the infiltration (Q_{inf}) over the area with radius r.

(Note that the flow from the aquifer is also generated by the infiltration, but, for the purpose of deriving the differential equation, it is considered simply as a flow through porous medium due to the existing hydraulic gradient.)

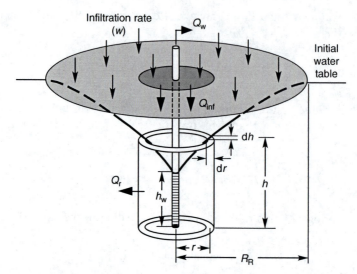

FIGURE 13.10 Scheme for deriving the differential equation of groundwater flow toward a fully penetrating well in an unconfined aquifer under the influence of infiltration.

According to Dupuit's assumptions, the flow through the porous medium at distance r from the well is described with the following equation:

$$Q_r = 2r\pi hK\frac{dh}{dr} \tag{13.38}$$

where h is the hydraulic head at distance r. This head is also the height of the cylinder, i.e., the thickness of the saturated zone if the reference level is placed at the base of the aquifer.

The flow generated by the infiltration over the area with radius r is

$$Q_{inf} = r^2 \cdot \pi \cdot w \tag{13.39}$$

where w is the infiltration rate.

As already mentioned, the well pumping rate is the sum of two flows:

$$Q_w = Q_r + Q_{inf} \tag{13.40}$$

$$Q_w = 2\pi rhK\frac{dh}{dr} + r^2\pi w \tag{13.41}$$

In order to solve this differential equation, the unknowns have to be separated:

$$h\,dh = \frac{Q_w}{\pi K}\frac{dr}{r} - \frac{w}{2K}r\,dr \tag{13.42}$$

and then integrated for the given boundaries. In our case (Figure 13.10) the boundary conditions are

- At distance r_w (which is the well radius) the hydraulic head is h_w
- At distance r the hydraulic head is h, i.e.,

$$\int_{h_w}^{h} h \, dh = \frac{Q_w}{\pi K} \int_{r_w}^{r} \frac{dr}{r} - \frac{w}{2K} \int_{r_w}^{r} r \, dr \tag{13.43}$$

All the integrals in the above equation are elementary and are easily solved (see Appendix B):

$$\frac{1}{2} h^2]_{h_w}^{h} = \frac{Q_w}{2\pi K} \times \ln r]_{r_w}^{r} - \frac{w}{2K} \times \frac{1}{2} r^2]_{r_w}^{r} \tag{13.44}$$

or

$$h^2 - h_w^2 = \frac{Q_w}{\pi K} \times \ln \frac{r}{r_w} - \frac{w}{2K} \times (r^2 - r_w^2) \tag{13.44a}$$

This is the final form of the equation describing flow toward a fully penetrating well in an unconfined aquifer under the influence of infiltration.

For the following boundaries:

- At distance R_R (which is the real radius of well influence) the hydraulic head is H (which is the initial head before pumping).
- At distance r_w the hydraulic head is h_w.

Equation 13.42 is solved as

$$H^2 - h_w^2 = \frac{Q_w}{\pi K} \ln \frac{R_R}{r_w} - \frac{w}{2K} (R_R^2 - r_w^2) \tag{13.45}$$

Since the squared well radius (r_w^2) is much smaller than the squared radius of well influence (R_R^2), it can be ignored and Equation 13.45 becomes

$$H^2 - h_w^2 = \frac{Q_w}{\pi K} \cdot \ln \frac{R_R}{r_w} - \frac{w}{2K} \cdot R_R^2 \tag{13.45a}$$

It can be seen that if there is no infiltration, the above equation is the same as Equation 13.33.

13.3.2 REAL RADIUS OF WELL INFLUENCE

For the equilibrium conditions assumed in our case, the flow at the well is generated entirely by the infiltration:

$$Q_w = R_R^2 \pi w \tag{13.46}$$

As shown in Figure 13.10, this flow is the product of the area exposed to infiltration and the infiltration rate that has dimensions of velocity. The real radius of well influence is then

$$R_R = \sqrt{\frac{Q_w}{\pi w}} \tag{13.47}$$

If the radius of well influence in Equation 13.45a is replaced by Equation 13.47, it gives

$$H^2 - h_w^2 = \frac{Q_w}{\pi K} \ln \frac{\sqrt{Q_w/\pi w}}{r_w} - \frac{w}{2K} \frac{Q_w}{\pi w} \qquad (13.48)$$

or, after simplification

$$H^2 - h_w^2 = \frac{Q_w}{\pi K} \times \ln \left(\frac{\sqrt{Q_w/\pi w}}{r_w} \right) - \frac{Q_w}{2\pi K} \qquad (13.49)$$

The above equation can also be written as

$$H^2 - h_w^2 = \frac{Q_w}{\pi K} \times \left[\ln \left(\frac{\sqrt{Q_w/\pi w}}{r_w} \right) - \frac{1}{2} \right] \qquad (13.50)$$

Knowing that

$$\ln e = 1 \quad \text{and} \quad 1/2 = \ln e^{1/2} = \ln \sqrt{e}$$

Equation 13.50 becomes

$$H^2 - h_w^2 = \frac{Q_w}{\pi K} \left(\ln \frac{\sqrt{Q_w/\pi we}}{r_w} \right) \qquad (13.51)$$

When the above equation is compared with Equation 13.33, it is apparent that, for the equivalent pumping rate and drawdown, the radius of well influence according to Dupuit (Dupuit's radius of well influence) is expressed as

$$R_D = \sqrt{\frac{Q_w}{\pi we}} \qquad (13.52)$$

As shown earlier, the real radius of well influence is given as

$$R_R = \sqrt{\frac{Q_w}{\pi w}}$$

which means that the two radii are connected through the following equation:

$$R_R = R_D \sqrt{e} \qquad (13.53)$$

Dupuit's radius of well influence determined graphoanalytically (see Figure 13.9) is 460 m and the real radius of well influence is

$$R_R = 460 \text{ m} \times 1.6487 = 758 \text{ m}$$

13.3.3 RATE OF INFILTRATION

The expression for the infiltration rate (w) is obtained by combining Equation 13.45 and Equation 13.46:

$$w = \frac{K}{R_R^2} \frac{(H^2 - h_w^2)}{\ln(R_R/r_w \sqrt{e})} \qquad (13.54)$$

which gives the following value when all the known values are inserted:

$$w = \frac{9.3 \times 10^{-5} \text{ m/s}}{(758 \text{ m})^2} \times \frac{(21.30 \text{ m})^2 - (15.50 \text{ m})^2}{\ln(758 \text{ m}/(0.2 \text{ m} \times 1.6487))}$$

$$w = 4.4635 \times 10^{-9} \text{ m/s} = 0.141 \text{ m/y}$$

The average annual infiltration is calculated from the infiltration rate and the average annual precipitation in the area (note that 29 in. of rain equals 0.737 m):

$$I = \frac{0.141 \text{ m}}{0.737 \text{ m}} \times 100\% = 19.1\%$$

13.3.4 WELL PUMPING RATE

The well pumping rate can be calculated from Equation 13.45a by solving for Q_w, or using Equation 13.46 directly:

$$Q_w = R_R^2 \pi w$$

$$Q_w = (758 \text{ m})^2 \cdot 3.14 \cdot 4.4635 \times 10^{-9} \text{ m/s}$$

$$Q_w = 8.05 \times 10^{-3} \text{ m}^3/\text{s}$$

which gives the total of 695.5 m³ or 183,730 gal of pumped water per day. This calculated value of the withdrawn groundwater is considerably larger than the amount of water used in the production (136,500 gal/d).

The efficiency of the water supply system is

$$\text{Efficiency} = \frac{\text{Used volume}}{\text{Withdrawn volume}} \times 100\% = \frac{136,500 \text{ gal}}{183,730 \text{ gal}} \times 100\% = 74.3\%$$

which indicates that there indeed may be a significant loss of water in the system requiring a thorough analysis. This analysis should, in addition to possible mechanical or distribution failures, include testing of the well efficiency.

13.4 CONFINED–UNCONFINED AQUIFER TRANSITION

An extraction well on a large river island (Figure 13.11) has been considered for a temporary lowering of the piezometric surface for construction purposes. The island's sandy aquifer, which is naturally confined, will have to undergo a change to unconfined conditions at the construction site because of the requirement that the piezometric surface must be kept at 1.5 m below the aquifer top during the construction (see Figure 13.12). Find the pumping rate at the well that will satisfy this requirement for safe construction. The well diameter is 500 mm. The hydraulic conductivity of the aquifer is 3.73×10^{-5} m/s. Other elements needed for the calculation are shown in Figure 13.12 through Figure 13.14.

The differential equation of groundwater flow toward a fully penetrating well in a confined aquifer, in steady-state conditions, is

$$Q = 2\pi r b K \frac{dh}{dr} \tag{13.55}$$

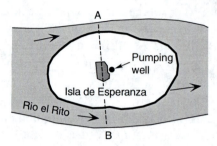

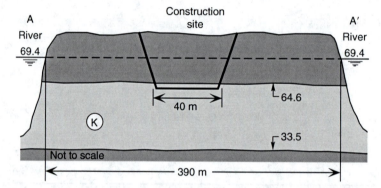

FIGURE 13.11 Map view of the location of the construction site and the dewatering well on an island (Isla de Esperanza) in a large river (Rio el Rito).

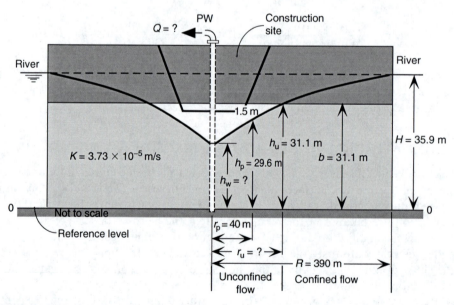

FIGURE 13.12 Hydrogeologic cross section of the confined aquifer with the outline of the excavation pit.

FIGURE 13.13 Calculation scheme for the problem.

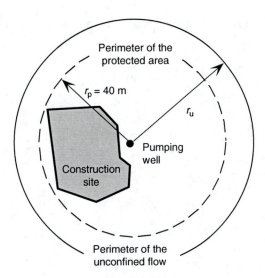

FIGURE 13.14 Map view of the construction site with the location of the extraction well. The dashed circle is the area of the drainage requirement (area to be protected), and the solid circle is the area of the aquifer with unconfined flow conditions during the well operation.

where Q is the well pumping rate, r is the radial distance from the well, K is the hydraulic conductivity, and h is the hydraulic head at distance r.

In our case the confined flow is in part of the aquifer with the following boundary conditions (Figure 13.13):

- At distance r_u (which is the radial distance from the well at which the flow changes from confined to unconfined), the hydraulic head is h_u and is equal to the aquifer thickness b.
- At distance R (which is the radius of well influence, i.e., the distance between the well and the river) the hydraulic head is H.

When integrated with these boundary conditions Equation 13.55 is solved as

$$Q = \frac{2\pi Kb(H - b)}{\ln(R/r_u)} \tag{13.56}$$

The differential equation of groundwater flow toward a fully penetrating well in an unconfined aquifer is

$$h \, dh = \frac{Q}{2\pi K} \frac{dr}{r} \tag{13.57}$$

In our case, the unconfined flow is described with the following boundary conditions:

- At distance r_p (which is the protected area—see Figure 13.14), the hydraulic head is h_p.
- At distance r_u (which is the radial distance from the well at which the flow changes from confined to unconfined), the hydraulic head is h_u and is equal to the aquifer thickness b.

The solution of Equation 13.57 for these boundary conditions is

$$Q = \frac{\pi K(b^2 - h_p^2)}{\ln(r_u/r_p)} \tag{13.58}$$

Equation 13.56 can be written as

$$\ln R - \ln r_u = \frac{2\pi Kb(H - b)}{Q} \tag{13.59}$$

Similarly, Equation 13.58 is

$$\ln r_u - \ln r_p = \frac{\pi K(b^2 - h_p^2)}{Q} \tag{13.60}$$

$$\ln r_u = \frac{\pi K(b^2 - h_p^2)}{Q} - \ln r_p \tag{13.60a}$$

Because of the principle of continuity of flow, Equation 13.56 and Equation 13.58 are equal, i.e., they both give the same pumping rate at the well. This means that Equation 13.60a can be inserted into Equation 13.59, which leads to the following:

$$\ln R - \ln r_p = \frac{2\pi Kb(H - b)}{Q} + \frac{\pi K(b^2 - h_p^2)}{Q} \tag{13.61}$$

or, after solving for Q:

$$Q = \frac{\pi K(2Hb - b^2 - h_p^2)}{\ln(R/r_p)} \tag{13.62}$$

As seen in Figure 13.13, the well pumping rate can be calculated directly as all the elements in Equation 13.62 are known. The hydraulic head at the perimeter of the protected area, h_p, is found from the requirement that the piezometric surface must be at least 1.5 m below the aquifer top within the excavation pit:

$$h_p = b - 1.5 \text{ m} = 31.1 \text{ m} - 1.5 \text{ m} = 29.6 \text{ m}$$

The well pumping rate that will meet the requirement for safe construction is

$$Q = \frac{3.14 \times 3.73 \times 10^{-5} \text{ m/s } [2 \times 35.9 \text{ m} \times 31.1 \text{ m} - (31.1 \text{ m})^2 - (29.6 \text{ m})^2]}{\ln(390 \text{ m}/40 \text{ m})}$$

$$Q = 0.02004 \text{ m}^3/\text{s} = 20.04 \text{ L/s}$$

13.4.1 RADIUS OF UNCONFINED FLOW

Any of Equation 13.59 through Equation 13.60a can be used to find the radius of the area with unconfined flow conditions during the well operation. For example, from Equation 13.60 it follows that

$$\ln\frac{r_u}{r_p} = \frac{\pi K(b^2 - h_p^2)}{Q} \tag{13.63}$$

$$\frac{r_u}{r_p} = e^{(\pi K(b^2 - h_p^2)/Q)} \tag{13.64}$$

which gives the following expression for the radius of unconfined flow:

$$r_u = r_p \, e^{(\pi K(b^2 - h_p^2))/Q} \tag{13.65}$$

and, after inserting all known values (note that $e = 2.7183$):

$$r_u = 40 \text{ m} \times e^{\frac{3.14 \times 3.73 \times 10^{-5} \text{ m/s}[(31.1 \text{ m})^2 - (29.6 \text{ m})^2]}{0.02 \text{ m}^3/\text{s}}} = 68.2 \text{ m}$$

13.4.2 DRAWDOWN IN THE WELL

The hydraulic head at the well, which is needed to calculate the drawdown, can be found by solving Equation 13.57 for the following boundary conditions:

- at distance r_w (which is the well radius) the hydraulic head is h_w;
- at distance r_p (which is the radius of the protected area), the hydraulic head is h_p

which gives

$$Q = \frac{\pi K(h_p^2 - h_w^2)}{\ln(r_p/r_w)} \tag{13.66}$$

Solved for the hydraulic head, the above equation becomes

$$h_w = \sqrt{\frac{Q}{\pi K} \ln\frac{r_p}{r_w} - h_p^2} \tag{13.67}$$

$$h_w = \sqrt{\frac{0.02 \text{ m}^3/\text{s}}{3.14 \times 3.73 \times 10^{-5} \text{ m/s}} \times \ln\frac{40 \text{ m}}{0.25 \text{ m}} - (29.6 \text{ m})^2} = 2.80 \text{ m}$$

The calculated drawdown in the pumping well is the difference between the initial hydraulic head (before pumping) and the hydraulic head during pumping:

$$s = H - h_w = 35.9 \text{ m} - 2.80 \text{ m} = 33.1 \text{ m}$$

Such a large drawdown (92.2% of the initial head) is not feasible because it does not leave enough saturated thickness for the screen and pump placement. (Note that the actual drawdown in the well would be higher than the calculated one because of the inevitable well loss.) This means that drainage of the construction site would have to be performed with two or more wells placed around the excavation perimeter. This would result in a smaller drawdown in individual wells and safer operation and maintenance of the dewatering system.

13.5   WELLS NEAR BOUNDARIES

A proposed system for the containment and cleanup of contaminated groundwater consists of a fully penetrating grout curtain and an extraction well as shown in Figure 13.15. The plume is in the confined aquifer that is in direct hydraulic connection with the Black River, a large perennial stream. The aquifer transmissivity is 5.04×10^{-4} m/s, and the average water table elevation is at 87 m above sea level. From the aquifer test at RW (real extraction well within the plume), it was determined that the radius of well influence is $R = 75$ m (this is an equivalent steady-state hydraulic radius which is very close to the radius of the river bend around RW). From the available data find the pumping rate at RW that would create a 5 m drawdown at the piezometer P1. The theoretical background on wells near boundaries is given in Section 2.5.11.

Figure 13.16 shows the real extraction well (RW) and its images that simulate the aquifer boundaries. The image wells IW1 and IW2 are created over the impermeable boundary and are also extracting water from the aquifer. The image well IW3 is a recharge well, i.e., with the opposite flow rate sign than the real well because it is created over the equipotential boundary. Well IW4 is the image of the image well IW3 and is also a recharge well—it is created over the extension of the impermeable boundary thus keeping its sign.

Drawdown at piezometer P1 is the algebraic sum of the individual drawdowns caused by the real well (s_r) and four image wells (s_{i2} through s_{i4}) shown in Figure 13.16:

$$s_P = s_r + s_{i1} + s_{i2} + s_{i3} + s_{i4} \tag{13.68}$$

or

$$s_P = \frac{Q}{2\pi T} \ln \frac{R}{r_r} + \frac{(+Q)}{2\pi T} \ln \frac{R}{r_{i1}} + \frac{(+Q)}{2\pi T} \ln \frac{R}{r_{i2}} + \frac{(-Q)}{2\pi T} \ln \frac{R}{r_{i3}} + \frac{(-Q)}{2\pi T} \ln \frac{R}{r_{i4}} \tag{13.69}$$

where the distances between P1 and all the image wells simulating the boundaries are denoted with their respective (r_i) and shown in Figure 13.17. R is the radius of well influence determined from a pumping test to be approximately 75 m. Distances between the piezometer

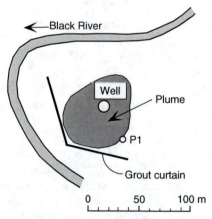

FIGURE 13.15 Map view of the proposed remediation of contaminated groundwater in the confined aquifer, which is in direct hydraulic connection with the Black River.

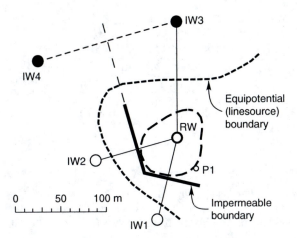

FIGURE 13.16 The real extraction-well (RW) and its images (IW). Note that the recharge image well IW3 has its own image well IW4 created over the extended portion of the impermeable boundary (grout curtain).

and the wells, measured on the map in Figure 13.17, are: $r_r = 40$ m; $r_{i1} = 68$ m; $r_{i2} = 110$ m; $r_{i3} = 155$ m; and $r_{i4} = 174$ m. Equation 13.69 can be written as

$$s_P = \frac{Q}{2\pi T} \ln \frac{R^3}{r_r r_{i1} r_{i2}} - \frac{Q}{2\pi T} \ln \frac{R^2}{r_{i3} r_{i4}} \tag{13.70}$$

or, after further simplifying it:

$$s_P = \frac{Q}{2\pi T} \ln \frac{R r_{i3} r_{i4}}{r_r r_{i1} r_{i2}} \tag{13.71}$$

Solved for Q (the pumping rate needed to create the drawdown of 5 m at P1) it is

$$Q = \frac{s_P 2\pi T}{\ln \left(\dfrac{R r_{i3} r_{i4}}{r_r r_{i1} r_{i2}} \right)} \tag{13.72}$$

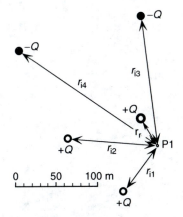

FIGURE 13.17 Calculation scheme for the problem.

or, after inserting all known values

$$Q = \frac{5\ m \times 2 \times 3.14 \times 5.04 \times 10^{-4}\ m^2/s}{\ln\left(\dfrac{75\ m\ \times\ 155\ m\ \times\ 174\ m}{40\ m\ \times 68\ m\ \times\ 110\ m}\right)}$$

$$Q = 8.28 \times 10^{-3}\ m^3/s = 8.3\ L/s$$

REFERENCES

Kashef, A.A.I., 1987. *Groundwater Engineering*. McGraw Hill, Singapore, 512 p.
Vukovic, M. and A. Soro, 1984. *Dinamika Podzemnih Voda* (*Dynamics of Groundwater*, in Serbo-Croatian). Posebna Izdanja, Knjiga 25, Institut Jaroslav Cerni, Beograd, 500 p.

14 Transient Flow to Water Wells

14.1 CONFINED, NONLEAKY AQUIFER

A 24 h well pumping test in a confined aquifer was conducted to determine its hydrogeologic parameters. Use information from the schematic hydrogeologic cross section in Figure 14.1 and pumping data from Table 14.1 to calculate aquifer transmissivity, hydraulic conductivity, and storage coefficient. Apply several methods of pumping test analysis. Also determine well loss and its acceptability (i.e., efficiency of the well). The pumping rate was kept constant at 8 L/s (0.008 m³/s) during all 24 h of the test. The drilling diameter of the well screen portion is 350 mm.

Section 2.5.3 provides a detailed derivation of the Theis equation, which describes groundwater flow toward a fully penetrating well in an "ideal" confined aquifer. One portion of the theoretical Theis curve, relating the Theis well function $W(u)$ and parameter u is shown in Figure 2.57. Field data of drawdown (s) versus time (t) are plotted separately for each observation-well on field data graphs with the same log–log scale as the Theis function graph (see Figure 14.2). Field curve for one observation-well at a time is superimposed on the Theis theoretical curve, also called type curve. It is essential that both graphs and curves have identical logarithmic scales and cycles as shown in Figure 14.3. Keeping coordinate axes of the curves parallel, field data for P3 are matched to the type curve. Figure 2.58 shows the same procedure for the P2 data.

Once a satisfactory match is found, a match point on the overlapping graphs is selected. The match point is defined by four coordinates, the values of which are read on two graphs: $W(u)$ and $1/u$ on the type curve graph, s and t on the field graph. The match point can be any point on the overlapping graphs, i.e., it does not have to be on the matching curves. Figure 14.3 shows two possibilities of choosing the match point: one is on the theoretical curve or measured data (which looks somewhat more logical), and one is chosen outside to obtain convenient values of $W(u)$ and $1/u$: 1 and 100, respectively. Both matching points should lead to the same values of T and S if the coordinates are read carefully.

The transmissivity is calculated using the following equation and the match point coordinates s and $W(u)$:

$$T = \frac{Q}{4\pi s} W(u) \qquad (14.1)$$

The storage coefficient (S) is calculated using the transmissivity already determined in the prior step, and the match point coordinates $1/u$ and t, with the following equation:

$$S = \frac{4Ttu}{r^2} \qquad (14.2)$$

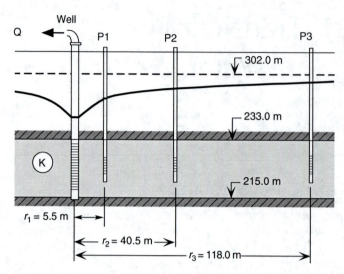

FIGURE 14.1 Schematic hydrogeologic cross section of the well pumping test conditions.

The match point coordinates are

$$W(u) = 2.35; \quad 1/u = 18.5; \quad u = 0.054; \quad s = 1.0 \text{ m}; \quad t = 37.5 \text{ min} = 2250$$

The transmissivity is calculated using Equation 14.1:

$$T = \frac{0.008 \text{ m}^3/\text{s}}{4\pi 1.0 \text{ m}} \times 2.35 = 1.50 \times 10^{-3} \text{ m}^2/\text{s}$$

The storage coefficient is calculated using Equation 14.2:

$$S = \frac{4 \cdot 1.50 \times 10^{-3} \text{ m}^2/\text{s} \times 2250 \times 0.054}{(118 \text{ m})^2} = 5.2 \times 10^{-5}$$

The hydraulic conductivity is calculated from the transmissivity and the aquifer thickness:

$$K = \frac{T}{b} = \frac{1.50 \times 10^{-3} \text{ m}^2/\text{s}}{18 \text{ m}} = 8.33 \times 10^{-5} \text{ m/s}$$

14.1.1 THE COOPER–JACOB STRAIGHT-LINE METHOD

When $u \leq 0.05$, the well function expressed in the following form is considered acceptable for practical purposes:

$$W(u) = \ln \frac{2.25Tt}{r^2 S} \tag{14.3}$$

and the Theis Equation 14.1 becomes

$$s = \frac{Q}{4\pi T} \ln \frac{2.25Tt}{r^2 S} \tag{14.4}$$

TABLE 14.1
Drawdown Observed at Piezometers and Pumping Well in a Homogeneous, Nonleaky Confined Aquifer

Time		Drawdown (in m)				Time		Drawdown (in m)			
min	s	P1	P2	P3	Well	min	s	P1	P2	P3	Well
1	60	1.875	0.412	0.041	6.407	120	7200	4.115	2.437	1.507	8.658
2	120	2.288	0.684	0.105	6.831	150	9000	4.175	2.513	1.593	8.717
3	180	2.495	0.862	0.175	7.038	180	10,800	4.254	2.59	1.667	8.795
4	240	2.651	0.992	0.241	7.197	210	12,600	4.319	2.655	1.731	8.861
5	300	2.757	1.092	0.302	7.302	240	14,400	4.408	2.73	1.792	8.95
6	360	2.839	1.172	0.357	7.384	270	16,200	4.429	2.765	1.839	8.971
7	420	2.906	1.24	0.407	7.451	300	18,000	4.498	2.824	1.887	9.04
8	480	2.965	1.298	0.451	7.509	330	19,800	4.537	2.862	1.926	9.08
9	540	3.016	1.348	0.492	7.56	360	21,600	4.574	2.989	1.963	9.117
10	600	3.06	1.394	0.53	7.605	390	24,400	4.607	2.931	1.996	9.151
12	720	3.146	1.471	0.594	7.691	420	25,200	4.637	2.961	2.027	9.181
14	840	3.212	1.535	0.65	7.757	450	27,000	4.666	2.989	2.056	9.21
16	960	3.267	1.592	0.7	7.812	480	28,800	4.692	3.016	2.083	9.245
18	1080	3.316	1.641	0.745	7.861	540	32,400	4.718	3.054	2.128	9.259
20	1200	3.36	1.686	0.786	7.904	600	36,000	4.791	3.177	2.178	9.332
25	1500	3.436	1.771	0.87	7.979	660	39,600	4.822	3.151	2.217	9.364
30	1800	3.534	1.854	0.94	8.078	720	43,200	4.861	3.188	2.253	9.403
35	2100	3.579	1.913	1.002	8.121	780	46,800	4.894	3.22	2.285	9.437
40	2400	3.651	1.974	1.005	8.194	840	50,400	4.924	3.25	2.315	9.467
45	2700	3.702	2.024	1.104	8.247	900	54,000	4.952	3.277	2.343	9.495
50	3000	3.73	2.064	1.147	8.273	960	57,600	4.977	3.303	2.37	9.521
55	3300	3.782	2.108	1.186	8.325	1020	61,200	5.002	3.328	2.394	9.545
60	3600	3.821	2.145	1.222	8.364	1080	64,800	5.024	3.351	2.418	9.568
70	4200	3.868	2.202	1.283	8.41	1140	68,400	5.046	3.372	2.44	9.59
80	4800	3.945	2.268	1.338	8.489	1200	72,000	5.066	3.393	2.461	9.61
90	5400	3.975	2.309	1.387	8.517	1260	75,600	5.085	3.412	2.481	9.63
100	6000	4.036	2.361	1.431	8.579	1320	79,200	5.105	3.434	2.501	9.647
110	6600	4.076	2.401	1.471	8.621	1380	82,800	5.124	3.453	2.52	9.666

or, when the logarithm to the base e or "natural logarithm" (ln) is replaced by the logarithm to the base 10 (log):

$$s = \frac{0.183Q}{T} \log \frac{2.25Tt}{r^2S} \tag{14.5}$$

The logarithmic form of Theis equation, known as the Cooper–Jacob equation, is much easier to work with and enables the usage of semilogarithmic graphs for well pumping test analyses. It is valid for small values of parameter u, i.e., sufficiently large values of pumping time t, and small values of r (i.e., when a monitoring well is close to the pumping well).

Figure 14.4 shows graph with plotted drawdown versus time data for all piezometers and the pumping well. There is a deviation from the straight line for P3 for the initial period of pumping which indicates that the parameter u is still greater than 0.05. This is because piezometer P3 is relatively far from the pumping well ($r = 118$ m) compared to P1 and P2

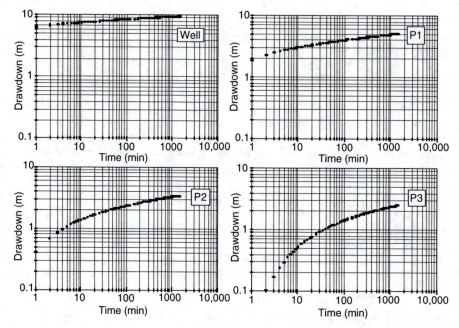

FIGURE 14.2 Field data of drawdown versus time for the pumping well and piezometers P1, P2, and P3 plotted on a log–log graph paper.

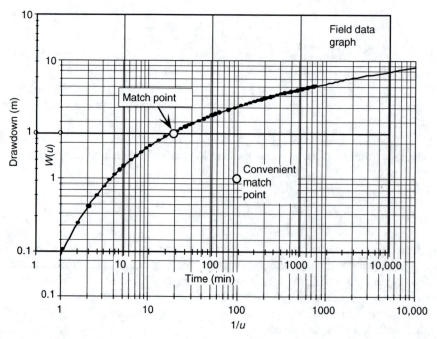

FIGURE 14.3 Curve matching for the piezometer P3. Coordinates of the match point are $W(u) = 2.35$; $1/u = 18.5$; $s = 1.0$ m; $t = 37.5$ min $= 2250$ s.

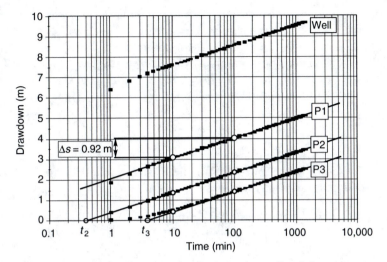

FIGURE 14.4 Semilog graph drawdown versus time for pumping well and three piezometers. The graph is used for the calculation of aquifer transmissivity and storage coefficient by the Cooper–Jacob straight-line method.

and more time is needed for the condition regarding parameter u to be satisfied. This time can be determined using the following equation:

$$t = \frac{r^2 S}{4Tu} \tag{14.6}$$

$$t = \frac{(118 \text{ m})^2 \times 5.2 \times 10^{-5}}{4 \times 1.5 \times 10^{-3} \text{ m}^2/\text{s} \times 0.05} = 2413 \text{ s} \approx 40 \text{ min}$$

In other words, data recorded in P3 form a straight line on the graph drawdown (s) versus logarithm of time (t) after approximately 40 min of pumping (see Figure 14.4).

If two points on the straight line for any of the piezometers are chosen to be one log cycle apart (i.e., ratio of time coordinates t_2 and t_1 is 10), the transmissivity is found using the following equation:

$$T = \frac{0.183Q}{\Delta s} \tag{14.7}$$

where Δs is the difference in drawdown between the two points as shown in Figure 14.4 for the P1 line. Knowing the pumping rate value (0.008 m^3/s) and using $\Delta s = 0.92$ m the transmissivity is

$$T = \frac{0.183 \times 8.0 \times 10^{-3} \text{ m}^3/\text{s}}{0.92 \text{ m}} = 1.59 \times 10^{-3} \text{ m}^2/\text{s}$$

The straight lines representing recorded data in our example are all parallel, i.e., with the same slope (Δs), which confirms the assumption that the aquifer is homogeneous and isotropic. Therefore any of the piezometers or the pumping well itself can be used to determine aquifer transmissivity as the result would be the same. If, however, the lines were with significantly different or varying slopes it would indicate that the aquifer may be heterogeneous or with some other disturbance (e.g., source of recharge or discharge or a boundary).

In addition to transmissivity, the Cooper–Jacob method is used to determine the storage coefficient of the aquifer. This is done graphically using the same graph shown in Figure 14.4. The straight line for any of the piezometers is extended to intercept the zero-drawdown (t_0) and the storage coefficient is found from

$$S = \frac{2.25 T t_0}{r^2} \tag{14.8}$$

For example, the zero-drawdown intercept for the P3 line is $t_3 = 3.9$ min or 244 s and the distance between the pumping well and P3 is $r = 118$ m:

$$S = \frac{2.25 \times 1.59 \times 10^{-3} \text{ m}^2/\text{s} \times 234 \text{ s}}{(118 \text{ m})^2} = 0.00006 = 6 \times 10^{-5}$$

Only data for the piezometers can be used to determine the storage coefficient because the actual drawdown measured in the pumping well includes a well loss. This additional drawdown is not known beforehand and neither is the theoretical position of the pumping well data (without the well loss).

14.1.2 Drawdown versus Ratio of Time and Square Distance (t/r^2)

This method is an obligatory part of every well pumping test analysis in homogeneous confined aquifers because, in addition to aquifer transmissivity and storage coefficient, it allows the determination of well loss and efficiency. It is also based on the application of the Cooper–Jacob straight-line method. Theoretical background of the method is given in Section 2.5.5.

When plotted together, as shown in Figure 14.5, data for all piezometers form a single straight line indicating that, for practical purposes, the aquifer can be considered as homogeneous and isotropic, without recharge, and without boundaries influenced by the

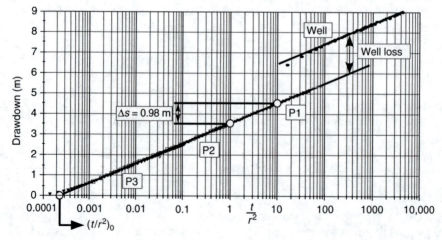

FIGURE 14.5 Semilog graph drawdown versus ratio t/r^2 used for determination of aquifer transmissivity, storage coefficient, and pumping-well loss.

pumping. Coordinates of any two points on the straight line can be used to determine aquifer transmissivity:

$$s_2 - s_1 = \Delta s = \frac{Q}{4\pi T} \left[\ln\left(\frac{t}{r^2}\right)_2 - \ln\left(\frac{t}{r^2}\right)_1 \right] \tag{14.9}$$

Solving Equation 14.9 for transmissivity, and introducing logarithm to the base 10, gives

$$T = \frac{0.183Q}{\Delta s} \left[\log\left(\frac{t}{r^2}\right)_2 - \log\left(\frac{t}{r^2}\right)_1 \right] \tag{14.10}$$

If two points on the straight line are chosen to be one log cycle apart (i.e., ratio of coordinates $(t/r^2)_2$ and $(t/r^2)_1$ is 10), Equation 14.10 reduces to a simple-to-use form:

$$T = \frac{0.183Q}{\Delta s} \tag{14.11}$$

In our example (Figure 14.5), the difference in drawdown for the log cycle 1–10 is 0.98 m. The transmissivity is then easily calculated knowing that the pumping rate is 8 L/s (8×10^{-3} m^3/s):

$$T = \frac{0.183 \times 8 \times 10^{-3} \text{ m}^3/\text{s}}{0.98 \text{ m}}$$

$$T = 0.001494 \text{ m}^2/\text{s} = 1.49 \times 10^{-3} \text{ m}^2/\text{s}$$

The storage coefficient is, similar to the Cooper–Jacob method, determined by substituting the previously calculated transmissivity and the zero-drawdown intercept of the straight line into the following equation:

$$S = 2.25T \left(\frac{t}{r^2}\right)_0 \tag{14.12}$$

In Figure 14.5, time, in the ratio (t/r^2), is given in minutes and the zero-drawdown intercept is 0.00025 min/m^2. When this time is expressed in seconds (which is needed for the consistency of units as T is given in m^2/s), this value is 0.00025 min/m$^2 \times 60 = 0.015$ s/m^2.

$$S = 2.25 \times 1.49 \times 10^{-3} \text{ m}^2/\text{s} \times 0.015 \text{ s/m}^2$$
$$S = 0.0000503 = 5 \times 10^{-5}$$

Well loss is the difference between the actual measured drawdown in the pumping well and the theoretical drawdown, which is expressed by the Theis equation and is the result of groundwater flow through the aquifer undisturbed zone only. The additional drawdown, or well loss, which is always present in pumping wells, is created by a combination of various factors such as improper well development (drilling fluid is left in the formation, mud cake along the borehole walls is not removed, and fines from the formation are not removed), and a poorly designed gravel pack or well screen. Well loss is determined directly from the graph in Figure 14.5. It is the vertical distance, in meters of drawdown, between the line of piezometers and the pumping-well line (if there were not any well losses, the well data would fall along the line of piezometers). In our case the well loss is 1.9 m.

Well efficiency, expressed in percent, is the ratio between the theoretical drawdown and the actual drawdown measured in the well:

$$\text{Well efficiency} = \frac{\text{Theoretical drawdown}}{\text{Measured drawdown}} \times 100\%$$

Values of theoretical and measured drawdown, as depicted from the graph in Figure 14.5, are 6 and 7.9 m, respectively, and the well efficiency is

$$\text{Well efficiency} = \frac{6 \text{ m}}{7.9 \text{ m}} \times 100\% = 76\%$$

More on well loss and efficiency, including other methods for determining these two important well characteristics, are given in Section 15.2.

14.2 APPLICATION OF THEIS EQUATION

A homogeneous confined aquifer is tapped by a fully penetrating well whose radius is 225 mm. The aquifer transmissivity is 8.35×10^{-4} m^2/s and the storage coefficient is 0.00065. For the given parameters determine the following:

1. Drawdown at 50 m from the pumping well 10 min and 24 h after the beginning of pumping with the rate of 12 L/s (0.012 m^3/s)
2. Time after which the drawdown at 30 m from the pumping well would be 2.5 m if the pumping rate were 9 L/s
3. Pumping rate needed to produce 1.5 m drawdown at 28 m from the pumping well after 10 d of pumping
4. Distance at which the drawdown is 0.85 m after one week of pumping with the rate of 5 L/s

14.2.1 DRAWDOWN

Drawdown at any time and at any distance from the pumping well is calculated using the well-known Theis equation:

$$s = \frac{Q}{4\pi T} W(u) \tag{14.13}$$

where Q is well pumping rate, T is aquifer transmissivity, and u is dimensionless parameter of the Theis (well) function given as

$$u = \frac{r^2 S}{4Tt} \tag{14.14}$$

where r is distance from the pumping well, S is storage coefficient, and t is time since pumping started.

The value of $W(u)$ can be found either in tables for the known parameter u (see Appendix D) or calculated using the Cooper–Jacob approximation and assuming that $u < 0.05$ (which is commonly the case after 110–120 min of pumping for distances that are not very far from the pumping well). In our case, parameter u after 10 min and at 50 m from the pumping well is

$$u = \frac{(50 \text{ m})^2 \times 0.00065}{4 \times 8.35 \times 10^{-4} \text{ m}^2/\text{s} \times 10 \times 60 \text{ s}} = 0.811$$

and the value of corresponding well function is (Appendix D)

$$W(u) = 0.305$$

The drawdown is

$$s = \frac{0.012 \text{ m}^3/\text{s}}{4 \times 3.14 \times 8.35 \times 10^{-4} \text{ m}^2/\text{s}} \times 0.305 = 0.35 \text{ m}$$

Well function calculated using the Cooper–Jacob approximation is

$$W(u) = \ln \frac{2.25Tt}{r^2 S}$$

$$W(u) = \ln \frac{2.25 \times 8.35 \times 10^{-4} \text{ m}^2/\text{s} \times 600 \text{ s}}{(50 \text{ m})^2 \times 0.00065} = -0.366 \tag{14.15}$$

Because the value is negative (which would lead to a negative drawdown), it is obvious that Equation 14.15 cannot be used to find the well function after 10 min of pumping. The time after which this approximate function can be used (accepting an error of 2% or less), i.e., time at which parameter u is equal to 0.05, is found using the following equation:

$$t = \frac{r^2 S}{4Tu} \tag{14.16}$$

In our case, after substituting all known values into Equation 14.16, this time equals 9730 s (162 min or 2.7 h).

Drawdown after 24 h of pumping at 50 m from the well using the Theis equation and table $W(u)$ vs. u (Appendix D) is

$$s = \frac{Q}{4\pi T} \times W \left[\frac{(50 \text{ m})^2 \times 0.00065}{4 \times 8.35 \times 10^{-4} \text{ m}^2/\text{s} \times 86,400 \text{ s}} \right]$$

$$s = \frac{0.012 \text{ m}^3/\text{s}}{4 \times 3.14 \times 8.35 \times 10^{-4} \text{ m}^2/\text{s}} \times W(0.00563)$$

$$s = 1.14 \text{ m} \times 4.61 = 5.255 \text{ m}$$

When using the approximate logarithmic formula, the drawdown is

$$s = \frac{Q}{4\pi T} \ln \frac{2.25T \times 24 \text{ h}}{r^2 S} = 5.26 \text{ m}$$

which is practically identical to the result obtained by the Theis equation.

14.2.2 Time

From the Theis equation, well function for a given drawdown is

$$W(u) = \frac{4s\pi T}{Q} \tag{14.17}$$

or, in our case

$$W(u) = \frac{4 \cdot 2.5 \text{ m} \times 3.14 \times 8.35 \times 10^{-4} \text{ m}^2/\text{s}}{0.009 \text{ m}^3/\text{s}} = 2.913$$

For this value of well function, parameter u is 0.0315 as determined from the table in Appendix D. The corresponding time is found using Equation 14.16:

$$t = \frac{r^2 S}{4Tu}$$

$$t = \frac{(30 \text{ m})^2 \times 0.00065}{4 \times 8.35 \times 10^{-4} \text{ m}^2/\text{s} \times 0.0315} = 5560 \text{ s or approximately 1.5 h}$$

14.2.3 Pumping Rate

From the Theis equation, the pumping rate needed to produce the given drawdown is

$$Q = \frac{4\pi s T}{W(u)} \tag{14.18}$$

Well function for the given aquifer parameters, time and distance is

$$W(u) = W\left(\frac{r^2 S}{4Tt}\right) = W\left(\frac{(28 \text{ m})^2 \times 0.00065}{4 \times 8.35 \times 10^{-4} \text{ m}^2/\text{s} \times 86{,}4000 \text{ s}}\right) = W(0.0001766) = 8.0684$$

After inserting all known values into Equation 14.18, the pumping rate is

$$Q = \frac{4 \times 3.14 \times 1.5 \text{ m} \times 8.35 \times 10^{-4} \text{ m}^2/\text{s}}{8.0684} = 0.0013 \text{ m}^3/\text{s or 1.3 L/s}$$

14.2.4 Distance

The distance at which a given drawdown is measured is part of the expression for parameter u in the Theis equation, i.e., it is part of the well function:

$$W(u) = W\left(\frac{r^2 S}{4Tt}\right) \tag{14.19}$$

The well function is (from the Theis equation):

$$W(u) = \frac{4\pi s T}{Q} \tag{14.20}$$

For the given values this function is

$$W(u) = \frac{4 \times 3.14 \times 0.85 \text{ m} \times 8.35 \times 10^{-4} \text{ m}^2/\text{s}}{0.005 \text{ m}^3/\text{s}} = 1.783$$

The corresponding parameter u is found from the table in Appendix D (note that it will almost always be necessary to choose an interpolated value for u; in our case this value is between 0.1 and 0.11 for the corresponding table values of $W(u) = 1.8229$ and 1.7371, respectively):

$$u = 0.104$$

From the expression for parameter u (Equation 14.14), the distance at which the drawdown is measured at a given time is

$$r = \sqrt{\frac{4uTt}{S}} \qquad (14.21)$$

or in our case:

$$r = \sqrt{\frac{4 \times 1.04 \times 8.35 \times 10^{-4} \times 7 \times 86,400 \text{ s}}{0.00065}} = 1798 \text{ m}$$

14.3 METHOD OF RECOVERY

A three-step pumping test at a fully penetrating well in a confined aquifer was conducted to determine the characteristics of the well. The graph in Figure 14.6 shows the drawdown at the well versus the time for all three steps, and the pumping rate hydrograph. The observed residual drawdown during recovery after the last step of pumping at the test well is given in

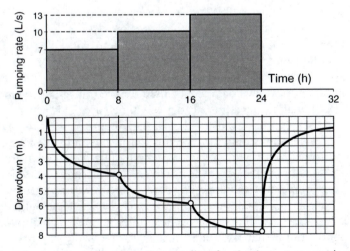

FIGURE 14.6 Drawdown at the pumping well versus time for the three-step pumping test.

TABLE 14.2
Residual Drawdown (s′) at the Pumping Well versus t/t′ Ratio

Time Since Stopped, t' (min)	Time Since Started, $t = t^* + t'$ (min)	Ratio t/t'	Residual Drawdown s' (m)
1	1109	1109	2.81
2	1110	555	2.67
3	1111	370	2.55
4	1112	278	2.49
5	1113	227	2.42
6	1114	186	2.36
7	1115	159	2.3
8	1116	139	2.26
9	1117	124	2.19
10	1118	112	2.14
15	1123	75	1.91
20	1128	56	1.78
25	1133	45	1.65
30	1138	38	1.56
40	1148	29	1.41
50	1158	23	1.29
60	1168	19	1.21
75	1183	16	1.11
90	1198	13	1.05
120	1228	10	0.89
150	1258	8.4	0.8
180	1288	7.2	0.68
240	1348	5.6	0.53
300	1408	4.7	0.44
360	1468	4.1	0.37
420	1528	3.6	0.32

Note: The time since pumping started is adjusted for the last pumping rate using Equation 14.22 (see also Figure 14.8), i.e., it is not the actual time since the beginning of the test.

Table 14.2. From the available data, find the aquifer transmissivity (well loss analysis for the same data is given in Section 15.2).

When the pumping rate varies during the test, which in the case of a step test is done deliberately, the residual drawdown data can be used for the transmissivity analysis after the pumping rate has been analytically adjusted to a new time base. This new pumping rate and the new time base preserve aquifer water balance: the volume of water actually withdrawn equals the volume of water that would have been withdrawn with the new constant pumping rate. It is important to realize that this adjusted pumping rate is not the average of the three rates. Rather, it is the pumping rate of the last step (13 L/s in our case), which has the most influence on the development of the cone of depression (drawdown) around the pumping well. The adjusted constant pumping rate is also the rate at which an imaginary recharge well simulating the recovery starts adding water to the aquifer after the actual pumping stopped.

During operation of the imaginary well, well loss is practically minimal as the pump is turned off. Although there will still be some linear resistance to the flow in the near-well zone due to aquifer disturbance during the well completion, turbulent well loss is absent. In addition, the water level recorded in the pumping well is much more accurate when the pump is not working. This means that both the test well and the monitoring wells' data can be used for transmissivity analysis. Practice shows that, if the pumping rate varies less than 10% –15% during the test, the residual drawdown method provides reasonable accuracy. However, in case of a three-step pumping test, this variation is commonly greater than 15% (often more than 100%), which means that the recovery data at the test well could be used only after the pumping rate were adjusted and the new time base for the recovery data determined.

Figure 14.7 shows hydrographs and the corresponding drawdown for an ideal pumping test. The first part of the test, which has three steps, is designed to determine the well characteristics such as well loss and an optimum pumping rate. The duration of each step should be the same, usually not more than 6–8 h. Drawdown recorded during the first step can also be used to find the transmissivity and the storage coefficient. The second part of the test should be performed after complete recovery and with a maximum feasible pumping rate. The duration of this part of the test, which is designed to determine the overall aquifer transmissivity for an extensive radius of influence, depends on specific project requirements and may vary from 24 h to several weeks in case of a major water-supply development. Long pumping with a maximum rate uncovers various aquifer irregularities that may be less obvious after a short test: distant boundaries, leakage, and changes in storage. Both drawdown and recovery data should be used to find the aquifer parameters. In case such an ideal test is not contemplated, the recovery data after the three-step pumping test can be used to gain more information on the expected aquifer transmissivity.

The adjusted duration of pumping that reflects the influence of the last-step pumping is found using the following formula (see Figure 14.8):

$$t^* = \frac{Q_1 t_1 + Q_2(t_2 - t_1) + Q_3(t_3 - t_2)}{Q_3} \tag{14.22}$$

$$t^* = \frac{0.007 \text{ m}^3/\text{s} \times 8 \text{ h} + 0.010 \text{ m}^3/\text{s} \times 8 \text{ h} + 0.013 \text{ m}^3/\text{s} \times 8 \text{ h}}{0.013 \text{ m}^3/\text{s}}$$

$$= 18.46 \text{ h} \approx 1108 \text{ min}$$

This new time base is then used to calculate the t/t' ratio for the recovery data; as shown in Figure 14.8, t is given as

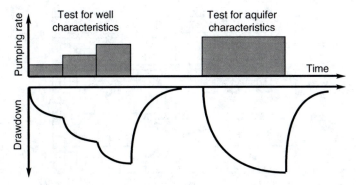

FIGURE 14.7 Pumping rate hydrographs and corresponding drawdown for an ideal pumping test designed to determine characteristics of both the test well and the aquifer.

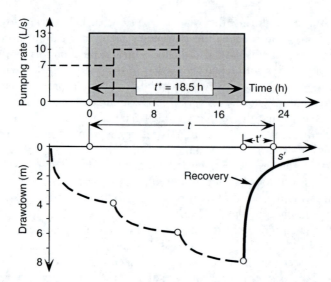

FIGURE 14.8 Hydrograph of the new adjusted constant pumping rate used to calculate the transmissivity from the residual drawdown data.

$$t = t^* + t'$$ (14.23)

where t' is the time since pumping stopped. Calculation of the t/t' ratio needed to plot the semilog graph shown in Figure 14.9 is given in Table 14.2. The transmissivity is found from

$$T = \frac{0.183Q}{\Delta s'}$$

$$T = \frac{0.183 \times 0.013 \text{ m}^3/\text{s}}{1.17 \text{ m}} = 2.03 \times 10^{-3} \text{ m}^2/\text{s}$$

14.4 LEAKY CONFINED AQUIFER

A pumping test was conducted at a fully penetrating well completed in a homogeneous confined aquifer, which, for practical purposes, can be considered to be of infinite extent. (It is assumed that the aquifer boundaries are distant and do not influence groundwater flow

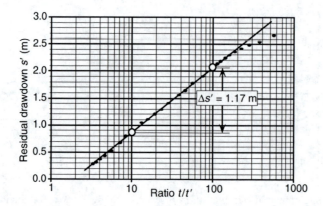

FIGURE 14.9 Residual drawdown at the test well versus ratio t/t'.

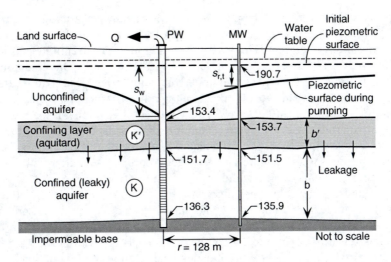

FIGURE 14.10 Simplified hydrogeologic cross section of the aquifer test at pumping well (PW) in a leaky confined aquifer.

for the duration of the test.) The aquifer is overlain by an aquitard, which in turn is overlain by an unconfined (water table) aquifer as shown in Figure 14.10. The hydraulic conductivity of the unconfined aquifer is 8.45×10^{-3} m²/s. The drawdown of the piezometric surface of the confined aquifer was measured at a monitoring well 128 m from the pumping well and the results are given in Table 14.3. From the available data find the hydrogeologic parameters of the confined aquifer and the aquitard. Calculate the drawdown at 10 m from the pumping well after 10 h of pumping.

TABLE 14.3
Time versus Drawdown (s, in m) Recorded at 128 m from the Pumping Well in a Leaky Confined Aquifer

T (min)	s (m)	T (min)	s (m)
1	0.11	120	1.46
2	0.52	150	1.47
3	0.75	180	1.48
4	0.86	240	1.49
5	0.96	300	1.5
6	1.01	360	1.5
7	1.05	420	1.51
8	1.08	480	1.51
9	1.11	600	1.51
10	1.14	720	1.51
15	1.24	840	1.52
20	1.28	960	1.52
25	1.33	1080	1.53
30	1.36	1200	1.53
45	1.39	1320	1.53
60	1.43	1440	1.53
90	1.45		

The Theis nonequilibrium equation describes flow toward a fully penetrating well in a confined aquifer and is based on numerous assumptions, one of which is that the aquifer receives no recharge for the duration of pumping. However, this condition is seldom entirely satisfied as most confined aquifers receive recharge, either continuously or intermittently. One common source of recharge is leakage through an overlying and underlying aquitard, which separates the pumped aquifer from another aquifer. The rate of leakage can significantly increase during pumping due to an increased hydraulic gradient between the two aquifers. Eventually, if the rate of leakage balances the pumping rate, the drawdown will stabilize at a certain level and the radius of well influence will cease to expand. Figure 14.11, which is a log–log plot of the recorded drawdown versus time in our case, shows this effect of leakage. Approximately after 100 min of initial pumping, the leakage balances the withdrawal of water from the confined aquifer and the drawdown remains practically unchanged for the duration of the test. (Note that the influence of an equipotential boundary would look very similar and interpreters should verify its absence before assuming leaky conditions.) Derivation of the equation describing groundwater flow toward well in a leaky confined aquifer is given in Section 2.5.9 together with the type curves, which are used for graphoanalytical solution of the equation.

The procedure for finding hydrogeologic parameters is as follows:

- Plot log–log graph time versus recorded drawdown at the same log-cycle scale as the graph with type curves (see Figure 14.11).
- Superimpose the field graph on the graph with type curves and, keeping the axes of both graphs parallel, match the field data with a type curve r/B (Figure 14.12).
- Choose a match point anywhere where the two graphs overlap and, noting the r/B value, read the four coordinates as follows: drawdown (s) and time (t) from the field graph, and $W(u,r/B)$ and $1/u$ from the type curve graph. It is recommended to choose rounded values for $1/u$ and $W(u,r/B)$ for easier calculations. In our case the coordinates of the match point are:

Well function $W(u,r/B)$	1
Ratio $1/u$	10 (and $u = 0.1$)
Drawdown s	0.32 m
Time t	2.5 min

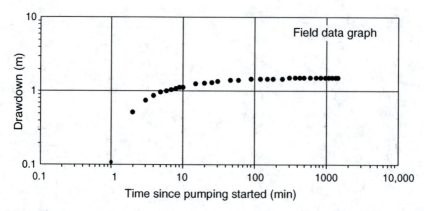

FIGURE 14.11 Time versus drawdown recorded at MW, 128 m from the pumping well (data in Table 14.3).

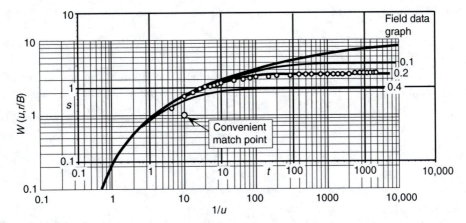

FIGURE 14.12 Matching field data to a type curve. The best fit is for the curve $r/B = 0.2$. The match point is chosen to have convenient coordinates $W(u,r/B) = 1$, and $1/u = 10$.

- Calculate transmissivity by substituting known values into equation:

$$T = \frac{Q}{4\pi s} W(u,r/B)$$

$$T = \frac{0.012 \text{ m}^3/\text{s}}{4 \times 3.14 \times 0.32 \text{ m}} \times 1 = 2.98 \times 10^{-3} \text{ m}^2/\text{s}$$

(14.24)

- Find the coefficient of storage using equation

$$S = \frac{4Tut}{r^2}$$

$$S = \frac{4 \times 2.98 \times 10^{-3} \text{ m}^2/\text{s} \times 0.1 \times 2.5 \times 60 \text{ s}}{(128 \text{ m})^2} = 1.09 \times 10^{-5}$$

(14.25)

- Find the hydraulic conductivity of the confining layer (aquitard) from

$$K' = \frac{Tb'(r/B)^2}{r^2}$$

$$K' = \frac{2.98 \times 10^{-3} \text{ m}^2/\text{s} \times 2 \text{ m} \times 0.2^2}{(128 \text{ m})^2} = 1.45 \times 10^{-8} \text{ m/s}$$

(14.26)

(Note that the average thickness of the aquitard is 2 m as determined from the cross section in Figure 14.10.)

It is important to remember that Equation 14.24 has limited application because of the following assumptions introduced as part of its solution (in addition to assumptions applied to the Theis equation):

- The water table in the unconfined aquifer does not change during pumping.
- There is no water released from the aquitard.
- The piezometric surface of the confined aquifer does not fall below top of the overlying aquitard.

The first assumption is valid if the following condition is satisfied:

$$T_{UA} = 100 \times T_{CA} \tag{14.27}$$

where T_{UA} is the transmissivity of the unconfined aquifer and T_{CA} is the transmissivity of the confined (pumped) aquifer. In our case the transmissivity of the unconfined aquifer is

$$T_{UA} = K_{UA} \times b_{UA} = 8.45 \times 10^{-3} \text{ m/s} \times 37 \text{ m} = 3.21 \times 10^{-1} \text{ m}^2/\text{s}$$

and the first assumption is valid because

$$3.21 \times 10^{-1} \text{ m}^2/\text{s} > 100 \times 2.98 \times 10^{-3} \text{ m}^2/\text{s}$$

The rate of flow attributed to the release of water from storage in the aquitard can be ignored if the following condition is satisfied:

$$t > \frac{0.036 b' S'}{K'} \tag{14.28}$$

where S' is storage coefficient of the aquitard. In our case, as this storage coefficient is not known, it has to be estimated. If S' is at least 10 times smaller than S ($S' = 1.1 \times 10^{-6}$), Equation 14.28 gives

$$36,000 \text{ s} > \frac{0.036 \times 2 \text{ m} \times 1.1 \times 10^{-6}}{1.45 \times 10^{-8} \text{ m/s}} > 5.5 \text{ s}$$

which means that the second assumption is valid and Equation 14.24 can be applied to our data.

14.4.1 DRAWDOWN AT GIVEN DISTANCE AND TIME

Drawdown at 50 m from the pumping well after 10 h of pumping is found using Equation 14.24, and table values or type curves of the well function—$W(u, r/B)$ after calculating u and r/B:

$$u = \frac{r^2 S}{4Tt}$$

$$u = \frac{(50 \text{ m})^2 \times 0.000011}{4 \times 2.98 \times 10^{-3} \text{ m}^2/\text{s} \times 36,000 \text{ s}} = 6.41 \times 10^{-5} \quad \text{and} \quad 1/u = 15,600$$

$$\frac{r}{B} = \frac{r}{\sqrt{\dfrac{Tb'}{K'}}}$$

$$\frac{r}{B} = \frac{50 \text{ m}}{\sqrt{\dfrac{2.98 \times 10^{-3} \text{ m}^2/\text{s} \times 2 \text{ m}}{1.45 \times 10^{-8} \text{ m/s}}}} = 0.078$$

For these values of $1/u$ and r/B, the well function is approximately 5.3 as determined from Figure 2.69 (graph with type curves in Section 2.5.9). Finally, the drawdown is calculated by inserting all known values into Equation 14.24 solved for s:

$$s = \frac{Q}{4\pi T} W(u,r/B)$$

$$s = \frac{0.012 \text{ m}^3/\text{s}}{4 \times 3.14 \times 2.98 \times 10^{-3} \text{ m/s}} \times 5.3 = 1.70 \text{ m}$$

14.5 UNCONFINED AQUIFER WITH DELAYED GRAVITY RESPONSE

An unconfined aquifer, with the average initial saturated thickness of 24 m, was tested with a fully penetrating well pumping 35 L/s constantly for 24 h. The drawdown recorded at a monitoring well 31 m from the pumping well is given in Table 14.4. Find hydrogeologic parameters of the aquifer.

The response of unconfined aquifers to well pumping tests is often considerably different than the response of confined aquifers. One of the assumptions made by Theis in deriving the well equation is that water is withdrawn from storage instantaneously with a decrease in hydraulic head. Although this assumption, for practical purposes, may be considered "true enough" for many confined aquifers, it is not acceptable for most unconfined aquifers because of a delay (lag) in releasing water from storage. Derivation of the related equation is given in Section 2.5.10.

Our data plotted on the graph in Figure 14.13 show such a delayed response. This field graph is matched to the theoretical type curves shown in Figure 2.72 (Section 2.5.10) as follows:

- Plot log–log graph time versus recorded drawdown at the same log-cycle scale as the graph with type curves (see Figure 14.13).
- Superimpose the field curve (data) on the type B curves and, keeping the axes of both graphs parallel, match as much of the late drawdown data as possible to a particular β curve (Figure 14.14).
- Choose a match point anywhere where the two graphs overlap and read the four coordinates as follows: s_D and t_y from the type graph, and s and t from the field graph.

TABLE 14.4
Drawdown (s, in m) Recorded at Monitoring Well 31 m from the Pumping Well in an Unconfined Aquifer with Delayed Gravity Response

T (min)	s (m)	T (min)	s (m)	T (min)	s (m)
0.25	0	15	0.34	300	0.53
0.5	0.04	20	0.34	360	0.57
1	0.14	25	0.35	420	0.6
1.5	0.22	30	0.36	480	0.63
2	0.25	45	0.37	540	0.65
3	0.3	60	0.38	600	0.68
4	0.31	75	0.39	720	0.72
5	0.31	90	0.4	840	0.77
6	0.32	105	0.42	960	0.8
7	0.32	120	0.43	1080	0.86
8	0.33	150	0.45	1200	0.91
9	0.33	180	0.47	1320	0.99
10	0.33	240	0.5	1440	1.08

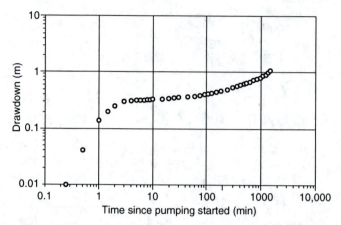

FIGURE 14.13 Log–log graph time versus drawdown recorded at 31 m from a pumping well in an unconfined aquifer.

It is recommended to choose rounded values for s_D and t_y (bottom scale) for easier calculations. In our case these B coordinates are:

Dimensionless drawdown s_D	10^0 i.e., 1
Dimensionless time t_y	10^0 i.e., 1
Drawdown s	0.85 m
Time t	1200 min

- Calculate the transmissivity using the following equation:

$$T = \frac{1}{4\pi} \times \frac{Q s_D}{s}$$ (14.29)

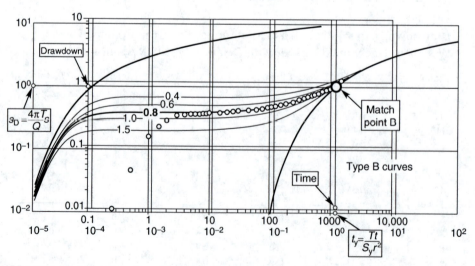

FIGURE 14.14 Matching late drawdown data to a type B curve. The best fit is for the curve $\beta = 0.8$. The match point is chosen conveniently ($s_D = 1$ and $t_y = 1$) where both graphs overlap. Note that the coordinate t_y is read on the lower axis of the type graph.

which in our case gives

$$T = 0.0796 \times \frac{0.035 \text{ m}^3/\text{s} \times 1}{0.85 \text{ m}} = 3.28 \times 10^{-3} \text{ m}^2/\text{s}$$

- Calculate the specific yield using equation:

$$S_y = \frac{Tt}{r^2 t_y} \qquad (14.30)$$

or, in our case

$$S_y = \frac{3.28 \times 10^{-3} \text{ m}^2/\text{s} \times 1200 \times 60 \text{ s}}{(31 \text{ m})^2 \times 1} = 0.246$$

- Overlap the early field data (curve) and the A type curve, which has the same value of β as the B curve (0.8 in our case). When overlapping try to match as much data as possible to the A type curve keeping the axes on both graphs parallel (see Figure 14.15). (Note that as data have to be matched to the same β curve, the graphs should move only in the horizontal direction.)
- Choose a match point anywhere where the two graphs overlap and read the four coordinates as follows: s_D and t_s (upper axis) from the type graph, and s and t from the field graph. Again, it is recommended to choose rounded values for s_D and t_s for easier calculations. These A coordinates are:

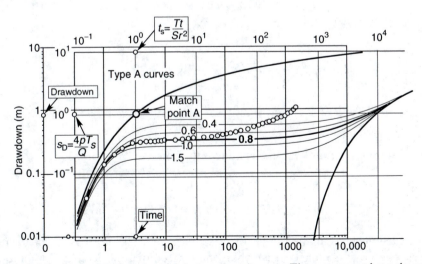

FIGURE 14.15 Matching early drawdown data to a type A curve. The curve must have the same value of β as the previously matched type B curve (0.8 in our case). The match point is again chosen conveniently ($s_D = 1$ and $t_s = 1$) where both graphs overlap. Note that the t_s coordinate is read on the upper axis of the type graph.

Dimensionless drawdown s_D	10^0 i.e., 1
Dimensionless time t_s	10^0 i.e., 1
Drawdown s	0.85 m
Time t	3.3 min

- Calculate the transmissivity again using Equation 14.29. The value should be the same as the one calculated for the late data:

$$T = 0.0796 \times \frac{0.035 \text{ m}^3/\text{s} \times 1}{0.85 \text{ m}} = 3.28 \times 10^{-3} \text{ m}^2/\text{s}$$

- Calculate the elastic storage coefficient using equation:

$$S = \frac{Tt}{r^2 t_s} \tag{14.32}$$

$$S = \frac{3.28 \times 10^{-3} \text{ m}^2/\text{s} \times 3.3 \times 60 \text{ s}}{(31 \text{ m})^2 \times 1} = 0.00068$$

- Find the horizontal hydraulic conductivity knowing the average aquifer thickness ($b = 24$ m) and the previously calculated transmissivity:

$$K_r = \frac{T}{b} \tag{14.32}$$

$$K_r = \frac{3.28 \times 10^{-3} \text{ m}^2/\text{s}}{24 \text{ m}} = 1.37 \times 10^{-4} \text{ m/s}$$

- Find the degree of anisotropy (K_D), which is needed to calculate the vertical hydraulic conductivity, using equation:

$$K_D = \frac{\beta b^2}{r^2} \tag{14.33}$$

$$K_D = \frac{0.8(24 \text{ m})^2}{(31 \text{ m})2} = 0.48$$

And finally, find the vertical hydraulic conductivity from

$$K_z = K_D \times K_r \tag{14.34}$$

$$K_z = 0.48 \times 1.37 \times 10^{-4} \text{ m/s} = 6.58 \times 10^{-4} \text{ m/s}$$

If the saturated thickness of the aquifer decreases more than 10% during the test, the recorded drawdown should be corrected using Jacobs formula (see Equation 2.120 in Section 2.5.10):

$$s' = s - \frac{s^2}{2h}$$

for the late data only. Correcting early drawdown would lead to erroneous results because the response of the aquifer to pumping is primarily due to the elastic properties of the porous media and water (i.e., the elastic storage is predominant).

14.6 SLUG TEST, BOUWER AND RICE METHOD

The results of a slug test performed in a monitoring well completed in an unconfined aquifer are given in form of graphs of displacement versus time in Figure 14.16. The test was for both rising and falling heads, measured with a pressure transducer and recorded with a digital data logger. The monitoring well is partially penetrating with the total depth of 25 ft. The water table in the well before the test was at 24.3 ft above the aquifer base, and 9.2 ft below the land surface. The well has a 4 in. diameter screen placed at the bottom 9 ft. The diameter of the borehole in the screen section is 9 in. Annular space between the borehole wall and the well screen is filled with gravel pack. From the available data, find the hydraulic conductivity of the aquifer.

As in our case, field data may often show the so-called double straight-line effect. The first straight-line portion is explained by a highly permeable zone around the well (gravel pack or developed zone), which quickly sends water into the well immediately after the water level in the well has been lowered. The second straight line is more indicative of the flow from the

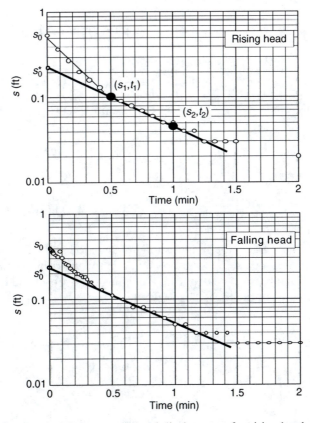

FIGURE 14.16 Semilog graphs of time versus head displacement for rising head and falling head slug tests in the same well.

undisturbed aquifer into the well and its slope should be used in the analysis. Derivation of equations and the graphs of the accompanying parameters used in the Bouwer and Rice method of slug test analysis are given in Section 2.4.2.2.

As the test well in our case is partially penetrating, the ratio $\ln(R_e/r_w)$ is found using Equation 2.7 and curves A and B in Figure 2.38. For our well, the ratio between the well depth (L_e) and the radius of developed zone (r_w, see Figure 2.37) is

$$\frac{L_e}{r_w} = \frac{9 \text{ ft}}{4.5 \text{ in.}} = \frac{2.74 \text{ m}}{0.114 \text{ m}} = 24$$

which gives the following values of the dimensionless parameters A and B (see Figure 2.38):

$$A = 2.45$$
$$B = 0.40$$

Inserting these values into Equation 2.7 gives

$$\ln \frac{R_e}{r_w} = \left[\frac{1.1}{\ln(L_w/r_w)} + \frac{A + B \times \ln[(H - L_w)/r_w]}{L_e/r_w} \right]^{-1}$$

$$\ln \frac{R_e}{r_w} = \left[\frac{1.1}{\ln(4.82 \text{ m}/0.114 \text{ m})} + \frac{2.45 + 0.40 \times \ln[(7.41 \text{ m} - 4.82 \text{ m})/0.114 \text{ m}]}{24} \right]^{-1} = 2.233$$

The slope of the second straight line for the rising head test (see Figure 14.16) is

$$\frac{\ln(s_1/s_2)}{(t_2 - t_1)} = \frac{\ln(0.1 \text{ ft}/0.045 \text{ ft})}{1 \text{ min} - 0.5 \text{ min}} = 1.597 \text{ min}^{-1}$$

After inserting this value and other known values into Equation 2.6, the hydraulic conductivity is found as

$$K = \frac{r_c^2 \ln \frac{R_e}{r_w}}{2L_e} \times \frac{1}{t} \times \ln \frac{s_0}{s_t}$$

$$K = \frac{(0.102 \text{ m})^2 \times 2.233}{2 \times 2.74 \text{ m}} \times 1.597 \times 60 \text{ s}^{-1} = 4.42 \times 10^{-5} \text{ m/s}$$

As can be seen in Figure 14.16, the second straight line in the graph for the falling head test has almost the identical slope as the corresponding line for the rising head test. This means that the hydraulic conductivity calculated from the falling head test would be almost the same as K already calculated from the rising head test.

15 Extraction Well Design and Analysis

15.1 GRAVEL PACK SIZE

Formulation of Problem: Determine the well gravel pack sizes for the two aquifers represented with the grain size curves shown in Figure 15.1. The wells tapping the aquifers have the same outside screen diameter of 273 mm (10 3/4 in.). The diameter of the boreholes in their screen sections is also the same—445 mm (17 1/2 in.). Note that the diameter of the borehole corresponds to the diameter of the drilling bit (drilling diameter), which is one of the main criteria for the well design.

Purpose of the gravel pack is to

- stabilize the formation;
- prevent or reduce sand pumping;
- enable larger screen openings; and
- establish transitional velocity and pressure fields between the formation and the well screen.

The placement of a filter pack makes the zone around the well screen more permeable and increases the effective hydraulic diameter of the well. The gravel pack allows the removal of some formation material during well development, and it retains most of the aquifer fine material during the well exploitation.

A gravel pack is particularly useful in fine-grained, uniformly graded formations, and in extensively laminated aquifers (e.g., alternating layers of silt, sand, and gravel) where it eliminates or reduces sand pumping and enables larger screen openings, thus increasing otherwise small well yields (see Section 3.1.1).

15.1.1 THICKNESS OF GRAVEL PACK

To successfully retain the formation particles, the thickness of the gravel (filter) pack in ideal conditions does not have to be more than 0.5 in. according to laboratory tests made by Johnson Division: "Filter-pack thickness does little to reduce the possibility of sand pumping, because the controlling factor is the ratio of the grain size of the pack material in relation to the formation material" (Driscoll, 1986). However, for practical purposes, the thickness of the gravel pack should be at least 3 in. to ensure its accurate placement and complete surrounding of the screen. On the other hand, a filter pack that is more than 8 in. (203 mm) thick can make final development of the well more difficult "... because the energy created by the development procedure must be able to penetrate the pack to repair the damage done by drilling,

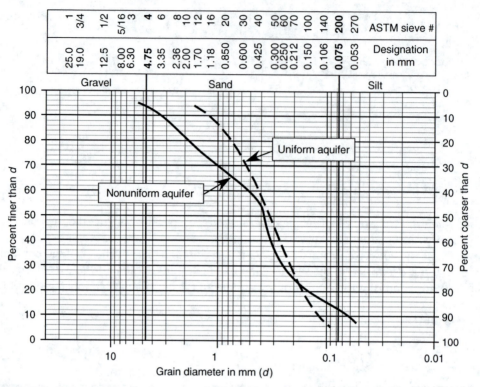

FIGURE 15.1 Grain size distribution curves for two samples.

break down any residual drilling fluid on the borehole wall, and remove fine particles (from the formation) near the borehole" (Driscoll, 1986).

The gravel pack thickness (GPT) corresponds to the thickness of the annulus between the well screen and the borehole wall:

$$\text{GPT} = \frac{D_b - D_s}{2} \tag{15.1}$$

where D_b is the diameter of the borehole in the screen section, and D_s is the outside diameter of the screen. In our case, this is

$$\text{GPT} = \frac{445 \text{ mm} - 273 \text{ mm}}{2} = 86 \text{ mm}$$

The GPT of 86 mm (~3.4 in.) satisfies the two criteria: it is more than 3 in. and less than 8 in. It is generally better to choose gravel packs of smaller thickness because drilling and design costs increase significantly with increasing well diameter. However, in some cases it may be necessary to place a thick gravel pack. This includes reverse rotary drilling, which is most efficient for large-diameter boreholes (>406 mm). The completed borehole may be significantly larger than an optimum screen diameter thus requiring stabilization of the aquifer material (the cost of a thicker gravel pack is incomparably lower than the cost of a larger screen).

15.1.2 SIZE OF GRAVEL PACK

The choice of the gravel pack size presented here is based on the recommendations of The Institution of Water Engineers and Scientists, London (Brandon, 1986). These recommendations are a compromise between numerous procedures that can roughly be divided into two groups:

- A more pragmatic approach followed mainly in the United States, which emphasizes simplicity of design (e.g., it is hard to find recommendations for design of a graded filter pack with the uniformity coefficient greater than 2.5)
- A strict approach that emphasizes the problems of choosing gravel packs for heterogeneous aquifers (the practice in most European countries) including design of filters with two or more layers

If the aquifer material is uniform and well sorted, a "uniform" gravel pack is needed. The grading of the filter pack is based on the grain size distribution curve of the finest aquifer material within the well screen section. A "graded" gravel pack should be considered for aquifers with a wide range of particle sizes and large uniformity coefficients. In this case the grading of the pack depends on both the coarsest and the finest aquifer materials.

15.1.3 UNIFORM GRAVEL PACK

Uniform gravel packs are designed for aquifers with uniformity coefficients less than 3. The first criterion for choosing the gravel pack is

$$D_{50} = 4 \text{ to } 6 \times d_{50} \tag{15.2}$$

where D_{50} is the sieve opening size that would allow 50% of the gravel pack material to pass, and d_{50} is the sieve opening size that would allow 50% of the aquifer material to pass. A greater multiplier is used for aquifers that have fines such as silt and clay.

The second criterion is

$$U_{\text{pack}} < 2.5 \tag{15.3}$$

where U_{pack} is the uniformity coefficient of the gravel pack. Figure 15.2 illustrates the procedure for choosing gravel pack characteristics for the aquifer material whose uniformity coefficient is less or close to 3:

$$U_{\text{aquifer}} = \frac{d_{60}}{d_{10}} = \frac{0.42 \text{ mm}}{0.18 \text{ mm}} = 2.3$$

First, an approximate "temporary" D_{50} is determined knowing that the d_{50} is 0.34 mm and applying Equation 15.2:

$$D_{50}^* = 4.5 \times 0.34 \text{ mm} = 1.53 \text{ mm}$$

Second, the grain diameter closest to a standard ASTM sieve is chosen for the final D_{50} (#12 or 1.70 mm in our case). A temporary gravel pack curve is then drawn through the D_{50} to roughly follow the shape of the aquifer curve. This temporary curve would usually have to be

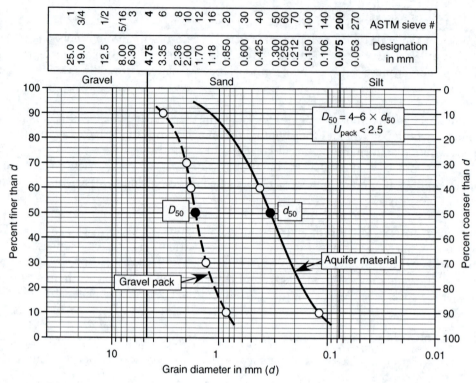

FIGURE 15.2 Grain size distribution curves for the uniform aquifer material and the corresponding gravel (filter) pack.

adjusted further as it lies in the coarser grain zone and, because of the logarithmic scale, the uniformity coefficient becomes larger than that of the aquifer material.

Third, the final gravel pack curve and its characteristic grain sizes (such as D_{10}, D_{30}, D_{50}, D_{70}, D_{90}) are chosen to match the closest standard ASTM sieves, and at the same time satisfy the requirement regarding the uniformity coefficient ($U < 2.5$). In our case this requirement is satisfied as $U = 2.1$ ($D_{60} = 1.80$ mm, and $D_{10} = 0.85$ mm). Choosing characteristic grain sizes that correspond to the standard sieves simplifies the process of constructing a suitable gravel pack if a proper commercial pack is not readily available. Table 15.1 shows the set of standard sieves that can be used to produce the gravel pack curve in Figure 15.2.

TABLE 15.1

Characteristic Grain Sizes of the Gravel Pack Curve in Figure 15.2 and the Corresponding Standard ASTM Sieves

	D_{10}	D_{30}	D_{50}	D_{70}	D_{90}
Grain size (mm)	0.850	1.40	1.70	2.00	3.35
ASTM sieve #	20	14	12	10	6

15.1.4 GRADED GRAVEL PACK

Graded gravel pack should satisfy the following criteria:

1. D_{15} of the pack is at least four times greater than d_{15} of the aquifer sample (if there is more than one aquifer sample, this criterion should be applied for the coarsest sample):

$$D_{15} \geq 4 \text{ to } 6 \times d_{15} \tag{15.4}$$

2. A greater multiplier is used for more heterogeneous aquifers that include silt and clay. D_{85} of the pack should be slightly less than four times the d_{85} (if there is more than one aquifer sample, this criterion should be applied for the finest sample):

$$D_{85} \leq 4 \times d_{85} \tag{15.5}$$

3. The curve of the gravel pack should match the shape of the aquifer sample (if there is more than one aquifer sample, the gravel pack curve should match the shape of the average sample):

$$U_{\text{pack}} \approx U_{\text{aquifer}} \tag{15.6}$$

Figure 15.3 shows a gravel pack curve that satisfies the above criteria. Two starting points for drawing an approximate temporary pack curve are determined knowing that d_{15} is 0.095 mm and d_{85} is 2.50 mm:

$$D^*_{15} = 6 \times 0.095 \text{ mm} = 0.57 \text{ mm}$$
$$D^*_{85} = 3.5 \times 2.50 \text{ mm} = 8.75 \text{ mm}$$

Then the final points are chosen to correspond to the closest standard sieves:

$$D_{15} = 0.60 \text{ mm} \quad \text{or} \quad \#30 \text{ sieve}$$
$$D_{85} = 8.00 \text{ mm} \quad \text{or} \quad 5/16 \text{ sieve}$$

and the curve is drawn to follow the shape of the aquifer curve, without trying to match minor "irregularities." Characteristic grain sizes on the curve (such as D_{10}, D_{30}, D_{50}, D_{70}, D_{90}) are also chosen to match the closest standard ASTM sieves, and at the same time satisfy the requirement that the uniformity coefficient of the gravel pack is similar to that of the aquifer.

TABLE 15.2
Characteristic Grain Sizes of the Gravel Pack Curve in Figure 15.3 and the Corresponding Standard ASTM Sieves

	D_{10}	D_{15}	D_{30}	D_{50}	D_{70}	D_{85}	D_{90}
Grain size (mm)	0.425	0.60	1.18	2.36	4.75	8.00	9.50
ASTM sieve #	40	30	16	8	4	5/16	3/8

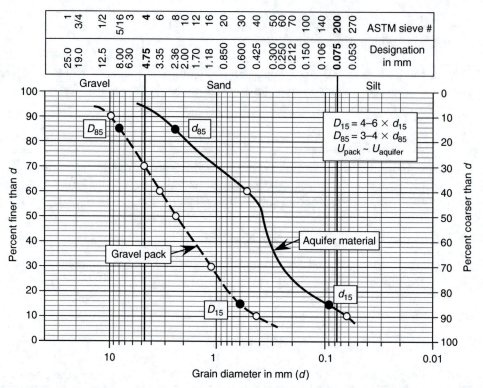

FIGURE 15.3 Grain size distribution curves for the nonuniform aquifer material and the corresponding gravel (filter) pack.

Table 15.2 shows the set of standard sieves that can be used to produce the gravel pack curve in Figure 15.3.

15.2 THREE-STEP PUMPING TEST, WELL SPECIFIC CAPACITY, LOSS, EFFICIENCY

Formulation of Problem: An extraction well and three monitoring wells were used to conduct a pumping test in a confined aquifer. The test was a three-step pumping test with each step lasting 6 h. The pumping rates were 10, 15, and 20 L/s. The radius of the pumping well in its screen section is 0.3 m. The location of the wells and a simplified hydrogeologic cross section are shown in Section 13.1. Table 15.3 shows measured drawdowns at all of the wells at the end of each step, and Figure 15.4 shows this graphically. Using the information provided, and assuming that the flow at the end of each step was in quasi–steady state (the drawdown "almost" stabilized), determine the well loss, efficiency, and specific capacity.

15.2.1 THREE-STEP PUMPING TEST

The first step in the analysis of well efficiency from a three-step pumping test is to plot a semilog graph of distance versus ratio S/Q for all of the wells, as shown in Figure 15.5. As the monitoring wells are not being pumped during the test, the S/Q ratio should remain the same for each step and all three steps should be represented with one data point per monitoring well. This is not the case with the S/Q ratios for the pumping well because of the inevitable well loss (see Section 3.1.2 for detailed explanation on the reasons for well loss).

TABLE 15.3
Drawdown Registered at the Pumping Well (PW) and Three
Monitoring Wells at the End of Each Step, for the Three
Respective Pumping Rates (Q, in L/s)

	Drawdown (m)		
Well	$Q = 10$ L/s	$Q = 15$ L/s	$Q = 20$ L/s
PW	1.10	1.84	2.72
MW1	0.54	0.80	1.08
MW2	0.32	0.48	0.65
MW3	0.17	0.24	0.33

Figure 15.5 shows that data for the monitoring wells MW1, MW2, and MW3 form straight lines for all three steps, while the three points for the pumping well are below this line.

Well loss is the difference between the actual measured drawdown in the pumping well and the theoretical drawdown expressed by the following equation (also called the formation loss equation):

$$s_0 = \frac{Q}{2\pi T} \ln \frac{R}{r_w} \tag{15.7}$$

where s_0 is the theoretical drawdown in the well (formation loss), Q is the pumping rate, T is the aquifer transmissivity, R is the steady-state radius of well influence, and r_w is the well radius.

The total measured drawdown in the well (s_w) is a combination of the linear losses and turbulent losses:

$$s_w = AQ + BQ^2 \tag{15.8}$$

where A is the coefficient of the linear losses, B is the coefficient of turbulent losses, and Q is the pumping rate. The turbulent losses are usually assumed to be quadratic, but other powers

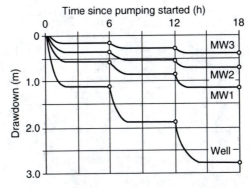

FIGURE 15.4 Drawdown registered at three monitoring wells and the pumping well during the three-step pumping test.

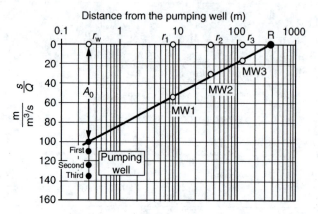

FIGURE 15.5 Graph of distance versus drawdown and pumping ratio for the three-step pumping test.

may be examined as well. The linear losses include both formation loss (Equation 15.7) and linear loss in the near-screen zone. Their respective coefficients are A, A_0, and A_1:

$$A = A_0 + A_1 \qquad (15.9)$$

The coefficient of linear formation loss (A_0) can be calculated as

$$A_0 = \frac{1}{2\pi T} \ln \frac{R}{r_\mathrm{w}} \qquad (15.10)$$

or determined graphically as shown in Figure 15.5: it is the point at distance r_w on the straight line formed by the monitoring wells (distance r_w is the radius of the pumping well).

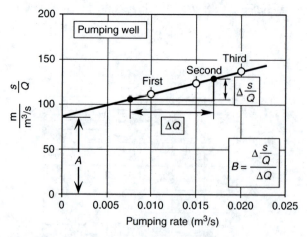

FIGURE 15.6 Graph of pumping rate versus drawdown–pumping rate ratio used to determine the components of the drawdown in the pumping well PW. A: coefficient of the linear loss, B: coefficient of the turbulent loss.

The coefficients of the total linear loss (A) and the quadratic loss (B) can be determined from a graph of pumping rate (Q) versus drawdown–pumping rate ratio (s/Q) as shown in Figure 15.6. The graph is a straight line of the following form:

$$\frac{s_w}{Q} = A + BQ \qquad (15.11)$$

where A is the intercept and B is the slope of the best-fit straight line drawn through the experimental data (in our case all the points fall exactly along the straight line). After substituting values of A and B determined from the graph into Equation 15.8, it is possible to calculate the total (i.e., expected to be actually recorded) drawdown in the pumping well for any pumping rate:

$$s_w = 87 \text{ s/m}^2 \times Q + 2600 \text{ s}^2/\text{m}^5 \times Q^2 \qquad (15.12)$$

The coefficient of turbulent (quadratic) well loss smaller than 2500–3000 s^2/m^5 is usually considered acceptable. Larger coefficients may indicate potential problems with the well such as inadequate well design and development, clogging of the well screen, or other deteriorations of the well. However, there are no widely accepted criteria or standards in the U.S. industry that would address well losses quantitatively.

The coefficients of the linear losses for the monitoring wells can also be found by constructing a graph similar to the graph in Figure 15.7. As there is no pumping from the monitoring wells there are no turbulent losses. The total drawdown is equal to the linear loss (AQ), which in turn is equal to the theoretical loss of the hydraulic head due to groundwater flow through the porous medium (i.e., the formation loss):

$$s_{MW} = AQ = A_0 Q \qquad (15.13)$$

The drawdown in the monitoring wells, for any pumping rate, can be calculated using the coefficient A determined graphically as shown in Figure 15.7 and then applying Equation 15.13. This method is convenient as it is not necessary to know the aquifer transmissivity and the real radius of well influence, but it is applicable only to (quasi) steady-state conditions.

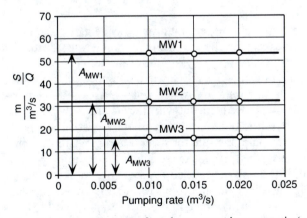

FIGURE 15.7 Graph of pumping rate (Q) versus drawdown–pumping rate ratio (s/Q) used to determine the coefficient of linear loss in the monitoring wells.

15.2.2 Well Efficiency

Well efficiency is the ratio between the theoretical drawdown and the actual drawdown measured in the well. It is expressed in percent:

$$\text{Well efficiency} = \frac{\text{Theoretical drawdown } (s_0)}{\text{Measured drawdown } (s_w)} \times 100\% \qquad (15.14)$$

The theoretical drawdown is easily determined from the graph in Figure 13.3, which is a plot of the same drawdown versus distance data for the three monitoring wells used in this problem. The theoretical drawdown for each step is the drawdown at distance r_w (which is the pumping well radius) read from the straight lines formed by the monitoring wells for each step. The values for the first, second, and third steps are 0.98, 1.45, and 1.96 m, respectively. As can be seen from the graph, the difference between the theoretical drawdown and the measured drawdown increases for each step, i.e., it increases with the increasing pumping rate. Consequently, the well efficiency decreases with the increasing pumping rate. Its values for the first, second, and third steps are 89%, 79%, and 72%, respectively. The example below illustrates determination of the well efficiency for the first step:

$$\text{Well efficiency} = \frac{0.98 \text{ m}}{1.10 \text{ m}} \times 100\% = 89.1\%$$

Note that a well efficiency of 70% or more is usually considered acceptable. If a newly developed well has less than 65% efficiency, there may be a problem with the well development or design, and the possible underlying reasons should be evaluated.

15.2.3 Well Specific Capacity and Optimum Pumping Rate

The specific capacity of the well is given as

$$\text{Well specific capacity} = \frac{\text{Pumping rate } (Q)}{\text{Measured drawdown } (s_w)} \qquad (15.15)$$

and is expressed as the pumping rate per unit drawdown (e.g., L/s/m of drawdown). Similar to the well loss, it also decreases with the increasing pumping rate: 9.1, 8.1 and 7.3 L/s/m for the first, second, and third steps, respectively, in our example.

Choosing an optimum rate at which a well will be pumped is a decision based on numerous factors. For example, if the well (or wells) will be used for a short-term construction dewatering, maintaining a desired drawdown may be the only relevant criterion. In some cases, where there are no alternatives, all that matters is a certain pumping rate. A graph showing drawdown versus pumping rate for several pumping steps, such as the one shown in Figure 15.8 for our well, is a useful tool for selecting an optimum pumping rate.

On the other hand, if the well is designed for a long-term exploitation, in addition to the energy cost of pumping, the hydraulic criteria are the most important in deciding the optimum pumping rate. This includes a comparative study of well losses and well efficiency. Graphs in Figure 15.9 and Figure 15.10 are the final results of such analyses for our pumping well. The procedure consists of the following steps:

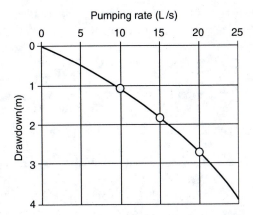

FIGURE 15.8 Graph of pumping rate versus drawdown for our well. Coordinates of the data points are the pumping rates for three steps and the drawdown registered at the end of each step (see Figure 15.4 and Table 15.3).

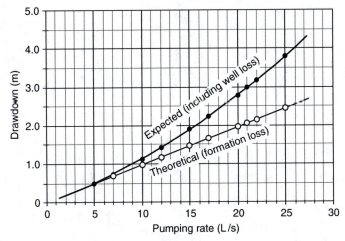

FIGURE 15.9 Pumping rate at PW versus the formation loss calculated using Equation 15.7, and the expected total drawdown calculated using Equation 15.8.

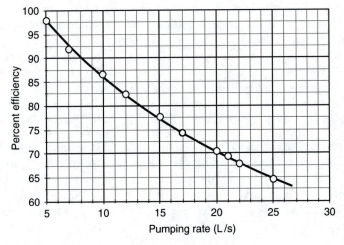

FIGURE 15.10 Pumping rate at PW versus the well efficiency.

- For up to 10 different pumping rates find drawdown that is expected to be actually recorded in the well (i.e., the drawdown that includes well losses) using Equation 15.8:

$$s_w = AQ + BQ^2$$

As determined from the graph in Figure 15.6, this equation for our pumping well (PW) is

$$s_w = 87 \text{ s/m}^2 \times Q + 2600 \text{ s}^2/\text{m}^5 \times Q^2$$

For example, the total expected drawdown in PW for the pumping rate of 25 L/s is

$$s_w = 87 \text{ s/m}^2 \times 0.025 \text{ m}^3/\text{s} + 2600 \text{ s}^2/\text{m}^5 \times (0.025 \text{ m}^3/\text{s})^2 = 3.80 \text{ m}$$

- For the same pumping rates find theoretical drawdown which is the result of groundwater flow through the undisturbed portion of the aquifer only (the formation loss) using Equation 15.7:

$$s_0 = \frac{Q}{2\pi T} \ln \frac{R}{r_w}$$

Substituting previously determined values of the transmissivity ($T = 1.16 \times 10^{-2} \text{ m}^2/\text{s}$), and the radius of influence ($R = 375$ m; see Section 13.1), this theoretical drawdown in PW (whose radius is 0.3 m) for the pumping rate of 25 L/s is

$$s_0 = \frac{0.025 \text{ m}^3/\text{s}}{2 \times 3.14 \times 1.16 \times 10^{-2} \text{ m}^2/\text{s}} \ln \frac{375 \text{ m}}{0.3 \text{ m}} = 2.45 \text{ m}$$

- Find the well efficiency for all chosen pumping rates using Equation 15.14. In our example for PW and the pumping rate of 25 L/s, the well efficiency is

$$\text{Well efficiency} = \frac{2.45 \text{ m}}{3.80 \text{ m}} \times 100\% = 64.5\%$$

The graph in Figure 15.10 is drawn after repeating the described procedure for 10 different pumping rates. Accepting that the well efficiency of 70% is the minimum for an optimal long-term exploitation, the pumping rate of 20 L/s is recommended.

Note that due to normal aging of pumping wells, the well efficiency is expected to decrease in time. Thus, the step-pumping tests should be performed every few years to adjust the exploitation (pumping) regime and apply well rehabilitation measures if necessary.

15.2.4 Transient Well Losses

When the drawdown at the end of each step is not stabilized, the above-described procedures for analyzing performance of pumping wells can still be applied with the adjusted drawdowns. Figure 3.13 shows the error that would occur if the drawdown registered at the end of each step were not corrected. Equation 3.14 derived in Section 3.1.2 describes Dupuit's radius of well influence in transient conditions as follows:

$$R_D = 1.5 \times \sqrt{Tt/S} \qquad (15.16)$$

Theoretically, for an infinite confined aquifer, the groundwater flow forms in infinity and reaches the well pumping rate at the well perimeter (r_0). The corresponding radius of well influence also approaches infinity for a long pumping period, which means that Dupuit's radius of well influence does not have a real physical meaning. For most practical purposes, however, Dupuit's radius of well influence given with Equation 15.16 will yield satisfactory results in various analytical calculations involving the Theis equation.

Again, it should be noted that a definite real radius of well influence cannot be formed in a homogeneous confined aquifer unless there is a source of recharge, such as from a boundary or leakage.

Using the expression for Dupuit's radius of well influence, the coefficient of the linear formation loss is

$$A_0 = \frac{1}{2\pi T} \ln \frac{1.5 \times \sqrt{Tt/S}}{r_w} \tag{15.17}$$

Similarly to steady-state conditions, the coefficients of linear and turbulent losses in transient conditions are found graphoanalytically from a graph of corrected drawdown/Q ratio versus Q. Table 15.4 lists drawdown versus pumping rate recorded during a three-step pumping test, which clearly showed no stabilization of drawdown at the end of any of the steps. The recovery phase of this test is analyzed in Section 14.3 and used to determine aquifer parameters.

The drawdown recorded at the end of the first step is:

$$s_1 = \frac{Q_1}{2\pi T} \ln \frac{1.5 \times \sqrt{T \times t_1/S}}{r_w} (+\text{well loss}) \tag{15.18}$$

or, after inserting all known values

$$3.87\,\text{m} = \frac{0.007\,\text{m}^3/\text{s}}{2 \times 3.14 \times 2.725 \times 10^{-3}\,\text{m}^2/\text{s}} \ln \frac{1.5 \times \sqrt{(2.725 \times 10^{-3}\,\text{m}^2/\text{s} \times 8\,\text{h})/0.0005}}{0.3\,\text{m}} (+\text{well loss})$$

$$3.87\,\text{m} = 3.10\,\text{m} + \text{well loss}$$

This drawdown of 3.87 m does not need adjustments and can be used directly to draw the first point on the graph with the coordinates s_1/Q_1 and Q_1 (553 s/m^2 and 0.007 m^3/s).

The drawdown at the end of the second step (s_2) has two components: the drawdown caused by the increase in pumping rate (s_2') and the drawdown caused by the extended

TABLE 15.4
Drawdown Registered at the Pumping Well (PW) at the End of Each Step for the Three Respective Pumping Rates (Q, in L/s)

	Drawdown (m)		
Well	$Q = 7$ L/s	$Q = 10$ L/s	$Q = 13$ L/s
PW	3.87	5.84	7.91

influence of the first pumping rate or step (s_1'). From the principle of superposition of flows, it follows that

$$s_2 = s_1' + s_2'(+\text{well loss}) \tag{15.19}$$

or

$$s_2 = \frac{Q_1}{2\pi T} \ln \frac{1.5 \times \sqrt{T \times t_2/S}}{r_w} + \frac{(Q_2 - Q_1)}{2\pi T} \ln \frac{1.5 \times \sqrt{T \times (t_2 - t_1)/S}}{r_w} (+\text{well loss}) \tag{15.20}$$

The adjusted drawdown at the end of the second step, which excludes the error caused by the extended influence of the first step, is

$$s_2^* = s_2 - \text{error} \tag{15.21}$$

where error is the difference between the theoretical (calculated) drawdown at the end of the second step (t_2) caused by the first pumping rate (Q_1), and the theoretical drawdown at the end of the first step (t_1) caused by the same pumping rate (Q_1):

$$\text{error} = \frac{Q_1}{2\pi T} \ln \frac{1.5 \times \sqrt{(Tt_2)/S}}{r_w} - \frac{Q_1}{2\pi T} \ln \frac{1.5 \times \sqrt{(Tt_1)/S}}{r_w} \tag{15.22}$$

or, after simplifying

$$\text{error} = \frac{Q_1}{2\pi T} \ln \sqrt{\frac{t_2}{t_1}} \tag{15.23}$$

If the time intervals of each step are the same, as in our case, the above equation becomes

$$\text{error} = \frac{Q_1}{2\pi T} \ln \sqrt{2} \tag{15.24}$$

and the adjusted drawdown is

$$s_2^* = s_2 - \frac{Q_1}{2\pi T} \ln \sqrt{2} \tag{15.25}$$

After substituting all the known values, the adjusted drawdown for the second step is

$$s_2^* = 5.84 \text{ m} - \frac{0.007 \text{ m}^3/\text{s}}{2 \times 3.14 \times 2.725 \times 10^{-3} \text{ m}^2/\text{s}} \ln 1.4142 = 5.70 \text{ m}$$

This adjusted drawdown, which includes the well loss, is used to draw the second point on the graph of s/Q versus Q with the following coordinates: s_2^*/Q_2 and Q_2 (570 s/m^2 and 0.010 m^3/s).

The recorded drawdown at the end of the third step (s_3) has three components: s_3', which is the drawdown caused by the increase in pumping rate from Q_2 to Q_3, the influence of the first step (s_1''), and the influence of the second step (s_2''):

$$s_3 = s_1'' + s_2'' + s_3' \tag{15.26}$$

$$s_3 = \frac{Q_1}{2\pi T} \ln \frac{1.5 \times \sqrt{Tt_3/S}}{r_w} + \frac{(Q_2 - Q_1)}{2\pi T} \ln \frac{1.5 \times \sqrt{T(t_3 - t_1)/S}}{r_w} + \frac{(Q_3 - Q_2)}{2\pi T}$$

$$\times \ln \frac{1.5 \times \sqrt{T(t_3 - t_2)/S}}{r_w} (+\text{well loss}) \tag{15.27}$$

The adjusted drawdown at the end of the third step excludes errors introduced by the influences of the first and second steps (see Figure 3.13):

$$s_3^* = s_3 - \text{error}_\mathrm{I} - \text{error}_\mathrm{II} \tag{15.28}$$

The error from the influence of the first step is

$$\text{error}_\mathrm{I} = \frac{Q_1}{2\pi T} \ln \frac{1.5 \times \sqrt{(Tt_3)/S}}{r_w} - \frac{Q_1}{2\pi T} \ln \frac{1.5 \times \sqrt{(Tt_1)/S}}{r_w} \tag{15.29}$$

or, after simplifying (note that $t_3 = 3t_1$ in our case)

$$\text{error}_\mathrm{I} = \frac{Q_1}{2\pi T} \ln \sqrt{3} \tag{15.30}$$

The error from the influence of the second step is

$$\text{error}_\mathrm{II} = \frac{Q_2 - Q_1}{2\pi T} \ln \frac{1.5 \times \sqrt{T(t_3 - t_1)/S}}{r_w} - \frac{Q_2 - Q_1}{2\pi T} \ln \frac{1.5 \times \sqrt{T(t_2 - t_1)/S}}{r_w} \tag{15.31}$$

or, after simplifying (note that $t_3 = 3t_1$ and $t_2 = 2t_1$ in our case):

$$\text{error}_\mathrm{II} = \frac{Q_2 - Q_1}{2\pi T} \ln \sqrt{2} \tag{15.32}$$

The corrected drawdown at the end of the third step is then

$$s_3^* = s_3 - \frac{Q_1}{2\pi T} \ln \sqrt{3} - \frac{(Q_2 - Q_1)}{2\pi T} \ln \sqrt{2}$$
$$s_3^* = 7.91 \text{ m} - 0.22 \text{ m} - 0.06 \text{ m} = 7.75 \text{ m} \tag{15.33}$$

The coordinates of the third adjusted point on the graph are then s_3^*/Q_3 and Q_3 (596 s/m^2 and 0.013 m^3/s).

By correcting the recorded drawdowns, the time-dependent influence of the preceding steps is eliminated and transient flow is converted into quasi–steady state. This allows the use of the graphoanalytical method for determining well losses for steady-state flow as explained earlier.

Figure 15.11 shows the graph of s/Q versus Q for the three-step pumping test for both corrected and raw data. The intercept of the straight line drawn through the corrected points is the coefficient of the total linear loss at the well (A). It includes the coefficients of the linear formation loss (A_0) and the linear well loss (A_1):

$$A = A_0 + A_1 \tag{15.34}$$

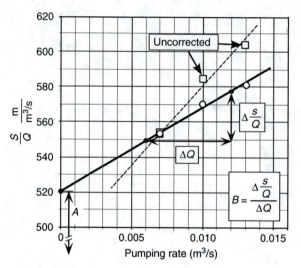

FIGURE 15.11 Graph of pumping rate versus drawdown–pumping rate ratio used to determine the components of the drawdown in the pumping well under transient conditions (A: coefficient of the linear loss, B: coefficient of the turbulent loss).

Unlike for steady-state flow, where these two components of linear loss can be separated, in transient conditions this separation is not possible as the coefficient of the linear formation loss is time dependent:

$$A = \frac{1}{2\pi T} \ln \frac{1.5 \times \sqrt{Tt/S}}{r_{\mathrm{w}}} + A_1 \qquad (15.35)$$

From the graph in Figure 15.11, the coefficient of the total linear loss determined with our three-step pumping test is 521 s/m^2. This value would change for other pumping rates and pumping times.

The coefficient of the turbulent well loss (B) is the slope of the straight line drawn through the corrected points. Theoretically, it is not time dependent and should remain the same for different pumping rates. A common exception is pumping from karst and fissured aquifers where turbulent well loss may increase with an increasing pumping rate. In such a case, the points on the graph of s/Q versus Q would form a parabola rather than a straight line.

From the two chosen points (filled black circles) on the straight line in our case, the coefficient of turbulent well loss is found as

$$B = \frac{\Delta(s/Q)}{\Delta Q}$$

$$B = \frac{29 \text{ s/m}^2}{0.006 \text{ m}^3/\text{s}} = 4833 \text{ s}^2/\text{m}^5 \qquad (15.36)$$

This value of B is high and indicates that the well should receive some corrective measures such as better development (if it is a new well), rehabilitation (if the well is old), or other appropriate treatment depending upon the cause of well loss. In general, if the coefficient of turbulent loss is less than 2500–3000 s^2/m^5, the well is considered to be in good condition.

15.3 RADIUS OF WELL INFLUENCE

Statement of Problem: A homogeneous confined aquifer is tapped by a fully penetrating well whose radius is 225 mm. The aquifer transmissivity is 8.35×10^{-4} m²/s and the storage coefficient is 0.00065. For the given parameters determine the radius of well influence after 6 months and 5 days of pumping with the rate of 15 L/s.

The radius of well influence in a homogeneous, confined aquifer of infinite extent and without recharge theoretically forms in infinity. However, it is often necessary, for practical purposes, to find some approximate "realistic" value for the radius of well influence. This radius is defined by a given small drawdown that can still be registered in the field (say, 5 cm or 0.05 m). Note that this radius of influence is not the same as the well capture zone; the well capture zone comprises the aquifer volume from which all the groundwater will eventually discharge at the well. This means that the gradient everywhere within the capture zone is toward the well. The radius of well influence is merely the distance at which the well pumpage is being registered, i.e., the hydraulic head is lowered. It does not necessarily mean that the hydraulic gradient everywhere within this distance is now being reversed toward the well from its original position.

The Theis equation for such a small drawdown (s_R) is

$$s_R = \frac{Q}{4\pi T} W(u_R) \tag{15.37}$$

where parameter u_R is defined by the distance from the pumping well where this small drawdown is registered, i.e., by the "realistic" radius of well influence (R_R):

$$u_R = \frac{R_R^2 S}{4Tt} \tag{15.38}$$

After finding the value for $W(u_R)$ from the following expression:

$$W(u_R) = \frac{4\pi T s_R}{Q} \tag{15.39}$$

and the corresponding value of u_R from the table in Appendix D, the radius of well influence is calculated by inserting all known values into

$$R_R = \sqrt{\frac{4Ttu_R}{S}} \tag{15.40}$$

The radius of well influence for sufficiently long pumping periods (i.e., when $u < 0.05$) can also be found from the Cooper–Jacob approximation of the This equation:

$$s_R = \frac{Q}{4\pi T} \ln \frac{2.25Tt}{R_R^2 S} \tag{15.41}$$

which gives

$$\frac{2.25Tt}{R_R^2 S} = e^{4\pi T s_R/Q} \tag{15.42}$$

$$R_R = \sqrt{\frac{2.25Tt}{S \times e^{4\pi T s_R/Q}}} \tag{15.43}$$

Note that (in general notation) if $\ln x = a$, then $e^a = x$. Assuming a small drawdown of 0.05 m at R_R, and inserting all known values into Equation 15.43, gives the following value for the realistic radius of well influence after 5 d of pumping:

$$R_R = \sqrt{\frac{2.25 \times 8.35 \times 10^{-4}\ \mathrm{m^2/s} \times 5 \times 86,400\ \mathrm{s}}{0.00065 e^{(4 \times 3.14 \times 8.35 \times 10^{-4} \times 0.05\ \mathrm{m})/0.015\ \mathrm{m^3/s}}}}$$

After 6 months of pumping with the same pumping rate of 15 L/s, the radius of well influence is 6643 m.

REFERENCES

Brandon, T.W., editor, 1986. *Groundwater: Occurrence, Development and Protection.* Water Practice Manuals, The Institution of Water Engineers and Scientists, London, 615 pp.

Driscoll, F.G., 1986. *Groundwater and Wells.* Johnson Filtration Systems Inc, St. Paul, MN, 1089 pp.

16 Spring Flow and Stream Base Flow

16.1 RECESSION (BASE FLOW) ANALYSIS

A portion of a spring hydrograph, which includes a recession period during months of June and July, is shown in Figure 16.5. Find the volume of water stored in the aquifer portion drained by the spring at the beginning of this recession period (first week of June) using available information from the hydrograph. Although the hydrograph shows influence of new precipitation at the end of July–beginning of August (i.e. the recession period has ended), assume that there was no rainfall in the spring drainage area for 3 months after the beginning of recession, and then calculate what the spring discharge would be at the end of this 3-month period.

The hydrograph of a spring is the final result of various processes that govern the transformation of precipitation and other water inputs in the spring drainage area into the flow at the point of discharge. In many cases, the discharge hydrograph of a spring closely resembles hydrographs of surface streams, particularly if the aquifer is unconfined and has a high transmissivity. In some low permeable media, such as silty and clayey sediments, or unaltered, unfractured metamorphic rocks, springs are weak and usually do not react visibly to daily, weekly, or even monthly (seasonal) water inputs. On the other hand, the reaction of large springs draining karst or intensely fractured aquifers to precipitation events is sometimes only a matter of hours. A thorough analysis of a spring discharge hydrograph provides useful information on aquifer characteristics such as the nature of its storage and transmissivity, and types and quantity of its groundwater reserves. Although the processes that generate hydrographs of springs and surface streams are quite different, there is much that is analogous between them, and the hydrograph terminology is the same (see Section 2.6.2.2).

Analysis of the falling hydrograph limb shown in Figure 16.1, which corresponds to a period without significant precipitation, is called the recession analysis. Knowing that the spring discharge is without disturbances caused by a rapid inflow of new water into the aquifer, the recession analysis provides good insight into the aquifer structure. By establishing an appropriate mathematical relationship between spring discharge and time, it is possible to predict the discharge rate after a given period without precipitation, and calculate the volume of discharged water. For these reasons, recession analysis has been a popular quantitative method in hydrogeological studies for a long time.

The shape and characteristics of a recession curve depend upon different factors such as type of aquifer porosity (the most important), position of piezometric surface, and recharge from other aquifers. The ideal recession conditions—a long period of several months without precipitation, are rare in moderate or humid climates. Consequently, summer and early fall storms can cause various disturbances in the recession curve that cannot be removed unambiguously during analysis. It is therefore desirable to analyze as many recession curves from different years as possible. Larger sample allows for the derivation of the average recession

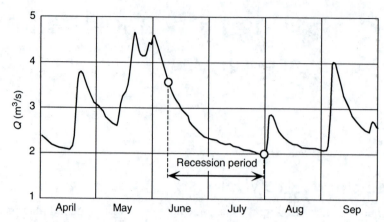

FIGURE 16.1 Part of the annual spring discharge hydrograph with the recession period.

curve as well as the envelope of minima (Figure 16.2). In addition, conclusions about the aquifer porosity structure, its accumulative ability, and expected long-term minimum discharge, are more accurate.

Boussinesq (1904) and Maillet (1905) proposed two well-known mathematical formulas that describe the falling limb of hydrographs and the base flow. Both equations give dependence of the flow at specified time (Q_t) on the flow at the beginning of recession (Q_0). The Boussinesq equation is of hyperbolic form:

$$Q_t = \frac{Q_0}{[1 + \alpha(t - t_0)]^2} \tag{16.1}$$

where t is time since the beginning of recession for which the flow rate is calculated and t_0 is time at the beginning of recession usually (but not necessarily) set equal to zero.

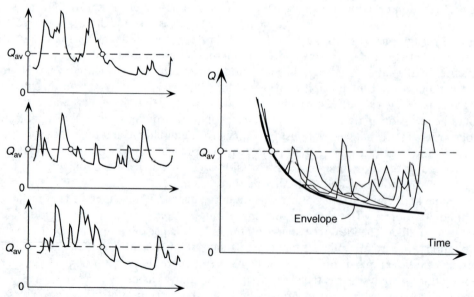

FIGURE 16.2 Envelope (bold line) of the minimum points on recession curves recorded at a spring over the period of record. The individual recession curves are overlaid at the point that equals the average discharge rate for the available period of record.

The Maillet equation, which is more commonly used, is an exponential function:

$$Q_t = Q_0 e^{-\alpha(t-t_0)} \tag{16.2}$$

The dimensionless parameter α in both equations represents the coefficient of discharge (or recession coefficient), which depends upon the aquifer's transmissivity and specific yield. The Maillet equation, when plotted on a semilog diagram, is a straight line with the coefficient of discharge (α) as its slope:

$$\log Q_t = \log Q_0 - 0.4343 \times \alpha \Delta t$$
$$\Delta t = t - t_0 \tag{16.3}$$

$$\alpha = \frac{\log Q_0 - \log Q_t}{0.4343(t - t_0)} \tag{16.4}$$

The introduction of the conversion factor (0.4343) in Equation 16.4 is a convenience for expressing discharge in cubic meters per second, and time in days. Dimension of α is day^{-1}.

Figure 16.3 is a semilog plot of time versus discharge rate for the recession period shown in Figure 16.1. The recorded daily discharges form three straight lines that mean that the recession curve can be approximated by three corresponding exponential functions with three different coefficients of discharge (α). The three lines correspond to three micro regimes of discharge during the recession. The coefficient of discharge for the first microregime, using Equation 16.4, is

$$\alpha_1 = \frac{\log Q_0 - \log Q_1}{0.4343(t_1 - t_0)}$$

$$\alpha_1 = \frac{\log(3.55 \text{ m}^3/\text{s}) - \log(2.25 \text{ m}^3/\text{s})}{0.4343 \times 24.5 \text{ d}} = 0.019$$

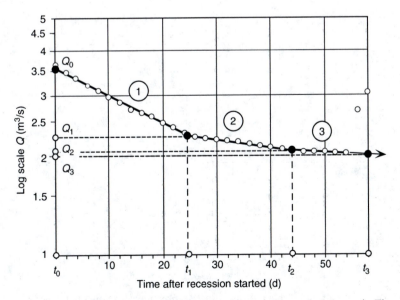

FIGURE 16.3 Semilog graph of time versus discharge for the recession period shown in Figure 16.1. The duration of the recession period is 54 d.

The coefficient of discharge for the second microregime is

$$\alpha_2 = \frac{\log(2.25 \text{ m}^3/\text{s}) - \log(2.06 \text{ m}^3/\text{s})}{0.4343(44 \text{ d} - 24.5 \text{ d})} = 0.0045$$

The third coefficient of discharge, or slope of the third straight line, is found by choosing a discharge rate anywhere on the line, including its extension if the actual line is short. In our case, the discharge rate after 60 d is 2.01 m³/s and α_3 is then

$$\alpha_3 = \frac{\log(2.06 \text{ m}^3/\text{s}) - \log(2.01 \text{ m}^3/\text{s})}{0.4343(60 \text{ d} - 44 \text{ d})} = 0.0015$$

After determining the coefficients of discharge, the flow rate at any given time after the beginning of recession can be calculated using the appropriate Maillet equation. For example, discharge of the spring 35 d after the recession started, when the second microregime is active, is calculated as

$$Q_{35} = Q_1 e^{-\alpha_2(35 \text{ d} - t_1)}$$
$$Q_{35} = 2.25 \text{ m}^3/\text{s} \times e^{-0.0045(35 \text{ d} - 24.5 \text{ d})} = 2.146 \text{ m}^3/\text{s}$$

Note that the initial discharge rate for the second microregime is 2.25 m³/s and the corresponding time is 24.5 d.

Alternatively, discharge after 35 d may be determined by extending the second straight line backward to where it intersects the Q axis and then using that value in the calculation as the initial discharge rate (see Figure 16.4). Then

$$Q_{35} = Q_2 e^{-\alpha_2(35 \text{ d} - t_0)}$$
$$Q_{35} = 2.55 \text{ m}^3/\text{s} \times e^{-0.0045 \cdot 35 \text{ d}} = 2.178 \text{ m}^3/\text{s}$$

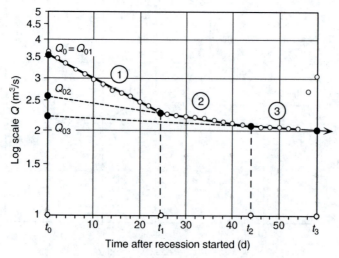

FIGURE 16.4 Alternative initial discharge rates for the three micro regimes shown in Figure 16.3.

Spring discharge 3 months after the beginning of recession, assuming no precipitation for the entire period, may be predicted by using the characteristic values for the third (or last-detected) microregime (see Figure 16.3):

$$Q_{90} = Q_2 e^{-\alpha_3(90\,d - t_2)}$$
$$Q_{90} = 2.06 \text{ m}^3/\text{s} \times e^{-0.0015(90\,d - 44\,d)} = 1.923 \text{ m}^3/\text{s}$$

or, alternatively (see Figure 16.4)

$$Q_{90} = Q_3 e^{-\alpha_3(90\,d - t_0)}$$
$$Q_{90} = 2.20 \text{ m}^3/\text{s} \times e^{-0.0015(90\,d)} = 1.922 \text{ m}^3/\text{s}$$

The variation of the coefficient of discharge has a physical explanation. It is accepted in practice that α on the order of 10^{-2} indicates rapid drainage of well-interconnected large fissures or fractures (or karst channels in case of karst aquifers), while milder slopes of the recession curve (α on the order of 10^{-3}) represent slow drainage of small voids, i.e., narrow fissures and aquifer matrix porosity. Accordingly, the main contribution to the spring discharge in our case is from storage in small voids.

The coefficient of discharge (α) and the volume of free gravitational groundwater stored in the aquifer above spring level (i.e., groundwater that contributes to spring discharge), are inversely proportional:

$$\alpha = \frac{Q_t}{V_t} \tag{16.5}$$

where Q_t is the discharge rate at time t, and V_t is the volume of water stored in the aquifer above the level of discharge (spring level). This relationship is valid only for descending gravitational springs. Equation 16.5 allows calculation of the volume of water accumulated in the aquifer at the beginning of recession, as well as the volume discharged during a given period of time. The calculated remaining volume of groundwater always refers to the reserves stored above the current level of discharge. The draining of an aquifer with three micro-regimes of discharge (as in our case), and the corresponding volumes of the discharged water are shown in Figure 16.5. The total initial volume of groundwater stored in the aquifer (above

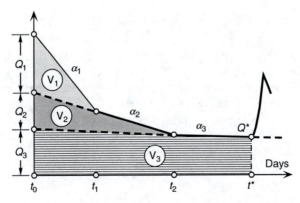

FIGURE 16.5 Schematic presentation of the recession with three microregimes of discharge, and the corresponding volumes of water discharged from the three different types of storage.

the level of discharge) at the beginning of the recession period is the sum of the three volumes that correspond to the three different types of storage (effective porosity):

$$V_0 = V_1 + V_2 + V_3 = \left[\frac{Q_1}{\alpha_1} + \frac{Q_2}{\alpha_2} + \frac{Q_3}{\alpha_3}\right] 86{,}400 \text{ s} \quad [\text{m}^3] \tag{16.6}$$

where discharge rates are given in m^3/s.

The volume of groundwater remaining in the aquifer at the end of the third microregime is the function of the discharge rate at time t^* and the coefficient of discharge α_3:

$$V^* = \frac{Q^*}{\alpha_3} \tag{16.7}$$

The difference between volumes V_0 and V^* is the volume of all groundwater discharged during the period $t^* - t_0$. In our case, the volume of groundwater stored in the aquifer at the beginning of recession is (see also Figure 16.4):

$$V_0 = \left[\frac{(Q_{01} - Q_{02})}{\alpha_1} + \frac{Q_{02} - Q_{03}}{\alpha_2} + \frac{Q_{03}}{\alpha_3}\right] \times 86{,}400 \text{ s} \quad [\text{m}^3]$$

$$V_0 = \left[\frac{(3.55 \text{ m}^3/\text{s} - 2.55 \text{ m}^3/\text{s})}{0.019} + \frac{(2.55 \text{ m}^3/\text{s} - 2.20 \text{ m}^3/\text{s})}{0.0045} + \frac{2.20 \text{ m}^3/\text{s}}{0.0015}\right] \times 86{,}400 \text{ s}$$

$$V_0 = 4.547 \times 10^6 \text{ m}^3 + 6.720 \times 10^6 \text{ m}^3 + 1.267 \times 10^8 \text{ m}^3 = 1.380 \times 10^8 \text{ m}^3$$

As can be seen, most water is stored in the fine storage, which is the main contributor to discharge during the recession. The volume of water remaining in the aquifer above the spring level, i.e., the volume of free gravitational water in the fine storage at the end of recession is

$$V^* = \frac{2.03 \text{ m}^3/\text{s}}{0.0015} \times 86{,}400 \text{ s} = 1.169 \times 10^8 \text{ m}^3$$

which gives the following volume of water discharged at the spring for the duration of recession (54 d):

$$V = V_0 - V^* = 1.380 \times 10^{-8} \text{ m}^3 - 1.169 \times 10^8 \text{ m}^3 = 21.1 \times 10^6 \text{ m}^3$$

Recession periods of large perennial karstic springs, or springs draining highly permeable fissured aquifers, often have two or three microregimes of discharge as in our example. However, the first microregime rarely corresponds to the simple exponential expression of the Maillet's type, and is better explained by hyperbolic functions. Deviations from exponential dependence can be easily detected if recorded data plotted on a semilog diagram do not form straight lines. Usually, the best approximation of the rapid (and often turbulent) drainage of large groundwater transmitters at the beginning of recession is the hyperbolic relation of the Boussinesq type. Its general form is

$$Q_t = \frac{Q_0}{(1 + \alpha t)^n} \tag{16.8}$$

In many cases this function correctly describes the entire recession curve. On the basis of 100 analyzed recession curves of karstic springs in France, Drogue (1972) concludes that among the six exponents studied, the best first approximations of exponent n are 1/2, 3/2, and 2.

The exact determination of exponent n and the discharge coefficient α for the function that best fits the measured data is performed graphically and by computation as follows:

- The minimum recorded discharge at the end of recession is noted ($Q_2 = 0.057 \text{ m}^3/\text{s}$ for the example in Figure 16.6).
- Any discharge Q_1 on the recession curve, which is not a consequence of (possible) recent rainfall, is chosen in the section between Q_2 and Q_0.
- The value of α that satisfies the following equation:

$$\frac{\log(Q_0/Q_1)}{\log(Q_0/Q_2)} = \frac{\log(1 + \alpha t_1)}{\log(1 + \alpha t_2)} \tag{16.9}$$

is determined by trial adopting an initial value for α (usually 0.5). The result can be graphically checked as shown in Figure 16.7: the correct coefficient of discharge gives the straight line through the points determined by Q_0, Q_1, Q_2, and the corresponding times t_0, t_1, t_2. The exact value of α for our example shown in Figure 16.6 and Figure 16.7 is 0.202.

$$\frac{\log(0.240 \text{ m}^3/\text{s}/0.105 \text{ m}^3/\text{s})}{\log(0.240 \text{ m}^3/\text{s}/0.060 \text{ m}^3/\text{s})} = \frac{\log(1 + 0.202 \times 15 \text{ d})}{\log(1 + 0.202 \times 47 \text{ d})}$$

$$0.5963 \approx 0.5929$$

The exponent n is calculated by substituting the determined value for α into either of the following two equations:

$$n = \frac{\log(Q_0/Q_1)}{\log(1 + \alpha t_1)} \tag{16.10}$$

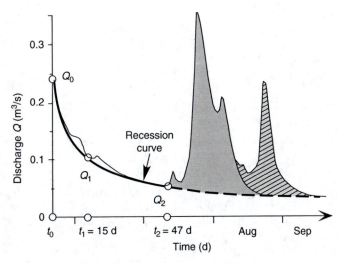

FIGURE 16.6 Recession curve of a spring with discharges used to determine the hyperbolic function. Hydrographs 1 and 2 are the result of summer storms.

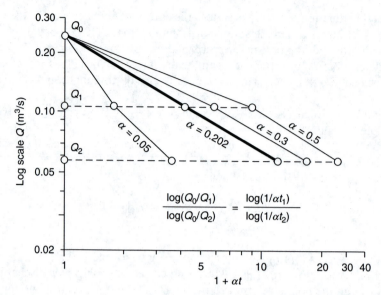

FIGURE 16.7 Graphical determination of discharge coefficient α for the example shown in Figure 16.6.

$$n = -\frac{\log(Q_1/Q_2)}{\log\left(\dfrac{1+\alpha t_1}{1+\alpha t_2}\right)} \tag{16.11}$$

For our example, Equation 16.10 gives

$$n = \frac{\log(0.240 \text{ m}^3/\text{s}/0.105 \text{ m}^3/\text{s})}{\log(1+0.202 \times 15 \text{ d})} = 0.593$$

and the recession discharge equation is then

$$Q_t = \frac{0.24 \text{ m}^3/\text{s}}{(1+0.202t)^{0.593}} \tag{16.12}$$

The coefficient of discharge and the exponent in Equation 16.8 have the following general relationship:

$$\alpha = \frac{\sqrt[n]{Q_0} - \sqrt[n]{Q_t}}{t \times \sqrt[n]{Q_t}} \tag{16.13}$$

As in the case of the Maillet equation, the determined hyperbolic function can be used to calculate the volume of free gravitational water stored in the aquifer above the spring level. In general, this volume at any time t since the beginning of recession is

$$V_t = \frac{Q_0}{\alpha(n-1)}\left[1 - \frac{1}{(1+\alpha t)^{n-1}}\right]86,400 \text{ s} \quad [\text{m}^3] \tag{16.14}$$

16.2   HYDROCHEMICAL SEPARATION OF SPRING HYDROGRAPH

Table 16.1 shows the results of daily monitoring of the spring discharge rate, the contents of the calcium ion in the spring water, and the precipitation in the drainage area of a permanent spring. Analyze the impact of precipitation on the spring discharge and find the percentage of newly infiltrated water in the overall spring discharge after major precipitation events.

The simultaneous recording of spring discharge and chemical constitution of the spring water allows for a fairly accurate separation into "old" (prestorm) and "new" (rain) water. This separation is based on the assumption that the constitution of the water entering the aquifer is considerably different than that already in it. When recharge by rain takes place, it is evident that the concentration of most cations characterizing groundwater, such as calcium

TABLE 16.1
Elements Monitored at the Spring during an 89 d Period with Two Major Precipitation Events

No.	Q (L/s)	Ca (mg/L)	P (mm)	No.	Q (L/s)	Ca (mg/L)	P (mm)	No.	Q (L/s)	Ca (mg/L)	P (mm)
1	35.2	95.8	0	31	74.4	97.0	0	60	135.0	98.6	0
2	35.9	96.2	0	32	70.6	97.8	0.2	61	139.7	99.8	0
3	36.7	97.0	0	33	66.9	97.8	0	62	131.5	97.4	0
4	35.9	97.0	0	34	63.3	96.2	2.9	63	128.1	97.4	0
5	37.4	96.2	0	35	65.1	95.0	0	64	121.3	97.4	0
6	35.9	96.2	0	36	63.3	94.6	0	65	105.9	96.6	0.9
7	36.7	97.0	0.3	37	57.9	95.8	0	66	122.4	96.6	0
8	34.4	96.6	2.55	38	57.1	96.2	1.2	67	119.0	96.2	0.4
9	34.4	96.6	0.3	39	55.3	97.0	1.1	68	110.2	96.2	2.75
10	33.7	96.2	0	40	51.0	97.0	0.9	69	114.6	95.8	5.55
11	34.4	97.0	0	41	55.3	96.6	6.8	70	100.6	96.2	1.55
12	34.4	97.0	0	42	161.4	96.2	16.1	71	112.4	96.2	4.8
13	33.7	96.2	0	43	301.0	95.4	10.6	72	114.6	95.8	0
14	31.5	96.2	0	44	280.0	95.4	0	73	112.4	95.4	0
15	30.8	97.0	0	45	283.0	72.7	25	74	105.9	95.0	0.5
16	31.5	97.4	0	46	346.2	69.3	11.6	75	102.7	96.2	2.2
17	68.8	97.8	17.1	47	330.4	66.5	0	76	103.8	94.2	0.2
18	55.3	97.0	6.1	48	268.2	73.7	0.9	77	97.4	93.0	0.6
19	93.3	96.3	11.8	49	260.9	76.5	2.6	78	83.2	93.8	0.9
20	97.4	95.4	4	50	243.7	85.0	0	79	101.6	93.8	19.1
21	437.4	91.4	18.8	51	202.5	85.8	0	80	137.3	95.0	7.8
22	330.4	79.4	0	52	73.5	90.6	0	81	110.2	96.2	0
23	232.5	83.0	0	53	70.6	92.8	0	82	91.2	95.4	0
24	87.2	85.0	0	54	62.4	95.8	0.9	83	91.2	95.8	0
25	87.2	85.8	0	55	71.6	96.6	0	85	93.3	95.8	0
26	51.0	87.4	0	56	151.6	96.6	0	86	95.3	95.8	1.2
27	64.2	88.6	0	57	149.2	96.6	0	87	104.8	96.2	10
28	66.0	91.4	1.5	58	156.5	97.0	8.3	88	99.5	96.2	11.1
29	114.6	95.4	14.5	59	140.9	97.8	8.7	89	99.5	97.0	2.1
30	79.2	96.2	0								

Note: Q, spring discharge; Ca, concentration of the calcium ion; P, precipitation in the spring drainage area.

and magnesium, is much lower in rainwater. Additional preconditions for the application of a hydrochemical hydrograph separation are (after Dreiss, 1989) that

- concentrations of the chemical constituents in the rainwater chosen for monitoring are uniform in both area and time;
- corresponding concentrations in the prestorm water are also uniform in area and time;
- the effects of other processes in the hydrologic cycle during the episode, including recharge by surface waters, are negligible; and
- the concentration and transport of elements are not changed by chemical reactions in the aquifer.

This last condition assumes a minor dissolution of carbonate rocks during the flow of new water through porous medium. The first two conditions regarding calcium ion, monitored in our case, are acceptable. From the graph in Figure 16.8, it is clearly visible that the calcium ion concentration in the spring water drops rapidly after the heavy rains that cause an increase in discharge rate. Assuming a simple mixing of old aquifer water (Q_{old}) and newly infiltrated rainwater (Q_{new}), the total recorded discharge of the spring is the sum of the two (after Dreiss, 1989):

$$Q_{total} = Q_{old} + Q_{new} \tag{16.15}$$

If chemical reactions in the aquifer do not cause significant and rapid changes in the concentration of calcium ion in the infiltrating rainwater (which is often true for unconfined karst and intensely fissured aquifers where the flow velocity is high), the calcium ion balance in the spring water is

$$Q_{total} \times C_{total} = Q_{old} \times C_{old} + Q_{new} \times C_{new} \tag{16.16}$$

where Q_{total} is the recorded spring discharge, C_{total} is the recorded concentration of calcium ion in the spring water, Q_{old} is the portion of the spring flow attributed to the "old" water (i.e., water already present in the aquifer before the rain event), C_{old} is the recorded concentration of calcium ion in the spring water before the rain event, Q_{new} is the portion of the

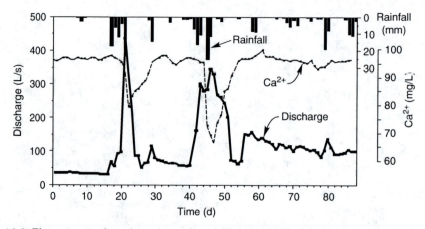

FIGURE 16.8 Elements monitored at the spring during an 89 d period. Plotted data are given in Table 16.1.

spring flow attributed to the "new" water, and C_{new} is the concentration of calcium ion in the new water.

If C_{new} is much smaller than C_{old} (which is correct in our case as the concentration of calcium ion in rainwater is usually less than 5 mg/L), the input mass of calcium ion is relatively small compared to its mass in the old aquifer water, i.e.,

$$Q_{new} \times C_{new} \ll Q_{old} \times C_{old} \tag{16.17}$$

From Equation 16.16 it follows, after excluding the (small) input mass, that

$$Q_{old} = \frac{Q_{total} \times C_{total}}{C_{old}} \tag{16.18}$$

Combining Equation 16.15 and Equation 16.18 gives

$$Q_{new} = Q_{total} - \frac{Q_{total} \times C_{total}}{C_{old}} \tag{16.19}$$

By applying Equation 16.19 it is possible to estimate the discharge component formed by the inflow of new rainwater if the spring discharge recordings and a continuous hydrochemical monitoring are performed before, during, and after the storm event.

The result of the hydrochemical hydrograph separation for the second major rainfall event, which started approximately 40 d after the beginning of monitoring (see Figure 16.8), is shown in Figure 16.8. For example, the flow rate attributed to "new water" 5 d after the first increase in spring discharge (day 45, see Table 16.1) is calculated by applying Equation 16.19 as

$$Q_{new} = 0.283 \text{ m}^3/\text{s} - \frac{0.283 \text{ m}^3/\text{s} \times 72.7 \text{ mg/L}}{96.4 \text{ mg/L}} = 0.070 \text{ m}^3/\text{s}$$

where 96.4 mg/L is the average concentration of calcium ion in the spring water before the rainfall event, i.e., during the recession period when the flow is generated by "old" water only (days 32–40; see Figure 16.8 and Figure 16.9).

As can be seen from the graph in Figure 16.9, only a minor amount of "new" water started discharging at the spring 1 d after the major increase in the flow rate. This lag is 3 d for the first major increase in the discharge of "new" water. The lag between the maximum spring discharge and the maximum discharge of "new" water is again 1 d. New water stopped discharging at the spring 14 d after the rainfall started influencing the overall flow rate. Figure 16.10 shows these three components in the overall spring discharge:

1. Flow that is the result of the rainfall event (the corresponding volume of the discharged water is $2.01 \times 10^5 \text{ m}^3$).
2. Flow that would have occurred if there were no precipitation. This component, equivalent to the base flow of surface streams, is separated by extending the prerainfall recession curve (the volume is $5.08 \times 10^4 \text{ m}^3$).
3. Flow due to discharge of newly infiltrated water ($3.67 \times 10^4 \text{ m}^3$).

A comparison of these three components and the corresponding volumes of discharged water show that "new" water contributes with only 18.3% to the overall flow increase at the spring (component 1). The percentage of new water in the total spring discharge for the 14 d

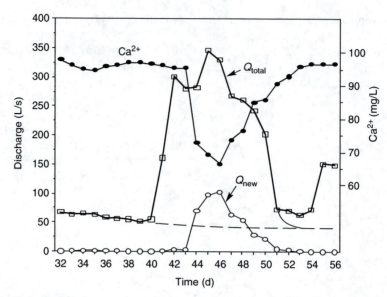

FIGURE 16.9 Hydrochemical separation of the newly infiltrated water discharging at the spring after a major rainfall event.

period is 14.6% (note that the total spring discharge includes components 1 and 2). This analysis shows that the main effect of the rainfall event and the related infiltration is the displacement of the "old" water, which is forced to discharge from the aquifer. This mechanism is attributed to the propagation of pressure through the well-interconnected conduits and fractures, which often causes a rapid increase in spring discharge. Only a small amount of "new" water, with a certain delay, actually reaches the spring during and immediately after the rainfall. Similar mechanisms of spring discharge are common in karst and intensely fissured aquifers (Dreiss, 1989; Kresic, 1991).

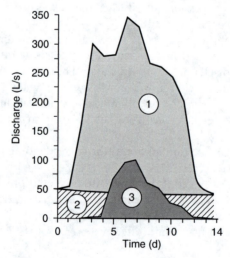

FIGURE 16.10 Components of flow at the spring during and after the second major rainfall event. 1: Overall increase in the discharge rate. 2: Discharge of prerainfall water that would have occurred if there was no precipitation. 3: Flow due to discharge of newly infiltrated water.

REFERENCES

Boussinesq, J., 1904. Recherches théoriques sur l'écoulement des nappes d'eau infiltrées dans le sol et sur les débits des sources. *Journal de Mathématiques Pures et Appliquées*, Paris, 10, 5–78.

Dreiss, S.J., 1989. Regional scale transport in a karst aquifer. 1. Component separation of spring flow hydrographs. *Water Resour. Res.*, 25(1): 117–125.

Drogue, C., 1972. Analyse statistique des hydrogrammes de decrues des sources karstiques. *J. Hydrol.*, 15: 49–68.

Kresic, N., 1991. *Kvantitativna hidrogeologija karsta sa elementima zastite podzemnih voda (Quantitative Karst Hydrogeology with Elements of Groundwater Protection*, in Serbo-Croatian). Naucna Knjiga, Belgrade, 192 p.

Maillet, E., editor, 1905. *Essais dihydraulique souterraine et fluviale.* Herman et Cie, Paris, Vol. 1, 218 p.

17 Groundwater Flow Modeling

17.1 THREE-LAYER MODEL

Groundwater contaminant plumes are delineated in the Monahne River alluvial aquifer and in the underlying confined aquifer as shown in Figure 17.1. The two aquifers are separated by a low-permeable, discontinuous silty clay layer (see Figure 17.2). Hydrogeologic assessment indicates that the contamination from the upper, unconfined aquifer enters the lower aquifer through the "window" in the silty clay layer due to the difference in hydraulic heads in the two aquifers. Develop a groundwater modeling concept and a model that will be used in the design of a well extraction system for containment of the two plumes. Hydraulic conductivities of the upper and lower aquifers, as determined from slug tests, are 2.6×10^{-4} and 3.4×10^{-3} m/s, respectively. Hydraulic conductivity of the confining silty clay layer, determined in the laboratory from Shelby tube samples, is 6.5×10^{-7} m/s. Average annual recharge rate in the alluvial aquifer is estimated at 2.5×10^{-9} m/s, which corresponds to 12% of the gross annual precipitation in the area (660 mm).

17.1.1 MODELING CONCEPT

Developing a modeling concept is the most important part of groundwater modeling and requires a thorough understanding of hydrogeology, hydrology, and groundwater dynamics. Often, as in our case, investigations are focused on a relatively small area and the collected data have to be supplemented with more regional information in order to develop a valid concept. As seen from Figure 17.1, the area of the two plumes, according to the information available from 25 monitoring wells, is approximately 250×250 m and has only one identified physical boundary of groundwater flow—the Monahne River. Based on water level measurements in the monitoring wells and one measurement of the river stage, this is the flow boundary for the alluvial (upper) aquifer only: the head in the confined (lower) aquifer is below the head in the upper aquifer and the flow directions in the two aquifers are slightly different.

Regional information shows that the Monahne River alluvial plain is approximately 4 miles wide and the stream is gaining water. There is no information on water table elevations as far as 1 mile from the plumes' area; there are four domestic wells within 500 m radius that were used to measure the hydraulic head in the confined aquifer. The monitoring well logs show a horizontal discontinuous base of the alluvial aquifer (the silty clay layer—see Figure 17.2). The regional pre-Quaternary base of the confined aquifer (dense clay) is known to be thick and continuous, and it dips toward northwest in the area of the two plumes. Water level measurements were collected only once, in late October, which is a month with approximately average precipitation on the annual basis. The following modeling concept is developed based on the above facts:

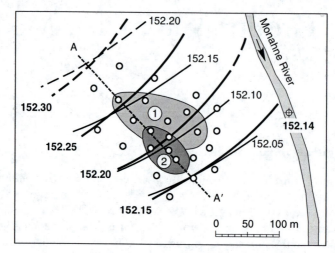

FIGURE 17.1 Water table contours (thin lines) for the unconfined alluvial aquifer, and the piezometric surface contours (thick lines) for the confined aquifer as determined from the monitoring wells (shown with circles) data. Contour interval is 0.05 m. 1: Contaminant plume in the unconfined aquifer. 2: Contaminant plume in the confined aquifer. The line AA' is the cross-section shown in Figure 17.2.

- The groundwater flow system consists of two aquifers separated by an aquitard, and will be modeled as a three-layer, full 3D system.
- Since the model might grow to be a fate and transport model for predicting the exact role of the aquitard, interactions between the two aquifers, concentrations of pollutants and cleanup times, the layers will be presented with their real thicknesses (top and bottom) and hydraulic conductivities rather than with the transmissivities only.
- There is only one physical (real) boundary for groundwater flow and that too only in the unconfined aquifer, which means that all other boundaries will have to be assumed from the flow nets in the two aquifers. These hydraulic (artificial) boundaries will have to be placed far enough in order not to influence the stress created by pumping wells for containment of the two plumes.
- The model will be calibrated in steady state as there is only one set of data available. The predictions will be made for transient conditions taking into account seasonal variations in the river stage, recharge, and groundwater levels.

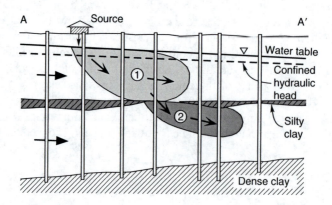

FIGURE 17.2 Simplified hydrogeologic cross section through the contaminant plumes in the upper (1) and lower (2) aquifers.

17.1.2 MODEL GEOMETRY

The model geometry defines the size and the shape of the model. In general, it consists of the model grid and the model boundaries, both external and internal, for all layers. Whenever possible, real physical boundaries should be chosen over artificial (hydraulic) boundaries. When the real boundaries are far from the area of interest and are impractical to model, as in our case (except for the Monahne River), it may be necessary to set hydraulic boundaries. Figure 7.7 (in Section 7.2.3) shows the general principle of setting such boundaries based on the flow net. Since the hydraulic boundaries have to be set far from the area of interest, a more regional flow net often has to be assumed because of lack of data beyond the area of site-specific hydrogeologic investigation. This is the case with the alluvial aquifer in our example. It is not rare in the modeling practice to assume that the rectangular model edges are far enough and there is no need to "bother" with setting some realistic flow boundaries. Note, however, that the model edges are by definition no-flow boundaries unless otherwise specified by the modeler. Such an approach is justified only if there is a strong physical internal boundary that will ensure formation of a realistic (natural) flow net in and around the area of interest. Figure 17.3 shows contours of the estimated hydraulic heads in the confined aquifer. In this case there are no identified physical boundaries around the area of interest (the Monahne River is not influencing the flow in the confined aquifer), and all model boundaries will have to be assumed.

17.1.3 GRID AND LAYERS

Once the possible external model boundaries are identified, the next step is to create the model grid. The overall dimensions of the regular rectangular grid should be large enough to

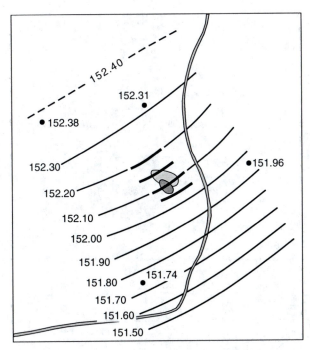

FIGURE 17.3 Contour lines of the hydraulic head surface in the confined (lower) aquifer. Contours drawn using monitoring well data in the area of the two plumes are shown with bold lines. Other contours are assumed based on just four measurements in domestic wells (shown with circles) and are therefore less accurate.

encompass all (usually) irregular boundaries of the flow fields in all layers. If the flow field areas in different layers are not of the same size, the grid is set for the largest combined area. The size and number of cells in the row and column directions are then set having the following in mind (see Figure 17.4 and Figure 17.5):

- Areas of interest, which will be exposed to various stresses (e.g., well pumpage), and areas with more data available will be covered with a finer mesh.
- Cell size along the physical boundaries should allow for their accurate representation.
- Length of successive cells in the row, column, and layer directions cannot increase by more than 1.5 times.

The model geometry along the vertical axis is controlled by the type, number, and thickness of the layers. In our case the alluvial aquifer is layer 1 of type 1 (strictly unconfined). The silty clay aquitard is layer 2 of type 0 (confined, the transmissivity is not expected to change), as is the confined aquifer (layer 3). This configuration can be changed later if it is concluded that the influence of future well pumpage on water levels should be modeled more precisely. If such be the case, more layers in each hydrostratigraphic unit can be easily added with most MODFLOW processors such as Groundwater Vistas and Processing Modflow. These new layers set in the same unit will have the same hydrogeologic parameters (hydraulic

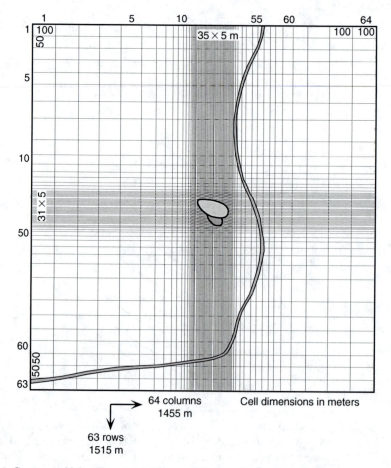

FIGURE 17.4 Custom grid for the model. The mesh is finer in the two-plume area.

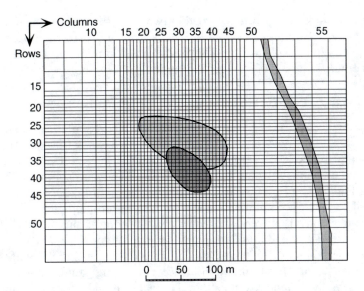

FIGURE 17.5 Fine grid spacing in the area of the two plumes. Note a gradual increase in cell dimensions beyond the area of interest following the rule of thumb in groundwater modeling—size of adjacent cells cannot increase (or decrease) by more than 1.5 times in row and column (and layer) directions.

conductivity, storage coefficient, porosity) as the original layer. Among other things, more layers set in the same aquifer or aquitard are useful when

- the hydraulic gradients created by pumping are steep;
- the pumping wells are partially penetrating and the well intake area has to be correctly simulated; and
- it is important to accurately simulate vertical migration (fate and transport) of pollutants.

A discontinuous aquitard layer in the area where it is not present can be modeled by adjusting its thickness and hydraulic conductivity as shown in Figure 7.13 (in Section 7.2.4). A realistic hydraulic connection between the two aquifers in the "window" area is simulated by assigning the hydrogeologic parameters (horizontal and vertical hydraulic conductivities, storage coefficient, effective porosity) of the layer above or below the aquitard. Alternatively, this portion of layer 2 (aquitard) can be given the average values of the parameters in layers 1 and 3. The top of layer 1 is the water table in the unconfined aquifer. It is interpolated from the monitoring well data in the area of the two plumes and then extrapolated to the rest of the model assuming a uniform gradient. (Note that the uniform gradient is assumed because there is no data on water table elevations or aquifer thickness beyond the two-plume area.) Water table contours (i.e., layer top) should be drawn to reflect the influence of the Monahne River, a gaining stream. Bottom of layer 1 (which is also top of layer 2) is assumed to be horizontal throughout the model and set at 142 m amsl, which is the average value derived from the monitoring well logs. The bottom of layer 2 (silty clay aquitard) is also relatively horizontal and set at 138 m, which gives its average thickness of 4 m. This thickness gradually decreases toward the "window" area where the bottom elevation is set at 141 m, i.e., the thickness of layer 2 is reduced to 1 m only.

Elevation of the bottom of layer 3 linearly decreases from 131 m in the southeast to 128 m in the northwest. This increase in thickness is estimated from the slope determined from the well logs and the piezometric surface recorded in two domestic wells in the north and north-east (see Figure 17.3). Note, however, that the wider spacing between contour lines in Figure 17.3 could be a consequence of two things: (1) an increase in thickness and (2) an increase in the hydraulic conductivity.

17.1.4   BOUNDARY AND INITIAL CONDITIONS

All boundaries in layers 2 and 3 are artificial and derived from the contour map of the hydraulic head shown in Figure 17.3. Prescribed head (or constant head) boundaries in the northwest and the southeast correspond to equipotential lines and provide for a realistic groundwater flow in the confined aquifer from northwest toward southeast. This flow field can be laterally limited by two boundary flowlines, as shown in Figure 17.6, in an attempt to accurately simulate the flow net in the confined aquifer. A more pragmatic (but less natural) way to model is to assume that the model edges are far enough from the area of interest and let the two equipotential boundaries "take care" of the flow field.

Figure 17.7 shows boundaries of the model for layer 1. Although the Monahne River is the only physical boundary, its position and influence as a strong gaining stream are certain to create a flow pattern very similar to the natural one even if additional hydraulic boundaries were not defined. However, in order to create as realistic a flow net as possible, and achieve better computational stability, one constant–head boundary simulating an equipotential line is set in the northwestern part of the model. Northern and eastern model edges coincide with the assumed streamlines. Note that all cells east of the river are inactive because the river is a complete physical boundary for the flow in the alluvial aquifer. The boundary condition

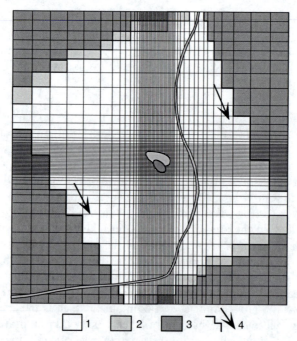

FIGURE 17.6 Artificial (hydraulic) boundaries for the confined aquifer (layer 3) and the silty clay aquitard (layer 2) based on the estimated contour map of the hydraulic head surface in layer 3 (see Figure 17.3). Prescribed–head boundaries correspond to two contour lines and no-flow boundaries are placed along two flowlines. 1: Active cell. 2: Prescribed (known) head cell. 3: Inactive cell. 4: Direction of groundwater flow and flowline.

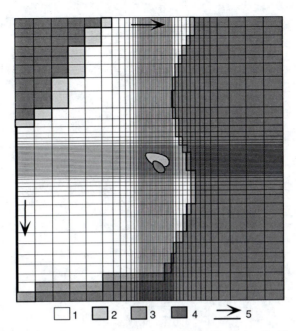

FIGURE 17.7 Model boundaries for the unconfined alluvial aquifer (layer 1). The Monahne River is the major physical boundary that can be simulated with various MODFLOW packages (see text). The northwestern boundary is along an assumed equipotential line placed far from the area of interest. Northern and western boundaries are along assumed flowlines, which coincide with the model edges. 1: Active cell. 2: Constant–head boundary (or head-dependent flux boundary). 3: River cell. 4: Inactive cell. 5: Direction of groundwater flow and flowline.

along the river can be simulated with constant heads (which is the simplest way), the Modflow River package or the General-Head Boundary package (see Section 7.2.5). In any case, the hydraulic head in the river cells should closely resemble the hydraulic gradient in the river, which means that the head should decrease in each cell in the downstream direction. This gradient usually has to be assumed and adjusted during model calibration. In our case, only one measurement of the river elevation was available near the two-plume area and the gradient was linearly interpolated between this and two other values that were estimated as shown in Figure 17.8. Simulation of known head boundaries with constant–head cells should be performed with care as the constant–head cells act as an inexhaustible source of water. Also, the hydraulic head in the constant–head cells cannot change during the model simulation and they are therefore not suitable for transient calibration or prediction. The time-variant specified-head package (CHD1) allows constant head cells to take on different user-specified head values for each simulation time-period and each time step within one period. The head at the boundary can also be specified by the user for different time periods in the General-Head Boundary and River packages. These three packages are therefore more appropriate for transient modeling of known-head or head-dependent flux boundaries than simple simulation with the constant–head cells.

17.1.5 MODEL PARAMETERS

As explained in Section 7.2.4, model parameters are divided into three groups: (1) time, (2) space, and (3) hydrogeologic characteristics. Time parameters are irrelevant for steady-state

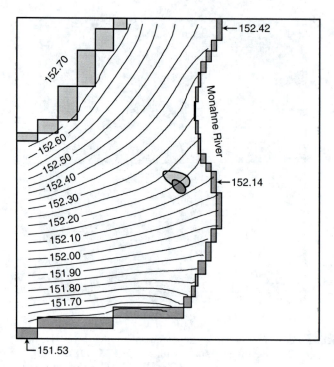

FIGURE 17.8 Contours of the calibrated water table in the unconfined aquifer. Contour interval is 0.05 m. The model check is done by comparing measured and calculated heads in the area of the two plumes (see Figure 17.9). The calculated heads cannot be checked beyond this area because of lack of data. They are controlled mainly by the assigned hydraulic gradient along the river to follow some logical flow pattern (three head values in the river are shown for illustration).

simulations. They will be important for the transient prediction, allowing for seasonal variations in aquifer recharge and boundary conditions. A common approach, in case of humid temperate climates, is to simulate one hydrologic year with four seasons or modeling time periods. If more accurate predictions are needed, 1 year can be simulated with 12 time periods, provided that there are enough data available to represent average monthly hydrologic conditions in the area. Space parameters, i.e., layer top and bottom, are needed when the user lets the model calculate transmissivity for each layer from its thickness and the hydraulic conductivity; this approach is taken in our example. Hydrogeologic parameters are transmissivity, horizontal and vertical hydraulic conductivities, leakance, storage parameters (specific yield for unconfined aquifers and storage coefficient for confined aquifers), and effective porosity. As mentioned earlier, most MODFLOW processors can calculate transmissivity for a layer of any type from its thickness and the horizontal hydraulic conductivity. Vertical leakance between layers, required by MODFLOW, can also be calculated by the model from the thicknesses of adjacent layers and their vertical hydraulic conductivities. This is recommended over often lengthy manual calculations. The vertical hydraulic conductivity, unless determined from aquifer pumping tests or estimated in some other way, is commonly assumed to be ten times smaller than the horizontal hydraulic conductivity. This assumption should be applied with care because many geologic settings do not "behave" accordingly. Storage properties are important for transient simulations, while effective porosity is required only if the model will be used in conjunction with a particle tracking code, in which case it is needed for calculations of linear flow velocities.

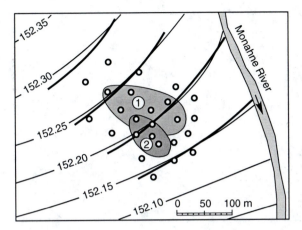

FIGURE 17.9 Contours of the calibrated (thin lines) and the measured (heavy lines) water table in the unconfined aquifer in the area of the two plumes. Contour interval is 0.05 m. Contours are strongly controlled by the assigned hydraulic heads in the cells simulating the river.

17.1.6 CALIBRATED MODEL

Figure 17.8 through Figure 17.11 show calibrated vs. measured heads in the alluvial and confined aquifers. Assumptions on relatively homogeneous hydraulic conductivities in all three layers proved to be valid as the initial K values were only slightly changed during the model calibration. The calibrated hydraulic conductivities in layers 1, 2, and 3 are 2.3×10^{-4}, 6.5×10^{-7}, and 3.7×10^{-4} m/s, respectively. The vertical hydraulic conductivities are ten times lower than the horizontal ones in all three layers. The recharge rate applied to layer 1

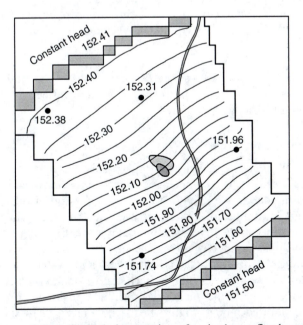

FIGURE 17.10 Contours of the calibrated piezometric surface in the confined aquifer. Contour interval is 0.05 m. The model check is done by comparing measured and calculated heads in the area of the two plumes (see Figure 17.11) and with field measurements in four domestic wells shown with circles.

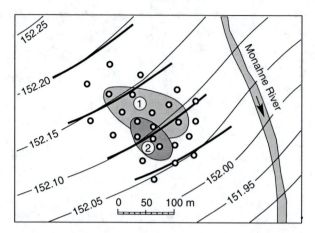

FIGURE 17.11 Contours of the calibrated (thin lines) and the measured (heavy lines) piezometric surface in the confined aquifer in the area of the two plumes. Contour interval is 0.05 m. Note that the contours are not influenced by the river which is simulated only in layer 1 (the aquifer is separated from the river by the silty clay aquitard).

(unconfined aquifer) is calibrated at 2.9×10^{-9} m/s, which is a bit more than the assumed value of 2.5×10^{-9} m/s and corresponds to approximately 14% of the average annual gross precipitation of 660 mm.

As expected, two major conclusions were made during the hydraulic head calibration in the alluvial (unconfined) aquifer:

- Adjusting distribution of the hydraulic head along the river cells is a very efficient way to closely match the calculated and the measured water table.
- Horizontal hydraulic conductivity and recharge are two parameters with equivalent quality—an increase in hydraulic conductivity creates the same effect as a decrease in recharge.

The adjustments of the hydraulic head at the two constant–head boundaries has the maximum effect on the calculated hydraulic head surface in the confined aquifer, in addition to changing the horizontal hydraulic conductivity.

17.1.7 CONTAINMENT OF THE PLUMES

The final test before using the model for various predictions is to check if the artificial boundaries were set correctly. It is recommended that the modeler performs a test prediction during early stages of model development rather than at the end of the calibration process. This is done by placing a number of anticipated wells in the area of interest and pumping them heavily in order to check if the hydraulic boundaries are far enough. Figure 17.12 shows that pumping in the upper aquifer is not influenced by the northwestern constant–head boundary. The pumping also does not reach the northern and western model edges representing two flowlines (no-flow boundaries). This check can be done visually, by comparing the position of the contours closest to the boundaries with their position before the pumping, i.e., for the steady-state calibrated model. Figure 17.13 and Figure 17.14 show containment of both plumes with three wells pumping the upper aquifer and two wells simultaneously pumping the lower aquifer. The main advantage of the numeric model is that it can be used to quickly analyze different pumping scenarios, including varying pumping locations,

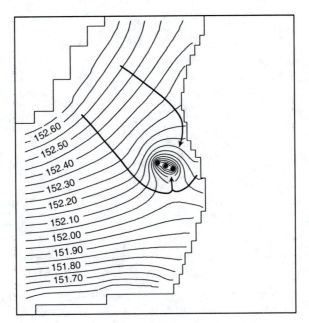

FIGURE 17.12 Water table contours in the unconfined (upper) aquifer as the result of pumpage with three wells containing the plume. Contour interval is 0.05 m. Note that this pumping stress does not reach the constant–head boundaries which are placed far enough (compare with Figure 17.8). The capture zone of the three wells is shown with the bounding flowlines. The wells receive portion of their pumpage from the infiltrating river water.

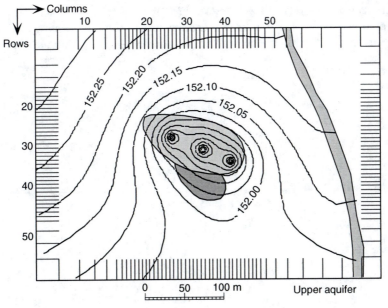

FIGURE 17.13 Containment of the plume in the upper aquifer with three wells pumping 0.0015 m³/s each. Water table contours are in meters amsl.

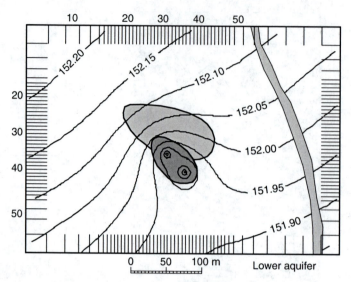

FIGURE 17.14 The plume in the lower (confined) aquifer contained by two wells pumping 0.001 m^3/s each. Contours of the piezometric surface are in meters amsl.

schedule, and rates. The model is irreplaceable when designing the system that would satisfy the following important requirements for an adequate containment of the two plumes in our example:

- First lower the water table in the upper aquifer below the hydraulic head surface in the lower aquifer to prevent further downward migration of the contaminant.
- Start pumping the lower aquifer, and at the same time increase the pumping in the upper aquifer to keep the water table slightly below the hydraulic head surface in the lower aquifer (compare Figure 17.13 and Figure 17.14).
- Adjust pumping rates for the seasonal changes in the aquifer recharge, the river stage, and the position of the water levels in the two aquifers (i.e., elevation of the known-head boundaries).

18 One-Dimensional Analytical Fate and Transport Models

18.1 DISSOLVED CONTAMINANT CONCENTRATIONS IN TIME AND SPACE

Find the time of arrival of the maximum concentration of dissolved tetrachloroethene (PCE) 30 ft downgradient from the monitoring well MW1, which has the highest recorded PCE concentration at the Chemical Company Site. Assume that the present-day conditions at the site represent an instantaneous source model (the actual source of PCE contamination has been removed). The concentration of PCE at MW1 is 0.053 mg/L. The average hydraulic conductivity of the porous media is 0.46 ft/d, and the total porosity is assumed to be 20%. The hydraulic gradient between MW1 and the property boundary, 30 ft downgradient, is 0.0062. There is no site-specific information on PCE fate and transport parameters available. After completing this calculation, assume that a source of PCE in the aquifer will maintain concentration recorded at MW1 (0.053 mg/L) constant for years to come, i.e., instead of the instantaneous source, assume a continuous source.

18.1.1 DISCONTINUOUS (INSTANTANEOUS) SOURCE

The maximum concentration of PCE recorded at the Site (well MW1) is used as a point source for contaminant fate and transport calculations assuming instantaneous contaminant loading in the dissolved phase. This situation is analytically described with the following equation known as Baetsle's model (Domenico and Schwartz, 1990, p. 650; see also Section 6.2):

$$C(x,y,z,t) = \left[\frac{C_0 V_0}{8(\pi t)^{3/2}(D_x D_y D_z)^{1/2}} \right] \exp \left[-\frac{X - v_c t}{4D_x t} - \frac{Y^2}{4D_y t} - \frac{Z^2}{4D_z t} - \lambda t \right] \tag{18.1}$$

where X, Y, Z are the distances in the x, y, z directions from the center of the gravity of the contaminant mass, t is time, C_0 is initial concentration at the source, V_0 is original volume of contaminant released so that $C_0 V_0$ is the mass of the release, D_x, D_y, D_z are coefficients of hydrodynamic dispersion ($D = \alpha v$; α is dispersivity), v_c is contaminant (retarded) velocity, and λ is the degradation constant for the contaminant.

The maximum contaminant concentration caused by a point source spill occurs at the center of the spreading plume, where $y = z = 0$, and $x = v_c t$:

$$C_{\max} = \frac{C_0 V_0 e^{-\lambda t}}{8(\pi t)^{3/2}(D_x D_y D_z)^{1/2}} \tag{18.2}$$

18.1.2 Contaminant (Retarded) Velocity

The retarded contaminant velocity for dissolved PCE is calculated as

$$v_c = \frac{v_L}{R} \tag{18.3}$$

where v_L is the linear velocity of groundwater and R is the retardation coefficient given as

$$R = 1 + \frac{\rho_b}{n} K_d \tag{18.4}$$

or

$$R = 1 + \left(\frac{1-n}{n}\right)\rho_s K_d \tag{18.5}$$

where ρ_b is soil bulk density, ρ_s is soil particle density (2.65 g/cm^3 for most mineral soils), n is total porosity, and K_d is the distribution (partitioning) coefficient given as

$$K_d = K_{oc} f_{oc} \tag{18.6}$$

where K_{oc} is organic carbon partition coefficient (values found in tables for various constituents) and f_{oc} is fraction of total organic carbon in the soil in terms of grams of organic carbon per gram of soil.

Linear groundwater velocity is found from the hydraulic conductivity of the porous medium (K), hydraulic gradient (i), and effective porosity of the aquifer material (n_{ef}):

$$v_L = \frac{Ki}{n_{ef}} \tag{18.7}$$

The site specific information for the Chemical Company Site is as follows:

- Hydraulic conductivity (geometric mean of the slug test data) $K = 0.46$ ft/d or 3.2×10^{-4} ft/min.
- Hydraulic gradient between the source area and the point of calculation (30 ft downgradient) is 0.0062.
- Effective porosity is assumed to be 15% based on the total porosity of 20%; $n_{ef} = 0.15$.

The linear velocity of groundwater is then

$$v_L = \frac{0.46 \text{ ft/d} \times 0.0062}{0.15} = 0.019 \text{ ft/d}$$

The distribution coefficient (K_d) for PCE is (Equation 18.6)

$$K_d = 263 \text{ mL/g} \times 0.0001 = 0.263 \text{ mL/g}$$

where 263 mL/g is the organic carbon partition coefficient for PCE (Cohen and Mercer, 1993), and 0.001 is a conservative value for fraction organic carbon (f_{oc}) in shallow unconsolidated sediments (Cohen and Mercer, 1993).

The retardation coefficient for PCE is (Equation 18.5)

$$R = 1 + \left(\frac{1 - 0.20}{0.20}\right) \times 2.65 \text{ g/cm}^3 \times 0.263 \text{ mL/g} = 3.79$$

where 0.20 is the estimated total porosity.
   The retarded velocity of dissolved PCE is (Equation 18.3)

$$v_c = \frac{0.019 \text{ ft/d}}{3.79} = 0.005 \text{ ft/d}$$

The time of travel of the center of the contaminant mass from the source area 30 ft downgradient is

$$t = \frac{L}{v_c} = \frac{30 \text{ ft}}{0.005 \text{ ft/d}} = 6000 \text{ d}$$

The numeric values for calculations of dissolved PCE concentration in groundwater at the Chemical Company Site using Equation 18.1 (Baetsle's model) are as follows:

$C_0 = 0.053$ mg/L is the highest concentration of PCE at the site (MW1);
$t_{1/2} = 720$ d is the most conservative value for PCE half-life found in literature (Howard et al., 1991, p. 502; Cohen and Mercer, 1993, p. A-8);
$\lambda = 9.63 \times 10^{-4} \text{d}^{-1}$ is the degradation constant calculated using a half-life of 720 d and the following equation:

$$\lambda = \frac{\ln 2}{t_{1/2}} \qquad (18.8)$$

$t = 6000$ d is the time of travel of the center of the mass of contaminant plume; from the source area to the property boundary 30 ft downgradient;
$v_L = 0.019$ ft/d is the linear velocity of groundwater;
$v_c = 0.005$ ft/d is the retarded velocity of dissolved PCE;
$D_x = 0.19$ ft^2/d is the longitudinal dispersion coefficient (assumed longitudinal dispersivity is $\alpha_x = 10$ ft);
$D_y = 0.019$ ft^2/d is the transverse dispersion coefficient (assumed transverse dispersivity is $\alpha_y = 1$ ft);
$D_z = 0.0019$ ft^2/d is the vertical dispersion coefficient (assumed vertical dispersivity is $\alpha_z = 0.1$ ft).

For the unit volume ($V_0 = 1$ ft^3, the maximum dissolved concentration of PCE at the time of arrival (6000 d) of the center of contamination 30 ft downgradient from the source area will be (Equation 18.2)

$$C_{max} = \frac{0.053 \text{ mg/L} \times 1 \text{ ft}^3 \times e^{-9.63 \times 10^{-4}/\text{d} \times 6000 \text{ d}}}{8 \times (3.14 \times 6000 \text{ d})^{3/2} \times (0.19 \text{ ft}^2/\text{d} \times 0.019 \text{ ft}^2/\text{d} \times 0.0019 \text{ ft}^2/\text{d})^{1/2}}$$

$$C_{max} = \frac{0.053 \text{ mg/L} \times e^{-5.778}}{8 \times 2,585,857 \times 2.619^{-3}}$$

$$C_{max} = \frac{0.053 \text{ mg/L} \times 0.0031}{54,181}$$

$$C_{\text{max}} = 3.03 \times 10^{-9} \text{ mg/L per unit aquifer volume of } 1 \text{ ft}^3$$

It is assumed that the dissolved contaminant with concentration 0.053 mg/L occupies an estimated 4375 ft³ of the porous medium (aquifer) voids, i.e.,

$$V_0 = 35 \text{ ft} \times 25 \text{ ft} \times 25 \text{ ft} \times 0.20 = 4375 \text{ ft}^3$$

where 35 ft is the estimated thickness of the contamination in the saturated zone around MW1, 25 ft × 25 ft is the surface area of contamination represented by 0.053 mg/L, and 0.20 is the porosity. The maximum concentration will then be (note that 1 ft³ = 28.57 L):

$$C_{\text{max}} = 3.03 \times 10^{-9} \text{ mg/L/ft}^3 \times 4375 \text{ ft}^3 \times 28.57 = 3.79 \times 10^{-4} \text{ mg/L}$$

i.e., PCE concentration will be 0.4 ppb due to its retardation and degradation.

In a more conservative case, i.e., if there is no biochemical degradation of PCE, the concentration at the site boundary is calculated using the following modified form of Equation 18.2; note that $\lambda = 0$:

$$C_{\text{max}} = \frac{C_0 V_0}{8(\pi t)^{3/2}(D_x D_y D_z)^{1/2}}$$

$$C_{\text{max}} = \frac{0.053 \text{ mg/L} \times 4375 \text{ ft}^3 \times 28.57}{8 \times (3.14 \times 6000 \text{ d})^{3/2} \times (0.19 \text{ ft}^2/\text{d} \times 0.019 \text{ ft}^2/\text{d} \times 0.0019 \text{ ft}^2/\text{d})^{1/2}}$$

$$C_{\text{max}} = 2.12 \times 10^{-3} \text{ mg/L} \quad \text{or} \quad 2.12 \text{ ppb}$$

18.1.3 CONTINUOUS SOURCE AND NO DEGRADATION

Domenico and Schwartz (1990, p. 647) analytical model of one-dimensional advective transport with three-dimensional dispersion of contaminants and including sorption (but not reaction or degradation) relates the initial concentration of the dissolved contaminant (C_0) and the calculated concentration (C) at distance (x) from the source, for the plane of symmetry $y = z = 0$:

$$C(x,0,0,t) = \frac{C_0}{2} \text{erfc}\left[\frac{x - v_c t}{2(\alpha_x v_c t)^{1/2}}\right]\left\{\text{erf}\left[\frac{Y}{4(\alpha_y x)^{1/2}}\right]\text{erf}\left[\frac{Z}{2(\alpha_z x)^{1/2}}\right]\right\} \qquad (18.9)$$

where

t is time, C_0 is initial concentration at the source, α_x, α_y, α_z are longitudinal, transverse, and vertical dispersivity, respectively, v_c is contaminant (retarded) velocity, x is distance along x axis from the release, for which the concentration is calculated, Y is the width of the planar source, Z is the height of the planar source, erfc is complimentary error function (see Table D.2), and erf is error function (see Table D.1).

The assumed geometry of the planar source around the highest recorded concentration of PCE (monitoring well MW1) is

$Y = 25$ ft (width of the planar source around MW1 perpendicular to the direction of groundwater flow);

$Z = 35$ ft (thickness of contamination).

All other numeric values needed for insertion into Equation 18.9 have been already calculated in the case of the assumed instantaneous source. At the time of arrival of the center of the plume 30 ft downgradient, the PCE concentration will be

$$C = \frac{0.053 \text{ mg/L}}{2} \text{ erfc}\left[\frac{30 \text{ ft} - 0.005 \text{ ft/d} \times 6000 \text{ d}}{2(10 \text{ ft} \times 0.005 \text{ ft/d} \times 6000 \text{ d})^{1/2}}\right]$$

$$\times \left\{ \text{erf}\left[\frac{25 \text{ ft}}{4(1 \text{ ft} \times 30 \text{ ft})^{1/2}}\right] \times \text{erf}\left[\frac{35 \text{ ft}}{2(0.1 \text{ ft} \times 30 \text{ ft})^{1/2}}\right] \right\}$$

$$C = \frac{0.053 \text{ mg/L}}{2} \text{erfc}(0) \times \text{erf}(2.28) \times \text{erf}(10.1)$$
$$C = 0.0265 \text{ mg/L} \times 1 \times 0.9986 \times 1 = 0.0265 \text{ mg/L}$$

which is exactly one-half of the initial concentration at the source. For other times and distances of travel Equation 18.9 would yield different results.

REFERENCES

Cohen, R.M. and J.W. Mercer, 1993. *DNAPL Site Evaluation*. C.K. Smoley. Boca Raton, FL.

Domenico, P.A. and F.W. Schwartz, 1990. *Physical and Chemical Hydrogeology*. John Wiley and Sons, New York, 824 p.

Howard, P., er al., 1991. *Environmental Degradation Rates*. Lewis Publishers, Chelsea, M1, 725 p.

19 Contaminant Fate and Transport Parameters

19.1 USE OF DISPERSION

Using site-specific groundwater flow parameters at the Chemical Company Site (Section 18.1), the recorded concentration at well MW1, and other estimated fate and transport parameters, analyze the influence of dispersivity on the calculated dissolved PCE concentrations at various distances downgradient from MW1, and for various times of travel.

Two user-friendly analytical programs in public domain, BIOSCREEN and BIOCHLOR, can be used to quickly analyze sensitivity of various fate and transport parameters included in most analytical equations such as those applied in Chapter 18. The programs were developed for the USEPA and the Technology Transfer Division of Air Force Center for Environmental Excellence (AFCEE) at Brooks Air Force Base by Groundwater Services, Inc., Houston, Texas. BIOSCREEN (Newell et al., 1996) simulates remediation through natural attenuation of dissolved hydrocarbons at petroleum fuel release sites. The software, programed in the Microsoft Excel spreadsheet environment and based on the Domenico analytical solute transport model, has the ability to simulate advection, dispersion, adsorption, and aerobic decay as well as anaerobic reactions that have been shown to be the dominant biodegradation processes at many petroleum release sites. BIOCHLOR (Aziz et al., 2000, 2002) simulates remediation by natural attenuation of dissolved solvents at chlorinated solvent release sites. This software is also programed in the Microsoft Excel spreadsheet environment and based on the Domenico analytical solute transport model. It has the ability to simulate 1-D advection, 3-D dispersion, linear adsorption, and biotransformation via reductive dechlorination (the dominant biotransformation process at most chlorinated solvent sites). Reductive dechlorination is assumed to occur under anaerobic conditions and dissolved solvent degradation is assumed to follow a sequential first-order decay process. BIOCHLOR includes three different model types:

1. Solute transport with or without first-order decay, for any dissolved constituent
2. Solute transport with biotransformation modeled as a sequential first-order decay process
3. Solute transport with biotransformation modeled as a sequential first-order decay process with two different reaction zones (i.e., each zone has a different set of rate coefficient values)
4. Decaying source, which allows the simulation of both plume expansion and contraction

BIOCHLOR, developed primarily as a quick screening tool, has the following limitations:

1. As an analytical model, BIOCHLOR assumes simple groundwater flow conditions. The model should not be applied where pumping systems create a complicated flow field. In

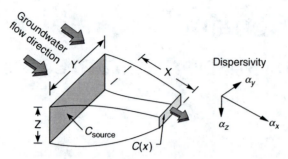

FIGURE 19.1 Source area and flow geometry implemented in the analytical solution by BIOCHLOR. (From Aziz, C.E., Newell, C.J., and J.R. Gonzales, 2002. BIOCHLOR; Natural attenuation decision support system; Version 2.2; March 2002; User's manual addendum. U.S. Environmental Protection Agency, Subsurface Protection and Remediation Division, National Risk Management Research Laboratory, Ada, Oklahoma, 10 p.)

addition, the model should not be applied where vertical flow gradients affect contaminant transport, and where recharge from the land surface or laterally from adjacent aquifers plays a significant role.

2. As a screening tool, BIOCHLOR assumes uniform hydrogeologic and environmental conditions over the entire model area (it is a typical "sandbox" model). For this reason, the model should not be applied where detailed, accurate results that closely match site conditions are required. More comprehensive numerical models should be applied in such cases.

3. When in biotransformation mode, BIOCHLOR is primarily designed for simulating the sequential reductive dechlorination of chlorinated ethanes (TCA) and ethenes (PCE, TCE).

To model a decaying source in BIOCHLOR, the Domenico (1987) semianalytical solution for reactive transport with first-order biological decay was modified to incorporate a decaying source (boundary) condition. The revised model assumes that the source decays exponentially via a first-order expression (i.e., $C_o \exp(-k_s t)$). Prior to using BIOCHLOR, the user must determine the source decay constant, k_s. The model includes option for a continuous source term as well. The modification of the Domenico solution was accomplished by extending a 1-D solution to the advection–dispersion equation that incorporated a decaying boundary condition to a 3-D solution by analogy (Van Genuchten and Alves, 1982). BIOCHLOR evaluates centerline concentrations at $y = 0$, $z = 0$, and the 2-D array at $z = 0$ (see Figure 19.1) by solving the following equation (Aziz et al., 2002):

$$C(x, y, z, t) = \frac{C_0}{8} e^{-k_s t} f_x f_y f_z \tag{19.1}$$

$$
f_x = \exp\left(\frac{x[1 - (1 + 4\alpha_x(\lambda - k_s R)/v_s]^{0.5}}{2\alpha_x}\right) \text{erfc}\left(\frac{x - v_c t(1 + 4\alpha_x(\lambda - k_s R)/v_s)^{0.5}}{2(\alpha_x v_c t)^{0.5}}\right)
$$

$$
+ \exp\left(\frac{x[1 + (1 + 4\alpha_x(\lambda - k_s R)/v_s]^{0.5}}{2\alpha_x}\right) \text{erfc}\left(\frac{x + v_c t(1 + 4\alpha_x(\lambda - k_s R)/v_s)^{0.5}}{2(\alpha_x v_c t)^{0.5}}\right)
$$

$$\tag{19.1a}$$

$$f_y = \text{erf}\left(\frac{(y + Y/2)}{2(\alpha_y x)^{0.5}}\right) - \text{erf}\left(\frac{y - Y/2}{2(\alpha_y x)^{0.5}}\right) \tag{19.1b}$$

$$f_z = \text{erf}\left(\frac{z + Z}{2(\alpha_z x)^{0.5}}\right) - \text{erf}\left(\frac{z - Z}{2(\alpha_z x)^{0.5}}\right) \tag{19.1c}$$

where $C(x,y,z,t)$ is the concentration at distance x downstream of source and distance y off centerline of plume at time t (mg/L)

C_0 is the concentration in source area at $t = 0$ (mg/L);

x is the distance downgradient of source (ft);

y is the distance from plume centerline of source (ft);

z is the distance from top of saturated zone to measurement point (assumed to be 0; concentration is always given at top of saturated zone);

α_x is the longitudinal groundwater dispersivity (ft);

α_y is the transverse groundwater dispersivity (ft);

α_z is the vertical groundwater dispersivity (ft);

n_{ef} is the effective porosity, needed for calculating linear (seepage) groundwater velocity;

λ is the first-order degradation rate coefficient (y^{-1});

v_s is the (seepage) linear groundwater velocity (ft/y); $v_s = Ki/n_{ef}$;

v_c is the contaminant velocity (ft/y); $v_c = v_s/R$;

K is the hydraulic conductivity (ft/y);

R is the contaminant retardation coefficient;

i is the hydraulic gradient;

Y is the source width (ft);

Z is the source thickness (ft);

k_s is the source decay constant (y^{-1}).

The following numeric values from Section 18.1 were used in BIOCHLOR to simulate PCE concentrations at the Chemical Company Site:

$C_0 = 0.053$ mg/L is the concentration of PCE in the source area;

$K = 0.46$ ft/d or 1.6×10^{-4} cm/s is the hydraulic conductivity;

$i = 0.0062$ is the hydraulic gradient;

$n = 0.2$ is the total porosity;

$n_{ef} = 0.15$ is the effective porosity;

$v_L = 0.019$ ft/d or 6.93 ft/y is the linear velocity of groundwater;

$R = 3.79$ is the retardation coefficient of dissolved PCE;

$\alpha_x = 10$ ft is the longitudinal dispersivity;

$\alpha_y = 1$ ft is the transverse dispersivity;

$\alpha_z = 0.1$ ft is the vertical dispersivity;

$Z = 35$ ft is the source thickness in the saturated zone;

$Y = 25$ ft is the source width.

Graph in Figure 19.2a shows the results of the calculations performed by BIOCHLOR for the above input parameters assuming no biodegradation of PCE. For comparison, the graph in Figure 19.2b shows the results for longitudinal, transverse, and vertical dispersivities of 3, 0.3, and 0.03 ft, respectively. The calculation times in both cases are 10, 15, and 30 y.

19.2 USE OF SORPTION (RETARDATION) AND SEQUENTIAL DEGRADATION

For the same Chemical Company Site, analyze the influence of changing sorption (retardation) on the dissolved PCE concentrations downgradient from the source zone after 5 and 30 y.

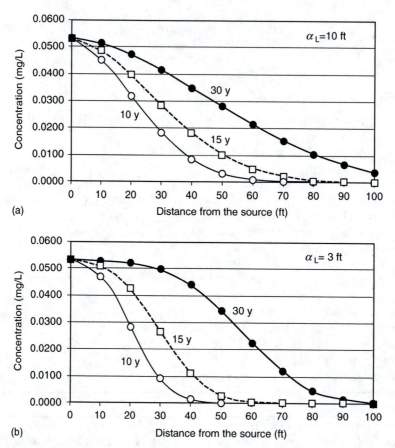

FIGURE 19.2 Concentration of dissolved PCE versus distance downgradient from the source area, calculated for two different longitudinal dispersivities and three time intervals. Transverse and vertical dispersivities are smaller than the longitudinal dispersivity by 10 and 100 times, respectively.

Using groundwater flow parameters and estimated fate and transport parameters from Section 18.1 and Section 19.1, test the following values of fraction organic carbon in the porous media: 0.5% and 1%. For one of the more conservative PCE degradation rates, analyzed in Section 18.1 (half-life of 2 y), and the minimal sorption caused by the assumed organic carbon content of 0.01% (0.0001), analyze concentrations of the PCE daughter products using "very conservative" degradation rates (half-lives). For these conservative half-lives, test the influence of a decaying source term on dissolved contaminant concentrations downgradient.

Percent organic carbon (f_{oc}) present in the aquifer porous media will influence the calculated value of the partitioning (distribution) coefficient (K_d) which, in turn, will change the retardation coefficient (R). These two parameters are calculated using the following two equations:

$$K_d = K_{oc} f_{oc} \qquad (19.2)$$

$$R = 1 + \left(\frac{1 - n}{n} \right) \rho_s K_d \qquad (19.3)$$

where n is total porosity, K_{oc} is PCE partitioning coefficient with respect to organic carbon, and ρ_s is soil particle density, usually estimated at 2.65 g/cm^3. One often used value of K_{oc} for PCE is 263 mL/g (Cohen and Mercer, 1993).

Calculation for 0.5% organic carbon:

$$K_d = 263 \text{ mL/g} \times 0.005 = 1.315 \text{ mL/g}$$

$$R = 1 + \left(\frac{1 - 0.20}{0.20}\right) \times 2.65 \text{ g/cm}^3 \times 1.315 \text{ mL/g} = 14.92$$

Calculation for 1% organic carbon:

$$K_d = 263 \text{ mL/g} \times 0.01 = 2.63 \text{ mL/g}$$

$$R = 1 + \left(\frac{1 - 0.20}{0.20}\right) \times 2.65 \text{ g/cm}^3 \times 2.63 \text{ mL/g} = 28.9$$

It is obvious that both values of the retardation coefficient will result in very low concentrations of dissolved PCE downgradient from the source, compared with the results in Section 19.1. Table 19.1 shows calculated PCE concentrations assuming no biodegradation, and using fate and transport parameters from Section 18.1 (longitudinal, transverse, and vertical dispersivities are 10, 0.1, and 0.01 ft, respectively).

Conservative half-lives for PCE and its degradation products are selected as follows:

PCE half-life 2 y
TCE half-life 3 y
DCE half-life 4 y
VC half-life 5 y

The results of BIOCHLOR calculations for these half-lives and the parameters used in Section 18.1 are shown in Figure 19.3 and Figure 19.4. Figure 19.3 shows a comparison of the calculated PCE concentrations with and without biodegradation, and Figure 19.4 shows concentrations of PCE and its degradation products when assuming the above "very conservative" half-lives.

TABLE 19.1
Calculated PCE Concentrations Downgradient from the Source for Two Different Values of the Retardation Coefficient (R) and Two Time Periods of Calculation (5 and 30 y)

Distance from Source (ft)	5 y		30 y	
	$R = 14.92$	$R = 28.9$	$R = 14.92$	$R = 28.9$
0	0.0530	0.0530	0.0530	0.0530
10	0.0119	0.0035	0.0421	0.0324
20	0.0004	0.0000	0.0259	0.0116
30	0.0000	0.0000	0.0120	0.0023
40	0.0000	0.0000	0.0041	0.0002
50	0.0000	0.0000	0.0010	0.0000
60	0.0000	0.0000	0.0002	0.0000
70	0.0000	0.0000	0.0000	0.0000

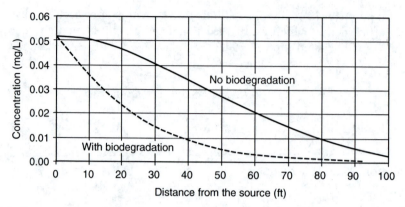

FIGURE 19.3 Concentration of PCE downgradient from the source area after 30 y, with and without biodegradation, assuming constant source at 0.053 mg/L.

The influence of a decaying source on dissolved PCE concentrations is simulated with the following function:

$$C = C_0 e^{-k_s t} \qquad (19.4)$$

where k_s is the assumed source decay constant in y^{-1}. The results are shown in Figure 19.5. This source decay constant can be determined if a time series of recorded dissolved concentrations in the source zone is available, showing a generally decreasing trend. It is the slope of the straight line drawn through the concentration data when they are plotted on a semilog graph concentration versus time, as shown in Figure 19.6. BIOCHLOR provides guidance for selecting and using both the contaminant degradation rates (λ), and the source decaying term (k_s).

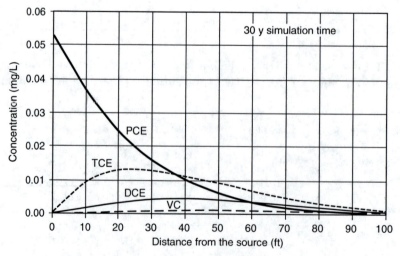

FIGURE 19.4 Concentration of PCE and its daughter products downgradient from the source area after 30 y, assuming constant source at 0.053 mg/L.

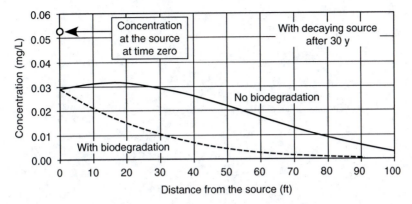

FIGURE 19.5 Concentration of PCE downgradient from the source area after 30 y, assuming a decaying source with $k_s = 0.02$ y^{-1}.

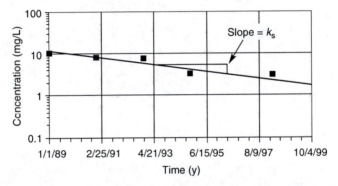

FIGURE 19.6 Determination of source decay constant, k_s using aqueous concentrations in source area wells. (Modified from Aziz, C.E., Newell, C.J., and J.R. Gonzales, 2002. BIOCHLOR; Natural attenuation decision support system; Version 2.2; March 2002; User's manual addendum. U.S. Environmental Protection Agency, Subsurface Protection and Remediation Division, National Risk Management Research Laboratory, Ada, Oklahoma, 10 p.)

REFERENCES

Aziz, C.E., et al., 2000. BIOCHLOR; Natural attenuation decision support system; User's manual, Version 1.0. EPA/600/R-00/008, U.S. Environmental Protection Agency, Office of Research and Development, Washington, D.C., 46 p.

Aziz, C.E., Newell, C.J., and J.R. Gonzales, 2002. BIOCHLOR; Natural attenuation decision support system; Version 2.2; March 2002; User's manual addendum. U.S. Environmental Protection Agency, Subsurface Protection and Remediation Division, National Risk Management Research Laboratory, Ada, Oklahoma, 10 p.

Cohen, R.M. and J.W. Mercer, 1993. DNAPL site evaluation. C.K. Smoley, Boca Raton, FL.

Domenico, P.A., 1987. An analytical model for multidimensional transport of a decaying contaminant species. *J. Hydrol.*, 91: 49–58.

Newell, C.J., McLeod, R.K., and J.R. Gonzales, 1996. BIOSCREEN; Natural attenuation decision support system; User's manual, Version 1.3. EPA/600/R-96/087, U.S. Environmental Protection Agency, Office of Research and Development, Cincinnati, Ohio, 65 p.

Van Genuchten, M.Th. and W.J. Alves, 1982. Analytical solutions of the one-dimensional convective–dispersive solute transport equation. U.S. Department of Agriculture, Technical Bulletin No. 1661.

20 Contaminant Fate and Transport Modeling

20.1 PLUME MIGRATION AND CONTAINMENT

Figure 20.1 shows an initial model setup and the flow model results for a portion of a confined aquifer between two equipotential boundaries, a lake and a river, which are 1000 m apart. The aquifer uniform thickness is 30 m and the aquifer does not receive any arealy distributed recharge, and it does not lose water to the underlying or overlying strata. The groundwater flow in the aquifer is entirely controlled by the two surface water features, which fully penetrate the aquifer and are in direct hydraulic connection with it. The elevation of the lake surface is 81 m asl, and the river has uniform gradient from the south to the north; in the south, the elevation of water surface in the river is 76 m asl, and in the north it is 71.5 m asl. For the initial analysis, assume no-flow boundaries along the southern and northern edges of the model domain shown in Figure 20.1. The effective porosity of the aquifer material is 25%, and the storage coefficient is estimated to be 0.001. The hydraulic conductivity is initially estimated at 50 m/d throughout the aquifer. There is a fully penetrating water supply well located near the river as shown in Figure 20.1. This well, when operating, is pumping 60 L/s or 5000 m^3/d. Cell size of the model is 10×10 m everywhere except in the far corners where the largest cell has dimension of 20×20 m. After setting up and running the flow model with the above parameters and boundary conditions in steady state, analyze potential migration of a dissolved contaminant plume that has just started to develop from a continuous source shown in Figure 20.1 for the next 10 y. Analyze two scenarios: when the well is not pumping and when it is pumping continuously. For the later case, run the model in transient conditions. The constant dissolved concentration of the contaminant in the source zone will be kept at 1000 mg/L for the entire period of 10 y. The contaminant does not sorb to the aquifer material and it does not degrade. After analyzing the contaminant fate and transport with the assumed uniform hydraulic conductivity in the aquifer, run the model with a varying hydraulic conductivity as shown later in this chapter.

The results of the groundwater flow, and fate and transport models are shown in Figure 20.1 through Figure 20.6, illustrating answers to the above tasks. There are several main conclusions that can be drawn from this modeling exercise: (1) The heterogeneity of the aquifer plays the most important role for both the groundwater flow and the contaminant transport; (2) Since the contaminant is conservative ("recalcitrant"), i.e., it does not sorb or degrade, dispersion is the most important parameter controlling contaminant concentrations downgradient from theo source and the overall expansion of the plume; (3) Particle tracking analysis alone is not sufficient to evaluate if there would be any impact on the pumping well from the contaminant plume.

This modeling example reveals one of the best kept secrets in hydrogeology: because of the dispersion phenomena, dissolved contaminants can cross the hydrodynamic flowlines

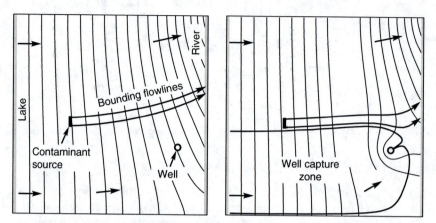

FIGURE 20.1 Hydraulic head in the confined aquifer assuming uniform hydraulic conductivity of 50 m/day. Based on this flow field, there is no direct pathway between the source zone and the pumping well: the well capture zone does not intersect with the bounding flowlines from the source area.

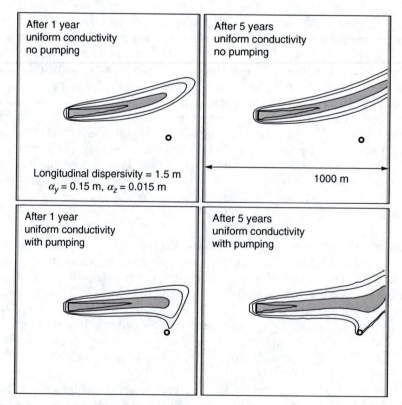

FIGURE 20.2 Spread of the contaminant plume 1 and 5 years following the release for two cases: without well pumpage (top) and with the continuous pumping at 60 L/s (bottom).

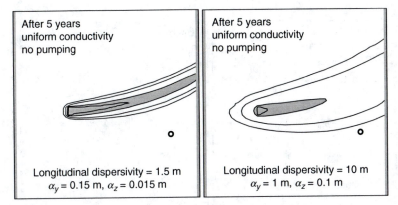

FIGURE 20.3 Influence of varying longitudinal dispersivity on the plume expansion. Even though the longitudinal dispersivity of 10 m may be considered "reasonable" based on various rules of thumb (e.g., it is 1/10 of the plume length) and in this case it satisfies the Pecklet number criterion, the map on the right shows spread of contamination upgradient from the source ("salmon effect") and a much wider plume downgradient.

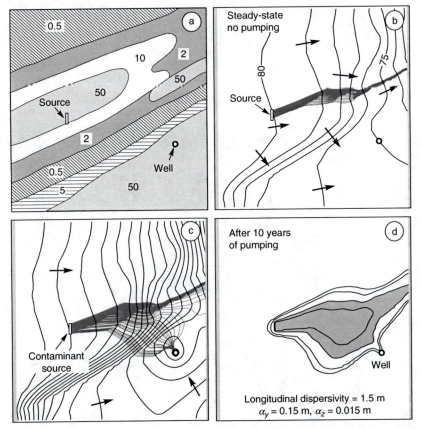

FIGURE 20.4 The influence of varying hydraulic conductivity on contaminant fate and transport. a: Estimated distribution of the hydraulic conductivity in the aquifer based on additional field information (values are in ft/d); b: Forward particle tracking from the source shows a very narrow steady-state flowpath between the source area and the river. c: Well pumpage causes wide spreading of particles leaving the source zone and traveling through several zones of varying hydraulic conductivity; refraction at the hydraulic conductivity boundaries causes additional spreading of the particles; d: Dissolved contaminant concentrations are starting to be detected in the well between 9 and 10 y following the release.

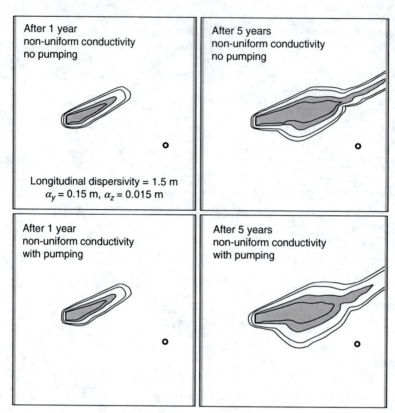

FIGURE 20.5 Spread of the contaminant plume for the heterogeneous hydraulic conductivity within first 5 years is not influenced by the well pumpage. The contaminant migration is influenced primarily by the spatial orientation of different hydraulic conductivity zones.

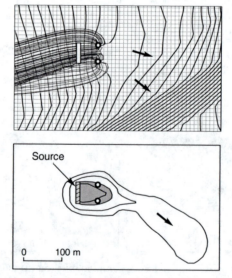

FIGURE 20.6 Particle tracking shows that two extraction wells located approximately 40 m downgradient from the source zone are containing the groundwater flux through the zone ("the plume is contained based on the hydraulics"). However, the fate and transport model predicts that some contamination will migrate downgradient from the pump-and-treat system at low concentrations, i.e., there is contaminant breakthrough.

(streamlines) that are estimated (calculated) from the hydraulic gradients and groundwater velocities. Flowlines in the flownet may act as no-flow boundaries for the volumetric (bulk) groundwater flow in the aquifer, but they are not necessarily the boundaries for contaminant transport. As illustrated with Figure 20.6, this is of particular importance when evaluating the effects of hydraulic plume containment. Pumping wells often create stagnant zones from which dissolved contaminants can "escape" due to dispersion. Although these concentrations are usually much lower than in the contained portion of the plume, they may be of importance if maximum contaminant level (MCL) for a particular contaminant is quite low.

Appendix A

TABLE A.1
Conversion Table for Units of Length

To Convert from	To	Multiply by
Meters (m)	Feet	3.281
Meters	Inches	39.37
Meters	Yards	1.094
Meters	Centimeters	100
Meters	Millimeters	1000
Meters	Kilometers	1.0×10^{-3}
Meters	Miles	6.2×10^{-4}
Feet (ft)	Meters	0.3048
Feet	Centimeters	30.48
Feet	Inches	12
Feet	Yards	0.333
Kilometers (km)	Miles	0.621
Kilometers	Meters	1.0×10^{3}
Kilometers	Feet	3.281×10^{3}
Miles	Feet	5280
Miles	Yards	1760
Miles	Kilometers	1.609

To Convert from	To	Multiply by
Square meters (m²)	Square feet	10.764
m²	Acres	2.47×10^{-4}
m²	Hectares	9.29×10^{-6}
m²	Square yards	1.196
Square feet (ft²)	Square meters	0.0929
ft²	Hectares	9.29×10^{-6}
ft²	Square inches	144
ft²	Acres	2.07×10^{-4}
Square km (km²)	Square meters	1.0×10^{6}
km²	Hectares	100
km²	Square feet	107.6×10^{5}
km²	Acres	247.105
km²	Square miles	0.3861
Square miles	Square meters	259×10^{4}
Square miles	Square km	2.59
Square miles	Square feet	2.788×10^{7}
Square miles	Acres	640

To Convert from	To	Multiply by
Cubic cm (cm³)	Liters	1.0×10^{-3}
cm³	Milliliters (ml)	1.0
cm³	Cubic inches	61.02×10^{-3}
cm³	Fluid ounce (fl. oz.)	33.81×10^{-3}
Liters (L)	Cubic meters	1.0×10^{-3}
L	Cubic feet	0.035
L	Gallons	0.264
L	Milliliters (ml)	1000
L	Fluid ounce (fl. oz.)	33.814
Gallons	Liters	3.785
Gallons	Cubic meters	3.785×10^{-3}
Gallons	Cubic feet	0.134
Gallons	Fluid ounce (fl. oz.)	128
Cubic feet (ft³)	Liters	28.317
ft³	Cubic meters	28.317×10^{-3}
ft³	Gallons	7.48
ft³	Acre-feet	22.957×10^{-6}

TABLE A.2
Conversion Table for Units of Flow

To Convert from	To	Multiply by	To Convert from	To	Multiply by	To Convert from	To	Multiply by
Gallons per minute (gpm)	Liters per second (L/s)	0.0631	Liters per second (L/s)	Gallons per minute (gpm)	15.85	Cubic feet per day (cfd)	Cubic meters per day (m³/d)	0.02832
gpm	Cubic meters per second (m³/s)	6.3×10^{-5}	L/s	Cubic meters per second (m³/s)	0.001	cfd	Cubic meters per second (m³/s)	3.28×10^{-7}
gpm	Cubic feet per day (m³/d)	5.451	L/s	Cubic feet per day (ft³/d)	3051.2	cfd	Liters per second (L/s)	3.28×10^{-4}
gpm	Cubic feet per second (ft³/s)	2.23×10^{-3}	Cubic feet per second (cfs)	Gallons per minute (gpm)	448.8	cfd	Gallons per minute (gpm)	5.19×10^{-3}
gpm	Cubic feet per day (ft³/d)	192.5	cfs	Liters per second (L/s)	28.32	cfd	Gallons per day	7.48
gpm	Acre-feet per day	4.42×10^{-3}	cfs	Cubic meters per second (m³/s)	0.0283	cfd	Acre-feet per day	2.3×10^{-5}

TABLE A.3
Conversion Table for Units of Hydraulic Conductivity

To Convert from	To	Multiply by
cm/s	m/d	864
cm/s	ft/d	2835
m/d	ft/d	3.28
m/d	cm/s	1.16×10^{-3}
ft/d	m/d	0.3049
ft/d	cm/s	3.53×10^{-4}

Appendix B

B.1 ALGEBRA

B.1.1 EXPONENTS AND ROOTS

If $a = x^b$ then $x = \sqrt[b]{a}$

$$x^a x^b = x^{a+b}$$

$$\frac{x^a}{x^b} = x^{a-b}$$

$$(x^a)^b = x^{ab}$$

$$x^{-a} = \frac{1}{x^a}$$

$$x^{\frac{1}{a}} = \sqrt[a]{x}$$

$$x^a y^a = (xy)^a$$

$$\frac{x^a}{y^a} = \left(\frac{x}{y}\right)^a$$

$$x^{\frac{a}{b}} = \sqrt[b]{x^a}$$

$$x^{-(1/a)} = \sqrt[a]{1/x}$$

$$(\sqrt[a]{x})^a = x$$

$$\sqrt[a]{xy} = \sqrt[a]{x} \times \sqrt[a]{y}$$

B.1.2 LOGARITHMS

If $a^x = b$

then x is the logarithm to the base a of the number b, i.e.,

$$x = \log_a b$$

For example:

$$2^3 = 8$$

$$\log_2 8 = 3$$

Logarithm to the base e, or natural logarithm, or Napierian logarithm is written as

$$\log_e x \quad \text{or} \quad \ln x$$

where $e = 1 + \dfrac{1}{1!} + \dfrac{1}{2!} + \dfrac{1}{3!} + \dfrac{1}{4!} + \ldots \cong 2.71828$

Logarithm to the base 10, or common logarithm, or Briggsian logarithm is written as

$$\log_{10} x \quad \text{or} \quad \log x$$

Relationships between natural and common logarithm

$$\log_{10} x = 0.4343 \log_e x \qquad (\log x = 0.4343 \ln x)$$

$$\log_e 10 = 2.302585 \qquad (\ln 10 = 2.3026)$$

$$\log_e x = 2.3026 \log_{10} x \qquad (\ln x = 2.3026 \log x)$$

$$\log_{10} e = 0.434294 \qquad (\log e = 0.4343)$$

Relations valid for any logarithmic base

$$\log ab = \log a + \log b$$

$$\log \frac{a}{b} = \log a - \log b$$

$$\log \frac{1}{a} = -\log a$$

$$\log 1 = 0$$

$$\log a^b = b \log a$$

$$\log_a a = 1$$

$$b^{\log_a b} - a$$

$$\log \sqrt[b]{a} = \frac{1}{b} \log a$$

$$\log \sqrt[b]{a^c} = \frac{c}{b} \log a$$

If $a^x = b$

then $\log a^x = \log b \quad \text{or} \quad x \log a = \log b$

and $\quad x = \dfrac{\log b}{\log a}$

B.1.3 THE QUADRATIC EQUATION

$$ax^2 + bx + c = 0 \quad (a, b, c \neq 0)$$

Roots of the equation: $x_{1,2} = \dfrac{-b \pm \sqrt{b^2 - 4ac}}{2a}$

If:
$b^2 - 4ac > 0$ the roots are real and not equal $x_1 \neq x_2$
$b^2 - 4ac = 0$ the roots are real and equal $x_1 = x_2$
$b^2 - 4ac < 0$ the roots are imaginary numbers called conjugate–complex numbers with the
same real part and the imaginary part having the opposite sign.
   $ax^2 + bx + c = a(x - x_1)(x - x_2)$ where x_1 and x_2 are the solutions

B.1.4 EXPANSIONS AND SPLITTING INTO FACTORS (FACTORING)

$$(x + y)^2 = x^2 + 2xy + y^2$$

$$(x - y)^2 = x^2 - 2xy + y^2$$

$$(x + y)^3 = x^3 + 3x^2y + 3xy^2 + y^3$$

$$(x - y)^3 = x^3 - 3x^2y + 3xy^2 - y^3$$

$$x^2 - y^2 = (x - y)(x + y)$$

$$a_1 x + b_1 y = c_1$$

$$x^2 + y^2 = (x + y - \sqrt{2xy})(x + y + \sqrt{2xy})$$

$$x^3 - y^3 = (x - y)(x^2 + xy + y^2)$$

$$x^3 + y^3 = (x + y)(x^2 - xy + y^2)$$

B.1.5 SYSTEM OF TWO LINEAR EQUATIONS WITH TWO UNKNOWNS

$$a_1 x + b_1 y = c_1 \tag{B.1}$$

$$a_2 x + b_2 y = c_2 \tag{B.2}$$

where
   $a_1, b_1, c_1, a_2, b_2, c_2$ are the coefficients, and x, y are the unknowns.

Solution by method of comparison
From Equation B.1 it follows that

$$y = \frac{c_1 - a_1 x}{b_1} \tag{B.3}$$

From Equation B.2 it follows that

$$y = \frac{c_2 - a_2 x}{b_2} \tag{B.4}$$

Since the left sides of Equation B.3 and Equation B.4 are equal, the right sides must also be equal:

$$\frac{c_1 - a_1 x}{b_1} = \frac{c_2 - a_2 x}{b_2}$$

It then follows that

$$x = \frac{b_2 c_1 - b_1 c_2}{a_1 b_2 - a_2 b_1}$$

The other unknown (y) is found by substituting the value of x into either Equation B.3 or Equation B.4:

$$y = \frac{a_1 c_2 - a_2 c_1}{a_1 b_2 - a_2 b_1}$$

Solution by method of substitution
From Equation B.1 it follows that

$$y = \frac{c_1 - a_1 x}{b_1} \tag{B.5}$$

Substituting Equation B.5 into Equation B.2 gives equation with only one unknown (x):

$$a_2 x + b_2 \cdot \frac{c_1 - a_1 x}{b_1} = c_2$$

which, solved for x, gives

$$x = \frac{b_2 c_1 - b_1 c_2}{a_1 b_2 - a_2 b_1}$$

Substituting the value of x into Equation B.5 gives the solution for y:

$$y = \frac{a_1 c_2 - a_2 c_1}{a_1 b_2 - a_2 b_1}$$

Solution by method of determinant
Matrix of the system is

$$\left(\begin{array}{cc|c} a_1 & b_1 & c_1 \\ a_2 & b_2 & c_2 \end{array} \right)$$

and the solution is

$$x = \frac{D_1}{D} \qquad y = \frac{D_2}{D}$$

where the determinants are

$$D = \begin{vmatrix} a_1 & b_1 \\ a_2 & b_2 \end{vmatrix} = a_1 b_2 - a_2 b_1$$

$$D_1 = \begin{vmatrix} c_1 & b_1 \\ c_2 & b_2 \end{vmatrix} = b_2 c_1 - b_1 c_2$$

$$D_2 = \begin{vmatrix} a_1 & c_1 \\ a_2 & c_2 \end{vmatrix} = a_1 c_2 - a_2 c_1$$

Note that the elements in the determinant are multiplied diagonally starting from the top of each column. The sign is positive for the multiplication from upper left to lower right, and it is negative for the multiplication from upper right to lower left.

B.1.6 SYSTEM OF THREE LINEAR EQUATIONS WITH THREE UNKNOWNS

$$a_1 x + b_1 x + c_1 x = d_1 \tag{B.6}$$

$$a_2 x + b_2 x + c_2 x = d_2 \tag{B.7}$$

$$a_3 x + b_3 x + c_3 x = d_3 \tag{B.8}$$

Solution by method of determinant
Matrix of the system is

$$\begin{pmatrix} a_1 & b_1 & c_1 & d_1 \\ a_2 & b_2 & c_2 & d_2 \\ a_3 & b_3 & c_3 & d_3 \end{pmatrix}$$

and the solution is

$$x = \frac{D_1}{D} \qquad y = \frac{D_2}{D} \qquad z = \frac{D_3}{D}$$

where the determinants are given as

$$D = \begin{vmatrix} a_1 & b_1 & c_1 \\ a_2 & b_2 & c_2 \\ a_3 & b_3 & c_3 \end{vmatrix}$$

$$D_1 = \begin{vmatrix} d_1 & b_1 & c_1 \\ d_2 & b_2 & c_2 \\ d_3 & b_3 & c_3 \end{vmatrix}$$

$$D_2 = \begin{vmatrix} a_1 & d_1 & c_1 \\ a_2 & d_2 & c_2 \\ a_3 & d_3 & c_3 \end{vmatrix}$$

$$D_3 = \begin{vmatrix} a_1 & b_1 & d_1 \\ a_2 & b_2 & d_2 \\ a_3 & b_3 & d_3 \end{vmatrix}$$

Note that determinant $D_1(D_2, D_3)$ is obtained from the basic determinant (D) in which the first (second, third) column is replaced by the free coefficients (d_1, d_2, d_3). The determinant of the third order is solved by applying Saruss's rule:

1. Add the first two columns to the right of the determinant.
2. Multiply elements in the "new" determinant diagonally, starting at the top of each column.
3. Sum the products following the rule of signs: multiplication from upper right to lower left is positive, and multiplication from upper right to lower left is negative. For example:

$$\begin{vmatrix} a_1 & b_1 & c_1 \\ a_2 & b_2 & c_2 \\ a_3 & b_3 & c_3 \end{vmatrix} \begin{matrix} a_1 & b_1 \\ a_2 & b_2 \\ a_3 & b_3 \end{matrix}$$

$$D = a_1b_2c_3 + b_1c_2a_3 + c_1a_2b_3 - c_1b_2a_3 - a_1c_2b_3 - b_1a_2c_3$$

B.2 ANALYTICAL GEOMETRY

B.2.1 ORTHOGONAL (CARTESIAN) COORDINATE SYSTEM

Orthogonal (Cartesian) coordinate system in a plane is determined by the following:

Horizontal (X) axis
Vertical (Y) axis
Origin (intersection of X and Y axes)

Position of point P is defined by its coordinates (see Figure B.1): (1) x coordinate (or abscissa), the orthogonal distance of the point from the vertical (Y) axis, and (2) y coordinate (or ordinate), the orthogonal distance of the point from the horizontal (X) axis.
Distance between two points (Figure B.2):

$$d = \sqrt{(x_2 - x_1)^2 + (y_2 - y_1)^2}$$

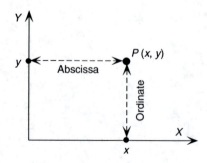

FIGURE B.1

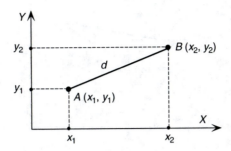

FIGURE B.2

B.2.2   POLAR COORDINATE SYSTEM

Polar coordinate system is determined by the following (see Figure B.3): (1) pole O, and (2) polar axis OX. Position of point P is defined by its polar coordinates:

(1) Radius vector r $(\overline{OP})$
(2) Angle φ (angle between axis OX and r)

Distance between two points (Figure B.4):

$$d = \sqrt{r_1^2 + r_2^2 - 2r_1 r_2 \cos(\varphi_2 - \varphi_1)}$$

B.2.3   CONNECTION BETWEEN CARTESIAN AND POLAR COORDINATES

If Cartesian coordinates are given (x, y), the polar coordinates are

$$r = \sqrt{x^2 + y^2} \quad \text{and} \quad \varphi = \arctan\frac{y}{x}$$

If polar coordinates are given (r, φ), the Cartesian coordinates are

$$x = r\cos\varphi \quad \text{and} \quad y = r\sin\varphi$$

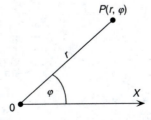

FIGURE B.3

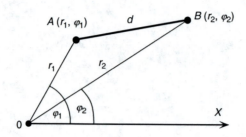

FIGURE B.4

B.3   TRIGONOMETRY

B.3.1   DEFINITIONS OF TRIGONOMETRIC FUNCTIONS

The trigonometric functions are arrived at from Figure B.5

$$\sin \alpha = \frac{a}{c}$$

$$\cos \alpha = \frac{b}{c}$$

$$\tan \alpha = \frac{a}{b}$$

$$\cot \alpha = \frac{b}{a}$$

B.3.2   TRIGONOMETRIC FUNCTIONS OF COMPLETE ANGLES

$$\sin (90° - \alpha) = \cos \alpha$$

$$\cos (90° - \alpha) = \sin \alpha$$

$$\tan (90° - \alpha) = \cot \alpha$$

$$\cot (90° - \alpha) = \tan \alpha$$

$$\sin^2 \alpha + \cos^2 \alpha = 1$$

B.3.3   BASIC RELATIONS

$$\tan^2 \alpha + 1 = \frac{1}{\cos^2 \alpha}$$

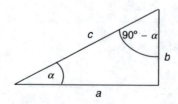

FIGURE B.5

$$1 + \cot^2 \alpha = \frac{1}{\sin^2 \alpha}$$

$$\tan \alpha = \frac{\sin \alpha}{\cos \alpha}$$

$$\cot \alpha = \frac{\cos \alpha}{\sin \alpha}$$

$$\tan \alpha \cot \alpha = 1$$

B.3.4 ADDING AND SUBTRACTING ANGLES

$$\sin(\alpha + \beta) = \sin \alpha \cos \beta + \cos \alpha \sin \beta$$

$$\sin(\alpha - \beta) = \sin \alpha \cos \beta - \cos \alpha \sin \beta$$

$$\cos(\alpha + \beta) = \cos \alpha \cos \beta + \sin \alpha \sin \beta$$

$$\cos(\alpha - \beta) = \cos \alpha \cos \beta - \sin \alpha \sin \beta$$

$$\tan(\alpha + \beta) = \frac{\tan \alpha + \tan \beta}{1 - \tan \alpha \tan \beta}$$

$$\tan(\alpha + \beta) = \frac{\tan \alpha - \tan \beta}{1 \mid \tan \alpha \tan \beta}$$

$$\sin \alpha + \sin \beta = 2 \sin\left(\frac{\alpha + \beta}{2}\right) \cos\left(\frac{\alpha - \beta}{2}\right)$$

$$\sin \alpha - \sin \beta = 2 \sin\left(\frac{\alpha + \beta}{2}\right) \sin\left(\frac{\alpha - \beta}{2}\right)$$

$$\cos \alpha + \cos \beta = 2 \cos\left(\frac{\alpha + \beta}{2}\right) \cos\left(\frac{\alpha - \beta}{2}\right)$$

$$\cos \alpha + \cos \beta = -2 \sin\left(\frac{\alpha + \beta}{2}\right) \sin\left(\frac{\alpha - \beta}{2}\right)$$

$$\tan \alpha + \tan \beta = \frac{\sin(\alpha + \beta)}{\cos \alpha \cos \beta}$$

$$\tan \alpha - \tan \beta = \frac{\sin(\alpha - \beta)}{\cos \alpha \cos \beta}$$

B.4 DERIVATIVES

B.4.1 DEFINITION

The derivative of function $f(x)$ at point $x = a$ is called the final limit value at point a of the ratio between the increment of the function and the increment of the argument, when the increment of the argument approaches zero (when $x \to a$):

$$f'(a) = \lim_{x \to a} \frac{f(x) - f(a)}{x - a}$$

where $f'(a)$ is the first derivative of the function at point a.
Derivative of function $f(x)$ at point x is

$$f'(x) = \lim_{\Delta x \to 0} \frac{f(x + \Delta x) - f(x)}{\Delta x}$$

which is also written as

$$f'(x) = \lim_{\Delta x \to 0} \frac{\Delta y}{\Delta x} \quad \text{or} \quad f'(x) = \frac{dy}{dx}$$

B.4.2 GEOMETRIC INTERPRETATION OF THE DERIVATIVE

The first derivative is the tangent of the angle between the horizontal (X-axis) and the line that touches the curve at a given point as shown in Figure B.6 (note that this line is also called tangent):

$$\frac{f(x) - f(a)}{x - a} = \tan \alpha'$$

Note that

$$\lim_{x \to a} \frac{f(x) - f(a)}{x - a} = f'(a) = \tan \alpha$$

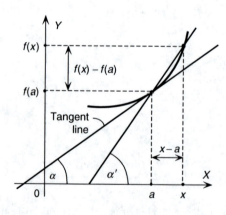

FIGURE B.6

B.4.3 ELEMENTARY DERIVATIVES

Function	First Derivative
$y = c$	$y' = 0$
$y = x$	$y' = 1$
$y = ax + b$	$y' = a$
$y = x^n$	$y' = nx^{n-1}$
$y = a^n$	$y' = a^x \ln a$
$y = e^x$	$y' = e^x$
$y = \log_a x$	$y' = \dfrac{1}{x} \log_a e$
$y = \ln x$	$y' = \dfrac{1}{x}$
$y = \log_x$	$y' = \dfrac{1}{x} 0.4343$
$y = \sin x$	$y' = \cos x$
$y = \cos x$	$y' = -\sin x$
$y = \tan x$	$y' = \dfrac{1}{\cos^2 x}$
$y = c \tan x$	$y' = -\dfrac{1}{\sin^2 x}$
$y = \arcsin x$	$y' = -\dfrac{1}{\sqrt{1 - x^2}}$ for $(|x| < 1)$
$y = \arccos x$	$y' = -\dfrac{1}{\sqrt{1 - x^2}}$ for $(|x| < 1)$
$y = \arctan x$	$y' = \dfrac{1}{1 + x^2}$
$y = \text{arcctan} \, x$	$y' = -\dfrac{1}{1 + x^2}$

B.4.4 BASIC RULES OF DERIVATION

If $f(x)$ and $g(x)$ are functions of an independent variable x, and if they have first derivatives $f'(x)$ and $g'(x)$, then it follows:

$$[f(x) \pm g(x)]' = f'(x) \pm g'(x)$$

$$[f(x)g(x)]' = f'(x)g(x) \pm f(x)g'(x)$$

$$\left[\frac{f(x)}{g(x)}\right]' = \frac{f'(x)g(x) \pm f(x)g'(x)}{[g(x)]^2} \quad \text{for } g(x) \neq 0$$

$$[c \times f(x)]' = c \times f'(x)$$

B.4.5 DERIVATIVES OF COMPLEX FUNCTIONS

Function $y = y(u)$, where $u = u(x)$, has the first derivative with respect to x as

$$y'(x) = y'(u) \times u'(x)$$

Examples

Function	First Derivative
$y = [f(x)]^n$	$y' = n \cdot [f(x)]^{n-1} \cdot f'(x)$
$y = a^{f(x)}$	$y' = a^{f(x)} \ln a \times f'(x)$ for a > 0
$y = e^{f(x)}$	$y = e^{f(x)} f'(x)$
$y = \ln[f(x)]$	$y' = \dfrac{f'(x)}{f(x)}$
$y = \sin[f(x)]$	$y' = f'(x) \cos[f(x)]$
$y = \cos[f(x)]$	$y' = -f'(x) \sin[f(x)]$
$y = \tan[f(x)]$	$y' = \dfrac{f'(x)}{\cos^2[f(x)]}$
$y = c\tan[f(x)]$	$y' = -\dfrac{f'(x)}{\sin^2[f(x)]}$

B.4.6 DERIVATIVES OF HIGHER ORDER

Derivative of the second order (second derivative) of function $f(x)$ is the derivative of its first derivative:

$$y''(x) = [y'(x)]'$$

Derivative of the n-th order is:

$$y^{(n)}(x) = [y^{(n-1)}(x)]'$$

B.5 INDEFINITE INTEGRAL

B.5.1 DEFINITION

A primitive function of function $f(x)$ is the function $F(x)$ such that its first derivative equals $f(x)$:

$$F'(x) = f(x)$$

Since

$$[F(x) + c]' = F'(x) = f(x) \quad \text{(because } c' = 0)$$

it follows that if the function $f(x)$ has a primitive function, then there exist an indefinite number of primitive functions of $f(x)$ that differ for the constant c. The array of all possible primitive functions of the continuous function $f(x)$ is called the indefinite integral of that function:

$$\int f(x) dx = F(x) + c$$

where $F(x)$ is the primitive function of the function $f(x)$, and c is an arbitrary constant of integration.

B.5.2 ELEMENTARY INDEFINITE INTEGRALS

$$\int dx = x + c$$

$$\int x \, dx = \frac{x^2}{2} + c$$

$$\int x^n dx = \frac{x^{n+1}}{n + 1} + c, \quad n \neq -1$$

$$\int \frac{1}{x} dx = \ln x + c$$

$$\int e^x dx = e^x + c$$

$$\int a^x dx = \frac{a^x}{\ln a} + c$$

$$\int \cos x \, dx = \sin x + c$$

$$\int \sin x \, dx = -\cos x + c$$

$$\int \frac{1}{\cos^2 x} dx = \tan x + c$$

$$\int \frac{1}{\sin^2 x} dx = -c \tan x + c$$

$$\int \frac{1}{\sqrt{1 - x^2}} dx = -\arcsin x + c$$

$$\int \frac{1}{1 + x^2} dx = -\arctan x + c$$

$$\int \frac{dx}{\sqrt{x^2 + 1}} = \ln\left(x + \sqrt{x^2 + 1}\right) + c$$

$$\int \frac{dx}{\sqrt{x^2 - 1}} = \ln\left(x + \sqrt{x^2 - 1}\right) + c \quad (\text{for } |x| > 1)$$

$$\int \frac{dx}{1 - x^2} dx = \frac{1}{2} \ln\left(\frac{1 + x}{1 - x}\right) + c \quad (\text{for } |x| < 1)$$

$$\int \frac{dx}{x^2 - 1} dx = \frac{1}{2} \ln\left(\frac{x - 1}{x + 1}\right) + c \quad (\text{for } |x| > 1)$$

B.5.3 INTEGRATION RULES

$$\int [f(x) \pm \varphi(x)] \, dx = \int f(x) \, dx \pm \int \varphi(x) \, dx$$

$$\int c \times f(x) \, dx = c \int f(x) \, dx$$

$$\int f(x) \, dx = \int f[\varphi(u)] \varphi'(u) du$$

where u is a new variable (method of substitution).

$$\int u \, dv = uv - \int v \, du$$

where u and v are the functions of the independent variable x (partial integration).

B.6. DEFINITE INTEGRAL

B.6.1 DEFINITION

The value of the definite integral is

$$\int_a^b f(x) \, dx = [F(x)]_a^b = F(b) - F(a)$$

where $F(x)$ is the primitive function of the function $f(x)$, i.e.,

$$F'(x) = f(x)$$

For example:

$$\int_5^9 5x^2 \, dx = 5 \int_4^9 x^2 \, dx = 5 \left[\frac{x^3}{3}\right]_4^9 = 5 \left[\frac{9^3}{3} - \frac{4^3}{3}\right] = 1108.3$$

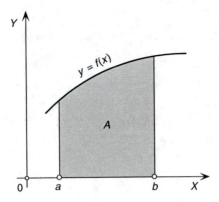

FIGURE B.7

B.6.2 Geometric Interpretation of the Definite Integral

The definite integral is arrived at from Figure B.7

$$\text{Area} = \int_a^b f(x)\,dx$$

B.6.3 Properties of the Definite Integral (Rules for Calculations with Definite Integrals)

See Figure B.8.

$$\int_a^b f(x)\,dx = -\int_b^a f(x)\,dx$$

$$\int_a^a f(x)\,dx = 0$$

$$\int_a^b [f(x) + g(x) - h(x)]\,dx = \int_a^b f(x)\,dx + \int_a^b g(x)\,dx - \int_a^b h(x)\,dx$$

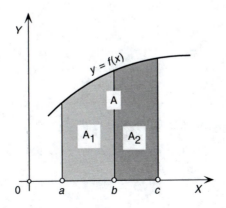

FIGURE B.8

$$\int_a^b f(x)\,dx + \int_b^c f(x)\,dx = \int_a^c f(x)\,dx$$

$$\int_a^b cf(x)\,dx = c\int_a^b f(x)\,dx$$

B.6.4 Rule of the Average Value in the Integral Calculations

$F(x)$ is a primitive function (indefinite integral) of the function $f(x)$.

$$\int_a^b f(x)\,dx = [F(x)]_a^b = F(b) - F(a)$$

and

$$\int_b^a f(x)\,dx = [F(x)]_b^a = F(a) - F(b)$$

give

$$\int_a^b f(x)\,dx = -\int_b^a f(x)\,dx$$

B.6.5 Examples of the Definite Integral Calculations

See Figure B.9 through Figure B.11.

1. Calculate the area bound by curve $y = \log x$, axis OX, and ordinate of the point $x = 4$

$$A = \int_1^4 \ln x\,dx = [x\log x - x]_1^4 = (4\log 4 - 4) - (\log 1 - 1) = 8\log 2 - 3$$

2. Find the area bound by curves $y = f(x)$, $y = \varphi(x)$, and lines $x = a$, and $x = b$.

The area A is found as the difference between the area under curve $y = f(x)$:

$$A_1 = \int_a^b f(x)\,dx$$

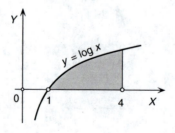

FIGURE B.9

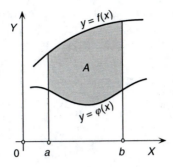

FIGURE B.10

and the area under curve $y = \varphi(x)$:

$$A_2 = \int_a^b \varphi(x)\,\mathrm{d}x$$

which gives

$$A = \int_a^b f(x)\,\mathrm{d}x - \int_a^b \varphi(x)\,\mathrm{d}x$$

This equation can also be written as

$$A = \int_a^b [f(x) - \varphi(x)]\,\mathrm{d}x$$

B.6.6 BASIC RULE OF THE INTEGRAL CALCULUS (CONNECTION BETWEEN DEFINITE AND INDEFINITE INTEGRALS)

The area $abAB$ in Figure B.11 is given as

$$A = \int_a^b f(x)\,\mathrm{d}x$$

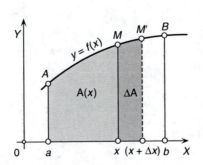

FIGURE B.11

If point x is an arbitrarily chosen point in the interval (a, b), then the area $axMA$ is a function of x:

$$A(x) = \int_a^x f(x)\,dx$$

If x is increased by Δx, then

$$A(x + \Delta x) - A(x) = \int_x^{x+\Delta x} f(x)\,dx$$

$$[F(x) + c]_b^a = [F(x)]_a^b$$

(c is a constant that drops out during the calculation of definite integrals).

Appendix C
Porosity of Rocks

TABLE C.1
Porosity of Clastic Sedimentary Rocks

Clastic (Indurated) Sedimentary Rocks	No. of Samples	Min	Max	Mean
Chert				
Keewatin, Precambrian	1			0.10
Onverwacht Gr., Swaziland, Precambrian	4	0.03	0.72	0.33
Chert	2			4.3
Chert	2			6.9
Claystone				
Claystone	12	22.1	32.3	29.0
Claystone	4	41.2	45.2	43.0
Greywacke	2	0.4	4.2	2.3
Quartzite				
Sioux Quartzite, Jasper, NM	1			0.000
Carroll & Frederick Counties	3			0.46
Cheshire, Rutland, VT				0.6
Quartzite	1			0.8
Globe-Miami, AZ	1			1.38
Johnson Camp, AZ	3	2.53	5.35	4.0
Quartzite	21	0.2	10.0	4.1
Chickies Fm., PA	5	3.8	7.8	5.4
Sandstone				
Mottled (Germany)			15.9	3.2
Stockton Fm., NJ & PA	5	1.3	7.9	4.0
Silurian	13	0.5	17.4	4.9
Clayey, Carroll & Frederick Counties, MD	1			6.10
Berea Sandstone, Lorain County, OH	1			6.400
Southern Italy	16	0.7	15.3	7.4
Carroll & Frederick Counties, MD	6	1.62	26.40	9.25
Permian	334	0.5	24.8	9.6
Sandstone		3.46	22.8	10.22
Bradford Fm., (low rank greywacke)	5	9.0	11.6	10.7
Cambrian	70	0.2	28.3	11.1
Clarendon Fm., PA	6	9.9	12.2	11.5
3d. Venango Fm., PA	16	7.5	17.8	14.0
3d. Bradford Fm., PA	10	12.2	15.6	14.0
Devonian	785	0.5	25.6	14.2

continued

TABLE C.1 (continued)
Porosity of Clastic Sedimentary Rocks

Clastic (Indurated) Sedimentary Rocks	No. of Samples	Min	Max	Mean
Ordovician	134	3.6	30.3	14.3
Weir Fm., (low rank greywacke)	6	12.9	16.6	14.5
Bradford Fm., PA	5	13.0	17.4	14.8
Kirkwood Fm., ("fairly clean orthoquartzite")	5	12.6	18.7	15.2
Sandstone	16	4.81	28.28	15.89
Pennsylvanian	6040	0.4	38.7	16.5
Cretaceous	1264	0.4	51.2	16.5
Triassic	303	0.0	35.0	16.9
Mississipian	375	3.8	27.6	17.6
Miocene	many	0.4	50.1	18.0
Assorted	23	6.8	25.4	18.3
Oligocene	many	0.8	45.0	22.0
Eocene	344	7.0	46.5	22.4
Miocene, Tunisia	2	20.0	37.0	28.5
Paleocene	18	9.4	53.6	30.8
Fine grained	55	13.7	49.3	33.0
Medium grained	10	29.7	43.6	37.0
Pliocene–Pleistocene	7	38.1	39.7	39.0

Source: Values processed from Wolff, R.G. Physical Properties of Rocks—Porosity, Permeability, Distribution Coefficient, and Dispersivity. U.S. Geological Survey Open-File Report 1982. 82–116, 118 pp.

TABLE C.2
Porosity of Carbonate Rocks

Carbonate Rocks	No. of Samples	Min	Max	Mean
Chalk				
Cemented, Northern France	2	7.7	8.3	8.0
Northern France	16	22.2	37.2	29.2
Chalk	3	45.9	46.8	46.2
Chalk				53.0
Coquina	1			56.7
Dolomite				
Martinsburg, WV				0.0
Webatuck, NY				0.4
Rustler, West Texas	2	0.41	1.37	0.89
Dolomite	27	0.8	12.4	4.5
Dolomite	5	3.0	8.6	5.5
Dolomite	2	19.1	32.7	26.0
Dolomite				1.0–22.2
Limestone				
Oak Hall, PA				0.0
Bone Springs, West Texas	1			0.44
Carroll & Frederick Counties, MD	7	0.27	4.36	1.70
Dolomitic, C. & Frederick Counties, MD	2			2.08

continued

TABLE C.2 (continued)
Porosity of Carbonate Rocks

Carbonate Rocks	No. of Samples	Min	Max	Mean
Chino, NM	3	0.366	4.38	2.44
Dolomite—Lower Ordovician		0.1	12.6	2.5
Johnson Camp, AZ	4	2.35	3.88	3.20
(Salem), Bedford, IN	1			3.63
Dolomite, Devonian	92	0.6	12.9	4.4
Dense, Southern Italy	24	0.3	14.0	4.5
Marble, dolomite	11	0.53	13.36	4.85
Dolomite, Ordovician	216	0.07	22.3	5.4
Carboniferous	29	0.6	14.9	5.5
Dolomite, Silurian	31	0.5	15.9	5.5
Dolomite, Cambrian				5.8
Pennsylvanian	2117	0.0	31.6	6.3
Triassic	37	0.4	36.5	9.3
Dolomite, Permian	56	3.2	27.1	9.8
Germany	6	3.1	28.4	10.4
Assorted	10	1.6	36.5	10.6
Mississipian	226	0.9	25.9	11.3
Oolite, Triassic	2109	0.0	34.4	13.4
Chalk, Cretaceous	601	0.2	42.8	17.5
Limestone	74	6.6	55.7	30.00
Dolomitic marine, Italy	12	10.5	66.6	34.7
Marble				
Marble				0.19
White, Portugal				0.26–27
Danbury, VT				0.30
Grenville Complex	9	0.01	1.06	0.35
Holston Marble, Knoxville, TN	1			0.52
White, Tyrol				0.59
Eastern US	100	0.4	0.8	0.60
Dolomitic, Carroll & Frederick Co., MD	2			0.60
Carroll & Frederick Counties, MD	7	0.31	2.02	0.62
Marble	6	0.7	1.1	0.9
Johnson Camp, AZ	1			2.6
Carrara, Italy				0.11–0.22

Source: Values processed from Wolff, R.G. Physical Properties of Rocks—Porosity, Permeability, Distribution Coefficient, and Dispersivity. U.S. Geological Survey Open-File Report 1982. 82–116, 118 pp.

TABLE C.3
Porosity of Evaporates

Evaporites	No. of Samples	Min	Max	Mean
Anhydrite				
Anhydrite	14	0.3	4.4	1.9
Anhydrite	21	0.3	40.8	10.3

continued

TABLE C.3 (continued)
Porosity of Evaporates

Evaporites	No. of Samples	Min	Max	Mean
Gypsum				
Assorted	2	1.3	4.0	2.6
Gypsum		1.32	3.96	2.64
Paris Basin				4.8
Castile, West Texas	4	3.42	6.43	4.81
Gypsum	3	27.2	30.6	28.4
Halite				
Halite	22	0.5	5.1	2.0
Halite	9	0.8	7.1	2.9
Salt				
(From 610–640 m below surface)		0.1	0.8	0.4
(From 790–823 m below surface)		0.3	0.7	0.5
Bedded, Hutchinson Salt, KS				0.59
Winnfield dome				1.28
Grand Saline dome, Texas				1.71
Bedded	17	0.62	7.17	2.10
Salt	11	1.5	8.6	3.7

Source: Values processed from Wolff, R.G. Physical Properties of Rocks—Porosity, Permeability, Distribution Coefficient, and Dispersivity. U.S. Geological Survey Open-File Report 1982. 82–116, 118 pp.

TABLE C.4
Porosity of Magmatic (Igneous) Rocks

Holocrystalline Rocks (Magmatic Rocks)	Samples	Min	Max	Mean
Dacite	3	3.5	16.0	9.0
Dacite, porphyry	101	2.0	29.9	10.4
Diabase, Frederick, MD				0.1
Diabase, Carroll & Frederick Counties, MD	2	0.47	1.0	0.58
Diabase	2	0.90	1.13	1.01
Diorite	38			0.0
Diorite				0.25
Diorite, quartz				0.6
Diorite, quartz (effective flow porosity)				0.2–0.003
Diorite, quartz, Sierrita–Esperanza, AZ	1			2.90
Diorite	1			3.0
Gabbro, Carroll & Frederick Counties, MD	3	0.00	0.62	0.29
Gabbro	1			0.84
Gabbro				0.6–0.7
Granite, Sherman (effective flow porosity)				0.002
Granite, Barre, VT	1			0.079
Granite, Westerly, RI	1			0.106
Granite, Stone Mt., GA				0.3
Granite	45			0.4
Granite, Tucson, AZ	1			0.611
Granite	451			0.7

continued

TABLE C.4 (continued)
Porosity of Magmatic (Igneous) Rocks

Holocrystalline Rocks (Magmatic Rocks)	Samples	Min	Max	Mean
Granite	26	0.4	3.0	0.9
Granite, Laramie, WY	1			1.08
Granite, Westerly, RI				1.1
Granite, Carroll & Frederick Counties, MD	17	0.44	3.98	1.11
Granite, Troy, AZ	1			1.36
Granite	322	0.1	11.2	1.4
Granite, equigranular Globe-Miami, AZ	1			1.77
Granite, Texas Canyon, AZ	1			2.96
Granite	9	0.7	5.5	3.0
Granite, porphyritic (alt.), Globe-Miami, AZ	1			5.35
Granite, porphyry, Bingham, UT	1			6.11
Granodiorite, St. Cloud, MN	1			0.76
Granodiorite, Carroll & Frederick Co., MD	1			0.50
Griesen	13			5.5
Latite, quartz porphyry, Bingham, UT	1			2.64
Latite, quartz	2	4.0	5.2	4.6
Latite, quartz porphyry, Silver City, NM	1			5.3
Latite, dike, Bingham, UT				12.5
Monzonite, quartz (altered), S. Esperanza, AZ	4	1.96	2.96	2.51
Monzonite, quartz porphyry (alt.), Chino, NM	1			2.60
Monzonite, quartz (some altered), Butte, MT	6	0.075	6.35	3.03
Monzonite, quartz	21	0.1	7.0	4.0
Monzonite, quartz (altered), San Manuel, AZ	7	1.5	4.1	4.0
Pegmatite	3			0.8
Pegmatite	4			0.9
Quartz monzonite	25			1.3
Quartz, monzonite	42	0.2	10.1	1.7
Quartz, monzonite	90	0.0	35.0	23.8
Seynites			1.38	0.5–0.6

Source: Values processed from Wolff, R.G. Physical Properties of Rocks—Porosity, Permeability, Distribution Coefficient, and Dispersivity. U.S. Geological Survey Open-File Report 1982. 82–116, 118 pp.

TABLE C.5
Porosity of Volcanic Rocks

Volcanic Rocks	No. of Samples	Min	Max	Mean
Basalt				
Dresser, WI	1			0.047
assorted	8	0.1	2.9	0.80
Germany	5	0.27	3.37	1.49
Basalt	20	1.2	18.7	9.4
Basalt	41	1.4	32.7	15.0
Basalt				4.40–5.60
Dacite				
Dike, Troy, AZ	1			3.08

continued

TABLE C.5 (continued)
Porosity of Volcanic Rocks

Volcanic Rocks	No. of Samples	Min	Max	Mean
Flow Troy, AZ	1			15.7
Obsidian				0.521
Phonolite			4.50	2.0–3.50
Porphyries, Germany	10	0.4	15.5	5.48
Pumice				
Pumice, from Champs Phlegreens				87.3
Pumice				50–75
Rhyodacite, dike, Troy, AZ	1			7.52
Rhyolite				
subvolcanic, Chino, NM	1			6.74
Altered, Sierrita Esperanza, AZ	1			7.48
Rhyolite	3	7.0	21.0	12.0
Rhyolite	6	10.2	17.9	14.6
Tuff				
Welded				14.1
Altered Red Mtn., AZ	1			21.5
Tuff	84	7.3	47.5	28.5
Zeolitized	23	15.8	37.7	29.4
Volcanic, Southern Italy	8	6.0	58.4	31.0
Zeolitized	28	23.2	39.5	31.1
Tuff	165	15.5	44.2	31.7
Volcanic, Rhine Valley	4	24.74	45.14	32.01
Tuff	15	29.3	40.0	33.5
Pumice	31	25.2	46.1	35.3
Friable				35.5
Pumice	16	31.5	43.3	36.2
Pumice	27	28.4	47.8	38.6
Bedded, Nevada				38.8
Bedded (pumiceous)				40.2
Tuff	180	7.2	54.7	41.0

Source: Values processed from Wolff, R.G. Physical Properties of Rocks—Porosity, Permeability, Distribution Coefficient, and Dispersivity. U.S. Geological Survey Open-File Report 1982. 82–116, 118 pp.

TABLE C.6
Porosity of Metamorphic Rocks

Metamorphic Rocks	No. of Samples	Min	Max	Mean
Gneiss				
Carroll & Frederick Counties, MD	5	0.30	2.23	0.78
Gneiss	56	0.7	1.8	1.2
Gneiss	30	0.3	4.1	1.6
Greenstone, Silver City, NM	1			0.669
Schist				
Argillaceous				0.62

continued

TABLE C.6 (continued)
Porosity of Metamorphic Rocks

Metamorphic Rocks	No. of Samples	Min	Max	Mean
Gneiss and granite	36	0.02	1.85	0.80
Siliceous				0.88
Pinal Schist, Globe-Miami, AZ	1			1.30
Pinal Schist, Johnson Camp, AZ	1			1.54
Slate, gneiss	6	0.2	8	2
Schist	39	0.6	6.0	2.6
Hornfels, gneiss, metap., Globe-Miami, AZ	5	0.66	8.42	3.12
Some weathered	18	4.4	49.3	38
Quartz-mica, some weathered	21	30.7	58.4	46.9
Serpentine	10	0.6	8.5	2.4
Shale				
Bangor, PA, Ordovician				1.0
Nonesuch Fm., Precambrian	6	1.5	1.7	1.6
Ophir Fm., UT, Cambrian	4			0.75
Johnson Camp, AZ	1			2.12
Devonian-Mississipian	5	1.6	7.6	5.9
Shale	20	1.4	9.7	6
Silurian	7			6.6
Pennsylvanian and Permian	5			8.8
Clay, Pennsylvanian	23	7.1	17.2	9.9
Mississipian	2	9.7	11.0	10.4
Wellington Fm., Selma, KS, Permian	2	15.3	15.5	15.4
Clays and mudstones, Cretaceous	34	0.8	42.3	18.8
Clay, Jurassic	11	8.8	30.7	20.2
Eastern Venezuela, Oligocene and Miocene	40	9.1	35.8	21.7
Shale	29	6.2	42.2	23.8
Ft. Union Fm., MT, Paleocene	3	21.2	36.9	27.2
Clays, Miocene	8			31.9
Shale				20–40
Scarn				
Silver City, NM	2	3.96	5.24	4.60
Chino, NM	8	0.73	9.43	4.65
Johnson Camp, AZ	1			14.7
Slate				
Carroll & Frederick Counties, MD	3	0.00	1.06	—
Black	3	0.40	0.50	0.49
Negaunee Iron Fm., Precambrian	2			0.6
Globe-Miami, AZ	1			0.73
Slate	76	0.1	4.3	0.8
Slate	6	1.91	5.66	3.12
Devonian	21	1.3	13.0	3.3
Slate-Shale	2	0.49	7.55	3.95
Slates, silts, and clays, Carboniferous		1.2	14.3	5.7

Source: Values processed from Wolff, R.G. Physical Properties of Rocks—Porosity, Permeability, Distribution Coefficient, and Dispersivity. U.S. Geological Survey Open-File Report 1982. 82–116, 118 pp.

TABLE C.7
Porosity of Unconsolidated Sediments

Unconsolidated Sediments	No. of Samples	Min	Max	Mean
Sand				
Upper Miocene	78	17.4	31.9	27.1
With gravel—glacial	8	18.6	37.6	28.1
Dune, France	3			31.5
Marine, Oligocene-Pliocene, Tunisia	25	23.2	41.5	31.5
Continental, Quarternary, Tunisia	6	32.0	34.0	33.1
Dune, Sahara	4	34.3	36.8	35.2
Fluvial	25	28.8	39.5	35.3
Silty, central CA	92	28.4	50.2	38.1
Medium (0.5–0.25 mm)	127	28.5	48.9	39.0
Coarse (1.0–0.5 mm)	26	30.9	46.4	39.0
With silt and clay, central CA	132	30.6	61.2	40.4
Beach, Quarternary	25	38.7	44.8	41.2
Beach accretion, Holocene, Galveston Island	17			42.2
Central CA	54	35.4	50.0	42.4
Fine ($\leq$0.25 mm)	243	26.0	53.3	43.0
Beach accretion, Holocene, N. Orleans Island	8			43.1
Clayey, central CA	13	28.0	52.8	44.4
Eolian	6	39.9	50.7	45.0
Silty, subaqueous, Holocene				49.4
Silt				
Sandy, central CA	36	33.9	55.6	40.9
Clayey, central CA	120	31.4	61.0	41.8
0.062–0.004 mm	281	33.9	61.1	46.0
Central CA	2	50.4	52.2	51.3
Loam soils	87			52.2
Loess	5	44.0	57.2	49.0
Clay				
Plastic, Oligocene	1			26.0
Plastic, Austria, Pliocene	2	26.0	26.1	26.1
Loam soils, Sacramento Valley, CA	43			37.3
Arlington, VA				40.1
($\leq$0.0045 mm)	74	34.2	56.9	42.0
Silty, central CA	72	35.6	53.3	43.1
Sandy, central CA	2	38.4	49.6	44.0
Boston blue clay				44.4
Clay		44.0	50.0	45.0
Wealden, Alamosa, CA	7	41.5	57.9	48.2
Blue marine clay, (42% < 2 μm)				49.2
Loam soils, Sacramento Valley, CA	148			50.1
Kaolin, Cornwall, residual (45% < 2 μm)				51.2
Gosport, Holocene estuarine (43% < 2 μm)				52.6
Ganges Delta, Holocene (48% < 2 μm)				55.0
London blue clay, Eocene (46% < 2 μm)				56.8
Paris Basin, Eocene (79% < 2 μm)				64.5
Kleinbelt ton, Denmark, Eocene (77% < 2 μm)				68.5

continued

TABLE C.7 (continued)
Porosity of Unconsolidated Sediments

Unconsolidated Sediments	No. of Samples	Min	Max	Mean
Glacial				
Drift, washed primarily sand size	3	34.6	41.5	39.0
Drift, washed primarily silt size	31	36.2	47.6	44.0
Drift, washed primarily clay size	5	38.4	59.3	49.0
Till	6	11.5	21.0	14.7
Till, primarily sand sized	10	22.1	36.7	31.0
Till, primarily silt sized	15	29.5	40.6	34.0

Source: Values processed from Wolff, R.G. Physical Properties of Rocks—Porosity, Permeability, Distribution Coefficient, and Dispersivity. U.S. Geological Survey Open-File Report 1982. 82–116, 118 pp.

Appendix D

TABLE D.1
Values of Error Function

λ	erf(λ)	λ	erf(λ)	λ	erf(λ)
0.00	0.000	0.42	0.448	0.84	0.765
0.01	0.011	0.43	0.457	0.85	0.771
0.02	0.023	0.44	0.466	0.86	0.776
0.03	0.034	0.45	0.476	0.87	0.781
0.04	0.045	0.46	0.485	0.88	0.787
0.05	0.056	0.47	0.494	0.89	0.792
0.06	0.067	0.48	0.503	0.90	0.797
0.07	0.079	0.49	0.512	0.91	0.802
0.08	0.090	0.50	0.520	0.92	0.807
0.09	0.101	0.51	0.529	0.93	0.812
0.10	0.113	0.52	0.538	0.94	0.816
0.11	0.124	0.53	0.547	0.95	0.821
0.12	0.135	0.54	0.555	0.96	0.825
0.13	0.146	0.55	0.563	0.97	0.830
0.14	0.157	0.56	0.572	0.98	0.834
0.15	0.168	0.57	0.580	0.99	0.839
0.16	0.171	0.58	0.588	1.00	0.843
0.17	0.190	0.59	0.596	1.05	0.862
0.18	0.201	0.60	0.604	1.10	0.880
0.19	0.211	0.61	0.612	1.15	0.897
0.20	0.223	0.62	0.619	1.20	0.910
0.21	0.234	0.63	0.627	1.25	0.928
0.22	0.244	0.64	0.635	1.30	0.934
0.23	0.255	0.65	0.642	1.35	0.944
0.24	0.266	0.66	0.649	1.40	0.952
0.25	0.277	0.67	0.657	1.45	0.960
0.26	0.287	0.68	0.664	1.50	0.966
0.27	0.297	0.69	0.671	1.60	0.976
0.28	0.308	0.70	0.678	1.70	0.984
0.29	0.318	0.71	0.685	1.80	0.989
0.30	0.329	0.72	0.691	1.90	0.993
0.31	0.339	0.73	0.698	2.00	0.9952
0.32	0.349	0.74	0.705	2.10	0.9970
0.33	0.359	0.75	0.711	2.20	0.9981
0.34	0.369	0.76	0.718	2.30	0.99883

continued

TABLE D.1 (continued)
Values of Error Function

λ	erf(λ)	λ	erf(λ)	λ	erf(λ)
0.35	0.379	0.77	0.724	2.40	0.99930
0.36	0.389	0.78	0.730	2.50	0.99958
0.37	0.399	0.79	0.736	2.60	0.99976
0.38	0.409	0.80	0.742	2.70	0.99987
0.39	0.419	0.81	0.748	2.80	0.99992
0.40	0.428	0.82	0.748	2.90	0.99996
0.41	0.438	0.83	0.760	3.00	0.99998

TABLE D.2
Values of Complimentary Error Function

λ	erfc(λ)	λ	erfc(λ)	λ	erfc(λ)
0.00	1.000	0.42	0.552	0.84	0.235
0.01	0.989	0.43	0.543	0.85	0.229
0.02	0.977	0.44	0.534	0.86	0.224
0.03	0.966	0.45	0.524	0.87	0.219
0.04	0.955	0.46	0.515	0.88	0.213
0.05	0.944	0.47	0.506	0.89	0.208
0.06	0.933	0.48	0.497	0.90	0.203
0.07	0.921	0.49	0.488	0.91	0.198
0.08	0.910	0.50	0.480	0.92	0.193
0.09	0.899	0.51	0.471	0.93	0.188
0.10	0.887	0.52	0.462	0.94	0.184
0.11	0.876	0.53	0.453	0.95	0.179
0.12	0.865	0.54	0.445	0.96	0.175
0.13	0.854	0.55	0.437	0.97	0.170
0.14	0.843	0.56	0.428	0.98	0.166
0.15	0.832	0.57	0.420	0.99	0.161
0.16	0.829	0.58	0.412	1.00	0.157
0.17	0.810	0.59	0.404	1.05	0.138
0.18	0.799	0.60	0.396	1.10	0.120
0.19	0.789	0.61	0.388	1.15	0.103
0.20	0.777	0.62	0.381	1.20	0.090
0.21	0.766	0.63	0.373	1.25	0.072
0.22	0.756	0.64	0.365	1.30	0.066
0.23	0.745	0.65	0.358	1.35	0.056
0.24	0.734	0.66	0.351	1.40	0.048
0.25	0.723	0.67	0.343	1.45	0.040
0.26	0.713	0.68	0.336	1.50	0.034
0.27	0.703	0.69	0.329	1.60	0.024
0.28	0.692	0.70	0.322	1.70	0.016
0.29	0.682	0.71	0.315	1.80	0.011
0.30	0.671	0.72	0.309	1.90	0.007
0.31	0.661	0.73	0.302	2.00	0.0048
0.32	0.651	0.74	0.295	2.10	0.0030

continued

TABLE D.2 (continued)
Values of Complimentary Error Function

λ	erfc(λ)	λ	erfc(λ)	λ	erfc(λ)
0.33	0.641	0.75	0.289	2.20	0.0019
0.34	0.631	0.76	0.282	2.30	0.00117
0.35	0.621	0.77	0.276	2.40	0.00070
0.36	0.611	0.78	0.270	2.50	0.00042
0.37	0.601	0.79	0.264	2.60	0.00024
0.38	0.591	0.80	0.258	2.70	0.00013
0.39	0.581	0.81	0.252	2.80	0.00008
0.40	0.572	0.82	0.252	2.90	0.00004
0.41	0.562	0.83	0.240	3.00	0.00002

TABLE D.3
Values of Function $R(\lambda)$ for Various Values of λ

λ	R(λ)	λ	R(λ)
0.00	1.0000	0.43	0.3408
0.01	0.9776	0.44	0.3312
0.02	0.9752	0.45	0.3224
0.03	0.9340	0.46	0.3132
0.04	0.9128	0.47	0.3048
0.05	0.8920	0.48	0.2960
0.06	0.8716	0.49	0.2880
0.07	0.8516	0.50	0.2800
0.08	0.8120	0.52	0.2644
0.09	0.8124	0.54	0.2492
0.10	0.7736	0.56	0.2332
0.11	0.7748	0.58	0.2220
0.12	0.7568	0.60	0.2092
0.13	0.7388	0.62	0.1968
0.14	0.7212	0.64	0.1852
0.15	0.7040	0.66	0.1744
0.16	0.6872	0.68	0.1640
0.17	0.6704	0.70	0.1528
0.18	0.6540	0.72	0.1448
0.19	0.6384	0.74	0.1360
0.20	0.6228	0.76	0.1276
0.21	0.6072	0.78	0.1196
0.22	0.5924	0.80	0.1120
0.23	0.5776	0.82	0.1048
0.24	0.5632	0.84	0.0984
0.25	0.5492	0.86	0.0920
0.26	0.5352	0.88	0.0860
0.27	0.5216	0.90	0.0804
0.28	0.5080	0.92	0.0748
0.29	0.4956	0.94	0.0700

continued

TABLE D.3 (continued)
Values of Function $R(\lambda)$ for Various Values of λ

λ	$R(\lambda)$	λ	$R(\lambda)$
0.30	0.4828	0.96	0.0652
0.31	0.4704	0.98	0.0608
0.32	0.4580	1.00	0.0568
0.33	0.4464	1.10	0.0396
0.34	0.4348	1.20	0.0272
0.35	0.4232	1.30	0.0184
0.36	0.4120	1.40	0.0120
0.37	0.3992	1.50	0.0080
0.38	0.3904	1.60	0.0052
0.39	0.3800	1.70	0.0032
0.40	0.3700	1.80	0.0020
0.41	0.3600	1.90	0.0012
0.42	0.3490	2.00	0.0008

Source: Adapted from Lebedev, A.V. Determination of Hydrogeologic Parameters by Means of Piezometric Data, in: Boreli, M. (ed.): *Seminar on Groundwater Balance* (in Serbo–Croatian), Yugoslav Committee for the UNESCO International Hydrologic Decade, Belgrade, 1968, 227 pp.

TABLE D.4
Values of $W(u)$ for Fully Penetrating Wells in a Confined, Isotropic Aquifer, for Values of u between 10^{-11} and 9.9

u	$W(u)$	u	$W(u)$	u	$W(u)$	u	$W(u)$	u	$W(u)$	u	$W(u)$
1.0E-11	24.7512	1.0E-09	20.1460	1.0E-07	15.5409	1.0E-05	10.9357	1.0E-03	6.3315	1.0E-01	1.8229
1.5E-11	24.3458	1.5E-09	19.7406	1.5E-07	15.1354	1.5E-05	10.5303	1.5E-03	5.9266	1.5E-01	1.4645
2.0E-11	24.0581	2.0E-09	19.4529	2.0E-07	14.8477	2.0E-05	10.2426	2.0E-03	5.6394	2.0E-01	1.2227
2.5E-11	23.8349	2.5E-09	19.2298	2.5E-07	14.6246	2.5E-05	10.0194	2.5E-03	5.4167	2.5E-01	1.0443
3.0E-11	23.6526	3.0E-09	19.0474	3.0E-07	14.4423	3.0E-05	9.8317	3.0E-03	5.2349	3.0E-01	0.9057
3.5E-11	23.4985	3.5E-09	18.8933	3.5E-07	14.2881	3.5E-05	9.6830	3.5E-03	5.0813	3.5E-01	0.7942
4.0E-11	23.3649	4.0E-09	18.7598	4.0E-07	14.1546	4.0E-05	9.5495	4.0E-03	4.9482	4.0E-01	0.7024
4.5E-11	23.2471	4.5E-09	18.6420	4.5E-07	14.0368	4.5E-05	9.4317	4.5E-03	4.8310	4.5E-01	0.6253
5.0E-11	23.1418	5.0E-09	18.5366	5.0E-07	13.9314	5.0E-05	9.3263	5.0E-03	4.7261	5.0E-01	0.5598
5.5E-11	23.0465	5.5E-09	18.4413	5.5E-07	13.8361	5.5E-05	9.2310	5.5E-03	4.6313	5.5E-01	0.5034
6.0E-11	22.9595	6.0E-09	18.3543	6.0E-07	13.7491	6.0E-05	9.1440	6.0E-03	4.5448	6.0E-01	0.4544
6.5E-11	22.8794	6.5E-09	18.2742	6.5E-07	13.6691	6.5E-05	9.0640	6.5E-03	4.4652	6.5E-01	0.4115
7.0E-11	22.8053	7.0E-09	18.2001	7.0E-07	13.5950	7.0E-05	8.9899	7.0E-03	4.3916	7.0E-01	0.3738
7.5E-11	22.7363	7.5E-09	18.1311	7.5E-07	13.5260	7.5E-05	8.9209	7.5E-03	4.3231	7.5E-01	0.3403
8.0E-11	22.6718	8.0E-09	18.0666	8.0E-07	13.4614	8.0E-05	8.8563	8.0E-03	4.2591	8.0E-01	0.3106
8.5E-11	22.6112	8.5E-09	18.0060	8.5E-07	13.4008	8.5E-05	8.7957	8.5E-03	4.1990	8.5E-01	0.2840
9.0E-11	22.5540	9.0E-09	17.9488	9.0E-07	13.3437	9.0E-05	8.7386	9.0E-03	4.1423	9.0E-01	0.2602
9.5E-11	22.4999	9.5E-09	17.8948	9.5E-07	13.2896	9.5E-05	8.6845	9.5E-03	4.0887	9.5E-01	0.2387
1.0E-10	22.4486	1.0E-08	17.8435	1.0E-06	13.2383	1.0E-04	8.6332	1.0E-02	4.0379	1.0	0.2194
1.5E-10	22.0432	1.5E-08	17.4380	1.5E-06	12.8328	1.5E-04	8.2278	1.5E-02	3.6374	1.5	0.1000
2.0E-10	21.7555	2.0E-08	17.1503	2.0E-06	12.5451	2.0E-04	7.9402	2.0E-02	3.3547	2.0	0.04890
2.5E-10	21.5323	2.5E-08	16.9272	2.5E-06	12.3220	2.5E-04	7.7172	2.5E-02	3.1365	2.5	0.02491
3.0E-10	21.3500	3.0E-08	16.7449	3.0E-06	12.1397	3.0E-04	7.5348	3.0E-02	2.9591	3.0	0.01305
3.5E-10	21.1959	3.5E-08	16.5591	3.5E-06	11.9855	3.5E-04	7.3807	3.5E-02	2.8099	3.5	0.00698
4.0E-10	21.0623	4.0E-08	16.4572	4.0E-06	11.8520	4.0E-04	7.2472	4.0E-02	2.6813	4.0	0.00378
4.5E-10	20.9446	4.5E-08	16.3394	4.5E-06	11.7342	4.5E-04	7.1295	4.5E-02	2.5684	4.5	0.00207
5.0E-10	20.8392	5.0E-08	16.2340	5.0E-06	11.6289	5.0E-04	7.0242	5.0E-02	2.4679	5.0	0.00115
5.5E-10	20.7439	5.5E-08	16.1387	5.5E-06	11.5336	5.5E-04	6.9289	5.5E-02	2.3775	5.5	0.000641
6.0E-10	20.6569	6.0E-08	16.0517	6.0E-06	11.4465	6.0E-04	6.8420	6.0E-02	2.2953	6.0	0.000360
6.5E-10	20.5768	6.5E-08	15.9717	6.5E-06	11.3665	6.5E-04	6.7620	6.5E-02	2.2201	6.5	0.000203
7.0E-10	20.5027	7.0E-08	15.8976	7.0E-06	11.2924	7.0E-04	6.6879	7.0E-02	2.1508	7.0	0.000116
7.5E-10	20.4337	7.5E-08	15.8286	7.5E-06	11.2234	7.5E-04	6.6190	7.5E-02	2.0867	7.5	6.58E-05
8.0E-10	20.3692	8.0E-08	15.7640	8.0E-06	11.1589	8.0E-04	6.5545	8.0E-02	2.0269	8.0	3.77E-05
8.5E-10	20.3086	8.5E-08	15.7034	8.5E-06	11.0982	8.5E-04	6.4939	8.5E-02	1.9711	8.5	2.16E-05
9.0E-10	20.2514	9.0E-08	15.6462	9.0E-06	11.0411	9.0E-04	6.4368	9.0E-02	1.9187	9.0	1.25E-05
9.5E-10	20.1973	9.5E-08	15.5922	9.5E-06	10.9870	9.5E-04	6.3828	9.5E-02	1.8695	9.5	7.19E-06

Source: Ferris, J.G., D.B. Knowles, R.H. Brown, and R.W. Stallman. *Geological Survey Water-Supply Paper 1536-E*, US Government Printing Office, Washington, 1962, 174 pp.

TABLE D.5
Values of Functions $W(u, r/b)$ for Various Values of u

Columns below are values of r/b.

u	0.002	0.004	0.006	0.008	0.01	0.02	0.04	0.06	0.08	0.1	0.2	0.4	0.6	0.8	1	2	4	6	8
0	12.7	11.3	10.5	9.89	9.44	8.06	6.67	5.87	5.29	4.85	3.51	2.23	1.55	1.13	0.842	0.228	0.0223	0.0025	0.0003
0.000002	12.1	11.2	10.5	9.89	9.44	8.06	6.67	5.87	5.29	4.85	3.51	2.23	1.55	1.13	0.842	0.228	0.0223	0.0025	0.0003
0.000004	11.6	11.1	10.4	9.88	9.44	8.06	6.67	5.87	5.29	4.85	3.51	2.23	1.55	1.13	0.842	0.228	0.0223	0.0025	0.0003
0.000006	11.3	10.9	10.4	9.87	9.44	8.06	6.67	5.87	5.29	4.85	3.51	2.23	1.55	1.13	0.842	0.228	0.0223	0.0025	0.0003
0.000008	11.0	10.7	10.3	9.84	9.43	8.06	6.67	5.87	5.29	4.85	3.51	2.23	1.55	1.13	0.842	0.228	0.0223	0.0025	0.0003
0.00001	10.8	10.6	10.2	9.8	9.42	8.06	6.67	5.87	5.29	4.85	3.51	2.23	1.55	1.13	0.842	0.228	0.0223	0.0025	0.0003
0.00002	10.2	10.1	9.84	9.58	9.30	8.06	6.67	5.87	5.29	4.85	3.51	2.23	1.55	1.13	0.842	0.228	0.0223	0.0025	0.0003
0.00004	9.52	9.45	9.34	9.19	9.01	8.03	6.67	5.87	5.29	4.85	3.51	2.23	1.55	1.13	0.842	0.228	0.0223	0.0025	0.0003
0.00006	9.13	9.08	9.00	8.89	8.77	7.98	6.67	5.87	5.29	4.85	3.51	2.23	1.55	1.13	0.842	0.228	0.0223	0.0025	0.0003
0.00008	8.84	8.81	8.75	8.67	8.57	7.91	6.67	5.87	5.29	4.85	3.51	2.23	1.55	1.13	0.842	0.228	0.0223	0.0025	0.0003
0.0001	8.62	8.59	8.55	8.48	8.40	7.84	6.67	5.87	5.29	4.85	3.51	2.23	1.55	1.13	0.842	0.228	0.0223	0.0025	0.0003
0.0002	7.94	7.92	7.90	7.86	7.82	7.50	6.62	5.86	5.29	4.85	3.51	2.23	1.55	1.13	0.842	0.228	0.0223	0.0025	0.0003
0.0004	7.24	7.24	7.22	7.21	7.19	7.01	6.45	5.83	5.29	4.85	3.51	2.23	1.55	1.13	0.842	0.228	0.0223	0.0025	0.0003
0.0006	6.84	6.84	6.83	6.82	6.80	6.68	6.27	5.77	5.27	4.85	3.51	2.23	1.55	1.13	0.842	0.228	0.0223	0.0025	0.0003
0.0008	6.55	6.55	6.54	6.53	6.52	6.43	6.11	5.69	5.25	4.84	3.51	2.23	1.55	1.13	0.842	0.228	0.0223	0.0025	0.0003
0.001	6.33	6.33	6.32	6.32	6.31	6.23	5.97	5.61	5.21	4.83	3.51	2.23	1.55	1.13	0.842	0.228	0.0223	0.0025	0.0003
0.002	5.64	5.64	5.63	5.63	5.63	5.59	5.45	5.24	4.98	4.71	3.50	2.23	1.55	1.13	0.842	0.228	0.0223	0.0025	0.0003
0.004	4.95	4.95	4.95	4.94	4.94	4.92	4.85	4.74	4.59	4.42	3.48	2.23	1.55	1.13	0.842	0.228	0.0223	0.0025	0.0003
0.006	4.54	4.54	4.54	4.54	4.54	4.53	4.48	4.41	4.30	4.18	3.43	2.23	1.55	1.13	0.842	0.228	0.0223	0.0025	0.0003
0.008	4.26	4.26	4.26	4.26	4.26	4.25	4.21	4.15	4.08	3.98	3.36	2.23	1.55	1.13	0.842	0.228	0.0223	0.0025	0.0003
0.01	4.04	4.04	4.04	4.04	4.04	4.03	4.00	3.95	3.89	3.81	3.29	2.23	1.55	1.13	0.842	0.228	0.0223	0.0025	0.0003
0.02	3.35	3.35	3.35	3.35	3.35	3.35	3.34	3.31	3.28	3.24	2.95	2.18	1.55	1.13	0.842	0.228	0.0223	0.0025	0.0003
0.04	2.68	2.68	2.68	2.68	2.68	2.68	2.67	2.66	2.65	2.63	2.48	2.02	1.52	1.13	0.842	0.228	0.0223	0.0025	0.0003
0.06	2.30	2.30	2.30	2.30	2.30	2.29	2.29	2.28	2.27	2.26	2.17	1.85	1.46	1.11	0.839	0.228	0.0223	0.0025	0.0003
0.08	2.03	2.03	2.03	2.03	2.03	2.03	2.02	2.02	2.01	2.00	1.94	1.69	1.39	1.08	0.832	0.228	0.0223	0.0025	0.0003
0.1	1.82	1.82	1.82	1.82	1.82	1.82	1.82	1.82	1.81	1.80	1.75	1.56	1.31	1.05	0.819	0.228	0.0223	0.0025	0.0003
0.2	1.22	1.22	1.22	1.22	1.22	1.22	1.22	1.22	1.22	1.22	1.19	1.11	0.996	0.857	0.715	0.227	0.0223	0.0025	0.0003
0.4	0.702	0.702	0.702	0.702	0.702	0.702	0.702	0.702	0.701	0.700	0.693	0.665	0.621	0.565	0.502	0.210	0.0223	0.0025	0.0003
0.6	0.454	0.454	0.454	0.454	0.454	0.454	0.454	0.454	0.454	0.453	0.45	0.436	0.415	0.387	0.354	0.177	0.0222	0.0025	0.0003
0.8	0.311	0.311	0.311	0.311	0.311	0.311	0.311	0.310	0.310	0.310	0.308	0.301	0.289	0.273	0.254	0.144	0.0218	0.0025	0.0003
1	0.219	0.219	0.219	0.219	0.219	0.219	0.219	0.219	0.219	0.219	0.218	0.213	0.206	0.197	0.185	0.114	0.0207	0.0025	0.0003
2	0.049	0.049	0.049	0.049	0.049	0.049	0.049	0.049	0.049	0.049	0.049	0.048	0.047	0.046	0.044	0.034	0.0110	0.0021	0.0003
4	0.0038	0.0038	0.0038	0.0038	0.0038	0.0038	0.0038	0.0038	0.0038	0.0038	0.0038	0.0038	0.0037	0.0037	0.0036	0.0031	0.0016	0.0006	0.0002
6	0.0004	0.0004	0.0004	0.0004	0.0004	0.0004	0.0004	0.0004	0.0004	0.0004	0.0004	0.0004	0.0004	0.0004	0.0004	0.0003	0.0002	0.0001	0
8	0	0	0	0	0	0	0	0	0	0	0	0	0	0	0	0	0	0	0

Source: Hantush, M.S., *Trans. Am. Geophys. Union*, 37(6):702–714, 1956. Copyright by the American Geophysical Union.

TABLE D.6
Values of s_D for the Construction of Type A Curves for Fully Penetrating Wells in Unconfined Aquifers with Delayed Response

t_s	$b = 0.001$	$b = 0.004$	$b = 0.01$	$b = 0.03$	$b = 0.06$	$b = 0.1$	$b = 0.2$	$b = 0.4$	$b = 0.6$	$b = 0.8$
1.00E−01	2.48E−02	2.43E−02	2.41E−02	2.35E−02	2.30E−02	2.24E−02	2.14E−02	1.99E−02	1.88E−02	1.70E−02
2.00E−01	1.45E−01	1.42E−01	1.40E−01	1.36E−01	1.31E−01	1.27E−01	1.19E−01	1.08E−01	9.88E−02	8.49E−02
3.50E−01	3.58E−01	3.52E−01	3.45E−01	3.31E−01	3.18E−01	3.04E−01	2.79E−01	2.44E−01	2.17E−01	1.75E−01
6.00E−01	6.62E−01	6.48E−01	6.33E−01	6.01E−01	5.70E−01	5.40E−01	4.83E−01	4.03E−01	3.43E−01	2.96E−01
1.00E+00	1.02E+00	9.92E−01	9.63E−01	9.05E−01	8.49E−01	7.92E−01	6.88E−01	5.42E−01	4.38E−01	3.60E−01
2.00E+00	1.57E+00	1.52E+00	1.46E+00	1.35E+00	1.23E+00	1.12E+00	9.18E−01	6.59E−01	4.97E−01	3.91E−01
3.50E+00	2.05E+00	1.97E+00	1.88E+00	1.70E+00	1.51E+00	1.34E+00	1.03E+00	6.90E−01	5.07E−01	3.94E−01
6.00E+00	2.52E+00	2.41E+00	2.27E+00	1.99E+00	1.73E+00	1.47E+00	1.07E+00	6.96E−01	5.07E−01	3.94E−01
1.00E+01	2.97E+00	2.80E+00	2.61E+00	2.22E+00	1.85E+00	1.53E+00	1.08E+00	6.96E−01	5.07E−01	3.94E−01
2.00E+01	3.56E+00	3.30E+00	3.00E+00	2.41E+00	1.92E+00	1.55E+00	1.08E+00	6.96E−01	5.07E−01	3.94E−01
3.50E+01	4.01E+00	3.65E+00	3.23E+00	2.48E+00	1.93E+00	1.55E+00	1.08E+00	6.96E−01	5.07E−01	3.94E−01
6.00E+01	4.42E+00	3.93E+00	3.37E+00	2.49E+00	1.94E+00	1.55E+00	1.08E+00	6.96E−01	5.07E−01	3.94E−01
1.00E+02	4.77E+00	4.12E+00	3.43E+00	2.50E+00	1.94E+00	1.55E+00	1.08E+00	6.96E−01	5.07E−01	3.94E−01
2.00E+02	5.16E+00	4.26E+00	3.45E+00	2.50E+00	1.94E+00	1.55E+00	1.08E+00	6.96E−01	5.07E−01	3.94E−01
3.50E+02	5.40E+00	4.29E+00	3.46E+00	2.50E+00	1.94E+00	1.55E+00	1.08E+00	6.96E−01	5.07E−01	3.94E−01
6.00E+02	5.54E+00	4.30E+00	3.46E+00	2.50E+00	1.94E+00	1.55E+00	1.08E+00	6.96E−01	5.07E−01	3.94E−01
1.00E+03	5.59E+00	4.30E+00	3.46E+00	2.50E+00	1.94E+00	1.55E+00	1.08E+00	6.96E−01	5.07E−01	3.94E−01
2.00E+03	5.59E+00	4.30E+00	3.46E+00	2.50E+00	1.94E+00	1.55E+00	1.08E+00	6.96E−01	5.07E−01	3.94E−01
3.50E+00	5.59E+00	4.30E+00	3.46E+00	2.50E+00	1.94E+00	1.55E+00	1.08E+00	6.96E−01	5.07E−01	3.94E−01

continued

TABLE D.6 (continued)
Values of s_D for the Construction of Type A Curves for Fully Penetrating Wells in Unconfined Aquifers with Delayed Response

t_s	$b = 1.0$	$b = 1.5$	$b = 2.0$	$b = 2.5$	$b = 3.0$	$b = 4.0$	$b = 5.0$	$b = 6.0$	$b = 7.0$
1.00E−01	1.70E−02	1.53E−02	1.38E−02	1.25E−02	1.13E−02	9.33E−03	7.72E−03	6.39E−03	5.30E−03
2.00E−01	8.49E−02	7.13E−02	6.03E−02	5.11E−02	4.35E−02	3.17E−02	2.34E−02	1.74E−02	1.31E−02
3.50E−01	1.75E−01	1.36E−01	1.07E−01	8.46E−02	6.78E−02	4.45E−02	3.02E−02	2.10E−02	1.51E−02
6.00E−01	2.56E−01	1.82E−01	1.33E−01	1.01E+00	7.67E−02	4.76E−02	3.13E−02	2.14E−02	1.52E−02
1.00E+00	3.00E−01	1.99E−01	1.40E−01	1.03E−01	7.79E−02	4.78E−02	3.13E−02	2.15E−02	1.52E−02
2.00E+00	3.17E−01	2.03E−01	1.41E−01	1.03E−01	7.79E−02	4.78E−02	3.13E−02	2.15E−02	1.52E−02
3.50E+00	3.17E−01	2.03E−01	1.41E−01	1.03E−01	7.79E−02	4.78E−02	3.13E−02	2.15E−02	1.52E−02
6.00E+00	3.17E−01	2.03E−01	1.41E−01	1.03E−01	7.79E−02	4.78E−02	3.13E−02	2.15E−02	1.52E−02
1.00E+01	3.17E−01	2.03E−01	1.41E−01	1.03E−01	7.79E−02	4.78E−02	3.13E−02	2.15E−02	1.52E−02
2.00E+01	3.17E−01	2.03E−01	1.41E−01	1.03E−01	7.79E−02	4.78E−02	3.13E−02	2.15E−02	1.52E−02
3.50E+01	3.17E−01	2.03E−01	1.41E−01	1.03E−01	7.79E−02	4.78E−02	3.13E−02	2.15E−02	1.52E−02
6.00E+01	3.17E−01	2.03E−01	1.41E−01	1.03E−01	7.79E−02	4.78E−02	3.13E−02	2.15E−02	1.52E−02
1.00E+02	3.17E−01	2.03E−01	1.41E−01	1.03E−01	7.79E−02	4.78E−02	3.13E−02	2.15E−02	1.52E−02
2.00E+02	3.17E−01	2.03E−01	1.41E−01	1.03E−01	7.79E−02	4.78E−02	3.13E−02	2.15E−02	1.52E−02
3.50E+02	3.17E−01	2.03E−01	1.41E−01	1.03E−01	7.79E−02	4.78E−02	3.13E−02	2.15E−02	1.52E−02
6.00E+02	3.17E−01	2.03E−01	1.41E−01	1.03E−01	7.79E−02	4.78E−02	3.13E−02	2.15E−02	1.52E−02
1.00E+03	3.17E−01	2.03E−01	1.41E−01	1.03E−01	7.79E−02	4.78E−02	3.13E−02	2.15E−02	1.52E−02
2.00E+03	3.17E−01	2.03E−01	1.41E−01	1.03E−01	7.79E−02	4.78E−02	3.13E−02	2.15E−02	1.52E−02
3.50E+00	3.17E−01	2.03E−01	1.41E−01	1.03E−01	7.79E−02	4.78E−02	3.13E−02	2.15E−02	1.52E−02

Source: Neuman, S.P., *Water Resour. Res.* 11(2):329–42, 1975. Copyright by the American Geophysical Union.

TABLE D.7
Values of s_D for the Construction of Type B Curves for Fully Penetrating Wells in Unconfined Aquifers with Delayed Response

t_y	$b = 0.001$	$b = 0.004$	$b = 0.01$	$b = 0.03$	$b = 0.06$	$b = 0.1$	$b = 0.2$	$b = 0.4$	$b = 0.6$	$b = 0.8$
1.00E−04	5.62E+00	4.30E+00	3.46E+00	2.50E+00	1.94E+00	1.56E+00	1.09E+00	6.97E−01	5.08E−01	3.95E−01
2.00E−04	5.62E+00	4.30E+00	3.46E+00	2.50E+00	1.94E+00	1.56E+00	1.09E+00	6.97E−01	5.08E−01	3.95E−01
3.50E−04	5.62E+00	4.30E+00	3.46E+00	2.50E+00	1.94E+00	1.56E+00	1.09E+00	6.97E−01	5.08E−01	3.95E−01
6.00E−04	5.62E+00	4.30E+00	3.46E+00	2.50E+00	1.94E+00	1.56E+00	1.09E+00	6.97E−01	5.08E−01	3.95E−01
1.00E−03	5.62E+00	4.30E+00	3.46E+00	2.50E+00	1.94E+00	1.56E+00	1.09E+00	6.97E−01	5.08E−01	3.95E−01
2.00E−03	5.62E+00	4.30E+00	3.46E+00	2.50E+00	1.94E+00	1.56E+00	1.09E+00	6.97E−01	5.09E−01	3.96E−01
3.50E−03	5.62E+00	4.30E+00	3.46E+00	2.50E+00	1.94E+00	1.56E+00	1.09E+00	6.98E−01	5.10E−01	3.97E−01
6.00E−03	5.62E+00	4.30E+00	3.46E+00	2.50E+00	1.94E+00	1.56E+00	1.09E+00	7.00E−01	5.12E−01	3.99E−01
1.00E−02	5.62E+00	4.30E+00	3.46E+00	2.50E+00	1.94E+00	1.56E+00	1.09E+00	7.03E−01	5.16E−01	4.03E−01
2.00E−02	5.62E+00	4.30E+00	3.46E+00	2.50E+00	1.94E+00	1.56E+00	1.09E+00	7.10E−01	5.24E−01	4.12E−01
3.50E−02	5.62E+00	4.30E+00	3.46E+00	2.50E+00	1.94E+00	1.56E+00	1.10E+00	7.20E−01	5.37E−01	4.25E−01
6.00E−02	5.62E+00	4.30E+00	3.46E+00	2.50E+00	1.95E+00	1.57E+00	1.11E+00	7.37E−01	5.57E−01	4.47E−01
1.00E−01	5.62E+00	4.30E+00	3.46E+00	2.51E+00	1.96E+00	1.58E+00	1.13E+00	7.63E−01	5.89E−01	4.83E−01
2.00E−01	5.62E+00	4.30E+00	3.46E+00	2.52E+00	1.98E+00	1.61E+00	1.18E+00	8.29E−01	6.67E−01	5.71E−01
3.50E−01	5.63E+00	4.31E+00	3.47E+00	2.54E+00	2.01E+00	1.66E+00	1.24E+00	9.22E−01	7.80E−01	6.97E−01
6.00E−01	5.63E+00	4.31E+00	3.49E+00	2.57E+00	2.06E+00	1.73E+00	1.35E+00	1.07E+00	9.54E−01	8.89E−01
1.00E+00	5.63E+00	4.32E+00	3.51E+00	2.62E+00	2.13E+00	1.83E+00	1.50E+00	1.29E+00	1.20E+00	1.16E+00
2.00E+00	5.64E+00	4.35E+00	3.56E+00	2.73E+00	2.31E+00	2.07E+00	1.85E+00	1.72E+00	1.68E+00	1.66E+00
3.50E+00	5.65E+00	4.38E+00	3.63E+00	2.88E+00	2.55E+00	2.37E+00	2.23E+00	2.17E+00	2.15E+00	2.15E+00
6.00E+00	5.67E+00	4.44E+00	3.74E+00	3.11E+00	2.86E+00	2.75E+00	2.68E+00	2.66E+00	2.65E+00	2.65E+00
1.00E+01	5.70E+00	4.52E+00	3.90E+00	3.40E+00	3.24E+00	3.18E+00	3.15E+00	3.14E+00	3.14E+00	3.14E+00
2.00E+01	5.76E+00	4.71E+00	4.22E+00	3.92E+00	3.85E+00	3.83E+00	3.82E+00	3.82E+00	3.82E+00	3.82E+00
3.50E+01	5.85E+00	4.94E+00	4.58E+00	4.40E+00	4.38E+00	4.58E+00	4.38E+00	4.58E+00	4.37E+00	4.37E+00
6.00E+01	5.99E+00	5.23E+00	5.00E+00	4.92E+00	4.91E+00	4.91E+00	4.91E+00	4.91E+00	4.91E+00	4.91E+00
1.00E+02	6.16E+00	5.59E+00	5.46E+00	5.42E+00	5.42E+00	5.42E+00	5.42E+00	5.42E+00	5.42E+00	5.42E+00

continued

TABLE D.7 (continued)
Values of s_D for the Construction of Type B Curves for Fully Penetrating Wells in Unconfined Aquifers with Delayed Response

t_r	b = 1.0	b = 1.5	b = 2.0	b = 2.5	b = 3.0	b = 4.0	b = 5.0	b = 6.0	b = 7.0
1.00E−04	3.18E−01	2.04E−01	1.42E−01	1.03E−01	7.80E−02	4.79E−02	3.14E−02	2.15E−02	1.53E−02
2.00E−04	3.18E−01	2.04E−01	1.42E−01	1.03E−01	7.81E−02	4.80E−02	3.15E−02	2.16E−02	1.53E−02
3.50E−04	3.18E−01	2.04E−01	1.42E−01	1.03E−01	7.83E−02	4.81E−02	3.16E−02	2.17E−02	1.54E−02
6.00E−04	3.18E−01	2.04E−01	1.42E−01	1.04E−01	7.85E−02	4.84E−02	3.18E−02	2.19E−02	1.58E−02
1.00E−03	3.18E−01	2.04E−01	1.42E−01	1.04E−01	7.89E−02	4.78E−02	3.21E−02	2.21E−02	1.58E−02
2.00E−03	3.19E−01	2.05E−01	1.43E−01	1.05E−01	7.99E−02	4.96E−02	3.29E−02	2.28E−01	1.64E−02
3.50E−03	3.21E−01	2.07E−01	1.45E−01	1.07E−01	8.14E−02	5.09E−02	3.41E−02	2.39E−02	1.73E−02
6.00E−03	3.23E−01	2.09E−01	1.47E−01	1.09E−01	8.38E−02	5.32E−02	3.61E−02	2.57E−02	1.89E−02
1.00E−02	3.27E−01	2.13E−01	1.52E−01	1.13E−01	8.79E−02	5.68E−02	3.93E−02	2.86E−02	2.15E−02
2.00E−02	3.37E−01	2.24E−01	1.62E−01	1.24E−01	9.80E−02	6.61E−02	4.78E−02	3.62E−02	2.84E−02
3.50E−02	3.50E−01	2.39E−01	1.78E−01	1.39E−01	1.13E−01	8.06E−02	6.12E−02	4.86E−02	3.98E−02
6.00E−02	3.74E−01	2.65E−01	2.05E−01	1.66E−01	1.40E−01	1.06E−01	8.53E−02	7.14E−02	6.14E−02
1.00E−01	4.12E−01	3.07E−01	2.48E−01	2.10E−01	1.84E−01	1.49E−01	1.28E−01	1.13E−01	1.02E−01
2.00E−01	5.06E−01	4.10E−01	3.57E−01	3.23E−01	2.98E−01	2.66E−01	2.45E−01	2.31E−01	2.20E−01
3.50E−01	6.42E−01	5.62E−01	5.17E−01	4.89E−01	4.70E−01	4.45E−01	4.30E−01	4.19E−01	4.11E−01
6.00E−01	8.50E−01	7.92E−01	7.63E−01	7.45E−01	7.33E−01	7.18E−01	7.09E−01	7.03E−01	6.99E−01
1.00E+00	1.13E+00	1.10E+00	1.08E+00	1.07E+00	1.07E+00	1.06E+00	1.06E+00	1.05E+00	1.05E+00
2.00E+00	1.65E+00	1.63E+00	1.63E+00	1.63E+00	1.63E+00	1.63E+00	1.63E+00	1.63E+00	1.63E+00
3.50E+00	2.14E+00	2.14E+00	2.14E+00	2.14E+00	2.14E+00	2.14E+00	2.14E+00	2.14E+00	2.14E+00
6.00E+00	2.65E+00	2.65E+00	2.64E+00	2.64E+00	2.64E+00	2.64E+00	2.64E+00	2.64E+00	2.64E+00
1.00E+01	3.14E+00	3.14E+00	3.14E+00	3.14E+00	3.14E+00	3.14E+00	3.14E+00	3.14E+00	3.14E+00
2.00E+01	3.82E+00	3.82E+00	3.82E+00	3.82E+00	3.82E+00	3.82E+00	3.82E+00	3.82E+00	3.82E+00
3.50E+01	4.37E+00	4.37E+00	4.37E+00	4.37E+00	4.37E+00	4.37E+00	4.37E+00	4.37E+00	4.37E+00
6.00E+01	4.91E+00	4.91E+00	4.91E+00	4.91E+00	4.91E+00	4.91E+00	4.91E+00	4.91E+00	4.91E+00
1.00E+02	5.42E+00	5.42E+00	5.42E+00	5.42E+00	5.42E+00	5.42E+00	5.42E+00	5.42E+00	5.42E+00

Source: Neuman, S.P., *Water Resour. Res.*, 11(2):329–42, 1975. Copyright by the American Geophysical Union.

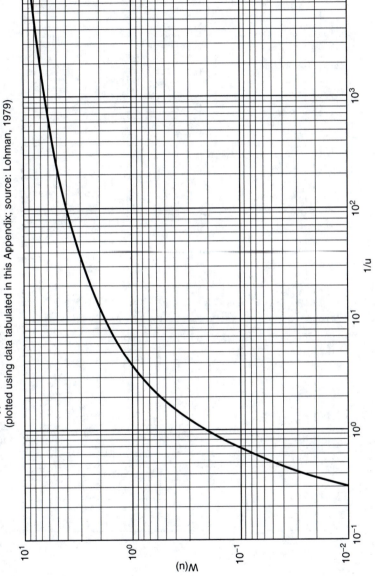

Theis type curve for fully penetrating wells in homogeneous confined aquifers (plotted using data tabulated in this Appendix; source: Lohman, 1979)

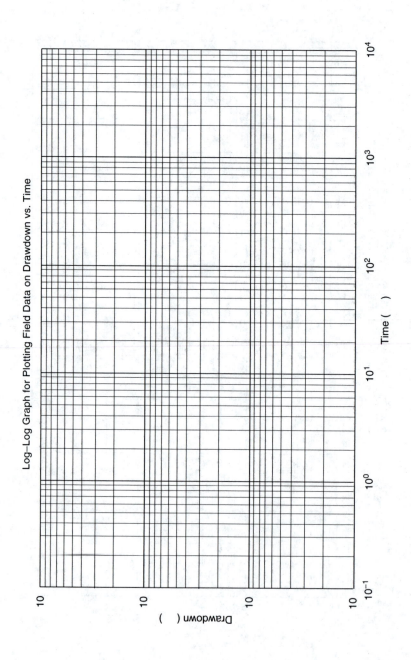

Log–Log Graph for Plotting Field Data on Drawdown vs. Time

WELL PUMPING TEST ANALYSIS
Cooper–Jacob Method
(Straight Line Method)

Well No.: _____   Location: _____   Pumping rate: _____

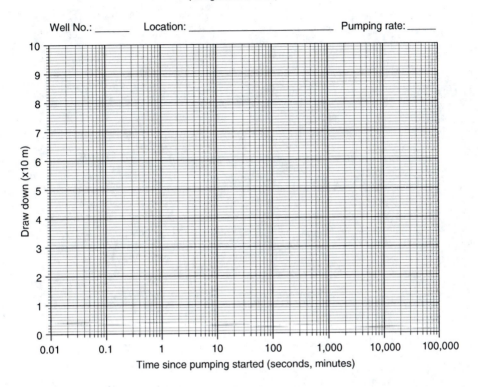

Draw down (x10 m)

Time since pumping started (seconds, minutes)

$$T = \frac{0.183\,Q}{\Delta s}$$

$$T = \quad \text{x10} \quad \text{m}^2/\text{s}$$

$$S = \frac{2.25\,T\,t_0}{r^2}$$

$$S =$$

Δs Difference in drawdown between two points that are one log cycle appart and lay on a stright line drawn through recorded data [in metric (SI) units].

Q Pumping rate in metric (SI) units

T Transmissivity

S Storage coefficient

t_0 Intercept of the straight line at zero drawdown (time on x axis)

r Distance between the pumping well and the observation well where the drawdown is recorded

Index